SI DERIVED UNITS

Quantity	Unit Name	Unit Symbol	Formula
force (F)	newton	N	$kg \cdot m/s^2$
pressure, stress (p)	pascal	Pa	N/m^2
energy, heat, work (W)	joule	J	$N \cdot m$
power (P)	watt	W	J/s
charge (Q)	coulomb	C	$A \cdot s$
electromotive force, voltage (V)	volt	V	W/A
resistance (R)	ohm	Ω	V/A
conductance (G)	siemens	S	A/V
capacitance (C)	farad	F	C/V
inductance (L)	henry	H	Wb/A
magnetic flux (Φ)	weber	Wb	$V \cdot s$
flux density (B)	tesla	T	Wb/m^2
frequency (f)	hertz	Hz	1/s
admittance (Y)	siemens	S	V/A
susceptance (B)	siemens	S	V/A

STANDARD DECIMAL PREFIXES

Multiplier	Prefix	Abbreviation
10^{12}	tera	T
10^{9}	giga	G
10^{6}	mega	M
10^{3}	kilo	k
10^{2}	hecto	h
10^{1}	deka	da
10^{-1}	deci	d
10^{-2}	centi	c
10^{-3}	milli	m
10^{-6}	micro	μ
10^{-9}	nano	n
10^{-12}	pico	p
10^{-15}	femto	f
10^{-18}	atto	a

ELECTRICAL AND ELECTRONICS FUNDAMENTALS

An Applied Survey of Electrical Engineering

V. A. Suprynowicz
University of Connecticut

WEST PUBLISHING COMPANY
St. Paul ▪ New York ▪ San Francisco ▪ Los Angeles

To Gina, Clark, Vin, and Wilma

Copyediting: James Hartman
Technical Illustrations: J & R Technical Services
Production Coordination: Technical Texts, Inc.
Cover image courtesy of AT&T Bell Laboratories

Library of Congress Cataloging-in-Publication Data

Suprynowicz, V. A. (Vincent A.)
Electrical and electronics fundamentals.

Includes index.
1. Electric engineering. 2. Electronics.
I. Title.
TK146.S84 1987 621.3 85-26510
ISBN 0-314-93520-7

Credits

Chapter 6: Problem 6.3 from Horowitz, Paul, and Winfield Hill, *The Art of Electronics* (New York: Cambridge University Press, 1980), Exercise 1.6, p. 7, reprinted with permission.

Chapter 11: Figures 11.62, 11.63, 11.64, and 11.65 from Hall, Jerry, and Charles Watts, *Learning to Work with Integrated Circuits* (Newington, Conn.: American Radio Relay League, 1977), reprinted with permission of the publishers, the American Radio Relay League.

Chapter 13: Figures 13.9, 13.10, 13.11, 13.12, and 13.16 from Shorter, Geoffrey, "Wireless World Dolby Noise Reducer," *Wireless World*, May 1975, reprinted with permission of the publishers of *Wireless World* and *High Fidelity Design*.

Chapter 15: Figures 15.17 and 15.18 from *Signetics Applications Handbook*, 1974, pp. 6.48 and 6.76, © Signetics Corp. 1974. All Rights Reserved, reprinted with permission; Figures 15.19 and 15.22 from Freimark, Ronald J., "Engine Controls Become Cost Effective," vol. 89, no. 8, pp. 28–35, © 1981 Society of Automotive Engineers, Inc., reprinted with permission; Figure 3 in *A Closer Look* and Figure 4 in *A Closer Look* from Heinen, Charles M., and Eldred W. Beckman, "Balancing Clean Air Against Good Mileage," *IEEE Spectrum*, November 1977, © 1977 IEEE, reprinted with permission.

Chapter 16: Figures 16.24, 16.25, 16.30, and 16.31 from Galloway, J.H., "Using the Triac for Control of AC Power," *Application Note 200.35*, pp. 7, 8, 14, reprinted with permission of General Electric Company; Figure 16.32 from *SCR Manual*, 5th Ed., 1972, p. 203, reprinted with permission of General Electric Company.

Chapter 19: Figures 19.38, 19.41, 19.42, and 19.43 from Dijken, R.H., "Designing a Small DC Motor," *Philips Technical Review*, vol. 35, 1975, pp. 96–103; Figure 3 in *A Closer Look* from Kamerbeek, E.M.H., "Electric Motors," vol. 33, 1973, pp. 215–234.

CONTENTS

PREFACE

This text is primarily intended to acquaint students in the various engineering disciplines with the fundamentals of electronics and electrical engineering in a most expeditious manner. It has long been recognized that the ever-increasing use of electrical and electronic equipment in all branches of engineering demands that students in the various engineering disciplines possess some degree of familiarity with their conceptual bases. In the past (and, unfortunately, in many instances to this day), the one or two introductory EE courses, primarily intended for electrical engineering students, were also required of all engineering students. From the viewpoint of such non-EE students, this procedure has two decided disadvantages:

1. A great deal of time is spent delving into fundamental electrical concepts to a depth not needed by the non-EE student.
2. This approach severely limits (if, in fact, not entirely eliminates) the time available to topics of much more utility to the non-EE student.

This text has been specifically written to rectify these objections. Its evolution over a seven-year period has been greeted with enthusiasm by non-EE students and faculty alike. The students being addressed herein will primarily end up being users of electronic equipment, not designers. They demand a rather different philosophic approach from that of a text intended primarily for electrical engineers. As a result, while some topics may not be developed as deeply, this approach has the advantage of allowing a much wider range of subject matter to be explored, with sufficient opportunity being available to illustrate the concepts in terms of applications.

Student enthusiasm is also periodically reinforced by the inclusion of short topics that would appear to have little to do with engineering objectives: Why ac public utilities, rather than the much simpler dc? Why high voltage transmission lines? How much does flashlight battery power cost per kilowatt-hour? Why FM? Classroom experience has shown that by relating material to matters with which the student may have had even a fleeting acquaintance proves vastly superior to its presentation in an experiential void.

This material was developed at the University of Connecticut primarily to satisfy requirements in the civil and mechanical engineering curricula. There is sufficient material for either a two-semester course, or for three quarters of coursework. At the end of one semester, the student, having advanced from a starting point of dc current obtained from a flashlight battery through to negative feedback stability in operational amplifiers, should have acquired sufficient knowledge and conversational proficiency to intelligently discuss professional electrical and electronic needs with instrumentation and sales engineers alike.

In keeping with the objective of presenting basic facts in each chapter, further elaboration of some topics is incorporated into the problems. Thus, as time and inclination of the instructor permit, topic matters may be expanded upon in this manner. Also, in many instances, the objectives of a problem are stated beforehand so that the student will know *why* a particular task has been assigned.

Problems that constitute extensions of text material are noted by a *Bourbaki Z* (△). Nicholas Bourbaki is the pseudonym of a group of French mathematicians who warn the reader of their papers of an approaching difficulty by means of a symbol resembling that of a French road sign indicating an S-curve. In this text, we adopt the symbol to indicate the incorporation into a problem of material not generally covered in the text.

As a starting point in the text, we consider the simplest source of electrical power, the battery. With the direct current (dc) available from batteries, we introduce its relation to voltage, resistance, and electrical power. We then turn to alternating current (ac) and consider why the added complication of ac circuits, in contrast to the much simpler dc ones, is justified. In addition to the resistors used in dc circuits, the topic of ac leads to the introduction of two new elements, the inductor and the capacitor. To show the utility of these concepts to the non–EE student, we consider automotive ignition, a system that makes use of all three elements. In fact, automotive ignition and engine controls constitute a continuing thread through a good part of the text. We also use them to demonstrate the application of transistors and microcomputers.

To assist non–EE engineers when working with either electronics engineers or with engineering sales personnel, in Chapter 4 we consider realistic values and specifications for resistors, capacitors, and inductors.

Prior coursework often leaves students with an erroneous view as to what constitutes electrical resistance (i.e., that it is due to the collision of charged carriers with the atoms making up the conductor). An understanding of the true nature of electrical conduction is necessary to comprehend current flow through semiconductors. Not only does this obviously arise in considering transistor operation, but semiconductors also find wide engineering application as temperature and exhaust sensors, in strain gages, and so forth, and their intelligent use demands some knowledge of their correct mode of operation.

Much of the mathematics employed in electrical engineering is common to all branches of engineering. The suppression and damping of mechanical vibration employs the same equation forms as those used in analyzing electrical filters and attenuators. In this text, these equations are applied to both types of circuits. Familiarity with the passage between mechanical and electrical domains can be used to improve upon mechanical designs.

All electronic devices, to varying degrees, distort the signals applied for processing. Minimizing such distortion constitutes a large part of electronic design, and even though the non–EE may not be called upon to undertake such designs, familiarity with the principles involved allows much more intelligent use of electronic apparatus. The Fourier analysis of distorted waveforms aids in this understanding, so we not only consider the mathematics involved but also review what is physically transpiring. This same analysis again proves invaluable in analyzing mechanical vibrations. The non-EE student might also be reminded that Fourier analysis was originally developed to solve heat conduction problems.

Having mastered the electrical fundamentals, we turn to some electronics ones: the diode, which can be used both to convert ac into dc and to take the received radio waves and convert them into intelligible sound waves; the transistor, which can be used to amplify electrical signals; the integrated circuit, which packages into an item smaller than a postage stamp an amplifier capable of voltage amplification in excess of 100,000; and pulse circuits such as used in TV sets, touch-system telephones, computers, and automotive engine controls.

Having introduced the elements of electrical power conversion (ac into dc) in an earlier chapter, these concepts are re-examined in the light of the subsequent consideration of a variety of semiconductor devices and integrated circuit amplifiers. In like manner, the dc measurements of an earlier chapter can now be extended into the ac realm. Amplifiers are also applied to two illustrative instrumentation areas, the measure of strains and the measurement of temperature.

In conclusion, there are now certain segments of electrical engineering that find a more receptive audience in the mechanical curriculum than in the electrical one, power control and electrical motors being cases in point. We consider both topics.

In the instructor's manual that accompanies this text, teaching notes are presented on a chapter-by-chapter basis to provide help for electrical engineering instructors who might approach the teaching of this course with a bit of timidity.

Acknowledgments

Numerous professors at other institutions were most helpful in providing review and criticism of the manuscript in its development stages. These include: Christopher Druzgalski, California State University, Long Beach; David Wade, University of Lowell; Alan Drake, University of New Hampshire; P.T. Hutchison, North Carolina State University, Raleigh; P. Banerjee, University of Rhode Island; Harold Boroson, University of Maryland; Herbert Hack, Duke University; H. Roland Zapp, Michigan State University; Robert Lee, Rochester Institute of Technology; Virginia McWhorter, Virginia Polytechnic Institute; Kenneth Johnson, Clarkson University; Tom Scott, Iowa State University; Michael S.P. Lucas, Kansas State University; Gregory E. Stillman, University of Illinois, Urbana; Charles Nunnally, Virginia Polytechnic Institute; Murray Sirkis, Arizona State University; John Truxal, State University of New York, Stony Brook; Octavio Salati, University of Pennsylvania; William Ross, Pennsylvania State University; George Webb, Tulane University; Dwayne Cooper, University of Illinois; Grant Myers, University of Nebraska, Lincoln. Acknowledgment is also due to that repository of device specifications, Bob Pease.

The author also wishes to express his gratitude and obligation to Pat Fitzgerald, Pam Rost McClanahan, Sylvia Dovner, Betty Slinger, and Jim Hartman.

PART I

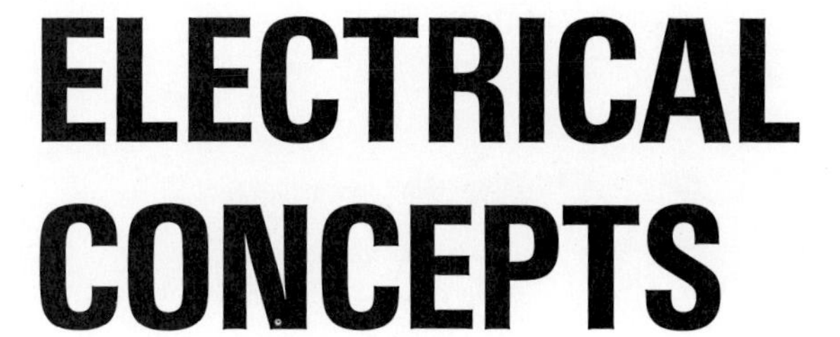

DIRECT CURRENT AND ALTERNATING CURRENT

OVERVIEW

A simple source of electrical current arises from the chemical reaction between the dissimilar metals of a battery. Such direct current is most easily generated, is easy to comprehend, and mathematically leads to the simple relations that constitute Ohm's law. Alternating current, on the other hand, is somewhat more involved in that it requires the manipulation of more than simple algebra in its analytical dissection. Alternating current is much more adaptable than direct current, particularly with regard to transmission over long distances with minimal loss.

In this chapter we consider one electrical element, the resistor, and its physical interaction with both dc and ac, but only dc measuring methods will occupy our attention. The specific device introduced in this chapter, the transformer, makes long-distance power transmission possible. We will also consider the transformer in Chapter 3 in conjunction with an automotive ignition system. In Chapter 4 we will analyze its operation in a more rigorous manner, and in Chapter 7 we will consider it as the means for efficiently extracting power from a transistor amplifier.

OUTLINE

1.1 ■ INTRODUCTION

The most elemental way of creating an electrical current is by means of a chemical reaction between dissimilar metals. This combination of two metals immersed in an electrolyte solution constitutes a **cell**. A **battery** consists of two or more cells arranged in series and/or parallel combinations. In some instances, however, a single cell is called a battery. Such is the case, for example, of a flashlight battery.

Secondary cells (such as automotive and many hand-calculator types) are rechargeable; **primary cells** (such as the disposable flashlight type) are not. In either case, because of chemical reactions, an excess of electrons at one terminal and a deficiency at the other can lead to an electron flow through electrical circuits connected between them. The charges at one electrode represent a higher potential energy than those at the other, and this potential difference (voltage) as well as the opposition to the flow (resistance) determine the magnitude of the flow (current). These three quantities are related through Ohm's law, which we will consider in Section 1.3.

Current through a resistance produces heat. Such heat production constitutes power dissipation whose unit of measure is the watt (to be considered in Section 1.3.3).

Electronic circuits generally employ **direct current** (dc)—current that is unidirectional and of rather constant magnitude (usually)—as the energy source. Today, with the almost universal use of transistorized electronics, direct current is most easily furnished by batteries. When an electronic circuit is employed within a home, laboratory, or an industrial establishment, it may prove more convenient and economical to employ current furnished by a public utility company. This current, **alternating current** (ac), periodically reverses direction, and its magnitude is continually changing. Alternating current must be converted into direct current prior to its being used to energize most electronic circuits, and a discussion of the methods employed to achieve such conversion will occupy us in Chapters 6 and 16, but initially we will address ourselves only to the question of why it is that public utilities furnish us with ac rather than dc.

The final segment of this chapter considers current measurement. Current through a conductor creates a surrounding magnetic field whose intensity is proportional to the magnitude of the current. The interaction between such current-induced magnetic fields and some external fixed magnets can serve as a measure of the current magnitude. Such is the basic principle involved in the operation of a current-measuring device, the **ammeter**. This basic instrument may also be easily converted to measure voltage **(voltmeter)** and resistance **(ohmmeter)**. With one exception, however, in this chapter we will confine our attention to the measurement of dc; measurement of ac is somewhat more involved, and its consideration will be deferred to Chapter 17.

The power needed to actuate the simple ammeter and voltmeter comes at the expense of the circuit being measured and can lead to significant inaccuracies. In Chapter 17 we will also consider methods that will obviate this difficulty.

1.2 ■ BATTERIES

1.2.1 Secondary Cells

Lead-Acid Battery. In the lead-acid storage battery used in automobiles the negative electrode is made of sponge lead (Pb) and the positive of lead peroxide (PbO_2), and both are immersed in sulfuric acid solution consisting of H^+ and OH^- ions (arising from water) and H^+ and SO_4^{2-} ions (from the acid).

Referring to Figure 1.1, we see that at the negative terminal, the lead goes into solution as Pb^{2+}, leaving the negative electrode with an excess of electrons. The Pb^{2+} ions combine with the SO_4^{2-} ions from the solution, forming $PbSO_4$, which, being insoluble, forms a deposit on the negative electrode. Through an external circuit connected between the electrodes, the electrons pass to the positive terminal where the four-valent lead (in PbO_2) is reduced to divalent form. This process also leads to insoluble $PbSO_4$, with the released oxygen combining with the hydrogen ions and going into solution as hydroxyl radicals. Thus, during discharge, lead sulfate is deposited on both electrodes. With the removal of the sulfate ions, the solution becomes progressively more aqueous. For this reason a check of the specific gravity may be used as an indication of the state of charge of a lead battery.

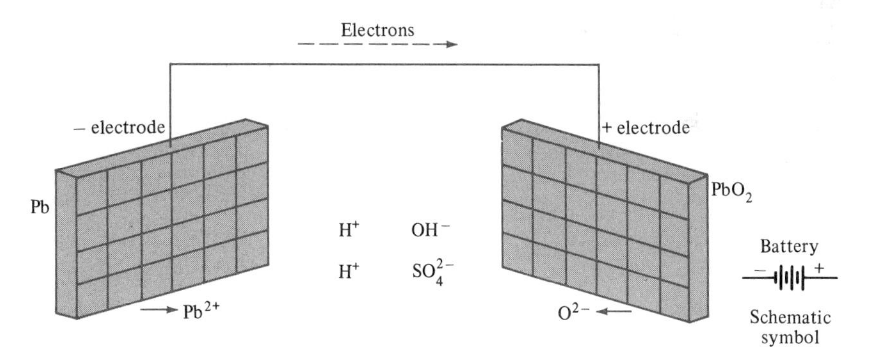

FIGURE 1.1. **Lead-Acid Battery** Solution of lead as ions at the left plate create an excess of electrons that move to the right through the external circuit. At the right plate a diminution of electrons has resulted in the *reduction* of PbO_2.

The excess of electrons at one terminal relative to the other creates an electrical pressure that gives rise to an **electromotive force** (emf) that is often called the *potential difference* or, more simply, *voltage*. The unit of measure is the volt. The unit of charge is the coulomb (C), and the charge carried by an electron is 1.6×10^{-19} C. Since an *ampere* constitutes a coulomb per second flowing past a given point, an ampere comprises 6.25×10^{18} electrons per second. The electrons flowing from the negative terminal of the source encounter a degree of opposition that depends on the material through which they are passing. Such opposition is termed the *resistance*. Resistance is measured in ohms. (The mathematical relationship between these quantities will be discussed in Section 1.3.)

When a battery is being charged, electrons are forced to flow in the opposite direction, and the reactions are reversed. Lead is created at the negative electrode and lead peroxide, at the positive one. In the process of charging, however, electrolysis of the water may create hydrogen and oxygen gas, which necessitates provision for venting the cells. Venting, however, leads to the ever-present hazard of spillage. Thus, sealed lead-acid batteries resulted from a study of the nature of gas evolution during charging.

In a sealed battery, during charging, at the positive electrode SO_4^{2-} ions are replaced by O_2^{2-}, some of which come off as a gas, O_2. Upon coming into contact with the negative electrode, this evolved oxygen can be reduced (becoming OH^-). Any hydrogen gas evolved, however, reacts very slowly, and its formation must be prevented. Prevention is achieved by making the negative electrode much larger than the positive one and keeping it at all times a partially charged state. The negative electrode is discharged by the oxygen atoms rather than by the creation of hydrogen gas. Thus, the charging equipment designed for one type of battery should not be used on any other type to avoid upsetting the charge balance.

The active material in a lead-acid battery is held in position by a grid structure. As the battery charges and discharges, there is considerable volumetric change in the electrodes, and considerable rigidity of the grid structure is desired. It had long been the practice to alloy this lead structure with antimony (Sb), thereby increasing the tensile strength from 1780 to 7280 psi. In the mid-1930s the deleterious effect of antimony on battery performance was recognized; solution of antimony takes place at the positive grid and by transport is deposited on the negative plate. There it facilitates hydrogen evolution (called depolarization) during charging and causes self-discharge in the form of an irreversible process that can ultimately lead to a failure of the negative plate. Depolarization may be corrected either by preventing such transport from taking place (by utilizing physical barriers) or by using a calcium (rather than an antimony) alloy.

An automotive battery is required to provide a few hundreds of amperes for a very short period of time (upon starting) and is otherwise generally kept in a fully charged condition. (Note that retention in a fully charged condition minimizes volume changes of the electrodes.) But a battery intended to operate a fork-lift truck, for example, will be subjected to a deep discharge over a period of hours, then will be recharged, and this cycle may be repeated frequently. In such batteries the positive plates must be protected from loss of active material, and this necessarily leads to a more expensive battery. Such batteries will last 3–6 years; conventional automotive ones would last about 6 months in this type of operation.

The Edison Battery. At the turn of the century Edison was manufacturing a nonrechargeable (primary) battery that the railroads were using extensively for signaling purposes. It consisted of zinc and copper oxide in a potassium hydroxide solution and was known as the Lalande cell. Edison particularly favored the use of alkaline solutions, which allowed metal containers to be employed. After a great deal of screening, he concluded that nickel oxide for the positive electrode and iron for the negative one would make a good secondary cell. He also found that cadmium would make an even better negative electrode, but ruled out its use primarily because of cost. (Today, rechargeable batteries, such as used in many hand calculators, are of the Ni-Cd type. Since very little hydrogen or oxygen is evolved during charge and discharge operations, they may be hermetically sealed.)

Unlike the lead-acid battery, the electrolyte of Edison's nickel-iron battery undergoes no net change during either charging or discharging and, therefore, maintains its voltage. At the time it was first put into extensive production (1909), it appeared that electric cars would be furnishing the prime motive power, at least in cities. The alkaline battery was more durable and delivered more energy per unit weight than the lead-acid battery. (After all, lead is heavy!) Today, the two are probably comparable in this respect, although on a volume basis the lead cell is probably superior.

The nickel-iron (Edison) battery continues to be manufactured today. It is made by the Electric Storage Battery Co.—E.S.B., Inc. They also make Exide acid batteries.

The advantages of the Edison battery are the following:

1. It is rugged.
2. It has a long life.
3. It is virtually electrical proof against overcharging, discharging, and reverse charging.
4. It can be stored indefinitely without deterioration.
5. It operates at higher temperatures than a lead-acid battery.

Its disadvantages are the following:

1. It provides lower voltage (1.10–1.30 volts) than the lead-acid battery (1.95–2.05 volts).
2. It gives poor low-temperature performance and is not usable below 0°F.
3. It is less efficient than the lead-acid battery, with a smaller fraction of the charging power being recoverable.
4. It gives poor performance at high discharge rates.
5. There is considerable O_2 and H_2 evolution during charging and upon standing idle after charging.
6. It is more expensive than the lead-acid battery.

Lead-acid cells can last up to 15–20 years in standby operation (such as needed by the telephone company). Nickel-iron cells have lasted up to 45 years.

1.2.2 Primary Cells

The simplest (and oldest) form of primary cell still in use is that named after Leclanché and consists of a zinc cylinder (the negative electrode) with a concentric carbon rod in contact with manganese dioxide (MnO_2) mixed with carbon (Figure 1.2). While the central carbon rod acts as the positive terminal, the cell action actually takes place between the zinc and the MnO_2, the two being electrically separated by a starch–flour gel that also contains the

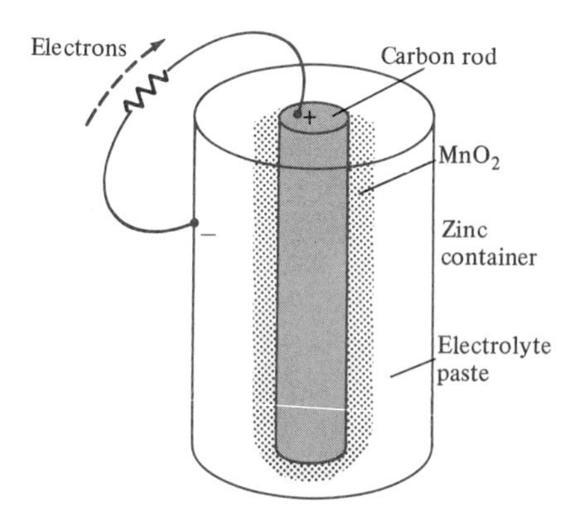

FIGURE 1.2. "Flashlight" Battery Zn^{2+} goes into solution leaving an excess of electrons that pass through the external circuit to replace a deficiency in MnO_2. The MnO_2 is in contact with the carbon rod that constitutes the positive electrode.

electrolyte. The entire assembly in turn may be steel-jacketed to prevent electrolyte leakage since the zinc electrode becomes perforated as the battery discharges. Another attempt at solving the leakage problem has the zinc inside, with MnO_2 surrounding it—the so-called inside-out cell.

The reaction that takes place is

$$2MnO_2 + Zn \rightarrow ZnO + Mn_2O_3$$

As the Zn^{2+} goes into solution, the excess of electrons that is left behind charges the metal negatively. At the MnO_2 electrode, electrons are accepted from the electrical circuit, and Mn^{4+} is reduced in the solid phase. Charge interchange takes place via electrolyte diffusion.

The electrolyte may be either Mn_4Cl or $ZnCl_2$, but for the sake of leakage reduction, the latter is preferred. The cheapest cells use natural MnO_2 and NH_4Cl, although better performance is obtained using synthetic manganese dioxide.

For continuous heavy discharge a KOH electrolyte is used, creating the so-called alkaline battery. Under continuous discharge conditions, the lifetime of a D-size alkaline cell is almost 8 times that of a standard D-size battery. Under intermittent conditions, this advantage is reduced to about $2\frac{1}{2}$.

The original Leclanché cell was developed in 1868 and held a spillable electrolyte. The present form appeared in the 1880s.

Silver oxide-zinc batteries (the button type used in hearing aids, electric watches, and so on) have the highest watt-hour output rating per unit weight (as well as volume) of the primary cells. The high cost of silver becomes less important as the unit size is reduced.

Having the highest negative potential on the scale of electropotentials and being very light, lithium represents another interesting cell material. One form of lithium cell is often used to energize heart pacemakers, providing a voltage of 3 volts and a life expectancy of 6–8 years.

1.3 ▪ DIRECT CURRENT

1.3.1 Ohm's Law

Whatever the resistance (R), the application of an electromotive force (V) to a material will result in the passage of an electrical current (I). In the simplest cases the current is proportional to the voltage, and the relationship among the three quantities (R, V, and I) constitutes **Ohm's law**. Qualitatively, it should be apparent that the greater the voltage, the greater the current, and the greater the resistance, the smaller the current. Thus,

$$I = \frac{V}{R} \tag{1.1}$$

with V being measured in volts (V), I in amperes (A), and R in ohms (Ω). Depending on which two of the three quantities are known, by permutation two other forms of Ohm's law prevail. In summary,

$$I = \frac{V}{R} \tag{1.2a}$$

$$V = IR \tag{1.2b}$$

$$R = \frac{V}{I} \tag{1.2c}$$

EXAMPLE 1.1

A 12 V battery is applied across a 47 Ω resistance. What is the resulting value of current?

Solution:

$$I = \frac{V}{R} = \frac{12\ \text{V}}{47\ \Omega} = 0.26\ \text{A}$$

It should be mentioned that the current through some devices may not be linearly related to the applied voltage. A light bulb is one example of such a nonlinear resistance; as the bulb's temperature rises, the resistance increases, and the current diminishes.

Early in the eighteenth century electrical experimenters, not knowing the actual nature of electricity but recognizing the existence of positive and negative charges, were faced with the problem of deciding whether the current carriers were positive or negative. They presumed them (erroneously) to be positive and thus had the current leaving the positive terminal of the battery. (The positive terminal can be looked upon as pushing away the positive current carriers into the connecting circuit.) We now know that the actual carriers in a metal wire are negative electrons, but since the *classical current* is so universally used, we will also adopt it in this text. When, at times, it becomes necessary to consider the true nature of metallic conduction, it will be termed *electron flow* and in figures will be distinguished by a dashed line in contrast to the solid line used to indicate classical current.

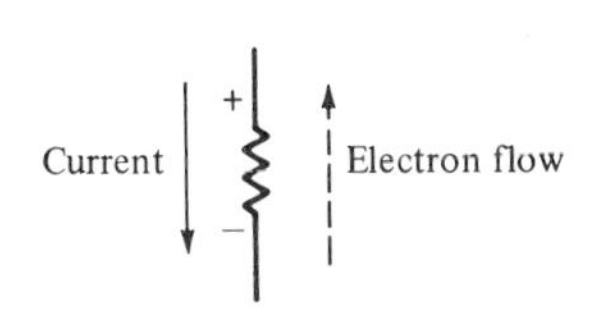

Resistor

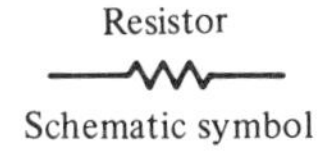

Schematic symbol

FIGURE 1.3. Voltage Polarity Depicted is the polarity of a voltage drop due to the passage of current or, alternatively, electron flow through a resistance.

Either classical current or electron flow, when used in a similar manner, will lead to like results, as can be illustrated by considering the voltage polarity across a resistance (Figure 1.3). When classical flow is used, the resistance terminal through which the current enters is taken to be positive. For electron flow, the entering terminal is taken to be negative. In either case, the polarity of the voltage drop across the resistance remains the same.

1.3.2 Kirchhoff's Laws

Figure 1.4 illustrates a simple electrical circuit: a battery in series with two resistors (R_1 and R_2). The junctures between the various elements of a circuit, however complex, are designated as the *nodes*, and it is common practice to select one node as a reference, generally termed the *ground*. The reference terminal may actually be connected to an external ground, or the ground symbol may merely be used to indicate the node that is being used as the reference. In Figure 1.4 we have three nodes (a, b, c), with node a being the reference.

As the circulating charges pass through the battery, their potential energy is increased. In the case of Figure 1.4, the potential at node b is +6 V relative

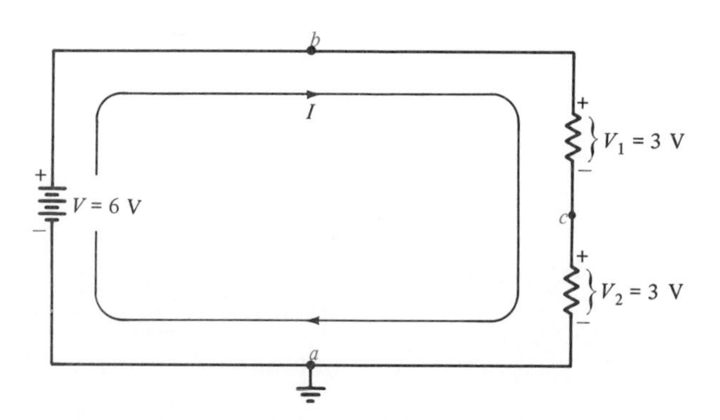

FIGURE 1.4. **Polarity of Voltage Drops across Series Resistors**

to that at a—that is, $V_{ba} = +6$ V. The first subscript indicates the node at which the potential is being considered and the second subscript, the reference node. The *potential difference* between a and b is said to be 6 V. If the reference node is ground, the second subscript may be omitted, or $V_a = +6$ V.

While the potential of the charged carriers rises as they go through the battery, the potential diminishes as they pass through the resistances, becoming $+3$ V at c and zero again at node a. The potential rise in going from a to b must then equal the potential drop (or voltage drop) in going from b to a.

Kirchhoff's voltage law (KVL) states that around any closed-loop path the sum of the voltage sources[1] and voltage drops is equal to zero. Considering Figure 1.4 and proceeding in a clockwise direction commencing at point a (ground), we see that a charge moving through the battery gains energy and that V may be taken to be positive. According to KVL, the charge's energy is diminished as it passes through the resistances.

$$V - V_1 - V_2 = 0$$
$$V - IR_1 - IR_2 = 0$$
$$V = I(R_1 + R_2) = I(R_{equivalent})$$
$$\frac{V}{I} = R_1 + R_2 = R_{equivalent}$$

From this result we see that the resistances in series are to be added to find their equivalent (R_{equiv}). Thus, in general,

$$R_{equiv} = R_1 + R_2 + R_3 + \cdots \tag{1.3}$$

EXAMPLE 1.2 A 12 V battery is applied to three resistors in series: $R_1 = 5\ \Omega$, $R_2 = 15\ \Omega$, and $R_3 = 20\ \Omega$. (See Figure 1.5.) What is the voltage drop across each resistor? What is the voltage at point a relative to ground? At point b? At point c?

[1] One sometimes finds E being used to designate emf, with V being reserved for voltage drops. We will usually follow the most general practice and use V for both, although in the chapter on motors we will employ E for emf.

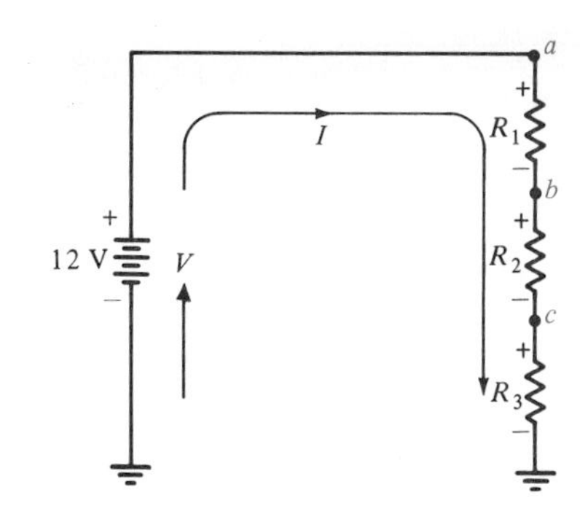

FIGURE 1.5

Solution:

$$I = \frac{V}{R_1 + R_2 + R_3}$$

$$I = \frac{12\text{ V}}{5\ \Omega + 15\ \Omega + 20\ \Omega} = 0.3\text{ A}$$

$$V_{R_1} = IR_1 = (0.3\text{ A})(5\ \Omega) = 1.5\text{ V}$$

$$V_{R_2} = IR_2 = (0.3\text{ A})(15\ \Omega) = 4.5\text{ V}$$

$$V_{R_3} = IR_3 = (0.3\text{ A})(20\ \Omega) = 6.0\text{ V}$$

$$V_a = +12\text{ V}$$

$$V_b = 12\text{ V} - 1.5\text{ V} = +10.5\text{ V}$$

$$V_c = 10.5\text{ V} - 4.5\text{ V} = +6.0\text{ V}$$

What is the equivalent (single) resistance if resistors are placed in parallel, as in Figure 1.6? In this case the total current furnished by the battery is I, and

$$I = I_1 + I_2 \tag{1.4}$$

It should be apparent that $V = V_1 = V_2$. Therefore,

$$\frac{V}{I} = R_{equiv} = \frac{V}{I_1 + I_2} = \frac{V}{(V/R_1) + (V/R_2)} \tag{1.5}$$

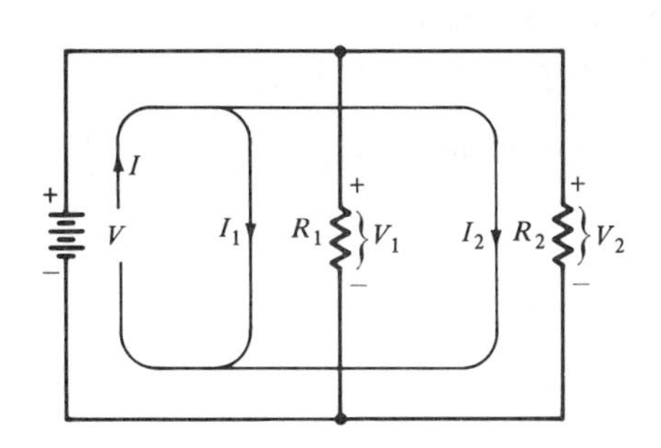

FIGURE 1.6. Current Through Parallel Resistors

from which it can be seen that

$$\frac{1}{R_{equiv}} = \frac{1}{R_1} + \frac{1}{R_2}$$

$$R_{equiv} = \frac{R_1 R_2}{R_1 + R_2} \tag{1.6}$$

and, in general, for resistances in parallel,

$$\frac{1}{R_{equiv}} = \frac{1}{R_1} + \frac{1}{R_2} + \frac{1}{R_3} + \cdots \tag{1.7}$$

In the latter case it proves convenient to define the quantity called **conductance,**[2] customarily represented by the letter G, as the reciprocal of resistance. Thus, for parallel circuits, conductances may be added directly, and the equivalent resistance is the reciprocal of the total conductance:

$$\frac{1}{R_{equiv}} = G_{equiv} = G_1 + G_2 + G_3 + \cdots \tag{1.8}$$

[2] The unit of conductance used to be the *mho*, which is ohm spelled backwards. Ampere per volt (as well as mA/V) was also used, particularly in Europe where it was considered disrespectful to have the name of Professor Ohm distorted in this manner. Today, the accepted SI unit is the siemens (abbreviated S); 1 S equals 1 A/V. While the symbol (as well as the abbreviation) for ohm is Ω (capital omega), that for the conductance is the inverted version, ℧, with S the abbreviation.

FOCUS ON PROBLEM SOLVING: Current Division Rule

When two resistors appear in parallel and the current into the circuit is known (Figure 1.42), the current through either one of the parallel branches may be easily determined by the following *current division rule*: Take the value of the *opposite* resistance, divide by the sum of the resistances, and multiply by the current into the parallel circuit. Proof of this useful rule is assigned as Problem 1.5.

EXAMPLE 1.3

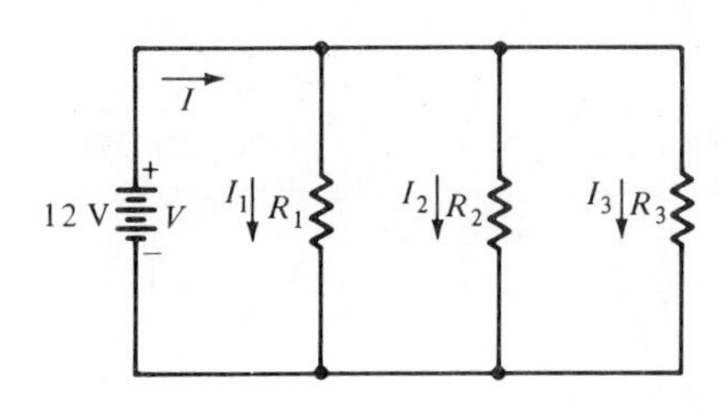

FIGURE 1.7

A 12 V battery is used to energize the following parallel combination of resistors: $R_1 = 5\ \Omega$, $R_2 = 12\ \Omega$, and $R_3 = 20\ \Omega$. (See Figure 1.7.) What value of current flows through each resistance? Using conductances, what is the single equivalent resistance that will lead to the same value of current?

Solution:

$$I = \frac{V}{R} = VG \qquad \text{since } G = \frac{1}{R}$$

$$I_1 = \frac{12\text{ V}}{5\ \Omega} = 2.4\text{ A}$$

$$I_2 = \frac{12\text{ V}}{10\ \Omega} = 1.2\text{ A}$$

$$I_3 = \frac{12\text{ V}}{20\ \Omega} = 0.6\text{ A}$$

$$G = G_1 + G_2 + G_3 = \frac{1}{5\ \Omega} + \frac{1}{10\ \Omega} + \frac{1}{20\ \Omega}$$

$$= 0.2\text{ S} + 0.1\text{ S} + 0.05\text{ S} = 0.35\text{ S}$$

$$I = VG = (12\text{ V})(0.35\text{ S}) = 4.2\text{ A}$$

$$R = \frac{1}{G} = \frac{1}{0.35\text{ S}} = 2.857\ \Omega$$

Check:

$$\frac{12\text{ V}}{2.857\ \Omega} = 4.2\text{ A}$$

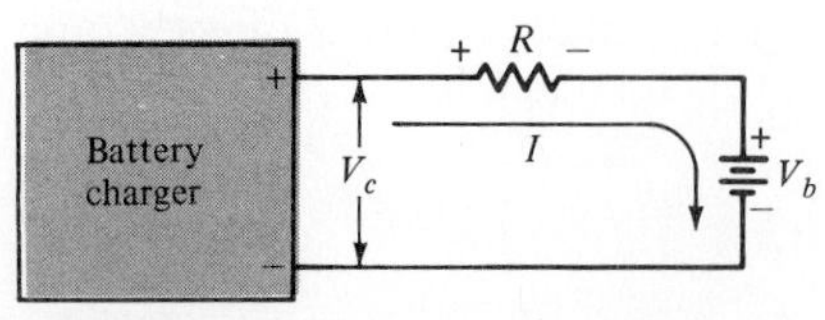

FIGURE 1.8. Multisource Circuit Current leaving the positive terminal of a device (battery charger) represents energy emission; current entering the positive terminal of a device (resistor and battery) represents energy absorption.

In a multisource circuit it is possible that a given source is absorbing rather than relinquishing energy. Such might be the case of a charging battery (Figure 1.8). In this instance, since I enters the positive terminal of the battery, the latter must be taken to represent a voltage drop. The KVL for this circuit would then take the form

$$+V_c - IR - V_b = 0$$

EXAMPLE 1.4

With regard to the circuit in Figure 1.9, determine the current using KVL.

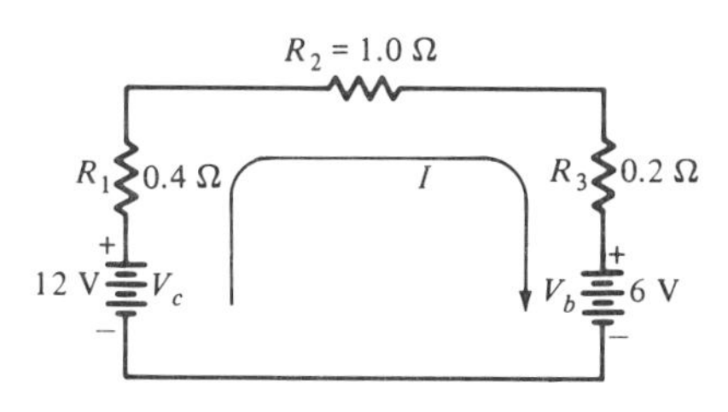

FIGURE 1.9

Solution:

$$+V_c - IR_1 - IR_2 - IR_3 - V_b = 0$$

$$+12 - 0.4I - 1.0I - 0.2I - 6 = 0$$

$$+12 - 6 - 1.6I = 0$$

$$1.6I = 6$$

$$I = \frac{6}{1.6} = 3.75\ \text{A}$$

In addition to Kirchhoff's voltage law (KVL), the solution of circuit problems is often aided by **Kirchhoff's current law** (KCL) to the effect that at every connecting point (node), the sum of the entering currents must equal the sum of the exiting currents. In solving a circuit problem, one *assumes* the currents to flow in arbitrarily chosen directions. If, upon solving the problem, it is found that a given current has a negative sign, the flow in that particular case is opposite to that assumed. The numerical value, however, remains the same.

The designations *circuit* and *network* are sometimes used interchangeably, but the latter usually implies a somewhat more complex interconnection than the former. Also, the word "network" carries with it an implication of generality such as when applied to a network theorem or network analysis.

FOCUS ON PROBLEM SOLVING: Resistive Network Problems

1. Designate currents through each network branch and arbitrarily assign a direction to each.
2. By using KCL, the number of unknown currents may possibly be reduced.
3. In using KVL, the sum of voltage sources and voltage drops about each closed loop must equal zero.
4. If the current is assumed to leave the positive terminal of a battery, the battery is to be considered a source of energy for that loop; if it enters the positive terminal, it is to be considered a drop. If a battery is common to two loops, it may be considered a drop for one loop and a source for the other (depending on the polarity encountered) if the current directions have been assumed to be opposed to one another. In a like manner, if a resistor is common to two loops with assumed currents being oppositely directed, for the loop under consideration it represents a drop proportional to that loop current and a source proportional to the opposing current.
5. The minimum number of independent loop currents equals the number of inner loops. An *inner loop* is one that does not contain another loop within itself. The resulting number of loop equations will then equal the number of unknowns, and their simultaneous solution will lead to values for all the unknowns.

6. If a particular solution results in a negative value for the unknown, the actual current is opposed to the direction assumed, but the computed magnitude is correct.

EXAMPLE 1.5

Referring to Figure 1.10, use loop equations to compute the current in each resistance.

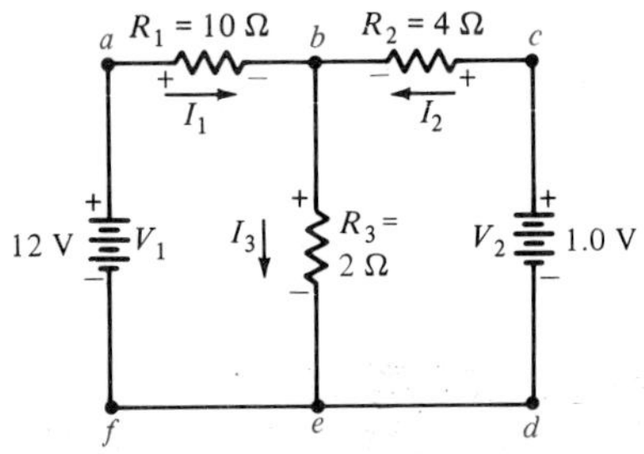

FIGURE 1.10

Solution:

1. Currents I_1, I_2, and I_3 are assumed to flow as shown.
2. Using KCL, we see that $I_3 = I_1 + I_2$.
3. Using KVL, we will have two loop equations.
4. Applying KVL to loop *fabe*, we have

$$+V_1 - I_1R_1 - I_3R_3 = 0$$
$$+V_1 - I_1R_1 - (I_1 + I_2)R_3 = 0$$
$$+12 - I_1(10) - (I_1 + I_2)(2) = 0$$
$$+12 - 12I_1 - 2I_2 = 0 \qquad \text{(I)}$$

For loop *dcbe*,

$$+V_2 - I_2R_2 - I_3R_3 = 0$$
$$+V_2 - I_2R_2 - (I_1 + I_2)R_3 = 0$$
$$+1 - I_2(4) - (I_1 + I_2)(2) = 0$$
$$+1 - 2I_1 - 6I_2 = 0 \qquad \text{(II)}$$

5. Multiplying Eq. I by 3 and subtracting Eq. II, we get

$$\begin{array}{r} 36 - 36I_1 - 6I_2 = 0 \\ -(1 - 2I_1 - 6I_2) = 0 \\ \hline 35 - 34I_1 \qquad = 0 \end{array}$$

$$I_1 = \frac{35}{34} = 1.029\ \text{A}$$

Substituting this solution into Eq. II, we obtain

$$1 - 2(1.029) - 6I_2 = 0$$

6. $$I_2 = \frac{1 - 2(1.029)}{6} = -0.176\ \text{A}$$

Thus, current I_2 flows in the opposite direction from that assumed. Rather than having the 1.0 V battery supply current, it is being charged instead.

We thus have

$$I_3 = I_1 + I_2 = 1.029\ \text{A} - 0.176\ \text{A} = 0.853\ \text{A}$$

EXAMPLE 1.6 Referring to Figure 1.11 use loop equations to compute the current in each resistance.

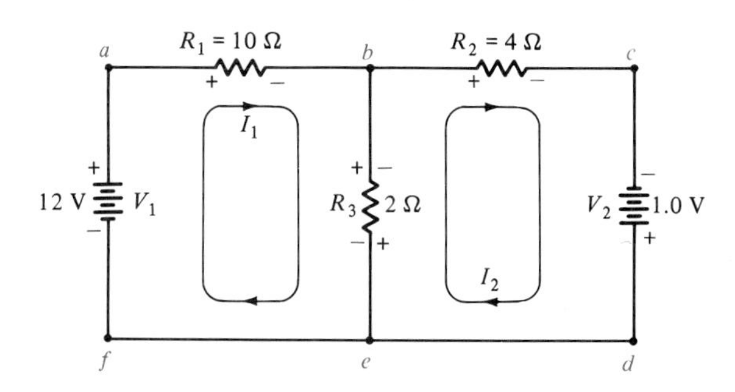

FIGURE 1.11

Solution:

1. Currents I_1 and I_2 are assumed to flow as shown.
2. The current through R_3 will be $I_1 - I_2$.
3. Using KVL, we will have two loop equations.
4. Applying KVL to loop *fabe*, we obtain

$$+V_1 - I_1R_1 - I_1R_3 + I_2R_3 = 0$$

$$+12 - I_1(10) - I_1(2) + I_2(2) = 0$$

$$+12 - 12I_1 + 2I_2 = 0 \qquad \text{(I)}$$

Applying KVL to loop *cdeb*, we obtain

$$+V_2 - I_2R_3 + I_1R_3 - I_2R_2 = 0$$

$$+1 - I_2(2) + I_1(2) - I_2(4) = 0$$

$$+1 + 2I_1 - 6I_2 = 0 \qquad \text{(II)}$$

5. Multiplying Eq. I by 3 and Eq. II, we obtain

$$\begin{array}{l} 36 - 36I_1 + 6I_2 = 0 \\ \underline{\ \ 1 + \ \ 2I_1 - 6I_2 = 0} \\ 37 - 34I_1 \qquad\ \ = 0 \end{array}$$

$$I_1 = \frac{37}{34} = 1.088 \text{ A}$$

Substituting this solution into Eq. I, we obtain

$$2I_2 = 12I_1 - 12 = 12(I_1 - 1)$$

6. $I_2 = 6(0.088) = 0.528$ A

$I_1 - I_2 = 1.088 \text{ A} - 0.528 \text{ A} = 0.56 \text{ A}$

1.3.3 Electrical Power

The passage of current through a resistance gives rise to a heating effect which is measured in terms of **power** (P) whose unit is the watt (W), that being equal to 1 joule/second (J/s). The joule, of course, is a unit of energy. The greater the voltage drop across a resistance, the greater will be the power dissipation (P_{dis}). Similarly, the greater the current through a resistance, the greater will be the dissipation. Therefore, assuming the current direction and polarity as in Figure 1.3, we have

$$P_{dis} = VI \tag{1.9a}$$

Using the Ohm's law relationships, we can derive two other expressions for the power dissipation:

$$P_{dis} = VI = (IR)I = I^2R \tag{1.9b}$$

$$P_{dis} = V\left(\frac{V}{R}\right) = \frac{V^2}{R} \tag{1.9c}$$

These can also be used to determine the magnitude of the power furnished by the source (P_s):

$$P_s = P_{dis} \tag{1.10}$$

EXAMPLE 1.7 Compute the power dissipation in a 10 Ω resistance connected to a 12 V battery. What is the current drawn from the battery? (Compute the power dissipation using the voltage drop and the current.)

Solution:

$$P = \frac{V^2}{R} = \frac{12^2}{10} = 14.4 \text{ W}$$

$$I = \frac{V}{R} = \frac{12}{10} = 1.2 \text{ A}$$

$$P = IV = 12 \times 1.2 = 14.4 \text{ W}$$

Whether a particular device dissipates power or is a source of power depends on the current direction and the polarity of the voltage. We have seen that current entering the positive terminal of a battery causes power to be withdrawn from the charging source; current leaving the positive terminal of a battery constitutes a source of power.

1.4 ■ ALTERNATING CURRENT

1.4.1 Principles

In the mid-nineteenth century electricity was a technological development looking for a use; coal furnished heat, waterfalls furnished power, and gas provided light. The first significant encroachment upon this triumvirate arose when Edison developed the electric light bulb. He used current obtained by rotating coils of wire within the fields furnished by magnets.

For example, consider Figure 1.12. If a strip of wire moves through the magnetic field in the direction away from the forefinger, it will cause positive charges to move upward. If the wire is oppositely directed, the induced voltage will be such the charges will move downward. If, rather than a single segment of wire, we now consider a single turn rotating about its axis (Figure 1.13), the current is seen to move from right to left in the upper loop and left to right in the lower one. After a turn through 180° the current directions through segments 1 and 2 will have been reversed. We may depict the current graphically as in Figure 1.14. The current periodically reverses direction, and the transition between the two states is rather gradual. In other words we have an alternating current, which is the "natural" output of a rotating electrical generator.

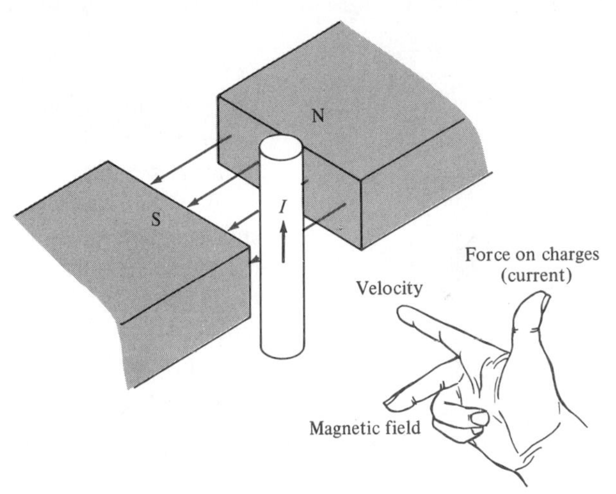

FIGURE 1.12. Right-Hand Rule Moving through a perpendicular magnetic field (directivity taken from N to S), a segment of wire (assumed part of an externally closed circuit) will have induced a voltage that gives rise to a current.

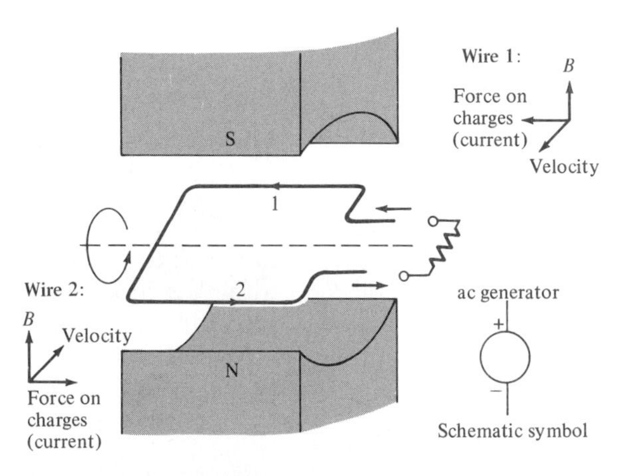

FIGURE 1.13. Elementary Generator Construction Rotating coil in a magnetic field will give rise to a current in a completed circuit.

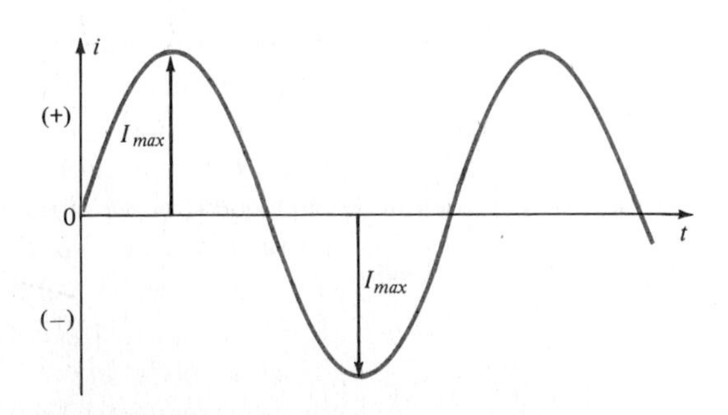

FIGURE 1.14. **Alternating Current as a Function of Time**

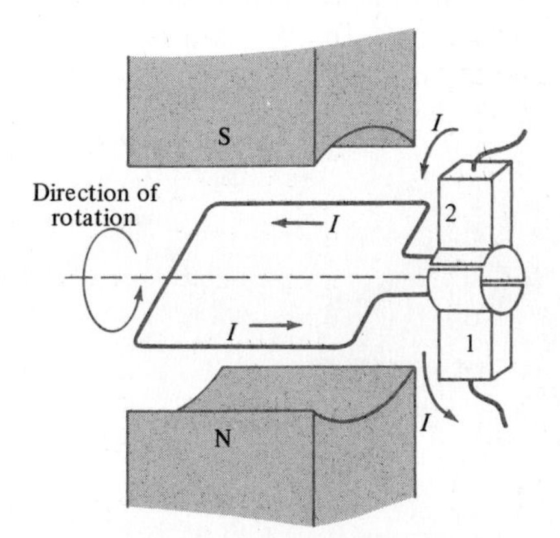

FIGURE 1.15. **Split Ring Commutator** Alternating current is converted into pulsating dc.

As a function of time, the instantaneous value of the voltage (v), in terms of the maximum voltage (V_{max}), is expressed in the form[3]

$$v = V_{max} \sin \omega t \qquad (1.11)$$

The angular frequency (ω) is expressed in radians per second and equals $2\pi f$, where f is the rotational frequency expressed in hertz (Hz), which equals 1 cycle per second.

If the connection to the two ends of the winding were made through a split ring commutator (Figure 1.15) with carbon "brushes" riding atop them, the current would always exit from brush 1 and enter through brush 2. In other words, the alternating current would have been changed to direct current—but not a *constant* direct current since (in an idealized case with a multiturn coil) the magnitude would vary in the manner shown in Figure 1.16. It was this type of **pulsating direct current** that Edison employed.

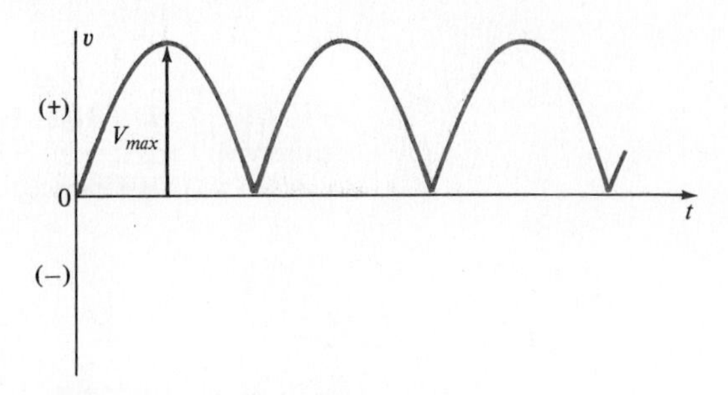

FIGURE 1.16. **Pulsating dc as a Function of Time**

The current was generated at a power station and transmitted along wires to the place where it was to be utilized. Some of the power was dissipated within the resistance of the power lines, constituting a loss. Since $P_{source} = IV$ and $P_{loss} = I^2R$, note that the loss is diminished as the line resistance is made smaller. Having made R as small as practicable, the loss goes up as I^2. Thus, to diminish the losses, we also wish I to be small. But to transmit a given

[3] In general, lowercase letters are used to indicate instantaneous values, and uppercase letters, to indicate constant values.

amount of power at small values of I, we must make the voltage (V) very large.

When using dc, a practical upper limit on the transmitted voltage at that time was set (by questions such as safety) at about 100 V, and if the voltage drop between the generating plant and the user was to be minimized, a generating plant was limited to a distribution range of about 1 mile. Thus, Edison's power grid required numerous power stations.

If ac is utilized, one has access to the use of transformers (which will not operate on dc). By using a modest generator voltage, the ac may be stepped up to a very large value, transmitted over large distances with minimized losses, and then stepped down to a safe level before entering home or factory. Thus, while a dc system required a generating plant within 1 mile of the user, an ac system allowed power to be generated at such distant points as Niagara Falls and "shipped" overland to New York City with relatively little loss.

To show mathematically the advantage of power transmission at high voltages, consider the following:

$$P_{generated} = VI$$

$$P_{loss} = I^2R$$

To minimize the ratio, we have

$$\frac{P_{loss}}{P_{generated}} = \frac{I^2R}{VI} = \frac{I}{V}R$$

from which we see that the larger we make V and the smaller I, the smaller will be the transmission loss.[4]

The development of dc and ac electricity furnishes excellent examples of two types of technical people: inventors, such as Thomas Edison, and engineers, such as the Serbian-born Nikola Tesla. Edison was not overly impressed with Tesla; a trial-and-error man like Edison had little confidence in the calculations of an academic type like Tesla. But Edison did employ him, after he rescued Edison from an embarrassing predicament by repairing the burned-out generator aboard the *S.S. Oregon*. They parted, however, when Edison refused to pay him a promised $50,000, dismissing him with the following: "Tesla, you don't understand our American humor."

Edison was a dc man; Tesla favored ac. The General Electric Company, a successor to the Edison Electric Company, was a dc company, while Westinghouse, a company that was the result of cooperation between Tesla and George Westinghouse, was an ac establishment. By way of supporting its program to prove ac to be more dangerous than dc, General Electric prevailed upon its home state of New York to use ac in the operation of its electric chair.

[4]One might justifiably ask why the transmission loss is I^2R rather than V^2/R. While the former implies increased loss at higher currents, the latter implies increased loss at higher voltages. The V that appears in this case is *not* the voltage applied to the transmission line but rather the voltage drop between the ends of the transmission lines. Therefore, $V = IR$, and we again have $P = I^2R$.

Electrical power systems in the nineteenth century utilized dc primarily because such concepts were much easier to understand and manipulate than ac. It was Tesla who first found the means by which ac could be used to drive motors without the bothersome commutators. But "informed" opinion of people like Edison refused to be budged from their antagonism toward ac. It was the much greater versatility of ac that finally prevailed, however.

The standardization of power systems at 60 Hz is also due to Tesla. He found that lower frequencies required the use of more iron in electrical machinery, which proved to be a large cost factor; higher frequencies required less of the costly iron, but there was a drop in efficiency. He settled on 60 Hz as an optimum compromise. (In much of Europe 50 Hz is used, and corresponding transformers tend to be physically larger for the same electrical capacity.)

1.4.2 Voltage and Current Sources

Some of the most troublesome aspects of learning a new field of endeavor arise from a nomenclature that is not to be interpreted literally. One such concept has to do with the classification of ac (as well as dc) sources as being either a *voltage* or a *current* source. Offhand, one might expect that on connecting a load across the output terminals of an electrical source, a voltage will give rise to a current. However, consider a common source of direct current, the battery. Since any source of emf may be depicted as an ideal source in series with a resistance, a flashlight cell may take the form shown in Figure 1.17.

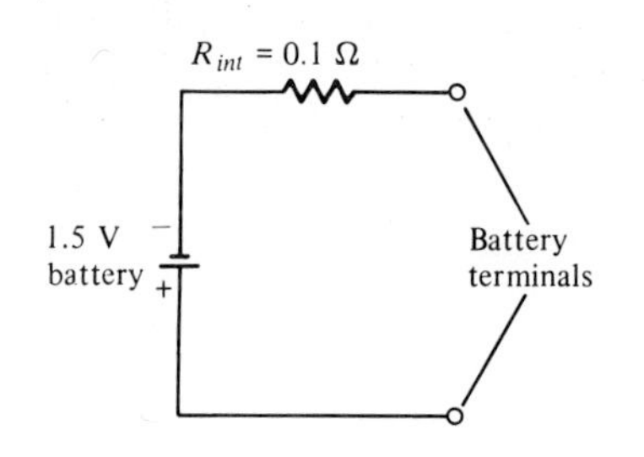

FIGURE 1.17. Flashlight Cell Circuit Equipment

Let us presume that the internal resistance (R_{int}) is 0.1 Ω. If we connect a light bulb whose internal resistance is 50 Ω across the battery, the total current withdrawn from the idealized battery will be about 0.03 A. The power expended within the load will be 0.045 W. As the battery becomes depleted, it remains a 1.5 V source; what is altered is the internal resistance. Thus if R_{int} has increased to 50 Ω, the total current withdrawn is 0.015 A, the voltage across the load is 0.75 V, and the power supplied to the load will be 0.0112 W. This is why one is always admonished to measure a battery's voltage *under load* conditions. If little or no current is being drawn from the battery, there is but a small drop across R_{int}, and the terminal voltage of a "spent" battery is about the same as that of a "good" battery.

Now, rather than considering R_{int} to be changing, let us consider the influence of a changing *load resistance*. Also, rather than a dc source, we consider an ac one. (The same concept, however, also holds for dc sources.)

When an alternating voltage is applied to a resistance, the latter will be heated due to current dissipation. While the value of the applied voltage is continually changing its value, we may conveniently specify the magnitude of an alternating voltage by determining what dc voltage, applied to the same resistor, will yield a like amount of heating. We may then specify that the alternating voltage has an *effective value* equal to that constant dc voltage. This value is also called **rms,**—root, mean, square—a term that arises from the mathematical procedure used in its determination. (See Eq. 1.17.)

In Figure 1.18 we consider an alternating source whose effective value (V_{eff}) is 10 V. The internal resistance of this generator is taken to be 10 Ω.

Across this generator we place a load resistor whose value we may vary. Let us compute the current *and* voltage delivered to the load when the load resistance is changed from a value of 1 kΩ to a value of 2 kΩ:

Current through load	Voltage across load
At 1000 Ω, 10 V/1010 Ω = 0.0099 A	9.90 V
At 2000 Ω, 10 V/2010 Ω = 0.00498 A	9.96 V

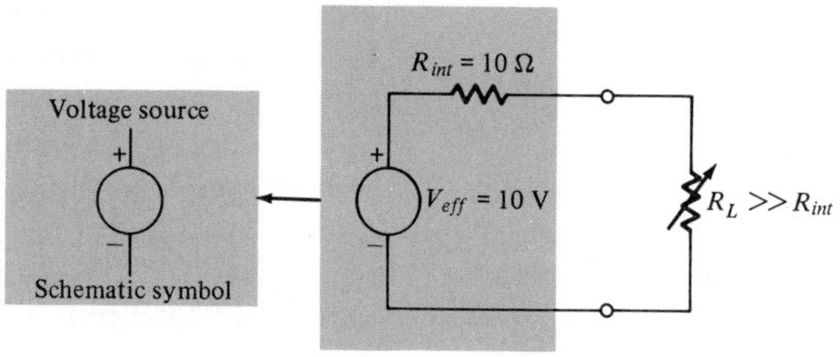

FIGURE 1.18. Voltage Source $R_L >> R_{int}$

We see that while there is a dramatic change in the delivered current, the voltage across the load shows very little change. We would speak of such a source as being a *constant voltage source*. Note the rather restrictive use of the word *constant* in this connection; since we are dealing with an alternating voltage, the voltage *is* changing its value from moment to moment, as with any alternating voltage.

By constant voltage source (or, simply, voltage source), we mean that as the value of the load changes, this type of source shows very little variation in the *delivered* effective voltage. In general, whenever the value of the load resistance is much larger than the internal resistance of the generator being used, we speak of such a source as being a **voltage source**.

Let us now go to the opposite extreme and consider an ac generator capable of producing V_{eff} = 10 V, but having an internal resistance of 100 kΩ. Again we compute the delivered current and voltage when the load resistance changed from 1 to 2 kΩ (Figure 1.19):

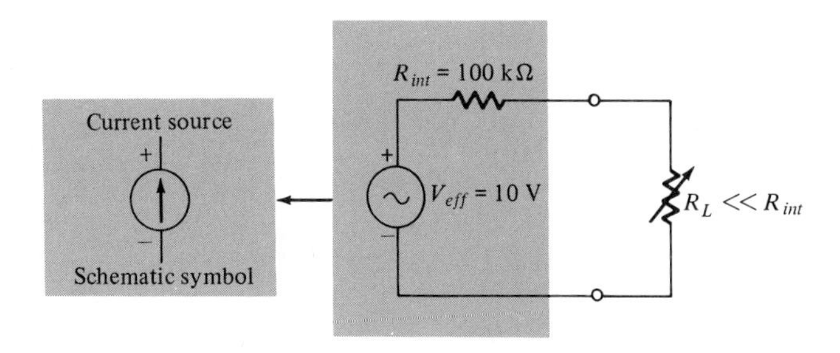

FIGURE 1.19. Current Source $R_L << R_{int}$

Current through load	Voltage across load
At 1000 Ω, 10 V/101,000 Ω = 9.9×10^{-5} A	Approx. 0.1 V
At 2000 Ω, 10 V/102,000 Ω = 9.8×10^{-5} A	Approx. 0.2 V

We now see that the situation is reversed; the current shows a relatively small variation under changing load conditions (~1%), and the voltage shows a significant change (100%). We now speak of this source as being a **current source**—that is, a generator is considered a current source when the internal resistance is much greater than the load resistance.

The generator we utilized in this example is not necessarily to be looked upon as some piece of rotating electrical machinery; in electronics work it usually is not. Most frequently, it will be some electronic generator that, as we have seen, may be broadly placed into one of two categories: It is a voltage source if its internal resistance is small compared to the load that is placed upon it, and it is a current source if its internal resistance is large compared to the load resistance. An *ideal* voltage source has zero internal resistance, while an *ideal* current source has infinite internal resistance.

EXAMPLE 1.8

A 1 kΩ resistor is connected to a 10 V ideal voltage source. (See Figure 1.20.)

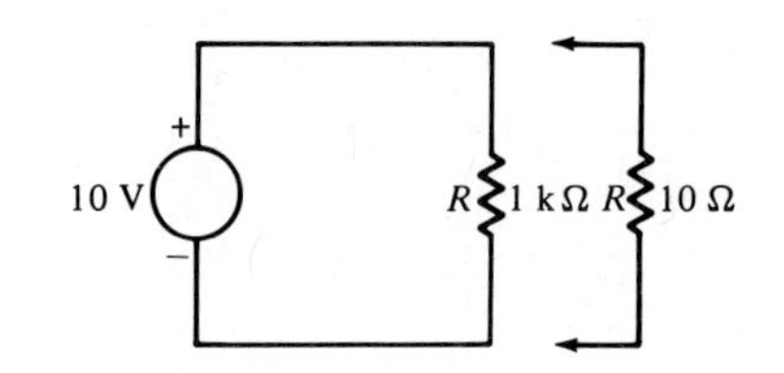

FIGURE 1.20

a. What is the current?
b. If an additional 10 Ω resistor is connected in parallel, what is the total current drawn from the source?

Solution:

a. $$I = \frac{V}{R} = \frac{10\ \text{V}}{10^3\ \Omega} = 0.01\ \text{A} = 10\ \text{mA}$$

b. For parallel resistances (designated as $R_{||}$), we have

$$R_{||} = \frac{1}{(1/R_1) + (1/R_2)} = \frac{R_1 R_2}{R_1 + R_2} = \frac{(10^3)(10)}{10^3 + 10} = 9.9\ \Omega$$

$$I = \frac{V}{R_{||}} = \frac{10\ \text{V}}{9.9\ \Omega} = 1.01\ \text{A}$$

An ideal voltage source, with its zero internal resistance, can furnish a limitless amount of current without the terminal voltage changing.

EXAMPLE 1.9

A 1 kΩ load is connected to a 1 mA ideal current source. (See Figure 1.21.)

a. What is the voltage across the load?
b. If an additional 2 kΩ resistance is placed in series with the first resistance, what is the voltage across each resistance?

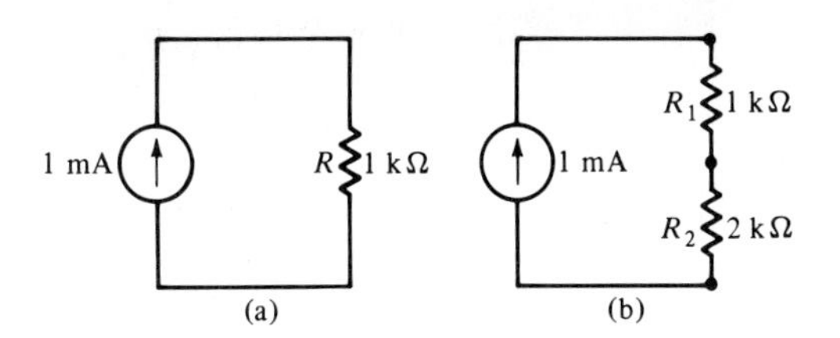

FIGURE 1.21

Solution:

a. $V = IR = (10^{-3}\ \text{A})(1.0 \times 10^3\ \Omega) = 1.0\ \text{V}$

b. $V_{1\,\text{k}\Omega} = (10^{-3}\ \text{A})(1.0 \times 10^3\ \Omega) = 1.0\ \text{V}$

$V_{2\,\text{k}\Omega} = (10^{-3}\ \text{A})(2.0 \times 10^3\ \Omega) = 2.0\ \text{V}$

$V_{terminal} = 1.0\ \text{V} + 2.0\ \text{V} = 3.0\ \text{V}$

An ideal current source furnishes the stated current irrespective of the magnitude of the load.

1.4.3 Superposition Principle

In a linear electrical network, the voltage across any element and the current through it can be regarded as being due to the current and voltage sources of the network acting independently. The same principle is applicable to other disciplines such as mechanical vibrations and fluid motion wherein the excitations x_1 and x_2 produce responses y_1 and y_2, respectively, and the application of $x_1 + x_2$ produces a response $y_1 + y_2$. Conversely, systems that behave in this manner are said to be linear.

The *superposition principle* can be used to solve complex network problems. The response is obtained by having each independent source act separately, while the other independent sources are suppressed. The suppression of a voltage source arises when it is replaced by a short circuit; an open circuit constitutes the suppression of a current source.

EXAMPLE 1.10

Referring to Figure 1.22, determine the current through R_3 using the superposition principle. What is the voltage across this resistor?

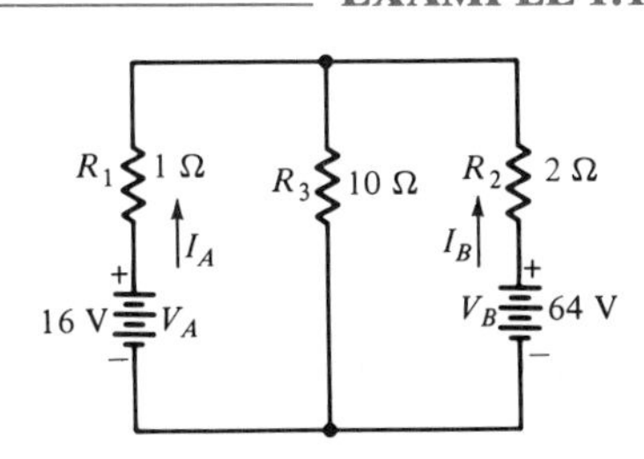

FIGURE 1.22

Solution: By setting $V_B = 0$, R_2 and R_3 appear in parallel. Using Eq. 1.6, we obtain

$$R_{||} = \frac{R_2 R_3}{R_2 + R_3} = \frac{(2)(10)}{2 + 10} = \frac{20}{12}\ \Omega$$

By using the current division rule, the current through R_3 due to V_A is

$$I'_A = \left(\frac{V_A}{R_{||} + R_1}\right)\left(\frac{R_2}{R_2 + R_3}\right) = \left[\frac{16}{(20/12) + 1}\right]\left(\frac{2}{2 + 10}\right)$$
$$= \left[\frac{(16)(12)}{20 + 12}\right]\left(\frac{2}{12}\right) = 1\ \text{A}$$

By setting $V_A = 0$, R_1 and R_3 appear in parallel:

$$R_{||} = \frac{R_1 R_3}{R_1 + R_3} = \frac{(1)(10)}{1 + 10} = \frac{10}{11}\ \Omega$$

By again using the current division rule, the current through R_3 due to V_B is

$$I'_B = \left(\frac{V_B}{R_{||} + R_2}\right)\left(\frac{R_1}{R_1 + R_3}\right) = \left[\frac{64}{(10/11) + 2}\right]\left(\frac{1}{1 + 10}\right)$$
$$= \left[\frac{(64)(11)}{10 + 22}\right]\left(\frac{1}{11}\right) = 2\ \text{A}$$

The total current through the 10 Ω resistor is 1 A + 2 A = 3 A. The voltage drop across the 10 Ω resistor is 3 A × 10 Ω = 30 V.

EXAMPLE 1.11

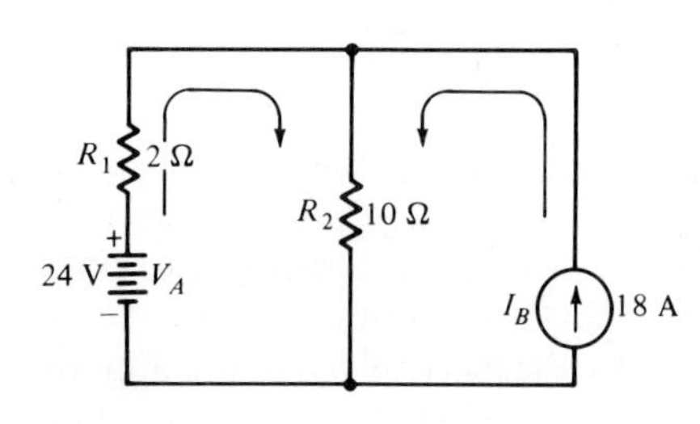

FIGURE 1.23

Referring to Figure 1.23, determine the current through the 10 Ω resistor. What is the voltage across this resistor? What is the current through the 2 Ω resistance?

Solution: Suppress I_B by opening the current generator branch:

$$I_{10} = \frac{V_A}{R_1 + R_2} = \frac{24}{2 + 10} = \frac{24}{12} = 2\ \text{A}$$

Suppress V_A by shorting out the voltage source, leaving 2 Ω in parallel with 10 Ω. Knowing the current into the parallel branch (18 A), we may immediately apply the current division rule to obtain the current in the 10 Ω resistance:

$$I_{10} = \frac{R_1}{R_1 + R_2} I_B = \frac{2}{2 + 10} 18 = \frac{18}{6} = 3\ \text{A}$$

The total current in the 10 Ω resistor is 2 A + 3 A = 5 A. The voltage drop across this same resistor is 5 A × 10 Ω = 50 V.

Using KVL for the left-hand loop, we obtain

$$+V_A - 2I_A - 50 = 0$$
$$+24 - 2I_A - 50 = 0$$
$$I_A = \frac{24 - 50}{2} = -13\ \text{A}$$

The battery is being charged by the current source.

FOCUS ON PROBLEM SOLVING: Superposition Theorem

Suppress all but one excitation source in the network and compute the circuit response with just that one active source. Ideal voltage sources are suppressed by replacing them with a short, while ideal current sources are suppressed by replacing them with an open circuit. The internal resistances of real energy sources must always be retained in the circuit. The actual circuit response is equal to the summation of the individual responses obtained in the above manner.

The superposition theorem is applicable only to linear relationships and thus cannot be used to determine power. For example,

$$(I_1 + I_2)^2R \neq I_1^2R + I_2^2R$$

1.4.4 Load Matching

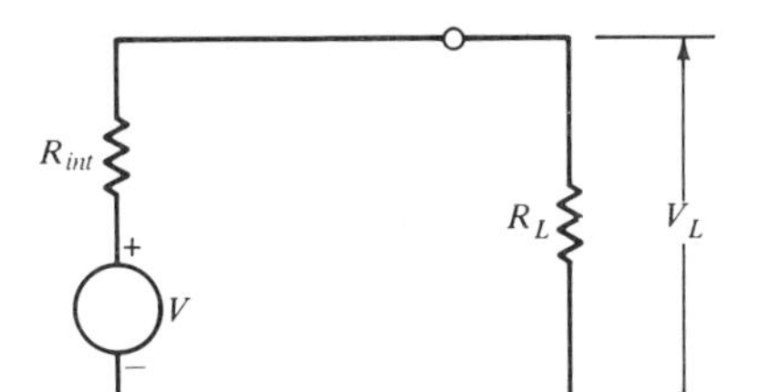

FIGURE 1.24. **Optimized Power Source** $R_L = R_{int}$

We have seen two extreme conditions, one in which the current shows little change with load resistance variation and another that has the voltage changing but little with load resistance. Another related problem has to do with that relationship between internal generator resistance and that load resistance which leads to a maximum *power* transfer between load and source.

For example, if the internal resistance of a generator is R_{int}, what should be the value of the load resistance in order that maximum power be delivered to the load? Referring to Figure 1.24, we take the source voltage to be V and the load resistance to be R_L. The current through the load is

$$I = \frac{V}{R_{int} + R_L} \tag{1.12}$$

The voltage across the load is

$$V_L = IR_L = \frac{VR_L}{R_{int} + R_L} \tag{1.13}$$

Calling the delivered[5] power P, we have

$$P = IV_L = \frac{V^2R_L}{(R_{int} + R_L)^2} \tag{1.14}$$

To maximize this result relative to R_L, we have a max-min problem. Taking the derivative of P relative to R_L and setting the result equal to zero, we solve for the necessary R_L:

$$\frac{dP}{dR_L} \equiv 0 = \frac{V^2}{(R_{int} + R_L)^2} - 2\frac{V^2R_L}{(R_{int} + R_L)^3} \tag{1.15}$$

[5] The delivered power is that which is delivered at the generator terminals. R_{int} lies within the generator and cannot be separated from the source, V.

Solving this equation shows that the condition is met when $R_L = R_{int}$—that is, the maximum transfer of power takes place when the load resistance equals the internal resistance of the generator.

We have mentioned in the previous section that the magnitude of the alternating source can be specified in terms of that value of constant dc which will give the same heating effect. In terms of I_{max}, the peak value of the alternating current, the corresponding direct current would satisfy the equation

$$I^2R = \frac{I_{max}^2 R}{2} \tag{1.16}$$

We called this dc equivalent the *effective* value of the ac. Thus,

$$I_{eff} = \frac{I_{max}}{\sqrt{2}} = \frac{I_{max}}{1.414} = 0.707 I_{max} \tag{1.17}$$

The mathematical reasoning used to derive this result proceeds in the following manner: Since the heating effect is independent of the current's direction, we discount the periodic change in direction by squaring the instantaneous value, compute its average over one cycle, and then take the square root. Thus, the effective value is the rms value:

$$i = I_m \cos \omega t$$

where $2\pi f = \omega = 2\pi/T$

I_m = maximum value of the ac waveform

f = frequency

T = period $(1/f = T)$

The instantaneous power (p) in R is

$$p = i^2R$$

Since the averages of the square for each quarter-cycle are identical, we integrate and average over $T/4$:

$$P = \frac{1}{T/4}\int_0^{T/4} I_m^2 R \cos^2 \omega t \, dt = \left(\frac{4I_m^2 R}{T}\right)\int_0^{T/4} \cos^2 \omega t \, dt$$

$$= \left(\frac{4I_m^2 R}{T}\right)\left(\frac{t}{2} + \frac{\sin 2\omega t}{4\omega}\right)_0^{T/4} = \left(\frac{4I_m^2 R}{T}\right)\left(\frac{T}{8}\right) = \frac{I_m^2 R}{2} \equiv I_{eff}^2 R$$

$$I_{eff} = I_{rms} = \frac{I_m}{\sqrt{2}} = 0.707 I_m$$

This same numerical relationship holds for alternating voltage. Thus, the ordinary house voltage, specified as being 120V, represents its effective value. A dc voltage of this value would have the same heating effect. The maximum value of the household voltage must then be $V_{eff}/0.707$, or about 170 V.

It is important to note that this unique numerical relationship between I_{eff} and I_{max} (as well as between V_{eff} and V_{max}) holds *only* for sinusoidal waveforms. Other waveforms will be subject to different numerical values.

There is also another rather useful number associated with sinusoidal waveforms. The average value of a sine wave, of course, is zero; the integral of its excursion in the negative direction equals that in the positive direction. But in the case of pulsating dc, such as shown in Figure 1.16, the average value is finite and (not surprisingly) is called the **average value**—that is, the value of dc voltage (and current) that is recoverable when the pulsations are removed.

If each pulse is half a sinusoid (note the wave as a whole is not a sine wave because it does not change its sign), the relation between I_{max} and I_{av} is

$$I_{av} = 0.637 I_{max} \tag{1.18}$$

and, similarly,

$$V_{av} = 0.637 V_{max} \tag{1.19}$$

(See Eq. 1.26 for a justification of this numerical result.)

We will subsequently see that when the pulsations are smoothed out, V_{av} is the value of the dc that is recoverable from the pulsating dc. Why, in this case, does the average differ from the effective (rms) value? It differs because the effective value is related to the heating effect, which is proportional to i^2 (or v^2). Therefore, the time that the waveform spends above I_{av} (or V_{av}) makes a larger contribution to the heating than does the portion below I_{av} (or V_{av}). Hence, the effective value is larger than the average value.

Power in *resistive* ac circuits is given by the expression

$$P = V_{rms} I_{rms} \tag{1.20}$$

or by either of the two alternative forms, involving V_{rms} and R in one case and I_{rms} and R in the other.

1.4.5 Transformers

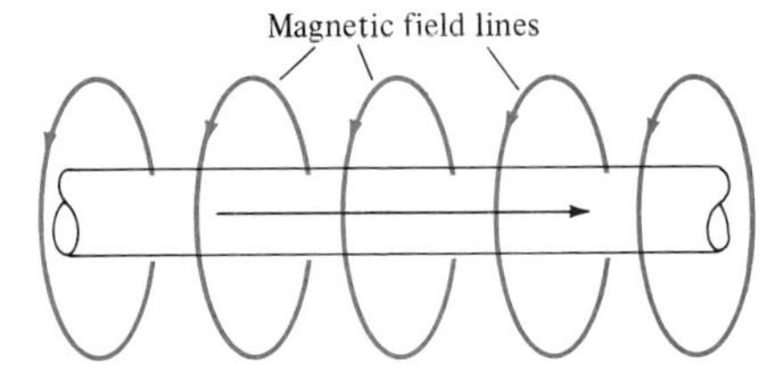

FIGURE 1.25. Magnetic Field Lines Surrounding Current-Carrying Wire

The fact that current through a wire creates about that wire a magnetic field constitutes an early discovery in electrical theory (Figure 1.25). The inverse effect, the creation of current by means of a magnetic field, has a more recent history. Early workers would wrap a coil around a permanent magnet and look for an induced voltage in the coil. Subsequently, it was discovered that such a voltage would only be generated if there was a *changing* magnetic field. Thus, like an old-time magician, if you "wave" a permanent magnet past a closed electrical conductor, you may detect an electrical current. Such "magnet waving" is in essence what is being performed in a rotating electrical generator.

Since it is a changing magnetic field that is necessary to create an induced voltage in a coil, rather than mechanical motion, one can instead pass a *changing current* through the coil of an electromagnet. Another coil (the secondary coil) in the vicinity will be subject to the resulting change in magnetic field (originating as a result of the primary coil) and have induced within it a voltage that, if one provides a closed secondary circuit, will give rise to a current. We may schematically depict this situation as in Figure 1.26. The magnetic field created by the changing primary current will be spread over a considerable volume of space, and since only the fraction that is coupled to the secondary coil will be effective, the "transformer" depicted in Figure 1.26 will be a very inefficient one. If, on the other hand, we start with an iron core (Figure 1.27) and wrap both windings about this core, the magnetic field will be confined to a small volume and result in tight coupling and efficient transformer action. (What we mean by efficient action is the presence of electrical power in the secondary circuit closely equal to that in the primary.)

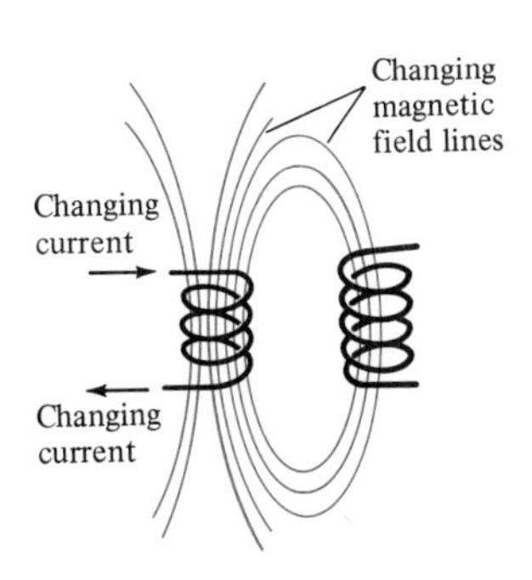

FIGURE 1.26. Transformer Operation A changing current through the primary coil creates a changing magnetic field that, upon threading through the secondary coil, creates an emf in the latter.

The magnetic field *intensity* depends on the magnitude of the changing

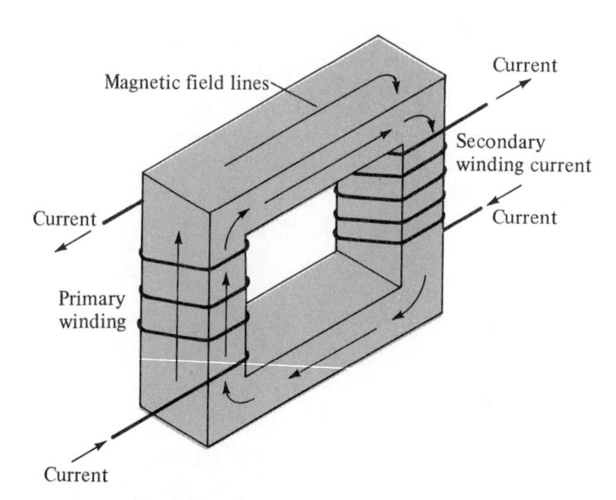

FIGURE 1.27. High-Efficiency Transformer Magnetic flux due to changing primary current is conducted via the iron core transformer to the secondary coil.

primary current *and* on the number of turns constituting the primary winding. If the number of turns constituting the secondary winding is twice the number of primary turns, the *voltage variation* across the secondary winding (ideally) will be equal to twice the variation applied to the primary winding.

Have we gotten something for nothing? Not at all; recall that power is equal to *voltage* times *current*, and although we have doubled the voltage, the current variation in a completed secondary circuit will be found to have a value about half that of the primary current variation. If N_1 and N_2 are, respectively, the number of primary and secondary turns on the transformer, then, as we have said, the voltage ratio is equal to N_2/N_1. Thus

$$V_{eff_2} = \frac{N_2}{N_1} V_{eff_1} \tag{1.21}$$

$$I_{eff_2} = \frac{N_1}{N_2} I_{eff_1} \tag{1.22}$$

But Eq. 1.22 may seem like a strange result; one would expect that the secondary current would depend on the load resistance (R_L) one connects across the secondary. It does, but R_L also determines what the primary current will be, and therein lies another useful aspect of transformer action.

In addition to altering the voltage (or the current), transformers may also be used to optimize the transfer of power between a source and a load. Recall from Section 1.4.4 that this transfer is most efficiently accomplished by having the source resistance (R_{int}) equal the load resistance (R_L). Often one is faced with a situation in which the two differ significantly.

Since

$$V_2 = \frac{N_2}{N_1} V_1$$

and

$$I_2 = \frac{N_1}{N_2} I_1$$

we have

$$R_L = \frac{V_2}{I_2} = \frac{N_2}{N_1} V_1 \left(\frac{N_2}{N_1 I_1}\right) = \left(\frac{N_2}{N_1}\right)^2 \frac{V_1}{I_1}$$

but $V_1/I_1 = R_{int}$, and

$$R_{int} = \left(\frac{N_1}{N_2}\right)^2 R_L \qquad (1.23)$$

By properly adjusting the transformer's turns ratio, R_L may be made to "look" to the generator as a resistance equal to R_{int} and vice versa.

EXAMPLE 1.12 A 10 V rms ac source has an internal resistance of 100 Ω and is connected to a 10 Ω load resistance. (See Figure 1.28.)

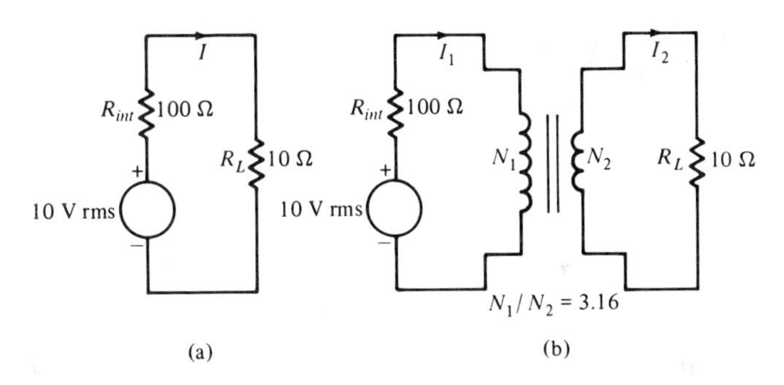

FIGURE 1.28

a. What would be the power delivered to the load?
b. What would be the power dissipated within the generator?
c. If a transformer with a primary to secondary turns ratio of 3.16 were connected between source and load, what would be the power delivered to the load and the power dissipated within the generator? (Assume an ideal transformer.)

Solution: For Figure 1.28a, we have

$$I = \frac{V}{R_{int} + R_L} = \frac{10 \text{ V}}{(100 + 10)\ \Omega} = 0.091 \text{ A}$$

a. $$P_L = I^2 R_L = \left(\frac{10}{110}\right)^2 (10) = 0.0826 \text{ W}$$

b. $$P_{int} = I^2 R_{int} = \left(\frac{10}{110}\right)^2 (100) = 0.826 \text{ W}$$

c. In Figure 1.28b, the generator "sees" a *total* resistance of

$$R_{tot} = R_{int} + \left(\frac{N_1}{N_2}\right)^2 R_L = 100 + (3.16)^2(10) \simeq 200\ \Omega$$

$$I_1 = \frac{V}{R_{tot}} = \frac{10\text{ V}}{200\ \Omega} = 0.05\text{ A}$$

$$P_{int} = I^2 R_{int} = (0.05)^2(100) = 0.25\text{ W}$$

$$P_{tot} = I^2 R_{tot} = (0.05)^2(200) = 0.5\text{ W}$$

$$P_L = P_{tot} - P_{int} = 0.5\text{ W} - 0.25\text{ W} = 0.25\text{ W}$$

Check:

$$I_1 = 0.05\text{ A}$$

$$I_2 = \frac{N_1}{N_2} I_1 = (3.16)\ (0.05) = 0.158\text{ A}$$

$$V_1 = \frac{(N_1/N_2)^2 R_L}{R_{int} + (N_1/N_2)^2 R_L} V = \frac{100}{200}(10) = 5\text{ V} \quad \text{(voltage applied to the primary)}$$

$$V_2 = \frac{N_2}{N_1} V_1 = \frac{1}{3.16}(5) = 1.58\text{ V} \quad \text{(secondary voltage)}$$

$$P_1 = V_1 I_1 = 5(0.05) = 0.25\text{ W} \quad \text{(power applied to the primary)}$$

$$P_2 = V_2 I_2 = 1.58(0.158) = 0.25\text{ W} \quad \text{(power in the secondary circuit)}$$

Note that while the voltage and current are modified by the turns ratio, resistances are modified by the *square* of the turns ratio.

FOCUS ON PROBLEM SOLVING: Transformer Ratios

Taking N_1 to be the number of primary turns and N_2 to be the number included in the secondary, currents and voltages are transformed as follows:

$$\frac{V_2}{V_1} = \frac{N_2}{N_1}$$

$$\frac{I_2}{I_1} = \frac{N_1}{N_2}$$

A load (R_L) connected across the secondary appears across the primary to have a resistance (R_{in}) of

$$R_{in} = \left(\frac{N_1}{N_2}\right)^2 R_L$$

A step-down transformer ($N_1 > N_2$) makes the resistive load look bigger across the primary; a step-up transformer ($N_1 < N_2$) makes it look smaller.

In the idealized case the secondary power equals the primary power. Of course, in practice there are always some losses, such as those due to flux leakage, and the secondary power is somewhat less than that in the primary. But the efficiency of transformers is generally in excess of 90%.

When dealing with a transformer, current passes into one primary terminal and out the other. At the same time (presuming a closed circuit across the secondary), current passes into one secondary terminal and out the other. It is often important to identify these respective terminals. In Figure 1.29, the corresponding terminals may be identified by dots as shown.

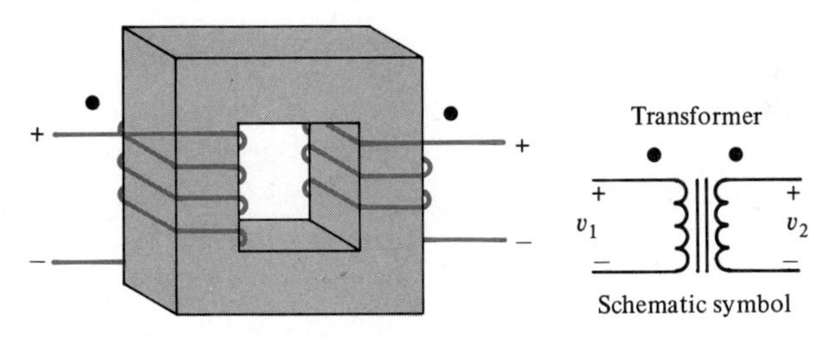

FIGURE 1.29. Polarity Relation Between Transformer Leads The instantaneous polarity on the transformer's primary leads to the indicated voltage polarity on the secondary winding. The corresponding polarities are identified by dots.

Often one terminal of the primary is grounded, and the secondary voltage can be made to be either 180° in phase or out of phase with the voltage across the primary, depending on which of the secondary terminals is also grounded. (See Figures 1.30a and 1.30b.) We will make use of this concept when discussing an automotive ignition system in a later chapter.

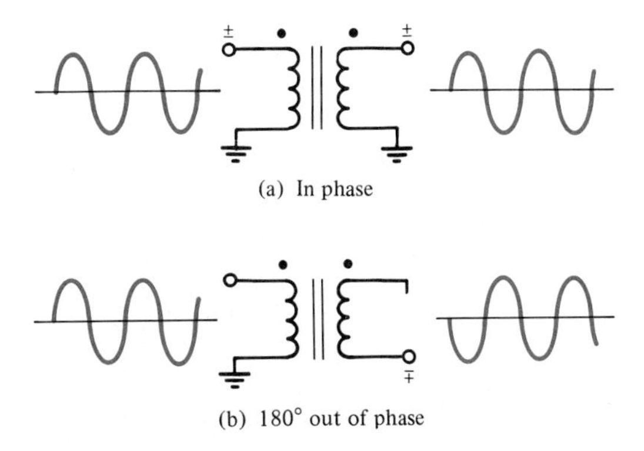

FIGURE 1.30. Phase Transformation (a) Polarity of the left terminal (measured relative to ground) always corresponds to the polarity of the right terminal (also measured relative to ground). (b) Polarity of the left terminal is always opposite to that of the right terminal.

1.5 ■ ELECTRICAL MEASUREMENTS

1.5.1 d'Arsonval Meters

Current through a conductor creates a surrounding magnetic field whose magnitude is proportional to the magnitude of the current. The interaction between current-induced magnetic fields and some external fixed magnets can serve as an indicator of the magnitude of current flow and can be measured by an **ammeter**, the basic instrument used to measure current.

The operating mechanism of one type of current-reading meter—a *d'Arsonval meter*—is shown in Figure 1.31. We see a current-carrying wire wound about a pivoted suspension, bracketed on one side by a north magnetic pole (N) and a south pole (S) on the other. Current flow through the coil will create a north pole perpendicular to the plane of the coil, as indicated by the dashed arrow. This north pole will be repelled by the fixed north pole and attracted by the fixed south pole, and the pivoted suspension will respond to these forces by turning. This torque, however, is opposed by the spring suspension. Therefore, there will be a rotation until the two torques balance; increasing the current will increase the angular deflection. One can, therefore, calibrate the needle position in terms of the current necessary to cause a particular deflection.

The strength of the magnetic field generated by the coil depends on the number of turns and the current passing through the coil. Therefore, the sen-

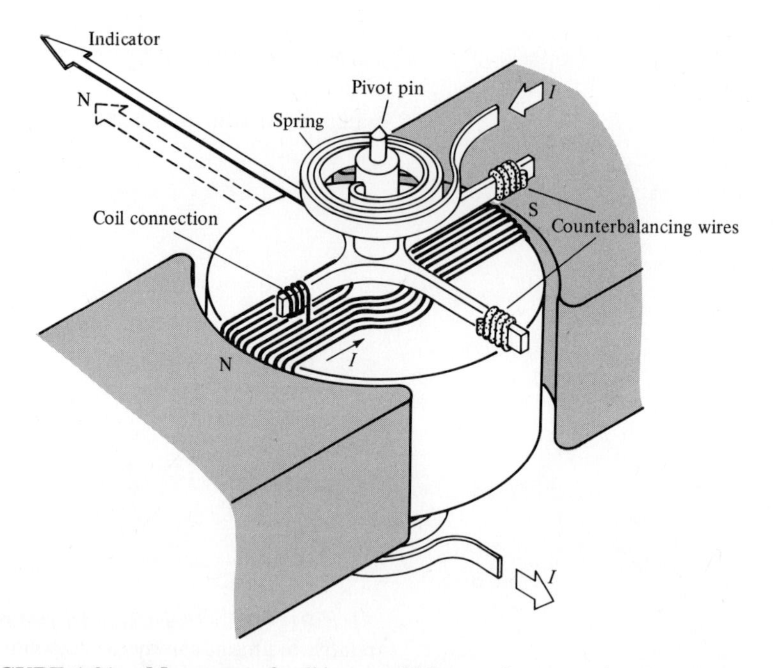

FIGURE 1.31. **Movement of a d'Arsonval Meter** Current to be measured passes through a movable coil, creating a proportional magnetic field which interacts with the fixed poles, N and S.

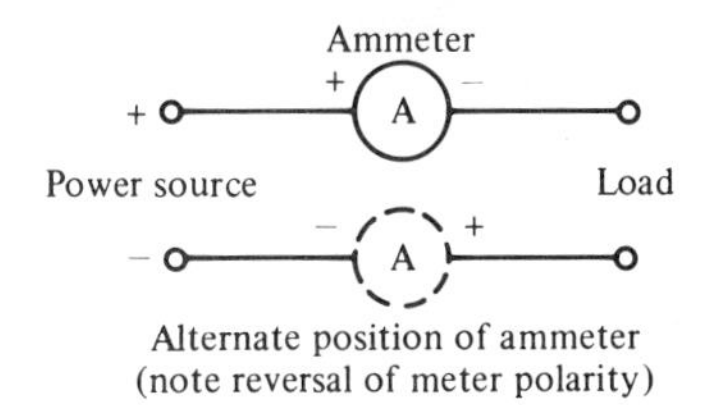

(a) Placement of an ammeter between power source and load

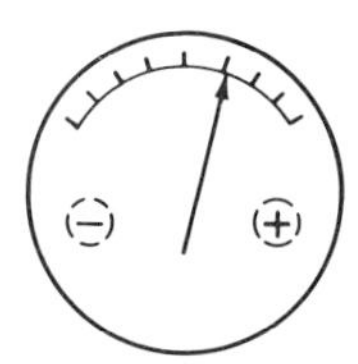

(b) Observed polarity of ammeter terminals

FIGURE 1.32. Current-Reading Meter in Series with Power Lead

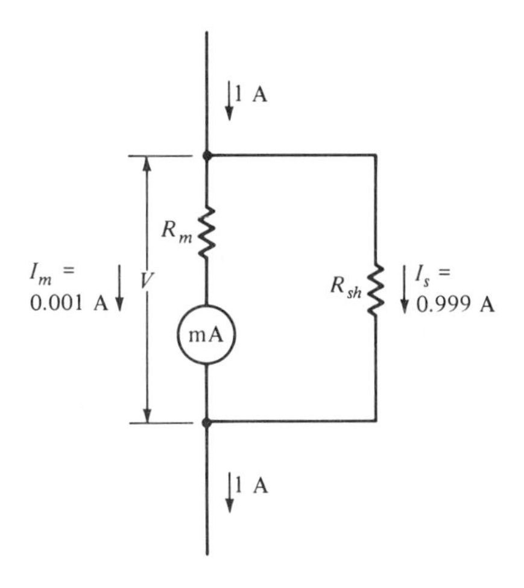

FIGURE 1.33. Ammeter Scale Expansion Shunt resistance expands the current range of a current-reading meter.

sitivity of a d'Arsonval movement (as it is called) can be increased by increasing the number of turns since less current will be required to achieve a given deflection. When the volume available for the winding is limited, a finer wire must be used, with an attendant increase in meter resistance.

The sensitivity also depends on the spring tension, and therein lies a precaution. If such a meter is overloaded, the winding may rupture, and for this reason, fuses are often employed. It may also happen that the meter may be overloaded without such a catastrophic failure and give rise to a rather insidious condition: The overload may have heated the suspension springs (through which the current passes) and permanently changed their temper. This change may grossly affect the accuracy of the meter. A periodic meter calibration is the only assurance that the meter has not been subjected to misuse.

One should also be cautious about accepting the accuracy of d'Arsonval meters when mounted on an iron relay-rack panel. The meter accuracy also depends on the magnetic field, which may be considerably altered because of the proximity of so much iron. Some meters are provided with shielding that minimizes such interaction.

The most common current-reading meters are designed to read full scale when 1 mA (10^{-3} A) flows through the coil. The second most common movement that is generally available requires 50 μA (50×10^{-6} A) for full-scale deflection. The winding resistance in the first case may be of the order of 50 to 100 Ω and of the order of a 1000 Ω in the second case. As shown in Figure 1.32, current meters are inserted in series with one or the other of the power leads activating the device. There is a polarity to be observed with such meters. When facing the front of the meter, the terminal on the left is the negative one, and the one on the right should thus be connected to the positive source potential.

The basic meter movements may easily be modified to measure larger values of current. Referring to Fig 1.33, for example, if we desire to use a 1 mA movement to measure current in the range 0–1 A, a shunt resistance (R_{sh}), whose value will pass the excess current, is placed in parallel with the meter. Thus, at full-scale reading, 0.001 A is passed through the meter and 0.999 A through the shunt.

To determine the required value of R_{sh}, one must know the value of the meter resistance (R_m). Since one always has the same voltage across R_m and R_{sh}, we have

$$\frac{V}{R_m} = I_m \qquad \frac{V}{R_{sh}} = I_{sh}$$

$$I_m R_m = I_{sh} R_{sh}$$

$$R_{sh} = \frac{I_m}{I_{sh}} R_m \tag{1.24}$$

If one desires to extend the meter's full-scale reading by a factor n (for example, for a 1 mA meter to read 1 A, $n = 1000$), $I_{sh} = (n - 1)I_m$. Then

$$R_{sh} = \frac{R_m}{n - 1} \tag{1.25}$$

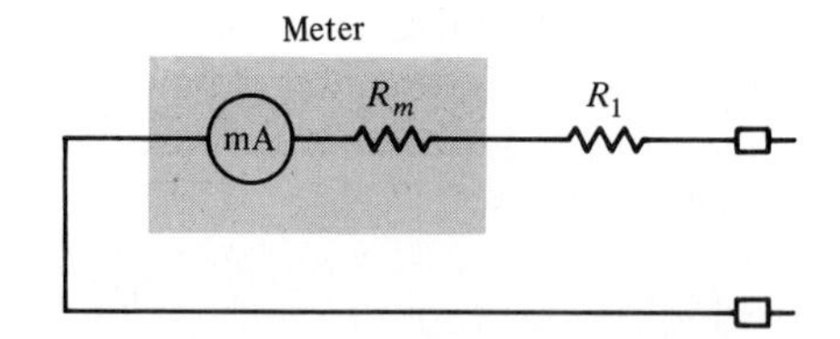

FIGURE 1.34. **Voltmeter** A milliammeter is converted into a voltmeter by the addition of a large series resistance.

1.5.2 Voltmeters

Basic current-measuring devices may easily be converted into voltage-reading meters, or *voltmeters*, by placing a large resistance in series with the meter, as shown in Figure 1.34. By way of a specific example, suppose we wish to convert a 1 mA meter into a 0–1 V meter. With 1 V applied to the terminals, we want 1 mA to flow through the meter:

$$R_m + R_1 = \frac{1\ \text{V}}{10^{-3}\ \text{A}} = 1000\ \Omega$$

$$R_1 = 1000 - R_m$$

If R_m is of the order of tens of ohms, the accuracy is not greatly affected by taking R_1 to be 1000 Ω. In fact, the 1 mA movement is often referred to as having a sensitivity of 1000 Ω/V.

The sensitivity of the 50 μA movement is 20,000 Ω/V, and we might choose to use it in place of the 1 mA meter to measure voltages. Since with proper design each might be made to read 100 V full scale, for example, is there any difference in their performance? There is, and it is a very important point to realize.

The series resistance for a 100 V meter using a 1 mA movement is 100,000 Ω or 2 MΩ with a 50 μA movement. In Figure 1.35, we have 1 mA passing through a 100 kΩ resistor, which gives rise to a 100 V drop across the resistor. We wish to measure this voltage with each of the voltmeters. When we place the 1 mA meter across the 100 kΩ resistance, we effectively have two 100 kΩ resistances in parallel; with 1 mA passing through the equivalent 50 kΩ, the *measured* voltage will be 50 V!

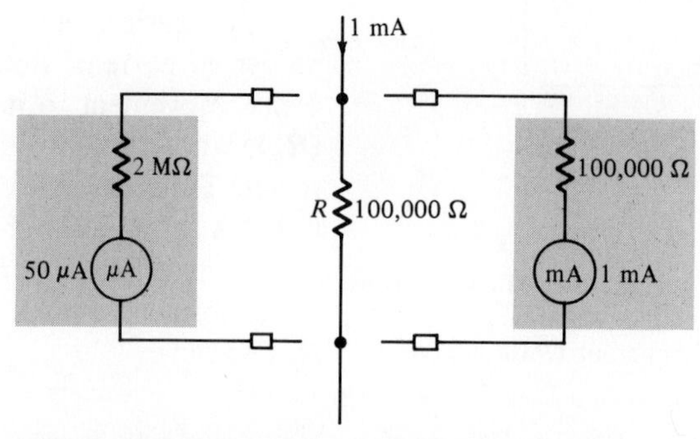

FIGURE 1.35. **Effect of Meter Loading on Voltage Measurement** Using a high-sensitivity meter (at the left) and a lower-sensitivity one (at the right) to measure a voltage drop across the center resistor.

Making the measurement with the 50 μA meter, we have $2 \times 10^6\ \Omega$ in parallel with 100 kΩ. The equivalent parallel resistance is about 95 kΩ. With 1 mA passing through 95 kΩ, we obtain a reading to 95 V.

In each of these cases an error arises because the measuring circuit disturbs the circuit being measured. This disturbance can be minimized by making the input resistance of the measuring circuit large by comparison to the impedence of the circuit being measured.

It takes electrical energy to set the meter movement into motion. It would help greatly if this electrical energy were supplied by an independent source rather than at the expense of the circuit being measured. Such supplementary energy is available if the meters are used in conjunction with transistors. We will defer a consideration of such electronic voltmeters until a later chapter.

EXAMPLE 1.13

A 1 mA basic movement is to be made into a 0–100 V voltmeter. What should be the value of the series resistance?

Solution:

$$R_1 = \text{sensitivity [ohms/volt]} \times \text{desired full-scale reading [volts]}$$
$$R_1 = 1000\ \Omega/\text{V} \times 100\ \text{V} = 100{,}000\ \Omega = 100\ \text{k}\Omega$$

1.5.3 Ohmmeters

Having looked at how current and voltage might be measured, we turn to a consideration of resistance measurement. Of course, we could put an ammeter in series with the resistance and a voltmeter to measure the drop across the resistance and determine R by use of Ohm's law, but there are more convenient ways of measuring resistance. Referring to Figure 1.36, we see the simplest form of *ohmmeter*, which consists of a battery, a milliammeter, and a resistor, connected in series with the unknown resistance (R_x).[6] The resistor (R_1) is made variable so that the meter can be made to read full scale when the input terminals are shorted—that is, when the unknown resistance is equal to zero. If a 100 Ω, 1 mA movement is used, together with a 4.5 V source (which could conveniently consist of three flashlight cells), R_1 should be 4400 Ω. This resistance can be made up of a 3 kΩ fixed resistor and a 1500 or 2000 Ω adjustable resistance. (The meter resistance has been acknowledged in computing R_1.) If we insert the unknown R_x and the meter reads half scale, then $R_x = 4.5$ kΩ. (The total is now 9 kΩ, and with $V = 4.5$ V, $I = 0.5$ mA.) A resistance lower than 4500 Ω will produce a reading greater than half scale; R_x greater than 4500 Ω will produce a reading less than half scale.

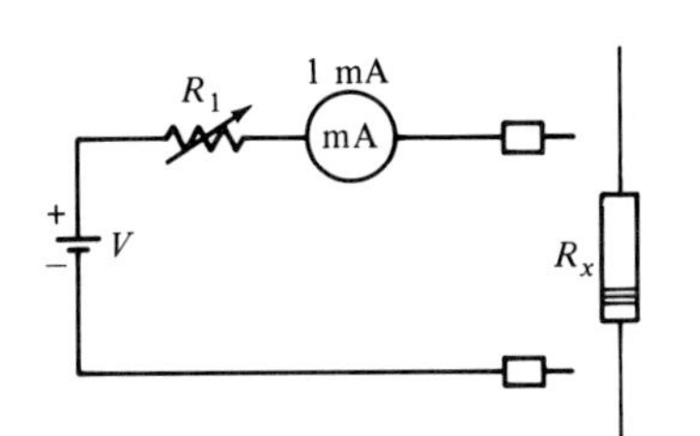

FIGURE 1.36. **Ohmmeter** A milliammeter in series with battery V and an adjustable resistor R_1 constitutes a simple ohmmeter.

Figure 1.37 shows what the ohmmeter calibration might look like. Note that the scale is reversed from what one normally expects: 0 Ω is at the extreme right and infinite resistance at the extreme left. (There are circuits, however, that invert the scale.) The resistance values get compressed at the right and left, and, for example, with the scale as designed, resistance values of tens or hundreds of ohms would give readings close to zero. The designed scale might be expected to give fairly accurate resistance readings in the range 1000 to, perhaps, 100,000 Ω. There are modifications of this simple circuitry that allow accurate ohmmeter measurements at both high and low values of resistance.

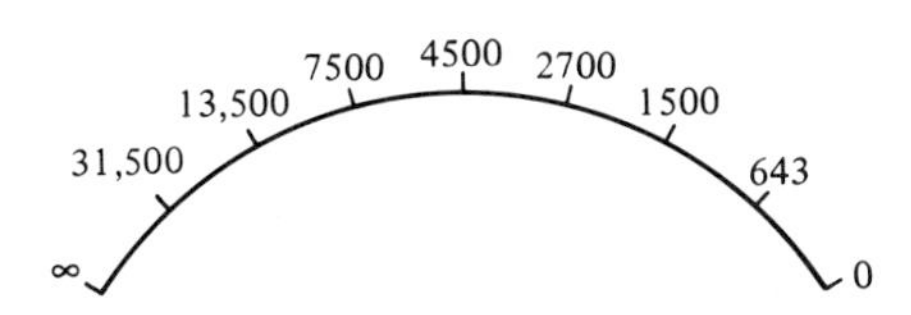

FIGURE 1.37. **Example of Nonlinear Scale of a Simple Ohmmeter**

It will have been noted that one and the same meter movement has been used to measure current, voltage, and resistance. By the use of appropriate switches and resistances, all these functions can be combined into one instrument, the **multimeter.**

[6] When making resistance measurements, the circuit should not be energized; you will probably damage the ohmmeter if it is. Also, the use of an ohmmeter to measure the continuity of devices such as sensitive meter fuses should be employed with caution.

1.5.4 Electrodynamometers

The application of sinusoidal ac to a meter of the d'Arsonval type will yield a reading of zero since such meters measure the *average* value. However, the application of a pulsating dc of the form obtained from a generator employing a commutator (Figure 1.38a) will yield the average value, which is 0.637 times the peak value:

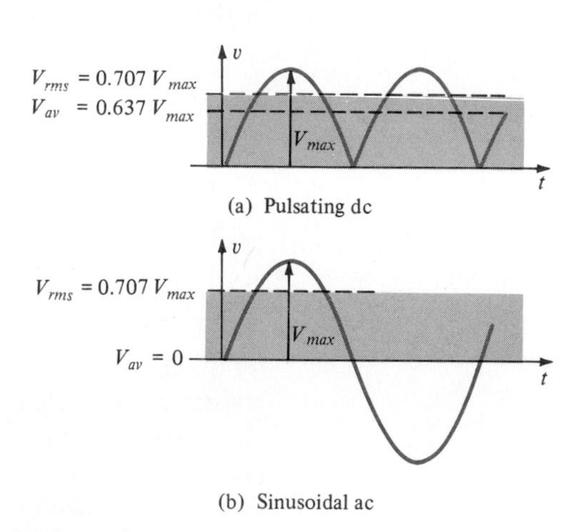

FIGURE 1.38. RMS and Average Values

$$\begin{aligned} V_{av} &= 4\frac{1}{T}\int_0^{T/4} V_{max} \sin \omega t \, dt = -\frac{4V_{max}}{\omega T} \cos \omega t \Big|_0^{T/4} \\ &= -\frac{4V_{max}}{2\pi}\left[\cos\left(\frac{2\pi}{T}\right)\left(\frac{T}{4}\right) - \cos 0\right] \\ &= -\frac{4V_{max}}{2\pi}(0 - 1) = \frac{2}{\pi} V_{max} = 0.637 V_{max} \end{aligned} \tag{1.26}$$

This average value is to be compared with the effective (rms) value of a sinusoidal wave:

$$V_{rms} = 0.707 V_{max} \tag{1.27}$$

As previously noted, the average value represents the constant dc value equivalent to the given waveform, and the rms value represents that value of dc that yields the same heating effect as the given waveform. Refer again to Figure 1.38; since the second pulse in the pulsating dc waveform is merely the inversion of the negative pulse in the sinusoidal waveform, the rms value of these two waveforms is the same. (The heating effect does not depend on the direction of the current.) That the average value is not the same resides in the fact that since power is proportional to i^2, the segment above the average makes a disproportionate contribution compared to the segment below.

If the permanent magnet pole pieces of the d'Arsonval movement are replaced by *electromagnets* and the current is made to pass through both the stationary and moving coils, the developed torque will be proportional to the square of the current. If the meter deflections are then calibrated in terms of the square root of the current squared, we have a meter whose deflection yields the rms value of the applied current. (Note that the stationary and rotating coils may both be made to reverse their polarity simultaneously with ac applied; hence the torque direction remains unchanged.) Such a meter is called an *electrodynamometer*. It can be used to measure either ac or dc (Figure 1.39).

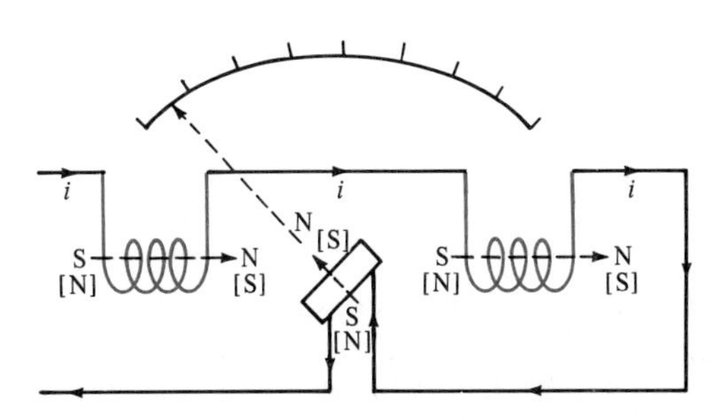

FIGURE 1.39. Dynamometer Magnetic polarities are shown as they prevail with indicated current direction. When current reverses, magnetic polarities are as indicated by [ ]; note that the torque remains directed clockwise irrespective of current direction.

Electrodynamometers use air-core coils, which is one of their major disadvantages: high power consumption. The current to be measured must not only pass through the movable coil but must also provide the field flux. To get sufficiently strong fields, either the source must provide significant current or the coil must contain a large number of turns.

As with the d'Arsonval movement, a series resistor converts the electrodynamometer into a voltmeter. But the sensitivity might be much less—that is, 10–30 Ω/V compared to 1000–20,000 Ω/V for the d'Arsonval movements.

The electrodynamometer can be used to measure power by applying the current to be measured to the fixed coils and a current proportional to the voltage to the moving coil. In a subsequent chapter we will consider other methods used in making ac measurements.

SUMMARY

1.2 Batteries

- A battery consists of two or more cells arranged in series and/or parallel combinations.
- Secondary cells are rechargeable; primary cells are not.

1.3 Direct Current

- Voltage, current, and resistance are related through Ohm's law: $V = IR, I = V/R$, or $R = V/I$.
- Classical current is directed from the positive terminal of a source; electron flow is directed from the negative terminal of a source.
- According to Kirchhoff's voltage law, around any closed loop the sum of voltage sources (taken to be positive) and voltage drops (taken to be negative) must equal zero.
- According to Kirchhoff's current law, at any circuit juncture point, termed a node, the sum of the incoming current must equal the outflow.
- For resistors in series, $R_{equiv} = R_1 + R_2 + R_3 + \cdots$. For resistors in parallel,

$$\frac{1}{R_{equiv}} = \frac{1}{R_1} + \frac{1}{R_2} + \frac{1}{R_3} + \cdots$$

- Current times voltage equals power: $P = IV$. Ohm's law also leads to two other expressions for power: $P = V^2/R$ and $P = I^2R$.

1.4 Alternating Current

- A coil rotating in a uniform magnetic field will give rise to a sinusoidally varying current—known as alternating current—in a circuit connected to the ends of the coil.
- The use of a split ring commutator in conjunction with a coil rotating in a magnetic field leads to a current that varies in magnitude but does not reverse direction. Such is the nature of pulsating dc.
- An ideal voltage source has zero internal resistance.
- An ideal current source has infinite internal resistance.
- According to the superposition principle, the total response of a linear electrical network is the summation of the responses due to each source acting independently.
- Transformers make possible a much more versatile and economical utility power system. Their proper operation requires changing currents. For this reason ac, despite its greater degree of mathematical complexity and more involved implementation, is preferred over the simpler dc system.
- The rms (effective) value represents the ac equivalent to the corresponding numerical value of dc in terms of power. For sinusoidal waveforms,

$$V_{rms} = 0.707 V_{max}$$

- Disregarding the periodic changes in direction, $V_{av} = 0.637\ V_{max}$ for sinusoidal waveforms.
- For transformers,

$$\frac{V_2}{V_1} = \frac{N_2}{N_1} \qquad \frac{I_2}{I_1} = \frac{N_1}{N_2} \qquad R_{in} = \left(\frac{N_1}{N_2}\right)^2 R_L$$

where subscript 1 refers to the primary (input) and 2 to the secondary (output).

1.5 Electrical Measurements

- The basic electrical meter measures current and is placed in series with the measured circuit.
- The inclusion of a substantial resistance in series with a current-reading meter converts it into a voltmeter. Voltmeters are placed in parallel with the circuit whose voltage is to be measured.
- An appropriate combination of dc voltage, a current-reading meter, and resistances can be made to serve as an ohmmeter.
- An electrodynamometer may be used to measure either ac or dc.

KEY TERMS

alternating current (ac): current that periodically reverses its direction

ammeter: a current-measuring device

average value: dc equivalent of an ac voltage or current

battery: a combination of electrical cells

cell: two dissimilar metals immersed in an electrolyte; a fundamental source of current

conductance: reciprocal of resistance (Unit is the siemens, formerly the mho.)

current source: an electrical source whose internal resistance is greater than the load resistance

direct current (dc): uni-directional current

electromotive force (emf): (alternatively, potential difference or voltage) a source of energy that creates a separation between the two polarities of charged carriers (The chemical separation in a battery and the separation by the motion of a conductor through a magnetic field represent two possible mechanisms that may be employed.)

Kirchhoff's current law (KCL): law of electricity stating that the sum of all currents entering and leaving a circuit node is zero

Kirchhoff's voltage law (KVL): law of electricity stating that the sum of voltage sources and voltage drops around any closed-loop circuit is zero

multimeter: a current-measuring instrument that, by means of switches and a variety of resistances, is made to measure voltage and resistance as well as current

Ohm's law: mathematical relationship between voltage, current, and resistance

ohmmeter: a resistance-measuring device

power: energy per unit time (Unit is the watt.)

primary cell: an electrical cell with consumable materials; not rechargeable

pulsating dc: unidirectional current, cyclically varying in magnitude

rms: root, mean, square value; representing the effective current or voltage; numerically equal to the respective dc values that would yield the same heating effect

secondary cell: a rechargeable electric cell

transformer: device for altering ac voltage, current, and resistance

voltage source: an electrical source whose internal resistance is less than the load resistance

voltmeter: a voltage-measuring device

SUGGESTED READINGS

Kordesch, Karl V., and Klaus Tomantschger, "Primary Batteries," THE PHYSICS TEACHER, vol. 19, no. 1, pp. 12–21, Jan. 1981. *Further information on primary cells, including a short history.*

Morehouse, C. K., R. Glicksman, and G. S. Lozier, "Batteries," PROC. IRE, vol. 46, no. 8, pp. 1462–1483, Aug. 1958. *A review of batteries.*

Cheney, Margaret, TESLA: MAN OUT OF TIME. New York: Dell, 1981. *A very readable account of the early controversy between ac and dc proponents.*

Bowers, Brian, A HISTORY OF ELECTRIC LIGHT AND POWER. London: Peter Peregrinus, Ltd., 1982. *Development of electric power, with a rather British slant.*

Monopoli, R. V., "Fundamentals of Electrical Networks," in INTRODUCTION TO ENGINEERING, M. Glorioso and Francis Hill, Jr. (eds.), Chap. 7. Englewood Cliffs, N.J.: Prentice-Hall, 1975. *Electrical fundamentals.*

Cooper, Wm. David, ELECTRONIC INSTRUMENTATION AND MEASURING TECHNIQUES. Englewood Cliffs, N.J.: Prentice-Hall, 1970. Chap. 4, *dc-indicating instruments, including details of the mechanical construction and suspension systems of a moving-coil meter,* and pp. 78–84, *discussion of the variety of ohmmeter types, including numerical examples.*

Colles, G. W., "The Metric vs. the Duodecimal System," TRANS. ASME, vol. 18, pp. 492–611, 1897. *A historical review of the pathological opposition hoisted against the adoption of the metric system (first conceived in 1791), including a photograph (page 557) comparing a fabric woven using English units and the apparently defective fabric that resulted from the use of metric units.*

Mechtly, E. A., "The International System of Units, Physical Constants and Conversion Factors," 2nd rev. NASA PUBL. SP-7012. Washington, D.C.: U.S. Government Printing Office, 1973. *A detailed listing of the SI units.*

A CLOSER LOOK: Units of Measure

The recommended units of measure to be used in electrical engineering are those of the SI system (Système International) proposed in 1960. It uses mks (meter, kilogram, second) units but additionally spells out usage rules. In one instance these include standardization of decimal abbreviations and in another the naming (or renaming) of units.

Whenever an SI unit that is named after an individual—for example, Volta (the volt), Siemens, Watt, or Coulomb—is written out, it is never capitalized. Thus, we might have a volt or an ampere. When expressed as an abbreviated unit, it is capitalized—0.5 V and 1.0 A, respectively—with no period appearing after the abbreviation. When accompanied by a numerical value, the spelled out form is used if the number is written out and the abbreviation is used in conjunction with the numeric form.

An s is never added to a symbol to denote the plural form, but when the unit is written out, the plural is formed in the usual manner—for example, volts, amperes. Notable exceptions to this rule are hertz and siemens, which assume the same form whether singular or plural.

While to a large extent in this text we use SI units and rules, we at times will also use the English engineering and cgs (centimeter, gram, second) units prevalent in other engineering fields.

Quantity Symbols and SI Units

Quantity		SI Unit	
Symbol	Name	Name	Abbreviation
v	Velocity	meter per second	m/s
F	Force	newton	N
Q	Charge	coulomb	C
ϵ	Permittivity	farad per meter	F/m
W	Work, energy	joule	J
L	Length	meter	m
V	Potential	volt	V
I	Current	ampere	A
R	Resistance	ohm	Ω
C	Capacitance	farad	F
H	Magnetic field intensity	ampere(-turn) per meter*	A(t)/m
B	Magnetic flux density	weber per meter squared	Wb/m^2
μ	Magnetic permeability	henry per meter	H/m
Φ	Magnetic flux	weber	Wb
T	Torque	newton-meter	N·m
M	Magnetization	ampere per meter	A/m
$\mathcal{R}$	Reluctance	ampere(-turn) per weber*	A(t)/Wb
L	Inductance	henry	H
ω	Radian frequency	radian per second	rad/s
f	Frequency	hertz	Hz
P	Power	watt	W
G	Conductance	siemens	S
$\mathcal{F}$	Magnetomotive force	ampere(-turn)*	A(t)

*The designated unit of magnetomotive force is the ampere, while in practice it often is stated as the ampere-turn; *turn,* however, is a dimensionless quantity.

Commonly Used Standard Decimal Prefixes

Multiplier	Prefix	Meaning	Abbreviation
10^{12}	tera	trillion (1 000 000 000 000)	T
10^{9}	giga	billion (1 000 000 000)	G
10^{6}	mega	million (1 000 000)	M
10^{3}	kilo	thousand (1000)	k
10^{-1}	deci	tenth (0.1)	d
10^{-2}	centi	hundredth (0.01)	c
10^{-3}	milli	thousandth (0.001)	m
10^{-6}	micro	millionth (0.000 001)	μ
10^{-9}	nano	billionth (0.000 000 001)	n
10^{-12}	pico	trillionth (0.000 000 000 001)	p

EXERCISES

1. Indicate the unit and symbol associated with each of the following quantities: voltage, resistance, charge, energy, power, conductance, current, time.

2. Express the following in their correct form utilizing abbreviated decimal units:

   a. 3.6×10^{-6} amperes

   b. 5.1×10^{3} volts

   c. 5.8×10^{-3} siemens

   d. 8.55 watts

   e. 5.1×10^{-6} coulombs

   f. 1.55×10^{-2} meters

   g. 6000 meters

   h. 1,000,000 ohms

   i. 2000 ohms

   j. 6×10^{-6} seconds

   k. 2×10^{-3} seconds

3. Express the following in terms of powers of 10 and the appropriate unit; note the conventions associated with SI units: 4.7 kΩ, 1.0 MΩ, 0.9 V, 15.0 mA, 220 kV, 2 μs, 18 ms, 18 mS.

4. Complete the following, including the appropriate unit:

$\frac{5 \text{ J}}{2.5 \text{ s}} =$ ______ $\frac{(8 \text{ V})^2}{2\ \Omega} =$ ______

$\frac{2 \text{ V}}{8 \text{ A}} =$ ______ $(3 \text{ A})^2 \times 2\ \Omega =$ ______

$2 \text{ W} \times 5 \text{ s} =$ ______

$\frac{2 \text{ C}}{5 \text{ s}} =$ ______ $2 \text{ A} \times 5 \text{ s} =$ ______

$8 \text{ A} \times 3\ \Omega =$ ______ $\frac{3 \text{ A}}{2 \text{ V}} =$ ______

$\frac{1}{8\ \Omega} =$ ______

5. Electromotive force (emf) alternatively is called either ______ or ______.

6. An ______ of current constitutes a ______ per second flowing past a given point.

7. **a.** List the three alternative forms of Ohm's law.

   b. Also list the three alternative forms of the power equation.

8. That the sum of voltage sources and voltage drops around any closed circuit is zero is a statement of ______ law.

9. **a.** The polarity where electron flow enters a resistance is taken to be ______ and ______ where it leaves.

   b. The polarity where classical current enters a resistance is taken to be ______ and ______ where it leaves.

10. That the sum of the entering currents must equal the sum leaving every circuit node is a statement of ______ law.

11. **a.** What is the resistance equivalent to the following placed in series: 47 Ω, 56 Ω, 82 Ω, 100 Ω?

    b. When placed in parallel?

12. What value of resistance should be placed in parallel with 56 Ω in order to get a 50 Ω equivalent?

13. **a.** What is the resistance equivalent of two parallel 47 Ω resistors placed in series with a 47 Ω resistor?

    b. Of four parallel 47 Ω resistors in series with a 47 Ω resistor?

14. **a.** What is the power dissipation when 120 V is applied across a 5 Ω resistor?

    b. When 20 A passes through the same 5 Ω resistor?

15. What is the resistance of a 100 W light bulb with 120 V applied?

16. A 40 W light bulb intended for 12 V trailer operation is accidentally placed in a 120 V socket. What would be the approximate power drawn? (Assume hot and cold resistances to be of the same order.)

17. What current corresponds to 130 C/min?

18. **a.** A ruby laser emits 0.05 J in a pulse whose duration is 20 ns. What is the power in the pulse?

    b. If the same energy is emitted on a continuous basis from an argon laser, what is the power?

19. Indicate the peak values corresponding to these common rms utility voltages: 120 V, 240 V, 440 V.

20. What is the internal resistance of a 12 V battery that delivers 100 A into a 0.1 Ω resistance? Is this a voltage or a current source?

21. The output of a transistor amplifier can be represented as a generator in series with an internal resistance. If this internal resistance is 1 kΩ and it is operating into a 10 kΩ load, what is the nature of this transistor amplifier?

22. If the internal resistance of a 10 V source is 1 Ω, compute the current, voltage, and power delivered successively to loads whose values are 0.1 Ω, 0.5 Ω, 1 Ω, 5 Ω, 10 Ω.

23. If 1 kW is equivalent to 1.341 horsepower (hp), what is the electrical equivalent of 1 hp?

24. **a.** If 1 J is equal to 0.73756 ft-lb, in terms of foot-pounds, what is the equivalent of 1 kWh?

b. Since 1 J = 1 N·m, what is the equivalent of 1 kWh in units of newton-meter?

25. What is the SI equivalent of 1 hp?

26. A TV antenna can be represented as a generator with an internal resistance of 300 Ω. What should be the input resistance of the TV receiver for maximum power transfer?

27. If 1000 W is delivered to a motor drawing 5 A and whose resistance is 3 Ω, what is the actual power converted into mechanical power? What is the motor efficiency? What is the horsepower rating of the motor?

28. **a.** An ac source applied to a resistive heater generates heat at the same rate as a 240 V dc source. What is the rms value of the sinusoidal ac source? What is the peak value of the ac source?

b. What would be the percentage increase in heat evolved if a dc source equal to the peak ac value computed in part a were applied to the same load?

c. Show that this is a general result in terms of the relationship between V_{rms} and V_{max} for sinusoidal voltage.

29. **a.** For a 60 Hz power source, what is the time interval between successive peak voltages—that is, the period of the waveform? (Express your result in terms of milliseconds.)

b. In Europe 50 Hz is employed by utility companies; what is the period of this waveform?

30. Is it correct to speak of a generator as a *source* of electricity?

31. A 120 V, 100 A generator is 85% efficient. What is the required power input? How much horsepower does this represent?

32. A 12 V storage battery delivers 250 A·h. How many joules does this represent?

33. It is desired to reduce the 120 V ac line voltage to a value of 12 V rms. What should be the transformer's turns ratio?

34. If the power applied to the primary of a transformer with a 10:1 turns ratio is 120 V at 1 A, what is the ideal secondary current available? Why might the delivered current be less?

35. **a.** The secondary of a transformer is to deliver 12 V at 1 A rms. What should be the turns ratio for operation using a 120 V ac utility?

b. What will be the idealized primary current?

36. A constant voltage of 100 V is applied to a transformer with a primary to secondary turns ratio of 1:5. What is the secondary voltage?

37. Electrical meters of the d'Arsonval type, whether used as a voltmeter or an ammeter, actually measure ______.

38. Moving-coil meters commonly have full-scale movements which are either ______ or ______. Alternatively these quantities may be expressed in terms of sensitivities—that is, as ______ ohms per volt, and ______ ohms per volt, respectively.

39. The full-scale reading of current-reading instruments can be extended by ______.

40. Moving-coil meters can be converted to voltmeters by placing a resistor ______.

41. Ammeters should be placed ______ relative to the circuit whose current is to be measured; voltmeters should be placed ______ relative to the circuit whose voltage is to be measured.

42. The ______ the meter resistance of an ammeter, the smaller its loading effect on the circuit being measured.

43. The ______ the meter resistance of a voltmeter, the smaller the loading effect on the circuit being measured.

44. A battery, an ammeter, and a variable resistor, all placed in series, may be used to serve as an ______ meter.

45. An instrument which, by an appropriate combination of resistances and switches, can be made to measure voltage, current, and resistance is termed a ______ meter.

PROBLEMS

1.1

a. A 12 V battery is placed in series with a 1 kΩ resistor. (See Figure 1.40.) What is the magnitude of the resultant current? (Express your answer in terms of amperes as well as milliamperes.)

b. The unit current charge is 1.6×10^{-19} in the circuit in Figure 1.40. How many charges pass a given point per second?

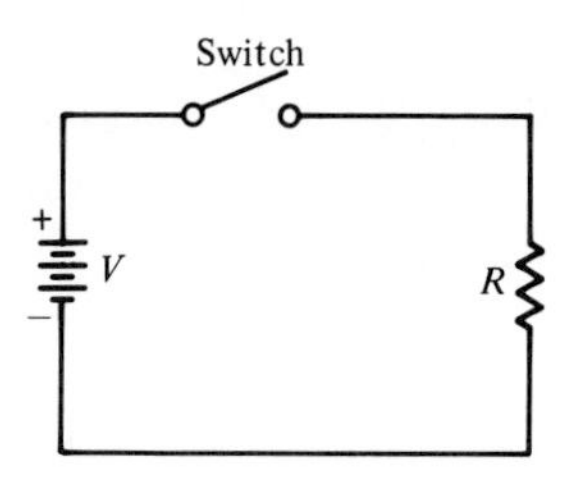

FIGURE 1.40

1.2

a. It is difficult for most of us to visualize the magnitudes of exponential numbers. By way of illustration, compute the duration of a human lifetime in terms of seconds, taking the average human lifetime to be 75 years.

b. An ampere consists of 6.25×10^{18} electrons/s flowing past a given point. If this number represented seconds of human life, to how many years would it correspond?

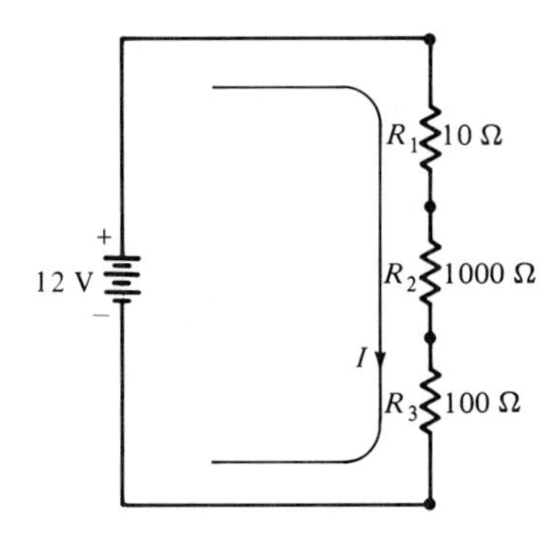

FIGURE 1.41

1.3

a. Calculate the current in Figure 1.41.

b. Determine the voltage drop across each resistance, and indicate the polarity of the drop across each resistance.

c. Verify Kirchhoff's voltage law (KVL).

1.4

a. Placing the three resistors of Problem 1.3 in parallel, what would be the magnitude of the current?

b. What would be the voltage drop across each resistance?

c. What would be the current through each resistance?

1.5

When two resistors appear in parallel and the current into the circuit is known, the current through either one of the parallel branches may easily be determined by using the current division rule. Verify this rule, considering I_2 (Figure 1.42) to be the desired current.

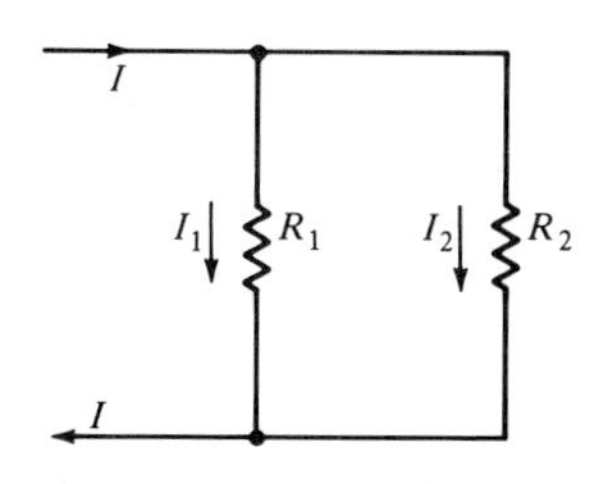

FIGURE 1.42

1.6

a. Determine the battery current in Figure 1.43.

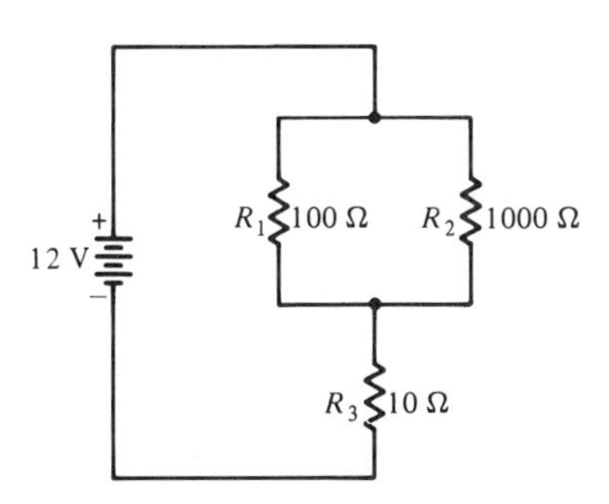

FIGURE 1.43

b. Determine the current through each resistance.

c. What is the voltage drop across each resistance?

1.7

a. Determine the power dissipated in each of the resistances in Problem 1.6.

b. What is the total power furnished by the battery?

1.8

State the resulting conductance upon placing the following resistances in parallel: 520 Ω, 4700 Ω, 5200 Ω, 1200 Ω. (Express your answer in the appropriate conductance unit.)

1.9

a. Express the power equations in terms of conductance (G) rather than resistance (R).

b. What power is represented by 1 mA of current passing through a conductance of 1 mS? By 1 V applied across a conductance of 1 mS? What is the current in the latter case?

1.10

Determine the voltage at the output terminals in the following circuits and indicate the polarity. (You should be able to solve them by inspection.)

a. Figure 1.44a.

b. What would be the output voltage if the polarity of the 10 V battery in part a were reversed?

c. Figure 1.44b.

d. Figure 1.44c.

e. Figure 1.44d.

1.11

With respect to each resistance in the circuit in Figure 1.45, determine the voltage drop and the current.

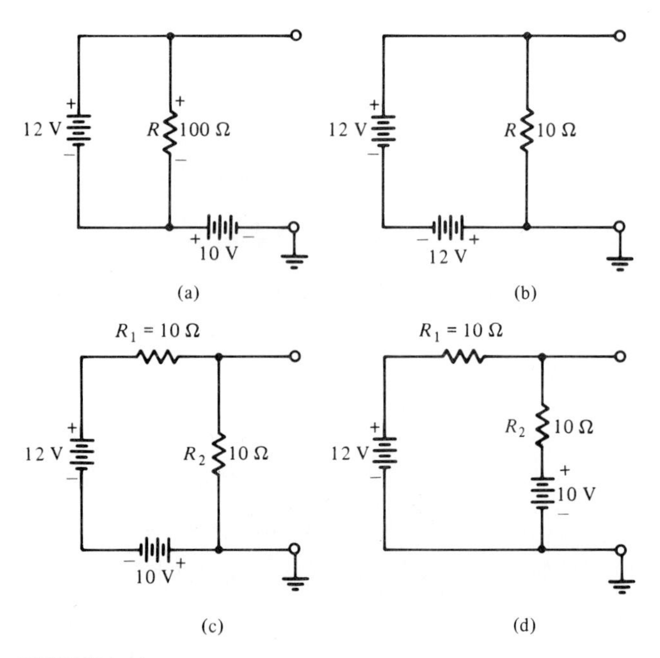

FIGURE 1.44

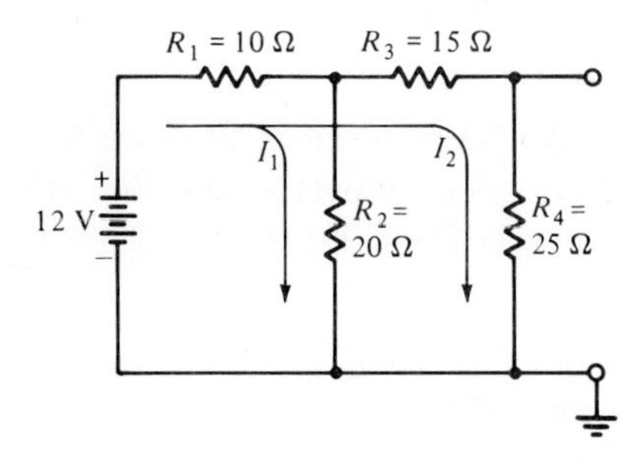

FIGURE 1.45

1.12

Calculate the current from each battery in the circuit in Figure 1.46. What power is furnished by each battery?

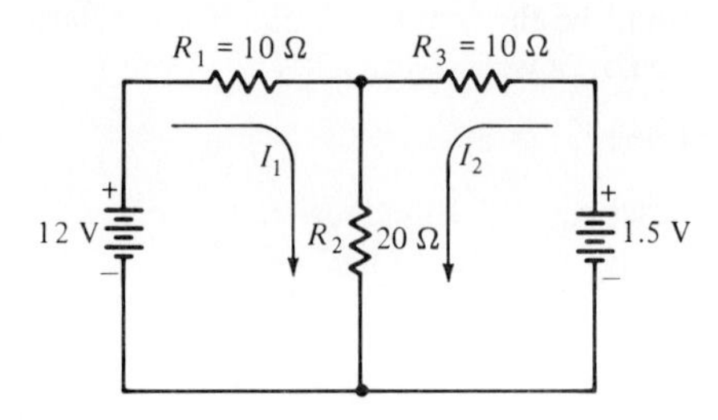

FIGURE 1.46

1.13

In simulating the use of a standard D-size flashlight battery, it is discharged into a 4 Ω resistance for 4 min periods each hour for 8 h a day (there are 16 h rest periods after each 8 h). From an initial voltage of 1.5 V the final useful voltage is taken to be 0.9 V. A typical lifetime is 480 min. (See Figure 1.47.)

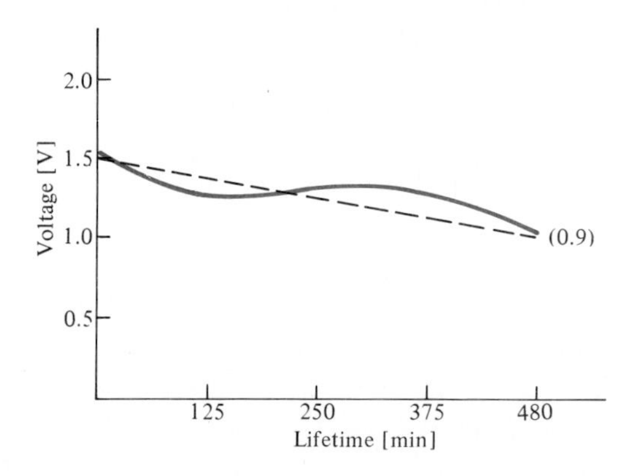

FIGURE 1.47

a. What is the average value of current during the battery's lifetime?

b. What is the average power supplied?

c. What is the power supplied in watt-hours?

d. Considering the battery to cost one dollar, compute the cost on the basis of kilowatt-hours, and compare this with the utility rate of about 8 cents/kWh.

COMMENT: Rather than watt-hours, it is more usual to specify battery lifetimes in terms of ampere-hours. For an initial current drain of 500 mA, a D-size flashlight battery has a capacity of 1.0 A·h in going from 1.5 to 0.9 V. The corresponding alkaline battery has a yield of about 7 A·h.

1.14

For a type AA cell (termed a penlight battery), operated for 5 min/day, 7 days/week, with an end point of 0.9 V (with an initial value of 1.5 V), the lifetime is about 45 min. Compute the cost per kilowatt-hour. The discharge was into a 4 Ω load. Take the unit cost to be 35 cents.

1.15

a. Most electric stoves use 240 V ac. What is the corresponding peak voltage?

b. Industrial wiring often makes use of 440 V ac. What is the peak value of the voltage?

1.16

A 120 V ac source has an internal resistance of 0.1 Ω.

a. Compute and plot the output voltage and current when operating successively into the following resistive loads (R_L): 10, 5, 1, 0.5, 0.1, 0.05, 0.01, 0.005, 0.001 (all in ohms).

b. Compute and plot the power delivered as a function of R_L.

c. Compute the power dissipated within the internal resistance of the generator as a function of R_L.

1.17
A 120 V generator delivers 2 A into a 10 Ω load. What should be the minimum generator internal resistance if the current in the load is not to change by more than 1% when R_L is increased to 50 Ω?

1.18
With a peak sinusoid voltage of 3 V and a peak current sinusoid of 2 A, compute and plot the instantaneous power during one cycle at 22.5° intervals.

1.19
In a three-phase power system there are three pairs of coils in a generator (or motor). Each output (or input) represents a sinusoidal variation with a 120° difference between them (termed the phase difference). The time they reach corresponding points in their cycles is shown in Figure 1.48. Mathematically,

$$v_A = V_m \sin \theta$$
$$v_B = V_m \sin (\theta + 120°)$$
$$v_C = V_m \sin (\theta + 240°)$$

For a peak voltage of 3 V and a peak current of 2 A, compute the power sum from the three phases for values of θ from 0° to 360° at 45° intervals. Plot the results.

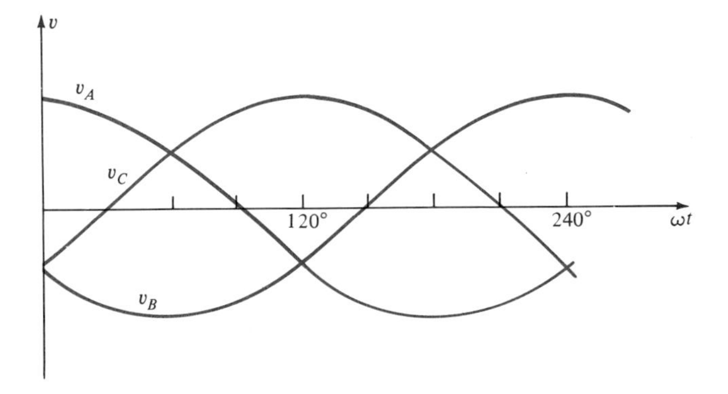

FIGURE 1.48

1.20
For a sinusoidally varying waveform, the instantaneous value (v) in terms of the maximum value (V_{max}) is

$$v = V_{max} \cos \omega t$$

where $\omega = 2\pi f = 2\pi/T$, f being the frequency
T = period

a. Mathematically show that the average value of v is equal to zero.

b. With the negative half-cycles positive, as in Figure 1.16, compute the average value. (*Hint*: Take the average over a quarter-cycle.)

c. What would be the average of only the positive half-cycles? (You should be able to do this by inspection after doing part b.)

1.21
The effective value of sinusoidal ac is also called the rms value. Show that $\sqrt{\overline{v^2}}$—that is, the square root of the mean of v^2—yields the numerical relationship

$$V_{rms} = 0.707 V_{max}$$

where $v = V_{max} \cos \omega t$. (*Hint*: Take the average over a quarter-cycle.)

1.22
What is the rms value of the positive half-cycles of a sinusoidal ac?

1.23
Compute the delivered current through each resistor in the circuit in Figure 1.49.

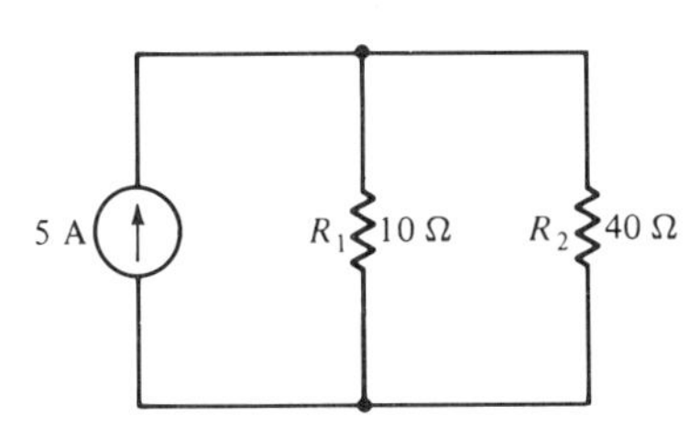

FIGURE 1.49

1.24
A 1 V ideal voltage source is applied to the following series-connected resistances: 2.2 kΩ, 4.7 kΩ, 10.0 kΩ. State the voltage across each resistance.

1.25
An ideal 1 mA current source furnished current to the following parallel resistances: 2.2 kΩ, 4.7 kΩ, 10.00 kΩ. What is the current through each resistance?

1.26
Ten thousand watts of power may alternatively be transmitted as 100 A at 100 V or as 50 A at 200 V. What is the relative transmission line loss in the two cases?

1.27 △

a. The watt-second is equal to 0.24 calorie (cal), the calorie being the quantity of heat necessary to raise 1 gram (g) of water by 1°C. If a 120 V electric heater in a coffee machine is to deliver 90°C hot water in 5 s (starting at 10°C), what is the required heater resistance? (Consider that a cup contains 0.25 kg.) What is the value of the current? What is the power?

b. Why might this result have to be modified in an actual design?

1.28

a. A 240 V power source, with an internal resistance of 1.5 Ω, is used to operate a motor. If the connecting lines each constitute a resistance of 2.0 Ω, what is the power delivered to the motor if I = 2 A?

b. If the source voltage is reduced to 120 V, what would be the power delivered to the motor?

c. How much power is delivered in each case by the source?

1.29

Over a 1 h period a motor utilizes 0.6714 MJ of energy. Since 746 W = 1 hp, what is the power rating of the motor?

1.30

a. A 100 W bulb, operating from a 120 V source, represents what operating resistance?

b. What would be the nature of the bulb resistance when first turned on?

1.31

The ac voltage and current applied to a circuit varies in the following manner:

$$v = 170 \sin 2\pi(60)t \quad [\text{V}]$$
$$i = 20 \sin 2\pi(60)\,t \quad [\text{A}]$$

What is the power drawn by the circuit?

1.32

A 1 mA movement with a 100 Ω internal resistance is to be converted into a 0–100 mA ammeter. Calculate the required shunt.

1.33 △

A multirange ammeter may be constructed by allowing different values of shunts to be switched into the circuit as shown in Figure 1.50. To prevent damage to the meter, the switch should be of the type known as a *make before break*—that is, the next contact is made before the previous one disconnects. Thus, at no time will the large shunting current be allowed to pass through the meter. Another solution to this problem is provided by the shunt shown in Figure 1.51, known as an Ayrton shunt. This assures that the meter is never without a shunt, at the expense of a slightly higher overall meter resistance. (Note that in the 10 A position, for example, $R_1 + R_2$ appears in series with R_m.) Design an Ayrton shunt which will provide ammeter ranges of 1 A, 5 A, and 10 A. Assume a 1 mA movement with R_m = 100 Ω.

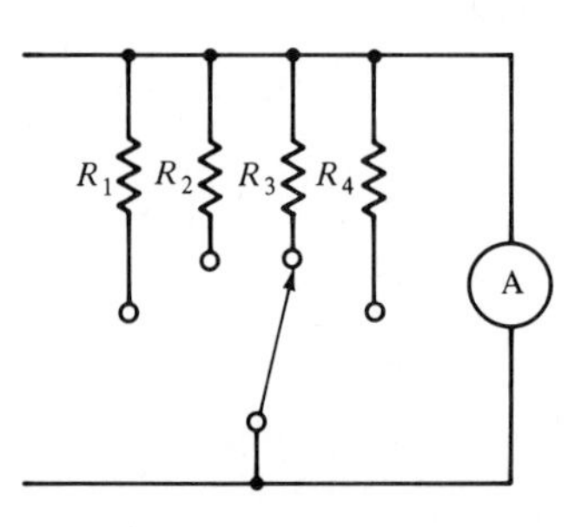

FIGURE 1.50

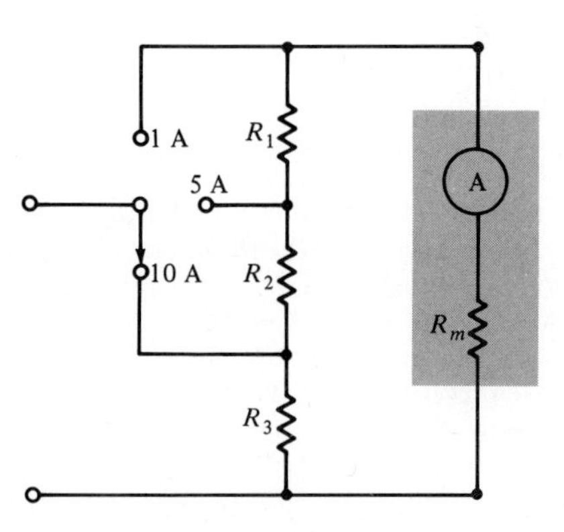

FIGURE 1.51

1.34

Using a 20,000 Ω/V meter, design a multirange voltmeter with the following scale readings: 10, 50, 100, and 500 V. Disregard the meter resistance.

1.35

Without any multiplier being used, what full-scale voltage could be indicated by a 1 mA meter whose R_m = 100 Ω? By a 50 μA meter whose R_m = 1000 Ω?

1.36

a. If a 1000 Ω/V meter is used as a 0–1 V voltmeter with a 1000 Ω multiplier, having disregarded the 100 Ω meter resistance, what is the percentage inaccuracy introduced at full scale?

b. If it is used as a 0–100 V voltmeter, what is the inaccuracy?

1.37

A 1 mA meter has a resistance of 40 Ω. What is the value of the shunt resistance necessary to convert it into a 10 mA full-scale instrument?

1.38
If a current meter is being used to measure current in a circuit whose resistance is comparable to the meter resistance, a significant error is incurred. What is the relationship between the circuit and meter resistance if the error is not to exceed 1%? 0.1%?

1.39 △
With regard to the series ohmmeter shown in Figure 1.36 as the battery ages and V diminishes, R_1 is diminished in order to maintain a full-scale current with $R_x = 0$. But this also changes the scale calibration. With $V = 4.5$ V and $R_1 + R_m = 4500\ \Omega$, if the battery voltage changes by 10%, what is the percent error in the midscale reading (4500 Ω)? The meter is a 1 mA movement with $R_m = 100\ \Omega$.

1.40 △
The changing calibration of a series ohmmeter as the battery ages may be minimized by using a shunt adjustment rather than a series adjustment. (See Figure 1.52.) As the battery voltage diminishes, R_2 is increased, forcing more of the battery current through the meter, thereby maintaining full-scale deflection when $R_x = 0$.

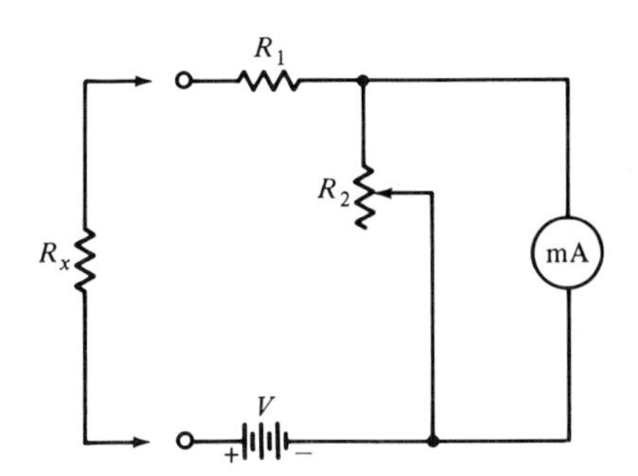

FIGURE 1.52

a. Determine the values of R_1 and R_2 for a 1 mA movement, with $R_m = 100\ \Omega$ and $V = 4.5$ V. It is desired to have a half-scale deflection of 2000 Ω.

b. What must be the maximum value of R_2 to take into account a 10% drop in the battery voltage?

c. How much of an error, at midscale, is introduced by a 10% drop in battery voltage? Compare this to the series ohmmeter of Problem 1.39.

1.41 △
Rather than placing the unknown resistance in series with the meter and voltage source (series ohmmeter), it may be placed in parallel with the meter leading to a shunt ohmmeter. This arrangement is particularly suitable for measuring low values of resistance. (See Figure 1.53.) In this case the meter reads zero when $R_x = 0$ and full scale when $R_x = \infty$—that is, when the test leads are open-circuited. For a 1 mA movement, $R_m = 50\ \Omega$ and $V = 4.5$ V, compute the midscale reading of the resulting ohmmeter.

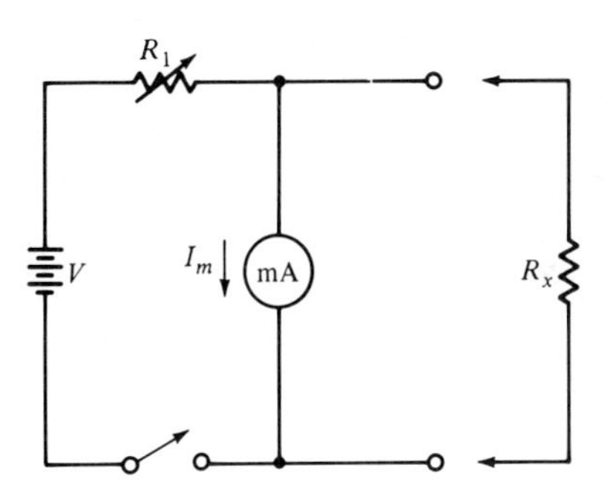

FIGURE 1.53

1.42
An ohmmeter of the form shown in Figure 1.36 utilizes a 4.5 V battery and a 1 mA meter in series with a 4400 Ω resistor. $R_m = 100\ \Omega$. Derive the general expression for the current in terms of the unknown resistance R_x. Compute the resistance for the following deflections: $\frac{1}{8}$, $\frac{1}{4}$, $\frac{3}{8}$, $\frac{1}{2}$, $\frac{5}{8}$, $\frac{3}{4}$, $\frac{7}{8}$, 1.0 (full scale). Sketch the resulting scale.

1.43 △
For an electrodynamometer, we have the following:

$$\theta_{av} = K\frac{1}{T}\int_0^T i\frac{v}{R_v}\,dt = K_2\frac{1}{T}\int_0^T iv\,dt$$

$$P_{av} = \frac{1}{T}\int_0^T vi\,dt$$

where θ_{av} = average angular deflection
K = instrument constant
i = instantaneous current in field coils
v/R_v = instantaneous current in voltage coils

Assume v and i are not in phase—that is, $v = V_{max}\sin\omega t$, and $i = I_{max}\sin(\omega t + \theta)$. Derive an expression for the deflection in terms of rms values of I and V and the cosine between the two quantities. R_v is the resistance of the voltage coils.

2 RESISTANCE, REACTANCE, AND IMPEDANCE

OVERVIEW

In Chapter 1 we considered a single electrical element, the resistor, and the nature of its interaction with ac and dc. Two additional electrical elements, the capacitor and the inductor, are introduced in this chapter, together with the mathematical methods needed to analyze their interaction with electrical currents. The necessary mathematics is that of complex numbers, which, as specifically adapted to electrical considerations, constitute the realm of phasors.

Both the inductor and the capacitor exhibit opposition to ac but in such a manner as to lead to mutual neutralization. Such mutual cancellation is known as the resonance condition, whose prime function is to isolate specific ac frequencies for the purpose of either discriminating against them or accentuating them.

OUTLINE

2.1 ■ INTRODUCTION

Whether employing dc or ac, resistors *impede* current and *dissipate* power. Thus, **resistance** is opposition to current that also involves power dissipation. In ac circuits two additional elements impede current but do *not* dissipate power: the capacitor (sometimes still called a condenser) and the inductor (often called a coil). Dissipationless opposition to current is called reactance.

In the latter part of this chapter, we will consider various combinations of resistances and reactances. Of particular interest to us will be their series and parallel arrangements when the magnitudes of inductive and capacitive reactances are equal. Such equality constitutes the resonant condition of an *RLC* circuit and may be variously used either to discriminate against a specific frequency or to accentuate it in preference to adjacent frequencies.

2.2 ■ INDUCTIVE REACTANCE

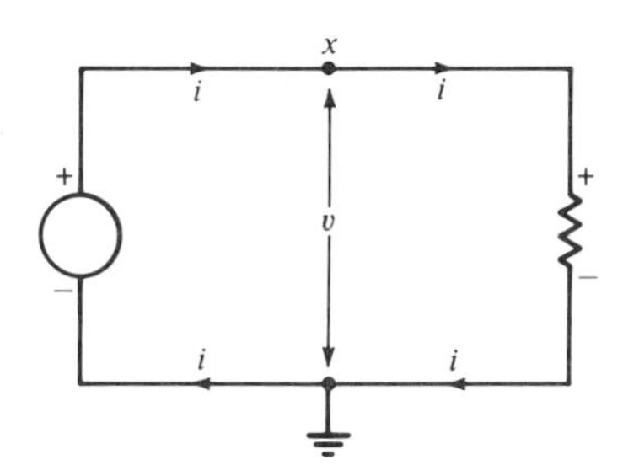

FIGURE 2.1. **Equality of Potential Rise and Fall** The ground symbol indicates the reference node. The potential at x is measured relative to ground.

Consider the ac power dissipation in a resistance. Since both voltage and current are continually changing, we must deal with the instantaneous power dissipation:

$$p = vi \tag{2.1}$$

(Again, lowercase letters are generally used to represent instantaneous values, and uppercase letters stand for either constant values, such as encountered in dc circuits, or fixed values, such as rms or maximum values in ac circuits.)

Consider Figure 2.1. The voltage is to be measured relative to the grounded terminal. Thus, while v is continually changing, the instantaneous output voltage of the generator is at all times equal to the voltage drop across the resistor. We may plot the instantaneous power variation in the resistor as in Figure 2.2. The power pulses in the resistor always remain positive.

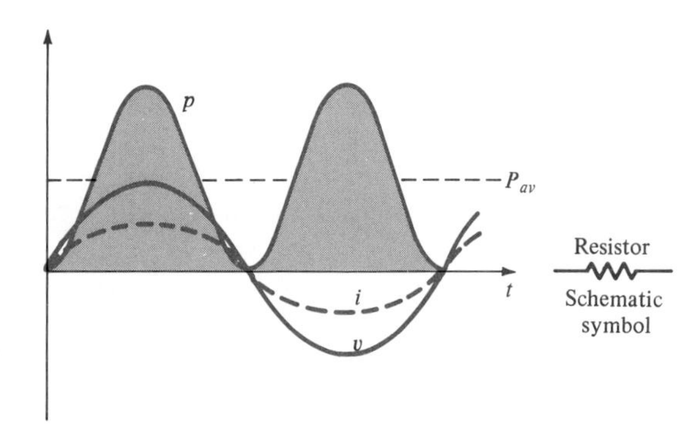

FIGURE 2.2. **Power Dissipation in a Resistance** Sinusoidal current and voltage variations in a resistance lead to positive power pulses, representing dissipation.

Relative to the current, let us now displace the voltage waveform to the left by a quarter-cycle, calling it v_L, without, for the time being, inquiring how or why one might want to do so. The resulting plot (Figure 2.3) shows that during a quarter-cycle the power is positive and that during the next quarter-cycle it is equally negative.

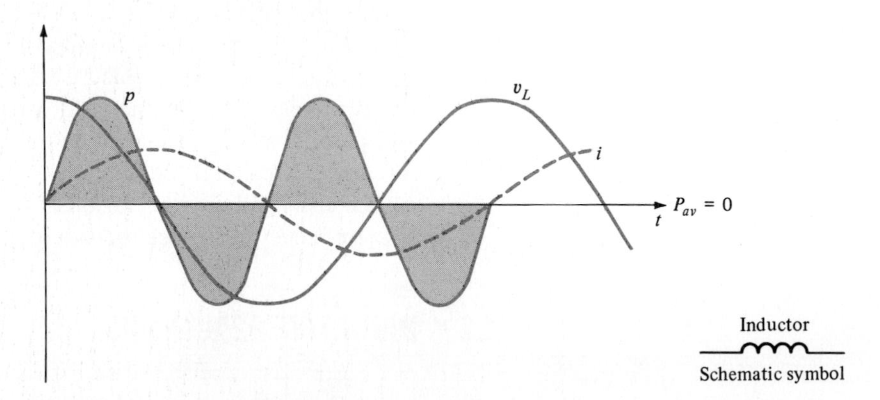

FIGURE 2.3. **Power Pulses in a Pure Inductor** ac voltage across an inductor leads current by 90° and gives rise to successive positive and negative power pulses, resulting in zero power dissipation in a pure inductor.

What is this device? It is a coil, also known as an inductor.

Why does an inductor cause this type of voltage displacement relative to the current? It does so because anytime the current changes through a coil, a counter-emf is produced.[1] Referring to Figure 2.4, note that at *a* the current is changing at its greatest rate. Therefore, the counter-emf (v_L) has a maximum value (a'). At *b* the current is not changing, and v_L is zero (b'). At *c*, again, the current is changing at its greatest rate, but it is now decreasing. The counter-emf assumes its greatest value but has a polarity opposite to that at a'.

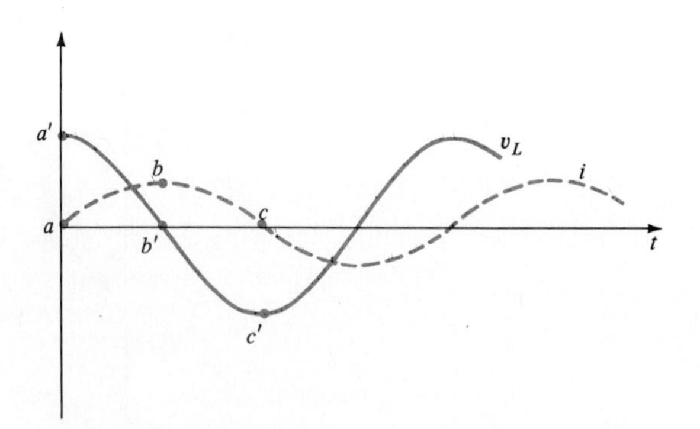

FIGURE 2.4. **Counter-emf Developed across an Inductor as a Function of the Instantaneous Current**

[1]This counter-emf is a result of Lenz's law. The induced voltage is always in a direction such as to oppose the current change.

Since in an ideal inductor (Figure 2.3) positive and negative pulses of equal magnitude alternate, there is no power dissipation. However, an inductor does oppose current or voltage variations (via the counter-emf). This dissipationless opposition is the **reactance,** and we designate it by the letter X. Reactance associated with inductors is termed **inductive reactance** (X_L).

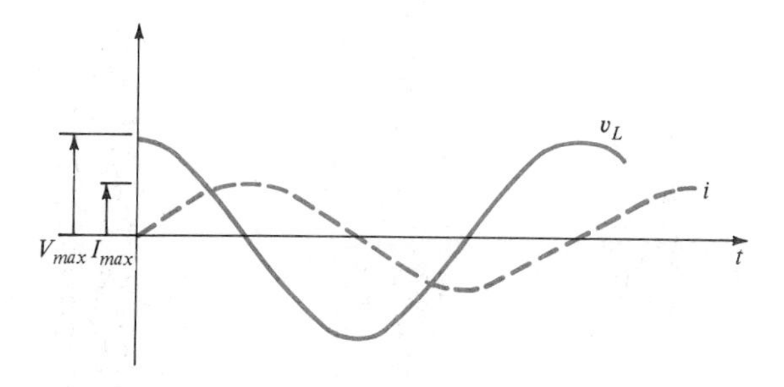

(a) Counter-emf vs. current through an inductor

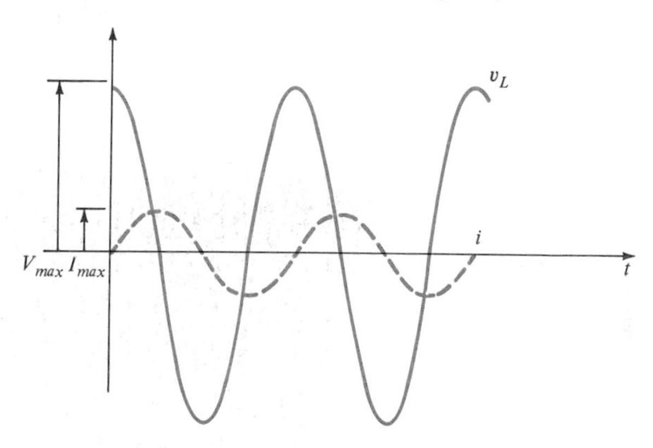

(b) Counter-emf vs. higher frequency current through an inductor

FIGURE 2.5. **Effect of Frequency on the Magnitude of Counter-emf**

If the frequency of the applied current variation is increased, the counter-emf is greater and gives rise to a greater reactance (Figure 2.5). Note that while I_{max} is the same in both cases, the higher frequency in the second case leads to a greater magnitude of counter-emf and hence a greater reactance. We may say that the inductive reactance X_L is proportional to the frequency f:

$$X_L \propto f \propto \omega \tag{2.2}$$

where ω = radian frequency $2\pi f$.

The inductance of a coil may be defined in the following manner: If a current change of 1 ampere per second (A/s) creates a counter-emf of 1 volt (V), the inductance is said to have a value of 1 **henry** (H). Thus, in general, $v_L = L(di/dt)$.

For the same current variation through a coil, the insertion of an iron core will lead to a larger counter-emf because the flux lines are more concentrated. Thus, a coil with an iron core has a larger inductance than the same coil with an air core.

Since counter-emf increases with L (the inductance), we have

$$X_L \propto L$$

The full expression for inductive reactance is

$$X_L = \omega L = 2\pi f L \tag{2.3}$$

The unit of reactance is also the ohm (Ω), since multiplying the current by reactance also equals voltage. We then have Ohm's law for reactances:

$$\begin{aligned} I &= \frac{V_L}{X_L} \\ V_L &= I X_L \\ X_L &= \frac{V_L}{I} \end{aligned} \tag{2.4}$$

The last version can be derived using $v_L = L(di/dt)$. Since $i = I_{max} \sin \omega t$, we have

$$v_L = L\left(\frac{di}{dt}\right) = \omega L I_{max} \cos \omega t$$

$$\frac{V_L}{I} = \omega L = X_L$$

where v_L = instantaneous value of voltage across coil
V_L = rms value
I = rms current (equal to $I_{max}/\sqrt{2}$)

In Eqs. 2.4, I and V_L are corresponding quantities—that is, both rms, both peak, and so forth.

When inductors are placed in series, the counter-emfs are additive, and the total inductance is the sum of the individual inductances. That is,

$$L_{tot} = L_1 + L_2 + L_3 + \cdots \tag{2.5}$$

Inductors in parallel are subjected to a common voltage, and since $v_L = L(di/dt)$, $i = (1/L)\int v_L\, dt$. Because the total current is the sum of the individual currents, we have

$$i_{tot} = i_1 + i_2 + i_3 + \cdots$$

$$\frac{1}{L_{tot}}\int v_L\, dt = \frac{1}{L_1}\int v_L\, dt + \frac{1}{L_2}\int v_L\, dt + \frac{1}{L_3}\int v_L\, dt + \cdots$$

$$\frac{1}{L_{tot}} = \frac{1}{L_1} + \frac{1}{L_2} + \frac{1}{L_3} + \cdots \tag{2.6}$$

Therefore, inductors in series and in parallel are additive, respectively, in the same manner as resistors *assuming* that the fields of the individual inductors do not interact with one another.

2.3 ■ CAPACITIVE REACTANCE

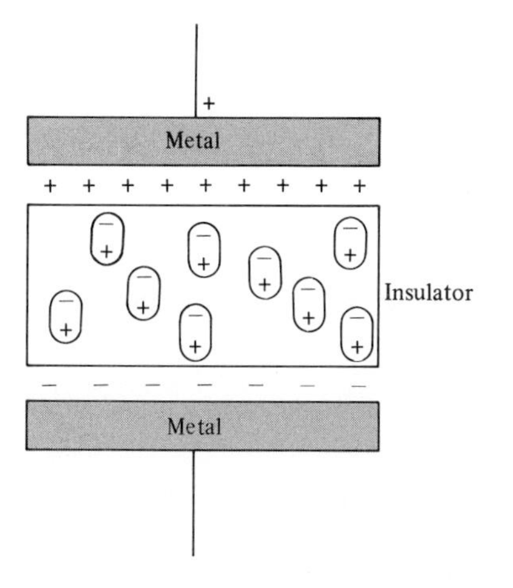

FIGURE 2.6. **Charge Distribution in a Capacitor with Voltage Applied**

Now consider a capacitor, which in its simplest form is composed of two metallic plates separated by an insulator. A voltage applied between the conducting plates (Figure 2.6) gives rise to plus (+) and minus (−) charges at the plate surfaces, and the positive and negative charges making up the insulation atoms are thereby displaced but not separated. (That, in fact, is why it is an insulator.) The insulation is thereby stressed, and the greater the charge difference applied to the plates, the greater the stress. Such stress represents stored energy.

Upon the application of a voltage V_C, it should easily be seen that the magnitude of the charge depends, for example, on the plate area. It will also depend on the nature of the insulation (how easily the charges are displaced) and on the separation distance between the plates. The ratio of charge to applied voltage is called the *capacitance* (C):

$$\frac{Q}{V_C} = C$$

or

$$Q = CV_C \tag{2.7}$$

where Q = charge in coulombs
C = capacitance in farads
V_C = voltage

A capacitor has a capacity of one **farad** if one volt applied across its plates results in a charge deposition of one coulomb.

Capacitors in parallel are additive since the capacitors are being subjected to the same voltage. Thus,

$$\begin{aligned} Q_{tot} &= C_{tot}V_{appl} = Q_1 + Q_2 + Q_3 + \cdots \\ &= C_1V_{appl} + C_2V_{appl} + C_3V_{appl} + \cdots \\ C_{tot} &= C_1 + C_2 + C_3 + \cdots \end{aligned} \tag{2.8}$$

When capacitors are placed in series, the total reactance is equal to the sum of the individual reactances. The equivalent capacitance is given by the relation

$$\begin{aligned} X_{C_{tot}} &= X_{C_1} + X_{C_2} + X_{C_3} + \cdots \\ \frac{1}{2\pi fC_{tot}} &= \frac{1}{2\pi fC_1} + \frac{1}{2\pi fC_2} + \frac{1}{2\pi fC_3} + \cdots \\ \frac{1}{C_{tot}} &= \frac{1}{C_1} + \frac{1}{C_2} + \frac{1}{C_3} + \cdots \end{aligned} \tag{2.9}$$

Therefore, capacitors in series add like resistors in parallel and vice versa.

What happens if such a capacitor is attached to a source of ac? A current will arise between the source and the capacitor as the latter charges, discharges, and then recharges in the opposite direction. Let us represent the instantaneous applied voltage as

$$v_C = V_C \cos \omega t \tag{2.10}$$

Then, since

$$i = \frac{dq}{dt}$$

that is, the time rate of change of electrical charge [coulombs per second] equals the current [amperes], from Eq. 2.7 we obtain

$$q = Cv_C \tag{2.11}$$

$$\frac{dq}{dt} = C\left(\frac{dv_C}{dt}\right) \tag{2.12}$$

$$i = C\left(\frac{dv_C}{dt}\right) = -\omega C V_C \sin \omega t \tag{2.13}$$

Therefore, v_C and i are 90° out of phase: While v_C varies as $\cos \omega t$, i varies as $\sin \omega t$.

But what is the significance of the negative sign in the expression for i? When $t = 0$, v_C has its maximum value, V_C. At this time, $i = 0$. A quarter-cycle later, $v_C = 0$, and i assumes its maximum *negative* value. We say, therefore, that in a capacitor the alternating current *leads* the applied voltage by 90°. (Recall that in an inductor the current *lags* behind the voltage by 90°.)

The relationship between v_C and i is shown in Figure 2.7. Note again that the power pulsations are alternately positive and negative and of equal magnitude. Therefore, no power is dissipated in an ideal capacitor.

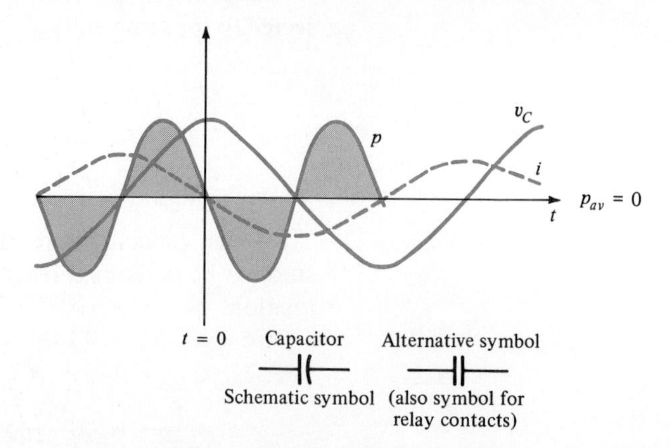

FIGURE 2.7. Power Pulses in a Pure Capacitor ac voltage across a capacitor lags the current by 90° and gives rise to successive positive and negative power pulses, resulting in zero power dissipation in a pure capacitor.

Let us stop a moment and summarize what is happening physically in an inductor and a capacitor. In an inductor, during one quarter-cycle power is withdrawn from the source and used to create a magnetic field; during the next quarter-cycle, this field collapses, and an equal amount of energy is returned to the source. In a capacitor, power is withdrawn from the source during one quarter-cycle and is used to create an electric field (a stressed dielectric); during the next quarter-cycle this energy is returned to the source. Thus, the capacitor, like the inductor, constitutes a reactance. The reactance associated with a capacitor is termed **capacitive reactance**.

How is the capacitive reactance (X_C) related to capacitance (C) and the applied frequency (ω)? The greater the capacitance, the greater the charge it can accommodate and hence the greater the current ($i = dq/dt$). Thus, we expect the following: the greater the C, the smaller the capacitive reactance. By analogy to Eq. 2.4, we have $I = V_C/X_C$. Hence,

$$X_C \propto \frac{1}{C} \tag{2.14}$$

Also,

$$X_C \propto \frac{1}{f} \tag{2.15}$$

The latter can be justified by noting that the higher the frequency, the less time there is during the cycle for the charge (hence, the voltage) to build up on the capacitor.

The full expression for capacitive reactance is

$$X_C = -\frac{1}{2\pi fC} = -\frac{1}{\omega C} \tag{2.16}$$

From Eq. 2.13, we see that X_C is negative.

2.4 ■ PHASORS

In alternating current circuits we are particularly interested in the phase relationship between the current and the voltage. Rather than drawing them as functions of time as we've done thus far (Figure 2.8a), it proves convenient to draw them as in Figure 2.8b. The radial lines representing **I** and **V** can each be considered as rotating about the origin with a frequency ω, which characterizes the applied frequency, but always maintaining the same phase relationship between them. A counterclockwise rotation of $\mathbf{V}_L$ and **I** (as well as $\mathbf{V}_C$ and **I**) with rotational frequency ω will cause their projections on the y-axis to trace out the time variations shown in Figure 2.8a. Analytically, we can express these variations as the real parts of complex numbers (complex numbers are discussed in Section 2.6):

$$\begin{aligned} i(t) &= Ie^{j\omega t} \\ v_L(t) &= V_L e^{j[\omega t + (\pi/2)]} \\ v_C(t) &= V_C e^{j[\omega t - (\pi/2)]} \end{aligned} \tag{2.17}$$

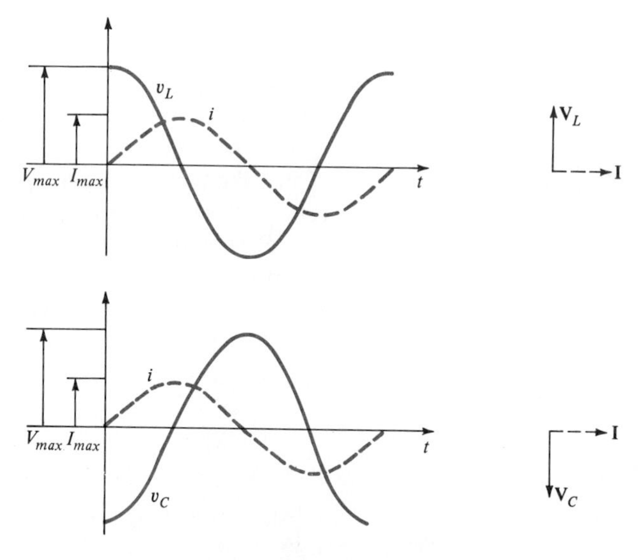

(a) Temporal representation (b) Phasor representation

FIGURE 2.8. Current and Voltage Variations in Reactances

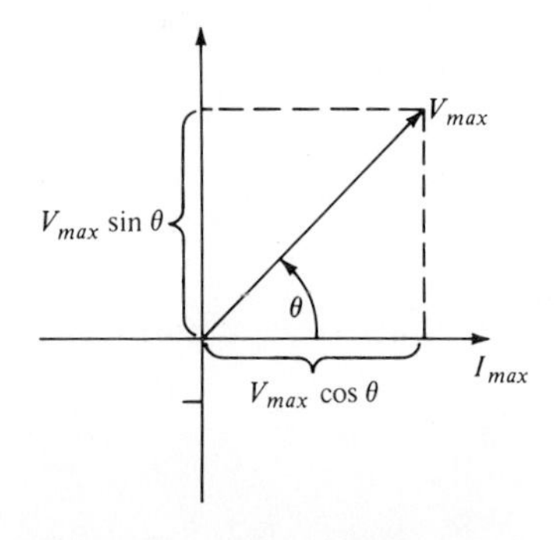

FIGURE 2.9. General Relationship Between Voltage and Current, with Corresponding Phase Angle θ

Since it is the phase angle between I and V that is of interest to us and it does not change with time, we can disregard the graphical rotation and, equivalently, delete the $e^{j\omega t}$ factor from each expression in Eqs. 2.17. Current is generally used as a reference and is drawn in the $+x$-direction. Voltage across a pure inductor has a $+y$-directivity and that for a capacitor, $-y$-directivity.

Let us consider a somewhat more general problem and take the voltage direction to differ from that of the current by an angle θ (Figure 2.9). With the instantaneous voltage varying sinusoidally as a function of time, we have

$$v(t) = V_{max} \cos(\omega t + \theta) \tag{2.18}$$

Since $e^{j(\omega t + \theta)} = \cos(\omega t + \theta) + j \sin(\omega t + \theta)$, $v(t)$ can be interpreted as the real part of a complex function:

$$v(t) = \text{Re}(V_{max} e^{j(\omega t + \theta)}) = \text{Re}[(V_{max} e^{j\theta})(e^{j\omega t})] \tag{2.19}$$

Dropping the $e^{j\omega t}$ factor, we are left with the following representation of voltage:

$$\mathbf{V} = V_{max} e^{j\theta} = V_{max} \cos\theta + jV_{max} \sin\theta \tag{2.20}$$

If, additionally, V_{max} and I_{max} (the peak values) are divided by $\sqrt{2}$, the resulting magnitudes represent rms values. Such expressions are termed **phasors** and are generally printed in boldface type to indicate their complex nature. Thus, typically,

$$\begin{aligned} \mathbf{I} &= Ie^{j\phi} \\ \mathbf{V} &= Ve^{j\phi} \end{aligned} \tag{2.21}$$

EXAMPLE 2.1

a. Express the waveform in Figure 2.10 as a function of time.
b. Express the waveform in Figure 2.10 as a phasor.

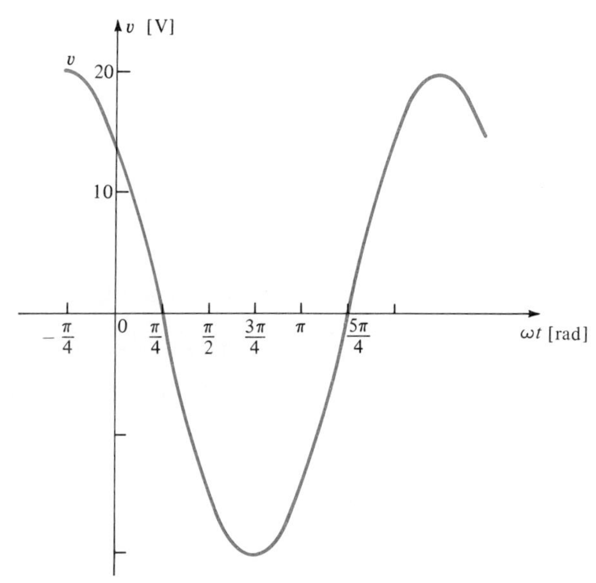

FIGURE 2.10

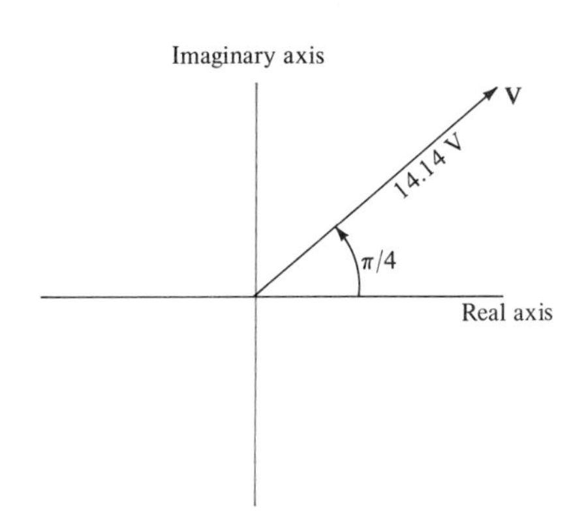

FIGURE 2.11

Solution:

a. The voltage reaches a maximum of 20 V at $-\pi/4$ before $\omega t = 0$. Therefore,

$$v(t) = 20\cos\left(\omega t + \frac{\pi}{4}\right)[\text{V}]$$

Alternatively, we can express the variation as

$$v(t) = \text{Re}\left(20e^{j[\omega t + (\pi/4)]}\right) = \text{Re}(14.14e^{j(\pi/4)}\sqrt{2}\, e^{j\omega t})\ [\text{V}]$$

b. The phasor voltage is

$$\mathbf{V} = 14.14e^{j(\pi/4)} = 14.14\angle\frac{\pi}{4}$$

The amplitude is the rms voltage, and the exponent represents the phase angle relative to the real axis. (See Figure 2.11.)

EXAMPLE 2.2

Express the current $i(t) = 5\sqrt{2}\sin[\omega t + (\pi/4)]$ as a phasor.

Solution: (See Figure 2.12.)

A phasor is represented as the real part of a complex function—that is, as the cosine—and

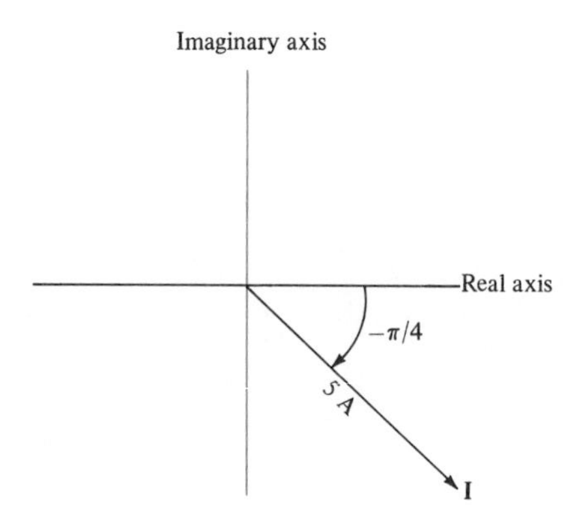

FIGURE 2.12

$$\cos\left(\theta - \frac{\pi}{2}\right) = \cos\theta\cos\frac{\pi}{2} + \sin\theta\sin\frac{\pi}{2} = \sin\theta$$

$$\sin\left(\omega t + \frac{\pi}{4}\right) = \cos\left(\omega t + \frac{\pi}{4} - \frac{\pi}{2}\right) = \cos\left(\omega t - \frac{\pi}{4}\right)$$

$$i(t) = 5\sqrt{2}\cos\left(\omega t - \frac{\pi}{4}\right)$$

$$\mathbf{I} = 5e^{j(-\pi/4)} = 5\angle -\frac{\pi}{4} = 5\angle -45^\circ$$

FOCUS ON PROBLEM SOLVING: Phasor Manipulation

Given a time-varying quantity such as

$$v(t) = V_{max}\cos(\omega t + \theta)$$

where V_{max} is the peak value, the phasor representation consists of the rms value ($V = V_{max}/\sqrt{2}$) times $e^{j\theta}$:

$$\mathbf{V} = Ve^{j\theta}$$

If the time variation is of the form

$$v(t) = V_{max}\sin(\omega t + \theta)$$

since $\cos[\theta - (\pi/2)] = \sin\theta$, then

$$v(t) = V_{max}\cos\left(\omega t + \theta - \frac{\pi}{2}\right)$$

$$\mathbf{V} = \frac{V_{max}}{\sqrt{2}}e^{j[\theta - (\pi/2)]} = Ve^{j[\theta - (\pi/2)]} = -jVe^{j\theta}$$

Addition and Subtraction. Because they are complex numbers, phasors must be added (or subtracted) by adding (or subtracting) their real and imaginary parts separately.

Multiplication. The product of phasors is equal to the product of their magnitudes and the sum of their angles.

Division. The division of phasors is accomplished by dividing their magnitudes and subtracting the divisor angle from that of the dividend.

Powers and Roots. When raising a phasor to a power n, the magnitude of the phasor is raised to the power n and the angle is multiplied by a factor n. When extracting the nth root of a phasor, the magnitude is obtained by extracting its nth root while the angle is divided by n.

$$\mathbf{V}^n = (Ve^{j\theta})^n = V^n\angle n\theta \qquad \mathbf{V}^{1/n} = V^{1/n}\angle\frac{\theta}{n}$$

EXAMPLE 2.3 If $v_1 = 120\sqrt{2}\cos[\omega t - (\pi/4)]$ and $v_2 = 120\sqrt{2}\cos[\omega t + (\pi/6)]$, express $v = v_1 + v_2$ both as a phasor and as a function of time. (See Figure 2.13.)

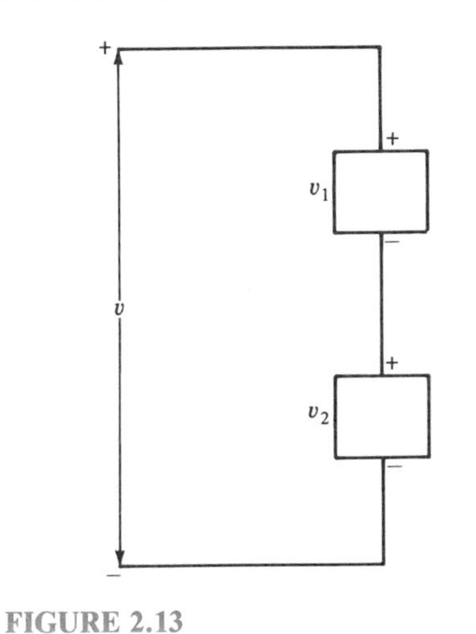

FIGURE 2.13

Solution: The phasor representation is obtained in the following manner:

$$\mathbf{V}_1 = 120e^{-j(\pi/4)} = 120\cos\left(-\frac{\pi}{4}\right) + j120\sin\left(-\frac{\pi}{4}\right)$$

$$\mathbf{V}_2 = 120e^{+j(\pi/6)} = 120\cos\frac{\pi}{6} + j120\sin\frac{\pi}{6}$$

$$\mathbf{V}_1 = 84.85 - j84.85$$

$$\mathbf{V}_2 = 103.92 + j60.00$$

$$\mathbf{V} = 188.8 - j24.85 = 190.4\angle -7.5° = 190.4e^{-j(0.0417\pi)}$$

where $7.5° = 0.0417\pi$.

The corresponding temporal variation is

$$v(t) = 190.4\sqrt{2}\cos(\omega t - 0.0417\pi)$$

2.5 ■ IMPEDANCE

In any practical case there is always some resistance that accompanies a reactance—particularly so in the case of an inductor. In Figure 2.14, while R and L are depicted separately, in actuality they are not distinct entities. The same current passes through L and R. This current and the voltage drop across the resistance are in phase, but the voltage across the inductance leads the current by 90°.

Figure 2.15 shows the variation of v_L and v_R as a function of time as well as the variation of the combined voltage. The addition of the phasors $\mathbf{V}_L$ and $\mathbf{V}_R$ yields $\mathbf{V}$ (Figure 2.16). It also proves informative to compute the power variation that arises in Figure 2.15. Figure 2.17 shows that the positive power pulses are greater than the negative ones, indicating a net power dissipation. Thus is what one would expect from a circuit containing a resistance.

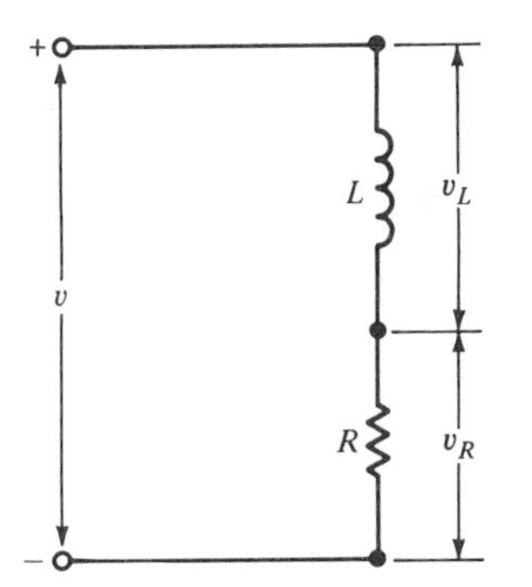

FIGURE 2.14. **Practical Inductor Showing Accompanying Resistance**

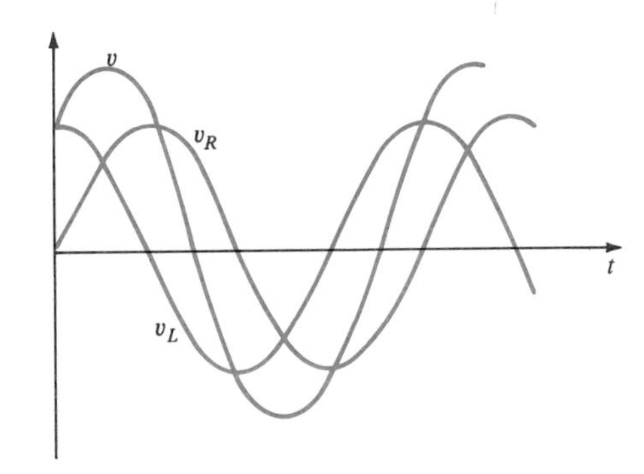

FIGURE 2.15. **Resistive (v_R) and Inductive (v_L) Temporal Variations of Voltage and the Combined Voltage Variation (v)**

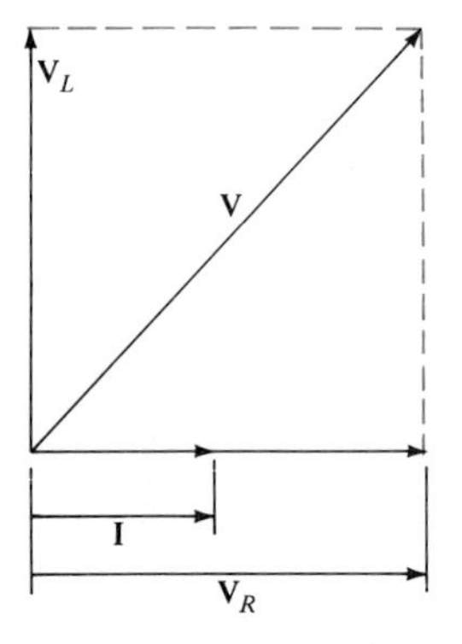

FIGURE 2.16. **Phasor Diagram of Voltage across a Series Combination of Inductance and Resistance**

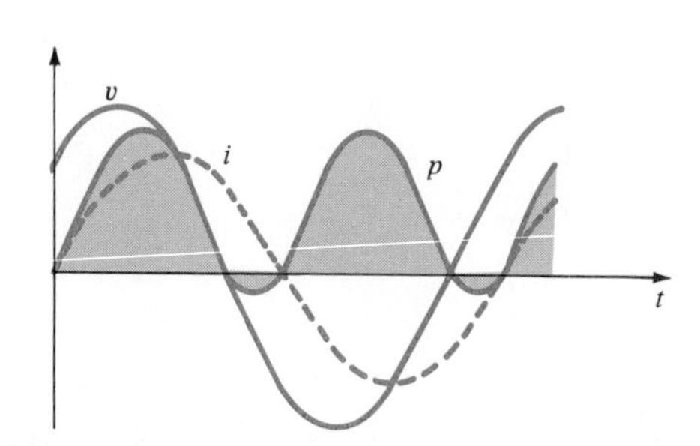

FIGURE 2.17. **Current, Voltage, and Power Variation in a Practical Inductor** Large positive power pulses indicate a net power dissipation due to the presence of the resistive component.

In phasor terms, this is shown in Figure 2.18, where we have made use of Ohm's law for ac circuits—that is, $\mathbf{V}_L = \mathbf{I}X_L$ and $\mathbf{V}_R = \mathbf{I}R$. When a circuit contains both reactance *and* resistance, the combination is termed **impedance** (**Z**), which is also measured in ohms. Figure 2.18 shows the graphical combination of R and X to yield **Z**. Mathematically, we have

$$\mathbf{Z} = R + jX_L \tag{2.22}$$

$$|\mathbf{Z}| = \sqrt{R^2 + X_L^2} \tag{2.23}$$

Thus, the general forms of Ohm's law for ac circuits are

$$\begin{aligned} \mathbf{V} &= \mathbf{IZ} \\ \frac{\mathbf{V}}{\mathbf{Z}} &= \mathbf{I} \\ \frac{\mathbf{V}}{\mathbf{I}} &= \mathbf{Z} \end{aligned} \tag{2.24}$$

While **Z** is a complex quantity, unlike **V** and **I** it is *not* a phasor. The designation phasor is applicable only to sinusoidally varying functions of time.

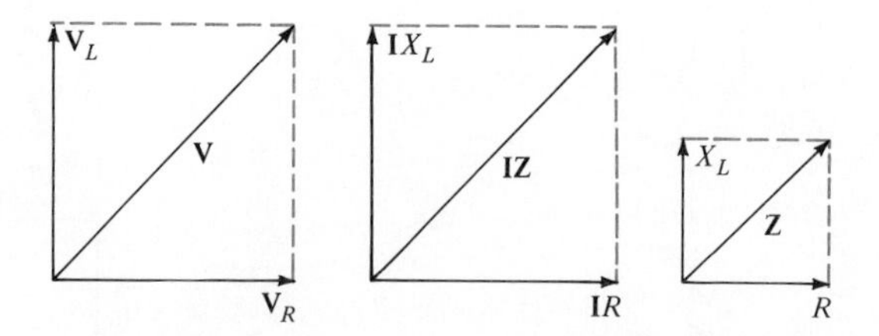

FIGURE 2.18. **Impedance Derived from Resistive and Inductive Voltage Components**

EXAMPLE 2.4

a. For $L = 1$ mH and $R = 5\ \Omega$, compute the impedance at 1000 Hz.
b. What value of capacitance at this same frequency (with $R = 0$) will yield the same magnitude of impedance?

Solution:

a. $X_L = 2\pi fL = 6.28(10^3)(10^{-3}) = 6.28\ \Omega$

$$\mathbf{Z} = R + jX_L = \sqrt{R^2 + X_L^2} \angle \tan^{-1}\frac{X_L}{R} = \sqrt{5^2 + 6.28^2} \angle 51.5°$$

$$= 8.03 \angle 51.5°\ \Omega$$

$$|\mathbf{Z}| = 8.03 = \frac{1}{2\pi fC}$$

b. $C = \dfrac{1}{2\pi(10^3)(8.03)} = 1.98 \times 10^{-5}\ \text{F} = 0.198\ \mu\text{F}$

For capacitive circuits also containing a resistance, we proceed in an analogous manner, as shown in Figure 2.19. We have

$$\mathbf{Z} = R + j\left(\frac{-1}{\omega C}\right) \tag{2.25}$$

$$|\mathbf{Z}| = \sqrt{R^2 + X_C^2} \tag{2.26}$$

The fact that the capacitive reactance is negative does not affect Eq. 2.26, but it does make a significant difference if there is also some L in the circuit (in addition to R and C).

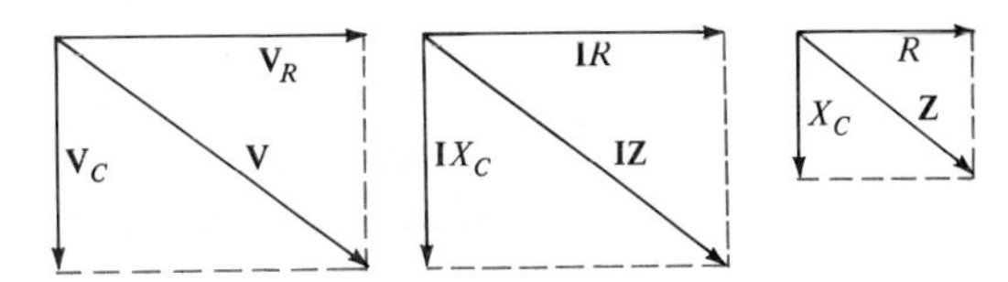

FIGURE 2.19. Impedance Derived from Resistive and Capacitive Voltage Components

Referring to Figure 2.20, we have

$$\mathbf{Z} = R + jX \tag{2.27}$$

where $X = \omega L - (1/\omega C)$. What happens when $|X_L| = |X_C|$? Since

$$|\mathbf{I}| = \left|\frac{\mathbf{V}}{\mathbf{Z}}\right| = \frac{V}{\sqrt{R^2 + X^2}} \tag{2.28}$$

X will equal zero, and I will have its maximum value. But if $|X_L| = |X_C|$, we have

$$2\pi fL = \frac{1}{2\pi fC} \tag{2.29}$$

$$f = \frac{1}{2\pi\sqrt{LC}} \equiv f_0 \tag{2.30}$$

where f_0 is termed the *resonant frequency* of the *RCL* circuit.

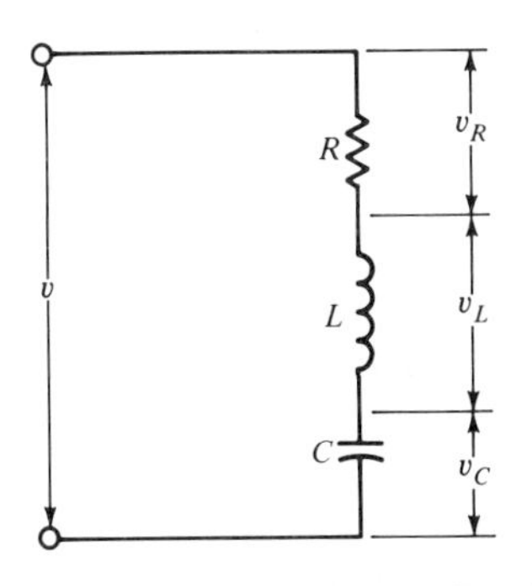

FIGURE 2.20. Series Circuit Consisting of Resistance, Inductance, and Capacitance

EXAMPLE 2.5 What is the series resonant frequency for $L = 1$ mH and $C = 0.2\ \mu$F?

Solution:

$$f_0 = \frac{1}{2\pi\sqrt{LC}} = \frac{1}{2\pi\sqrt{10^{-3} \times 2 \times 10^{-7}}} = 1.125 \times 10^4 \text{ Hz}$$

2.6 ■ COMPLEX NUMBERS

When we considered resistance and reactance, it proved convenient to use complex numbers, such as $a + jb$, where a and b are real numbers and j is equal to $\sqrt{-1}$. This *j operator* is a term used to indicate that an electrical quantity has gone through a 90° rotation in either a counterclockwise ($+j$) or clockwise ($-j$) direction.[2]

For example, if a numerical value, such as 3, is to be plotted along the real x-axis and a numerical value (say 4) along the y-axis, a j is placed before the number 4. The j means rotate the vectorial representation by 90° counterclockwise. Thus, we plot 3 units along the $+x$-axis and 4 units along the $+y$-axis. A vector addition (according to the Pythagorean theorem) yields the vector sum whose magnitude is $\sqrt{3^2 + 4^2} = 5$, and whose orientation is at an angle (with respect to the $+x$-axis) of $\theta = \tan^{-1}(\frac{4}{3}) = 53.1°$.

If we continue to apply the j operator, $j(j4)$ means rotate the four-unit vector through a total rotation of 180° [thereby yielding a value of -4, or $\sqrt{-1}\,(\sqrt{-1})4 = -4$]. Further, $jjj4 = -j4$ (three 90° rotations), and $jjjj4 = +4$ (four 90° rotations, bringing us back to where we started).

With these concepts in mind, we can show that X_C is a negative quantity by the following: If X_C were taken to be positive, jX_C would have to be plotted in the $+y$-direction. With X_C taken to be negative, as a real number it would have to be plotted in the $-x$-direction, and application of j would turn it into the $-y$-direction, which is correct.

There are a number of alternative ways of expressing impedance (and other ac quantities) analytically:

$$R + jX = Z\angle\theta = Ze^{j\theta} = Z\cos\theta + jZ\sin\theta$$

where

$$Z = \sqrt{R^2 + X^2}$$

$$\theta = \tan^{-1}\frac{X}{R}$$

$$R = Z\cos\theta$$

$$X = Z\sin\theta$$

Expressing complex numbers in terms of real and imaginary components (for example, $a + jb$) constitutes their *rectangular* or *Cartesian* form. Ex-

[2]In mathematics i is used rather than j, but since in electrical engineering i is used to represent current, the indicated substitution is made.

pressing complex numbers as a magnitude times an exponential (for example, $Me^{j\theta}$) constitutes their *polar* form (also expressible symbolically as $M \angle \theta$).

The angle θ equals the angular difference between the current and voltage waveforms and is termed the **phase angle**. Using the current as a reference, we find that in a pure inductor the voltage leads by 90°, and hence $\theta = +90°$. In a capacitor, the voltage lags by 90°, and $\theta = -90°$.

If we encounter a complex number as a denominator, it is usually desirable to separate the complex expression into a real and an imaginary term. An example arises when a voltage V is applied to a circuit impedance $R + jX$ and we want to determine the resulting current:

$$\mathbf{I} = \frac{V}{R + jX}$$

One method of evaluation proceeds by multiplying both numerator and denominator by the complex conjugate of the denominator. (The complex conjugate of a quantity is obtained by changing the signs on all imaginary terms.) Thus, we have

$$\mathbf{I} = \left(\frac{V}{R + jX}\right)\left(\frac{R - jX}{R - jX}\right) = \frac{VR - jVX}{R^2 + X^2}$$

$$= \frac{VR}{R^2 + X^2} - j\left(\frac{VX}{R^2 + X^2}\right)$$

With these real and imaginary terms, the current magnitude and phase may be determined by the usual methods.

Alternatively, we can handle the equation in the following manner:

$$\mathbf{I} = \frac{V}{R + jX} = \frac{V}{\sqrt{R^2 + X^2} \angle \theta} = \frac{V}{\sqrt{R^2 + X^2}} \angle -\theta$$

$$= \frac{V}{\sqrt{R^2 + X^2}} e^{-j\theta}$$

where $\theta = \tan^{-1}(X/R)$. Since $e^{-j\theta} = \cos\theta - j\sin\theta$, we have

$$\cos\theta = \frac{R}{\sqrt{R^2 + X^2}} \qquad \sin\theta = \frac{X}{\sqrt{R^2 + X^2}}$$

Thus,

$$\mathbf{I} = \frac{V}{\sqrt{R^2 + X^2}}\left(\frac{R}{\sqrt{R^2 + X^2}} - j\frac{X}{\sqrt{R^2 + X^2}}\right)$$

and we see that the two results are identical.

EXAMPLE 2.6

a. For an effective voltage of 1 V, applied to a circuit consisting of $X_C = -100\ \Omega$ and $R = 50\ \Omega$, compute the real and reactive parts of the current.

b. Determine the magnitude of the current and the phase relationship between current and voltage.

Solution:

a. $\mathbf{Z} = R - jX_C = 50 - j100 = 112\angle -63.4°\ \Omega$

$$\mathbf{I} = \frac{V}{\mathbf{Z}} = \frac{1}{112\angle -63.4°} = 0.0089\angle 63.4° = (0.004 + j0.008)\text{ A}$$

The real part of the current is 4 mA, and reactive part is 8 mA.

b. The current magnitude is 8.9 mA. The current leads the voltage by 63.4°.

An alternative solution would proceed as follows:

$$\mathbf{I} = \frac{1}{R - jX} = \left(\frac{1}{50 - j100}\right)\left(\frac{50 + j100}{50 + j100}\right)$$

$$= \frac{50}{50^2 + 100^2} + j\left(\frac{100}{50^2 + 100^2}\right) = (0.004 + j0.008)\text{ A}$$

Note: A reference angle of 0° was selected for the voltage—that is, $V = 1\angle 0°$. In most cases such a reference selection is arbitrary, but once made applies to all phasors in a given problem.

EXAMPLE 2.7 $\mathbf{V} = 120\angle 38°$ V is applied to an impedance $\mathbf{Z} = 3 + j5$. Compute the resulting current.

Solution:

$$\mathbf{I} = \frac{\mathbf{V}}{\mathbf{Z}} = \frac{|\mathbf{V}|\angle\theta}{R + jX_L} = \frac{|\mathbf{V}|\angle\theta}{\sqrt{R^2 + X_L^2}\angle\tan^{-1}(X_L/R)} = \frac{|\mathbf{V}|\angle\theta}{|\mathbf{Z}|\angle\phi}$$

$$= \frac{120\angle 38°}{3 + j5} = \frac{120\angle 38°}{5.83\angle 59°} = 20.6\ \angle 38° - 59°$$

$$= 20.6\angle -21°\text{ A}$$

In this case, relative to the positive real axis, we would graphically show **V** with a positive angle of 38° and **I** with a negative angle of −21°. We could then say that the voltage was leading the current by 59°—that is, by 38° − (−21°). In keeping with the convention that **I** be drawn along the positive real axis, we could rotate both phasors in the positive direction and represent **V** at a positive angle of 59°.

2.7 ■ REAL AND REACTIVE POWER

Series *RL* circuits are frequently encountered in industrial practice. For example, the wire used as the heater element in a resistance furnace is generally wound in the form of a coil, giving rise to a significant inductance. In such cases the power dissipation is *not* equal to the product of voltage (V) and current (I) since the two are not in phase. In a *purely* inductive circuit,

$$i = I_{max}\sin\omega t \tag{2.31}$$

$$v = V_{max} \sin\left(\omega t + \frac{\pi}{2}\right) \tag{2.32}$$

The latter indicates the 90° phase difference between current and voltage. To compute the instantaneous power (p), since

$$\sin\left(\omega t + \frac{\pi}{2}\right) = \cos \omega t$$

and

$$2 \sin \omega t \cos \omega t = \sin 2\omega t$$

we have

$$p = \tfrac{1}{2} V_{max} I_{max} \sin 2\omega t \tag{2.33}$$

Thus, we have verified our prior graphical result showing that the power pulses vary at twice the applied frequency, with equal positive and negative pulses yielding zero average power dissipation.

With only resistance present,

$$p = vi = V_{max} I_{max} \sin^2 \omega t \tag{2.34}$$

and since $\sin^2 \omega t = \frac{1}{2}(1 - \cos 2\omega t)$, we have

$$p = \tfrac{1}{2} V_{max} I_{max}(1 - \cos 2\omega t) \tag{2.35}$$

Not unexpectedly, the power remains positive, varying between zero and the value $I_{max}V_{max}$, with an average value of $\frac{1}{2} V_{max} I_{max}$. (The latter, of course, is $V_{rms} I_{rms}$.)

With both resistance and inductance present, the voltage leads the current by some angle that is less than 90°:

$$i = I_{max} \sin \omega t$$
$$v = V_{max} \sin(\omega t + \theta)$$

Thus,

$$p = vi = V_{max} I_{max} \sin \omega t \sin(\omega t + \theta) \tag{2.36}$$

Since, in general, $\sin \alpha \sin \beta = \frac{1}{2}[\cos(\alpha - \beta) - \cos(\alpha + \beta)]$ and $\cos(-\alpha) = \cos \alpha$, we have

$$\begin{aligned} p &= \tfrac{1}{2} V_{max} I_{max}[\cos(\omega t - \omega t - \theta) - \cos(\omega t + \omega t + \theta)] \\ &= \tfrac{1}{2} V_{max} I_{max}[\cos \theta - \cos(2\omega t + \theta)] \end{aligned} \tag{2.37}$$

Thus, the instantaneous power consists of a cosine term that will average out to zero and a constant term, $\frac{1}{2} V_{max} I_{max} \cos \theta$. The average value of p is then

$$P = \tfrac{1}{2} V_{max} I_{max} \cos \theta = VI \cos \theta \tag{2.38}$$

where V and I are the rms values and $\cos \theta$ is termed the **power factor** (*pf*). Note that a unity power factor means the voltage and current are in phase. A lagging power factor means the current lags the voltage (as in an inductive circuit), and a leading power factor has the current leading the voltage (as in a capacitive circuit).

The product VI represents **apparent power**—that is, the product of the rms voltage and rms current with an angular difference between them being disregarded. It is measured in either volt-amperes (VA) or kilovolt-amperes (kVA) to distinguish it from real power, which is measured in either watts (W) or kilowatts (kW). **Real power** is equal to $VI \cos \theta$, with θ being the angular difference between the voltage and current waveforms. A particular utility line might be specified as capable of furnishing a stated kVA. Since the available voltage is fixed, this specification is an indication of the maximum current permitted.

The use of a power triangle is instructive in understanding these concepts. In Figure 2.21a, the voltage is shown leading the current in an RL load. In Figure 2.21b, we see the redrawn in-phase ($I \cos \theta$) and quadrature ($I \sin \theta$) components of the current. Finally, in Figure 2.21c, we have multiplied each component by V. Note that the hypotenuse represents the apparent power (volt-amperes) composed of the real power ($VI \cos \theta$) (in watts) and the reactive power, $VI \sin \theta$ (in volt-amperes reactive, VAR). The **reactive power** ($VI \sin \theta$) represents that portion of the apparent power that is exchanged between source and load without dissipation.

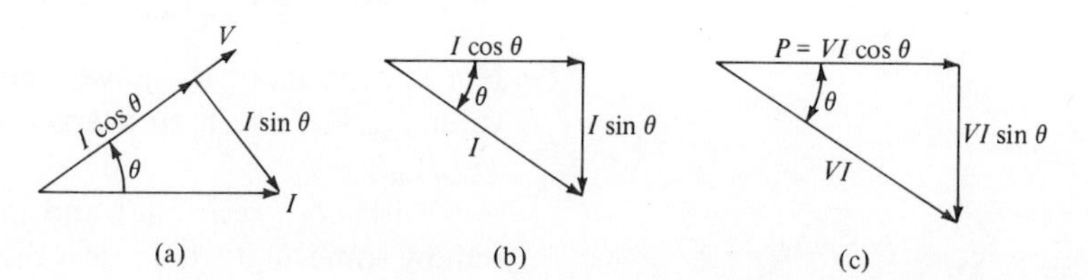

FIGURE 2.21. **Power Triangle Showing Apparent (VI), Real ($VI \cos \theta$), and Reactive ($VI \sin \theta$) Power**

FOCUS ON PROBLEM SOLVING: Real and Reactive Power

With the phase angle θ between voltage and current,

$$\text{power factor} = \cos \theta = \frac{R}{Z}$$

$$\text{real power } VI \cos \theta = I^2R = \frac{V_R^2}{R}$$

$$\text{reactive power} = VI \sin \theta = I^2X = \frac{V_X^2}{X}$$

$$\text{apparent power} = VI = I^2Z = \frac{V^2}{Z}$$

where $V_R = VR/Z$ and $V_X = VX/Z$ and where V and I are rms values.

For a **lagging power factor**, the reactive power should be depicted as being negative as in Figure 2.21c. However, it is customary in power applications to consider a lagging power factor as giving rise to positive reactive power, and negative when a **leading power factor** prevails.

EXAMPLE 2.8

A motor rated at 5.5 kVA operates from a 240 V line with a lagging power factor of 0.8. What are the real and reactive components of power? What is the value of the current? The value of R? The value of X?

Solution:

$$5.5 \text{ kVA} = 5500 \text{ VA}$$

$$\cos\theta = 0.8 = \frac{R}{Z} = \text{power factor}$$

$$\theta = 36.9°$$

$$\text{apparent power} = VI = 5.5 \text{ kVA} = \frac{V^2}{Z} = \frac{240^2}{Z}$$

$$Z = \frac{240^2}{5500} = 10.5\ \Omega$$

$$R = 0.8Z = 0.8(10.5) = 8.4\ \Omega$$

$$I = \frac{VI}{V} = \frac{5500}{240} = 22.9 \text{ A}$$

$$Z^2 = R^2 + X^2$$

$$X = \sqrt{Z^2 - R^2} = \sqrt{10.5^2 - 8.4^2} = 6.3\ \Omega$$

$$\text{real power} = VI\cos\theta = I^2R = 22.9^2(8.4) = 4400 \text{ W}$$

$$\text{reactive power} = VI\sin\theta = I^2X = 22.9^2(6.3) = 3300 \text{ VAR}$$

2.8 ▪ RESONANCE AND CIRCUIT Q

2.8.1 Series Resonance

We have seen that for a series RLC circuit (Figure 2.20) the current will be at a maximum when $\omega_0 L = 1/\omega_0 C$. Such equality between the magnitudes of inductive and capacitive reactance constitutes the **resonant frequency**. The resonant frequency is related to L and C through the equation

$$2\pi f_0 = \omega_0 = \frac{1}{\sqrt{LC}} \tag{2.39}$$

In general,

$$|\mathbf{I}| = \left|\frac{\mathbf{V}}{\mathbf{Z}}\right| = \frac{V}{\sqrt{R^2 + [\omega L - (1/\omega C)]^2}} \tag{2.40}$$

and the current at resonance (I_0) is

$$I_0 = \frac{V}{R} \tag{2.41}$$

For example, if the applied voltage has an effective value of 1 mV, $R = 0.1\ \Omega$, $C = 0.1\ \mu\text{F}$, and $L = 1$ mH, then we have

$$f_0 = \frac{1}{2\pi\sqrt{LC}} = \frac{1}{2\pi\sqrt{10^{-3} \times 10^{-7}}} \simeq 15.9 \text{ kHz}$$

At resonance,

$$I_0 = \frac{10^{-3}}{10^{-1}} = 10^{-2} \text{ A} = 0.01 \text{ A}$$

With 0.01 A through the circuit at resonance, let us compute the effective voltage drop across each series element:

$$V_R = I_0 R = 10^{-2} \text{ A} \times 0.1\ \Omega = 10^{-3} \text{ V} = 1 \text{ mV}$$

$$V_L = I_0 X_L = I_0(2\pi f_0 L) = I_0\left[2\pi\left(\frac{1}{2\pi\sqrt{LC}}\right)L\right] = I_0\sqrt{\frac{L}{C}}$$

$$= 10^{-2}\sqrt{10^4} = 1 \text{ V!}$$

$$V_C = I_0 X_C = I_0\left(\frac{1}{2\pi f_0 C}\right) = I_0\left[\frac{1}{2\pi(1/2\pi\sqrt{LC})C}\right] = I_0\sqrt{\frac{L}{C}}$$

$$= 10^{-2}\sqrt{10^4} = 1 \text{ V!}$$

The question of how we can get a voltage across L and C that is larger than that being applied may most easily be answered graphically, as in Figure 2.22. The same alternating current passes through all three elements, but while the voltage drop across the resistance is in phase with this current, the voltage across the capacitor lags the current by 90° and across the inductor it leads the current by 90°. While the signal voltages across L and C *individually* may be large at any given time, the voltage across the L and C combination will be zero; hence the entire applied voltage appears across R. Thus, a series RLC circuit may be used to magnify a signal voltage. For example, a small radio signal applied to such a series circuit allows a much larger signal voltage to be made available across the capacitor (or the inductor).

Solving the problem in general terms, we obtain

$$\begin{aligned} V_L &= I_0 X_L = \frac{V}{R}\omega_0 L = \frac{\omega_0 L}{R} V \\ V_C &= I_0 X_C = \left(-\frac{V}{R}\right)\left(\frac{1}{\omega_0 C}\right) = -\frac{1}{\omega_0 CR} V \end{aligned} \tag{2.42}$$

Thus, the magnification depends on either $\omega_0 L/R$ or $1/\omega_0 CR$. Actually, these quantities equal each other. Since $\omega_0 = 1/\sqrt{LC}$, $L = 1/\omega_0^2 C$,

$$\frac{\omega_0 L}{F} = \left(\frac{\omega_0}{R}\right)\left(\frac{1}{\omega_0^2 C}\right) = \frac{1}{\omega_0 CR}$$

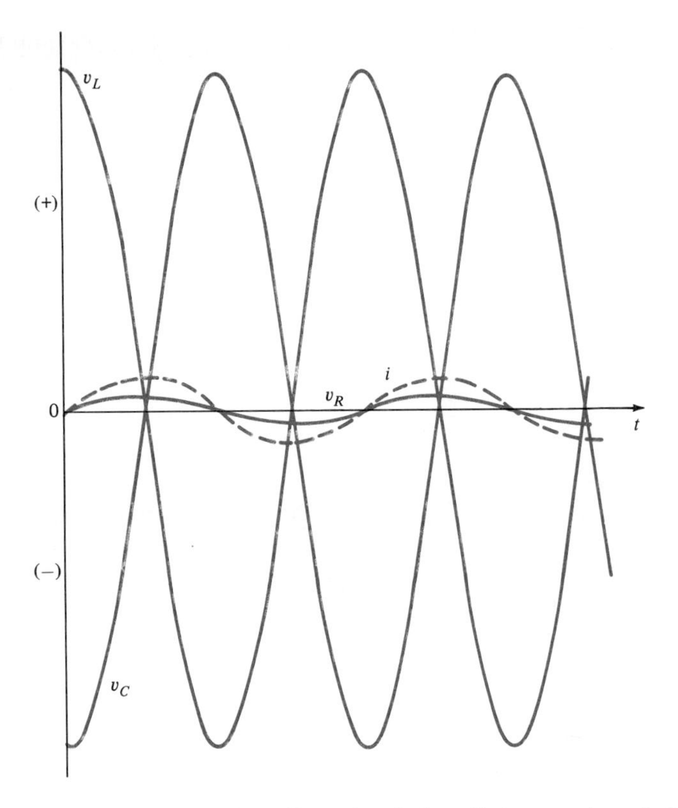

FIGURE 2.22. **Component Voltage Drops in a Series** Inductor voltage (v_L) and capacitor voltage (v_C) remain 180° out of phase in a series resonance circuit at resonance. Voltage across the resistance (v_R) remains in phase with current (i) through the circuit. (The magnitude of v_R is exaggerated.)

These quantities, termed the Q of the circuit, represent a measure of this magnification property[3]:

$$\frac{\omega_0 L}{R} = \frac{1}{\omega_0 CR} \equiv Q \tag{2.43}$$

As the alternating current passes through the circuit, the magnetic and electric fields alternately build up and then collapse. As can be seen from Figure 2.22, during the first quarter-cycle, v_L and i lead to a positive product ($v_L i$ being equal to power), while the product v_C times i is negative. As the magnetic field is building up, the electric field is collapsing and vice versa. If there were no resistance in the circuit, there would be no power dissipated, and the energy would simply be exchanged between the coil and capacitor ad infinitum. But, because there is resistance, some power is dissipated during each cycle. The following represents the most basic definition of **circuit Q**:

[3]This Q should not be confused with electrical charge, which is also represented by the same letter.

$$Q = 2\pi\left(\frac{\text{maximum energy stored}}{\text{energy dissipation per cycle}}\right) \tag{2.44}$$

Since we have taken I to be the rms value of the current, $I_{max} = \sqrt{2}\, I$, we have the following equation[4]:

$$W_{stored} = \tfrac{1}{2}LI^2_{max} = LI^2 \tag{2.45}$$

We know

$$\frac{\text{energy}}{\text{time}} = \text{power}$$

$$\text{dissipated energy} = \text{power dissipation} \times \text{time}$$

$$\frac{\text{time}}{\text{cycle}} = \frac{1}{\text{frequency}}$$

Therefore, the dissipated energy per cycle is

$$\frac{\text{dissipated energy}}{\text{cycle}} = \text{dissipated power} \times \frac{\text{time}}{\text{cycle}}$$

$$= \frac{\text{dissipated power}}{f_0}$$

Thus,

$$\frac{W_{dis}}{\text{cycle}} = \frac{P}{f_0} = \frac{I^2R}{f_0} \tag{2.46}$$

$$Q_{series\ circuit} = 2\pi\left(\frac{LI^2}{I^2R/f_0}\right) = 2\pi f_0\left(\frac{L}{R}\right) = \frac{\omega_0 L}{R} \tag{2.47}$$

which agrees with the prior result.

There is yet another expression for the circuit Q that is particularly informative, but may best be explored in terms of the parallel resonant circuit.

2.8.2 Parallel Resonance

In conjunction with the parallel resonant circuit (Figure 2.23), there are two currents to be considered: There is a large *circulating* current I that represents the exchanging of energy between the magnetic and electric fields. If there were no resistance (R_s), the magnitude of I would not change. To the extent that there is some energy dissipation in R_s, the external circuit must furnish an alternating current I', which makes up for this loss. (Here I and I' are to be interpreted as being rms values.) Note that the larger the value of R_s, the greater will be the loss per cycle, and the greater will be the value of I' that the external circuit will be called upon to furnish if I is to be maintained.

[4]At this point this equation may be justified as follows: We will show in Chapter 3 that the electrical inductance (L) corresponds to mechanical mass (m) and that electrical current (i) corresponds to mechanical velocity (v). In mechanics the kinetic energy is $\frac{1}{2}mv^2$; the corresponding electrical energy is $\frac{1}{2}LI^2_{max}$. (For dc circuits it is $\frac{1}{2}LI^2_{dc}$.)

Insofar as the external circuit is concerned, if one imagines that it doesn't know what appears between the terminals, when it is called upon to furnish a small I', it interprets the impedance between the terminals to be large. If it must furnish a large I', it interprets that impedance to be small. Thus, we have the rather peculiar condition that the bigger R_s, the smaller the impedance of the parallel circuit appears to be.

The parallel *RLC* circuit may be represented by the *equivalent form* shown in Figure 2.24. It should be noted that the R_p of the *equivalent* circuit is *not* equal to the R_s of the *actual* circuit.

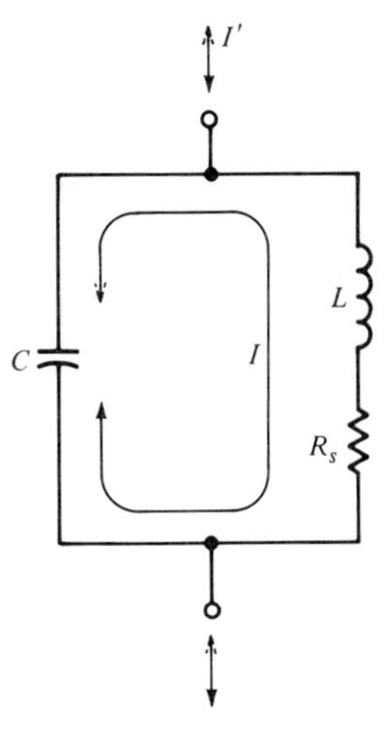

FIGURE 2.23. **Parallel Resonant Circuit** A large circulating current (I) is replenished by a much smaller current (I') necessitated by power dissipation in the series resistance R_s.

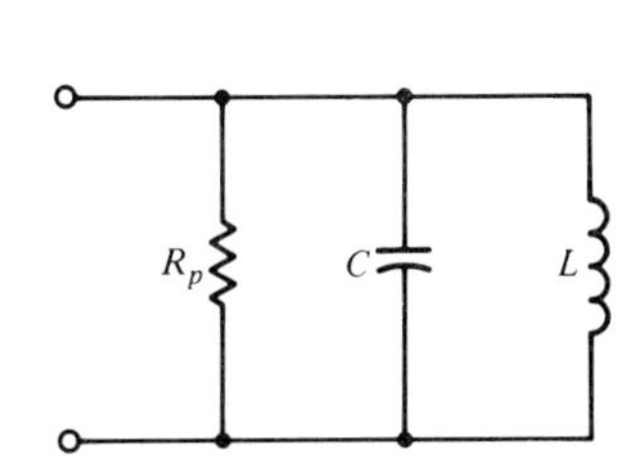

FIGURE 2.24. **Current Equivalent to Actual Parallel Circuit** At resonance C and L effectively cancel, and only R_p appears as the impedance.

Like most parallel circuits, Figure 2.23 may be solved most conveniently in terms of reciprocals. The reciprocal of reactance is termed **susceptance,** usually designated by the letter B (which should not be confused with magnetic flux density, which is also specified by the same letter):

$$B_C = \omega C \qquad \text{[siemens]} \tag{2.48}$$

The reciprocal of inductive reactance would be B_L, but in the inductive branch of Figure 2.23 we also have resistance. Thus, we must consider the **admittance Y**, the reciprocal of the impedance **Z**:

$$\begin{aligned}\mathbf{Y}_L &= \frac{1}{R_s + j\omega L} = \left(\frac{1}{R_s + j\omega L}\right)\left(\frac{R_s - j\omega L}{R_s - j\omega L}\right) = \frac{R_s - j\omega L}{R_s^2 + \omega^2 L^2} \\ &= \frac{R_s}{R_s^2 + \omega^2 L^2} - \frac{j\omega L}{R_s^2 + \omega^2 L^2} \equiv G_L - jB_L\end{aligned} \tag{2.49}$$

$$\mathbf{Y}_{tot} = \frac{R_s}{R_s^2 + \omega^2 L^2} - j\left(\frac{\omega L}{R_s^2 + \omega^2 L^2} - \omega C\right) \tag{2.50}$$

At resonance the impedance of the parallel circuit is purely resistive; hence, the resonance condition is one for which the coefficient of the j term is zero:

$$\frac{\omega_0 L}{R_s^2 + \omega_0^2 L^2} = \omega_0 C \tag{2.51}$$

$$\omega_0 = \sqrt{\frac{1}{LC} - \frac{R_s^2}{L^2}} \tag{2.52}$$

However, in computing the Q of a parallel resonant circuit, it is the *natural* resonance frequency ($\omega_0 = 1/\sqrt{LC}$) that must be used. This may be justified by observing that when the circuit is considered in its equivalent form (Figure 2.24), we have $X_C = X_L$, which leads to $\omega_o = 1/\sqrt{LC}$.

One generally strives to make R_s small, in which case the resonance condition for a parallel circuit is approximately the same as that for a series *RLC* circuit:

$$\omega_0 \approx \sqrt{\frac{1}{LC}} \tag{2.53}$$

For a parallel resonant circuit (using the equivalent in Figure 2.24), we have

$$W_{stored} = \tfrac{1}{2} C V_{max}^2 = CV^2 \tag{2.54}$$

where V is the rms voltage.[5] Also, we have

$$\frac{W_{dis}}{\text{cycle}} = \frac{P}{f_0} = \frac{V^2}{f_0 R_p}$$

$$Q_p = 2\pi\left(\frac{CV^2}{V^2/f_0 R_p}\right) = 2\pi f_0 R_p C = \omega_0 R_p C \tag{2.55}$$

Note that the *mathematical* expression for Q_p is the reciprocal of that for a series circuit, Q_s. Likewise, $Q_p = R_p/\omega_0 L$. *In the literature it is sometimes stated that $Q_p = 1/Q_s$. This statement should not be taken literally. The resistances in the two expressions may not be the same.*

At resonance, R_p represents the parallel circuit impedance. Note again that as R_s increases, the value of Q_s (and hence also Q_p) diminishes, and the value of R_p decreases, substantiating our prior statement.

EXAMPLE 2.9 Express the following in terms of their reciprocals:

$$\mathbf{Z} = (3 + j4)\ \Omega$$

$$X_C = -5\ \Omega$$

$$X_L = 10\ \Omega$$

[5] In an analogous fashion to what we encountered with an inductor, Chapter 3 will show that C corresponds to the reciprocal of a mechanical spring force constant (k) and that the voltage (v) corresponds to mechanical force (kx) (x being displacement). As a result, $\frac{1}{2}CV_{max}^2$ corresponds to potential energy in the mechanical case, $\frac{1}{2}kx^2$. We may conclude then that, just as in the mechanical case, energy is exchanged between potential and kinetic. In the electrical case this corresponds to an exchange between the magnetic and electric fields.

Solution:

$$\mathbf{Y} \equiv \text{admittance} = \frac{1}{\mathbf{Z}} = \frac{1}{3 + j4} = \frac{1}{5\angle 53.1^\circ} = 0.2\angle -53.1^\circ \text{ S}$$
$$= 0.2 \cos(-53.1^\circ) + j0.2 \sin(-53.1^\circ)$$
$$= 0.12 - j0.16 = (G - jB) \text{ S}$$
$$G \equiv \text{conductance} = 0.12 \text{ S}$$
$$B \equiv \text{susceptance} = -0.16 \text{ S}$$
(inductive susceptance is negative)
$$\mathbf{Y} = \frac{1}{0 - j5} = \frac{1}{5\angle -90^\circ}$$
$$= 0.2\angle 90^\circ = 0.2 \cos(90^\circ) + j0.2 \sin(90^\circ) = (0 + j0.2) \text{ S}$$
$$G = 0$$
$$B = 0.2 \text{ S} \qquad \text{(capacitive susceptance is positive)}$$
$$\mathbf{Y} = \frac{1}{0 + j10} = \frac{1}{10\angle 90^\circ} = 0.1\angle -90^\circ$$
$$= 0.1 \cos(-90^\circ) + j0.1 \sin(-90^\circ)$$
$$= (0 - j0.1) \text{ S}$$
$$G = 0$$
$$B = -0.1 \text{ S} \qquad \text{(negative)}$$

EXAMPLE 2.10 If the components of the series circuit in Figure 2.25 are placed in parallel, compute the equivalent parallel resistance (R_p) at resonance.

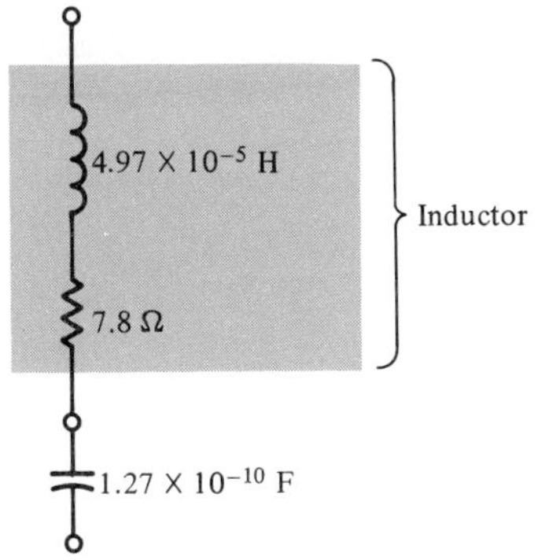

4.97 X 10⁻⁵ H
7.8 Ω
Inductor
1.27 X 10⁻¹⁰ F

FIGURE 2.25

Solution:

$$f_0 = \frac{1}{2\pi\sqrt{LC}} = \frac{1}{2\pi\sqrt{4.97 \times 10^{-5} \times 1.27 \times 10^{-10}}}$$
$$= 2.00 \times 10^6 \text{ Hz}$$
$$Q_s = \frac{\omega_0 L}{R_s} = \frac{2\pi \times 2 \times 10^6 \times 4.97 \times 10^{-5}}{7.8} = 80$$
$$Q_p = \frac{R_p}{\omega_0 L}$$
$$R_p = \omega_0 L Q_p = 2\pi(2 \times 10^6)(4.97 \times 10^{-5})(80) = 5 \times 10^4\ \Omega$$

FOCUS ON PROBLEM SOLVING: Parallel Equivalent of a Series Resonant Circuit

The resonant frequency of a series resonant LCR_s circuit is

$$\omega_0 = \frac{1}{\sqrt{LC}}$$

For a parallel LCR_s circuit, it is

$$\omega_0 = \sqrt{\frac{1}{LC} - \frac{R_s^2}{L^2}}$$

With R_s the series resistance, we have

$$Q_s = \frac{\omega_0 L}{R_s} = \frac{1}{\omega_0 C R_s}$$

Since Q_s is a measure of the energy dissipation per cycle, this does not change if these same elements are configured into a parallel circuit. Therefore, numerically, $Q_s = Q_p$.

Letting R_p represent the equivalent resistance (R_p) in parallel with C and L, we have

$$Q_p = \frac{R_p}{\omega_0 L} = \omega_0 C R_p$$

It must be remembered that in computing the parallel Q_p, the resonant frequency used is that of the series circuit because R, L, and C are in series around the loop involving I.

2.8.3 Response Curves of Resonant Circuits

Returning to the *series RLC*, we have

$$\mathbf{Z} = R + j\left(\omega L - \frac{1}{\omega C}\right) \tag{2.56}$$

At resonance, we have

$$\mathbf{Z}_0 = R \tag{2.57}$$

$$\frac{\mathbf{Z}_0}{\mathbf{Z}} = \frac{R}{R + j[\omega L - (1/\omega C)]} = \frac{1}{1 + j[(\omega L/R) - (1/\omega CR)]} \tag{2.58}$$

Multiplying the term in brackets by ω_0/ω_0, we get

$$\frac{\mathbf{Z}_0}{\mathbf{Z}} = \frac{1}{1 + j\{[(\omega_0/\omega_0)(\omega L/R)] - [(\omega_0/\omega_0)(1/\omega CR)]\}} \tag{2.59}$$

Letting $\omega_0 L/R = 1/\omega_0 CR \equiv Q$, we obtain

$$\frac{\mathbf{Z}_0}{\mathbf{Z}} = \frac{1}{1 + j[Q(\omega/\omega_0) - Q(\omega_0/\omega)]}$$

$$= \frac{1}{1 + jQ[(\omega/\omega_0) - (\omega_0/\omega)]} \tag{2.60}$$

The *reciprocal* of Eq. 2.60 can be used to describe the *parallel RLC* circuit:

$$\frac{\mathbf{Z}_0}{\mathbf{Z}} = 1 + jQ\left(\frac{\omega}{\omega_0} - \frac{\omega_0}{\omega}\right) \tag{2.61}$$

Figure 2.26 illustrates the variation of $|\mathbf{Z}/\mathbf{Z}_0|$ for a parallel *RLC* circuit.

Most radio and TV sets utilize a parallel resonant circuit to "tune in" the desired station by having either L or C variable. As one tunes through resonance, the impedance varies as in Figure 2.26. The "sharpness" of this resonant curve determines how well the circuit can discriminate against interference from stations at adjoining frequencies. At resonance, due to the large impedance, a large signal *voltage* is developed at the desired frequency.

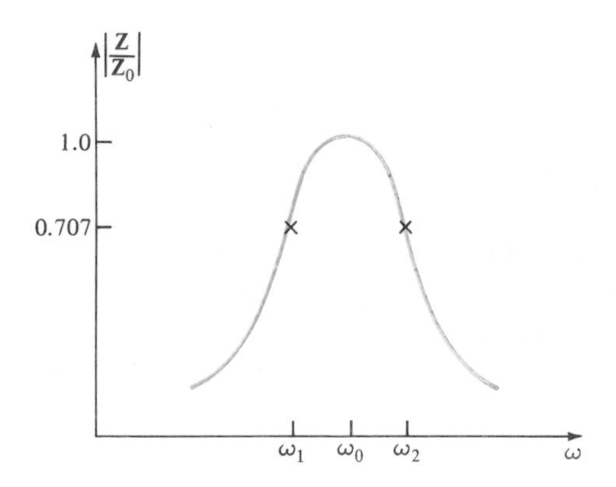

FIGURE 2.26. **Ratio of the Magnitude of Parallel Circuit Impedance (Z) at Frequency ω to the Magnitude of Impedance ($\mathbf{Z}_0$) at the Resonant Frequency ω_0** The frequency span, $\omega_2 - \omega_1$, defines the bandwidth.

The sharpness of the resonance is defined in terms of the **bandwidth**, which is the frequency interval between those frequencies at which the parallel circuit impedance has dropped to 0.707 of its resonant value. Calling these frequencies ω_1 and ω_2, respectively, we find that at these frequencies the expression in parentheses in Eq. 2.61 is equal to $\pm 1/Q$:

$$\frac{\mathbf{Z}_0}{\mathbf{Z}} = 1 \pm j1 = \sqrt{2} \angle \pm 45^\circ$$

$$|\mathbf{Z}| = 0.707\,|\mathbf{Z}_0|$$

We then have

$$\frac{\omega_1}{\omega_0} - \frac{\omega_0}{\omega_1} = -\frac{1}{Q}$$

$$\frac{\omega_2}{\omega_0} - \frac{\omega_0}{\omega_2} = +\frac{1}{Q}$$

$$\omega_1 = \omega_0\sqrt{1 + (\tfrac{1}{2}Q)^2} - \frac{\omega_0}{2Q}$$

$$\omega_2 = \omega_0\sqrt{1 + (\tfrac{1}{2}Q)^2} + \frac{\omega_0}{2Q}$$

$$\omega_2 - \omega_1 = \frac{\omega_0}{Q} \tag{2.62}$$

That is, the greater the circuit Q, the sharper the resonance curve. When used in a radio receiver, such circuits have good discrimination against interference from adjoining stations. Such receivers are said to have good **selectivity**. Inexpensive radio sets that exhibit considerable adjacent channel interference have low-Q resonant circuits.

SUMMARY

2.1 Introduction

- Resistors impede current and dissipate power.
- Reactances impede current but do not dissipate power.

2.2 Inductive Reactance

- Inductive reactance (X_L) equals $2\pi fL$, where f is the frequency of the ac (in hertz) and L the inductance (in units of henry).
- For inductors in series

$$L_{tot} = L_1 + L_2 + L_3 + \cdots$$

- For inductors in parallel,

$$\frac{1}{L_{tot}} = \frac{1}{L_1} + \frac{1}{L_2} + \frac{1}{L_3} + \cdots$$

2.3 Capacitive Reactance

- Capacitive reactance (X_C) equals $-1/2\pi fC$, where f is the ac frequency (in hertz) and C the capacitance (in units of farad).
- For capacitors in series,

$$\frac{1}{C_{tot}} = \frac{1}{C_1} + \frac{1}{C_2} + \frac{1}{C_3} + \cdots$$

- For capacitors in parallel,

$$C_{tot} = C_1 + C_2 + C_3 + \cdots$$

2.5 Impedance

- The combination of resistance and reactance constitutes impedance.

2.6 Complex Numbers

- Mathematically, resistance is represented by a real number, reactance is represented by an imaginary number, and impedance is represented by a complex number.

2.7 Real and Reactive Power

- Real power, expressed in watt units, represents dissipated power. Reactive power represents power exchanged between source and load without any dissipation and is expressed in units of volt-ampere. Apparent power represents real and reactive power and is also expressed in units of volt-ampere.

2.8 Resonance and Circuit Q

- In a series circuit, consisting of L, C, R, at resonance the inductive and capacitive reactances cancel, and the circuit behaves as if it consisted solely of resistance.
- The resonant frequency of a series circuit (ω_0) is given by the equation $\omega_0 = 1/\sqrt{LC}$.
- The Q_s of a series resonant circuit, alternatively, equals either X_L/R_s or X_C/R_s, where R_s is the series resistance.
- Q is also a measure of the rate at which energy is being dissipated in a resonant circuit, being 2π (maximum energy stored/energy dissipated per cycle). The concept is applicable to mechanical systems and electrical ones.
- Series resonant circuits present their lowest value of impedance at resonance. Parallel resonant circuits present their highest impedance at resonance.
- At resonance, a parallel circuit encompasses a large circulating current that passes back and forth between L and C. Usually, the much smaller current that enters from outside the parallel circuit merely replenishes the current that has been dissipated in the series resistance.
- A parallel resonant circuit can alternatively be represented by an ideal L in parallel with an ideal C, paralleled in turn by resistance R_p.
- In utilizing identical components in parallel and series arrangements, the Q of a parallel circuit (Q_p) equals that of a series circuit (Q_s), and this may be used to

obtain the relationship between R_s and R_p equivalents of the parallel circuit; namely,

$$\frac{\omega_0 L}{R_s} = \frac{R_p}{\omega_0 L}$$

or, alternatively,

$$\frac{1}{\omega_0 C R_s} = \omega_0 C R_p$$

- The resonant frequency of a parallel circuit is

$$\omega_0 = \sqrt{\frac{1}{LC} - \frac{R_s^2}{L^2}}$$

but in computing circuit Q, one always uses the "natural" frequency of resonance—that is, $\omega_0 = \sqrt{1/LC}$.

- The resonant sharpness of a tuned circuit is determined by its Q.

KEY TERMS

admittance (Y): reciprocal of impedance (Unit is the siemens.)

apparent power: the vectorial sum of reactive and real powers (Unit is the volt-ampere.)

bandwidth (BW): the frequency interval within which the voltage amplitude is above 0.707 of its midband value; corresponds to the frequency interval between the half-power points

capacitive reactance (X_C): reactance exhibited by a capacitor (Unit is the ohm.)

farad (F): the unit of capacitance (If the application of 1 V to a capacitor leads to a charge of 1 C, the value of the capacitance is 1 F.)

henry (H): the unit of inductance (If a change of 1 A/s creates a counter-emf of 1 V, the inductance is equal to 1 H.)

impedance (Z): the vectorial combination of resistance and reactance (Unit is the ohm.)

inductive reactance (X_L): reactance exhibited by an inductor (coil) (Unit is the ohm.)

lagging power factor: arises in an inductively dominant circuit (Current lags the voltage waveform.)

leading power factor: arises in a capacitively dominant circuit (Current leads the voltage waveform.)

phase angle: the angular difference between voltage and current waveforms

phasor: a representation of rms voltage and current magnitudes, together with their respective phase relationships, in terms of complex numbers

power factor (pf): the cosine of the angle between voltage and current waveforms in ac circuits (A unity pf means the voltage and current are in phase.)

***Q*:** a measure of the stored energy relative to the dissipation rate

reactance (X): opposition to ac but with no power dissipation in ideal elements (Unit is the ohm.)

reactive power: power that is periodically drawn from the source and then returned with no net consumption (Unit is the volt-ampere.)

real power: power consumed within a device (Unit is the watt.)

resistance (R): opposition to current; also involves power dissipation (Unit is the ohm.)

resonance: the mutual cancellation of reactances resulting in a circuit dominated by the resistance

selectivity: a measure of the sharpness of resonance; related to circuit Q

susceptance (B): reciprocal of reactance (Unit is the siemens.)

SUGGESTED READINGS

Green, Estill I., "History of Q," AMERICAN SCIENTIST, vol. 43, pp. 584–594, 1955. *The historical development of circuit Q.*

Cathey, J. J., and S. A. Nasar, SCHAUM'S THEORY AND PROBLEMS OF BASIC ELECTRICAL ENGINEERING, pp. 48–55. New York: McGraw-Hill, 1984. *Phasors and resonance, with many solved problems.*

Edminister, Joseph A., SCHAUM'S THEORY AND PROBLEMS OF ELECTRIC CIRCUITS, 2nd ed., pp. 285–287. New York: McGraw-Hill, 1983. *Complex number system,* pp. 285–287, and *power and power factors,* pp. 134–147.

EXERCISES

1. Compute the reactance values for the following:

   a. $L = 10\ \mu H$ at 1 MHz

   b. $L = 10\ \mu H$ at 1 kHz

   c. $L = 10\ \mu H$ at 1 Hz

   d. $L = 10$ mH at 1 MHz

   e. $L = 10$ mH at 1 kHz

   f. $L = 10$ mH at 1 Hz

   g. $L = 1$ H at 1 MHz

   h. $L = 1$ H at 1 kHz

   i. $L = 1$ H at 1 Hz

2. Compute the reactance values for the following:

   a. $C = 100$ pF at 1 MHz

   b. $C = 100$ pF at 1 kHz

   c. $C = 100$ pF at 1 Hz

   d. $C = 100\ \mu F$ at 1 MHz

   e. $C = 100\ \mu F$ at 1 kHz

   f. $C = 100\ \mu F$ at 1 Hz

   g. $C = 1\ \mu F$ at 1 MHz

   h. $C = 1\ \mu F$ at 1 kHz

   i. $C = 1\ \mu F$ at 1 Hz

3. What is the decimal equivalent of 100 pF in units of μF? Of 10,000 pF in terms of μF?

4. **a.** What is the impedance magnitude of a series circuit consisting of $L = 1$ H and $R = 400\ \Omega$ operating at 60 Hz?

   b. What is the phase difference between voltage and current?

   c. Is the voltage leading or lagging the current?

   d. What value of capacitance would yield the same reactance at 60 Hz as the inductor in part a? What would be the phase angle? Would the voltage be leading or lagging?

5. If a 7.02 μF capacitor and a 1 H inductor were placed in series with a 400 Ω resistor, what would be the impedance at 60 Hz? What would be the phase relationship between voltage and current?

6. What is the resonant frequency of a series circuit consisting of $L = 1$ H, $C = 7\ \mu F$, and $R = 2\ \Omega$?

7. The reactance of a capacitor at 1 MHz is 10 Ω. What is the value of the capacitor?

8. Ten volts dc appears across a 100 μF capacitor. What is the charge on the capacitor?

9. Express the following complex number in terms of its polar form and in terms of its Cartesian components: $5 + j10$.

10. Convert the following into a real and an imaginary term: $1/(5 + j10)$.

11. A series 120 V circuit draws 2 A of current with a lagging power factor of 0.9. What is the phase difference between the voltage and the current? Is the voltage leading or lagging the current?

12. A 120 V circuit draws 1 A with a leading power factor of 0.85. What is the real power? The reactive power? The apparent power?

13. If the real power in a circuit is 8000 W and the reactive component is reduced from 2 to 0.5 kVA, what is the magnitude of the power factor improvement?

14. What is the series resonant frequency of a 1.0 H inductor in series with a 1.0 μF capacitor? The inductor resistance is 100 Ω.

15. **a.** What value of capacitor should be used with a 1.0 H inductor to provide a series resonant frequency of 120 Hz?

    b. What is the circuit Q if the coil resistance is 100 Ω? (The resistance of the capacitor can be taken to be negligible.)

16. What is the Q of a 120 Hz series resonant circuit whose capacitance is 1.0 μF if the resistance is 100 Ω?

17. What is the Q of a circuit that stores a maximum energy of 0.5 J, with 0.05 J dissipated per cycle?

18. The rms current through a 1 mH inductor is 100 mA. If the resonant frequency is 10 kHz and the resistance is 10 Ω, what is the series circuit Q?

19. **a.** What is the susceptance of a 100 pF capacitor operating at 1000 kHz? (Include the appropriate abbreviation unit in your answer.)

    b. What is the necessary inductance for resonance at 1000 kHz?

20. **a.** What is the resonant frequency when a 1.0 H inductor, whose resistance is 100 Ω, is placed in parallel with a 1.8 μF capacitor?

    b. What is the circuit Q?

    c. At resonance, what is the parallel circuit resistance?

    d. What can be done to decrease the resistance of the inductor?

21. **a.** For a series circuit Q of 15, what is the bandwidth of the circuit with a resonant frequency of 1 MHz?

b. What is the relationship between the impedance at resonance and at 967 kHz?

22. At the band limits, what is the relationship between the impedance at resonance and that at the band limits for a parallel resonant circuit?

23. If it is desired to have a circuit resistance of 10 kΩ at resonance, arising from a parallel resonant circuit employing a 100 μH inductor at 1 MHz, what should be the resistance of the inductor?

24. At the band limits of a resonant RLC circuit, the phase shift between the current and the applied voltage is ______. At resonance the phase shift is ______.

25. The sharpness of a series RLC resonant circuit decreases as the series resistance ______.

26. If one considers the resistance of a parallel RLC circuit to arise because of the resistance of the inductor, decreasing the value of this resistance ______ the equivalent parallel resistance at resonance.

27. The Q_p of a parallel RLC circuit is related to the Q_s of a series circuit employing the same components in the following manner: ______.

28. At resonance the voltage across the capacitor of a series RLC circuit, with the same applied voltage, may be increased by ______ the resistance.

29. The ability of a tuned RLC circuit, used in a radio or TV receiver, to discriminate against adjoining channel interference is a measure of its ______.

PROBLEMS

2.1

a. Given a 10 H choke (inductor), compute the inductive reactance at 60 Hz.

b. What would be the inductive reactance at 1000 kHz, a frequency corresponding to that used in AM broadcasting? At 100 MHz, a frequency corresponding to FM broadcasting?

2.2

a. What value capacitance at 60 Hz will have the same *magnitude* of reactance as the inductor in Problem 2.1a?

b. What would be the reactance of that capacitor at 1000 kHz? At 100 MHz?

2.3

If the inductor in Problem 2.1 and the capacitor in Problem 2.2 are placed in series with a 4700 Ω resistor, what is the circuit impedance at 60 Hz? At 1000 kHz? (Indicate both magnitude and phase angle.)

2.4

The application of a 10 V rms sinusoidal voltage to a capacitor, at a frequency such that the capacitive reactance is −25 Ω, will yield what value of current? What will be the phase relationship between the current and voltage? If the capacitor is 0.01 μF, what is the applied frequency?

2.5

If a 5 Ω resistor is added in series with a capacitive reactance of −25 Ω, what will be the magnitude and phase of the current with 10 V rms applied?

2.6

Assume 5 Ω of resistance were added in series with an inductive reactance of 25 Ω. What would be the current magnitude and its relationship to the applied voltage of 10 V rms?

2.7

If 10 V rms were applied to a series circuit consisting of R = 5 Ω, $X_C = -25$ Ω, and $X_L = 25$ Ω, what would be the current magnitude and its phase relationship to the applied voltage?

2.8

A charge of 10^{-5} C is deposited on a capacitor as a result of applying a 10 V dc potential difference. What is the value of the capacitor?

2.9

a. What total charge passes through a 12 V, 30 W automobile bulb in 1 s?

b. What would be the value of capacitance needed to accommodate the same charge with 10 V applied? (The resulting capacitance is impractically large.)

2.10

The light output of a typical helium-neon laser is emitted in a continuous fashion and represents 10 mW of (optical) power. One may safely have this light beam strike a hand. How much energy (per second) does this represent? If this same amount of light energy is emitted from a pulsed ruby laser in 20 ns, how much power does this represent? (Incidentally, unless you want a hole poked through it, it is not recommended that you hold your hand in front of the beam in the latter instance.)

2.11

The impedance of a series RC circuit is 112 Ω, the phase angle is −63.5°, and $X_C = -100$ Ω. What is the value of R?

2.12

A 120 Hz source is connected in series with an RC circuit consisting of R = 10.0 kΩ, and C = 0.500 μF. Find

a. The magnitude of Z.

b. The phase angle between current and voltage across the combination.

2.13
A 120 Hz, 25 V peak source is connected in series with X_L = 50 Ω and R = 100Ω. Calculate

a. The impedance.

b. The impedance angle.

c. The peak voltage across R and L.

d. The phase angle between the current and voltage.

2.14
440 V rms is applied to an impedance $Z = 3 + j4$. Compute the real, reactive, and apparent values of the power. (Specify the appropriate units.) What is the value of the power factor?

2.15
The power factor of an inductive load may be improved by placing capacitors in parallel with the load. If the *pf* in Problem 2.14 is made to be 0.9 by the addition of such capacitance, what will be the new value of the apparent power? (Note that since the voltage across the load remains the same, the real power does not change.)

2.16
What would be the capacitance in Problem 2.15 if it were placed in series with the load? Compare this to the value of capacitance needed in Problem 2.15.

2.17
With regard to the circuit in Figure 2.27,

a. What is the resonant frequency?

b. With 1 mV rms impressed, what is the voltage across C?

c. What would be the "measured" voltage across L? (Remember, the 1 Ω resistance is part of the inductance.)

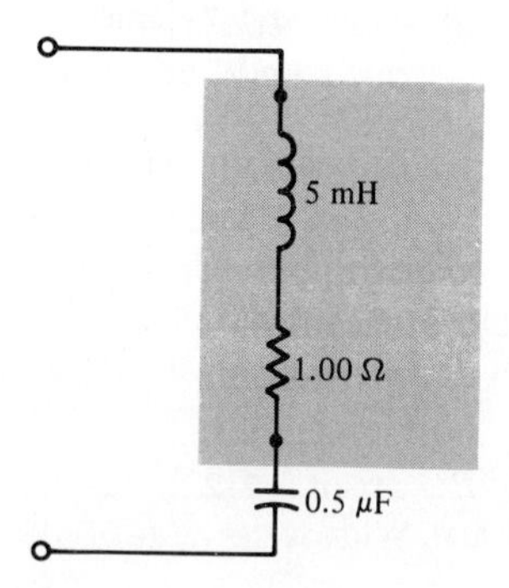

FIGURE 2.27

2.18
With regard to the circuit in Problem 2.17, if a 50.0 Ω resistance appears in place of the 1.00 Ω resistance,

a. With 1 mV impressed at the natural resonant frequency—that is, $f_0 = 1/(2\pi\sqrt{LC})$—what is the voltage across C?

b. What would be the "measured" voltage across the *actual* inductance—that is, including R?

c. Sketch the phasor relationship among V_L, V_C, and I.

2.19
With regard to the circuit in Figure 2.28, with 1 mV rms applied,

a. What single resistance can replace the parallel circuit at resonance?

b. Check your results using the circuit Q.

c. How can the magnitudes of I_1 and I_2 be much larger than I? (Sketch this using a phasor diagram for the currents.)

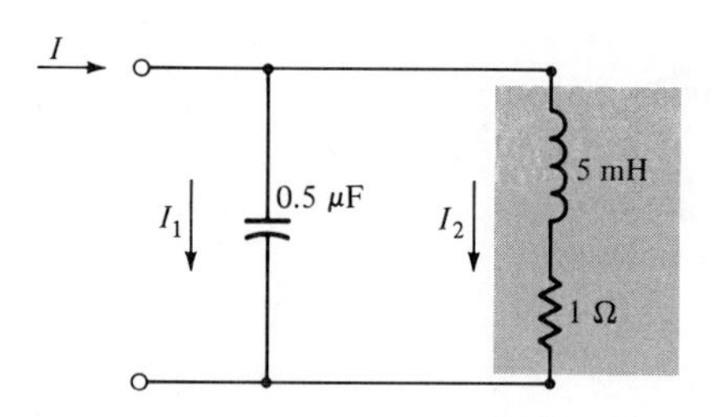

FIGURE 2.28

2.20
Rather than 1.00 Ω, the inductive reactance in Problem 2.19 is take to be 50.0 Ω, with 1 mV applied.

a. What is the resonant frequency?

b. What single resistance can replace the circuit at the natural resonant frequency?

c. Check your results using circuit Q. (When computing circuit Q, one always uses the "natural" frequency of resonance—that is, $\omega_0 = 1/\sqrt{LC}$.)

2.21
In an AM radio receiver a typical value for the Q of the tuning circuit might be 125.

a. For f_0 = 1000 kHz and C = 150 pF, compute the necessary value of inductance.

b. What is the resultant bandwidth?

2.22 ⚠
In a series RLC circuit, the voltage across the capacitor is

$$v_C = iX_C = \frac{VX_C}{\sqrt{R^2 + (X_L - X_C)^2}}$$

where V is the voltage applied at frequency f_0.

a. If R = 100 Ω, X_L = 200 Ω, and f_0 = 60 Hz, what is the value of C that will lead to the maximum capacitor voltage?

b. With a fixed frequency (f_0) applied and C made to be variable, v_C will be a maximum at a frequency lower than f_0. The reason for this is the following: With decreasing values of C, as f_0 is approached, the current rises and so does the value of X_C. At f_0, i is constant, but X_C is still increasing. v_C will reach a maximum at a smaller value of X_C than that which prevails at resonance. The peak value of v_C will be reached when the falling current balances the increase in X_C. Determine the value of C at which v_C is a maximum, if C is varied. (*Hint*: Set $dv_C/dX_C \equiv 0$ and solve for C.)

c. For the values given in part a, what is the percentage difference between the maximum v_C computed in part a and that computed as a result of part b?

d. Show that for high-Q circuits the two values become virtually the same.

2.23

In a series RLC circuit, the voltage across the inductor is

$$v_L = iX_L = \frac{VX_L}{\sqrt{R^2 + (X_L - X_C)^2}}$$

a. What is the value of L when v_L assumes its maximum value, with a fixed frequency (f_0) applied and L varied? (*Hint*: Set $dv_L/dX_L \equiv 0$ and solve for L.)

b. Using the values of Problem 2.22 (with $X_C = -200\ \Omega$ at 60 Hz), what is the percent change in the value of L between resonance and the value of L that leads to the largest value of v_L?

c. What would be the resonant frequency for that value of L that leads to the largest value of v_L in part a?

2.24

Use the results of Problem 2.23. At resonance a series LCR circuit has $|X_L| = |X_C| = 200\ \Omega$ and $R = 100\ \Omega$.

a. What is the voltage across L at resonance if 1000 V is being applied to the circuit?

b. What is the maximum voltage across L if L is varied?

2.25

With a signal being applied to a series circuit consisting of $R = 100\ \Omega$, $L = 5$ mH, and $C = 0.1\ \mu$F, compute and plot the following on a linear frequency scale: R, X_L, X_C, and Z, as a function of frequency. Use a frequency interval from $0.4f_0$ to $3.0f_0$.

2.26

Repeat Problem 2.25 with $R = 10\ \Omega$.

2.27

Repeat Problem 2.26 by plotting the results on semilog paper.

2.28

Using $f_0 = 5 \times 10^3$ Hz, $L = 5$ mH, and $R = 50\ \Omega$, compute the necessary value of C for series resonance. Calculate and plot Z, X_L, X_C, and R on semilog paper (magnitudes only) and compare the results with those for Problem 2.27.

2.29

A radio receiver invariably employs a parallel resonance to select the desired station. A typical circuit Q is 125. If the received signal voltages of a 1080 kHz station and that of an adjoining one at 1100 kHz are both 50 μV, what is the signal voltage developed by the 1100 kHz station when the resonant circuit is tuned to 1080 kHz?

2.30

A very useful laboratory instrument is the Q-meter, which allows one to experimentally determine the Q of a coil. See Figure 2.29. The ac source delivers current into a very low-resistance shunt R_{sh} across which the delivered voltage is measured. So that it does not significantly influence the Q, the value of R_{sh} is sufficiently low, typically of the order of 0.02 Ω. Across the variable capacitor is another voltmeter, this one calibrated directly in units of Q.

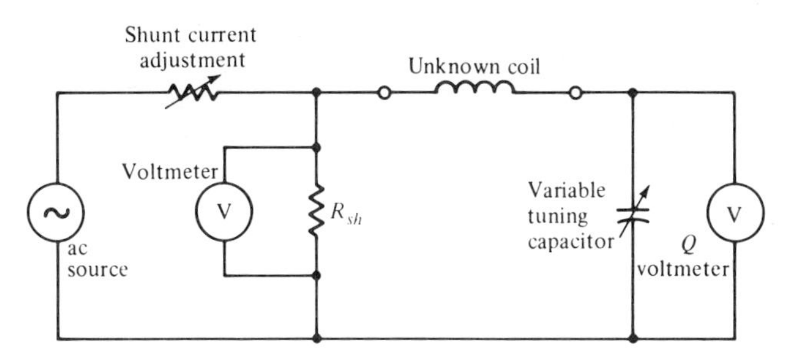

FIGURE 2.29

a. When a 1 MHz signal is used, a coil with a 10 Ω resistance resonates with a capacitance of 50 pF. Compute the percent error as a result of the insertion loss of 0.02 Ω.

b. Repeat the calculation for a coil with $R = 0.1\ \Omega$, at 40 MHz, which resonates with $C = 12.5$ pF.

2.31

a. What is the susceptance of the circuit elements in Figure 2.30? (Include units.)

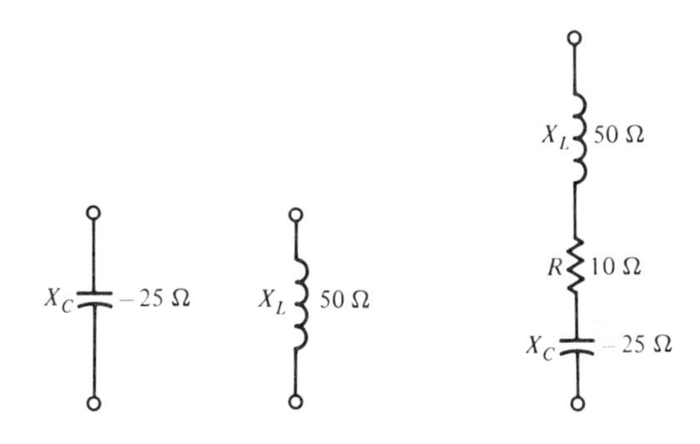

FIGURE 2.30 FIGURE 2.31

b. What is the value of the conductance and of the susceptance for the circuit in Figure 2.31?

3 RESISTANCE, INDUCTANCE, AND CAPACITANCE: PRACTICAL ASPECTS

OVERVIEW

In Chapter 2 we developed the mathematical methods needed to analyze circuits composed of resistors, inductors, and capacitors. In this chapter we consider the use that may be made of such combinations and some electrical theorems that may be used to simplify their mathematical treatment, specifically Thevenin's theorem and Norton's theorem.

The various combinations of R, L, and C are often used to discriminate against a specific frequency (or band of frequencies) much in the manner of a spring or a shock absorber used to minimize mechanical vibrations. It will be shown that the mathematical analyses in the mechanical and electrical cases are almost identical and that the application of one may be used to reinforce an understanding of the other.

The specific example used in this chapter to illustrate the combined application of resistance, inductance, and capacitance is the mechanical automotive ignition system. In Chapter 7 we will consider a transistorized ignition system, and in Chapter 15 we will study the use of a microcomputer for ignition and engine control.

OUTLINE

3.1 ■ INTRODUCTION

Rather than presenting the three fundamental quantities R, L, and C as rather sterile entities, we will undertake to consider them in terms of a practical engineering system that makes use of all three, the high-voltage automotive ignition system. The "coil" used with such a system is in reality a transformer, and this engineering application will also provide us with an opportunity to take a closer look at the operation of this electrical device.

For diagnostic purposes we generally wish to observe the temporal variation of the spark plug voltages. This is accomplished by means of an oscilloscope whose operation at this point we will consider only in the qualitative form sufficient to the task at hand. A more detailed consideration of oscilloscope performance will be deferred until after we have discussed the operation of electronic amplifiers.

The signal voltages developed in the automotive system are very high, of the order of tens of kilovolts. They cannot be applied directly to an oscilloscope but rather must first be **attenuated**—that is, diminished—to an amplitude of but a few volts while still retaining their waveform integrity. To accomplish this, we will make use of voltage dividers.

Divider networks, as well as more complex combinations of resistances and reactances, may be analyzed more simply by the use of two electrical network theorems, one named after Thevenin and the other, after Norton. They will constitute our next topic for discussion.

In the final sections of this chapter we will discuss resistance-reactance combinations that will allow the passage (or, alternatively, the rejection) of only a selected band of frequencies, and will conclude with a discussion of comparable tasks in the mechanical domain (for example, the use of springs and shock absorbers to restrict the transmission of mechanical vibrations). The objective of the latter discussion is to show the close relationship between some mechanical and electrical concepts.

3.2 ■ SIMPLIFIED IGNITION SYSTEM

Figure 3.1 illustrates the basic operation of a two-cycle, four-stroke internal combustion engine. In Figure 3.1a the piston is on its downward stroke; the fuel-air mixture is being drawn into the cylinder. In Figure 3.1b the compression stroke is illustrated; the fuel-air mixture is being compressed. In Figure 3.1c the spark plug is energized, and the fuel-air mixture is ignited; the resulting compressive force is applied to the piston and represents the power stroke. In Figure 3.1d the piston again starts upward, pushing out the exhaust gases, preparatory to initiating another cycle.

Since ignition takes place only on alternate cycles, the spark plug firings must also be made to take place at half the rate of the crankshaft revolutions (rpm). This is handled in the engine by the camshaft, whose rotational speed is half that of the engine.

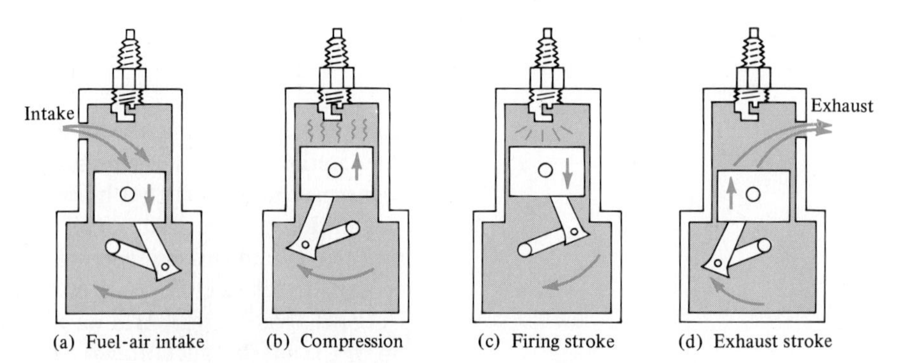

FIGURE 3.1. **Four-Stroke, Two Cycle Engine Sequence**

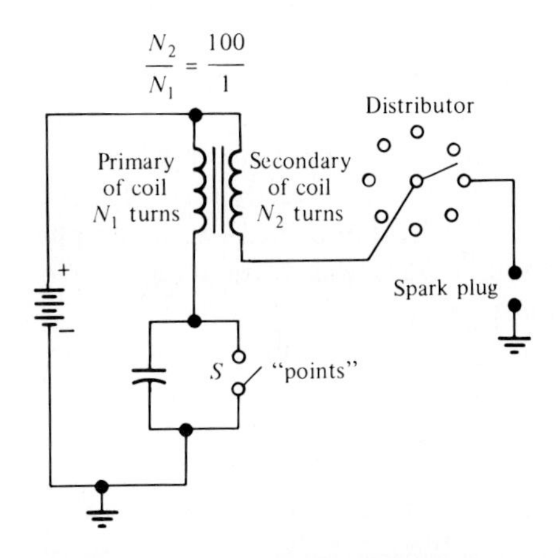

FIGURE 3.2. **Simplified Schematic for "Standard" Automotive Ignition System**

While the *rotating distributor arm* is enroute between the contacts (Figure 3.2), switch S_1 is closed. In automotive parlance, the switch contacts are termed the **points**. The closed switch allows a large current to flow from the battery through the low resistance of the primary coil (an inductance). This primary coil is physically wound in close proximity to the secondary coil, and together they constitute a transformer. One terminal of the primary winding is grounded (through the battery), and the secondary voltage can be made to be either in phase or out of phase with that across the primary, depending on which of the secondary terminals is also grounded. (See Figure 1.19.) With neither secondary terminal grounded, one may be connected to the ungrounded primary terminal as in Figure 3.3. In Figure 3.3a the secondary voltage can be seen to subtract from the primary as the polarity of the primary voltage changes. In Figure 3.3b the primary and secondary voltages are additive.

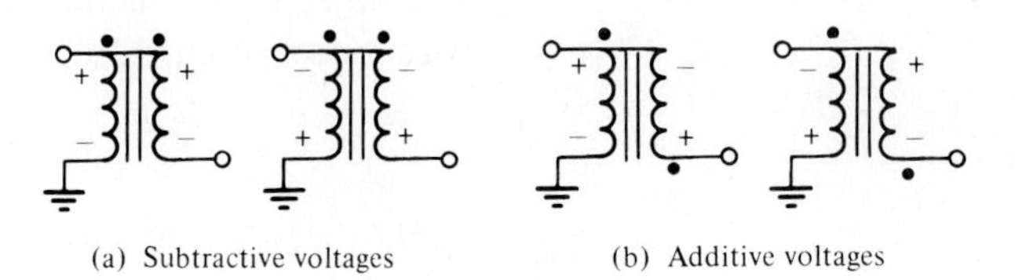

FIGURE 3.3. **Series-Connected Transformer Windings** (a) Alternate voltage cycles applied to the primary at the left subtract from the secondary output at the right. (b) Alternate cycles at the primary add to the output voltage at the secondary. Dots signify in-phase primary and secondary terminals.

Figures 3.4a and b show the type of situation we encounter in the ignition system. With a *constant* dc flowing through the primary (Figure 3.4a), the secondary voltage (measured relative to ground) is *essentially* zero since there is no changing magnetic field. However, if the switch (representing the points) is opened, the current should go to zero, but the primary inductance tries to maintain the current flow by assuming a counter-emf polarity that is additive to that of the battery. This creates a negative voltage across the secondary, and we will see that this is a desirable condition for spark plug operation.

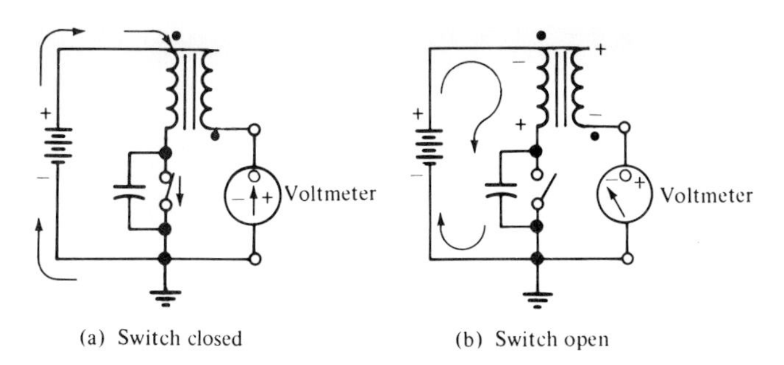

FIGURE 3.4. Action of Ignition "Points" Contact (a) With switch closed and constant dc, the output voltage is zero. (b) Upon opening the switch, the current decreases, the polarity of the voltage across the primary is as shown, and the output voltage at the secondary assumes a negative polarity relative to ground.

3.3 ▪ MATHEMATICS OF IGNITION

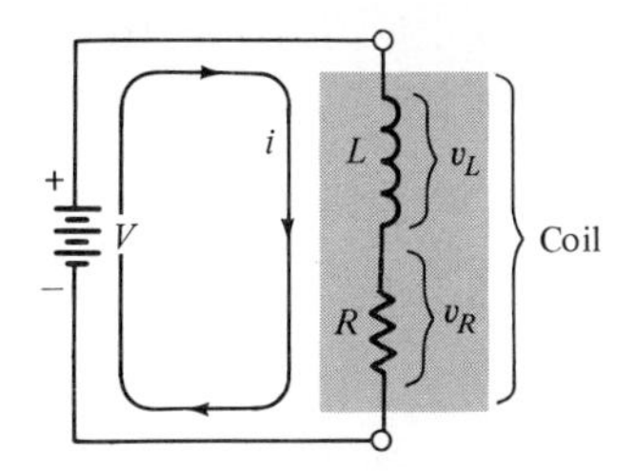

FIGURE 3.5. Charging Cycle of an Ignition Coil A coil's inductance and resistance are separately depicted. The source voltage must at all times equal the sum of the voltage drops across the inductance and the resistance.

When the points close, we are faced with the situation depicted in Figure 3.5. Using Kirchhoff's voltage law (KVL), we obtain

$$V - L\left(\frac{di}{dt}\right) - iR = 0$$

$$\frac{V}{R} - i = \left(\frac{L}{R}\right)\left(\frac{di}{dt}\right)$$

$$\int_{i=0}^{i} \frac{-di'}{(V/R) - i'} = -\frac{R}{L}\int_0^t dt'$$

$$\ln\left(\frac{V}{R} - i'\right)\Bigg|_{i=0}^{i} = -\frac{R}{L}t$$

$$\ln\left[\frac{(V/R) - i}{V/R}\right] = -\frac{R}{L}t$$

$$\frac{V}{R} - i = \frac{V}{R}e^{-Rt/L}$$

$$i = \frac{V}{R}(1 - e^{-Rt/L}) = \frac{V}{R}(1 - e^{-t/\tau}) \tag{3.1}$$

where τ is defined as the **time constant**—that is, the time interval required for the exponential factor to reach a value of $1/e$, where e is the base of the natural logarithms. In the preceding case the time constant is equal to L/R. The change in current with time is illustrated in Figure 3.6.

Since the time needed for the current to reach its final asymptotic value is infinite, the **rise time** (t_R) of the circuit is defined as the time interval between signal values of 10% and 90% of the asymptotic value. (A comparable behavior is experienced in an RC circuit; see "A Closer Look" at the end of this chapter.)

EXAMPLE 3.1

If $L = 5$ mH and $R = 2\ \Omega$, what will be the rise time of the current upon application of a 10 V step?

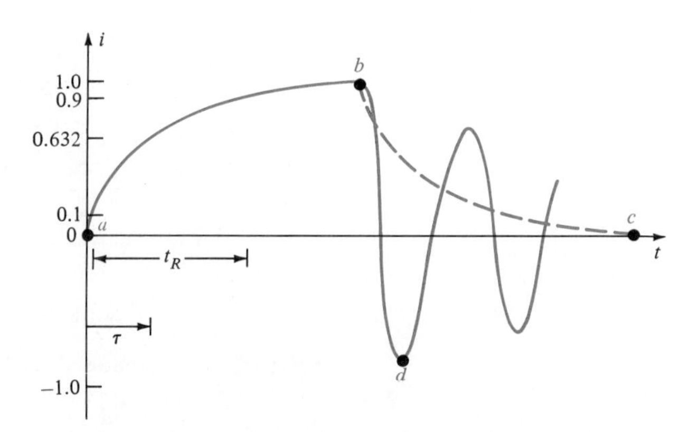

FIGURE 3.6. Charging and Discharging Curves of an Ignition Coil Curve *a* to *b*: Charging current for *RL* circuit. Curve *b* to *c*: Discharging current for *RL* circuit. Curve *b* to *d*: Discharging current for *RLC* circuit.

Solution: We proceed to compute the times at which i is equal to $0.1V/R$ (time t_1) and $0.9V/R$ (time t_2) to find the intervening time, which represents the rise time:

$$\tau = \frac{L}{R} = \frac{5\times 10^{-3}\text{ H}}{2\ \Omega} = 2.5 \times 10^{-3}\text{ s} = 2.5\text{ ms}$$

$$i_1 \equiv 0.1\left(\frac{V}{R}\right) = \frac{V}{R}(1 - e^{-t_1/\tau}) = \frac{V}{R}(1 - e^{-t_1/(2.5\times 10^{-3})})$$

$$-\frac{t_1}{2.5 \times 10^{-3}} = \ln(1 - 0.1)$$

$$t_1 = 2.63 \times 10^{-4}\text{ s}$$

$$i_2 \equiv 0.9\left(\frac{V}{R}\right) = \frac{V}{R}(1 - e^{-t_2/\tau}) = \frac{V}{R}(1 - e^{-t_2/(2.5 - 10^{-3})})$$

$$-\frac{t_2}{2.5 \times 10^{-3}} = \ln(1 - 0.9)$$

$$t_2 = 5.76 \times 10^{-3}\text{ s}$$

$$t_R = t_2 - t_1 = 5.50 \times 10^{-3}\text{ s}$$

It should be noted that the rise time (t_R) depends only on the time constant (τ) and is independent of the applied voltage (V).

With the current approaching its asymptotic value, if the inductor is short-circuited, the collapsing magnetic field will lead to an exponential decay with an identical time constant (curve *b-c* in Figure 3.6). The **fall time** of the circuit is defined in a manner analogous to that of the rise time—that is, the elapsed time interval during which the current value changes from 90% to 10% of its initial value. Thus,

$$i = \frac{V}{R}e^{-Rt/L} = \frac{V}{R}e^{-t/\tau} \tag{3.2}$$

If however, the decay process takes place with a capacitor (C) inserted in series with R and L, under appropriate circumstances one may achieve a decay time that is much shorter than would be obtained with the RL circuit alone. This will lead to a large primary voltage and a correspondingly large secondary voltage applied to the spark plug.

The general solution of the RLC circuit will again be of an exponential form, $i = Ae^{st}$, where A is the initial amplitude and s, in general, is a complex number to be represented by $\alpha + j\omega$. For the RL circuit, s was real, so that $\alpha = -R/L$, with $\omega = 0$. For the RLC circuit, using KVL, we have the following equations[1]:

$$v_L + v_R + v_C = 0 \tag{3.3}$$

$$L\left(\frac{di}{dt}\right) + iR + \frac{\int i\,dt}{C} = 0 \tag{3.4}$$

Differentiating this equation, we get

$$L\left(\frac{d^2i}{dt^2}\right) + R\left(\frac{di}{dt}\right) + \frac{i}{C} = 0 \tag{3.5}$$

and substituting into this equation the assumed solution, $i = Ae^{st}$, we obtain

$$s^2 + s\left(\frac{R}{L}\right) + \frac{1}{LC} = 0 \tag{3.6}$$

This quadratic equation has two solutions:

$$\begin{aligned} s_1 &= -\frac{R}{2L} + \sqrt{\left(\frac{R}{2L}\right)^2 - \frac{1}{LC}} \\ s_2 &= -\frac{R}{2L} - \sqrt{\left(\frac{R}{2L}\right)^2 - \frac{1}{LC}} \end{aligned} \tag{3.7}$$

There are three solution regimes to Eqs. 3.7, depending on the magnitudes of $R/2L$ and $1/LC$. Our interest at present resides with the solution for **damped oscillation**—that is, cyclical variation with diminishing magnitudes in successive cycles—when $(R/2L)^2 < 1/LC$. (The other two cases are assigned as problems.) In the damped oscillatory solution, both roots will be complex numbers and will be represented by $s_{1,2} = \alpha \pm j\omega$, where $\omega = \sqrt{(1/LC) - (R/2L)^2}$ represents the oscillation frequency.

The general solution will be the sum of the two solutions:

$$i = A_1e^{s_1t} + A_2e^{s_2t} = A_1e^{(\alpha_1 + j\omega)t} + A_2e^{(\alpha_2 - j\omega)t}$$

But $\alpha_1 = \alpha_2 = -R/2L$. Then we have

[1]From Eq. 2.9, $I = dq/dt$. Then $q = \int i\,dt$, and $v_C = q/C = \int i\,dt/C$.

$$i = e^{-Rt/2L}(A_1 e^{j\omega t} + A_2 e^{-j\omega t})$$

Since

$$e^{j\omega t} = \cos \omega t + j \sin \omega t$$
$$e^{-j\omega t} = \cos \omega t - j \sin \omega t$$

we have

$$\begin{aligned} i &= e^{-Rt/2L}[(A_1 + A_2) \cos \omega t + j(A_1 - A_2) \sin \omega t] \\ &= e^{-Rt/2L}(B_1 \cos \omega t + jB_2 \sin \omega t) \end{aligned} \quad (3.8)$$

In our specific case, at $t = 0$, $i = V/R$. Therefore, $B_1 = V/R$, and we have the result

$$i = \frac{V}{R} e^{-Rt/2L} \cos \omega t \quad (3.9)$$

that is, a damped sinusoid. The frequency of oscillation is represented by ω, and $(V/R)e^{-Rt/2L}$ defines the oscillation envelope. (See Figure 3.6.)

Now let's return to our ignition system. When it is time for a spark plug to fire, S_1 is made to open by means of the cam attached to the distributor rotor. Since, typically, the change in primary voltage (accompanying the current change) might be hundreds of volts and the step-up ratio of the transformer is of the order of 1:100, the secondary voltage is of the order of tens of kilovolts. This large voltage is applied to the spark plug where it causes an arc to develop and leads to fuel combustion. This arc will be sustained until the magnetic field energy is no longer sufficient to maintain it. When it ceases, the remainder of the magnetic field (which had been established about the transformer windings) is dissipated in the form of a circulating current between the capacitor and the inductance primary, these constituting a parallel *RLC* circuit. The distributor shaft has continued to rotate, and its cam arrangement will again close S_1, thereby allowing the field to build up within the transformer in preparation for a repeat performance when the distributor contact reaches the next terminal.

EXAMPLE 3.2 Combine the inductance and resistance of Example 3.1. With a capacitance of 0.25 μF, if the initial current is 5 A, determine the time for the current to fall to zero if the capacitor is suddenly inserted in series with the circuit. What would be the primary voltage accompanying this current drop (assuming the spark plug did not fire)?

Solution:

$$\omega_0 = \sqrt{\frac{1}{LC} - \frac{R^2}{4L^2}} = \sqrt{\frac{1}{5 \times 10^{-3} \times 2.5 \times 10^{-7}} - \frac{2.0^2}{4(5 \times 10^{-3})^2}}$$
$$= \frac{2\pi}{T} = 2.83 \times 10^4 \text{ rad/s}$$
$$T = 2.22 \times 10^{-4} \text{ s} = \text{period of oscillation}$$

The current will pass through zero at $T/4$—that is, at $t = 5.55 \times 10^{-5}$ s. (Compare this with the decay time of 5.50×10^{-3} s for the *LR* circuit alone.)

The accompanying voltage variation (providing the spark plug did not fire) would be

$$L\left(\frac{di}{dt}\right) = 5 \times 10^{-3}\left(\frac{5}{5.55 \times 10^{-5}}\right) = 450 \text{ V}$$

More accurately, one should also take into account that there is an exponential decay accompanying the oscillation, but its contribution to the result will be small due to the short time interval—that is,

$$e^{-Rt/2L} = e^{-2(5.55 \times 10-5)/2(5 \times 10^{-3})} = 0.989$$

3.4 ▪ IGNITION SIGNAL DISPLAY

An **oscilloscope** is an electronic device that can be used to observe the temporal variations of electrical signals. One of our ultimate objectives will be to study the operation of an oscilloscope, but for now we will simply depict what the oscilloscope will display for us with regard to the preceding events.

Figure 3.7 is meant to represent a typical sequence of events that each individual cylinder exhibits. When S_1 opens, we see the very sharp downward (negative) pulse, which represents the large secondary voltage pulse. The spark plug "fires," and this "short circuit" across the secondary winding drops the voltage to a rather small value. This portion of the display (while the plug is "firing") represents the portion labeled the *arc line*. When ignition ceases, we can see the "ringing" that characterizes the circulating current passing back and forth between the capacitor and inductance. Since each time the current circulates some energy is lost, we see that the amplitude of oscillation diminishes with time. This is labeled *dissipation oscillation*.

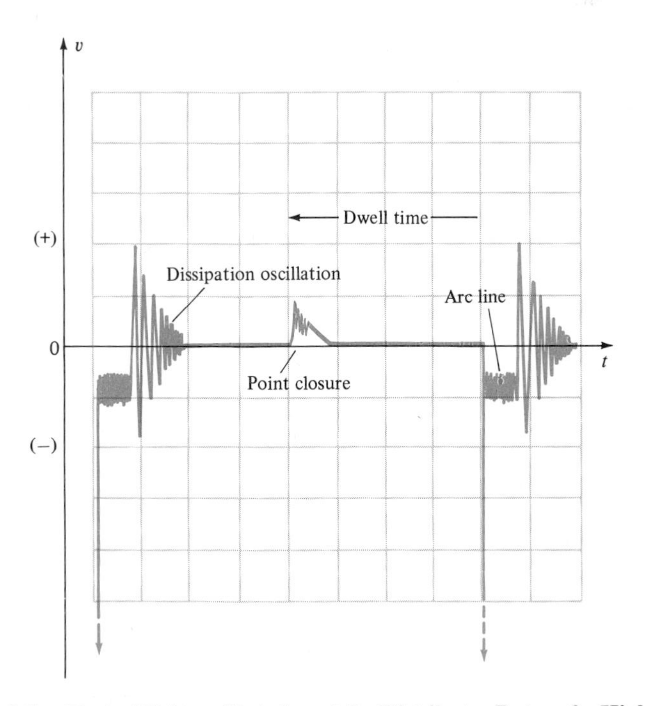

FIGURE 3.7. **Typical Voltage Variation at the Distributor Rotor of a High-Voltage Automotive Ignition System**

Now that the energy has been dissipated, the points are made to close. There is also some ringing that accompanies this closure. It is smaller in magnitude, rapidly diminishes as the current builds up, and has a significantly higher frequency than the dissipation oscillation. Remember that at this time only L and R are in the circuit; C has been shorted out. But between each turn of an inductor there is a small capacitance—that is, two conductors separated by insulation—and the cumulative effect leads to a very small capacitance that is in parallel with the inductor. This represents a high-frequency resonant circuit and accounts for the ringing at point closure.

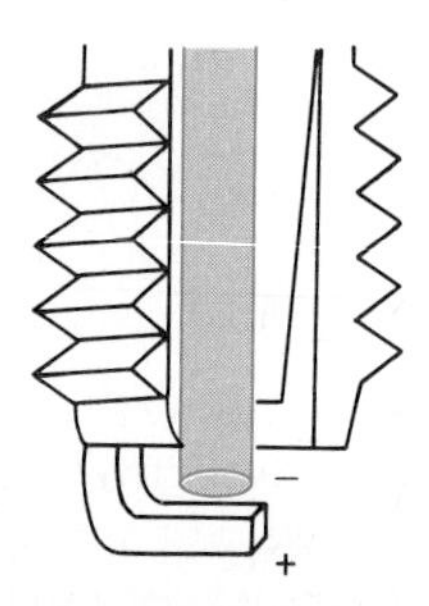

FIGURE 3.8. Cutaway View of Spark Plug The desired polarity between center and outer electrodes is indicated.

Note that shortly after S_1 closes and before it again opens (the interval labeled *dwell time* in Figure 3.7), there is a large current flowing through the primary winding. But there is no *changing* current, and, therefore, no voltage appears across the secondary coil until S_1 again is opened, at which time the sequence of events outlined here again commences.

Figure 3.8 depicts the lower portion of the spark plug. In contrast to "classical" current, as we have indicated, it is actually electrons that are in motion in metals, and they leave the negative terminal of the battery and reenter at the positive terminal. Also, if one has a heated metallic surface, electrons are rather readily "evaporated" from such surfaces.

If a negative voltage is applied to the spark plug, the outer spark plug electrode is positive, while the central—more confined and hence hotter—electrode is at a negative potential. Electrons leave this central terminal for two reasons:

1. The potential difference across the electrodes attracts them to the positive electrode.
2. The high temperature of the central electrode gives rise to thermionic emission.

The second contribution allows ionization to take place with 20–40% less voltage than would be the case if a positive pulse were applied to the spark plug.

Let's return to our ignition display and see what sort of ignition faults may be detected with the oscilloscope. Adjustment of the oscilloscope allows the display of all the spark plug firings in sequence (Figure 3.9). From such a display one may immediately note that one firing differs significantly from the others. We may then alter the oscilloscope operation and display that single waveform in an expanded version that fills the screen and allows a more detailed examination. For the case in point the disparity in Figure 3.9 might be recognized as showing cylinder 3 to have an arc line that is longer than normal.

There are two characteristics of the arc line that are of interest:

1. The amplitude
2. The duration

These two quantities are inversely related to one another—that is, the large amplitude will lead to a short duration and vice versa. A large amplitude means that the arc resistance is too high, which might be the result of a spark gap spacing being too large.

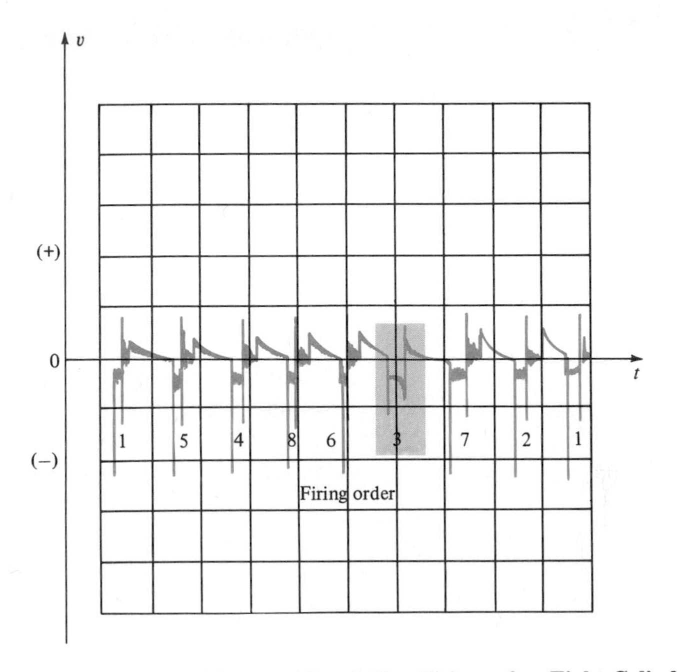

FIGURE 3.9. **Sequential Display of Spark Plug Firings of an Eight-Cylinder Engine** Note that the display of cylinder 3 differs significantly from others, indicative of a malfunction.

If the amplitude is initially correct but then increases as the plug warms up, there may be some sort of temperature-dependent fouling. A diminishing amplitude can indicate a deposit fouling. If the amplitude stays fairly constant but the arc line grows shorter with time, the available energy may be diminishing, which indicates an incipient failure of either the capacitor or the coil.

About 1955–1956, when automotive manufacturers changed from a 6 V to a 12 V ignition system, a ballast resistor was placed in series with the 12 V lead. This resistor reduces the coil voltage to between 6 and 9 V, depending on the car. During starting, the ignition switch momentarily shorts out the ballast resistor, allowing the full 12 V to be used for starting. (In actuality only about 10 V is applied because the heavy current drawn by the starter reduces the battery voltage to about that level.)

The lower voltage (and consequent current) keeps the points current within safe limits, but allows momentarily higher starting current by shorting out the ballast resistor. If the large starting current persisted continually, the points contacts would quickly oxidize (indicated by their blue color) and lead to poor performance.

3.5 ■ VOLTAGE DIVIDERS

A **voltage divider** consists of a series of electrical elements (usually resistors) whose individual values are adjusted in such a manner as to reduce a single voltage down to one or more requisite smaller values. In the early days of radio, because various magnitudes of voltage were required to energize circuits, one resorted to using an assortment of batteries: an A battery (ca. 1.5 V), a B battery (ca. 90–180 V), and a C battery (ca. 45 V). Ultimately, with the development of ac-operated power supplies, a single voltage was delivered, and by means of voltage dividers, the needed assortment of voltages could be obtained.

3.5.1 Resistive Voltage Dividers

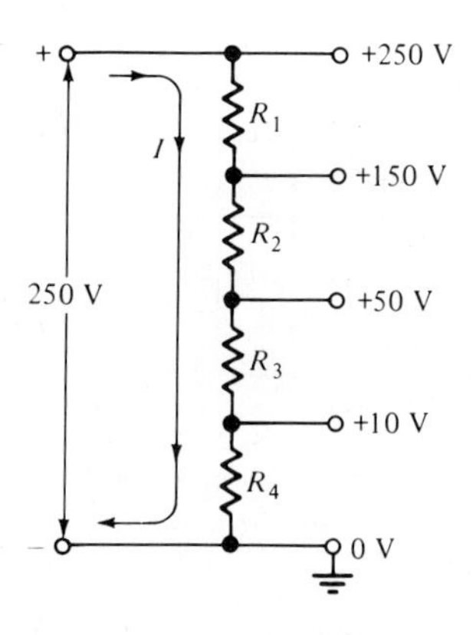

FIGURE 3.10. **Voltage Divider** Current I through the resistors gives rise to a variety of voltages, measured relative to ground.

Consider Figure 3.10. We presume that a 250 V supply is available. In addition to this voltage we also require the following: +150 V, +50 V, and +10 V. Due to current I through the resistors, we may achieve the desired values by selecting the values of R_1, R_2, R_3, and R_4. Thus, we have

$$\begin{aligned} IR_4 &= 10\text{ V} \\ I(R_3 + R_4) &= 50\text{ V} \\ I(R_2 + R_3 + R_4) &= 150\text{ V} \\ I(R_1 + R_2 + R_3 + R_4) &= 250\text{ V} \end{aligned}$$

Suppose we assume that $R_1 + R_2 + R_3 + R_4 = 100\text{ k}\Omega$. Using Ohm's law, we obtain

$$I = \frac{2.50 \times 10^2}{10^5} = 2.5 \times 10^{-3}\text{ A} = 2.5\text{ mA}$$

This current, through the voltage divider, is colloquially termed the **bleeder current**.

Then, successively, we have

$$R_4 = \frac{10}{2.5 \times 10^{-3}} = 4 \times 10^3\ \Omega = 4\text{ k}\Omega$$

$$R_3 = \frac{50 - 10}{2.5 \times 10^{-3}} = 16\text{ k}\Omega$$

$$R_2 = \frac{150 - 50}{2.5 \times 10^{-3}} = 40\text{ k}\Omega$$

$$R_1 = \frac{250 - 150}{2.5 \times 10^{-3}} = 40\text{ k}\Omega$$

More simply, there is no need to compute the current. Each resistor represents a proportional voltage drop. Thus, we have

$$\frac{R_4}{R_{tot}} \times 250\text{ V} = 10\text{ V}$$

$$\frac{R_3 + R_4}{R_{tot}} \times 250\text{ V} = 50\text{ V}$$

.
.
.

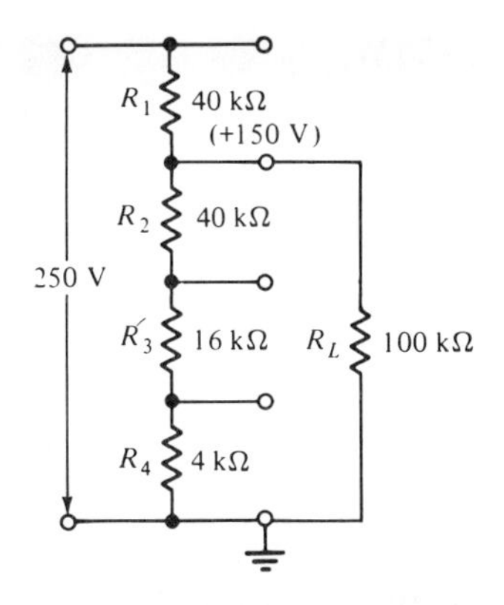

FIGURE 3.11. A Loaded Voltage Divider For proper operation, +150 V, required by R_L, is to be obtained from the voltage divider.

Why did we select the total resistance to be 100 kΩ? This was an arbitrary choice on our part. Some other value might be equally satisfactory, but let us consider first the consequences of our choice.

The voltages shown in Figure 3.10 may change once we start to draw power from these points, as we will see when we connect *loads* across these voltage sources. For example, suppose that the part of the circuit that requires +150 V draws 1.5 mA of current. By Ohm's law this load constitutes a resistance of 100 kΩ. Thus, our circuit, including the load, will look as shown in Figure 3.11. Since $R_2 + R_3 + R_4$ is 60 kΩ, this resistance in parallel with the 100 kΩ load will represent an equivalent resistance of 37.5 kΩ, and the output voltage at this terminal will now be

$$\frac{R_{equiv}}{R_1 + R_{equiv}} \times 250 = \frac{37.5}{40 + 37.5} \times 250 \simeq 121 \text{ V}$$

Therefore, rather than 150 V across the load, we have only 121 V. We say that such a voltage divider exhibits poor regulation in that the voltage varies significantly when load current is being withdrawn. **Regulation** is defined in the following manner:

$$\text{Regulation} = \frac{\text{no-load value} - \text{full-load value}}{\text{full-load value}}$$

If this ratio is multiplied by 100, the result constitutes the regulation percentage.

Let us now redesign this system by taking the total resistance to be 10 kΩ instead of 100 kΩ. We now have

$$I = \frac{250}{10^4} = 2.5 \times 10^{-2} \text{ A}$$

Proceeding as previously, we will find that

$$R_4 = 400 \ \Omega$$
$$R_3 = 1.6 \text{ k}\Omega$$
$$R_2 = 4.0 \text{ k}\Omega$$
$$R_1 = 4.0 \text{ k}\Omega$$

Now connect the same 100 kΩ load across the 150 V terminals. The 6 kΩ constituting $R_2 + R_3 + R_4$ now appears in parallel, with 100 kΩ yielding an equivalent resistance of 5.66 kΩ. The output voltage (Figure 3.12) is now

$$\frac{5.66}{5.66 + 4.0} \times 250 \simeq 146 \text{ V}$$

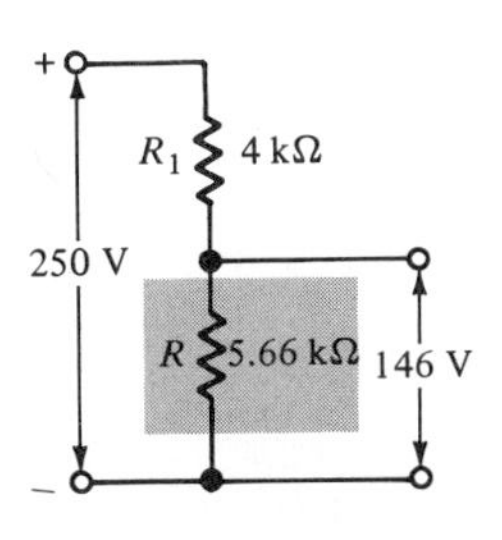

FIGURE 3.12. Circuit Equivalent of a Loaded Voltage Divider The parallel combination of modified bleeder resistance and the resistive load (Figure 3.11) leads to a diminished equivalent resistance.

Thus, with the load connected, where previously we had 121 V, we now have 146 V, and what we want is 150 V. The regulation is much better. It can still be further improved by making the total resistance even smaller. Should we do this? To answer this question, one must consider the price to be paid.

In the first design, *without any load connected*, the bleeder current was 2.5 mA. The power dissipated in the total bleeder resistance was

$$W = I^2R = 6.25 \times 10^{-6} \times 10^5 = 0.625 \text{ W}$$

In the second design the bleeder current was 25 mA:

$$W = 6.25 \times 10^{-4} \times 10^4 = 6.25 \text{W}$$

If we make R_{tot} smaller, we will improve the regulation still further but at the expense of greater power expenditure. If, however, the load resistance is much greater than the bleeder resistance across which it is connected, the effect of loading on the delivered voltage will be negligible.

3.5.2 Attenuators

What we are ultimately concerned with in this chapter is the means by which one can observe the ignition *signals*. Their magnitude of some tens of kilovolts precludes the direct application of such signals to measuring instruments, nor is there any need to do so. In such cases a voltage divider consisting of but two elements will suffice in remedying the difficulty. The term *voltage divider* is generally applicable to a series network used to diminish a *dc* voltage; a network used to diminish a *signal* voltage is termed an **attenuator.**

Presuming we have a 30 kV input pulse and desire, perhaps, a 3 V output pulse, we may conveniently use the simple attenuator shown in Figure 3.13, where

$$\frac{R_2}{R_1 + R_2} = \frac{3 \text{ V}}{30{,}000 \text{ V}}$$

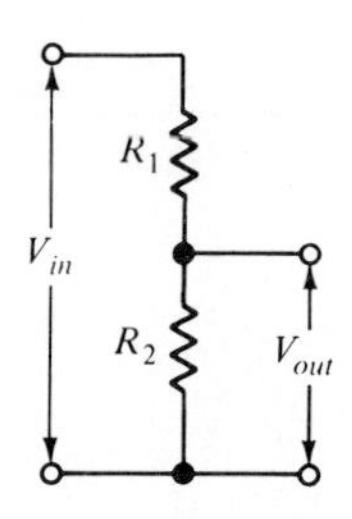

FIGURE 3.13. **Simple Resistive Voltage Divider (Attenuator)**

Since the resistance represented by the measuring device might typically be 10 MΩ, we might select R_2 to be 100 kΩ. With such a choice the loading effect of the measuring instrument will be small and the total attenuator resistance so large that the power dissipation due to the bleeder current will be negligible:

$$R_1 + R_2 = \frac{30{,}000}{3} R_2$$

R_1, therefore, should be set equal to about 1000 MΩ.

EXAMPLE 3.3

The input resistance of an oscilloscope is taken to be 10 MΩ. (See Figure 3.14.) If a 20 kV ignition pulse is to be observed, what values of resistance should be used in a resistive attenuator to reduce the output signal to 20V?

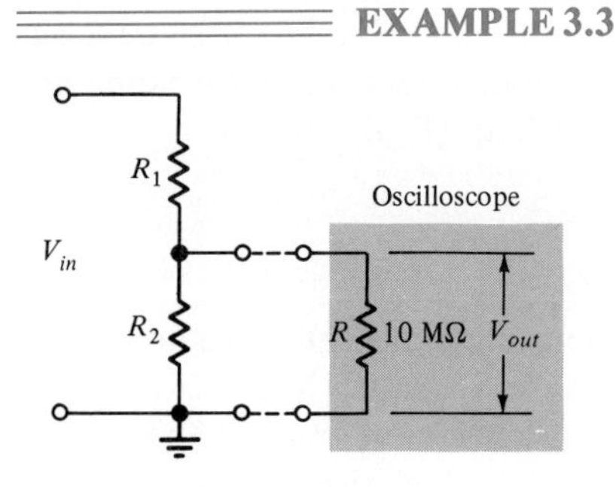

FIGURE 3.14

Solution: To minimize loading effects, R_2 should be no greater than $\frac{1}{100} R$:

$$R_2 = \frac{1}{100} \times 10 \text{ M}\Omega = 100 \text{ k}\Omega$$

$$\frac{V_{out}}{V_{in}} = \frac{20}{20{,}000} = 10^{-3}$$

$$\frac{V_{out}}{V_{in}} = \frac{R_2}{R_1 + R_2} = \frac{100 \text{ k}\Omega}{R_1 + 100 \text{ k}\Omega} = 10^{-3}$$

$$R_1 \simeq 100 \text{ M}\Omega$$

It is customary to refer to V_{out}/V_{in} as the **gain** of a circuit even if $V_{in} > V_{out}$, while V_{in}/V_{out} is termed the **attenuation**. Thus, in Example 3.3, by way of illustration, the gain is 10^{-3}, while the attenuation is 10^3.

A major difficulty with the attenuator in Example 3.3 stems from the requirement that the top of R_1 must make physical contact with the high-voltage ignition wires. In other words, we must penetrate through the ignition wire insulation. This proves to be an inconvenience that can be avoided in this case by resorting to another type of attenuator.

Consider Figure 3.15. Rather than using resistances, we use a capacitance divider. Suppose we proportion the divider so that $C_2 = 0.1\ \mu F$ and $C_1 = 10$ pF. Then the application of a voltage pulse of magnitude V_{in} will lead to an output pulse of magnitude $10^{-4}\ V_{in}$:

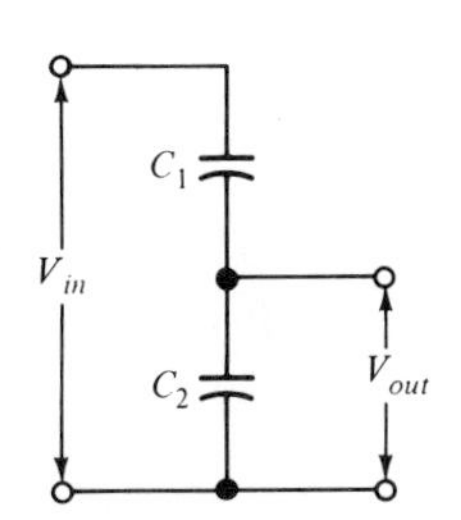

FIGURE 3.15. Capacitive Voltage Divider (Attenuator)

$$V_{out} = \frac{X_{C_2}}{X_{C_1} + X_{C_2}} V_{in} = \frac{1}{2\pi f C_2} \left[\frac{1}{(1/2\pi f C_1) + 1/2\pi f C_2)} \right] V_{in}$$

$$\frac{C_1}{C_1 + C_2} V_{in} = \frac{10^{-11}}{10^{-11} + 10^{-7}} V_{in} \simeq 10^{-4} V_{in}$$

We have thereby achieved the same type of divider action attained with resistors.[2]

Figure 3.16 shows the sampling lead being terminated by a metal spring clamp that is strapped around the ignition lead. This metallic clamp together with the central ignition wire constitute two electrodes which, together with the intervening cable insulation, give rise to a capacitance C_1. Within the shielded cable leading to the oscilloscope, between the central wire and the (outer) shielded braid of the cable, there is a discrete capacitor, C_2. The value of the latter capacitance can be changed as we require different magnitudes of attenuation.

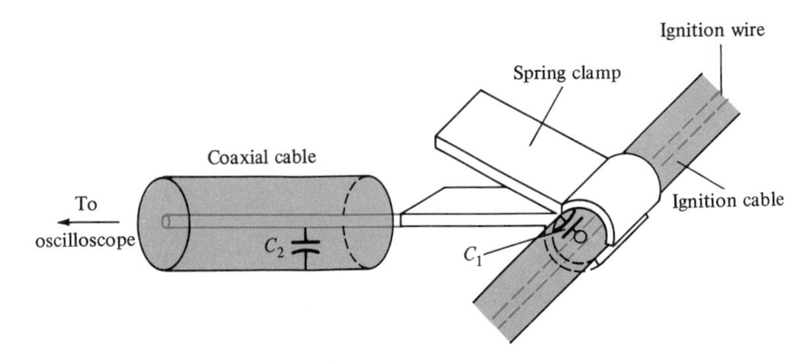

FIGURE 3.16. Oscilloscope Probe to Observe Ignition Signals Capacitance between metal clamp and ignition wire constitutes C_1; capacitor C_2 is an actual capacitor located within the probe lead.

[2] While the attenuation of a resistive divider is $R_2/(R_1 + R_2)$, that of a capacitive divider is $C_1/(C_1 + C_2)$.

Thus, we see that we do not need to make *physical* contact with the ignition wires. If we wish to monitor all the cylinders in succession, we place the clamp around the central lead to the distributor cap. If we wish to monitor an individual cylinder, we place the clamp around that individual ignition wire.

EXAMPLE 3.4

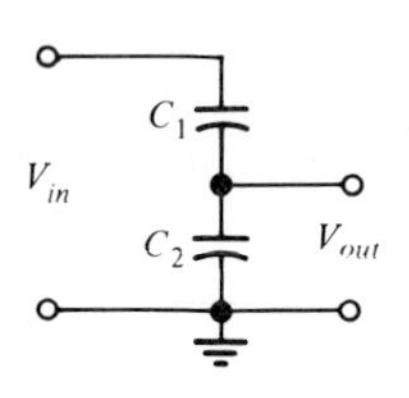

FIGURE 3.17

The pickup probe of an ignition monitoring system has a capacitance (C_1) of 6 pF (6×10^{-12} F). (See Figure 3.17.) What value of C_2 should be utilized in order to achieve an attenuation of 1000?

Solution:

$$\frac{C_1}{C_2 + C_1} = \frac{6 \times 10^{-12}}{C_2 + (6 \times 10^{-12})} \simeq \frac{C_1}{C_2} = 10^{-3}$$

$$C_2 = 6 \times 10^{-9} = 0.006\ \mu\text{F}$$

3.6 ▪ THEVENIN AND NORTON THEOREMS

3.6.1 Thevenin's Theorem

The solution of the loaded voltage divider can be solved in yet another manner. Use of the **Thevenin theorem** allows us to replace complex networks that may include many source voltages and complicated circuitry with a single equivalent source and a single series impedance. For the voltage divider shown in Figure 3.10, we wish to find what one equivalent resistance (R_T) and equivalent source voltage (V_T) between the output terminals will behave like the original circuit (as least insofar as the load is concerned).

Referring to Figure 3.18, we consider the connection between two networks, N_1 and N_2. To optimize performance, it is often desirable to first determine the effect of N_2 on N_1, for example, by obtaining the maximum transfer of power from N_1 to N_2.

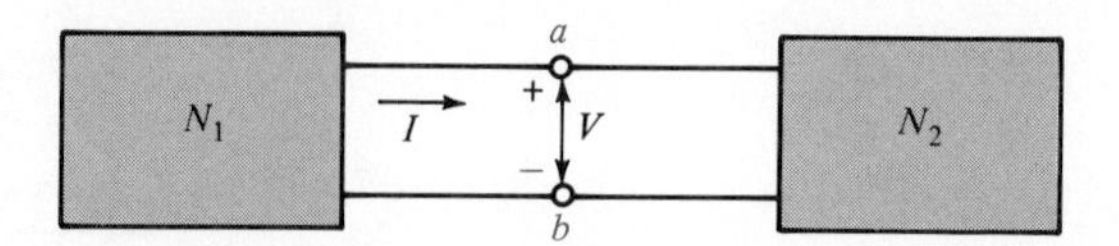

FIGURE 3.18. **Current Flow Between Networks N_1 and N_2, Together with Common Terminal Voltage V**

With terminals a and b short-circuited, the resultant current through the short will be termed I_{SC}, with $V = 0$. With the load N_2 and the short removed, the voltage between a and b will be termed V_{OC}—that is, the open-circuit voltage—and will be set equal to V_T.

Presume that N_1 is composed of voltage sources and a complex network of resistances. According to Thevenin's theorem, N_1 may be replaced by a single voltage source (V_T) in series with a single resistance (R_T) equal to V_{OC}/I_{SC}.

EXAMPLE 3.5

Solve the voltage divider problem shown in Figure 3.11 using Thevenin's theorem.

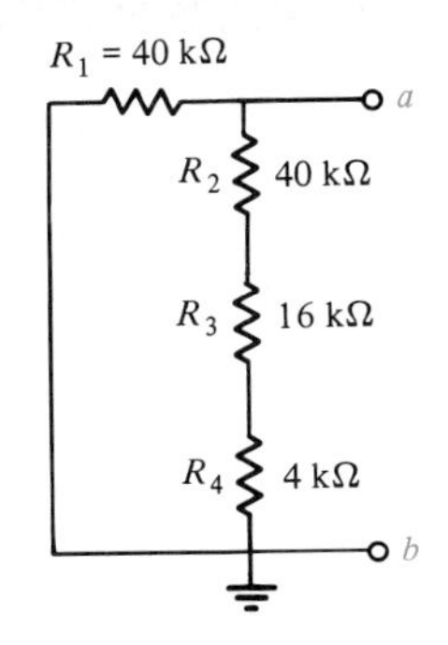

FIGURE 3.19

Solution:

$$I_{SC} = \frac{250}{R_1} = \frac{250}{4 \times 10^4} = 6.25 \times 10^{-3} \text{ A}$$

$$V_T = \frac{R_2 + R_3 + R_4}{R_{tot}} \times 250 = \frac{40 + 16 + 4 \text{ k}\Omega}{40 + 40 + 16 + 4 \text{ k}\Omega} \times 250$$
$$= 150 \text{ V}$$

$$R_T = \frac{V_{OC}}{I_{SC}} = \frac{150}{6.25 \times 10^{-3}} = 24 \text{ k}\Omega$$

With the 100 kΩ load (R_L) connected in series with the Thevenin resistance (R_T) and the Thevenin source (V_T), the voltage across the load is

$$V = \frac{R_L}{R_T + R_L} V_T = \frac{100 \text{ k}\Omega}{24 + 100 \text{ k}\Omega} \times 150 = 121 \text{ V}$$

which agrees with the prior calculation.

Alternatively, R_T could be obtained by shorting out the 250 V source and determining the resulting resistance between the output terminals. (See Figure 3.19.) This results in 40 kΩ (R_1) in parallel with the series combination ($R_2 + R_3 + R_4$), resulting in an equivalent resistance of 24 kΩ, agreeing with the preceding:

$$\frac{1}{R_T} = \frac{1}{R_1} + \frac{1}{R_2 + R_3 + R_4}$$

$$R_T = \frac{1}{(1/40) + [1/(40 + 16 + 4)]} = 24 \text{ k}\Omega$$

FOCUS ON PROBLEM SOLVING: Thevenin Theorem (Resistive Network)

Determine the open-circuit voltage V_{OC}—that is, the output voltage with the load disconnected. Determine the short-circuit current I_{SC} obtained by shorting the output terminals. The equivalent series resistance (R_T) is equal to V_{OC}/I_{SC}. The Thevenin equivalent is V_T (= V_{OC}) in series with R_T.

An alternative procedure again starts with a determination of the open-circuit voltage V_{OC}. All independent sources are then suppressed—that is, voltage sources are shorted out, and current sources are open-circuited. R_T is the resulting equivalent resistance between the output terminals. The Thevenin equivalent is V_T (= V_{OC}) in series with R_T.

3.6.2 Norton's Theorem

Rather than expressing a circuit equivalent in terms of a voltage generator in series with a resistance as a result of using Thevenin's theorem, we may instead depict it in terms of a current source. The procedure used to derive a

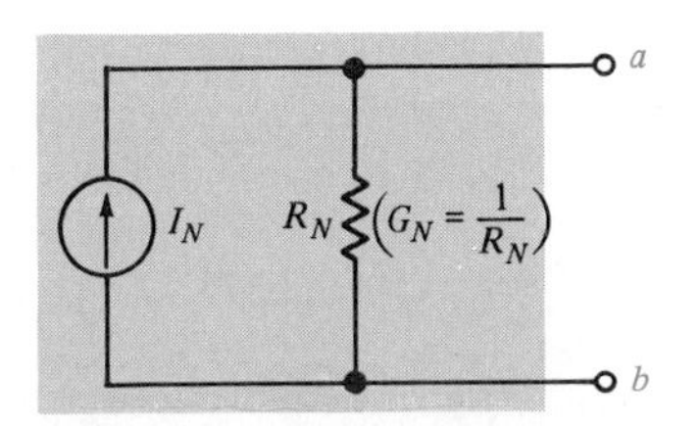

FIGURE 3.20. Norton Equivalent Circuit Consisting of a Current Source I_N in Parallel with Conductance G_N

circuit's current equivalent, in parallel with an equivalent resistance, constitutes **Norton's theorem** (Figure 3.20). In this case the current generator (I_N) is the current that flows when the output terminals are short-circuited. The equivalent parallel source resistance (R_N) is obtained by dividing the open-circuit output voltage by the short-circuit current. (Often the parallel resistance is labeled as a conductance—that is, $G_N = 1/R_N$.)

FOCUS ON PROBLEM SOLVING: Norton's Theorem (Resistive Network)

Determine the open-circuit voltage (V_{OC}) and the short-circuit current (I_{SC}). The Norton equivalent consists of a current source I_N (equal to I_{SC}) in parallel with conductance G_N $(1/R_N)$, which equals I_{SC}/V_{OC}.

EXAMPLE 3.6

Determine the Norton equivalent of the circuit considered in Figure 3.11 and show that voltage under load remains the same as previously calculated.

Solution: See Figure 3.21.

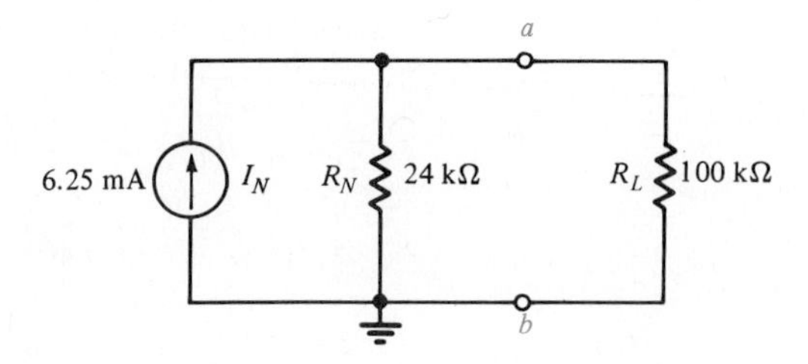

FIGURE 3.21

$$I_N = I_{SC} = \frac{250}{R_1} = \frac{250}{4 \times 10^4} = 6.25 \times 10^{-3} \text{ A}$$

$$V_{OC} = 150 \text{ V}$$

$$R_N = \frac{V_{OC}}{I_{SC}} = \frac{150}{6.25 \times 10^{-3}} = 24 \times 10^3 = 24 \text{ k}\Omega$$

The current (I_L) through the 100 kΩ load resistor (R_L) may be obtained by using the current division rule—that is,

$$I_L = \frac{R_N}{R_N + R_L} I_N = \frac{24}{24 + 100} \times 6.25 \times 10^{-3} = 1.21 \times 10^{-3} \text{ A}$$

The output voltage across the load is

$$V_L = I_L R_L = (1.21 \times 10^{-3})(100 \times 10^3) = 121 \text{ V}$$

3.6.3 Generalized Thevenin and Norton Theorems

The application of the Thevenin and Norton theorems is not limited to resistive networks. They may be applied to any *linear* network—that is, any combination of voltage (or current) sources involving resistors, capacitors, and inductors. Nor are they limited to dc sources. In the general case, for example, the Thevenin equivalent is a voltage source (which may include a phase angle) in series with an impedance, while the Norton equivalent in general is a current source with a phase angle in parallel with an admittance. The power loss in either a Thevenin or a Norton source, however, is not the same as in the original network except for the case of zero dissipation.

FOCUS ON PROBLEM SOLVING: Generalized Thevenin and Norton Theorems

These theorems are also applicable to any linear electrical network that is in a sinusoidal steady state. The basic procedures are the same as for strictly resistive circuits.

For the Thevenin equivalent, determine the open-circuit voltage (which may be a complex number) and the circuit *impedance* with all voltage and current sources suppressed—that is, with voltage sources shorted and current sources open. The Thevenin equivalent is the resulting impedance in series with the (complex) open-circuit voltage source.

The Norton equivalent is an admittance (rather than a conductance) that appears in parallel with an equivalent (complex) current source.

EXAMPLE 3.7 Determine the Thevenin equivalent for the circuit in Figure 3.22.

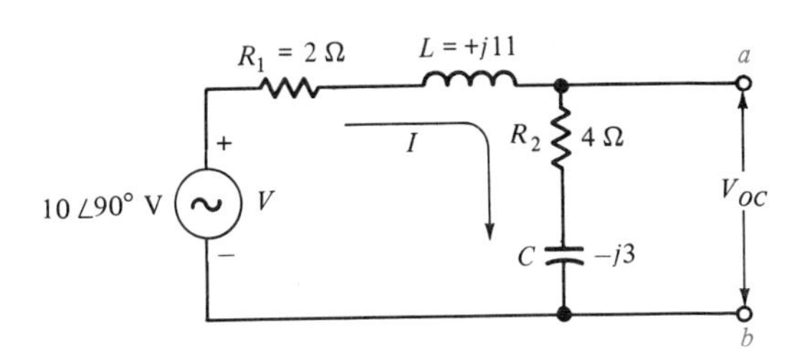

FIGURE 3.22

Solution:

$$\mathbf{I}(R_1 + jX_L + R_2 - jX_C) = \mathbf{V}$$
$$\mathbf{I}(2 + j11 + 4 - j3) = 10\angle 90^\circ$$
$$\mathbf{I}(6 + j8) = 10\angle 90^\circ$$
$$\mathbf{I} = \frac{10\angle 90^\circ}{6 + j8} = \frac{10\angle 90^\circ}{10\angle 53.1^\circ} = 1\angle 36.9^\circ\ \text{A}$$

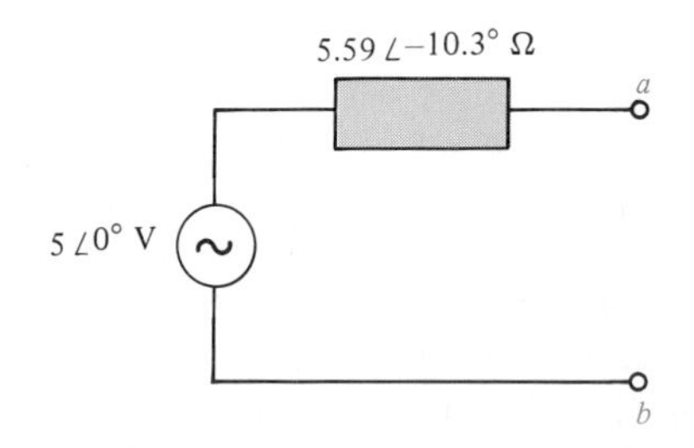

FIGURE 3.23

$$\mathbf{V}_{OC} = \mathbf{I}(R_2 - jX_C) = (1\angle 36.9°)(4 - j3)$$
$$= (1\angle 36.9°)(5\angle -36.9°) = 5\angle 0° \text{ V} = \mathbf{V}_T$$

To find $\mathbf{Z}_T$, the Thevenin impedance, short out the voltage source and determine the impedance between the output terminals (Figure 3.23):

$$\mathbf{Z}_T = \frac{1}{[1/(R_1 + jX_L)] + [1/(R_2 - jX_C)]}$$
$$= \frac{1}{[1/(2 + j11)] + [1/(4 - j3)]} = \frac{(2 + j11)(4 - j3)}{4 - j3 + 2 + j11}$$
$$\mathbf{Z}_T = \frac{41 + j38}{6 + j8} = \frac{55.9\angle 42.8°}{10\angle 53.1°} = 5.59\angle -10.3° \ \Omega$$

In the general case, the load itself may be an impedance (Z_L). We then have

$$\mathbf{Z}_T = R_T + jX_T = Z_T\angle\theta_T$$
$$\mathbf{Z}_L = R + jX = Z_L\angle\theta_L$$

The current magnitude is

$$I = \frac{V_T}{\sqrt{(R_T + R)^2 + (X_T + X)^2}}$$

and the power furnished to the load is

$$P_L = I^2R = \frac{V_T^2R}{(R_T + R)^2 + (X_T + X)^2}$$

We consider the values of the Thevenin source to be fixed, but the values of the load to be adjustable. If maximum power transfer is desired, from the power equation we see that this arises when $X_T = -X$, considering the reactance of the load to be adjustable.

If only R is adjustable, optimum power transfer takes place when

$$R = \sqrt{R_T^2 + (X_T + X)^2}$$

If both X and R are adjustable, the optimum power transfer arises when $R = R_T$ and $X = -X_T$.

Such impedance matching is important in electronic and communication circuits. It is not generally employed with electric power devices—motors, generators, power lines—because of the low efficiency that may result. Half the power would have to be dissipated at the source and connecting networks, not a very attractive aspect for the power company.

3.7 ■ FILTERS

Often in electronic circuits one encounters what looks like a voltage divider consisting of a capacitor *and* a resistor (Figure 3.24). Such a configuration of

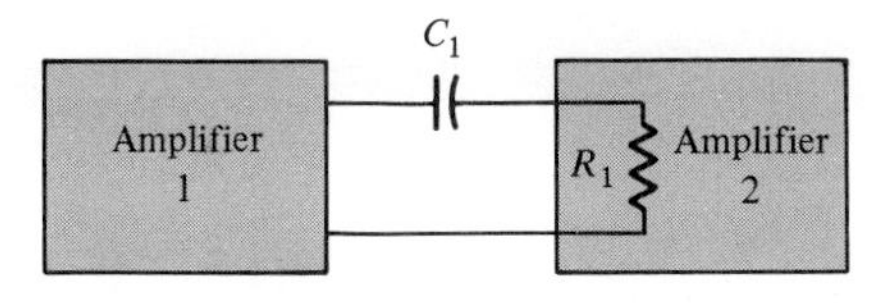

FIGURE 3.24. **Interstage Coupling Between Amplifiers** C_1 is needed to prevent dc interaction between amplifiers yet is chosen to be large enough so that ac may pass without significant attenuation. R_1 is the input resistance to the second stage of amplification.

resistance and capacitance constitutes a frequency-dependent attenuator called a **filter**. Specifically, the one shown in Figure 3.24 is a *high-pass filter*.

A single stage of electronic amplification usually is insufficient to the task at hand, and amplifier stages are cascaded (as in Figure 3.18). One wishes to transfer the ac output signal from one stage to the next where it will be subjected to additional amplification. But the dc voltage at the output of an amplifier may differ from the dc voltage at the input to the next stage, and therefore, they cannot be directly connected. Making the connection through a capacitor, however, prevents dc interaction yet allows the ac to pass between them.

In the present instance we wish to diminish as little as possible the signal passing through C_1—that is, we want X_{C_1} to be small compared to R. It would help to make R as large as possible, but generally there is a limit in this regard, set by the circuitry in the second amplifier stage. Therefore, this primarily leaves the selection of C_1 as a major means of reducing the attenuation of this *coupling circuit*. To make X_{C_1} small at frequency f, we wish to use a large capacitor. There are, however, a number of limitations in this regard: availability of specific values, physical size, and so on.

EXAMPLE 3.8

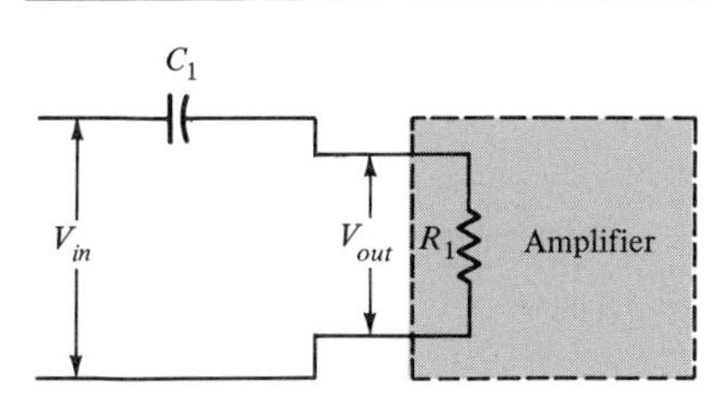

FIGURE 3.25

Relative to Figure 3.24, the input resistance of an amplifier is $10^5\ \Omega$. With 40 Hz as the low-frequency limit of the amplifier, what should be the value of C_1? (See Figure 3.25.)

Solution: A rule of thumb is to make $X_{C_1} \leq 0.1\ R_1$ at the low-frequency limit:

$$X_{C_1} \equiv 0.1R_1 = 0.1 \times 10^5\ \Omega = 10^4\ \Omega = \frac{1}{2\pi fC}$$

$$C_1 = \frac{1}{2\pi f X_{C_1}} = \frac{1}{2\pi(40)(10^4)} \simeq 4 \times 10^{-7}\ \text{F} = 0.4\ \mu\text{F}$$

For any frequency greater than 40 Hz, the reactance will be even smaller, which is desirable. Of course, if one wishes to make the signal drop even less, C_1 should be made even larger.

Interchanging R and C (Figure 3.26) leads to a low-pass filter. Low frequencies are readily passed through, but at high frequencies the capacitor acts progressively more as a short circuit across the output. The respective transmission of the two filters is shown in Figure 3.27.

By convention the *band limit* of filters is designated as being that frequency at which the output voltage (relative to the input) drops by a factor of 0.707—that is, $1/\sqrt{2}$. In a high-pass filter, this is designated as the lower band limit (f_L) and in the low-pass filter, as the upper band limit (f_H).

Considering a filter as a voltage divider, for the high-pass filter in Figure 3.25 we have

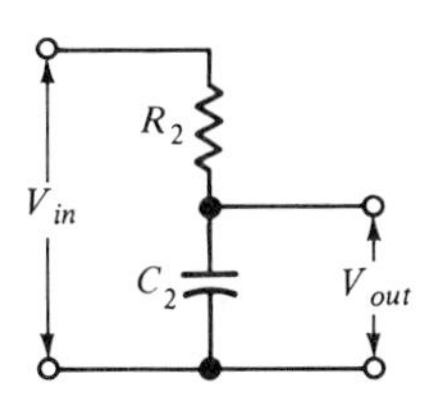

FIGURE 3.26. **Low-Pass Filter** Capacitor progressively shorts out higher frequencies.

$$\mathbf{V}_{out} = \frac{R_1}{R_1 - (j/\omega C_1)}\mathbf{V}_{in} = \frac{1}{1 - (j/\omega C_1 R_1)}\mathbf{V}_{in}$$

$$= \frac{1}{\sqrt{1 + (1/\omega^2 C_1^2 R_1^2)}} \angle \tan^{-1}\left(\frac{1}{\omega C_1 R_1}\mathbf{V}_{in}\right)$$

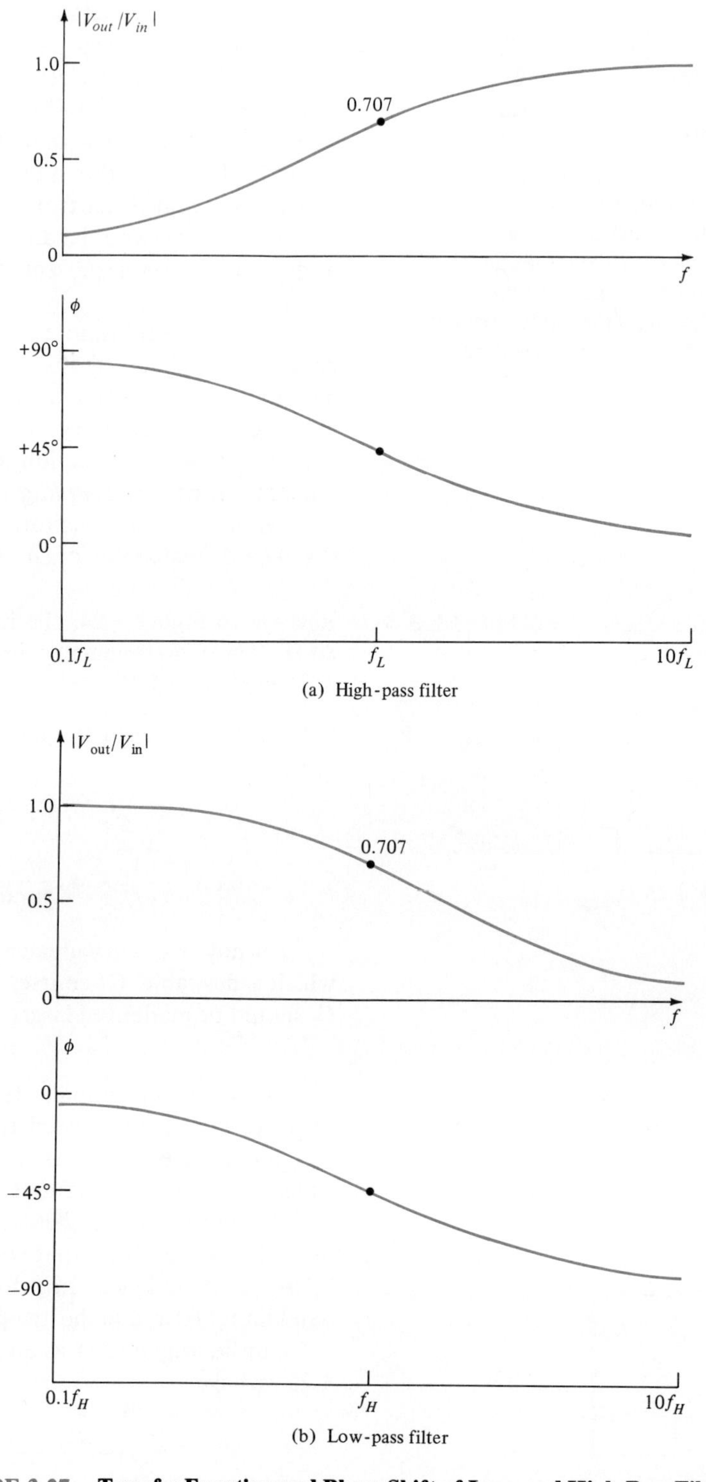

FIGURE 3.27. **Transfer Function and Phase Shift of Low- and High-Pass Filters**

$$V_{out} = \frac{1}{\sqrt{1 + (1/\omega^2 C_1^2 R_1^2)}} V_{in}$$

It can be seen that $V_{out} = 0.707 V_{in}$ when $\sqrt{2} = \sqrt{1 + (1/\omega^2 C_1^2 R_1^2)}$. Defining ω_L as the frequency that satisfies this condition, we obtain

$$1 = \frac{1}{\omega_L C_1 R_1}$$

$$\omega_L = 2\pi f_L = \frac{1}{R_1 C_1}$$

$$f_L = \frac{1}{2\pi R_1 C_1} \tag{3.10}$$

This constitutes the lower band limit for a simple *RC* high-pass filter. In a comparable manner the upper band limit for a simple *RC* low-pass filter takes the form

$$f_H = \frac{1}{2\pi R_2 C_2} \tag{3.11}$$

In general, the ratio of output signal to input signal is called the **transfer function**. The magnitude of the transfer function ($|V_{out}/V_{in}|$) is most conveniently expressed in decibels (see "A Closer Look" at the end of this chapter), calculated by the defining equation.

$$\text{dB} = 20 \log_{10} \frac{V_{out}}{V_{in}} \tag{3.12}$$

Thus, at the band limits, when $V_{out} = 0.707 V_{in}$, this represents a decibel change of

$$20 \log_{10} \frac{0.707 V_{in}}{V_{in}} \simeq -3 \text{ dB}$$

For this reason the band limits of a filter are often referred to as the 3 dB points.

EXAMPLE 3.9

a. The addition of a preamplifier to an audio amplifier increases the signal voltage by a factor of 10^4. To what does this correspond in terms of decibels?
b. How many dB are represented by each factor of 10 increase (or decrease) in signal voltage?

Solution:

$$\text{dB} = 20 \log_{10} \frac{V_{out}}{V_{in}}$$

a. $20 \log_{10}(10^4) = 20 \times 4 = 80$ dB
b. $20 \log_{10}(10) = 20 \times 1 = 20$ dB

Figure 3.27 also includes the phase change that accompanies the variation of the transfer function. Although of no immediate interest, it will prove useful in subsequent chapters.

3.8 ■ BRIDGE CIRCUITS

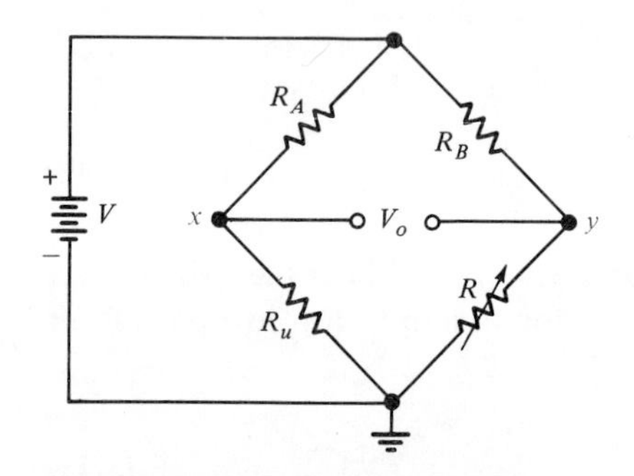

FIGURE 3.28. Wheatstone Bridge R_A and R_B are fixed resistors, R is a calibrated resistor, R_u is an unknown resistance.

Combinations of voltage dividers, such as the two parallel resistive dividers that constitute the Wheatstone bridge (Figure 3.28), can also be made to serve useful purposes. The voltages at x and y, respectively, will be

$$V_x = \frac{R_u}{R_A + R_u} V \quad \text{and} \quad V_y = \frac{R}{R_B + R} V$$

leading to an output voltage of

$$V_{out} = V_x - V_y = V\left(\frac{R_u}{R_A + R_u} - \frac{R}{R_B + R}\right) \tag{3.13}$$

Adjusting the variable R for $V_{out} = 0$, we obtain

$$\frac{R_u}{R_A + R_u} = \frac{R}{R_B + R}$$

$$R_u(R_B + R) = R(R_A + R_u)$$

$$R_u(R_B + R) - RR_u = RR_A$$

$$R_uR_B = RR_A$$

$$R_u = R\left(\frac{R_A}{R_B}\right) \tag{3.14}$$

While V_x and V_y may be finite when measured relative to ground, under balanced conditions—that is, $V_{out} = 0$—the potential difference between them is zero. Thus, if ac, rather than dc, is applied to the balanced bridge, while V_x relative to ground varies in a sinusoidal fashion, as does V_y, the *difference* between them always remains at zero.

As employed in a common physics experiment, the Wheatstone bridge is used to determine the value of an unknown resistor R_u. R_A and R_B are fixed precision resistors, while R is a calibrated variable resistor. Eq. 3.14, adjusted to yield $V_{out} = 0$, may be used to determine the value of R_u.

In Chapter 17 we will make use of the Wheatstone bridge in conjunction with strain gage measurements. The bridge is initially brought to a balance. The unknown resistance is then made to vary (as it is subjected to a stress), and the output voltage due to the bridge unbalance is used to measure the resulting strain. In such cases we are interested in the bridge sensitivity, that is, dV_{out}/dR_u:

$$\frac{dV_{out}}{dR_u} = \left[\frac{1}{R_A + R_u} - \frac{R_u}{(R_A + R_u)^2}\right]V = \frac{R_A}{(R_A + R_u)^2} V \tag{3.15}$$

EXAMPLE 3.10

Using a Wheatstone bridge with $R_A = 1000\ \Omega$ and $R_B = 100\ \Omega$, if the bridge is in balance when $R = 65.0\ \Omega$, find the value of R_u. If $V = 12$ V, what is the bridge sensitivity? What would it be if R_A were made equal to 10 kΩ?

Solution:

$$R_u = R\left(\frac{R_A}{R_B}\right) = 65.0\left(\frac{1000}{100}\right) = 650\ \Omega$$

$$\frac{dV_{out}}{dR_u} = \frac{R_A}{(R_A + R_u)^2}V = \frac{1000}{(1000 + 650)^2}12 = 4.4\ \text{mV}/\Omega$$

If R_A were 10 kΩ, we would obtain

$$\frac{dV_{out}}{dR_u} = \frac{10{,}000}{(10{,}000 + 650)^2}12 = 1.06\ \text{mV}/\Omega$$

Figure 3.29 represents another useful combination of voltage dividers, the Wien bridge. We will make use of it when discussing an electronic device (called the Wien bridge oscillator; Section 8.4.2) that generates ac whose frequency is controllable. The right branch is strictly resistive and point b is always at a potential equal to one third that of the input:

$$V_b = \frac{R_2'}{R_1' + R_2'}V_{in} \equiv \tfrac{1}{3}V_{in} \tag{3.16}$$

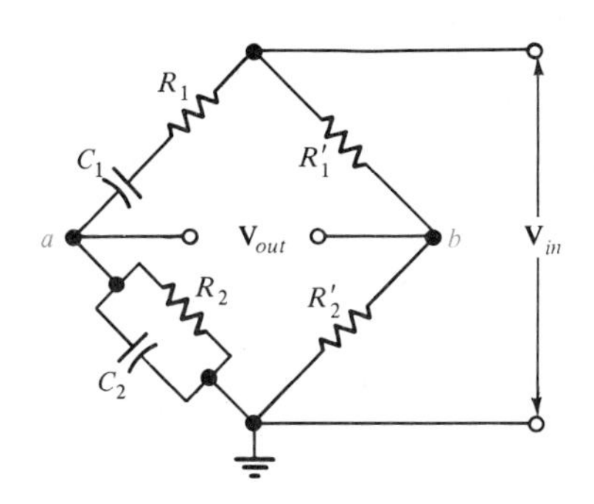

FIGURE 3.29. **Wien Bridge** By proper selection of Rs and Cs, a specific input frequency will be rejected and not appear at the output, while other frequencies are passed through with little or no attenuation.

To compute the potential at a, by designating the impedance of the parallel branch as $\mathbf{Z}_P$ and $\mathbf{Z}_S$ for the series branch, we have

$$\frac{\mathbf{V}_a}{\mathbf{V}_{in}} = \left(\frac{\mathbf{Z}_P}{\mathbf{Z}_S + \mathbf{Z}_P}\right)\left(\frac{\mathbf{Y}_P}{\mathbf{Y}_P}\right) = \frac{1}{1 + \mathbf{Z}_S\,\mathbf{Y}_P} \tag{3.17}$$

Since $\mathbf{Z}_S = R_1 - (j/\omega C_1)$ and $\mathbf{Y}_P = G_2 + jB_2 = (1/R_2) + j\omega C_2$, we have

$$\frac{\mathbf{V}_a}{\mathbf{V}_{in}} = \frac{1}{1 + [R_1 - (j/\omega C_1)][(1/R_2) + j\omega C_2]}$$

Setting $R_1 = R_2 \equiv R$ and $C_1 = C_2 \equiv C$ leads to

$$\frac{\mathbf{V}_a}{\mathbf{V}_{in}} = \frac{1}{3 + j[\omega CR - (1/\omega CR)]} \tag{3.18}$$

When the imaginary term is zero, $V_a = \frac{1}{3}V_{in}$. Under this condition, we obtain

$$V_a - V_b = \tfrac{1}{3}V_{in} - \tfrac{1}{3}V_{in} = 0 \tag{3.19}$$

Thus, the output signal is zero when $\omega^2C^2R^2 = 1$. Then we have

$$\omega_0 \equiv \frac{1}{RC} \tag{3.20}$$

The Wien bridge is a rejection filter at this frequency.

Neither the left branch alone (Figure 3.30) nor the right branch alone yields a particularly sharp response, but the two in combination do so (Figure 3.31).

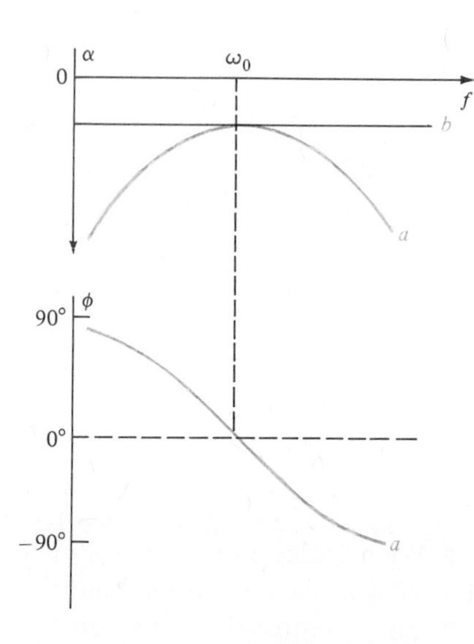

FIGURE 3.30. **Attenuation (α) vs. Frequency (*f*) at the Output of a Wein Bridge Measured Separately Relative to Ground and the Accompanying Phase Shift (φ) of the *a* Branch**

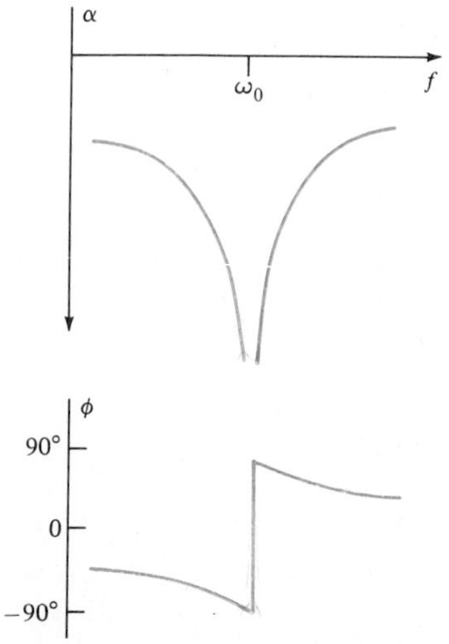

FIGURE 3.31 **Attenuation (α) vs. Frequency (*f*) at the Output of a Wien Bridge [Between Terminals *a* and *b* (Figure 3.29)] Together with Accompanying Phase Shift (φ)**

EXAMPLE 3.11

If a 3 V rms signal is applied to the input of a Wien bridge at the rejection frequency ($\omega_0 = 1/RC$), the voltage at point *a* has an rms value of 1 V. (See Figure 3.32.) If $R = 1\ \text{k}\Omega$ and $C = 0.159\ \mu\text{F}$, determine

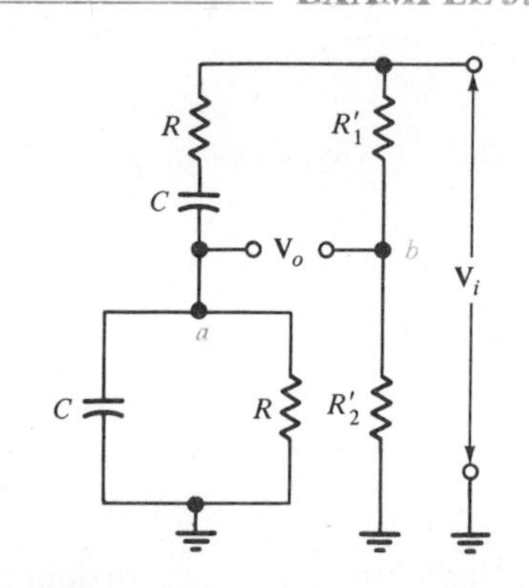

FIGURE 3.32

a. The rejection frequency.
b. The output voltage from the bridge under the following conditions:
   1. $R_2' = 1\ \text{k}\Omega$, $R_1' = 2\ \text{k}\Omega$
   2. $R_2' = 0.9\ \text{k}\Omega$, $R_1' = 2.1\ \text{k}\Omega$
   3. $R_2' = 1.1\ \text{k}\Omega$, $R_1' = 1.9\ \text{k}\Omega$

Solution:

a. $$\omega_0 = 2\pi f_0 = \frac{1}{RC}$$

$$f_0 = \frac{1}{2\pi RC} = \frac{1}{2\pi(10^3)(1.59 \times 10^{-7})} = 1\ \text{kHz}$$

b. Condition 1:

$$V_a = 1\ \text{V}$$

$$V_b = \left(\frac{R_2'}{R_1' + R_2'}\right)V_{in} = \left(\frac{1}{1 + 2}\right)3 = 1\ \text{V}$$

$$V_a - V_b = 0 \text{ V}$$

Condition 2:

$$V_a = 1 \text{ V}$$

$$V_b = \frac{0.9}{2.1 + 0.9} \times 3 = 0.9 \text{ V}$$

$$V_a - V_b = 0.1 \text{ V}$$

Condition 3:

$$V_a = 1 \text{ V}$$

$$V_b = \frac{1.1}{1.1 + 1.9} \times 3 = 1.1 \text{ V}$$

$$V_a - V_b = 1 - 1.1 = -0.1 \text{ V}$$

Note that in Condition 2 the output voltage is in phase with the input. In Condition 3, $V_b > V_a$, and the output is 180° out of phase with the input. Thus, when $R_1' = 2R_2'$, the output at frequency ω_0 is zero. When $R_1' > 2R_2'$, the output at ω_0 is finite and in phase with the input. When $R_1' < 2R_2'$, the output at ω_0 is finite and 180° out of phase with the input.

3.9 ▪ ELECTROMECHANICAL ANALOGS

3.9.1 Electrical and Mechanical Equivalents

If we look at the equations that govern the operation of a mechanical device and those for an electrical one, we can easily see that they are often identical in form. Accordingly, the operation of an electrical circuit can be studied via a mechanical analog and vice versa. Sometimes a device represents a combination of both elements, as in the case of a wind tunnel, which combines mechanical airflow generation with electrical controlling mechanisms. A more familiar example, however, is the phonograph pickup cartridge. Because of their easier simulation, it often helps to convert such electromechanical devices into an exclusively electrical equivalent circuit.

We start by considering the parallel *RLC* circuit shown in Figure 3.33a and its mechanical equivalent in Figure 3.33b. In the electrical case, we have an alternating voltage (whose instantaneous emf is v) and in the mechanical case an oscillating force (f). In the first instance, this leads us to an alternating current (i) and in the second case, to a varying velocity (u):

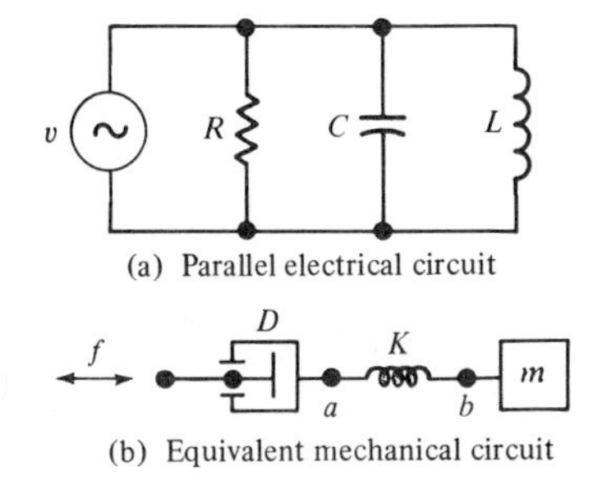

FIGURE 3.33. Example of Electrical and Mechanical Equivalent

$$\begin{aligned} &\text{voltage } (V) \leftrightarrow (F) \text{ force} \\ &\text{current } (I) \leftrightarrow (U) \text{ velocity} \\ &v = V_{peak} e^{j\omega t} \leftrightarrow f = F_{peak} e^{j\omega t} \\ &i = I_{peak} e^{j\omega t} \leftrightarrow u = U_{peak} e^{j\omega t} \end{aligned} \tag{3.21}$$

The various elements are arranged in corresponding order—that is,

electrical resistance (R) $\leftrightarrow$ (D) mechanical resistance
inductance (L) $\leftrightarrow$ (m) mass
capacitance (C) $\leftrightarrow$ (K) compliance

(Note that despite the *schematic symbol* similarity, the electrical inductor *does not* correspond to the mechanical spring.)

In the series electrical circuit, the applied voltage must equal the sum of the voltage drops:

$$V = V_R + V_L + V_C \tag{3.22}$$

Each of these terms might represent the corresponding effective values. We may again use the analogies to derive corresponding mechanical terms:

$$V_R = IR \leftrightarrow F_D = UD$$

$$V_L = X_L I = j\omega LI \leftrightarrow F_M = j\omega mU$$

$$V_C = X_C I = \frac{I}{j\omega C} \leftrightarrow F_K = \frac{U}{j\omega K}$$

Thus, summarizing, for Figures 3.33a and b, we have

$$V = V_R + V_L + V_C = I\left(\frac{R + j\omega L + 1}{j\omega C}\right)$$

$$F = F_D + F_M + F_K = U\left(\frac{D + j\omega m + 1}{j\omega K}\right)$$

EXAMPLE 3.12 An object weighing 2.5 lb is set on three rubber mounts whose compliance is $\frac{1}{16}$ in./lb.

a. Determine the natural resonant frequency.
b. What would be the corresponding values of L and C that would yield the same frequency? (k = spring constant = $1/K$, where K is the compliance.)

Solution:

a. $k = 1/K = 1/(\frac{1}{16}\text{ in./lb}) = 16\text{ lb/in./rubber mount:}$

$$k_{tot} = 3\text{ mounts} \times 16\text{ lb/in./mount} = 48\text{ lb/in.}$$

$$m = \frac{W}{g} = \frac{2.5\text{ lb}}{386\text{ in./s}^2}$$

$$f_c = \frac{1}{2\pi}\sqrt{\frac{k}{m}} = \frac{1}{2\pi}\sqrt{\frac{kg}{W}}$$

$$= \frac{1}{2\pi}\sqrt{\frac{48 \times 386}{2.5}} = 13.7\text{ Hz}$$

b. $\omega_0^2 = \dfrac{1}{LC}$

$$LC = \frac{1}{\omega_0^2} = \frac{1}{(2\pi \times 13.7)^2} = 1.35 \times 10^{-4}$$

By selecting $C = 1\ \mu F$ as a typical capacitor, the necessary L would be

$$L = \frac{1.35 \times 10^{-4}}{10^{-6}} = 1.35 \times 10^{2} = 135\ \text{H}$$

a very large value that, even if available, might have a large resistance and very low Q.

By selecting $C = 4700\ \mu\text{F}$, the necessary L would be 0.029 H, which is somewhat more practical.

EXAMPLE 3.13

Consider the two springs in Figure 3.34 subjected to an end force.

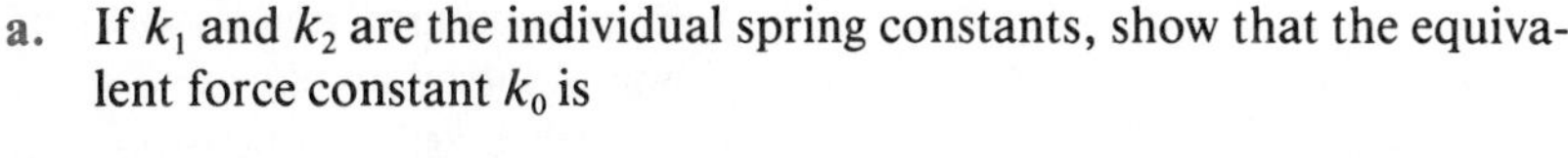

a. If k_1 and k_2 are the individual spring constants, show that the equivalent force constant k_0 is

$$k_0 = \frac{1}{(1/k_1) + (1/k_2)} = \frac{k_1 k_2}{k_1 + k_2}$$

b. What is the corresponding electrical equivalent?

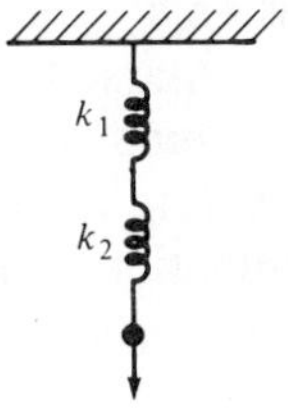

FIGURE 3.34

Solution:

a. Both springs are subjected to the same force. Their respective displacements are $F/k_1 = x_1$ and $F/k_2 = x_2$. The total displacement of the end is $x_1 + x_2$. We have

$$\frac{F}{k_1} + \frac{F}{k_2} = x_1 + x_2$$

$$F = k_0(x_1 + x_2)$$

$$k_0 = \frac{F}{x_1 + x_2} = \frac{F}{(F/k_1) + (F/k_2)}$$

$$= \frac{1}{(1/k_1) + (1/k_2)} = \frac{k_1 k_2}{k_1 + k_2}$$

b. Capacitors are analogous to the springs. Both capacitors have the same voltage applied (the two springs have the same force applied). We have

$$Q_{tot} = C_{equiv} V$$

$$Q_1 = C_1 V$$

$$Q_2 = C_2 V$$

$$Q_{tot} = Q_1 + Q_2 = C_1 V + C_2 V = (C_l + C_2)V = C_{equiv} V$$

$$C_{equiv} = C_1 + C_2$$

This is the equation for two capacitors in parallel. Therefore, the two springs in series correspond to two capacitors in parallel.

Since capacitance corresponds to compliance (K), which is the reciprocal of the spring constant (k), we have

$$k_1 = \frac{1}{K_1}$$

$$k_2 = \frac{1}{K_2}$$

$$\frac{1}{K_0} = \frac{(1/K_1)/(1/K_2)}{(1/K_1) + (1/K_2)} = \frac{1}{K_1 + K_2}$$

$$K_0 = K_1 + K_2$$

which is of the same form as capacitors in parallel.

3.9.2 Vibration Suppression

A very common and simple problem in the suppression of vibration transmission is solved by placing a spring beneath a vibrating mass. The effectiveness of such elastic suspension is measured by the *transmissivity*, which is the ratio of the disturbance transmitted through the spring to the disturbance when directly applied to the foundation. If the foundation is rigid, the latter is simply the reaction force. The function of the spring is not to pass less force than it receives but rather to reduce the force applied.

The equation of motion of mass (m) subjected to an oscillating force ($Fe^{j\omega t}$), with y the displacement and k the spring constant, is

$$m\left(\frac{d^2y}{dt^2}\right) + ky = Fe^{j\omega t} \tag{3.23}$$

In terms of an arbitrary constant (A) and forcing frequency (ω), the solution is $Ae^{j\omega t}$. This leads to

$$y(k - m\omega^2) = F \tag{3.24}$$

Therefore,

$$y = \frac{F}{k - m\omega^2} \tag{3.25}$$

This represents the forced solution. The transient solution that arises at the start of the excitation is assumed to have died away.

The force transmitted through the spring is

$$ky = \frac{kF}{k - m\omega^2} \tag{3.26}$$

and the transmissivity (ϵ) becomes

$$\epsilon = \left(\frac{kF}{k - m\omega^2}\right)\left(\frac{1}{F}\right) = \frac{k}{k - m\omega^2} = \frac{k/m}{(k/m) - \omega^2} = \frac{\omega_c^2}{\omega_c^2 - \omega^2}$$

$$= \frac{f_c^2}{f_c^2 - f^2} \tag{3.27}$$

where f_c is the natural resonant frequency. Defining the ratio of the exciting frequency (f) to the resonant frequency (f_c) as r, we obtain

$$\epsilon = \frac{1}{1 - r^2} \tag{3.28}$$

where $f/f_c \equiv r$. A plot of this function is shown as Figure 3.35. It is apparent that for reduction in ϵ the spring must be soft enough so that the resonant frequency lies below the frequency of excitation.

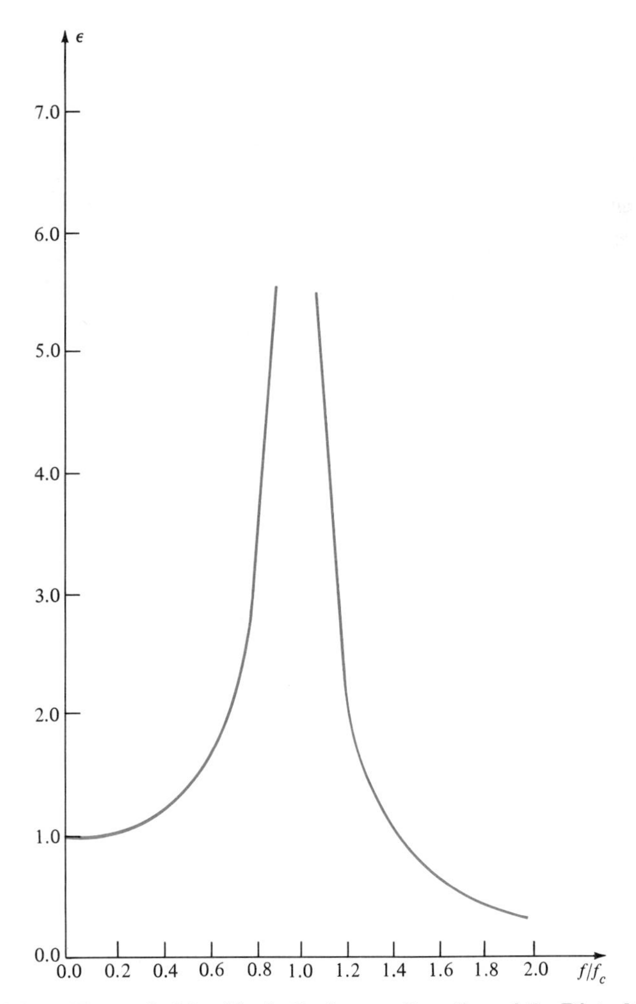

FIGURE 3.35. **Transmissivity (ϵ) of a Spring as a Function of the Disturbing Frequency (f) and the Resonant Frequency (f_c)**

Electrical engineers are generally interested in current (corresponding to velocity), while mechanical engineers are usually interested in displacement (corresponding to charge). In electrical engineering, we have

$$I = \frac{V}{Z} \tag{3.29}$$

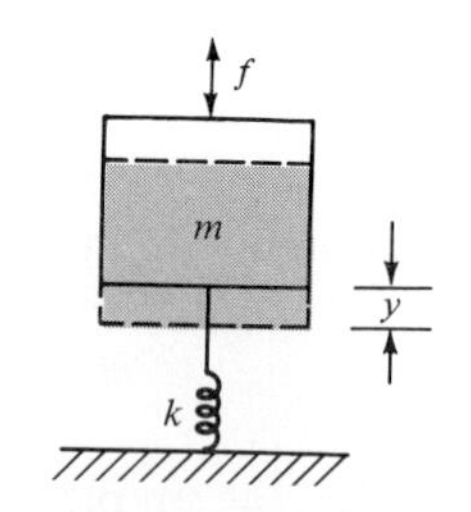

(a) Spring (k) used to reduce vibration communicated to foundation

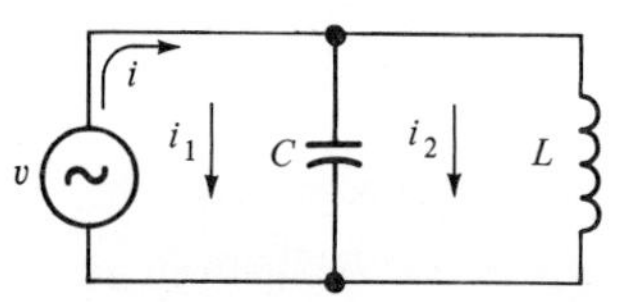

(b) Equivalent electrical circuit

FIGURE 3.36. Electrical Equivalent of a Spring

where Z is the **electrical impedance**. In mechanics, we have

$$v = \frac{F}{Z'} \tag{3.30}$$

where Z' is the **motional impedance**. From this we see that velocity (v) corresponds to current (i), in agreement with our system of analogies.

Figure 3.36a shows a vibrating mass above an elastic suspension while Figure 3.36b shows the equivalent electrical circuit. Being interested in the vibration of the mass, we apply the current division rule to the electrical equivalent circuit and obtain

$$i_2 = \frac{1/j\omega C}{(1/j\omega C) + j\omega L} i \tag{3.31}$$

Because the resonant frequency in the electrical case is $\omega_0 = 1/\sqrt{LC}$, we have

$$\frac{i_2}{i} = \frac{1}{1 - \omega^2 LC} = \frac{1}{1 - (\omega/\omega_0)^2} = \frac{1}{1 - r^2} \tag{3.32}$$

If $\omega < \omega_0$, $i_2/i > 1$. This is not a desirable situation in the mechanical case if we are attempting to attenuate vibration. In the electrical case, if we are faced with current pulsations that are superimposed on dc and wish to remove them, we will operate in the regime $\omega \gg \omega_0$ such that for ac $i_2/i < 1$.

While in the mechanical case it is the mass that is available in "pure" form, in the electrical case it is generally the capacitance. The mechanical spring usually incorporates some mechanical resistance, while in the electrical case it is difficult to obtain an inductance devoid of some electrical resistance.

Like the ideal inductor, the ideal spring would not dissipate power. Realistically, however, they both do since each incorporates some resistance. Additionally, in the mechanical case a resistive element is often purposely introduced in parallel with a spring—for example, a shock absorber placed in parallel with the springs of an automobile.

3.9.3 Vibration Damping

Refer to Figure 3.37. The equation of motion is

$$m\left(\frac{d^2y}{dt^2}\right) + D\left(\frac{dy}{dt}\right) + ky = Fe^{j\omega t} \tag{3.33}$$

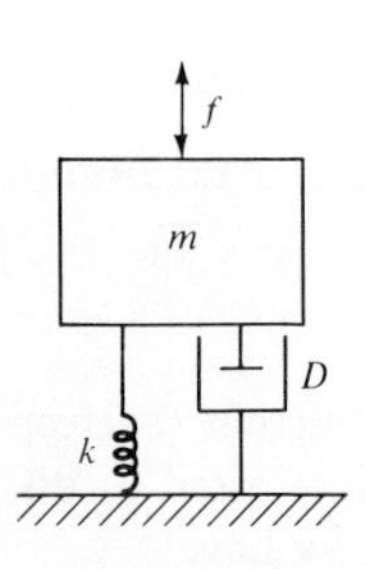

FIGURE 3.37. Vibration Suppression (k) and Damping (D) Elements Used to Reduce Vibration Communicated to Foundation

where the frictional force $[D(dy/dt)]$ is presumed to be proportional to the velocity (dy/dt). In the electrical case this corresponds to the voltage drop RI (or, more commonly, the IR drop). By again assuming the solution $Ae^{j\omega t}$, the forced vibration solution is

$$y(k - m\omega^2 + jD\omega) = f \tag{3.34}$$

where f represents the instantaneous force leading to displacement y, and

$$y = \frac{f}{k - m\omega^2 + jD\omega}$$

$$= \frac{f/k}{[1 - (\omega^2/\omega_c^2)] + j[(D\omega_c/k)(\omega/\omega_c)]} \tag{3.35}$$

$$\epsilon = \frac{y(k + jD\omega)}{f} = \frac{k + jD\omega}{k - m\omega^2 + jD\omega} \tag{3.36}$$

The electrical equivalent of this result may easily be derived by again using the current division rule as applied to the equivalent electrical circuit upon solving for the ratio i_2/i. (See Figure 3.39.)

Since we are only interested in the magnitude of ϵ, the simplest procedure is to divide the magnitude of the numerator by that of the denominator:

$$\epsilon = \sqrt{\frac{f_c^4 + (D^2f^2/4\pi^2m^2)}{(f_c^2 - f^2)^2 + (D^2f^2/4\pi^2m^2)}} \tag{3.37}$$

If we set $D = 0$, we obtain the prior result—that is, Eq. 3.27. The presence of damping can, under some circumstances, *increase* the transmissivity. Why, then, use dampers, such as automotive shock absorbers?

Springs do not damp vibration but rather prevent the transmission of vibration. Damping implies energy dissipation, and (ideal) springs do not dissipate energy [any more so than do (ideal) capacitors]. (See Figure 3.38.)

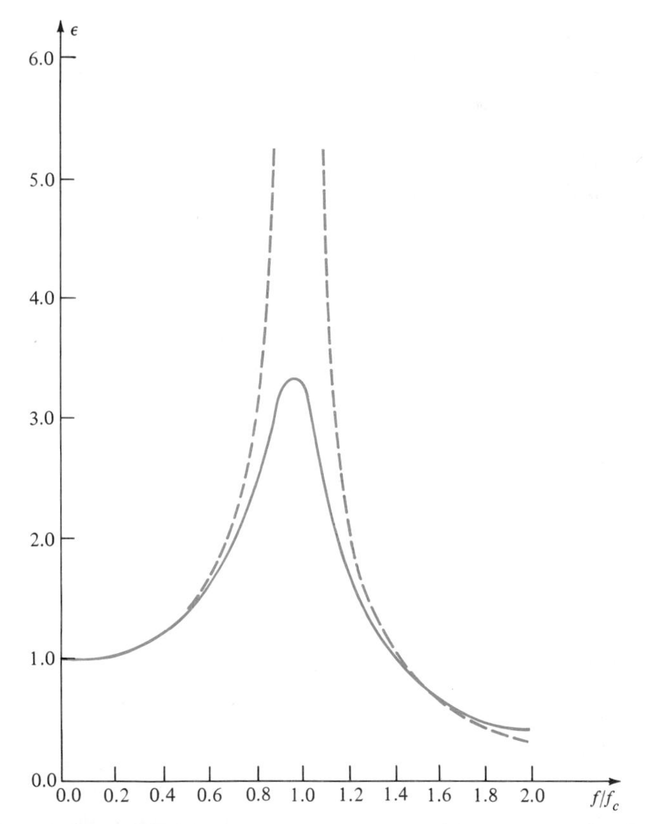

FIGURE 3.38. **Transmissivity (ϵ) of a Simple Spring with Accompanying Damper** Note that at frequencies significantly higher than resonance the damper actually increases transmissivity, but the resonance peak is diminished

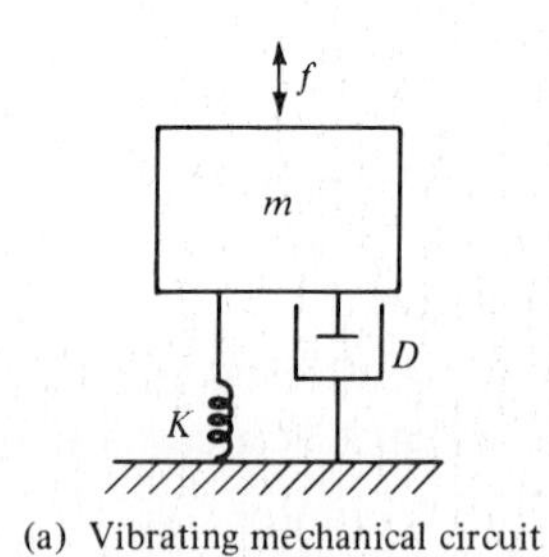

FIGURE 3.39. Electrical Equivalent of a Damped Spring

Dissipation reduces the amplitude of the *resonant* vibration and prevents the possible buildup of peak values that may destroy the springs. Figure 3.39 shows the electrical equivalent of the mechanical damping circuit. Whereas K and D appear in parallel, C and R appear in series. Why? Such topological substitutions represent a general rule: *A parallel mechanical circuit is represented by a series electrical circuit and vice versa.*[3]

Let us proceed to verify this rule by means of some simple examples: Consider two mechanical springs connected in series as in Figure 3.40. Here the displacements (hence the velocities) of the two are different, while each is subjected to the same force. This may most easily be seen by imagining K_2, for example, to be so stiff that it hardly shows any extension (or compression). The equivalent electrical circuit must then have different current flow through the two elements. Since the elements are presumed to differ in value, the equivalent electrical circuit must be a parallel arrangement.

Now consider two springs as shown in Figure 3.41. While the spring impedances are additive, they are connected in parallel. Are they both being subjected to the same force? No. They are both being subjected to the same displacement, but since their compliances are taken to be different (compliance ≡ displacement/force), there is a difference in the force experienced by each. Since their displacements are the same, so is the velocity of each.

What, then, is the electrical analog corresponding to Figure 3.41? In the electrical case we have current corresponding to velocity, and for two unequal

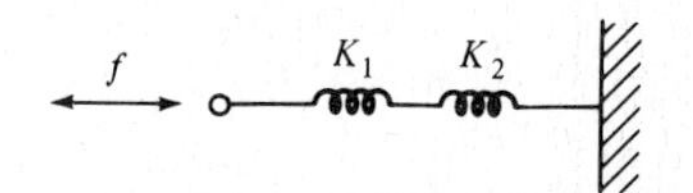

FIGURE 3.40. Series Spring Circuit Differing spring compliances subjected to the same force but different displacements correspond to electrical capacitances in parallel.

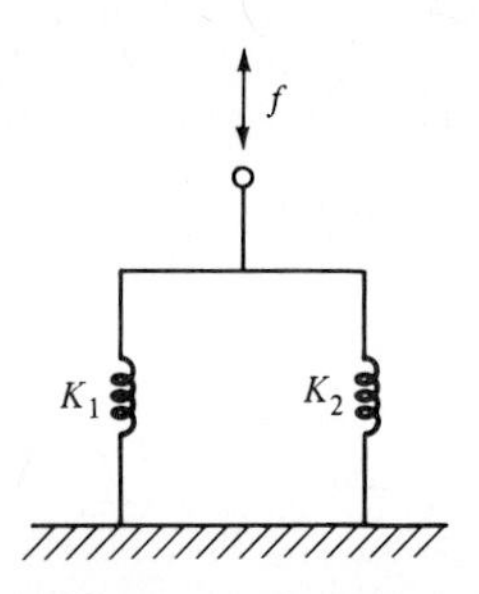

FIGURE 3.41. Parallel Spring Circuit Differing spring compliances subjected to the same displacement must be subject to different forces. The electrical equivalent is two capacitors in series.

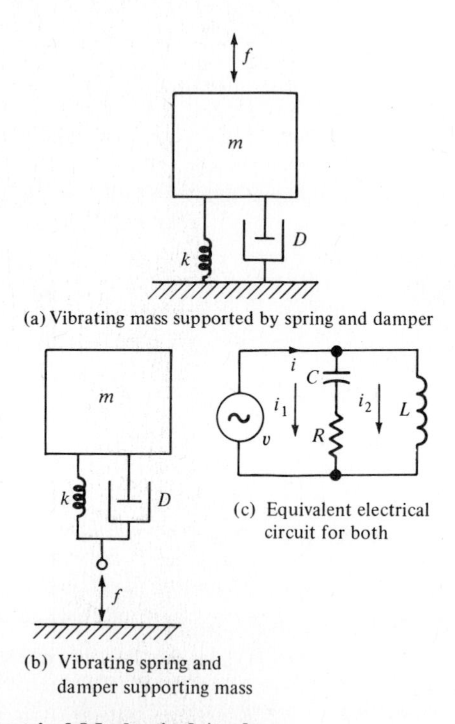

FIGURE 3.42. Electrical-Mechanical Analogs

[3] This arises because of the particular set of analogs that we have employed—that is, force corresponding to voltage, mass corresponding to inductance, and so forth. With a different set of conversions, termed the *duals*, such inversion would not be necessary. This dual set, for example, would have force corresponding to current, mass corresponding to capacitance, etc.

elements to have the same current, they must be in series. Thus, the equivalent electrical circuit will consist of two capacitors in *series*.

The case of a vibrating force applied to a mass suspended above a damped spring is shown again in Figure 3.42a and that of a mass suspended above a damped spring that is subjected to a vibrating force is shown in Figure 3.42b. (The latter case is comparable to that of an automotive suspension system.) The transmissivity in both cases is the same. While it might seem that in Figure 3.42a there is only one displacement (that is, at the juncture between the elements), one must also consider a finite displacement at the base. (Otherwise there would be no need for damping.) This leads to identical equivalent electrical circuits.

In the mechanical circuits the displacement of the mass is not necessarily the same as the displacement to which the spring and damper are subjected. (This may be verified by letting m in Figure 3.42b become very massive. Its displacement is then essentially zero, while the displacements of the spring and damper remain finite.) In the equivalent electrical circuit (Figure 3.42c), this must mean the elements are arranged in parallel and have differing currents. It also explains why in this case the transmissivity starts at unity; the capacitor does not pass dc current, and $i_2/i = 1$ (Figure 3.43). On the other hand, i_1/i starts out at zero, this being the transmission through the capacitor (Figure 3.44).

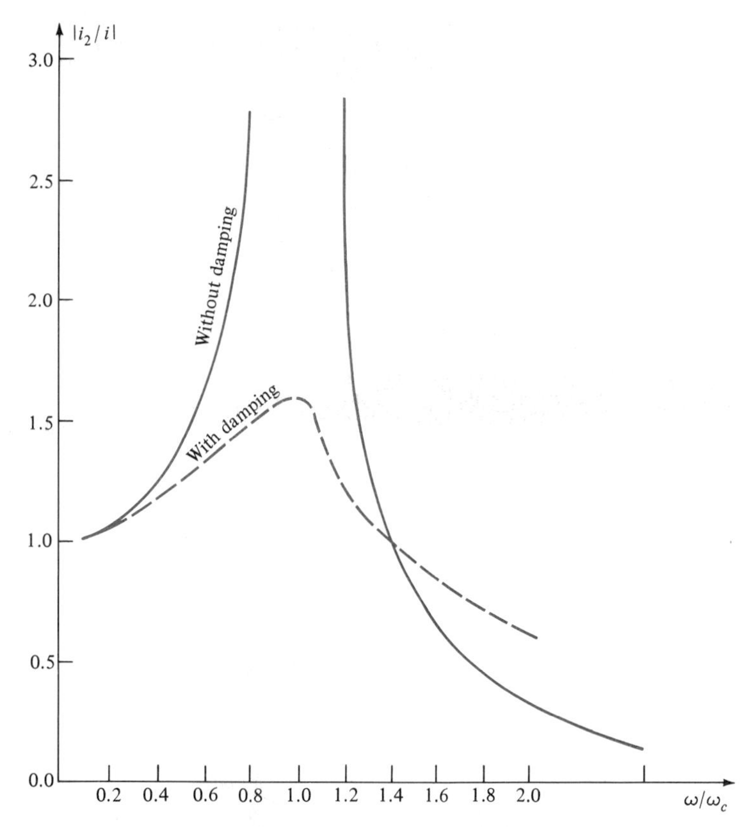

FIGURE 3.43. Transfer Function (i_2/i) Through the Inductor of Figure 3.42, with and Without Damping

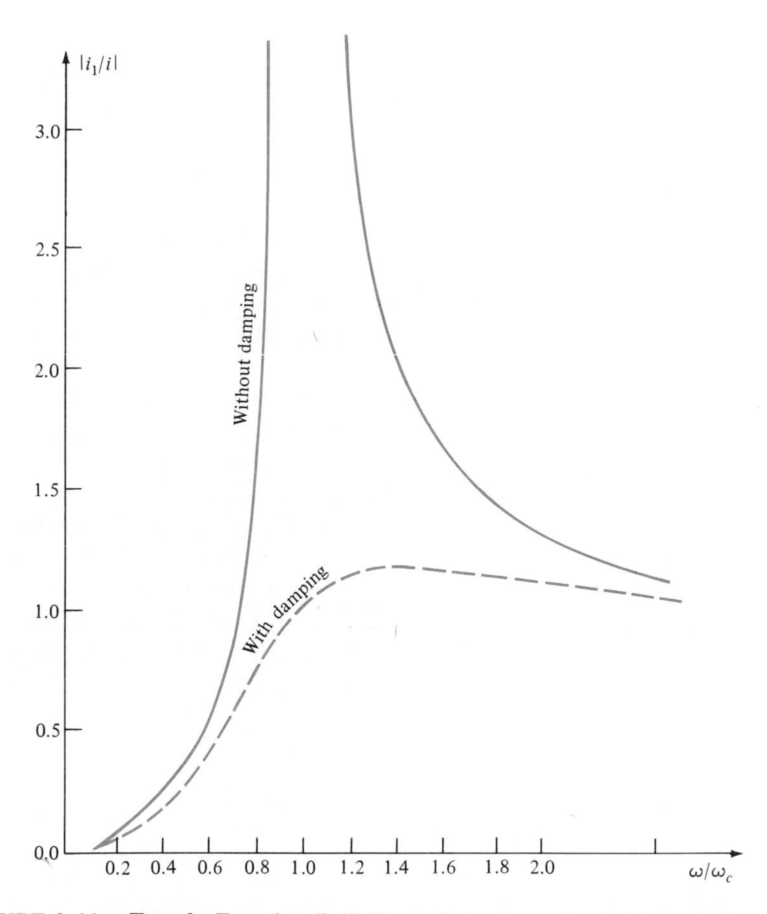

FIGURE 3.44. **Transfer Function (i_1/i) Through the Capacitor in Figure 3.42, with and Without Damping**

The electrical situation in Figure 3.42c does not generally arise in practice because the resistance is usually associated with the inductance and not the capacitance. In such cases $|i_2/i|$ does not exhibit the crossing characteristic at high frequencies.

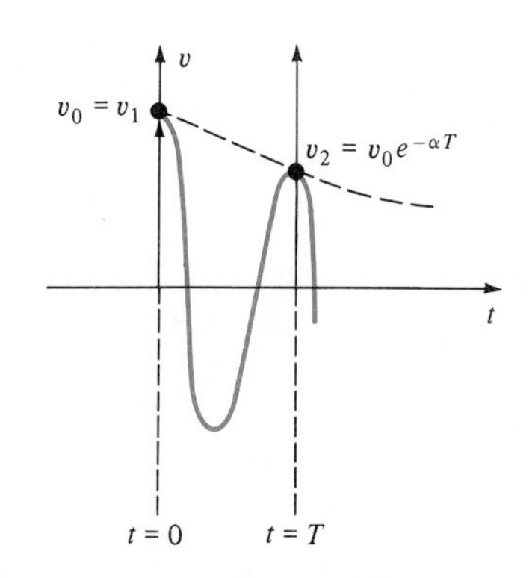

FIGURE 3.45. **Damped Wave in Terms of Amplitude (v_0), period (T), and Damping Constant (α)**

3.9.4 Transient Solutions

It might be recalled that when the point contacts of an automotive ignition system first opened there was an oscillatory response. (See, for example, Figure 3.7.) Such oscillation represents the transient solution whose real part constitutes the voltage variation:

$$v = v_0 e^{-\alpha t} \cos \omega_0 t \tag{3.38}$$

where v_0 = initial value

$\alpha = R/2L$

The natural resonant frequency is ω_0. The resulting damped oscillation is shown in Figure 3.45. The rapidity of the decay depends on the damping factor, the damping being due to energy dissipation. The natural logarithm of the ratio of successive amplitudes is termed the **logarithmic decrement** δ:

$$\delta \equiv \ln \frac{v_1}{v_2} \tag{3.39}$$

Since the time between successive peaks equals one oscillatory period (T), $f_0 = 1/T$, and with $v = v_0 \equiv v_1$ at $t = 0$ and $v = v_2 = v_0 e^{-\alpha T}$ at $t = T$, we have

$$\delta = \ln \frac{v_1}{v_2} = \ln e^{\alpha T} = \alpha T = \frac{\alpha}{f_0} = \frac{R}{2Lf_0} = \frac{\pi R}{\omega_0 L} \tag{3.40}$$

Or, in terms of circuit Q, we have

$$\delta = \frac{\pi}{Q} \tag{3.41}$$

Of course, the higher the circuit Q, the smaller is δ, and the more slowly the transient solution fades away. After the transient has died out, if the system is being subjected to a forced oscillation, that vibration will prevail.

The same type of performance arises in mechanical systems. When subjecting a mechanical resonant system to a sudden impulse (comparable to opening, or closing, the switch in the electrical case), there may arise a transient oscillation at or near the natural resonant frequency. In the mechanical case, by analogy, we have

$$Q = \sqrt{\frac{m}{KD^2}} = \sqrt{\frac{km}{D^2}} \tag{3.42}$$

3.9.5 Vibration with Resonant Supports

The next step in the complexity of a vibration attenuation problem is to consider the support itself to have a finite mass and elasticity. We then have the situation depicted in Figure 3.46. Spring k_1 is subjected to a displacement $(y_1 - y_2)$, and this leads to the equation

$$-y_1 m\omega^2 + k_1(y_1 - y_2) = f \tag{3.43}$$

while the second spring responds to the equation

$$-y_2 m\omega^2 + k_2 y_2 + k_1(y_2 - y_1) = 0 \tag{3.44}$$

Solving simultaneously for y_2, we obtain

$$y_2 = \frac{fk_1}{k_1k_2 - (k_1m_1 + k_1m_2 + k_2m_1)\omega^2 + m_1m_2\omega^4} \tag{3.45}$$

If k_1 is very large, resulting in f being applied directly to m_2, we obtain

$$y_2' = \frac{f}{k_2 - (m_1 + m_2)\omega^2} \tag{3.46}$$

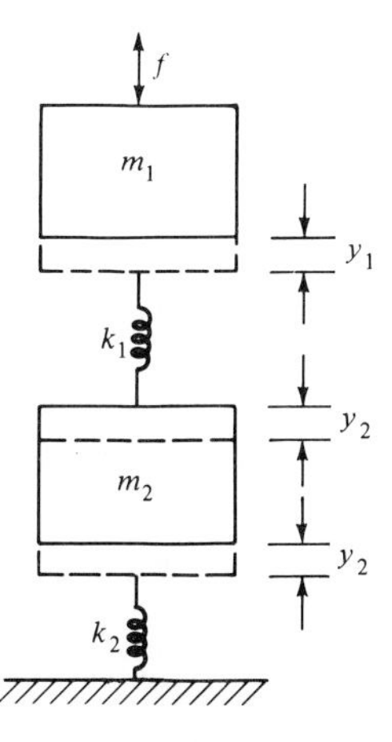

FIGURE 3.46. **Oscillating Masses with Spring Suspensions Intervening**

The transmissivity through k_1 is

$$\epsilon = \frac{y_2}{y_2'} = \frac{k_1k_2 - k_1(m_1 + m_2)\omega^2}{k_1k_2 - (k_1m_1 + k_2m_2 + k_2m_1)\omega^2 + m_1m_2\omega^4}$$
$$= \frac{1 - [(m_1 + m_2)/k_2]\omega^2}{1 - [(m_1/k_2) + (m_2/k_1) + (m_1/k_1)]\omega^2 + (m_1m_2/k_1k_2)\omega^4} \quad (3.47)$$

The denominator is a quadratic equation that leads to two solutions, ω_1^2 and ω_2^2. Therefore, we have

$$\epsilon = \frac{1 - [(m_1 + m_2)/k_2]\omega^2}{[1 - (\omega^2/\omega_1^2)][1 - (\omega^2/\omega_2^2)]} \quad (3.48)$$

Define $\omega_3 = k_2/(m_1 + m_2)$. While ω_1 and ω_2 are real resonant frequencies, ω_3 will be a resonant frequency if k_1 is rigid. (See Eq. 3.46.) In Eq. 3.48, however, it represents a frequency at which the transmissivity goes to zero. We will consider a practical example of this behavior in a moment.

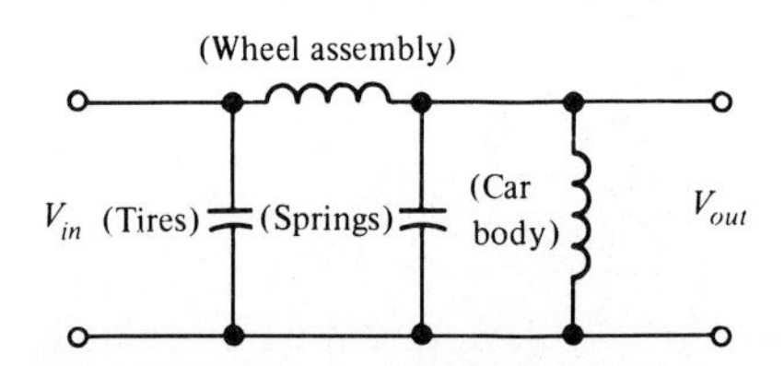

FIGURE 3.47. Electrical Equivalent of Figure 3.46 and of a Simplified Automotive Suspension System

The electrical equivalent circuit is shown in Figure 3.47. One may qualitatively justify this circuit in the following manner: m_1 and k_1 appear in series. Their electrical equivalents, therefore, appear in parallel. If k_1 were stiff, m_1 and m_2 would be combined into a single entity in series with k_2. In the electrical equivalent, k_1 would be an open circuit, and the two inductors would be in series across k_2.

Now consider Figure 3.48. This, in simplified form, typically represents an automotive suspension system. The displacement of the body (y_1) and the wheel assembly (y_2) are given by the expressions

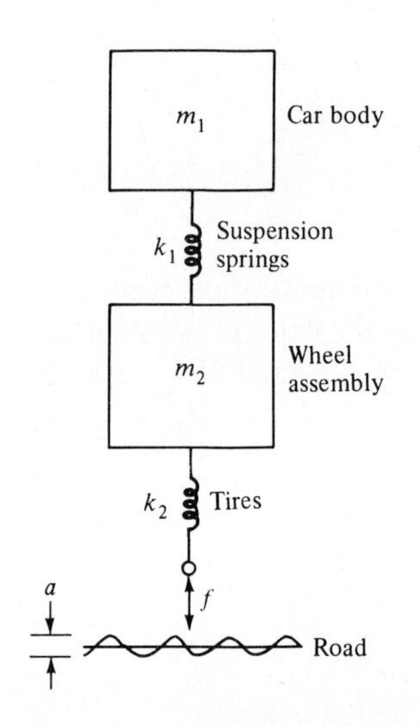

FIGURE 3.48. Simplified Mechanical Circuit of an Automotive Suspension System

$$y_1 = \frac{a}{[1 - (\omega^2/\omega_1^2)]\,[1 - (\omega^2/\omega_2^2)]} \cos \omega t \quad (3.49)$$

$$y_2 = \frac{a[1 - (m_1/k_1)\omega^2]}{[1 - (\omega^2/\omega_1^2)][1 - (\omega^2/\omega_2^2)]} \cos \omega t \quad (3.50)$$

The road's roughness (amplitude) is represented by a. Figure 3.49 represents y_1/a and y_2/a, the relative vibration of car body and wheels, respectively.

For the most part we have thus far disregarded phase relationships. In the case of Figure 3.49, however, they represent an interesting aspect of the problem. In ranges A and B, both masses move together, being in phase with the road's oscillation in range A and of opposite phase in range B. In ranges C and D, the two masses move out of phase with one another, with m_2 being in phase with the road oscillation in range C and m_1 being in phase with the road oscillation in range D.

With regard to the automobile, it should be noted that as long as both resonances are below the road frequencies, the wheels (m_2) are subjected to a greater degree of vibration than the body of the car (m_1), although there is a frequency between the two resonances at which the wheel vibration goes to zero.

SUMMARY

3.1 Introduction

- The "coil" of an automotive ignition system is actually a transformer.

3.2 Simplified Ignition System

- The ratio of secondary to primary voltage of a transformer is directly proportional to the turns ratio.

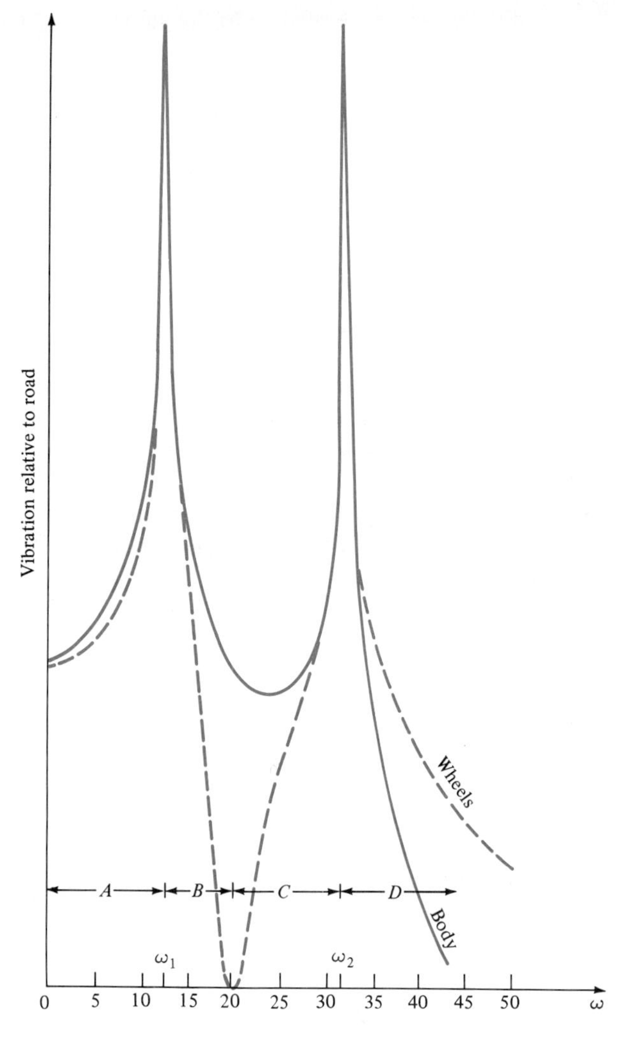

FIGURE 3.49. **Vibration Transmission of an Automotive Suspension System**

Current transformation is inversely proportional to this ratio.

3.3 Mathematics of Ignition

- The time constant (τ) of an RL circuit is L/R.
- With an increasing magnetic field,

$$i = \frac{V}{R}(1 - e^{-t/\tau})$$

With a collapsing magnetic field,

$$i = \frac{V}{R}e^{-t/\tau}$$

- The general solution of an RLC circuit can take on one of three different solutions depending on the degree of damping (that is, $R/2L$), being, respectively, the damped oscillatory solution, the overdamped solution, and the critically damped solution.
- The largest ignition voltage arises when we deal with the damped oscillatory solution of an RLC circuit.

3.5 Voltage Dividers

- Upon connecting the load, the degree of change in the output voltage of a divider circuit is a measure of the circuit's regulation ability.

- The smaller the divider resistance relative to the value of the load resistance, the better the circuit regulation.
- Good regulation of a voltage divider is obtained as a result of considerable power dissipation in the divider circuit.
- Pulsed signals, such as obtained from an ignition system, may be attenuated by using a capacitive voltage divider.

3.6 Thevenin and Norton Theorems

- The analysis of complex linear circuitry may be simplified by replacing the circuits with a single voltage source, in series with a single impedance. This constitutes Thevenin's theorem.
- Alternatively, linear complex circuitry may be replaced with a single current source in parallel with a single impedance. This constitutes Norton's theorem.
- With a fixed source resistance maximum power transfer arises when source resistance matches the load resistance.

3.7 Filters

- Combinations of R and C may be used to selectively pass only high or only low frequencies.
- The magnitude of the transfer function (that is, $|V_{out}/V_{in}|$) of the electronic and electrical circuits is conveniently expressed in decibels (dB). For voltage ratios, $\text{dB} = 20 \log_{10}(V_2/V_1)$. For power ratios, $\text{dB} = 10 \log_{10}(P_2/P_1)$.

3.8 Bridge Circuits

- The Wien bridge circuit is an example of a band rejection filter, passing higher and lower frequencies but discriminating against a selectable frequency range.

3.9 Electromechanical Analogs

- Each electrical quantity may be shown to correspond to a mechanical quantity, and the differential equations relating them are often identical.
- An LC circuit (without R) can merely act to exclude certain frequencies; it cannot dissipate their energy. In the same manner, a mechanical spring can be used to prevent transmission from a vibrating mass; it cannot attenuate those vibrations.
- The inclusion of resistance in the electrical case and of damping in the mechanical case introduces a mechanism that can actually dissipate signals.
- In the electromechanical analogs that we have adopted, a parallel mechanical system is represented by a series electrical analog and vice versa. In the dual system no such interchange would be necessary.
- Interchange of electrical elements does not change the performance of an electrical circuit. In the mechanical case the ordering of the elements can change the performance.
- The ratio of successive amplitudes gives rise to the logarithmic decrement, which is inversely proportional to the circuit Q.

KEY TERMS

attenuate: to reduce

attenuator: an electrical network used to reduce the magnitude of electrical signals

bleeder current: that portion of the current passing through a voltage divider that represents a power loss

damped oscillation: a cyclical variation with diminishing magnitudes in the successive cycles

decibel: a logarithmic unit used to measure either gain or attenuation

fall time: time interval involved in the parametric change from 90% to 10% of the initial value

filter: an electrical network used to selectively either pass or reject certain frequencies

gain: ratio of V_{out}/V_{in}, even if $V_{in} > V_{out}$ (Attenuation is V_{in}/V_{out}.)

impedance, electrical: voltage divided by current; in general, a complex quantity

impedance, motional: force divided by velocity

logarithmic decrement: the natural logarithm of the ratio of successive peak amplitudes of a damped sinusoid

Norton theorem: provides for the replacement of complex linear electrical networks with a single current source in parallel with a single admittance

oscilloscope: an electronic device generally used to observe the time variation of waveforms

points: electrical contacts in an automotive ignition system that, upon "making" and "breaking," cause the current to, respectively, either flow or not flow

regulation: as applied to a voltage source, the ability to maintain a constant voltage under changing source and/or load conditions

rise time: time interval involved in a parametric change from 10% of the final value to 90% of the final value

Thevenin theorem: provides for the replacement of complex linear electrical circuits by a single voltage source in series with a single impedance

time constant: time interval involved in a parametric change by a factor of $1/e$, where e is the base of the natural logarithms (2.718)

transfer function: relationship between the source and response of a network (It can be a dimensionless quantity—for example, V_{out}/V_{in}—I_{out}/I_{in}—or have the dimensions of either impedance or admittance—for example, V_{out}/I_{in}, I_{out}/V_{in}.)

voltage divider: electrical elements used to reduce voltage magnitude

SUGGESTED READINGS

Tepper, Marvin, ELECTRONIC IGNITION SYSTEMS, Chaps. 1 and 2. Rochelle Park, N.J.: Hayden Book Co., 1977. *A brief introduction to automotive ignition systems.*

Crede, Charles E., VIBRATION AND SHOCK ISOLATION. New York: Wiley, 1951. *Mechanical vibration, absorption, and isolation.*

Fitzgerald, A. E., D. E. Higginbotham, and A. Grabel, BASIC ELECTRICAL ENGINEERING, 5th ed., pp. 101–102. New York: McGraw-Hill, 1981. *In place of the electromechanical analog system which we considered (called the force-voltage analog), there is another system termed the force-current analog. A very brief consideration of such duals (as they are called) may be found in this work.*

Cathey, J. J., and S. A. Nasar, SCHAUM'S OUTLINE OF THEORY AND PROBLEMS OF BASIC ELECTRICAL ENGINEERING, Chaps. 1–5. New York: McGraw-Hill, 1984. *R, L, and C circuits, with many numerical examples.*

Fagan, M. D. (ed.), A HISTORY OF ENGINEERING AND SCIENCE IN THE BELL SYSTEM. THE EARLY YEARS (1875–1925), pp. 303–309. New York: Bell Telephone Laboratories, 1975. *A short history of the development of the decibel unit.*

A CLOSER LOOK: *RC* Circuit Response and Decibels

RC Circuit Response

See Figure 1. Upon switch closure, we have

$$V - iR - \frac{\int i\,dt}{C} = 0$$

$$R\left(\frac{di}{dt}\right) + \frac{i}{C} = 0$$

$$\int_{V/R}^{i} \frac{di'}{i'} = -\int_{0}^{t} \frac{dt'}{RC}$$

$$\ln i' \Big|_{V/R}^{i} = -\frac{t}{RC}$$

$$\ln \frac{i}{V/R} = -\frac{t}{RC}$$

$$i = \frac{V}{R} e^{-t/RC}$$

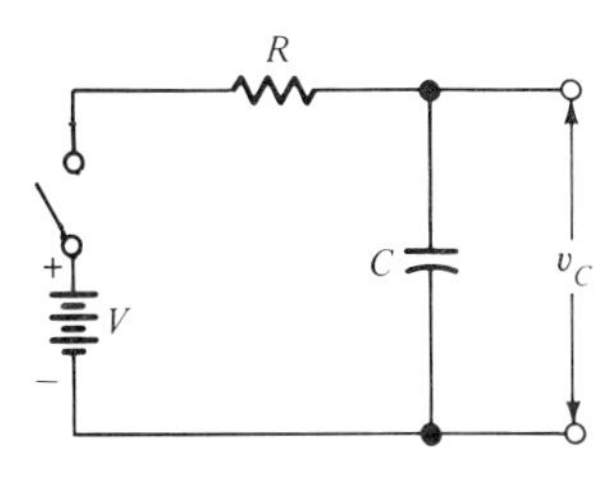

FIGURE 1

$$v_C = V - iR = V - \frac{V}{R} R e^{-t/RC} = V(1 - e^{-t/RC})$$
$$= V(1 - e^{-t/\tau})$$

where $\tau = RC \equiv$ time constant. It has the dimensions of time (second) and in the preceding case represents the time at which the current has dropped to 0.368 of its initial value and the capacitor voltage has reached a value equal to 0.632 times the impressed voltage.

Decibels

See Figure 2. The gain of a typical electronic amplifier, plotted in terms of frequency, utilizes a logarithmic scale along the x-axis. This procedure compresses a rather wide frequency range without incurring the loss of any information. Likewise, a rather wide range of gain can be accommodated by logarithmic means. But along the y-axis, one uses a logarithmic unit, the *decibel*, rather than a logarithmic scale.

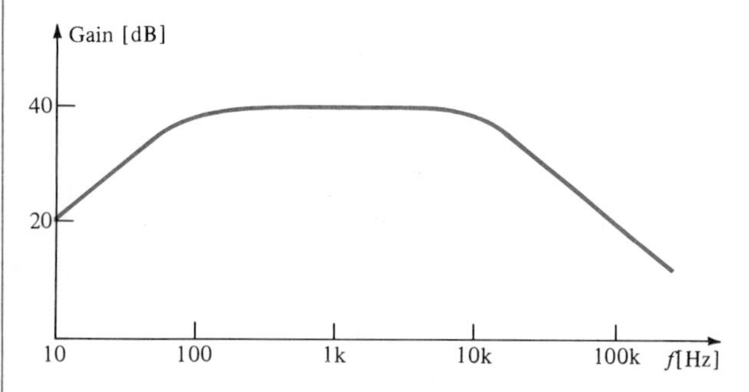

FIGURE 2

Actually the defined unit is the *bel*, which is equal to the base 10 logarithm of a power ratio:

$$\text{bel} = \log_{10} \frac{P_2}{P_1}$$

But it proves more convenient to use a unit one-tenth as large, the decibel (dB). A change of 1 dB is about the minimal change that a human ear can detect.

Since 10 dB = 1 bel, we have

$$\log_{10} \frac{P_2}{P_1} \text{[bels]} \times 10\text{[dB/bel]} = 10 \log_{10} \frac{P_2}{P_1} \text{[dB]}$$

Thus, an increase in power from 1 to 2 W represents an increase of 3 dB. Similarly, an increase from 10 to 20 W also represents a 3 dB increase.

Often, however, we express the output in terms of signal voltage rather than signal power. Since $P = V^2/R$, we have

$$10 \log_{10} \frac{P_2}{P_1} = 10 \log_{10} \frac{V_2^2 R}{RV_1^2} = 10 \log_{10} \left(\frac{V_2}{V_1}\right)^2$$
$$= 20 \log_{10} \frac{V_2}{V_1} \text{[dB]}$$

Note that if one measures the gain of an amplifier, that is, the input and output signal voltages, these voltages are generally developed across different values or R. Therefore, to compute dB, we have

$$10 \log_{10} \frac{V_2^2 R_1}{R_2 V_1^2} = 10 \log_{10} \left(\frac{V_2}{V_1}\right)^2 + 10 \log_{10} \frac{R_1}{R_2}$$
$$= 20 \log_{10} \frac{V_2}{V_1} + 10 \log_{10} \frac{R_1}{R_2}$$

The decibel is a measure of either signal power or signal voltage *change*. It may be made into an absolute unit if a reference level is adopted as a standard. Thus, as is often done in audio work, this could be 0.001 W (1 mW). With this as a reference, +dB represents powers greater than 1 mW; −dB represents power levels less than 1 mW. To indicate the use of such a reference, the unit is often designated as dBm, m standing for milliwatt. Since signal voltages in dB are dependent on the load resistance, 1 dBm corresponds to 1 mW across a 600 Ω load (600 Ω is one of the standard impedance levels utilized by the phone company and corresponds to about 0.78 V).

Consider a power level of 0.1 W:

$$10 \log_{10} \frac{0.1 \text{ W}}{0.001 \text{ W}} = +20 \text{ dBm}$$

For a power level of 1 μW, we have

$$10 \log_{10} \frac{10^{-6}}{10^{-3}} = -30 \text{ dBm}$$

The dBm is used to indicate a level at constant frequency. If, instead, one is dealing with an average sound level (such as in broadcasting), the volume unit VU is used instead.

Some other forms of dB in use are the following: (*a*) A 1 W reference is designated as dBW; (*b*) dBV uses 1 V as a standard; and (*c*) in acoustics, 0 dB SPL represents an rms pressure of 0.0002 μbar (a bar being 10^6 dyne/cm^2, or about 1 atm).

EXERCISES

1. **a.** A 12 V battery gives rise to a steady-state current of ______ A when applied to an ignition "coil" whose resistance is 2.5 Ω.

b. If the inductance is 5.0 mH, what is the rise time of the circuit?

2. What is the order of magnitude of the voltage developed across the secondary of an automotive ignition coil?

3. What is the electronic device used to visually observe the ignition waveforms?

4. What is the meaning of the dots that are sometimes found alongside the primary and secondary leads of a transformer's schematic figure?

5. What is the function of the capacitive voltage divider used in conjunction with an ignition probe unit?

6. Since a changing magnetic field about an ignition coil gives rise to a secondary voltage, why doesn't a spark plug fire when the points close upon completion of the antidwell time?

7. The current in a series RL circuit increases as

$$i = \frac{V}{R}(1 - e^{-t/\tau})$$

where τ = time constant = L/R. Show that at $t = \tau$ the current has reached about 0.632 of the ultimate value.

8. **a.** In discharging an RL circuit, what is the expression for the instantaneous current in terms of the initial current (I_0) and the time constant τ?

b. At $t = \tau$, what is the value of the current in terms of I_0?

9. Why does the exponential for current decay in Eq. 3.2 differ from that in Eq. 3.9?

10. What do metallic-blue relay contacts signify?

11. Why is there an oscillatory voltage when the point contacts close in an automotive ignition system?

12. Why is it desirable to have the "coil" polarity arranged in such a fashion that the center spark plug electrode is negative upon firing?

13. What is the function of the ballast resistor in an automotive ignition system?

14. **a.** Two 100 kΩ resistors are placed in series. If 10 V is applied to the combination, what is the voltage across one of them?

b. If the 10 V source has an internal resistance of 1 kΩ, what is the voltage available across one of the resistors?

c. If the source voltage drops to 9 V, what is the voltage across one of the resistors?

d. If the voltage divider consists of two 1 kΩ resistors and a 10 V source with internal resistance of 1 kΩ is applied across the combination, what is the voltage across one of the resistors?

e. If the same source as in part d drops to 9 V, what is the output voltage across one of the resistors?

f. What is the percentage change in cases involving parts b and c and between parts d and e?

15. A voltage divider consisting of two 1 kΩ resistors has 10 V applied across the combination. What is the power dissipation in each resistor?

16. State, in its two alternative forms, the procedure used to obtain an equivalent circuit using Thevenin's theorem.

17. State the procedure used to obtain a Norton equivalent circuit.

18. **a.** A voltage divider consists of a dc voltage V = 12 V, R_1 = 1 kΩ, and R_2 = 100 Ω. If a 1.5 kΩ load is placed across R_2, calculate the output voltage using Thevenin's theorem.

b. If an ac voltage of 12 V rms were used, what would be the output voltage?

19. With V = 12 V, R_1 = 100 kΩ, R_2 = 1 kΩ, and a load resistance R_L = 1.5 kΩ across R_2, using Thevenin's theorem, compute the loaded and unloaded output voltage.

20. It is desired to have the load voltage across a 1.5 kΩ resistor be within 0.3 V of 1.5 V. Using a source voltage of 15 V, design a voltage divider to satisfy this requirement.

21. What is the Norton equivalent of a 12 V source driving a series circuit consisting of 1 kΩ and a load of 10 kΩ? Show that the load currents are the same in both cases.

22. Find the Norton equivalent of the circuit in Exercise 18. Compare the Thevenin and Norton circuits. Does this comparison suggest an easy way of finding the Norton resistance once the Thevenin resistance is known?

23. Find the Norton equivalent of the circuit in Exercise 19.

24. **a.** A simple voltage divider consists of a battery V in series with two resistors, one of value R and the other with value $2R$. Determine the Thevenin equivalent of the circuit that acts as a source for $2R$.

b. What is the voltage across, and the current through, $2R$?

c. What is the Norton equivalent of the circuit driving $2R$?

d. Show that the voltage across and the current through $2R$ are the same as in part b.

25. **a.** A 12 V battery is used in conjunction with a voltage divider consisting of two 2 Ω resistors. A 6 Ω load is placed across one of them. Compute the total power dissipation in the potential divider and that in the load.

b. Determine the Thevenin equivalent of the driving circuit and compute the power dissipation in the 6 Ω load and in the Thevenin resistance.

c. Comparing the results of parts a and b, what do you conclude?

26. **a.** A 12 V source has an internal resistance of 2 Ω. Compute the voltage and power delivered to the following loads: 0.5 Ω, 1.0 Ω, 2.0 Ω, 4.0 Ω, 8.0 Ω.

b. In each of the preceding cases, compute the power dissipated in the internal resistance.

c. What is the efficiency of the power transfer from source to load in each case?

27. **a.** A Thevenin source with internal resistance R_T is connected across the primary of a transformer with a primary-to-secondary turns ratio of N_1:N_2. Determine the Thevenin equivalent of the secondary circuit with load resistance R_L.

b. In the secondary circuit, what is the condition for maximum load voltage? For maximum load power? What is the load voltage at maximum load power?

28. A step-down audio transformer has a number of taps on the secondary. With an unknown source resistance driving the primary, the optimum power coupling may be found by placing a voltmeter across the load and selecting the tap that leads to the highest load voltage. Why doesn't the voltage simply continue to increase as one taps off a larger number of turns across the secondary?

29. An audio transformer has a fixed number of primary turns but a number of taps on the secondary so that the turns ratio may be altered. The source driving the primary has a 2000 Ω resistance, while the resistive secondary load is 8 Ω. If a 10 V rms signal is impressed across the primary, what is the voltage available across the 8 Ω load as it is placed across the taps representing the following turns ratios: 31.6, 22.3, 15.8, 11.2? (Assume no losses in the transformer.) What is the optimum ratio? Check the impedance match for this ratio.

30. **a.** A 1 V rms signal at 10 Hz is applied to a series combination consisting of 0.1 μF and a 100 kΩ resistor. What is the signal voltage across the resistor?

b. What is the phase angle between the output voltage and the applied voltage?

31. **a.** A 1 V rms signal at 10 Hz is applied to a series combination of 0.1 μF and a 100 kΩ resistor. What is the signal voltage across the capacitor?

b. What is the phase angle between input and output voltages?

32. **a.** Compute the current flow in either Exercise 30 or Exercise 31.

b. Using this current (including the phase angle), compute the voltage drop across C and across R. Express the two results separately in terms of their real and imaginary parts.

c. Why doesn't the voltage across C plus the voltage across R equal the applied voltage?

d. Show that the two results when added vectorially *do* lead to an equality.

33. Show analytically that when written in terms of complex numbers the sum of the voltages across R and C of a series circuit equals the applied voltage.

34. For a high-pass filter, what should be the value of the RC product if the attenuation at 50 Hz is not to differ by more than 5% of the high-frequency value? If $C = 0.1\ \mu$F, what should be the value of R?

35. What should be the value of the RC product for a low-pass filter if the attenuation at 100 KHz is not to differ by more than 5% of its value at low frequencies?

36. **a.** The output voltage from an amplifier is increased from 10 to 12.5 V. What is the change in dB?

b. If the output is decreased from 12.5 to 10 V, what is the change in dB?

37. A 30 W amplifier is replaced by a 100 W amplifier. What is the increase in power output expressed in dB?

38. The output voltage of an amplifier, measured across an 8 Ω resistance, is 1.0 V. The input signal to the amplifier, measured across an input resistance of 5 kΩ, is also 1.0 V. What (if any) is the power gain of this amplifier, expressed in dB?

39. The ability of the output voltage of a voltage divider to resist changes under variations of the imposed load is a measure of its ______.

40. Indicate the corresponding electrical-mechanical equivalents in the following cases:

Inductance ______
Voltage ______
Charge ______
Compliance ______
Velocity ______

41. In the general case, force divided by displacement constitutes ______; in the general case, voltage divided by current represents ______.

42. Displacement divided by force represents ______; charge divided by voltage represents ______.

43. The rapidity of decay of a transient oscillation depends on the ______ dissipation. The natural log of the ratio of successive amplitudes is termed the ______. In terms of circuit Q, this equals ______.

44. **a.** The resonant frequency of a damped electrical oscillation is given by the equation

$$\omega_0 = \frac{1}{2\pi}\sqrt{\frac{1}{LC} - \left(\frac{R}{2L}\right)^2}$$

Write the corresponding equation in mechanical terms—that is, K, D, ω_c.

b. The solution for the damped mechanical oscillation is

$$y = Ae^{-(Dt/2m)} \sin(\omega_c t + \alpha)$$

where A and α are constants. If the effect of the damping on the value of ω_c is small, show that the number of cycles (n) before the amplitude falls to 50% of its original value is given by the expression

$$n = 0.22\left(\frac{m\omega_c}{D}\right)$$

45. A spring scale extends to 6 in. when subjected to its full-scale loading of 100 lb. What is the compliance of the spring? What is the spring constant?

46. Rework Exercise 45 in terms of SI units.

47. What is the SI unit for electrical impedance? For motional impedance? What is the electrical equivalent of motional impedance?

48. What is the function of shock absorbers in conjunction with automotive springs?

49. **a.** Two springs, each solidly anchored at one end and subjected to simultaneous equal oscillating displacment (hence velocity) at their opposite ends, are analogous to an electrical circuit consisting of two ______ in series and subjected to an equal oscillating ______.

b. Two springs connected in series are equivalent to ______ connected in ______.

50. Does the order in which mechanical elements (damper, spring, mass) are connected alter the overall mechanical performance? How about the ordering of corresponding electrical components?

51. What is the rule concerning the series and parallel representation of mechanical circuits by their electrical analogs?

52. **a.** What is the magnitude of the attenuation of the frequency-determining portion of a Wien bridge filter at the rejection frequency?

b. What is the phase shift of the filter at this frequency?

53. **a.** What is the rejection frequency of a Wien bridge if in the frequency-determining branch both resistors equal 100 kΩ and both capacitors equal 0.1 μF?

b. With $C = 0.1\ \mu$F, what should be the value of R in order to reject 60 Hz interference?

54. **a.** Indicate the arrangement of R and C in order to form a low-pass filter.

b. Indicate the arrangement of R and C in order to form a high-pass filter.

55. In the electrical case, an expression for circuit Q is $\omega_0 L/R$; in the analogous mechanical case, with $\omega_c = \sqrt{k/m}$, it equals ______.

56. What is the natural resonant frequency of a 5 lb weight suspended on a spring whose compliance is 0.25 in./lb?

PROBLEMS

3.1

a. A conventional ignition "coil" has an inductance of 5 mH and a resistance of 2.6 Ω and can safely handle 5 A. Compute the rise time.

b. A transistorized "coil" has an inductance of 1.25 mH and a resistance of 1.0 Ω (which includes the ballast resistance). Compute the rise time.

3.2

a. Estimate the ignition time of a system with $L = 5$ mH, $R = 2.5\Omega$, and $C = 0.25\ \mu$F. Take the initial current to be 3 A, the turns ratio to be 1:100, and the firing voltage of the spark plug to be 20 kV.

b. What would be the primary current flow at that time?

c. The greatest rate of change of current arises when the decaying current passes through zero. If there has not been a prior firing of the spark plug, what would be the developed voltage at that time?

d. What would be the secondary voltage if RL were shorted without C being present?

3.3

For a four-cylinder car each cylinder is allotted 90° of rotation of the camshaft. (Remember that the camshaft rotates at half the engine speed.) A typical dwell time is 53%; therefore, in terms of the camshaft rotation, each cylinder is allotted 47.7° (53% × 90°). Compute the energy stored in the magnetic field at 500 and 4000 rpm of engine speed. Take the maximum points current to be 3 A. A typical spark plug, depending on spacing, dissipates 10–20 mW·s. (The energy is generally specified in milliwatt-seconds, being equal to millijoules.) (*Hint*: Energy $= \int vi\,dt$.) (Also, for RL circuits, $\alpha = R/L$; for RLC circuits, $\alpha = R/2L$.) Use the values of R, L, and C as given in Problem 3.2.

3.4

We have shown that the general solution for current decay in a series RLC circuit is of the form

$$i = A_1e^{s_1t} + A_2e^{s_2t}$$

where

$$s_{1,2} = -\frac{R}{2L} \pm \sqrt{\left(\frac{R}{2L}\right)^2 - \frac{1}{LC}}$$

There are three solution regimes depending on the values under the square root sign. To demonstrate the physical difference with the least amount of confusion, we elect to deal with some rather unrealistic values for C. Take $R = 2\ \Omega$, $L = 1$ H, and C variously equal to

a. $\frac{25}{9}$F

b. 0.5 F

c. 1.0 F

Determine the solution in each case, taking the initial current to be 1 A. Plot the results.

COMMENT: In the critically damped case, with two equal roots, for a second-order differential equation there must be two arbitrary constants. As a result the solution takes the form

$$I = A_1te^{-t} + A_2e^{-t}$$

3.5
Derive the solution for the current in Figures 3.50 and 3.51. Note particularly the difference in the exponentials. (Use the oscillatory solution in Figure 3.51.)

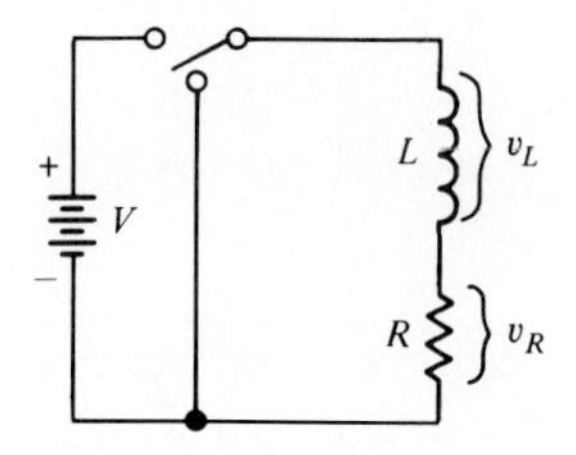

FIGURE 3.50

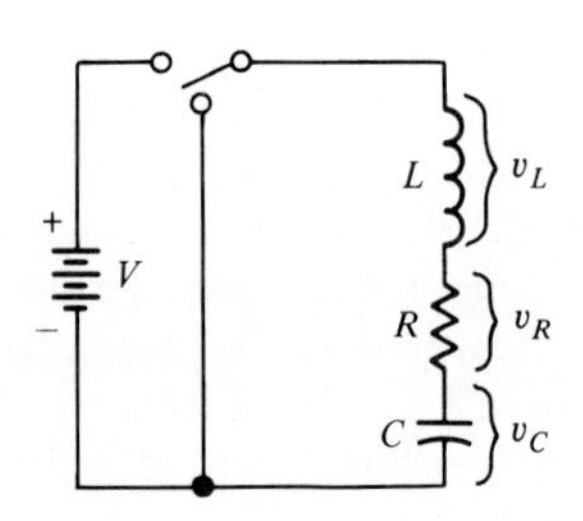

FIGURE 3.51

3.6
For the ignition coil whose specifications are given in Problem 3.2, what would be the value of C that would lead to critical damping?

3.7
For the voltage divider shown in Figure 3.13, the source voltage is taken to be 30 V, and both R_1 and R_2 are each equal to 10 kΩ.

a. What will be the unloaded output voltage?

b. What will be the output voltage with a 10 kΩ load connected across the lower resistance?

c. What will be the current drawn from the source without the load?

d. With the load connected, what will be the current drawn from the source?

3.8
Find the Thevenin equivalent circuit for each of the circuits in Figure 3.52.

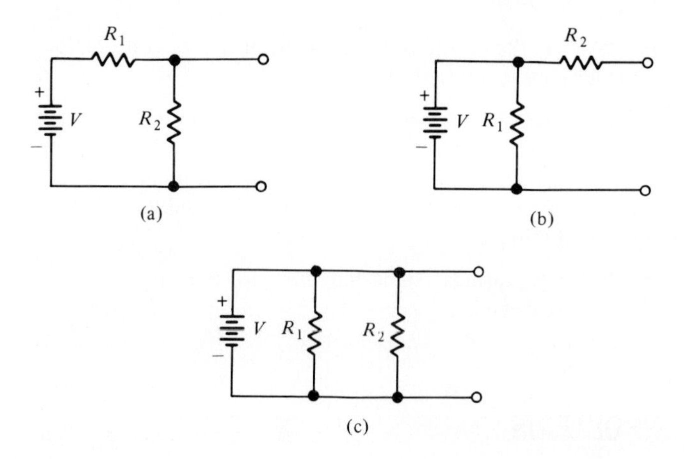

FIGURE 3.52

3.9

a. Determine the Thevenin equivalent for the circuit in Figure 3.53.

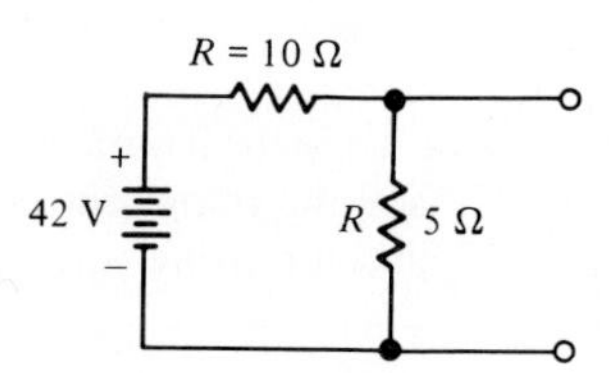

FIGURE 3.53

b. Determine the Norton equivalent circuit. (Express in terms of a conductance.)

c. With a 15 Ω load resistance, show that both circuits result in identical voltages at the load.

3.10
Find the Norton equivalents of the circuits shown in Problem 3.8. Why is there a difficulty with the circuit in Figure 3.52c?

3.11
Determine the current through the 15 Ω resistance in Figure 3.54.

a. By using loop currents

b. By Thevenin's theorem

c. By Norton's theorem

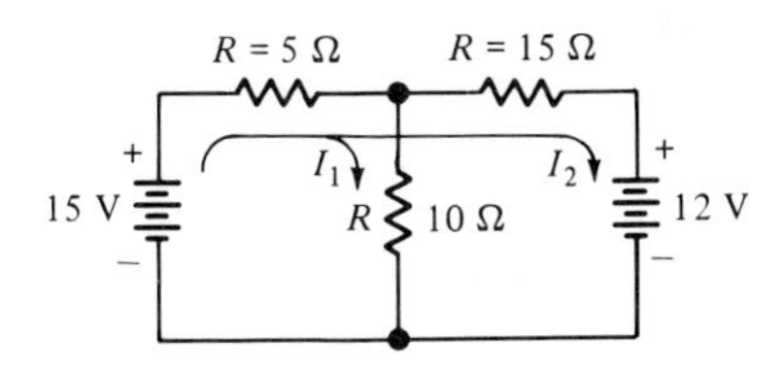

FIGURE 3.54

3.12
Determine the power dissipation in a load 3 + *j*4 connected across the output of the Thevenin equivalent developed in Example 3.7.

3.13
a. If a voice coil of a radio speaker has a resistance of 8 Ω, and recalling that maximum power transfer takes place when the load matches the generator's internal impedance, what should be the turns ratio if R_T of the driving transistor is 200 Ω?

b. If the measured rms voltage at the output of the amplifier is 2 V, what would be the signal voltage across the voice coil of the loudspeaker?

c. What would be the output power driving the loudspeaker?

3.14
Indicate the Thevenin equivalent "seen" by the load resistance R_L, with $i = Ie^{j\omega t}$, in the circuit in Figure 3.55.

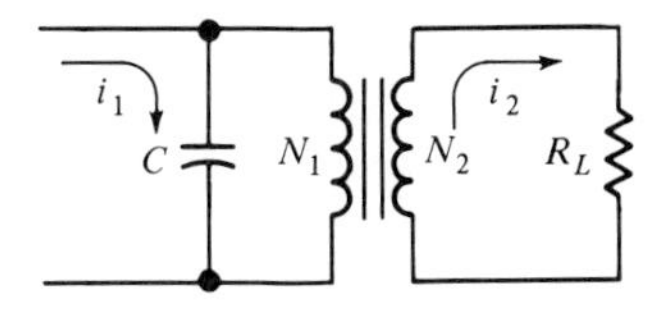

FIGURE 3.55

3.15
a. With a 1 V rms input signal, as the input frequency in the circuit in Figure 3.56 is varied, determine the output signal amplitudes at the following frequencies and plot the results on a linear scale: 100, 200, 500, 1000, 2000, 10,000, 100,000 Hz.

b. Replot the results of part a on semilog paper, expressing the linear scale in decibels attenuation. (Note particularly the frequency at which the tangent extension of the two segments of the curve intersect.)

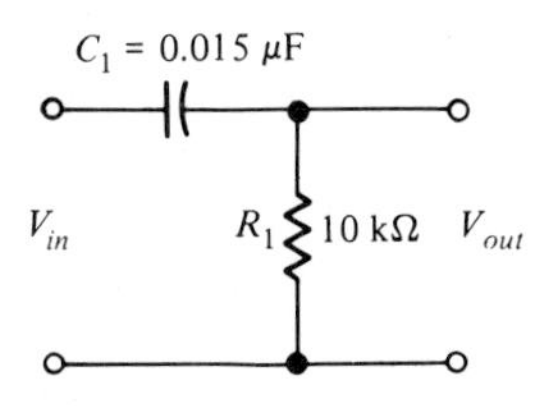

FIGURE 3.56

3.16
a. Reverse the positions of *R* and *C* in Problem 3.15, as in Figure 3.57, and plot the output signal amplitudes on a linear scale for the following frequencies: 100, 500, 1000, 2000, 5000, 10,000, 30,000, 100,000 Hz.

b. Replot the results of part a on semilog paper, expressing the linear scale in decibels attenuation. (Note particularly the frequency at which the tangent extension of the two segments of the curve intersect.)

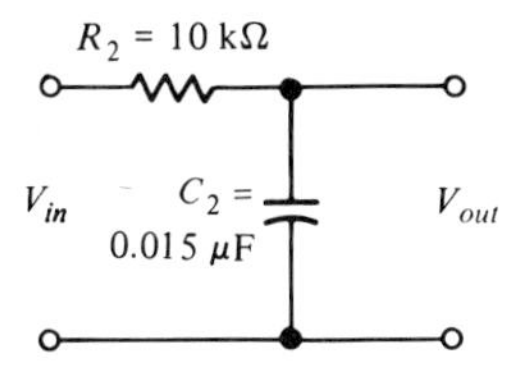

FIGURE 3.57

3.17
An easy way to determine that frequency at which the output of simple low- or high-pass *RC* filters drops to 0.707 of the input amplitude is to use the expression $f = 1/(2\pi RC)$. Verify that this is true for Problems 3.15 and 3.16.

3.18
A capacitor (of sufficient size for the frequency in question) has the effect of passing ac voltages between two points without affecting the dc voltages in the respective circuits. See Figure 3.58.

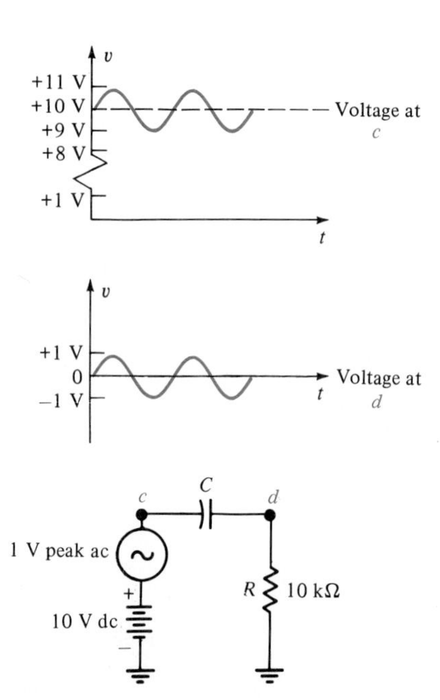

FIGURE 3.58

a. Sketch the voltage variation at c and d in the circuit in Figure 3.59.

b. If the ac frequency is 1000 Hz, what should be the value of C to assure *no* attenuation of the ac signal? (Generally one selects $X_C \leq \frac{1}{10}R$.)

c. What will be the value of the ac voltage at point d using the computed value of C?

d. What will be the value for a capacitor 10 times larger?

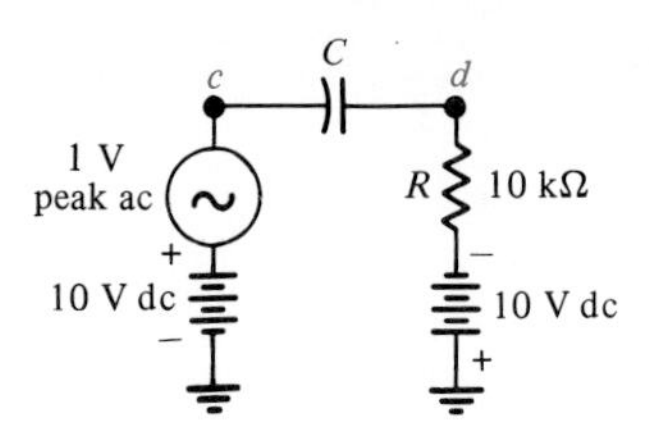

FIGURE 3.59

3.19

a. The signal output voltage of an amplifier at midband is 1000. Its output at the upper band limit is 707. In decibels, how much lower is the output at the band limit?

b. If the power output at the band limit is half what it is at midband, how much lower, in decibels, is the power output at the band limit?

3.20

a. The power output of an amplifier is increased from 1 to 2 W. Express the change in decibels.

b. The output is changed from 10 to 11 W. How many decibels does the change represent?

c. An output is increased from 10 to 20 W. How many decibels does this represent?

3.21

a. The input resistance of an amplifier is 100,000 Ω, and the output resistance is 2000 Ω. An input signal voltage of 1 μV leads to an output signal voltage of 1 V. What is the increase in signal, expressed in decibels? (*Note:* Solve in terms of a power ratio.)

b. Suppose the problem had been solved in terms of the voltage ratio. What would the answer have been? Why does the difference arise?

3.22

The minimal change in a signal level detectable by the human ear is generally taken to be 1 dB. To what does this correspond in terms of a voltage ratio? A power ratio?

3.23

The loudest one can shout is about 18 dB above the normal conversational level. What power ratio does this represent?

3.24

a. The midband voltage gain of an amplifier is 2×10^5. What is this in terms of decibels?

b. What power range does this represent?

3.25

a. The dynamic sound range of a concert can be as large as 90 dB. What power range does this represent?

b. A high-quality recording has a range of about 65 dB. What power range does this represent?

c. A broadcasting station is limited to about 40 dB dynamic range. What power range does this represent?

3.26

Let $F = F_0 \sin \omega_c t$ represent the force applied to a damped harmonic oscillator, resulting in a displacement $y = Y_0 \sin(\omega_c t - \phi)$, where ϕ represents the lag angle between force and displacement. Compute the work done per cycle, Verify that 2π times the maximum energy at resonance divided by the work done per cycle (at resonance) equals the Q of the circuit.

3.27

Repeat Problem 3.26 for the electrical case, with $q = Cv$, where q is the charge, C is the capacitance, and v is the applied voltage.

3.28 △

In an *RLC* circuit critical damping arises when

$$\frac{R_c}{2L} = \sqrt{\frac{1}{LC}}$$

where R_c is the value of the resistance that leads to critical damping:

$$R_c = \frac{2L}{\sqrt{LC}} = 2L\omega_0$$

Damping can be specified in terms of critical damping:

$$R = \zeta R_c$$

$$\zeta \equiv \text{damping ratio} = \frac{R}{R_c}$$

Express the roots of an *RLC* circuit solution and write the general solution of the circuit in terms of ζ. Note that if the damping is significant, the period of the oscillation changes, becoming zero for critical damping.

3.29
Using the expression for Q of a mechanical resonant system (Eq. 3.42), by analogy convert this into a recognizable form for an electrical circuit.

3.30
Using the equation $v = V_0 e^{-\alpha t} \cos \omega_0 t$, show that the number of cycles necessary to reduce a damped sinusoidal wave by a factor $e^{-\pi}$—that is, to approximately 4% of its initial value—is equal to the Q of the circuit.

3.31 △

a. The logarithmic decrement arises from energy loss per cycle. Derive the relationship between logarithmic decrement (δ) and Q (assuming small damping— using the expression for the potential energy of a spring; that is, $U = \frac{1}{2}kx^2$. (*Hint:* Use $e^{-\delta} \simeq 1 - \delta$.)

b. What value of x_2/x_1 will lead to an error of 10% in $\Delta U/U$ in using the approximation of part a?

c. What Q corresponds to this value of δ?

d. In the case of this value of δ, how much error is introduced by disregarding the third term in the exponential expansion?

3.32
With regard to Figure 3.37 and the plot in Figure 3.38, at what frequency (relative to f_c) does the damping increase the transmissivity?

3.33 △
Show that δ is given by the equation

$$\delta = \frac{1}{n} \ln \frac{x_0}{x_n}$$

where x_n = amplitude after n cycles
x_0 = initial amplitude

3.34

a. With the same instantaneous voltage (v) applied in both cases, derive the expression for $|i/i'|$ for a series electrical circuit under steady-state conditions, where i is the current for a series *RLC* circuit and i' is the current with only L present.

b. To what does this correspond in the mechanical case?

c. What is the significance of the current i through C in the mechanical case?

3.35

a. Derive the expression for $|i_2/i|$ for the electrical circuit of Figure 3.42c.

b. To what does this correspond in the mechanical analog?

c. What is the significance of the current i_1 through C in the mechanical case?

d. Derive the expression for $|i_1/i|$.

3.36
With regard to Figure 3.46, if the *weight* of each of the masses is taken to be 1500 lb, and each of the springs is taken to have a spring constant of 1500 lb/in.,

a. What are the values of ω_1 and ω_2?

b. Compute the ratios $|y_1/a|$ and $|y_2/a|$ for the following values of ω: 1, 5, 10, 12, 13, 14, 15, 19, 20, 21, 22, 23, 25, 30, 31, 32, 40, 45.

c. Plot the results using three-cycle semilog paper.

d. At what value of ω does the zero of y_2 arise?

3.37
In terms of $|i_2/i|$, show the distinction between the resistance of a parallel circuit being dominant in the capacitive branch and in the inductive branch. See Figure 3.60.

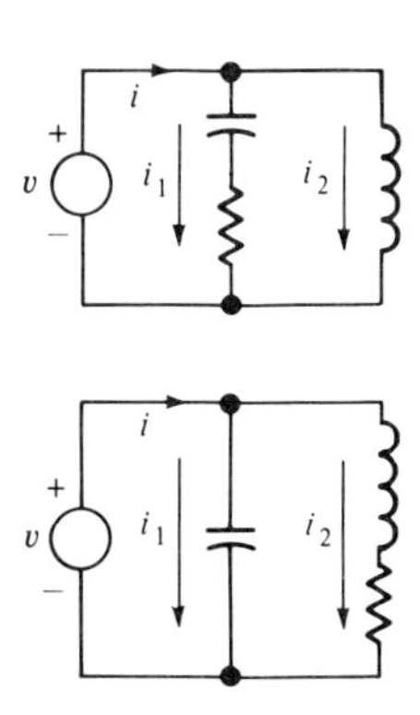

FIGURE 3.60

3.38

a. Rather than operating a Wheatstone bridge in a balanced condition, it is frequently desirable to have the output voltage vary linearly with the value of the unknown resistor R_u. Thus, an output meter may be directly calibrated in terms of resistance. Using the expression from Eq. 3.14, derive the condition for such linear operation.

b. What is the condition for maximum sensitivity of a Wheatstone bridge?

3.39

A 1 V rms signal is applied to a Wien bridge having $R_1 = 2R_2$. Utilizing Eqs. 3.16, 3.18, and 3.20 and a normalized frequency (ω/ω_0), in terms of the rejection frequency ω_0,

a. Compute and plot (using semilog paper) the output voltage at *a* (relative to ground) as a function of frequency.

b. Compute and plot the output and phase at *a* relative to *b* as a function of frequency. Peform the calculations at the following values of ω/ω_0: 0.2, 0.4, 0.5, 0.8, 1.25, 2.0, 2.5, 5.0.

3.40

See Figure 3.61. The capacitance of a cylindrical capacitor (in picofarads per foot) is given by the expression

$$C[\text{pF/ft}] = \frac{7.354K}{\log_{10}(D/d)}$$

where K = dielectric constant of the insulation
D = diameter of the outer conductor
d = diameter of the inner conductor

If the outer diameter of an ignition wire is $\frac{1}{4}$ in., the diameter of the central conductor is $\frac{1}{16}$ in., and an attentuation of $\frac{1}{1000}$ is desired from an ignition probe used with $C_2 = 0.01\ \mu\text{F}$, what should be the approximate length (l) of the probe clamp? For the insulation, take $K = 2.2$.

3.41

Using Thevenin's theorem, determine the current through the 5 Ω resistance in the circuit in Figure 3.62.

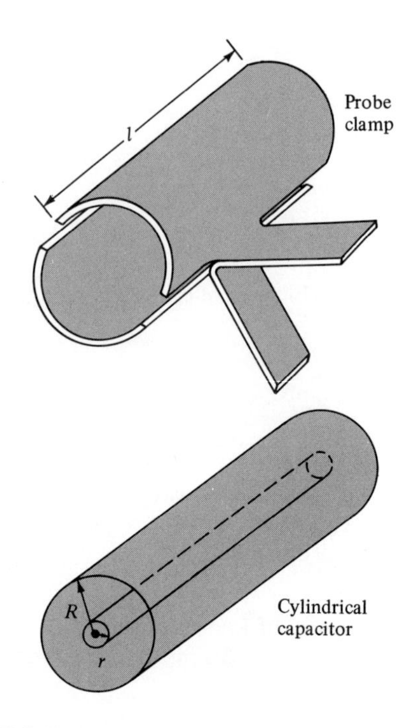

FIGURE 3.61

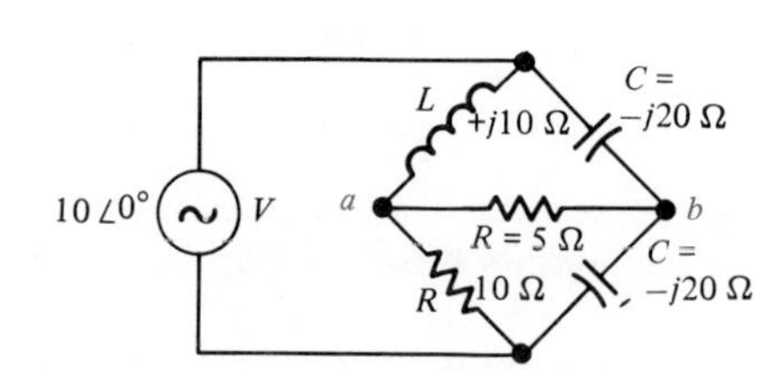

FIGURE 3.62

4 SIZES, DIMENSIONS, AND TOLERANCES

OVERVIEW

Having achieved some understanding of the function and performance of resistance, capacitance, and inductance in the previous chapters, we turn now to the practical matter of their physical appearance, specifications, deviations from ideal performance, and various methods of construction. A legitimate question might be raised as to the utility of this type of information for the user of electronic equipment in contrast to a designer. It is useful for two reasons.

First, while much of today's electronic instrumentation takes on the aspects of a *black box*—a signal goes in and an appropriately modified signal comes out—there is often some interfacing that needs to be employed. This interfacing must generally be provided by the user and usually consists of appropriate combinations of resistors, capacitors, and (less frequently) inductors.

Second, the quality of the performance provided by an electronic instrument depends on the quality of the components that make up the instrument. The user should be in a position to appreciate the distinction between quality components and inferior ones.

While most of the discussion in this chapter is of a rather qualitative nature, we will take the opportunity to elaborate in a more realistic manner on the operation of a transformer, in contrast to the operation of the idealized form we have heretofore considered.

OUTLINE

4.1 ■ INTRODUCTION

In our computation of the capacitance necessary for an interstage coupling, we arrived at a value of 0.4 μF (Example 3.8). As in so many technical areas, one handicap faced by the relative newcomer involves an appreciation for what represents a realizable component value and what does not. It will be the function of this chapter to furnish at least some information in this regard.

One might question the need for nonelectrical engineers to be concerned with the composition of various resistor, capacitor, and (to a lesser degree) inductor units such as represented in this chapter. But such engineers are often called upon to furnish appendages to be used in conjunction with existing equipment: as bypass capacitors to eliminate interference, as voltage dividers, and as analog components. They should certainly be made aware that one does not use a type II disc ceramic capacitor in conjunction with an *LC* resonant circuit. Additionally, the gain, stability, and interference rejection properties of integrated circuits are strongly dependent on the quality and precision of external components. And, finally, the electronic user, while not being involved in the design of the equipment, should appreciate the quality (and expense) of the components that make up a reliable, well-designed electronic device. Such components represent the difference in the performance of a high-quality oscilloscope in contrast to an inexpensive one, for example.

Here, however, inductors will get short *shrift* because they represent elements usually to be avoided since they do not lend themselves very well to integrated circuit techniques. We will, however, take this opportunity to deal with them in the form of transformers in a more realistic fashion.

4.2 ■ RESISTORS

4.2.1 Composition and Film Resistors

By far the most commonly encountered resistors are those classed as *carbon resistors*, available in values from a few ohms to 20 MΩ. The resistance of such units cannot be determined by their physical dimensions. A 47 Ω resistor may have the same dimensions as a 10 MΩ resistor. It is a second specification that influences the physical size, the power rating. Thus, a 2 W resistor is physically longer and thicker than a $\frac{1}{2}$ W resistor. Since the heat generated within a resistor leaves primarily because of radiation, the larger surface area of the 2 W resistor aids in its dissipation.

Carbon composition resistors and **carbon film resistors** represent the two basic categories of carbon resistor. The first is made by mixing carbon powder with a binder, and the density of the carbon employed determines the resistance. Although still available and encountered in large numbers, such composition resistors are obsolete primarily because they tend to generate considerable electrical noise. They are being replaced by the film resistor. The carbon film resistor is made by depositing a carbon film on an insulating cylindrical base and then cutting away a helical path in the carbon film. By virtue of the film thickness, and by altering the pitch of the helix, the length and breadth of the conducting path can be altered, and this determines the value of the resistance. This process can be much more closely controlled in manufacturing than can the mixing process used to make composition resistors. For a given power rating (commonly available in $\frac{1}{8}$, $\frac{1}{4}$, $\frac{1}{2}$, 1, and 2 W versions), the various resistance values have the same physical size.

The third specification of a resistor is its *tolerance value*. Because of manufacturing variations, the value of one resistor may differ somewhat from the value of the next one. In other words, a resistor specified as having a value of 470 Ω may actually have a value that is either somewhat more or somewhat less than this nominal value. A large sample of supposedly identical resistors will have values that represent a Gaussian distribution centered on the stated value.

In some applications the actual value of the resistance is not very important, and one may utilize anything within, say, 20% of the stated value. In other cases the value is much more critical, and one needs to use a resistor with a smaller tolerance. Carbon composition resistors, in addition to being available with a 20% tolerance, are also readily available with a value within 10% of that specified, as well as 5%. This accounts for the rather peculiar values assigned to resistors. Rather than having values such as 200 or 5000 Ω, they are labeled as 220 and 4700 Ω, respectively. Thus, 20% tolerance resistors have values that are multiples of

1.0 1.5 2.2 3.3 4.7 6.8

and 10% tolerance resistors have values that are multiples of

1.0 1.2 1.5 1.8 2.2 2.7
3.3 3.9 4.7 5.6 6.8 8.2

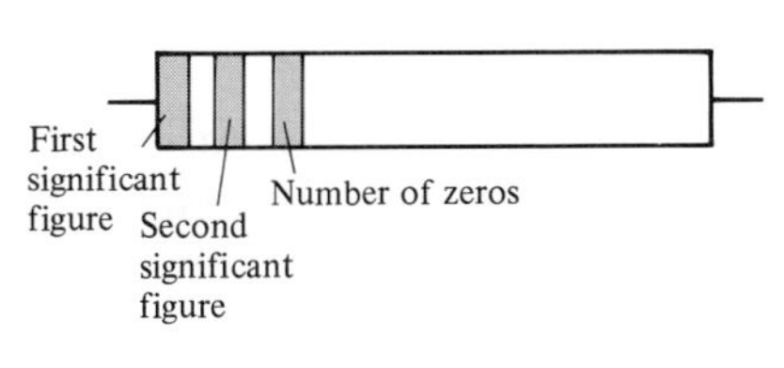

Black	0	Green	5
Brown	1	Blue	6
Red	2	Violet	7
Orange	3	Gray	8
Yellow	4	White	9

FIGURE 4.1. Basic Resistor Color Code

while 5% tolerance resistors have values that are multiples of

1.0 1.1 1.2 1.3 1.5 1.6 1.8 2.0
2.2 2.4 2.7 3.0 3.3 3.6 3.9 4.3
4.7 5.1 5.6 6.2 6.8 7.5 8.2 9.1

Such resistors are usually color-coded. A sketch of a carbon resistor is shown in Figure 4.1. Most of the color bands lie closer to one end. Starting at that end, note that the first two successive color bands indicate the two significant figures of the resistance, while the third color band indicates the number of zeros to follow.

If there are only three color bands, the resistor has a 20% tolerance. If the three color bands are followed by a silver band, the resistor has a 10% tolerance. A gold band following the initial three colors indicates a 5% tolerance, and a red band represents 2% tolerance.

EXAMPLE 4.1

Determine the values of the following resistor color codes:

a. Red, red, brown, silver
b. Yellow, violet, black, gold
c. Brown, black, green

Solution:

a. 220 Ω, ±10%
b. 47 Ω, ±5%
c. 1,000,000 Ω = 1 MΩ, ±20%

Because of the closer manufacturing tolerances that prevail when making carbon film resistors, they are also available with smaller tolerance ratings, including $\frac{1}{2}$% and 1%. Rather than carbon films, metal and metal oxide films may be used to create such resistors. Such precision resistors must have their values specified to three significant figures, and the color code must be expanded to accommodate them. The first three color bands represent the required three significant figures, while a fourth constitutes the multiplier. In such cases the tolerance band is moved to the other end of the resistor and given a width that is 50% greater. Additionally, in such cases, the tolerance code takes on the following values:

- brown ±1%
- red ±2%
- green ±0.5%
- blue ±0.25%
- violet ±0.1%
- gray ±0.05%

It should be assumed that a resistor will retain its value within the stated tolerance forever. That tolerance is termed the *purchase tolerance*—in contrast to the *design tolerance*, which takes into account estimated variations due to such factors as vibration, moisture, soldering, overload, and so forth.

Rather than being banded, precision resistors may have their values imprinted on their bodies in the form of a four-digit code, but the code is easily interpreted. Thus, 2373 means 237 kΩ. The last digit indicates the number of zeros that follow the first three significant figures. On the other hand, 1000 represents 100 Ω. For values too small to be stated in this manner, an R is used to indicate the location of the decimal point: 46R4 is 46.4 Ω, and 10R0 is 10.0 Ω.

Metal film resistors will retain their accuracy only within stated limits of temperature, humidity, and so forth. When greater stability is desired (better than 0.1%), ultra-precision **wire-wound resistors**—or some special metal film resistors—may be used. The objection to wire-wound resistors is the inductance that often accompanies them, limiting their high-frequency application. Whereas metal film resistors may be used up to 50 MHz before such effects become bothersome, a wire-wound resistor might be limited to 5 MHz.

Since the use of precision resistors in electronic circuitry is dramatically increasing, it helps to become familiar with at least some of the additional degrees of complexity that their specifications entail. Consider the widely used 1% precision metal film resistors. Their power ratings depend on the ambient temperature:

Resistor Designation	Power Rating at 70°C	Power Rating at 125°C
RN50	1/8 W	1/20 W
RN55	1/8	1/10
RN60	1/4	1/8
RN65	1/2	1/4
RN70	3/4	1/2
RN75	—	1
RN80	2	—

Their temperature coefficient of resistance (TCR) is specified in terms of resistance change in ppm/°C using the following designations:

Designation	TCR	Temperature Span
D	0 ± 100 ppm/°C	−55°C to +165°C
C	0 ± 50 ppm/°C	−55°C to +175°C
E	0 ± 25 ppm/°C	−55°C to +175°C

Thus, an RN60C is a 1% resistor rated at $\frac{1}{4}$ W at 70°C and $\frac{1}{8}$ W at 125°C and will show a TCR of 100 ppm/°C in the range −55°C to +165°C.

To determine the tolerance rating, one also has to consider the heating effect resulting from connecting the resistor into the circuit, whether done without heat, by soldering, or by welding. There are additional specifications that deal with failure rates (0.1%/1000 h, 0.01%/1000 h, and so forth), voltage coefficients, and so on. The main point of all this is to impress on the reader that if one wants circuit reliability, one must be willing to underwrite (that is, pay for) good design procedures.

4.2.2 Power Resistors

Where large amounts of power must be dissipated (for example, in the resistors making up a voltage divider with large amounts of bleeder current), wire-wound resistors are used. Their values may range from a few ohms to a megohm, with power ratings of a few watts to hundreds of watts. Again, their accompanying inductance limits their use to dc or power ac. They should be mounted in such a manner that their considerable heat may easily be either radiated or conducted away.

4.2.3 Variable Resistors

Variable resistors usually take the form shown in Figure 4.2. A carbon or a plastic film with contacts at each end is deposited on the inside of a circular form. A movable wiper arm makes contact with the film, thereby allowing for a variation in the resistance between either end and the center contact. Alternatively, the resistance may be provided by wire wound along the circular form.

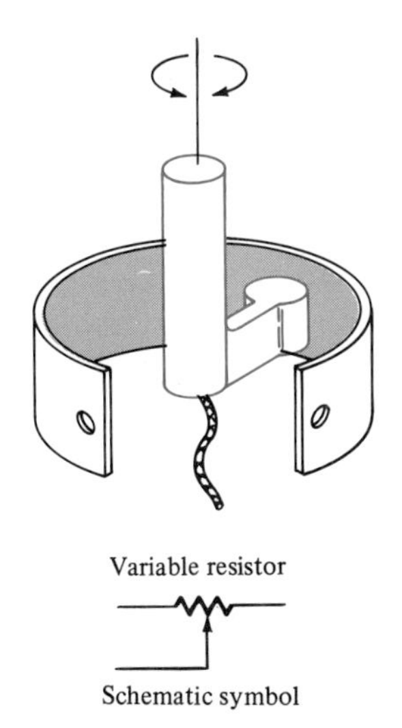

FIGURE 4.2. **Variable Resistor** A wiper arm makes contact with the resistive component that lines the inner circular surface.

Such a variable resistor is often called a **pot**, which is an abbreviation for *potentiometer* (a wire-wound pot is abbreviated on circuit diagrams as WW). A true potentiometer, however, is a rather elaborate electrical instrument used to measure potentials accurately, and purists reserve this name for such instruments. It should also be apparent that a pot represents a variable voltage divider and attenuator. Pots generally constitute the gain control on stereo sets and volume control on radio receivers and TV sets.

The lifetime of a pot is rated in terms of how many cycles (rotations) it can endure. For wire-wound pots this might be on the order of hundreds of thousands and for carbon pots, significantly less. When used as a gain or volume control, excessive wear may be recognized by noisy performance and uncertain contact, particularly as the control is adjusted. The unit should be replaced when such performance is noted.

The usable angular deflection of the pots so far discussed is about 270°. Some forms of variable resistor, termed **helipots**, utilize a movable contact that moves along a helically wound resistance, requiring as many as 25 turns to travel from one end of the resistance to the other. They provide rather precise control of resistance variation.

An important specification of a pot is its **resolution**, the smallest incremental travel of the wiper arm needed to produce an incremental change in the resistance. The resolution of wire-wound pots is poor since the output is "stepped" as it moves from one turn to the adjoining one. The resolution of the helipot, on the other hand, is excellent.

Another specification is the **taper:** a measure of how the resistance varies as a function of angular position. The most obvious taper is a linear one. Equal increments of angular position yield equal increments of resistance variation throughout the range. But there are other tapers that allow fine control over a certain range of angular position (the most used) and a gross control otherwise.

The resistance specification for a pot constitutes the total resistance between the fixed terminals. For carbon film types this ranges from 100 Ω to 10 MΩ. Wire-wound types span the range from about 1 Ω to 200 kΩ. The power rating of carbon pots is generally a fraction of a watt up to about 2 W; wire-wound pots go up to about 5 W. Where a wire-wound variable resistor is made to handle large amounts of power (such as in a light-dimming circuit), it is termed a **rheostat.** In such cases only one terminal end and the wiper arm are utilized. Rheostats are available with ratings in the hundreds of watts, but the maximum resistance value diminishes with increasing power rating. Thus, a 300 W unit is available up to 1500 Ω, while a 150 W unit is available up to 10 kΩ.

4.3 ■ CAPACITORS

4.3.1 Introduction

The basic capacitor configuration consists of two flat plates separated by an insulating layer (dielectric). If the insulation is vacuum (or air, which in this regard can be considered to be almost the same), we have

$$C = \frac{\epsilon_0 A}{d} \tag{4.1}$$

where C = capacitance in farads
A = area in square meters
d = separation in meters
ϵ_0 = permittivity of free space (8.854×10^{-12} farad per meter)

If the dielectric is other than vacuum (or air), the preceding expression must be multiplied by the dielectric constant of the material (also termed the relative permittivity).

While a significant resistance (and even some capacitance at times) is associated with most inductors, it is rather common practice to consider capacitors to be "pure." But under some circumstances one must also consider inductance and resistance to be associated with capacitors. Let's take a look at the associated resistance first.

No insulator is perfect, so, of course, there is some current leakage in a capacitor, and this leakage represents a resistance associated with the dielectric. Also, when a charge is placed upon the capacitor, it has already been indicated that there is a charge displacement in the dielectric, creating electric dipoles. (See Figure 2.6.) If it is an ac voltage that is impressed, the dipoles will be made to rotate, and the frictional forces involved in such rotation produce a loss. Since in a pure reactance there should be no energy dissipation, this loss means that there is a resistive component associated with such movement. We represent these losses schematically by means of a resistance R_p placed in parallel with the ideal capacitor. Additionally, the capacitor leads and the electrodes themselves represent some resistance. We show this resistance as a series r. And, finally, those same leads represent a small amount of inductance. The changing current creates a changing magnetic field, and as it cuts across the leads, it creates a counter-emf and, therefore, constitutes an inductance. Thus, a real capacitor should be represented by the equivalent circuit shown in Figure 4.3.

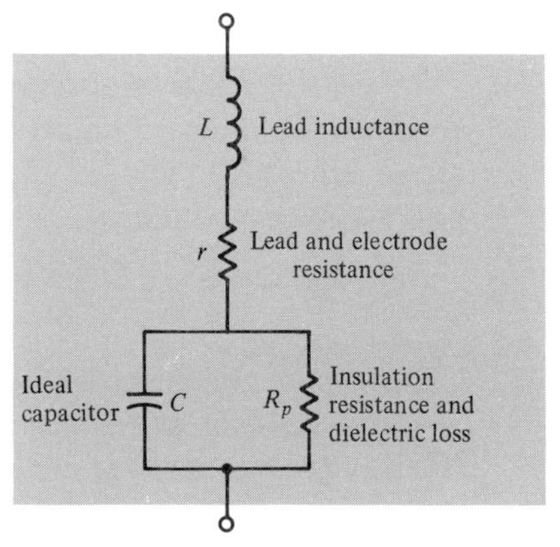

(a) Circuit equivalent of a capacitor

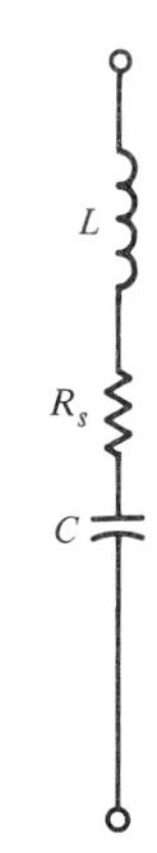

(b) Circuit equivalent of a capacitor, with series equivalent R_p added to r yielding the value R_s

FIGURE 4.3. **Nonideal Capacitor**

At low frequencies, L is small enough so that its effect can generally be disregarded. Also, the parallel resistance R_p can be replaced by its series equivalent, which, when added to r, leads to the *equivalent series resistance* (ESR) R_s. An actual capacitor represents an impedance:

$$Z = \sqrt{R_s^2 + X_C^2} \tag{4.2}$$

$$\cos\theta = \frac{R_s}{Z} \equiv \text{power factor} \tag{4.3}$$

$$\tan\delta = \frac{R_s}{X_C} = D \equiv \text{dissipation factor} \tag{4.4}$$

You may recognize D as the reciprocal of the expression for circuit Q.

EXAMPLE 4.2 Compute the expression for the equivalent series impedance (Figure 4.3b) taking into account that $R_p \gg X_C$ and, therefore, $1/R_p \ll \omega C$.

Solution:

$$\frac{1}{\mathbf{Z}_{\|}} = \frac{1}{X_C} + \frac{1}{R_p} = j\omega C + \frac{1}{R_p} = \frac{1 + j\omega CR_p}{R_p}$$

$$\mathbf{Z}_{\|} = \left(\frac{R_p}{1 + j\omega CR_p}\right)\left(\frac{1 - j\omega CR_p}{1 - j\omega CR_p}\right) = \frac{R_p - j\omega CR_p^2}{1 + \omega^2C^2R_p^2}$$

$$= \frac{R_p}{1 + \omega^2C^2R_p^2} - j\left(\frac{\omega CR_p^2}{1 + \omega^2C^2R_p^2}\right)$$

$$= \frac{1/R_p}{(1/R_p^2) + \omega^2C^2} - j\left[\frac{\omega C}{(1/R_p^2) + \omega^2C^2}\right] \simeq \frac{1}{\omega^2C^22R_p} - j\left(\frac{1}{\omega C}\right)$$

since $R_p \gg 1/\omega C$. Therefore,

$$\mathbf{Z} = r + \frac{1}{\omega^2C^2R_p} + j\left(-\frac{1}{\omega C} + \omega L\right) = R_s + j\left(-\frac{1}{\omega C} + \omega L\right)$$

where r = lead and electrode resistance

$1/\omega^2C^2R_p$ = dielectric loss

Note also that there is a resonant frequency at which the capacitor changes over and becomes an inductor. Because L is small, this resonant frequency has a rather high value.

4.3.2 Paper and Mylar Capacitors

One relatively simple means by which the electrode area may be increased makes use of two metal foils separated by a pair of impregnated paper insulators. This assemblage is then "rolled up," much in the manner of a jelly roll, thereby achieving a large area within a reasonable volume. However, such insulation has considerable limitations, and these **paper capacitors** find use primarily at low frequencies. More recently, polyester plastics, such as **mylar**, have been replacing paper since they can be made more homogeneous, with fewer imperfections. Paper and plastic metallized capacitors are also available. Rather than using metal foil electrodes, the metal is vacuum-deposited on the insulation, leading to a much smaller volume for a given value of capacitance.

Paper and plastic capacitors are available in values from about 0.001 to 1.0 μF, with voltage ratings up to about 600 V dc. The latter represents the rated voltage of the paper or plastic insulation.

The considerable leakage resistance of such capacitors (particularly the paper variety) limits their use. For example, neither a paper nor a plastic capacitor should ever be used as the resonating capacitor in a high-Q circuit. Their typical dissipation factor is 0.01 (with the factor for the plastic version being somewhat better).

Plastic capacitors with closer tolerances may be obtained by the use of **polycarbonates**. Such capacitors exhibit better stability and resistance to severe environmental conditions but are somewhat more expensive.

4.3.3 Mica Capacitors

At high frequencies mica dielectrics, with their much lower leakage resistance, show superior performance. In many cases their construction more closely approximates the classical one of two flat conductors separated by a layer of insulation (mica) except that the attainment of larger values of capacitance (up to 0.1 μF) is achieved by interleaving many such plates. Mica is available in extremely thin sheets (which allow smaller separations of the metal plates and thereby increase the capacitance) and exhibits the high insulation resistance previously mentioned.

For a given capacitance, mica tends to occupy a relatively small volume. Additionally, the metal plates may also be made by vacuum deposition on both sides of the mica dielectric, thereby diminishing the volume to an even greater degree. A typical dissipation factor for mica might be 0.005. Only polystyrene capacitors have a lower value.

4.3.4 Polystyrene Capacitors

There is an important electronic operation called *sample and hold*: A varying signal is sampled by having an instantaneous value of the signal voltage deposited as a proportional charge on the capacitor. The source is then removed, and the measurement of the voltage on the capacitor is made at a relatively slow pace. Such operations demand a capacitor that has a very small leakage resistance. The capacitor of choice for such operations utilizes **polystyrene** insulation. The insulation resistance is 10^6 MΩ, compared to 10^5 MΩ for mica and 10^4 MΩ for paper.

However, metallized coatings are difficult to apply to polystyrene. As a result such capacitors use relatively large physical dimensions. **Polypropylene film capacitors**, which can be metallized, represent their major competition, although the dissipation factor tends to be an order of magnitude greater (0.001 vs. 0.0003), while the insulation resistance is somewhat lower.

4.3.5 Electrolytic Capacitors

Since thin dielectric layers allow for the closer proximity of the metallic plates and thereby lead to significantly higher capacitance, one may achieve very large capacitance within a reasonable volume by employing a dielectric that consists of a thin molecular layer. By such means one may obtain a capacitance as large as 100,000 μF, in contrast to the fractional microfarads available in plastic or mica forms. Such **electrolytic capacitors**, as they are called, generally consist of aluminum foils on which thin insulating layers have been electrodeposited. The one precaution to be observed when using such devices is the maintenance of the same polarity as was employed in the formation of the dielectric. A reversal of polarity may permanently damage the capacitor. Obviously, therefore, they must be employed under dc conditions, with superimposed ac amplitudes insufficient to reverse the polarity. However, some electrolytic capacitors consist of two anodized electrodes separated by an electrolyte. Such **nonpolarized** electrolytics do not require a polarity distinction between their leads, but they tend to occupy a volume twice that of the **polarized** type for the same value of capacitance. Capacitances for aluminum electrolytics cover the range 0.5–140,000 μF with dc ratings up to about 600 V. Their leakage resistance tends to be low, yielding dissipation factors of about 0.08.

Tantalum electrolytic capacitors have much better dissipation factors, 0.005 to 0.02. Their voltage limitations, however, are somewhat more restrictive, being limited to the range 1–100 V, with capacitance values of 0.1–1000 μF.

4.3.6 Ceramic Capacitors

The dielectric constant for mica is 5.4; for impregnated paper, 3.7; for aluminum oxide, 8.4; and for tantalum oxide, 27.6. The value for ceramic materials such as titanium dioxide is in the range 80–120, while for barium titanate, it can run as high as 16,000. One might therefore expect that such **ceramic capacitors** make available a large value of capacitance within a small volume, and they do. Physically, disc ceramic capacitors exhibit a size and shape comparable to a penny. A typical value of capacitance might be 0.1 μF, with a voltage rating in the order of kilovolts. However, their tolerance rating can be quite large. The actual value of capacitance for a given unit may be off by as much as 100% from the indicated value, and the value depends on the voltage applied.

Barium titanate is classed as a type II ceramic, and such capacitors exhibit poor high-frequency characteristics. They should never be used in an LC resonant circuit because of the low Q they exhibit. Type I ceramics, using titanium oxide, on the other hand, have reasonably good stability and high-frequency characteristics.

4.3.7 Variable Capacitors

Adjustable capacitors, termed **variable capacitors**, often take the form of semicircular aluminum plates that interleave with a set of stationary plates (Figure 4.4). All the stators are connected in common, as are the rotor plates. This effectively causes the capacitance between each rotor plate face and the opposing stator plate face to appear in parallel with all other such pairings. The maximum capacitance (which arises when the rotor is fully interleaved with the stator) depends on the number of plates used. A typical variable capacitor used for station tuning in an AM radio might have a maximum value of 350 pF and in an FM radio, perhaps 15 pF.

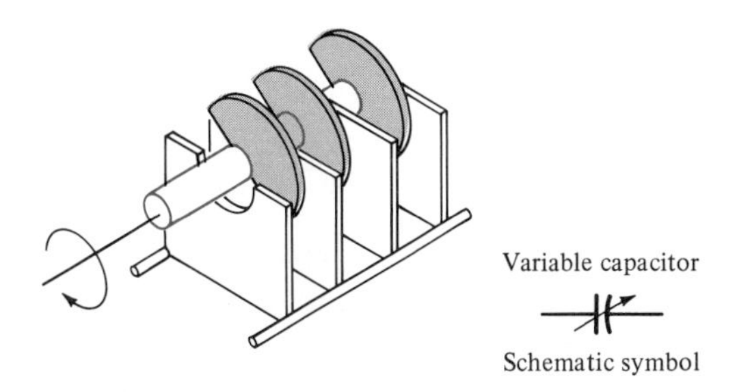

FIGURE 4.4. **Variable Capacitor** Capacitance depends on the magnitude of the interleaved area between stationary and rotary plates.

A ceramic variable capacitor (termed a ceramic trimmer) consists of a thin deposit of metal on a circular insulator below which is another semicircular metallic deposit which remains in a fixed position. By adjusting the overlap area, one may change the capacitance. The available range of variability is from 1 to 50 pF.

Another type of trimmer consists of a stack of mica plates placed between two outer metallic plates, one of which is spring-loaded. By altering the pressure between the plates, one changes the separation distance and thereby the capacitance. Values from 1 to 1000 pF are available.

4.4 ▪ INDUCTORS

4.4.1 Introduction

The third basic electrical element, the inductor, tends to be the most restrictive with respect to the values available. Also, unlike the resistor and the capacitor, both of which are available in rather "pure" form, the inductor inevitably exhibits considerable resistance (due to the wire) and capacitance (due to the capacitance between the turns). Also, while microcircuit techniques easily create resistors and capacitors, inductors are not easily formed. The inductor is frequently encountered in the form of a relay solenoid. In older vacuum tube circuits, it also served as a smoothing element in the power supplies where the inductance might typically have a value of some tens of henries. (The other main specification would be the maximum current that should be allowed to pass through the winding.) An inductor intended to attenuate ac components is termed a **choke**.

One other form of inductor might be encountered—the RF choke (radio frequency choke). Its function is to prevent the entrance of high-frequency signals into specific circuitry paths. Typical values of such inductors might be a few millihenries and, again, a specification of the maximum allowable current.

Inductors are, of course, to be found as part of *LC* resonant circuits in radio and TV circuits. Depending on the value of capacitance used and on the frequency, their values would be of the order of millihenries and microhenries. Providing their magnetic fields do not interact, inductors in series and parallel are additive in the same manner as resistors (Section 2.2).

4.4.2 Transformer Theory

There are, of course, situations wherein inductors do interact with one another, purposely so in the case of a transformer. We have thus far considered only idealized transformers (Sections 1.4.5 and 3.2), and this is an opportune time to extend our treatment into a more practical situation. We will subsequently also use such concepts when discussing electric motor operation.

We have seen that secondary circuit impedances appear in the primary circuit to have their respective values multiplied by the square of the turns ratio. As a first step, consider the resistances in Figure 4.5. Part of the input signal power is dissipated in the primary's resistance (R_p) and part in the secondary's resistance (n^2R_s). The remainder will appear in the load (n^2R_L).

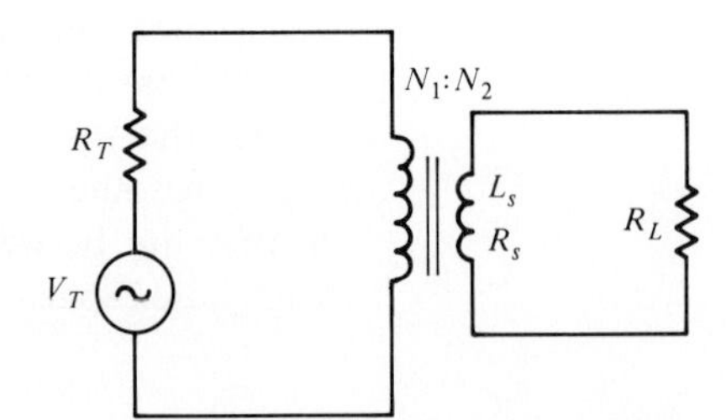

(a) Thevenin generator with intervening transformer used to match resistive load R_L

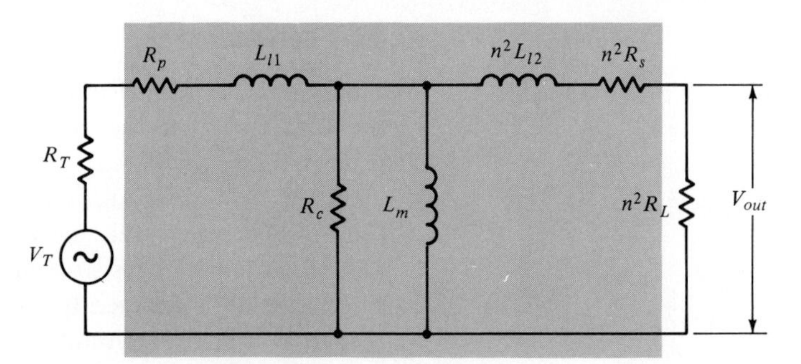

(b) Equivalent circuit with transformer's secondary parameters referred to the primary

FIGURE 4.5. Load Matching with a Transfer

Not all of the flux lines from the primary link the secondary. Those that do not lead to the **leakage inductance** of the primary, L_{l1}. Similarly, the flux lines from the secondary winding do not all link the primary, and this leads to a secondary leakage inductance (L_{l2}) that appears in the primary circuit as n^2L_{l2}. The flux that links the primary and secondary is associated with the **magnetizing inductance**.

There is also a power loss in the transformer's iron core, which is why transformers can get hot. In most models this power loss appears in the form of a shunt resistance (R_c). Some of the signal current passes through R_c (in our model) rather than reaching n^2R_L.

EXAMPLE 4.3

An audio transformer is represented by the following parameters: L_{l1} = 63.5 mH, L_{l2} = 0.159 mH, L_m (measured on the primary side) = 1.59 H. (See Figure 4.6.) It is driven by a Thevenin generator whose resistance is 2 kΩ and whose output is 5 V rms. Disregarding the winding resistances (R_p and R_s) as well as the core losses (R_c), compute the output voltage at 100 Hz, 1 kHz, and 5 kHz. What is the output bandwidth of the transformer? (Assume R_L to be 5 Ω.)

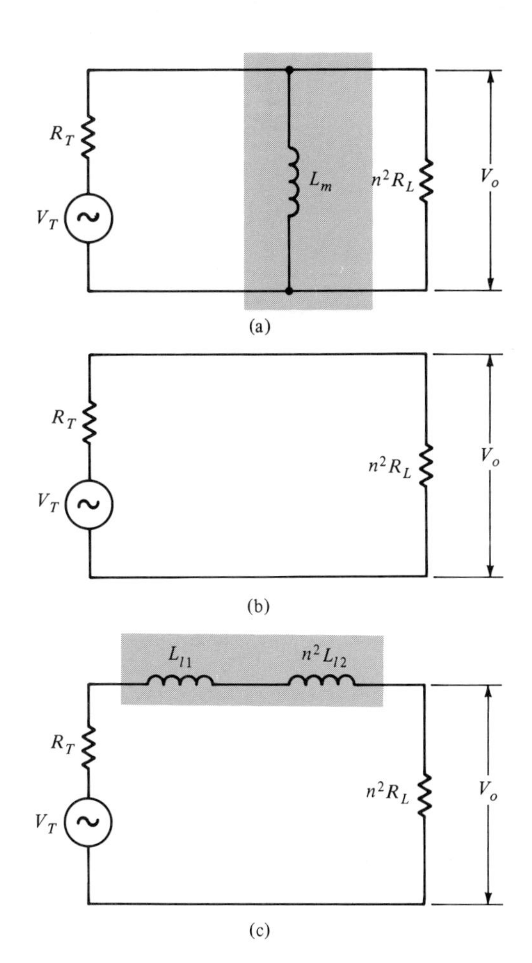

FIGURE 4.6

Solution: For maximum power transfer, we have

$$n \equiv \frac{N_1}{N_2} = \sqrt{\frac{R_T}{R_L}} = \sqrt{\frac{2000}{5}} = 20$$

To match the load to the source, we have

$$n^2 R_L = R_T = 2000\ \Omega$$

By using Eq. 2.3, the computed inductive reactances at the various frequencies are as follows:

	100 Hz	1 kHz	5 kHz
$X_{Ll} = X_{Ll1} + n^2 X_{Ll2}$	80	800	4,000
X_{Lm}	1000	10,000	50,000

The parameters with the most significant values at 100 Hz are shown in Figure 4.6a. X_{L_m} in parallel with n^2R_L constitutes a series equivalent impedance approximately equal to $400 + j800\ \Omega$:

$$\frac{2000(j1000)}{2000 + j1000} = \frac{2 \times 10^6 \angle 90°}{2.236 \times 10^3 \angle 26.6°}$$
$$= 0.89 \times 10^3 \angle 63.4° \simeq (400 + j800)\ \Omega$$

The output voltage at 100 Hz is

$$\frac{0.89 \times 10^3 \angle 63.4°}{2000 + 400 + j800} 5\ \text{V} = \frac{4.45 \times 10^3 \angle 63.4°}{2530 \angle 18.4°} = 1.76 \angle 45°\ \text{V}$$

The most significant parameters at 1 kHz are shown in Figure 4.6b. The output voltage at 5 kHz is

$$\frac{n^2 R_L}{R_T + n^2 R_L} 5 = \frac{2000}{2000 + 2000} 5 = 2.5\ \text{V}$$

Figure 4.6c shows the significant parameters at 5 kHz. The output voltage is

$$\frac{n^2 R_L}{R_T + n^2 R_L + jX_{L_l}} 5 = \frac{2000}{2000 + 2000 + j4000} 5$$
$$\simeq \frac{2000}{5660 \angle 45°} 5 = 1.76 \angle -45°\ \text{V}$$

The values at 100 Hz and 5 kHz represent the 3 dB points. That is,

$$20 \log_{10} \frac{1.76}{2.5} = -3\ \text{dB}$$

Therefore, the bandwidth is

$$5\ \text{kHz} - 100\ \text{Hz} = 4900\ \text{kHz}$$

SUMMARY

4.2 Resistors

- Resistor specifications include resistance value, tolerance (that is, accuracy of stated resistance value), and power rating.
- Composition resistor values are available within the approximate range of 10 Ω to 22 MΩ with tolerance values of 20%, 10%, and 5%.
- Carbon film resistors have values within the approximate range of 1 Ω to 10 MΩ with tolerance values of 5%, 2%, and 1%.
- Metal film resistors have a value range of 10 Ω to 1 MΩ with tolerance values of 5%, 1%, and less than 1%.
- Precision wire-wound resistors have values of 0.1 Ω to 1 MΩ and tolerance values of less than 1%.

4.3 Capacitors

- Capacitor specifications include capacitance value, tolerance, and voltage rating.
- The most inexpensive capacitors are of the paper and polyester varieties. Closer tolerances are available with polycarbonate types.
- At radio frequencies, mica capacitors are generally satisfactory.
- The highest insulation resistance is available with polystyrene capacitors. Polypropylene capacitors are their closest rivals in this regard.
- The largest values of capacitance are available with electrolytic capacitors.

4.4 Inductors

- Inductor specifications normally include inductance value, current rating, and resistance.
- The high-frequency performance of transformers is primarily dependent on the leakage inductance and the low-frequency performance, on the magnetizing inductance.

KEY TERMS

carbon composition resistor: a resistance composed of carbon granules held together by a binder

carbon film resistor: a thin resistive film deposited on an insulation form

ceramic capacitors:

- **type I:** reasonable stability and high-frequency characteristics
- **type II:** poor high-frequency characteristics and wide tolerance values

choke: an inductor intended to attenuate ac components

electrolytic capacitor: a type of capacitor that leads to the largest capacitance per unit volume

- **nonpolarized:** with no distinction between terminals
- **polarized:** with a dc polarity at its terminals

helipot: a variable resistance whose total variation extends over many turns of the rotor

inductance: magnetic flux divided by the current used to create the flux

leakage inductance: in the case of a transformer, the flux that links only the coil that created it, divided by the creating current (Primary and secondary windings each have individual values of leakage inductance.)

magnetizing inductance: in the case of a transformer, the inductance associated with the primary current

mica capacitor: a capacitor using a mica dielectric; gives superior performance at high frequencies

mylar capacitor: a capacitor using a polyester dielectric; a replacement for paper capacitors

paper capacitor: a capacitor using a paper dielectric (obsolete)

polycarbonate capacitor: a plastic capacitor with superior tolerance and stability ratings

polypropylene capacitor: a superior capacitor that is somewhat less expensive than a polystyrene capacitor

polystyrene capacitor: a capacitor containing the ultimate in insulation

pot: an abbreviation for potentiometer; a variable resistance with the total resistance range extending over a circular movement of less than 360° of arc

resolution: the smallest incremental travel needed to produce an incremental change in resistance

rheostat: a form of variable resistor

taper: a measure of the uniformity of resistance change with a change in the angular position of a pot

variable capacitor: an adjustable capacitor
wire-wound resistor: resistance wire wound in the form of a helix and, therefore, accompanied by some inductance

SUGGESTED READINGS

Mazda, F. F., DISCRETE ELECTRONIC COMPONENTS. New York: Cambridge University Press, 1981. *A detailed consideration of component values, tolerances, and stability.*

Mandl, Matthew, DIRECTORY OF ELECTRONIC CIRCUITS, Appendices F, G, H, I, J, and K. Englewood Cliffs, N.J.: Prentice-Hall, 1978. *An extensive treatment of color coding as applied to resistors, capacitors, diodes, and transformers.*

Brotherton, M., CAPACITORS, THEIR USE IN ELECTRONIC CIRCUITS. New York: Van Nostrand, 1946. *A short history of capacitor development may be found in the introduction to this book.*

Snow, C., "Formulas for Computing Capacitance and Inductance," NATIONAL BUREAU OF STANDARDS CIRCULAR 544. Washington, D.C.: U.S. Government Printing Office, 1955. *Inductance and capacitance calculations for various geometrical forms.*

EXERCISES

1. The minimal specification of a fixed resistor should include ______, ______, and ______.

2. With regard to variable resistors, distinguish between a pot and a rheostat.

3. The largest power rating available in the form of carbon resistors is ______ W.

4. Resistance tolerances of 1% require ______ significant figures for their proper resistance specification.

5. Compute the power dissipation in each of the following if the current is 1 mA: 100 kΩ, 1 MΩ, 10 MΩ, 100 MΩ.

6. What should be the limiting current if the power rating of a 1 W, 1 MΩ resistor is not to be exceeded?

7. What is the resistance equivalent to the following: 47 Ω, 52 Ω, 100 Ω

- **a.** When placed in parallel
- **b.** When placed in series

8. What is the capacitance equivalent to the following: 0.01 μF, 0.1 μF, 0.015 μF

- **a.** When placed in parallel
- **b.** When placed in series

9. If a charge is placed successively on a polystyrene capacitor (with leakage resistance equal to 10^6 MΩ), a mica capacitor (10^5 MΩ), and a paper capacitor (10^4 MΩ), compare the decay times for the charge to have diminished to $1/e$ of its initial value. (The leakage resistance can be taken to appear in parallel with C.) The capacitors have a value of 0.1 μF.

10. The smallest incremental travel of a pot's wiper arm needed to produce an incremental change in resistance constitutes its ______.

11. Specify the resistance indicated by the following coding:

- **a.** Red, red, orange
- **b.** Orange, white, red, silver
- **c.** Green, brown, yellow, gold
- **d.** Blue, gray, green, silver
- **e.** Brown, brown, red, gold
- **f.** Brown, black, green
- **g.** 49R9
- **h.** 10R0
- **i.** 1693

12. What type of variable resistor should be used in order to achieve a high degree of resolution?

13. What can be a major disadvantage of wire-wound pots?

14. How do electrolytics achieve a large value of capacitance within a relatively small volume?

15. Why are type II (barium titanate) ceramic capacitors not suitable for resonant circuit applications?

16. What is a major advantage offered by disc ceramic capacitors?

PROBLEMS

4.1
Specify the values of resistance indicated by the following coding:

a. Brown, black, orange
b. Yellow, violet, black
c. Yellow, violet, red, gold
d. Green, black, green, silver
e. 1020
f. 4423
g. 11R8
h. 10R0

4.2
What is the meaning of the following codes?

a. RN55D
b. RN60C

4.3
a. A 1 MΩ resistor passes 1 mA of current. What should be the power rating of the resistor used?

b. If the same 1 mA passes through a 4.7 kΩ resistor, what should be the power rating of the resistor used?

4.4
a. What is a typical value of capacitance of an electrolytic capacitor?

b. What is a typical value of capacitance of a paper- or plastic-wound capacitor?

c. What is a typical voltage rating of a tantalum electrolytic capacitor?

d. A typical value of capacitance used in an AM receiver for tuning is 100 pF. With this value of C, what value of L is necessary to tune the receiver to the middle of the broadcast band, taken to be 1000 kHz?

4.5
When a variable air capacitor is fully opened, there is still a slight capacitance between the two sets of plates. If a 300 pF (maximum value) variable capacitor is used to tune through the AM broadcast band (535–1605 kHz), what is its capacitance in the fully opened position?

4.6
With regard to a capacitor, what is the relationship between the magnitudes of the phase angle (θ) and the loss angle (δ)?

4.7
At 120 Hz a 100 μF aluminum electrolytic capacitor has a dissipation factor of 0.2. Compute the equivalent series resistance and sketch the impedance diagram showing the loss angle and the phase angle.

4.8
a. Four 0.01 μF capacitors are placed in series. Compute the equivalent capacitance.

b. If placed in parallel, what would be the equivalent capacitance?

4.9
Compute the equivalent capacitance if the following capacitors are placed in series: 0.01 μF, 0.5 μF, 2.0 μF.

4.10
A voltage of 150 V dc is applied to an 8 μF capacitor. What is the charge accumulated?

4.11
If two 1 H inductors are connected in series, and presuming their fields do not interact, what is the equivalent inductance? When placed in parallel?

4.12
An electronics supply house catalog lists a 680 pF disc ceramic capacitor with a dissipation factor of 0.02 at 1 kHz. On the other hand, a polystyrene capacitor, also 680 pF, is rated as having a power factor of 0.05% at 1 kHz. Determine the equivalent series resistance in each case.

4.13
For a plastic capacitor, $C = 1\ \mu$F, and $L = 0.2\ \mu$H. Compute the resonant frequency.

4.14
Taking into account the inductance of a capacitor, compute the equation that yields the effective value of the capacitance.

4.15
If the bandwidth in Example 4.3 were to be extended to 20 kHz, what parameter changes would have to be made?

5 FOURIER ANALYSIS

OVERVIEW

Most of the signals that have thus far occupied our attention have been of a sinusoidal nature. The notable exceptions were those encountered in automotive ignition systems and those involved with damped vibrations. But once passed through electronic circuits, even sinusoidal signals will be found to deviate from true sinusoids, for all electronic circuits exhibit some degree of nonlinearity, and a substantial part of electronic design involves an attempt to minimize such nonlinearities. In order to diminish distortion, one must be able to determine the degree to which it exists. Such is one of the functions of Fourier analysis.

Signals that are not initially sinusoidal (such as the ignition signals and damped oscillation) are also subject to alteration upon being passed through nonlinear circuits. Fourier analysis will be applied to study the nature of such transformations as well. The concepts developed in this chapter will find application in just about every chapter that follows.

OUTLINE

5.1 ▪ INTRODUCTION

Circuits consisting solely of ideal *R, L,* and *C* elements constitute linear electrical systems. They satisfy the superposition principle. Each source can be considered independently, and the linear response is the summation of the individual responses. There can be no alteration of the frequency applied to a linear system; a sinusoidal waveform remains a sinusoidal waveform of the same frequency.

We will shortly be considering devices that exhibit nonlinear behavior in varying degrees. One such device is the transistor, and a consequence of nonlinearity is the partial conversion of an applied frequency into other extraneous frequencies. The summation of the resulting frequencies constitutes an output that is nonsinusoidal. Alternatively, we may say that a nonsinusoidal response represents the summation of a number of different frequencies.

By way of preparing ourselves for such eventualities, we turn now to a consideration of the method used to analyze the results of such frequency transformations, Fourier analysis. Nonlinear analyses are not confined to electrical systems. Consider the jarring motion that a rough road communicates to a car. Such vibration includes a variety of frequencies, some high and some low.

Some commonly encountered electrical waveforms are illustrated in Figure 5.1. The most fundamental one is the sinusoid illustrated in Figure 5.1a. It also represents the simplest periodically repeating signal that may be produced by a string stretched between two stable supports (Figure 5.2). Such a vibrating string will produce a given frequency (pitch). Its magnitude of vibration will determine the intensity of the sound.

The frequency of vibration depends on the length and tension of the string, with the frequency being increased by increasing the tension. By such means, for example, the vibration might be made to take the form shown in Figure 5.3. The lowest possible frequency of vibration is termed the **fundamental.** Frequencies that are integral multiples of the fundamental are termed **harmonics.** Thus the form shown in Figure 5.3 represents the third harmonic of that shown in Figure 5.2. In comparable fashion we might obtain the fifth harmonic, the seventh, and so forth.

If we pluck a string indiscriminately, it will simultaneously vibrate in a number of these modes. It is the number and strength of the harmonics that accompany a fundamental that allow us to recognize the musical instrument that is sounding a given note.[1]

Consider two electronic amplifiers, *A* and *B*. On a piano we might play the tune "Mary had a little lamb . . . ," which, picked up by a microphone, is passed through the two amplifiers. The assumed listener cannot hear the original sound, but only the sound that comes to him or her via loudspeakers attached to the two amplifiers. If amplifier *A* reproduces only the fundamental tones of the tune, the listener will probably not be able to identify the musical instrument. If, however, amplifier *B* reproduces a wide range of

[1]Supposedly, the only musical instrument capable of emitting pure tones is the ocarina.

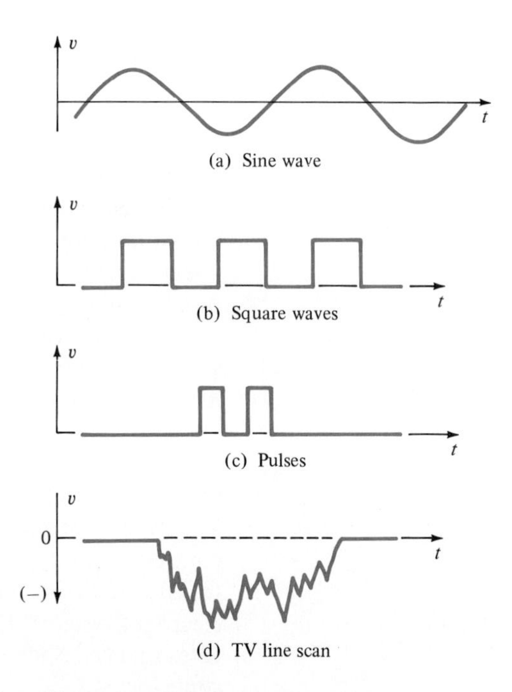

FIGURE 5.1. Some Examples of Electronic Waveforms

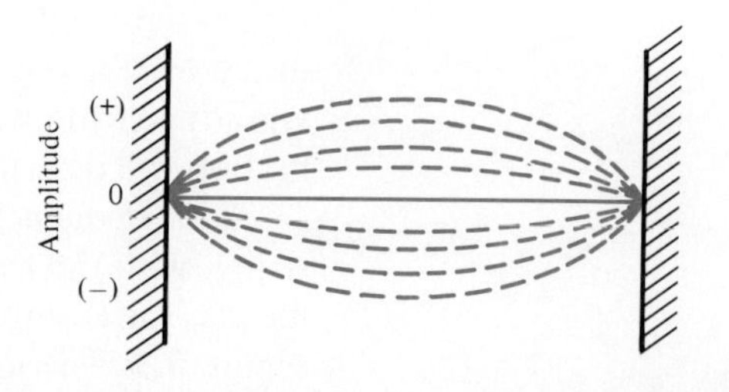

FIGURE 5.2. String Vibrating at Its Fundamental Frequency (ω_0)

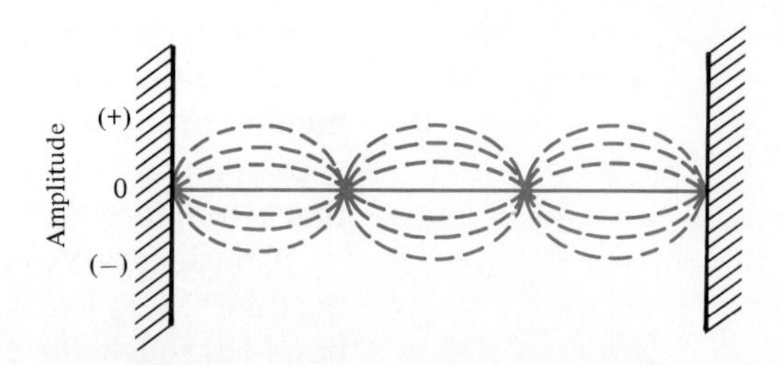

FIGURE 5.3. String Vibrating at Its Third Harmonic ($3\omega_0$)

frequencies—that is, not only the fundamental tones but also their accompanying harmonics—the listener should easily be able to identify the instrument since it is the number and loudness of the various harmonics that distinguish one type of instrument from another.

As a function of time, the fundamental and some of its harmonics can be depicted as in Figure 5.4. Let us add ω_0, $3\omega_0$, $5\omega_0$, and so on to see what we get for the resulting waveform, which is also shown in Figure 5.4. It approximates a square wave. Such additions of harmonically related sine waves constitute the **synthesis** of a periodically repeating waveform.

Note that while it is the fundamental and the lower-order harmonics that create the relatively flat horizontal portions of the resulting square wave, it is

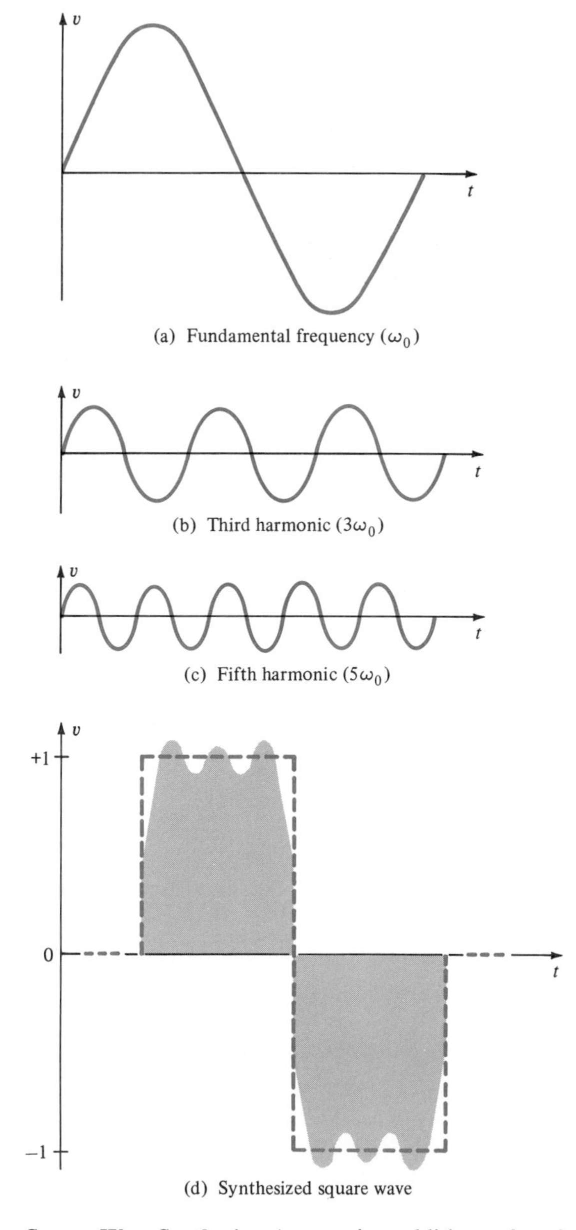

FIGURE 5.4. Square Wave Synthesis Appropriate additions of ω_0, $3\omega_0$, $5\omega_0$, and so on will synthesize a square wave (dotted).

the high-order harmonics that will lead to the sharp leading and trailing vertical edges. We thus see that appropriate combinations of the fundamental and harmonics may be used to synthesize a square wave. Or, alternatively, we may say that a square wave contains a fundamental and various harmonics.

Suppose we start with a square wave signal of small amplitude and wish to amplify it. Assume the period of the square wave to be 25 ms. This period means that the square wave repeats itself 40 times per second [$1/(25 \times 10^{-3}$ s) $= 40$ Hz] (Figure 5.5).

The instrument used to observe such waveforms, the oscilloscope, generally has a capacitor in series with the input terminals, as shown in Figure 5.6. At the discretion of the operator, this capacitor can be either switched into the circuit or shorted out. (At this time we need not go into why one might wish to do one or the other.) *Assume* the amplifier in the oscilloscope is capable of amplifying frequencies up to 300,000 Hz.

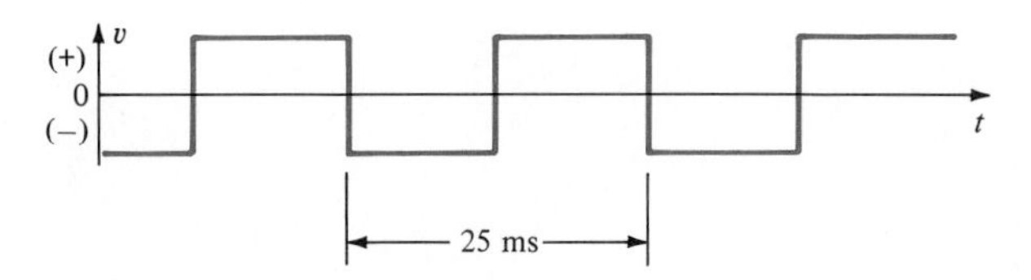

FIGURE 5.5. Square Wave with 25 ms Period

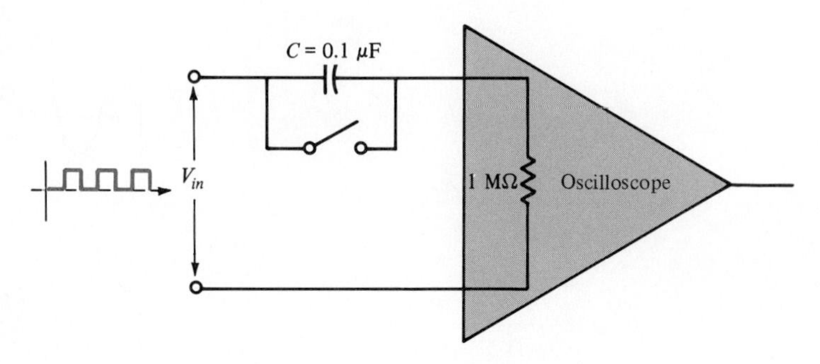

FIGURE 5.6. Oscilloscope Input Circuit The signal input is applied through a 0.1 μF capacitor that may be switched either in or out of the circuit.

With the capacitor shorted out, we apply our 40 Hz square wave to the oscilloscope and see a rather accurate reproduction of the waveform. But if we open the switch and thereby insert the capacitor in series with the input leads, while the vertical portions of the square wave are still accurately reproduced, we will find the horizontal portions to be slanted, as in Figure 5.7. Let us consider the reason for such behavior.

In Figure 5.6, with the switch open, we have a high-pass filter such as discussed in conjunction with Figure 3.25. We have noted that a square wave is made up of a fundamental and its odd harmonics. In the preceding case the fundamental represents a frequency of 40 Hz. At 40 Hz the capacitive reactance is about 40,000 Ω, and the amplitude and phase of the fundamental (as

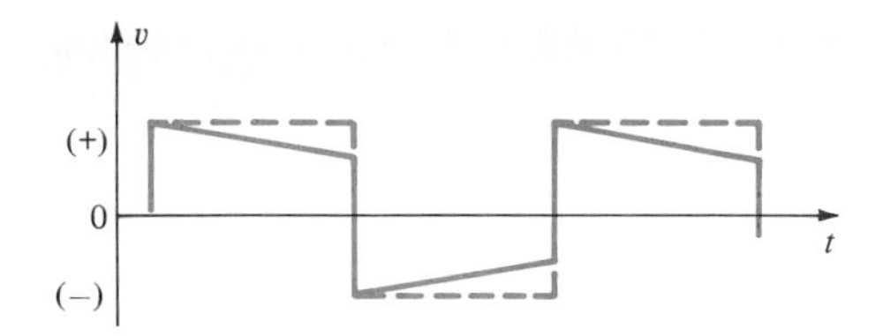

FIGURE 5.7. **Tilted Square Wave Resulting from Insufficient Low-Frequency Content**

presented for display) are thereby altered. The same holds for the low-order harmonics, but to a lesser extent as the order of the harmonic increases. It is this alteration that gives rise to the slanted portions. With the capacitor shorted out, all components of the square wave pass through essentially unaltered in relative magnitude, and the waveform is rather accurately displayed.

Let us now increase the frequency of the square wave to such a value that its period is 25 μs. Now the fundamental frequency is 40,000 Hz. With the capacitor shorted out, we observe something resembling Figure 5.8. With the capacitor in place, the display stays the same!

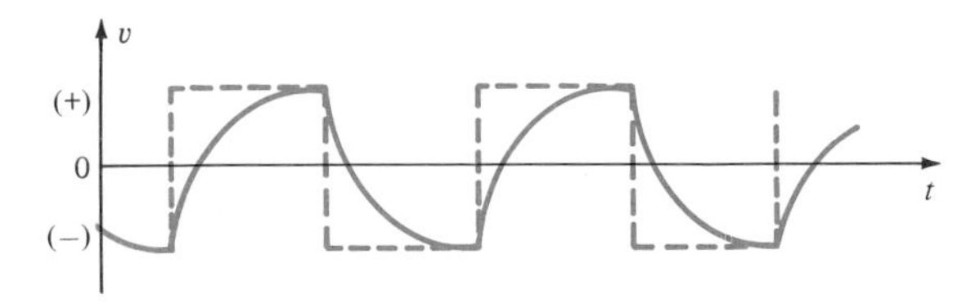

FIGURE 5.8. **Distorted Square Wave Resulting from Insufficient High-Frequency Content**

Since the lowest frequency that makes up the square wave in Figure 5.8 encounters a capacitive reactance that is only a few (negligible) ohms, the capacitor has little effect on the waveform's appearance. Thus, there is no change whether the capacitor is in or out of the circuit. The limitation in display integrity in this case is due to the limitation imposed by the amplifier itself. We assumed it had a bandwidth of 300,000 Hz, and hence it cannot amplify the higher harmonics of the square wave. Recall that it is these higher harmonics that are necessary to reproduce the sharp leading and trailing edges. To reproduce such a rapidly varying square wave, we would have to resort to an oscilloscope with a much wider bandwidth. Some oscilloscopes have bandwidths as large as 100 MHz (100,000,000 Hz).

Consider now a succession of two short-duration pulses such as shown in Figure 5.1c. Again, wide-band amplifiers are generally necessary to reproduce such pulses. If narrow-band amplifiers are used, the result might look as shown in Figure 5.9. We may, in fact, have difficulty in determining that there are two pulses present.

Figure 5.1d typifies a single line of scan on a TV picture tube. The peaks represent bright regions on the picture tube screen and the depressions, the

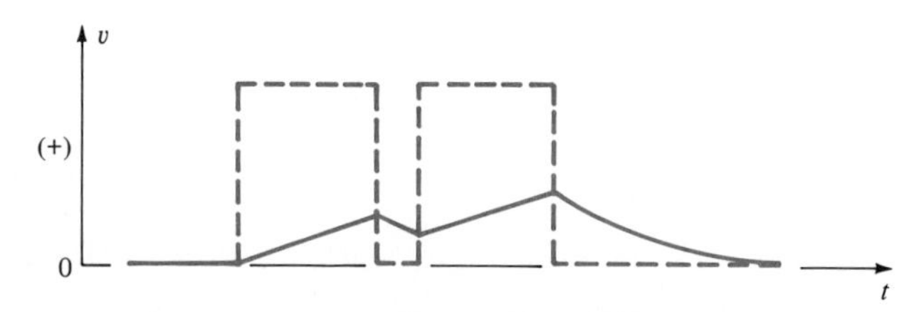

FIGURE 5.9. The Distortion (Solid Trace) of Two Successive Short-Duration Pulses (Dashed Trace) Because of Limited Amplifier Bandwidth

degree of darkness. To reproduce the sharp boundaries between such areas—that is, to obtain a sharply focused picture—wide-band amplifiers are necessary. Typical TV amplifiers have bandwidths that are 4.5 MHz.

How wide should the bandwidth be in order to reproduce a recognizable pulse? A rough rule of thumb is to make the bandwidth at least equal to the reciprocal of the pulse width.

EXAMPLE 5.1

a. A 1 μs pulse of sinusoidal frequency (f) is to be amplified. What should be the order of magnitude of the amplifier's bandwidth ($f_H - f_L$) in order to maintain the pulse integrity? (f_H and f_L are the upper and lower amplifier bandwidths, respectively.) (See Figure 5.10a.)

b. In what way would the reproduction of a 1 μs rectangular pulse (Figure 5.10b) differ from that of part a?

c. It is possible to generate optical pulses of the order of picoseconds (10^{-12} s) in duration. What amplifier bandwidth would be necessary to amplify the signal from a photoelectric device used for such detection?

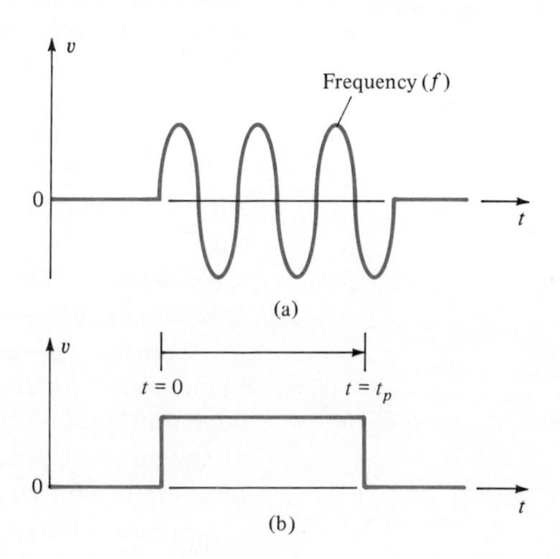

FIGURE 5.10

Solution:

a. $f_H - f_L \sim \frac{1}{t_p} = \frac{1}{10^{-6}\ \text{s}} = 10^6\ \text{Hz} = 1\ \text{MHz}$

We would need a bandwidth of 1 MHz centered on frequency f.

b. We would again need (at least) a 1 MHz bandwidth, but extending from $f_L = 0$ to $f_H \sim 1$ MHz. Actually, it should extend out at least another order of magnitude if significant rounding of pulse edges is to be avoided.

c. $f_H - f_L \sim \frac{1}{10^{-12}\ \text{s}} = 10^{12}\ \text{Hz} = 10^6\ \text{MHz}$

Since a typical visible optical frequency is of the order of 10^{14} Hz, we would need an amplifier with a bandwidth of 10^6 MHz centered on 10^8 MHz. This is much beyond the capabilities of conventional electronic amplifiers.

5.2 ■ TRIGONOMETRIC FOURIER SERIES

Any periodic function of time, $f(t)$—whose period is T—can be written in the form

$$\begin{aligned} f(t) = a_0 &+ a_1 \cos \omega t + b_1 \sin \omega t \\ &+ a_2 \cos 2\omega t + b_2 \sin 2\omega t \\ &+ a_3 \cos 3\omega t + b_3 \sin 3\omega t + \cdots \end{aligned} \tag{5.1}$$

where $\omega = 2\pi/T$. The as and bs are constants that represent weighting values and indicate the relative strengths of the fundamental and various harmonics. For the square wave we have been considering, with its amplitude centered on zero, we then have $a_0 = 0$. We may alternatively have a square wave elevated on a "pedestal" as in Figure 5.11. In this case a_0 represents the average value of the square wave. Its value can be determined by the equation

$$a_0 = \frac{1}{T}\int_0^T f(t)\, dt \tag{5.2}$$

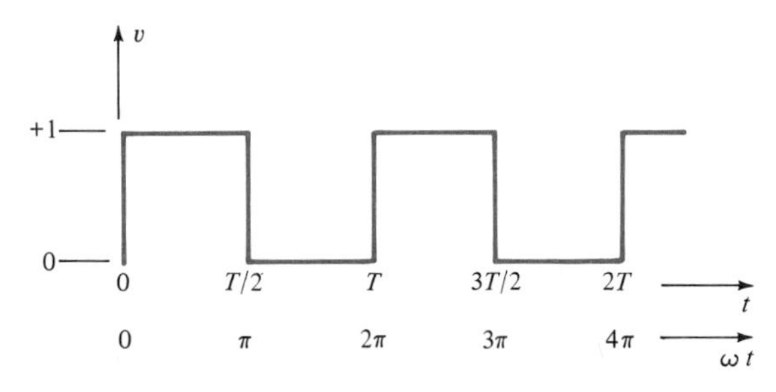

FIGURE 5.11. **Square Wave with $a_0 = \frac{1}{2}$**

In discussing the voltage and current waveforms applied to capacitors and inductors, we have seen that the 90° phase shift difference between them leads to a succession of positive and negative power pulses. Over one period (or even over one-half period), therefore, the integrated average of the product of sin ωt and cos ωt is equal to zero. Thus,

$$\frac{1}{T}\int_0^T \cos \omega t \sin \omega t \, dt = 0 \tag{5.3}$$

It may likewise be easily verified, both graphically and mathematically, that the integrated average of the sine and cosine product involving different integrally related frequencies will also yield a zero result. We have

$$\frac{1}{T}\int_0^T \cos n\omega t \sin m\omega t \, dt = 0 \tag{5.4}$$

where n and m = integers. In fact, the same type of average of the product between cosine functions of harmonically related frequencies, or between sine functions, also leads to a value of zero:

$$\left.\begin{aligned} \frac{1}{T}\int_0^T \cos n\omega t \cos m\omega t \, dt &= 0 \\ \frac{1}{T}\int_0^T \sin n\omega t \sin m\omega t \, dt &= 0 \end{aligned}\right\} \quad \text{if } m \neq n \tag{5.5}$$

It is only in the case of the last two integrals, and then only if $m = n$, that we get a finite result.

Equations 5.4 and 5.5, together with Eq. 5.3, may be used to obtain the frequency content of any periodic function of time. Let us illustrate the procedure by using the square wave of Figure 5.4. We easily determine that

$$\begin{aligned} a_0 &= \frac{1}{T/2}\int_0^{T/2} 1 \, dt + \frac{1}{T/2}\int_{T/2}^{T} (-1) \, dt \\ &= \frac{2}{T}\left(\frac{T}{2}\right) - \frac{2}{T}\left(T - \frac{T}{2}\right) = 0 \end{aligned}$$

Suppose we wish to determine the value of b_3 for this waveform. We multiply each term of Eq. 5.1 by sin $3\omega t$, obtaining

$$\begin{aligned} f(t) \sin 3\omega t = {} & a_0 \sin 3\omega t + a_1 \cos \omega t \sin 3\omega t + b_1 \sin \omega t \sin 3\omega t \\ & + a_2 \cos 2\omega t \sin 3\omega t + b_2 \sin 2\omega t \sin 3\omega t \\ & + a_3 \cos 3\omega t \sin 3\omega t + b_3 \sin 3\omega t \sin 3\omega t + \cdots \end{aligned}$$

From what we know concerning these products, only the b_3 term will lead to a finite value when we take an integrated average over one period of the function. Then we need to evaluate only

$$\begin{aligned} \frac{1}{T}\int_0^T f(t) \sin 3\omega t \, dt &= \frac{b_3}{T}\int_0^T \sin 3\omega t \, dt \\ &= \frac{b_3}{T}\int_0^T \left(\frac{1}{2} - \frac{\cos 6\omega t}{2}\right) dt = \frac{b_3}{2} \end{aligned}$$

Thus, we have

$$b_3 = \frac{2}{T}\int_0^T f(t)\sin 3\omega t\, dt$$

Substituting the appropriate values for our square wave—that is, +1 in the interval 0 to $T/2$ and −1 in the interval $T/2$ to T—we obtain

$$\begin{aligned} b_3 &= \frac{2}{T/2}\int_0^{T/2} (+1)\sin 3\omega t\, dt + \frac{2}{T/2}\int_{T/2}^{T} (-1)\sin 3\omega t\, dt \\ &= \frac{2}{3}\pi + \frac{2}{3}\pi \end{aligned}$$

and, therefore, $b_3 = \frac{4}{3}\pi$.

In general, we have

$$\begin{aligned} a_n &= \frac{2}{T}\int_0^T f(t)\cos n\omega t\, dt = \frac{\omega}{\pi}\int_0^{2\pi/\omega} f(t)\cos n\omega t\, dt \\ &= \frac{1}{\pi}\int_0^{2\pi} f(\omega t)\cos n\omega t\, d(\omega t) \\ b_n &= \frac{2}{T}\int_0^T f(t)\sin n\omega t\, dt = \frac{\omega}{\pi}\int_0^{2\pi/\omega} f(t)\sin n\omega t\, dt \\ &= \frac{1}{\pi}\int_0^{2\pi} f(\omega t)\sin n\omega t\, d(\omega t) \end{aligned} \tag{5.6}$$

For the square wave under consideration, we find that

$$a_1 = a_2 = a_3 = \cdots = 0$$
$$b_2 = b_4 = b_6 = \cdots = 0$$
$$b_1 = \frac{4}{\pi}, \qquad b_3 = \frac{4}{3\pi}, \qquad b_5 = \frac{4}{5\pi}, \qquad \cdots$$

Thus, for our square wave we have

$$f(t) = \frac{4}{\pi}\left(\sin \omega t + \frac{1}{3}\sin 3\omega t + \frac{1}{5}\sin 5\omega t + \cdots\right)$$

5.3 ■ GRAPHICAL FOURIER ANALYSIS

To better visualize what is being done in a Fourier analysis, we can consider Figure 5.12, which depicts what started out to be a sinusoidal alternating voltage but which has had the negative cycle excised. The resulting waveform represents a **half-wave rectified voltage** and constitutes the first step in converting alternating current into direct current. It is a waveform that we will encounter again in Chapter 6 (Section 6.2.3).

Figures 5.13 and 5.14 illustrate graphically the Fourier process: In Figure 5.13a, sin ωt is multiplied by the half-wave voltage (dashed curve), and the area that is shaded constitutes an evaluation of the coefficient b_1. This coefficient will have a finite positive value for the waveform under consideration.

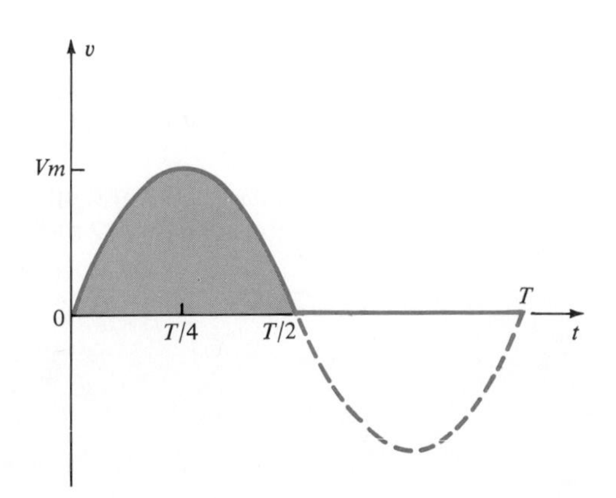

FIGURE 5.12. **Half-Wave Rectified Sine Wave (Solid) with Continuation of Sine Wave (Dotted)**

Figure 5.13b shows the result of evaluating the product of cos ωt and the rectified voltage. It can be assumed that a_1 for this waveform will equal zero. Likewise since the sin $2\omega t$ factor (Figure 5.14a) will lead to a zero average, b_2

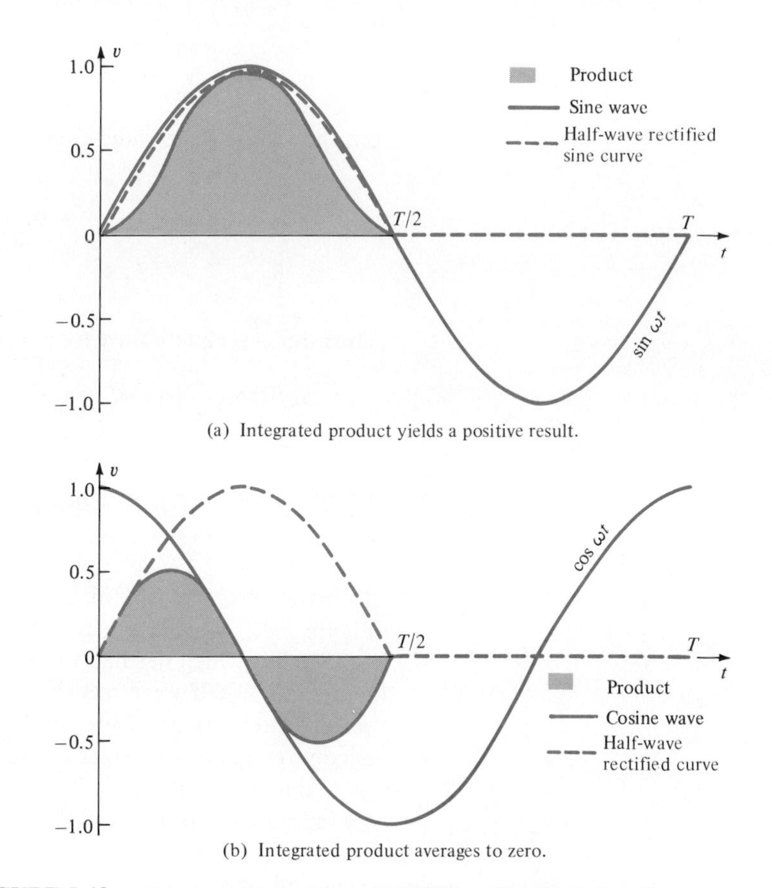

FIGURE 5.13. **Graphical Evaluation of Half-Wave Rectified Sine Wave**

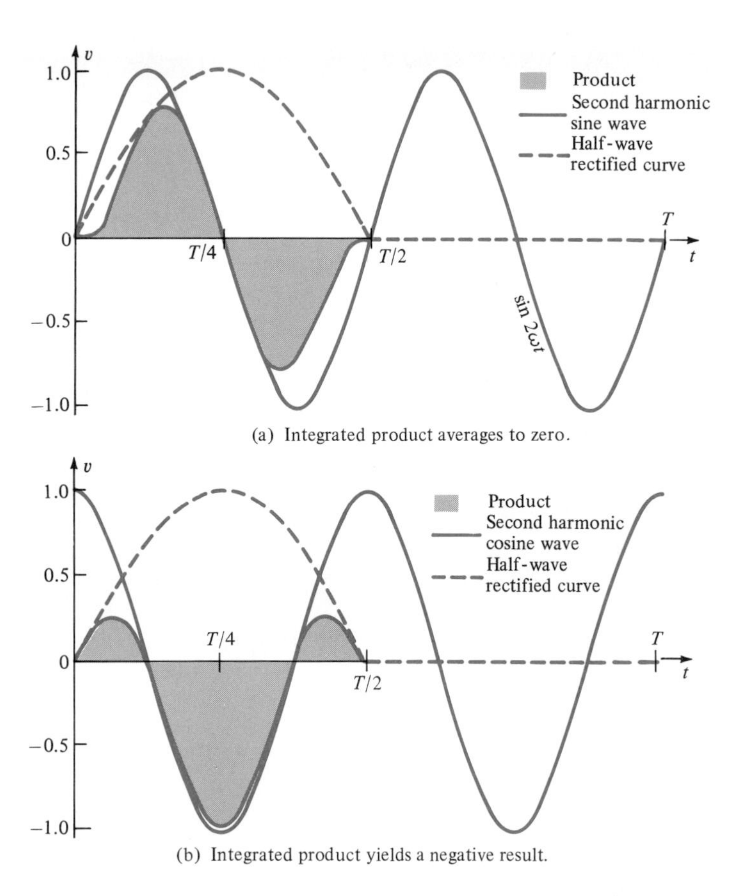

FIGURE 5.14. Additional Graphical Analysis of Half-Wave Rectified Sine Wave

may be presumed to equal zero. In Figure 5.14b, the cos $2\omega t$ factor leads to a large negative pulse whose area is larger than the sum of the two smaller positive excursions, and an evaluation of a_2 should lead to a finite result that is negative.

$$v = V_{max}\left(\underset{a_0}{\frac{1}{\pi}} + \underset{b_1}{\frac{1}{2}\sin \omega t} - \underset{a_2}{\frac{2}{3\pi}\cos 2\omega t} - \underset{a_4}{\frac{2}{15\pi}\cos 4\omega t} - \cdots\right) \tag{5.7}$$

EXAMPLE 5.2

Show graphically that there is no sin $4\omega t$ term in the Fourier analysis of a half-wave rectified signal (Figure 5.12) but that there is a cos $4\omega t$ term. (See Figures 5.15 and 5.16.)

Solution: The obvious equality of the paired pulses in the sin $4\omega t$ sketch (Figure 5.15) makes it obvious that there is no sin $4\omega t$ term. It is difficult to tell in the case of the cos $4\omega t$ sketch (Figure 5.16) whether the sum of the three positive pulses equals that of the two negative ones. They do not, but

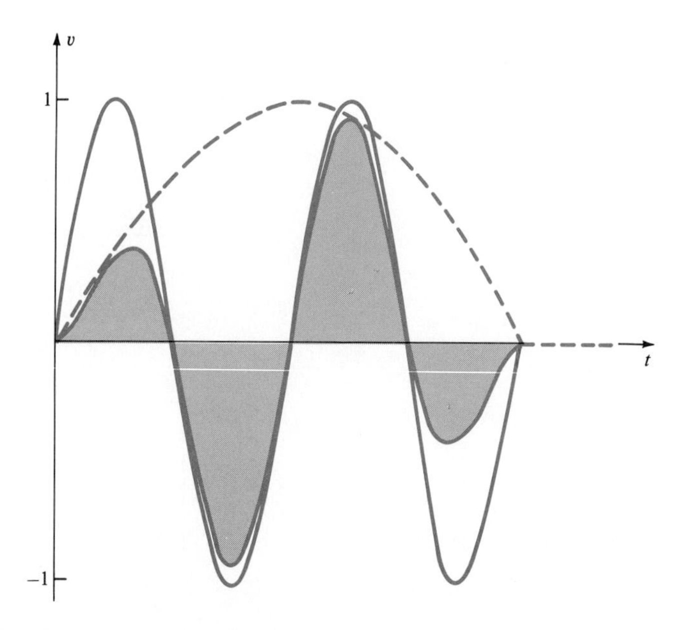

FIGURE 5.15

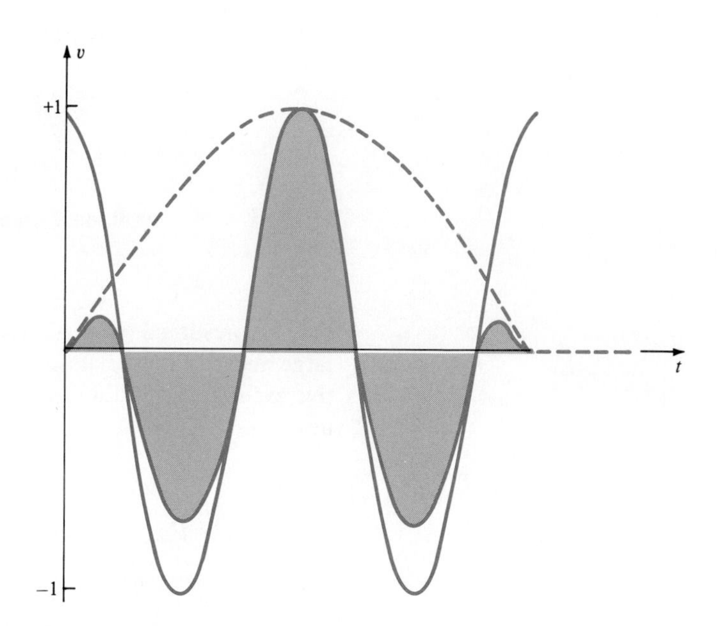

FIGURE 5.16

their difference is small, and one must resort to more accurate means of determination, such as waveform symmetry, discussed in Section 5.4.

As the first step toward converting ac into dc, rather than deleting alternate half-cycles, we may instead invert their polarity, thereby giving rise to a **full-wave rectified signal** and a higher conversion efficiency.

5.4 ▪ WAVEFORM SYMMETRY

Sketching the product of a waveform and the various harmonic components represents a very time-consuming method of determining what Fourier components need to be evaluated. This determination can be accomplished much more rapidly by analyzing the symmetry form of the waveshape.

A function is said to have **even symmetry** if $f(x) = f(-x)$. A few examples of this form are shown in Figure 5.17. All such waveforms will contain only cosine terms (plus a constant if there is a nonzero average). Additionally, the sum and the product of even functions are also even functions and will also contain only cosine terms—that is, only a_n terms need to be evaluated.

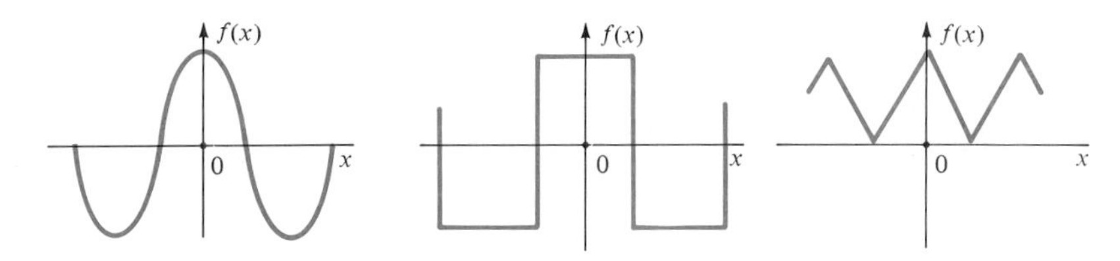

FIGURE 5.17. **Examples of Even Symmetry Waveforms**

A function is said to have **odd symmetry** if $f(x) = -f(-x)$. A few examples of odd functions are shown in Figure 5.18. The sum of two odd functions remains an odd function. The product of two odd functions is an even function. The addition of a constant removes the odd nature of a function.

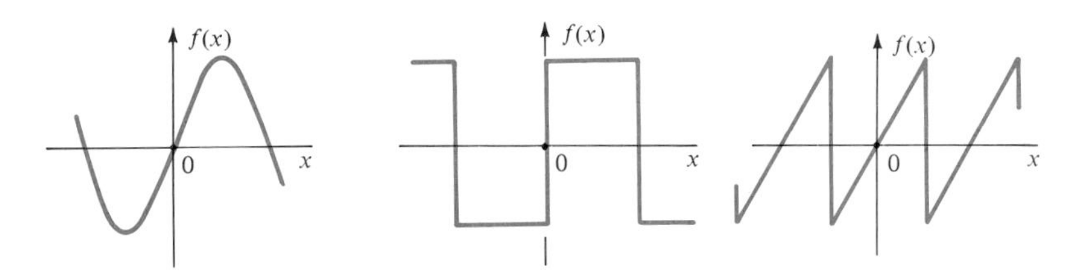

FIGURE 5.18. **Examples of Odd Symmetry Waveforms**

Odd waveforms contain only sine terms. The waveform may become odd upon subtracting out its average value in which case only the sine terms need to be evaluated.

A waveform is said to have **half-wave symmetry** if $f(x) = -f[x + (T/2)]$, where T is the period (Figure 5.19). For such waveforms only odd harmonics are present, but the series will contain both sine and cosine terms unless the function is also odd or even. In either case, a_n and b_n are equal to zero for $n = 2, 4, 6, \ldots$. Half-wave symmetry may arise upon subtracting the average value.

Note that in some cases we can change whether the function is odd or even by shifting the location of the vertical axis.

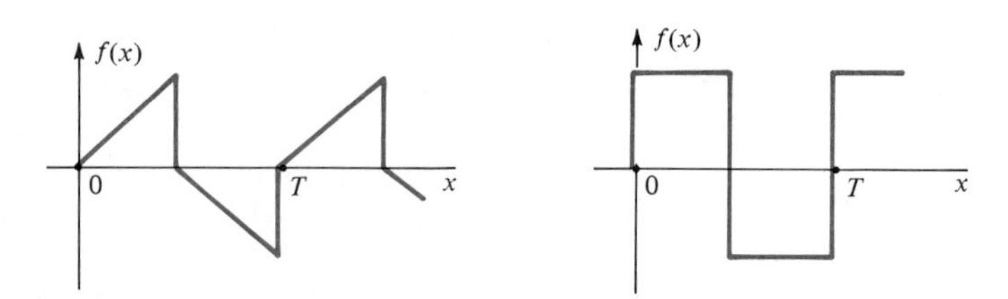

FIGURE 5.19. **Examples of Half-Wave Symmetry**

EXAMPLE 5.3 By means of symmetry conditions, determine the finite Fourier terms for the waveform in Figure 5.20a.

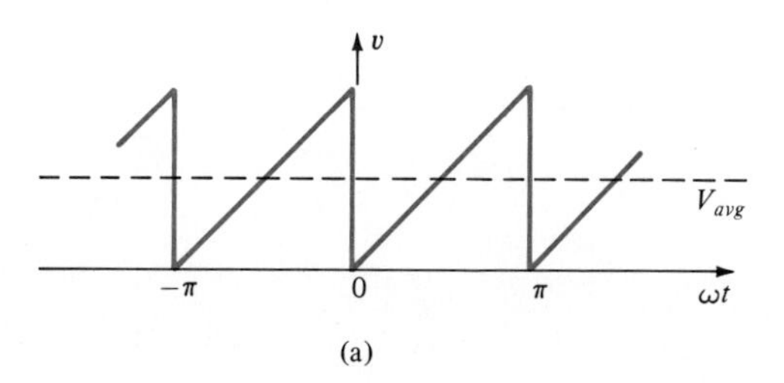

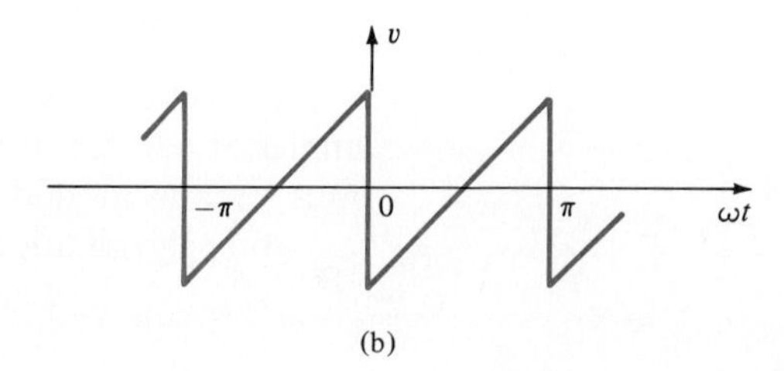

FIGURE 5.20

Solution: After subtracting out the average value, we have odd symmetry (Figure 5.20b)—that is, $f(t) = -f(-t)$. Therefore, only sine terms will arise.

FOCUS ON PROBLEM SOLVING: Fourier Analysis

1. Determine whether the periodic waveform has a zero average. If so, $a_0 = 0$.

2. Analyze the waveform for possible symmetry: Odd waveforms, where $f(x) = -f(-x)$, contain only sine terms. Even waveforms, where $f(x) = f(-x)$, contain only cosine terms (plus a constant if there is a nonzero average). Half-wave symmetry, where $f(x) = -f[x + (T/2)]$, will lead to the presence of odd harmonics only, but both sine and cosine terms will be present unless the function is also odd or even. Sometimes, the

waveform may be made odd by removing a constant. The location of the vertical axis can alter a waveform's being odd or even.

3. If the average is not zero, we have

$$a_0 = \frac{1}{T}\int_0^T f(t)\,dt \qquad \text{or} \qquad a_0 = \frac{1}{2\pi}\int_0^{2\pi} f(\omega t)\,d(\omega t)$$

(Sometimes the Fourier expansion is presented with the constant term being $a_0/2$, in which case the preceding integrals should be multiplied by 2. In Section 5.6 we define the constant in this manner because it leads to a simpler mathematical expression.)

4. The values of a_n and b_n (with $n \neq 0$) may be obtained by using the following:

$$a_n = \frac{2}{T}\int_0^T f(t)\cos n\omega t\,dt \qquad \text{or} \qquad a_n = \frac{1}{\pi}\int_0^{2\pi} f(\omega t)\cos n\omega t\,d(\omega t)$$

$$b_n = \frac{2}{T}\int_0^T f(t)\sin n\omega t\,dt \qquad \text{or} \qquad b_n = \frac{1}{\pi}\int_0^{2\pi} f(\omega t)\sin n\omega t\,d(\omega t)$$

The integration should encompass one full period, but the limits are not necessarily confined to being $0 \rightarrow T$ or $0 \rightarrow 2\pi$. They may be $(-T/2) \rightarrow (+T/2)$, $-\pi \rightarrow +\pi$, or any other full period that simplifies the integration.

EXAMPLE 5.4 Electronic circuits generally require dc power for proper operation. The first step in converting ac into dc is the elimination of alternate half-cycles of an ac sinusoidal waveform, as shown in Figure 5.12. Compute the Fourier series applicable to this waveform using a peak value of V_{max}.

Solution: Since the waveform shows no symmetry, both sine and cosine terms are to be expected. The value of a_0 is obviously not zero and must be evaluated. We have

$$a_0 = \frac{1}{2\pi}\int_0^{\pi} V_{max}\sin\omega t\,d(\omega t) = \frac{V_{max}}{2\pi}\left(-\cos\omega t\right)_0^{\pi} = \frac{V_{max}}{2\pi}\left(1 + 1\right)$$
$$= \frac{V_{max}}{\pi}$$

(The integration limit in this case extends from 0 to π rather than 2π because the function itself is zero between π and 2π.)

$$a_n = \frac{1}{\pi}\int_0^{\pi} V_{max}\sin\omega t\cos n\omega t\,d(\omega t)$$
$$= \frac{V_{max}}{\pi}\left\{-\frac{1}{2}\left[\frac{\cos(1-n)\omega t}{1-n} + \frac{\cos(1+n)\omega t}{1+n}\right]\right\}_0^{\pi}$$
$$= \frac{V_{max}}{\pi(1-n^2)}(\cos n\pi + 1)$$

When n is even, $a_n = 2V_{max}/[\pi(1 - n^2)]$. When n is odd, $a_n = 0$. a_1 must be evaluated separately because of the singularity in the denominator:

$$a_1 = \frac{1}{\pi}\int_0^{\pi} V_{max} \sin \omega t \cos \omega t \, d(\omega t) = \frac{V_{max}}{\pi}\int_0^{\pi} \frac{1}{2} \sin 2\omega t \, d(\omega t) = 0$$

$$\begin{aligned} b_n &= \frac{1}{\pi}\int_0^{\pi} V_{max} \sin \omega t \sin n\omega t \, d(\omega t) \\ &= \frac{V_{max}}{\pi}\left[\frac{\sin(1 - n)\omega t}{2\,(1 - n)} - \frac{\sin(1 + n)\omega t}{2(1 + n)}\right]_0^{\pi} \\ &= \frac{V_{max}}{\pi(1 - n^2)}(0) = 0 \end{aligned}$$

Here again, because of a singularity, b_1 must be evaluated separately. We have

$$\begin{aligned} b_1 &= \frac{1}{\pi}\int_0^{\pi} V_{max} \sin^2 \omega t \, d(\omega t) = \frac{V_{max}}{\pi}\left(\frac{\omega t}{2} - \frac{\sin 4\omega t}{4}\right)_0^{\pi} \\ &= \left(\frac{V_{max}}{\pi}\right)\left(\frac{\pi}{2}\right) = \frac{V_{max}}{2} \end{aligned}$$

Therefore, the Fourier expansion of a half-wave rectified sine wave takes the following form:

$$\begin{aligned} f(t) = \frac{V_{max}}{\pi}\Big(1 &+ \frac{\pi}{2} \sin \omega t - \frac{2}{3} \cos 2\omega t - \frac{2}{15} \cos 4\omega t \\ &- \frac{2}{35} \cos 6\omega t - \cdots\Big) \end{aligned}$$

The fourth harmonic term verifies the statement in Example 5.2.

5.5 ■ DISTORTED SINE WAVES

Electronic amplifiers are not perfectly linear devices. The application of a sinusoidal signal for amplification will invariably result in the appearance of some amplitude distortion in the output. Since it is usually only the first new harmonics that have a significant magnitude, we turn to a consideration of their computation. A sinusoidal variation of current is shown in Figure 5.21. (In actuality we depict it as a cosine.) We presume the distortion is such that the magnitude of the positive swing is not equal to that in the negative direction. Since this is an even function—that is, $f(t) = f(-t)$—all sine terms are zero, and we have

$$i = I_c + A_0 + A_1 \cos \omega t + A_2 \cos 2\omega t + A_3 \cos 3\omega t + A_4 \cos 4\omega t$$

At $\omega t = 0°$, we have

$$I_{max} = I_c + A_0 + A_1 + A_2 + A_3 + A_4$$

At $\omega t = 60°$, we have

$$I_x = I_c + A_0 + \frac{A_1}{2} - \frac{A_2}{2} - A_3 - \frac{A_4}{2}$$

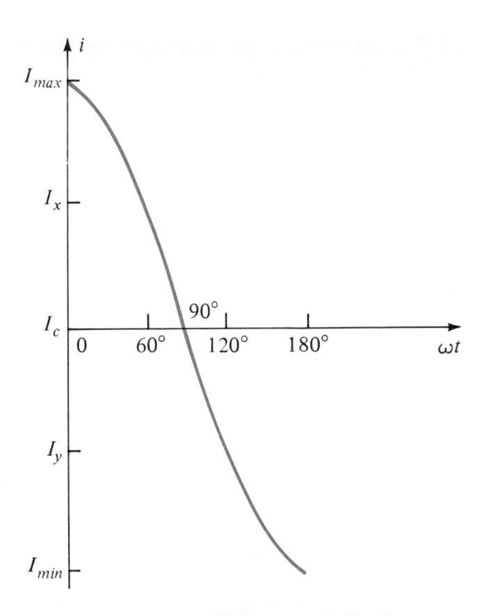

FIGURE 5.21. Segment of Cosine Current Waveform with Values Used in Distortion Computation

At $\omega t = 90°$, we have

$$I_b = I_c + A_0 - A_2 + A_4$$

At $\omega t = 120°$, we have

$$I_y = I_c + A_0 - \frac{A_1}{2} - \frac{A_2}{2} + A_3 - \frac{A_4}{2}$$

At $\omega t = 180°$, we have

$$I_{min} = I_c + A_0 - A_1 + A_2 - A_3 + A_4$$

Elimination among the last five equations can be used to obtain expressions for the various coefficients, and we may thereby determine the magnitudes of the fundamental and the various harmonics. Thus,

$$\begin{aligned} A_0 &= \tfrac{1}{6}(I_{max} + I_{min}) + \tfrac{1}{3}(I_x + I_y) - I_c \\ A_1 &= \tfrac{1}{3}(I_{max} - I_{min}) + \tfrac{1}{3}(I_x - I_y) \\ A_2 &= \tfrac{1}{4}(I_{max} + I_{min}) - \tfrac{1}{2}I_c \\ A_3 &= \tfrac{1}{6}(I_{max} - I_{min}) - \tfrac{1}{3}(I_x - I_y) \\ A_4 &= \tfrac{1}{12}(I_{max} + I_{min}) - \tfrac{1}{3}(I_x + I_y) + \tfrac{1}{2}I_c \end{aligned} \quad (5.8)$$

The distortion percentage due to a specific harmonic is obtained by using

$$D_2 = \frac{A_2}{A_1} \times 100\% \qquad D_3 = \frac{A_3}{A_1} \times 100\% \qquad D_4 = \frac{A_4}{A_1} \times 100\% \quad (5.9)$$

The total harmonic distortion is defined as the ratio

$$D = \frac{\sqrt{A_2^2 + A_3^2 + A_4^2 + \cdots}}{A_1} \times 100\% \quad (5.10)$$

EXAMPLE 5.5

Assume in the process of amplification that the positive half-cycle of a sine wave is amplified to a lesser degree than the negative half-cycle and leads to the following current values: $I_{max} = 3.9$ mA, $I_x = 2.95$ mA, $I_c = 2.0$ mA, $I_y = 1.0$ mA, and $I_{min} = 0$ mA. Compute the various distortion percentages.

Solution:

$$\begin{aligned} A_0 &= \tfrac{1}{6}(3.9 + 0.0) + \tfrac{1}{3}(2.95 + 1.0) - 2.0 = -0.033 \\ A_1 &= \tfrac{1}{3}(3.9 - 0.0) + \tfrac{1}{3}(2.95 - 1.0) = 1.95 \\ A_2 &= \tfrac{1}{4}(3.9 + 0.0) - \tfrac{1}{2}(2.0) = -0.025 \\ A_3 &= \tfrac{1}{6}(3.9 - 0.0) - \tfrac{1}{3}(2.95 - 1.0) = 0.0 \\ A_4 &= \tfrac{1}{12}(3.9 + 0.0) - \tfrac{1}{3}(2.95 + 1.0) + \tfrac{1}{2}(2) = 0.0083 \end{aligned}$$

$$D_2 = \frac{0.025}{1.95} \times 100\% = 1.28\%$$

$$D_3 = 0\%$$

$$D_4 = \frac{0.0083}{1.95} \times 100\% = 0.43\%$$

When subjected to a Fourier analysis, the asymmetric distortion of a sine wave, as illustrated in Example 5.5, leads to a major contribution in the second harmonic. Symmetric distortion of a sine wave makes a major contribution to the third harmonic.

5.6 ■ EXPONENTIAL FOURIER SERIES

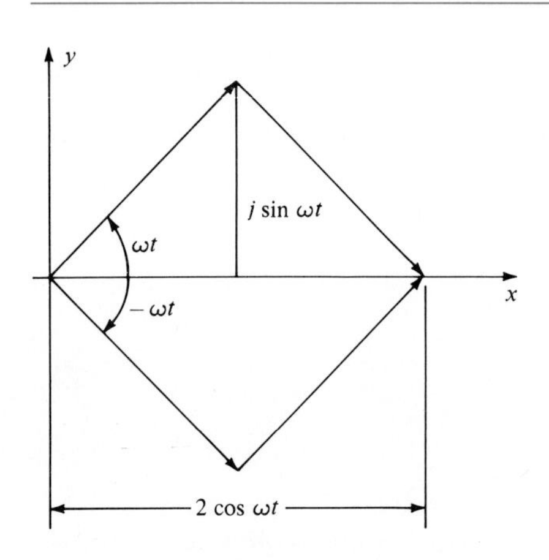

FIGURE 5.22. Real and Imaginary Phasor Components at Positive and Negative Angles

The exponential $e^{j\omega t}$ represents a unit phasor rotating in a counterclockwise direction with an angular velocity ω. At a particular time t, the phasor lies at an angle ωt relative to the $+x$-axis and can be resolved into real and imaginary components (Figure 5.22). We have

$$e^{j\omega t} = \cos \omega t + j \sin \omega t \tag{5.11}$$

It is also possible to have the unit phasor rotate in the clockwise direction, leading to a negative angle. We have

$$e^{-j\omega t} = \cos \omega t - j \sin \omega t \tag{5.12}$$

An alternating voltage, $v = V \cos \omega t$, can then be seen to be expressible as

$$v = V \cos \omega t = V\left(\frac{e^{j\omega t} + e^{-j\omega t}}{2}\right) \tag{5.13}$$

where adding Eqs. 5.11 and 5.12 and dividing by 2 leads to Eq. 5.13. A like result arises from the vector addition shown in Figure 5.22 and verifies that ac voltages incorporate both positive and negative frequencies. In a similar manner, mathematically, one may obtain the equivalence:

$$\sin \omega t = \frac{e^{j\omega t} - e^{-j\omega t}}{2j} \tag{5.14}$$

In exponential form the Fourier expansion of a function is

$$f(t) = \frac{a_0}{2} + a_1\left(\frac{e^{j\omega t} + e^{-j\omega t}}{2}\right) + a_2\left(\frac{e^{j2\omega t} + e^{-j2\omega t}}{2}\right) + \cdots$$
$$+ b_1\left(\frac{e^{j\omega t} - e^{-j\omega t}}{2j}\right) + b_2\left(\frac{e^{j2\omega t} - e^{-j2\omega t}}{2}\right) + \cdots$$

Upon rearranging, we obtain

$$f(t) = \cdots + \left(\frac{a_2}{2} - \frac{b_2}{2j}\right)e^{-j2\omega t} + \left(\frac{a_1}{2} - \frac{b_1}{2j}\right)e^{-j\omega t} + \frac{a_0}{2}$$
$$+ \left(\frac{a_1}{2} + \frac{b_1}{2j}\right)e^{j\omega t} + \left(\frac{a_2}{2} + \frac{b_2}{2j}\right)e^{j2\omega t} + \cdots \tag{5.15}$$

One may then define a series of complex coefficients as

$$c_n = \tfrac{1}{2}(a_n - jb_n) \qquad c_{-n} = \tfrac{1}{2}(a_n + jb_n) \tag{5.16}$$

$$f(t) = +c_{-2}e^{-j2\omega t} + c_{-1}e^{-j\omega t} + c_0 + c_1 e^{j\omega t} + c_2 e^{j2\omega t} + \cdots \tag{5.17}$$

Multiplication of the two sides of this equation by $e^{-jn\omega t}$, as well as integration over a full period, causes all definite integrals on the right to be zero except for $\int_0^{2\pi} c_n \, d(\omega t)$, which has the value $2\pi c_n$. Therefore,

$$c_n = \frac{1}{2\pi}\int_0^{2\pi} f(t)e^{-jn\omega t}\,d(\omega t) \tag{5.18}$$

or, alternatively,

$$c_n = \frac{1}{T}\int_0^{T} f(t)e^{-jn2\pi t/T}\,dt \tag{5.19}$$

$$f(t) = \sum_{n=-\infty}^{\infty} c_n e^{jn\omega t} \tag{5.20}$$

This equation constitutes the exponential Fourier series. It is compact and easily differentiated and integrated and yet contains all the prior information. Note that corresponding positive and negative *c*s are complex conjugates—that is,

$$c_n = c_{-n}^{*} \tag{5.21}$$

Since energy (or power) is proportional to the square of the amplitude, the relative energy associated with the *n*th harmonic is

$$a_n^2 + b_n^2 = 4c_n c_n^{*} \tag{5.22}$$

The Fourier series leads to discrete frequencies. The repetitive function $f(t)$ is broken down into its harmonics, and c_n gives the amplitude of the *n*th harmonic. As the repetition frequency of the waveform diminishes—that is, as T gets larger—the spacing between the harmonics gets smaller, ultimately becoming a continuum of frequencies. The function $f(t)$ in this case takes on the form of an integral. One may then deal with the spectrum associated with isolated pulses.

Rather than restricting the analysis to one cycle (or a fraction of a cycle), it often proves convenient to analyze a waveform extending over a number of cycles. This procedure is particularly useful under the following conditions: For a sinusoidal waveform to represent a single frequency (ω_0), in theory it must extend in time from $-\infty$ to $+\infty$. But in any practical situation, the waveform must have a starting time. Therefore, initially, there is a distribution of frequencies surrounding ω_0, and only after many cycles have transpired does the frequency become rather specifically ω_0. This initial distribution is that associated with the transient circuit behavior.

EXAMPLE 5.6

Consider a cosine waveform of amplitude unity, extending from $-t_1$ to t_1. (See Figure 5.23.) If $f' = 100$ Hz, plot the spectrum for $t_1 = 0.1$ s and for $t_1 = 0.25$ s.

Solution:

$$f(t) = \cos 2\pi f' t$$

$$F(f) = K\int_0^{t_1} f(t)\cos 2\pi f t\,dt$$

(*Note:* Since we are only interested in noting how the frequency distribution changes and not its absolute values, we simply indicate a constant K ahead of the integral rather than a specific value.)

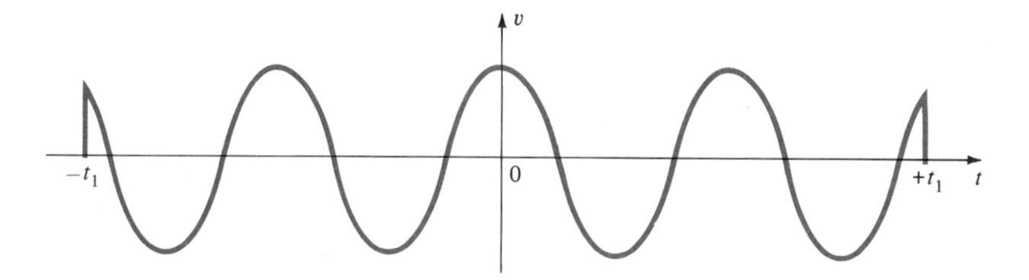

FIGURE 5.23

$$F(f) = K \int_0^{t_1} \cos \omega' t \cos \omega t \, dt$$

$$= K' \left[\frac{\sin(\omega' - \omega)t}{\omega' - \omega} + \frac{\sin(\omega' + \omega)t}{\omega' + \omega} \right] \Bigg|_0^{t_1}$$

The second term represents a very rapidly oscillating quantity whose integrated value will be very small compared to that from the first term. The evaluation, then, resolves itself into an evaluation of the first term:

$$F(f) \simeq K' \frac{\sin 2\pi(f' - f)t_1}{2\pi(f' - f)}$$

By setting $K' = 1$ and $f' = 100$ Hz and allowing f to take on a succession of values from 90 to 110 Hz with $t_1 = 0.1$ s, evaluation of $F(f)$ leads to the spectral distribution shown in Figure 5.24. A comparable evaluation for $t_1 = 0.25$ s leads to the narrower spectral distribution shown in the same figure. In the limit, when $t_1 = \infty$, the spectral distribution becomes a single frequency (100 Hz).

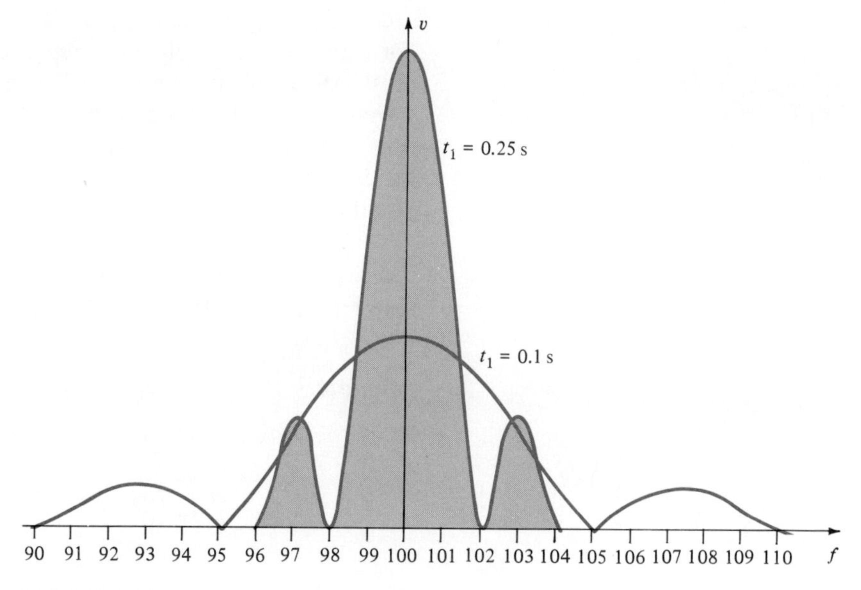

FIGURE 5.24

SUMMARY

5.1 Introduction

- The leading and trailing edges of a periodic signal make up the higher harmonic components. The low-frequency components go to make up the horizontal portions of the signal.

5.2 Trigonometric Fourier Series

- Any periodic signal can be resolved into a constant average plus a number of harmonically related frequencies termed the Fourier series.

5.4 Waveform Symmetry

- Fourier evaluation is simplified by making use of waveform symmetry. Odd functions contain only sine terms and even functions, only cosine terms.

5.5 Distorted Sine Waves

- The unequal reproduction of positive and negative half-cycles of a sinusoid gives rise to a dominant second harmonic distortion. Symmetrical distortion of positive and negative half-cycles leads to the dominance of the third harmonic.

5.6 Exponential Fourier Series

- The commencement of a sinusoidal waveform of ω_0 is accompanied by the presence of a number of frequencies adjoining ω_0. With the persistence of the waveform (assuming constant amplitude), the transients disappear and finally result in the exclusive presence of frequency ω_0.

KEY TERMS

even symmetry: $f(x) = f(-x)$

full-wave rectified signal: the periodic waveform resulting from the sign reversal of alternate half-cycles of a sine wave

fundamental: the lowest frequency present in a periodically repeating waveform

half-wave rectified signal: the periodically repeating waveform resulting from the excision of alternate half-cycles of a sine wave

half-wave symmetry: $f(x) = -f[x + (T/2)]$, where T is the period

harmonic: starting with $n = 2$, integral values times the fundamental frequency (In musical terms, the first overtone equals the second harmonic, and so on.)

odd symmetry: $f(x) = -f(-x)$

synthesis: the combining of various harmonically related waves to form a periodically repeating waveform

SUGGESTED READINGS

Van Vleck, E. B., "The Influence of Fourier's Series upon the Development of Mathematics," SCIENCE, vol. 39, pp. 113–124, 1914. *Some historical aspects of the development of Fourier analysis.*

Edminister, Joseph A., SCHAUM'S OUTLINE OF THEORY AND PROBLEMS OF ELECTRIC CIRCUITS, 2nd ed., Chap. 12. New York: McGraw-Hill, 1983. *Extensive series of worked examples involving Fourier waveform analysis.*

Boylestad, Robert, and Louis Nashelsky, ELECTRONIC DEVICES AND CIRCUIT THEORY, 3d ed., pp. 392–403. Englewood Cliffs, NJ: Prentice-Hall, 1982. *Distortion calculations, with numerical examples.*

Karlekar, B. V., and R. M. Desmond, HEAT TRANSFER, 2nd ed., pp. 152–163. St. Paul, Minn.: West Publishing Co., 1982. *Examples of Fourier analysis applied to the solution of heat conduction problems, which was the first application of this method.*

Bell, E. T., MEN OF MATHEMATICS, pp. 191–204. New York: Simon & Schuster, 1937. *Biographical material about Joseph Fourier.*

EXERCISES

1. A square wave consists of a fundamental and its ______ harmonics.

2. What is the fundamental frequency of a square wave whose period is 16.7 ms? What is its second harmonic and its third harmonic?

3. **a.** If a 16.7 ms square wave is applied to an amplifier with a 30 kHz bandwidth, approximately what is the highest harmonic passed without suffering significant reduction?

b. Would you expect a rather accurate square wave output?

c. What if the square wave period were 16.7 μs?

4. **a.** What is the value of a_0 in the Fourier analysis of a 60 Hz sinusoidal waveform?

b. What is the value of a_0 for a square wave whose limits are -1 and $+1$? With limits of 0 and $+1$?

c. If the value of a_0 for a half-wave rectified sinusoid was $1/\pi$, what is the value for a full-wave rectified sinusoid?

5. What do the various values of a and b in the Fourier analysis represent?

6. In general, ______ coefficients in Fourier analysis are associated with cosine terms and ______ coefficients with sine terms.

7. **a.** A function with $f(x) = -f(-x)$ is said to be an ______ function; such waveforms contain only ______ terms in the Fourier expansion.

b. A function with $f(x) = f(-x)$ is said to be an ______ function; such waveforms contain only ______ terms in their Fourier analysis.

8. Sketch some waveforms with even symmetry.

9. Sketch some waveforms with odd symmetry.

10. Certain waveforms can be either odd or even depending on the location of the vertical axis. Show how the symmetry of a square wave can be altered by shifting the vertical axis.

11. A waveform with neither odd nor even symmetry will contain ______ terms.

12. A sawtooth waveform with the horizontal axis connecting its lower extremities results in neither odd nor even symmetry. Show how it may be made to exhibit odd symmetry by subtracting out the average value.

13. The Fourier analysis of a half-wave rectified sinusoid, with the vertical axis at the leading edge of the sinusoid, exhibits the components indicated in Eq. 5.7. If the vertical axis were shifted to the peak of the sinusoid portion, how would this alter the component content?

14. **a.** The uneven amplification of positive and negative half-cycles of a sinusoid lead to distortion whose major contribution is at the ______ harmonic.

b. A sinusoid subjected to symmetrical distortion (for example, a nonlinear amplification of both of the peaks) will lead to a distortion whose major contribution is at the ______ harmonic.

15. For a sinusoidal waveform to represent a single frequency, in theory it must extend over an ______ time interval.

16. The initial distribution of frequencies surrounding what will, with the passage of time, turn out to be a sinusoid of frequency ω_0 represents the frequencies associated with ______ circuit behavior.

PROBLEMS

5.1

a. Rather than eliminating the negative half-cycles as is done with a half-wave rectified waveform, one may instead invert the negative half-cycles, resulting in the waveform in Figure 5.25, termed a full-wave rectified signal. This yields a higher average (hence greater dc) voltage. Calculate the Fourier components for this waveform.

b. Calculate the Fourier components for a full-wave rectified waveform with its peak centered at $\omega t = 0$ (Figure 5.26).

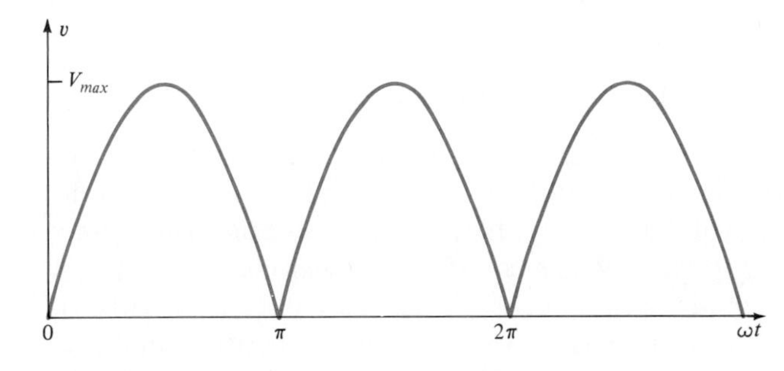

FIGURE 5.25

5.2

a. In the text the Fourier series for a square wave with $a_0 = 0$ was developed. What is the series for a square wave centered on the value $\frac{1}{2}$, as shown in Figure 5.11?

b. How would the series be altered if the square wave were symmetric about $\omega t = 0$?

5.3

a. Compute the Fourier series for a triangular wave of magnitude V_{max} (Figure 5.27). Draw a line spectrum and compare

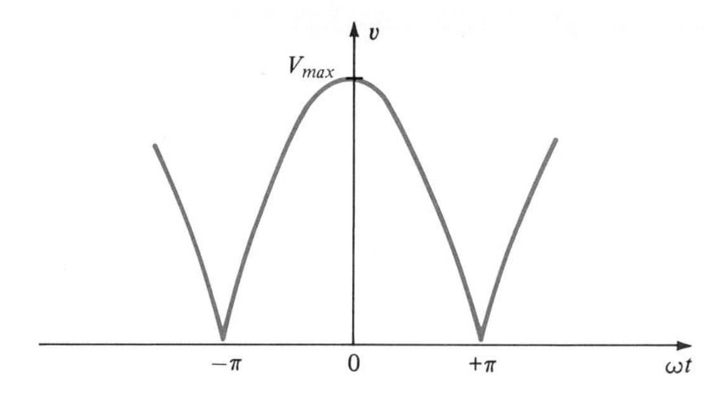

FIGURE 5.26

the results with a line spectrum of a square wave. (A line spectrum is a bar graph of the harmonics.)

b. Solve the same problem with the vertical axis moved to the left by $\omega t = \pi$.

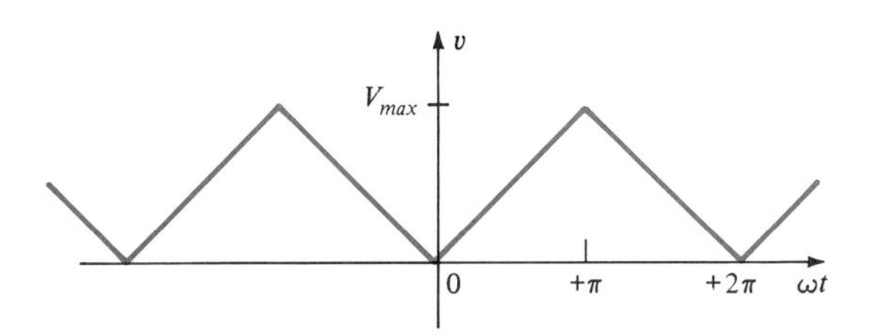

FIGURE 5.27

5.4

a. In Example 5.4 we developed the major distortion terms of an asymmetrically distorted sine wave. Calculate the first five terms of the Fourier analysis for a symmetrically distorted sine wave according to the following specifications: I_{max} = 3.9 mA, I_x = 3.0 mA, I_c = 2.0 mA, I_y = 1.0 mA, I_{min} = 0.1 mA. Compute the various distortion percentages.

b. Upon comparing the two cases, what general statement can you make concerning the distortion?

5.5

a. All the Fourier waveforms thus far considered seem to contain only every other harmonic. Are there any that can serve as a source that provides each harmonic? Such a waveform, for example, is needed in musical synthesizers, electronic organs, etc. Thus, a given generator can furnish the corresponding harmonic in each succeeding higher octave. A waveform that will do this is the sawtooth (Figure 5.28). This is also the waveshape that is used to provide the horizontal scan on a TV picture tube (the gradual sloping portion causes the beam to be pulled from left to right). At the right extremity the voltage rapidly drops to a negative value, and the beam returns to the left edge where it again starts its relatively slow journey to the right. If there are 525 scan lines to a TV picture (as there are in the United States) and the framing time—that is, the time to present one complete picture—is $\frac{1}{30}$ s, what is the repetition rate of the horizontal sawtooth in the TV set?

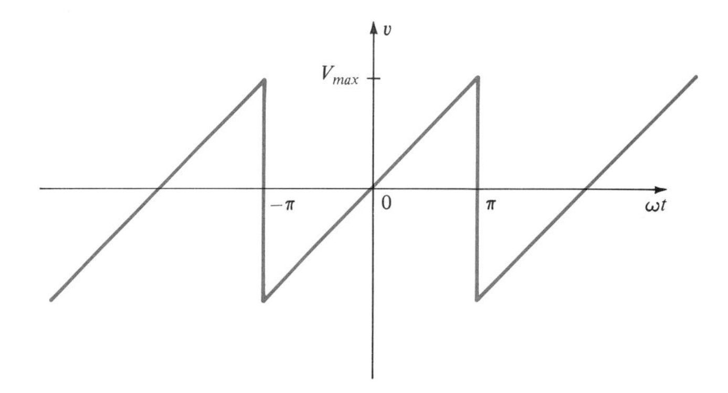

FIGURE 5.28

b. Show that the Fourier analysis of the sawtooth contains all the harmonics.

5.6

Write out the Fourier series as a summation for the following (with $a_0 = 0$):

a. A square wave

b. A triangular wave

Qualitatively explain the reason for the difference between parts a and b.

5.7

A square wave is applied to a low-pass filter of the form shown in Figure 3.26. Recognizing that the sharp leading and trailing edges are due to the higher harmonics that will effectively be "shorted out" at the output, sketch the expected appearance of the output waveform.

5.8

The output of a high-pass filter (Figure 3.25) to which a square wave is applied should have the fundamental and low-frequency harmonics attenuated to a considerable degree because of the voltage drops across the capacitor. Sketch the expected output under the circumstances.

5.9

A series of very short-duration voltage pulses with a repetition frequency f is applied to a series LCR circuit whose resonant frequency is also f. What is the appearance of the current waveform through the circuit? What is the nature of the frequency content of the impressed voltage?

5.10

The steady-state response of a network to which a nonsinusoidal periodic waveform is applied may be found by summing the individual responses to each of the Fourier components. Thus, the application of a triangular waveform (Figure 5.29) whose Fourier voltage components are

$$v(t) = \frac{8}{\pi^2}\left(\cos \omega t + \frac{1}{3^2} \cos 3\omega t + \frac{1}{5^2} \cos 5\omega t + \cdots\right)$$

may be used to obtain the Fourier current elements by dividing each voltage component by the circuit impedance. Determine the Fourier current waveform if the preceding voltage variation is applied to a series circuit consisting of a 1 Ω resistance and a 1 H inductance, taking $\omega = 1$ for the sake of simplicity. Sketch the result after summing the first few Fourier components.

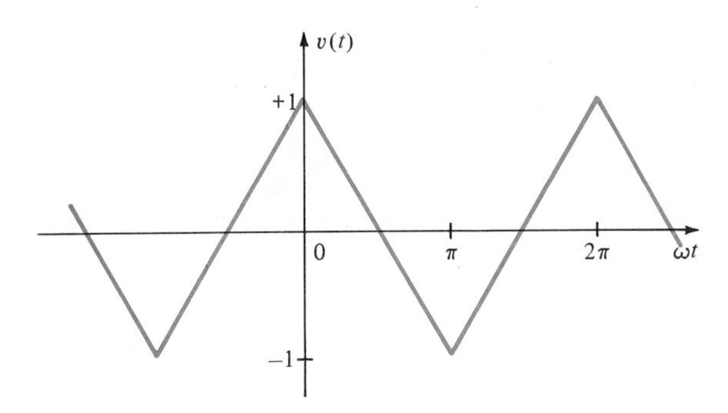

FIGURE 5.29

5.11

Evaluate the following at 22.5° intervals, sketch the results, and identify the limiting waveshape:

$$f(t) = -1 \sin \omega t - \tfrac{1}{2} \sin 2\omega t - \tfrac{1}{3} \sin 3\omega t - \cdots$$

5.12

Consider a function to be separable into even and odd parts:

$$f(t) = f_e(t) + f_0(t)$$

where $f_e(t) = f_e(-t)$

$$f_0(t) = -f_0(-t)$$

Adding yields

$$f(t) + f(-t) = f_e(t) + f_e(-t) + f_0(t) + f_0(-t) = 2f_e(t)$$

$$f_e(t) = \frac{f(t) + f(-t)}{2} \qquad \text{(I)}$$

Subtracting yields

$$f(t) - f(-t) = f_e(t) - f_e(-t) + f_0(t) - f_0(-t) = 2f_0(t)$$

$$f_0(t) = \frac{f(t) - f(-t)}{2} \qquad \text{(II)}$$

a. Identify the following function:

$$f(t) = 0 \qquad (-\pi < \omega t < 0)$$
$$f(t) = \sin \omega t \qquad (0 < \omega t < \pi)$$

b. Sketch the even part of this function using Eq. I in the preceding.

c. Sketch the odd part of this function using Eq. II in the preceding.

d. Combine the Fourier components for parts b and c and show that they equal that of the function in part a.

5.13

The Fourier components of two unit amplitude square waves, the first with unity amplitude centered at $\omega t = 0$ and the other with zero centered at $\omega t = 0$, may be used to explain the operation of stereo broadcasting (Figure 5.30). The first square wave (Figure 5.30a) periodically samples the left sound channel whose instantaneous voltage is v_L. The second square wave (Figure 5.30b) samples the right sound channel whose instantaneous value is v_R.

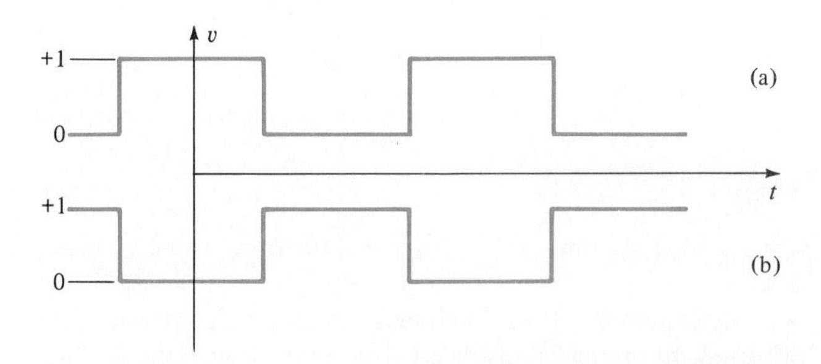

FIGURE 5.30

a. Show by addition and subtraction how the left *plus* right channel information becomes distinct from the left *minus* right channel information. (Only the dc and fundamental terms need to be retained.)

b. In terms of low- and high-pass filters, how may one discriminate between the passage of the $L + R$ and $L - R$ information?

c. Having separated $L + R$ and $L - R$, at the receiving end, how may one recover the L and R information separately?

5.14

The $L - R$ information of a stereo system (see Problem 5.13) is used to vary the amplitude of the cos ωt signal. In FM broadcasting the frequency corresponding to ω is 38 kHz—that is, $2\pi(38 \times 10^3) = \omega$. To receive the information at the receiver, one must have a 38 kHz signal that is accurately synchronized with that at the transmitter. But for technical reasons, what is actually transmitted is a 19 kHz signal. How may

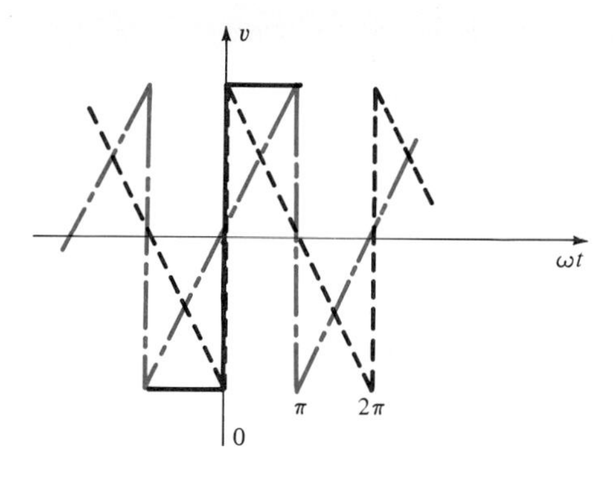

FIGURE 5.31

one convert this 19 kHz signal into the required 38 kHz synchronization?

5.15

a. With regard to the sawtooth depicted in Figure 5.28, how are the Fourier components altered if the sawtooth is one that is diminishing in magnitude as it passes through zero rather than increasing as in Figure 5.28?

b. If one adds the two sawtooth waveforms, the result should be a square wave (Figure 5.31). Show that adding the two sets of Fourier components leads to a series that represents a square wave.

PART

ELECTRONIC CONCEPTS: Analog Signals

6 ELECTRICAL CONDUCTION

OVERVIEW

Having examined linear electrical processes in some detail, we turn now to nonlinear considerations. In a linear electrical circuit, no change occurs in the applied frequency. It may be attenuated or subject to a phase change, but it remains the same. Such is not the case with nonlinear devices. In this chapter we make use of Fourier analysis to study the nature of such frequency changes.

Particularly interesting are the changes accompanying the creation of radio signals, and it is with that in mind that we undertake the study of amplitude modulation (AM), as used in broadcasting, as well as frequency modulation (FM). The techniques we will study will provide us with an opportunity to see how diodes may be used to extract information from radio waves in the form of sound, music, and so forth, and will also be found useful with regard to physical measurements—for example, with techniques that employ strain gages.

Conduction processes will play an important role in understanding the operation of bipolar junction transistors (Chapter 7), field effect transistors (Chapter 9), and integrated circuits (Chapter 10). Modulation methods will be encountered again in Chapter 15 (phase-locked loops) and Chapter 17 (measurements).

OUTLINE

6.1 ■ INTRODUCTION

What is electrical resistance? In the case of electrical conduction within a metallic wire, it is often visualized as arising from the collision of electrons with atoms. This definition is *not*, however, a correct depiction of the resistive process.

To isolate the mechanism involved, we first consider electron interaction with relatively rare atoms such as might exist in a *vacuum*. (Of course, even in the best of vacuums there are always some residual gas atoms.) Having recognized the true nature of resistance in this relatively simple case, we then proceed to extend the concept to a crystalline solid and finally to practical solids such as might be used to make up electronic devices. We conclude with a discussion of what is probably the simplest of such electronic devices, the semiconductor diode.

6.2 ■ THE PHYSICS OF ELECTRICAL RESISTANCE[1]

6.2.1 Evacuated Space Resistance

In Figure 6.1 we depict a few of the atoms that constitute a gas-filled differential volume whose frontal area is a^2 and that has a depth dx. We are interested in the possibilities of a collision between an incoming particle of radius r_1 and the gas atoms whose radii are taken to be r_2. About each atom we draw a (dashed) circle whose radius is $r_1 + r_2$. A collision arises whenever the incident particle strikes within these areas. We presume that there is no significant overlap between these areas, the thickness dx to be small enough. If n represents the number of atoms per unit volume, the number of gas atoms within the volume under consideration is N, where

$$N = na^2\, dx \tag{6.1}$$

The total area enclosed by the dashed circles is a':

$$a' = n\pi(r_1 + r_2)^2 a^2\, dx \tag{6.2}$$

Since the frontal area is a^2, the dashed circles represent a *fractional* area equal to $n\pi(r_1 + r_2)^2\, dx$. Each incoming particle that strikes within a dashed area is removed from the beam, and if the beam current is taken to be J [particles per square centimeter second],[2] within a distance dx the reduction in beam current dJ is:

$$\frac{dJ}{J} = -n\pi(r_1 + r_2)^2\, dx \tag{6.3}$$

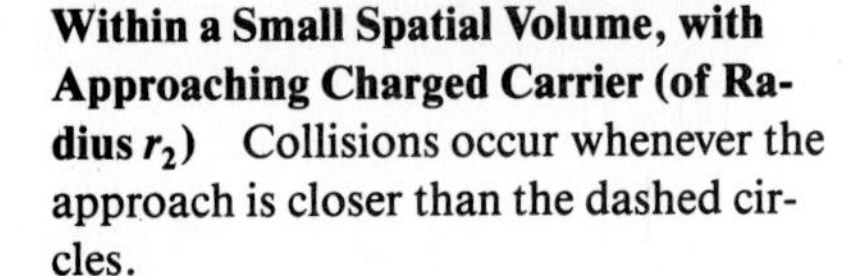

FIGURE 6.1. Atoms (of Radius r_2) Within a Small Spatial Volume, with Approaching Charged Carrier (of Radius r_2) Collisions occur whenever the approach is closer than the dashed circles.

[1]See R. A. Smith, "Physics and the New Electronics," *J. Sci. Inst.*, vol. 34 (1957), 377–382.

[2]We choose to use CGS units because gas pressures are usually expressed in terms of millimeters of mercury.

Upon integrating between 0 and some generalized depth x, we obtain

$$J = J_0 e^{-n\pi(r_1 + r_2)^2 x} \tag{6.4}$$

where J_0 is the incident current density.

When we encounter an exponential law of this sort, we generally select some characteristic distance within which the quantity in question drops to $1/e$ of its original value, where e is the base of the system of natural logarithms. Designating this distance as $\bar{l}$, from Eq. 6.4, we may determine that for the particle under consideration

$$\bar{l} = \frac{1}{n\pi(r_1 + r_2)^2} \tag{6.5}$$

The characteristic distance that leads to the particle density dropping to $1/e$ of its original value is termed the **mean free path.**

A mole of gas, at standard temperature and pressure, has about 2.69×10^{19} molecules/cm^3. According to the perfect gas law, the concentration varies directly with pressure and inversely with temperature. We thus have

$$n = 2.69\left(\frac{p}{760}\right)\left(\frac{273}{T}\right) = 9.68 \times 10^{18}\,\frac{p}{T}\ [\text{molecules/cm}^3] \tag{6.6}$$

where p = gas pressure in millimeters of mercury (mm Hg)

T = absolute temperature

If the incident particle is presumed to be an electron, $r_1 \ll r_2$, and from Eq. 6.5 we see that

$$\bar{l} \sim \frac{1}{n(\pi r_2^2)} \tag{6.7}$$

If the gas pressure is 10^{-3} mm Hg (not a particularly good vacuum but one that can easily be achieved with a mechanical pump), by taking the area of a gas atom to be 10^{-15} cm^2, at room temperature the value of $\bar{l}$ is about 30 cm. If electrons are to travel in an unimpeded manner between various electrodes within an evacuated space (such as prevailed in vacuum tubes), it is necessary that the probability of a collision with the residual air molecules be negligible. At a pressure of 10^{-3} mm Hg, while $\bar{l}$ is fairly large, we would still lose a significant number of electrons from the beam. Therefore, in vacuum tubes the vacuum was made significantly better, about 10^{-8} mm Hg. But for our purposes, let us stick with the 10^{-3} mm Hg figure.

6.2.2 Resistance in Solids

If the block in Figure 6.1 is now taken to represent a solid that we want to utilize in a transistor amplifier, we again need to have the conducting particles pass through the material with little chance of suffering a collision while traveling between the electrodes. Since in a solid there are about 10^{23} atoms/cm^3 and since, again, the atomic area is taken to be about 10^{-15} cm^2, we see that now $\bar{l}$ will be about 10^{-8} cm, which is of the same order as the distance between adjoining atoms. Clearly something is wrong with this approach since it would indicate that electrical conduction through solids, even such as copper, would be minimal.

As long ago as 1928, F. Bloch proposed that in a perfectly regular crystalline solid, with no impurities or defects and no lattice vibrations—that is, at a temperature of 0 K—the electrons should pass through the crystal in an unimpeded fashion. The basis for this statement lies in the fact that an electron exhibits wave-like properties in addition to its particle-like behavior.

When waves travel through an otherwise uniform medium and encounter a discontinuity, a reflected wave arises. What Bloch proved was that an electron wave, similarly, would continue to travel unimpeded through a homogeneous crystal. Only when it encountered impurity atoms, lattice defects, and so forth would its flow be impeded, leading to some of its energy being reflected (scattered), and it was such encounters—*not* collisions with lattice atoms—that constituted resistance to current. Let us go along with this idea and see whether it leads to reasonable results.

Let us return to the solid-state conductor. How pure must we make the crystal if we are to have $\bar{l} \sim 30$ cm as we did in the vacuum case? The answer: about $3 \times 10^{13}/\text{cm}^3$.

Because we have about 10^{23} atoms/cm^3, the maximum impurity concentration we can tolerate is about $3 \times 10^{13}/10^{23}$, or about one impurity atom per 10^{10} atoms. By chemical means we may achieve material purities of about 1 part per million (1 ppm). Thus, we see why the development of the transistor had to await the development of improved methods of crystalline purification. The material technology was not available to make a practical device.

In summary, then, from the point of view of electrical conduction, we can look upon a perfectly regular crystalline solid as if it were a vacuum. Electronic conduction is impeded only to the extent that the crystal's regularity is disturbed by imperfections, impurities, and lattice vibrations. Such defects correspond to residual gas atoms that impede the flow of current through a vacuum.

But it should be stressed that the copper making up the customary wire conductors used in electrical circuitry is far from being a regular crystalline solid. It is polycrystalline, and the grain boundaries cause a scattering of electrons and contribute a significant resistance to the current. On the other hand, materials such as silicon, which are used to fabricate transistors, do closely approximate regular crystalline solids.

6.3 ▪ METALLIC CONDUCTION

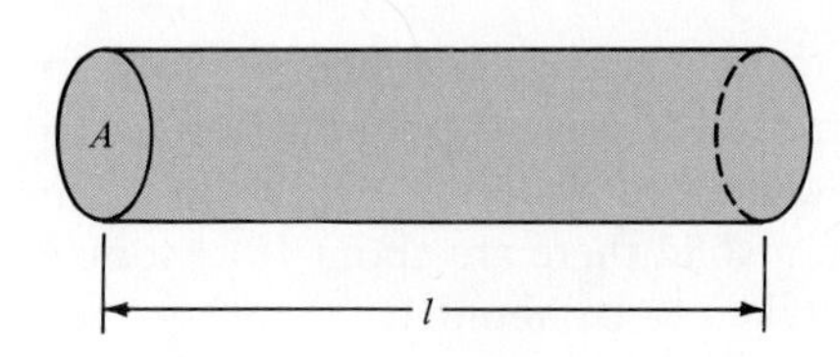

FIGURE 6.2. **Cylindrical Conductor of Length *l* and Cross-Sectional Area *A***

Consider a cylindrical conductor such as the one in Figure 6.2. The resistance is directly proportional to the length and inversely proportional to the area:

$$R = \frac{l}{\sigma A} \tag{6.8}$$

The resistance obviously depends on the material composition, which is represented by σ, the conductivity, whose dimensions are easily determined to be either $(\Omega\cdot\text{cm})^{-1}$ or $(\Omega\cdot\text{m})^{-1}$. Conductivity also depends on the concentration of carriers available for conduction and the ease with which they can move about in the material. In the case of metals, the former is represented by the conducting electron concentration (n) and the latter, by the **mobility** (μ):

$$\sigma = n\mu e \tag{6.9}$$

The quantity e is the charge per electron, numerically equal to 1.9×10^{-19} C.

While the **conductivity** (σ) is a measure of the ease with which a material conducts, the **resistivity** (ρ), the reciprocal of conductivity, is a measure of the opposition to current.

EXAMPLE 6.1

For copper, $n = 9.06 \times 10^{22}$ electrons/cm^3, $\sigma = 5.61 \times 10^5\ (\Omega\cdot\text{cm})^{-1}$, and $e = 1.6 \times 10^{-19}$. What is the value of the mobility? Justify the dimensions of μ being square centimeters per volt second. What is the value of the resistivity?

Solution: Since $\sigma = \mu ne$, we have

$$\mu = \frac{\sigma}{ne} = \left(\frac{5.61 \times 10^5}{\Omega\cdot\text{cm}}\right)\left(\frac{\text{cm}^3}{9.06 \times 10^{22}\ \text{electrons}}\right)\left(\frac{1}{1.6 \times 10^{-19}\ \text{C}}\right)$$

$$= 38.7\ \frac{\text{cm}^2}{\Omega\cdot\text{C}}$$

$$\frac{\text{coulomb}}{\text{second}} = \text{ampere}$$

$$\text{coulomb} = \text{ampere-second}$$

$$\text{ohm ampere} = \text{volt}$$

$$\text{ohm coulomb} = \text{ohm ampere-second} = \text{volt second}$$

$$\frac{\text{centimeters}^2}{\text{ohm coulomb}} = \frac{\text{centimeters}^2}{\text{volt second}}$$

$$\rho = \frac{1}{\sigma} = \frac{1}{5.61 \times 10^5\ (\Omega\cdot\text{cm})^{-1}} = 1.76 \times 10^{-6}\ \Omega\cdot\text{cm}$$

Copper is among the best electrical conductors, and its value of resistivity is frequently used as a basis for comparison to other materials (Table 6.1). While silver has the lowest resistivity in this list, its expense militates against its common usage in favor of copper. As the price of copper has risen, however, there has been a tendency to replace its use as a utility wire with the cheaper aluminum. The lower melting point of the latter, however, constitutes a more serious hazard under overload conditions than does copper. Additionally, the oxide coating that rapidly forms on aluminum and is responsible for its corrosion resistance also constitutes a rather effective insulation material, in contrast to a copper oxide surface, which is electrically conductive. Aluminum oxide may give rise to a serious resistive heating hazard when aluminum and copper are joined, such as at a junction box. Industrial installations of aluminum wiring result in substantial savings, but much greater care must be exercised in such installations in contrast to those employing copper.

TABLE 6.1 Relative Resistivity of Some Metals

Material	Resistivity Compared to Copper
Silver	0.94
Copper (annealed)	1.00
Copper (hard-drawn)	1.03
Gold	1.40
Aluminum (pure)	1.60
Brass	3.70–4.90
Iron (pure)	5.68
Steel	7.60–12.70
Lead	12.80

TABLE 6.2 Some Typical Values of Wire Gauges and Their Resistance

AWG	Diameter [mm]	Resistance [Ω · m]
.	.	.
.	.	.
.	.	.
22	0.644	0.0530
20	0.812	0.0333
18	1.024	0.0209
.	.	.
14	1.628	0.0083
12	2.052	0.0052
.	.	.
.	.	.
1	7.350	0.00041
.	.	.
.	.	.
0000*	11.700	0.00016

*Gauge sizes larger than gauge 1 are, successively, 0, 00, 000, 0000. The latter is the largest gauge size.

Copper wire is available in various diameters, termed *gauges*. The numbers assigned to the gauges in Table 6.2 represent their AWG (American Wire Gauge) values. The stranded wire is used where flexibility is desired. Electronic circuits are generally wired with #22 gauge stranded wire, and household lamp cords use #18 stranded wire. Household wiring employs either #12 or #14 solid wire.

Most materials exhibit a positive temperature coefficient. That is, their resistance increases with temperature (typically because of the more violent atomic vibrations that lead to an increase in electron scattering). This increase with temperature takes place in an approximately linear fashion:

$$R = R_0(1 + \alpha \, \Delta T) \tag{6.10}$$

where R is the resistance at a temperature ΔT degrees above the temperature at which the initial resistance was measured (R_0). (If the wire is cooled, ΔT is a negative number.)

EXAMPLE 6.2

If an iron wire has a resistance of 100 Ω at 25°C, with a temperature coefficient (α) at 25°C equal to 0.006 Ω/Ω·°C, what will be its resistance at 75°C?

Solution:

$$R = R_0(1 + \alpha \, \Delta T) = 100[1 + (0.006)(50)] = 130 \, \Omega$$

Some metallic alloys may be made to yield materials that have a very low temperature coefficient of resistance. One of the best in this regard is *manganin*, an alloy of copper, manganese, and nickel. Its resistivity reaches a peak value between 0°C and 100°C, so that over a range of temperatures near this peak its temperature coefficient is essentially zero. The peak may be moved by changing the composition. For resistance standards the peak is placed at 25°C. Thus, between 20°C and 30°C such resistances have a temperature coefficient that is essentially zero and ±0.00001 Ω/Ω·°C in the adjoining temperature range.

Another frequently encountered resistive material is *nichrome*, widely used for electrical heating. It is its nonoxidizing characteristic at high temperatures that makes it particularly suitable in this regard.

One precaution concerning the computation of metallic resistance should be mentioned: As the frequency of an alternating current is increased, the current progressively becomes more concentrated near the surface of the conductor. Thus, while dc utilizes the entire cross-sectional area of a conductor, ac makes use of only a fractional volume. The resistance of a wire, therefore, depends on the frequency of the applied current and becomes greater as the frequency is increased. This concentration of ac near the conductor's surface is known as the *skin effect*.

6.4 ▪ SEMICONDUCTORS

The mobility of charged carriers within a solid is a measure of the ease with which they move through the material. We might expect that a material such as silicon, which is classed as a **semiconductor**, would exhibit a smaller degree of mobility than a material such as copper, which is classed as a *conductor.* But the mobility of electrons in silicon is 1300 cm^2/V·s, which is much greater than the 32 cm^2/V·s that characterizes copper. Why then is the conductivity of silicon so much less than that of copper?

Conductivity (Eq. 6.9) involves the product of mobility and charge density, and while the electron mobility is high in silicon (because of crystalline regularity, and so on), there are many fewer charged carriers. Because of thermal vibrations, there are some free carriers in silicon that arise due to the rupture of crystalline bonds, but they are few in number. Additional carriers may be created by purposely introducing selected inpurity atoms, but the number of carriers that they introduce still does not equal the number available in a metal such as copper. Thus, while the mobility of carriers in copper is low, the large value of carrier density yields a large value for conductivity. The purposeful introduction of impurity atoms gives arise to **doped** semiconductors.

Unlike metallic conductors in which the number of carriers available for conduction remains fairly constant, as a function of temperature in a semiconductor the number increases in an exponential fashion. The resultant increase in conduction far outweighs the decrease due to carrier collisions at higher temperatures and leads to a negative temperature coefficient—that is, the resistance drops as the temperature increases. This may be approximated by the equation

$$R = Ae^{\beta/T} \tag{6.11}$$

where A = constant largely dependent on the shape of the material

β = constant determined by the material used

T = absolute temperature in K

(There is a temperature below which this equation does not apply. This limit is about $-100°C$.) The rate of change of resistance with temperature, termed the **temperature coefficient of resistance**, is about $-0.04\ \Omega/\Omega \cdot °C$ for semiconductors. A material whose resistance increases with increasing temperature represents a **positive** temperature coefficient, while a decrease in resistance represents a **negative** temperature coefficient.

Thermistors are semiconductor resistors that exhibit negative temperature coefficients, and they may be used to compensate for the more usual positive coefficients one encounters with ordinary resistors. Thus, a properly selected combination of positive and negative resistance can lead to a resistance that is essentially independent of temperature (within a range of temperatures). Commercial thermistors consist of mixtures of materials such as NiO, Mn_2O_3, and Co_2O_3. The large variation of resistance with temperature also makes thermistors useful for temperature measurement.

Heavily doped semiconductors, on the other hand, acquire a positive temperature coefficient because of the decrease in carrier mobility. Such resistors typically have positive coefficients of the order of $+0.007\ \Omega/\Omega \cdot °C$ and may be combined with negative coefficients for purposes of compensation. Semiconductors with positive temperature coefficients are called **sensistors.**

EXAMPLE 6.3 In a sample of silicon semiconductor, at room temperature, the concentration of electrons is $1.5 \times 10^{10}/cm^3$. With a mobility of $1300\ cm^2/V \cdot s$, what is the resistivity due to the electrons? Calculate the resistivity if the doping impurity provides an electron concentration of $2 \times 10^{15}/cm^3$. Compare the values with that for copper.

Solution:

$$\rho = \frac{1}{ne\mu} = \frac{1}{(1.5 \times 10^{10})(1300)(1.6 \times 10^{-19})} = 3.2 \times 10^5\ \Omega\cdot\text{cm}$$

$$= \frac{1}{(2.0 \times 10^{15})(1300)(1.6 \times 10^{-19})} = 2.4\ \Omega\cdot\text{cm}$$

The resistivity for copper is $1.76 \times 10^{-6}\ \Omega\cdot\text{cm}$.

EXAMPLE 6.4

With $\beta = 2500$ K, calculate the temperature coefficient of resistance (α) for a thermistor at 25°C. Compare the value with that for copper.

Solution:

$$R = Ae^{\beta/T}$$

$$\alpha \equiv \frac{dR/dT}{R} = \frac{-\beta}{T^2} = -\frac{2500}{298^2} = -0.028\ \Omega/\Omega\cdot°\text{C}$$

The value for copper is $0.0039\ \Omega/\Omega\cdot°\text{C}$.

6.4.1 Semiconductor Materials

Germanium and silicon are two semiconductor materials that can be made to approximate the flawless, crystalline structures conducting current in solids. While such materials exhibit little opposition to the flow of charged carriers, in their usual fabricated state (termed **intrinsic**) there are few charged carriers to do any conducting. In a vacuum tube a heated filament furnished the electron carriers. In the transistor we have done away with such power-consuming, heat-generating, inefficient producers of charged carriers by introducing controlled quantities of impurities. These impurities lead to the creation of charged carriers.

Both germanium and silicon are atoms with four valence electrons. In crystalline form each atom gives rise to a double bond with adjoining atoms. In Figure 6.3, in addition to such silicon atoms we also show a purposely introduced indium impurity. This is a three-valence atom. Therefore, there is one missing bond (shown as the dashed line). When a voltage is impressed across such a crystal, an electron from an adjoining complete bond will be attracted to fill the vacancy, creating in turn a vacancy to the right, and so forth. Such a vacancy (termed a *hole*) moves toward the negative electrode. It thus behaves as a classical charged carrier and can simply be looked upon as a positive carrier moving through the crystal. Such materials are termed *p-type semiconductors.*

If, rather than three-valence impurities, we introduced, in a controlled fashion, atoms with five valence electrons (such as antimony), there would be more electrons than needed to form the double bonds. As is the case with *p*-type materials, the *n*-type remains electrically neutral since the positive charge of the ion impurity equals the number of negative electrons that accompany it. Upon applying a voltage across the crystal, the rather mobile

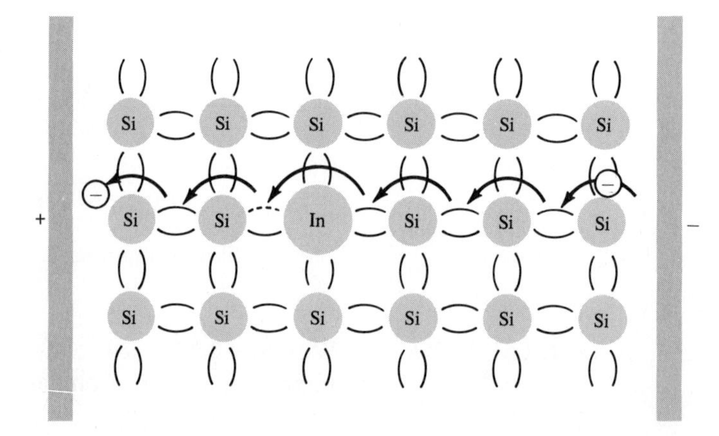

FIGURE 6.3. **Hole Conduction Within *p*-Type Semiconductor** The missing bond (dashed line) attracts an electron from the bond to its right, which in turn attracts another electron from still farther to the right, and so on. The broken bond thus moves toward the negative electrode where it is neutralized, at which time another hole is created at the positive electrode.

electrons will be attracted to the positive electrode. Such materials constitute *n-type semiconductors.*

In an intrinsic semiconductor each thermally induced bond fracture creates a pair of opposite charges. Thus, in intrinsic silicon at room temperature there are about 1.5×10^{10} free carriers *of each polarity* per cubic centimeter. The purposeful introduction of additional carriers by doping creates an **extrinsic** material, with the polarity of the dominant carrier (termed the **majority carrier**) determined by the nature of the dopant. In a doped semiconductor there still remain some carriers of the opposite polarity (**minority carriers**), and the factor by which the majority carriers have been increased by doping equals the factor by which the minority concentation has been decreased.

Five-valent atoms that provide excess electrons are termed **donors**, and three-valent atoms that create holes are termed **acceptors.**

6.4.2 Semiconductor Current

We have mentioned that in metallic conduction, negative electrons are responsible for current and that within semiconductors, depending on the fabrication techniques, the current carriers are either positive or negative, being, respectively, *p*-type and *n*-type semiconductors. We should pause a moment and consider the behavior of *positive* and *negative* current.

Refer to Figure 6.4. We wish to measure the electrical flow in the cylindrical conductor by means of an ammeter. When connecting a dc ammeter into such a circuit, we observe the polarity indicated. Thus, electrons from the battery enter the left terminal and leave at the right one, with the meter deflection being proportional to the magnitude of the flow. If the semiconductor were of the *p*-type, in the connecting wires, we would have electron flow.

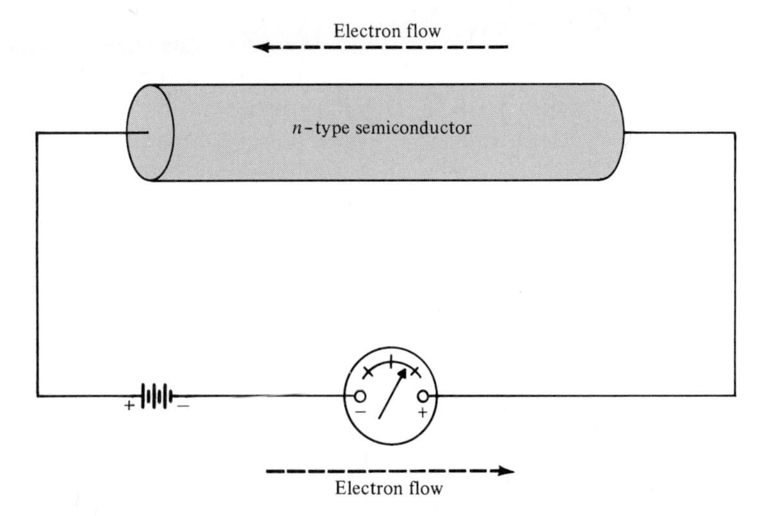

FIGURE 6.4. **Electron Conduction Within *n*-Type Semiconductor and a Metallic External Circuit**

At the left of the semiconductor (Figure 6.5), electrons are pulled out, which creates holes that travel to the right. At the right edge these holes are cancelled out by electrons that have arrived from the battery's negative terminal. For the sake of simplicity, however, we can continue to assume that we have classical current from the positive (+) terminal of the battery to the negative (−) terminal.

Note that in both cases the terminal connections to the meter stay the same. We conclude that insofar as the external circuit is concerned, negative carrier flow in one direction is equivalent to positive flow in the opposite direction.

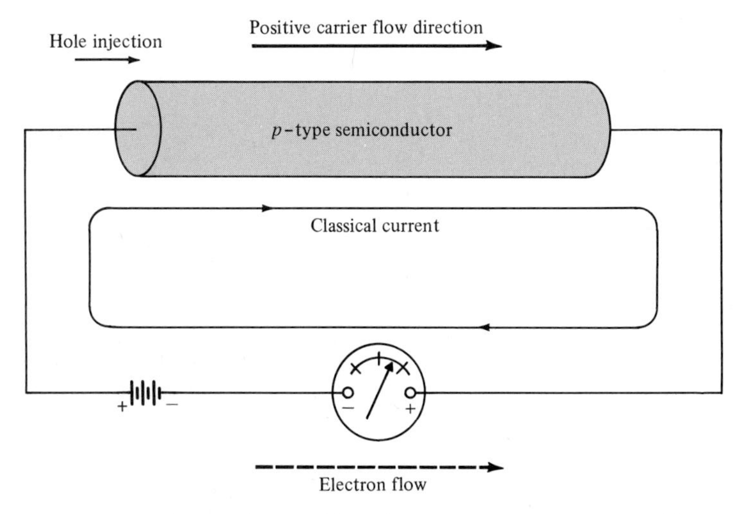

FIGURE 6.5. **Hole Conduction in *p*-Type Semiconductor and Electron Flow in Metallic External Conductor**

What happens when we simultaneously have both types of carriers in the same semiconductor, as in fact we do, since the dominant carrier in each is accompanied by minority carriers? To a large degree they may be looked upon as being independent of one another with each making a proportional contribution to the current in the external circuit. To the extent that they *do* interact (and one usually attempts to adjust conditions so as to minimize such interaction by making one type of carrier dominant), there is a mutual cancellation.

The application of a voltage between the ends of a section of material will create an electric field and will impart to any free charged carriers an average velocity (u), known as *drift velocity*, as they bump their way through the material. The *electric field intensity* ($\mathcal{E}$) is equal to the voltage applied, divided by the distance between the electrodes (l) or $\mathcal{E} = V/l$. The ratio of u to $\mathcal{E}$ is termed the *mobility* of the charged carriers—that is, $u = \mu\mathcal{E}$, where μ is the mobility.

The movement of charged carriers with drift velocity u constitutes a current (density). If the concentration of charged carriers is n [1/m^3], the current density J [A/m^2] is equal to neu, where e is the charge per carrier (1.60×10^{-19} C). We have

$$J = neu = ne\mu\mathcal{E} = \sigma\mathcal{E} \tag{6.12}$$

where σ = the conductivity.

In a semiconductor where both positive and negative carriers are simultaneously present, each contributes to the conductivity. We have

$$\sigma = (n\mu_n + p\mu_p)e \tag{6.13}$$

where n and p are, respectively, the carrier concentration of electrons and holes and μ_n and μ_p are their respective mobilities. In the intrinsic state there is an equality of electrons and holes, and for silicon at 300 K, we have approximately

$$n = p = 1.5 \times 10^{10}/\text{cm}^3$$

Also, for silicon at 300 K, we have, typically,

$$\mu_n = 0.1300 \text{ m}^2/\text{V}\cdot\text{s} \qquad \mu_p = 0.0500 \text{ m}^2/\text{V}\cdot\text{s}$$

6.5 ■ SEMICONDUCTOR DIODES

Figure 6.6 depicts the junction between two types of semiconductor materials, one p-type and one n-type. The p-type constitutes the *anode* and the n-type, the *cathode*. The resulting device is called a **diode.**

When the junction is formed, the excess holes in the p-material diffuse into the n-material, and the excess electrons in the n-material diffuse into the p-material. This diffusion leaves uncovered negative ions on the p-side of the junction and uncovered positive ions on the n-side (Figure 6.6). This junction

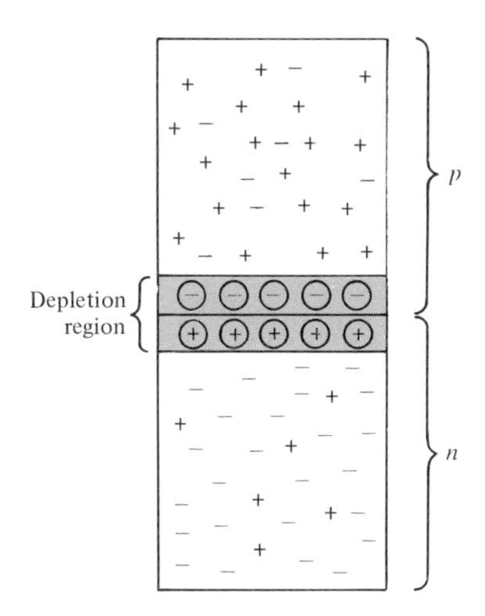

FIGURE 6.6. ***pn*** **Semiconductor Junction** Majority current diffusion across the junction leaves behind uncovered ions that constitute the depletion region.

region, with its mobile charges considerably depleted, constitutes the **depletion region**, which is also called the *space-charge region*.

As the number of uncovered ions on either side of the junction increases, the majority carriers of the *n*-type become progressively more repelled by the negative ions on the *p*-side of the junction, while the majority carriers of the *p*-type likewise become increasingly repelled by the positive ions on the *n*-side of the junction. As a result the majority carrier diffusion diminishes as the potential barrier builds up.

But due to the electric field that has been created by the uncovered charges, the minority carriers in each type of semiconductor drift into the opposing domains. Thus, the (diminishing) diffusion of the majority carriers is offset by the (increasing) drift of the minority carriers. Eventually an equilibrium will prevail, and the open-circuited diode will end up with a fixed voltage (termed the contact potential) across the junction region.

The application of a **forward bias**—that is, positive potential to the *p*-side and negative potential to the *n*-side—diminishes the potential barrier (Figure 6.7), and majority carriers from the *p*-side cross over into the *n*-region where they constitute minority carriers. We refer to this crossover as *hole injection* into the *n*-material. The resulting charge unbalance is remedied by having the external circuit provide compensating electrons via the negative electrode.

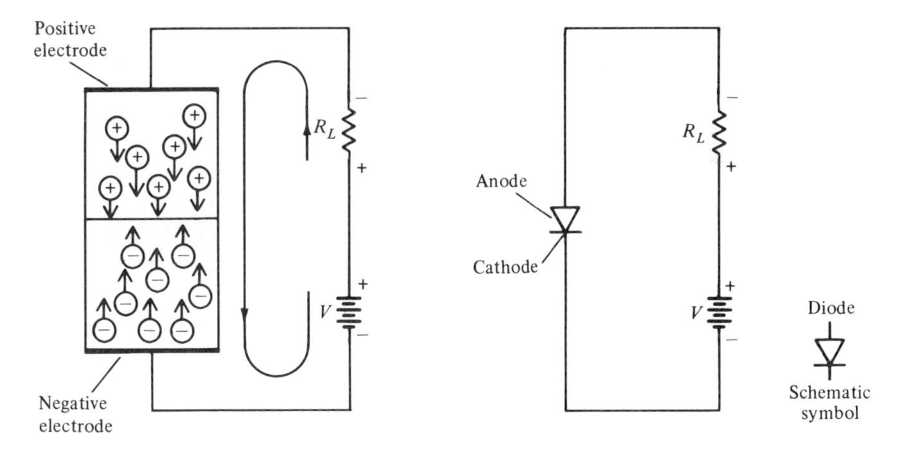

FIGURE 6.7. Conduction Through a Semiconductor Diode with Forward Bias Applied The arrow in the schematic symbol indicates the direction of current flow with forward bias.

Likewise, forward bias leads to the injection of electrons into the *p*-region through the junction, and the resulting charge unbalance is remedied by having electrons withdrawn at the positive electrode. The combined effect represents a significant current from the positive to the negative electrode through the diode.

If the **bias** is **reversed** (potentials applied to the diode are opposite those shown in Figure 6.7), the effect is to increase the width of the depletion region, create a greater potential barrier, and substantially diminish the circuit current. (The current does not go to zero since a reverse bias for the majority carriers constitutes a forward bias for the minority carriers. It does, however, become orders of magnitude smaller.)

6.5.1 The Diode Equation

Ideally there should be no current through a reverse-biased semiconductor diode. In practice there is. It arises because of the minority carriers. For them the reverse bias condition constitutes forward bias, and a small current (typically, for silicon diodes, 10^{-9} A or less) will be the result. This current is called the **reverse saturation current** (I_0), and despite its small magnitude, it plays an important role in the forward conduction (I) of a silicon semiconductor diode through the equation

$$I = I_0(e^{V/V_T} - 1) \tag{6.14}$$

where V = the voltage applied between the anode and cathode.

The quantity V_T is called the voltage equivalent of temperature and satisfies the relation

$$V_T \simeq \frac{T}{11.600} \tag{6.15}$$

where T = the absolute temperature of the semiconductor.

Figure 6.8 shows how the diode current varies under forward and reverse bias conditions. Note the scale change between $-I$ and $+I$ that is used to make the reverse current visible. Also, after but a small fraction of a volt, the reverse current reaches a constant value, but at a sufficiently large reverse bias (V_{BR}), there is a breakdown of the insulation, and a large current can suddenly arise. (We will defer further consideration of this phenomenon until Chapter 16.) The forward current shows but a small rise until the voltage reaches a value of about 0.6 V (for a silicon diode). Therefore, since the current growth is exponential, it rises rapidly, in keeping with Eq. 6.15. This voltage is given various names such as cut-in voltage and threshold voltage (V_γ). (For germanium the comparable value is about 0.2 V.)

We will consider two different resistances associated with the diode: The **dc resistance** is simply the voltage drop divided by the diode current, but superimposed on the dc there may be ac. This is subject to the **dynamic resistance** (r), which is equal to $\Delta v/\Delta i$ and whose value obviously depends on where one is operating along the curved characteristic. A very useful approximation for the dynamic resistance value is given by the following:

$$r\ [\Omega] \simeq \frac{26\ \text{mV}}{I\ [\text{mA}]} \tag{6.16}$$

EXAMPLE 6.5 In terms of the reverse saturation current (I_0), the silicon diode current varies with applied voltage (V) according to the equation

$$I = I_0(e^{qV/kT} - 1)$$

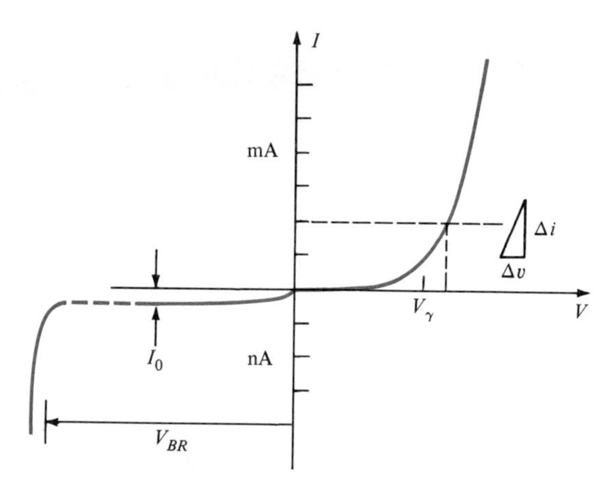

FIGURE 6.8. Semiconductor Diode Current under Forward and Reverse Bias Conditions I_0 is the reverse saturation current. V_γ is the cut-in voltage. Dynamic resistance is represented by $\Delta v/\Delta i$.

where q = electronic charge
k = Boltzmann's constant
T = temperature in K

Derive the expression for the dynamic resistance.

Solution:

$$\frac{dI}{dV} = \frac{q}{kT} I_0 e^{qV/kT} = \frac{q}{kT}(I + I_0) \simeq \frac{q}{kT} I$$

$$r = \frac{dV}{dI} = \frac{kT}{qI} = \frac{kT/q}{I}$$

$$\frac{kT}{q} \simeq 0.026 \text{ V}$$

$$r = \frac{0.026 \text{ V}}{I\text{ [A]}} = \frac{26 \text{ mV}}{I\text{ [mA]}}\ \Omega$$

To this, we should add a resistance representing the bulk of the semiconductor as well as the contact resistance of the metallic leads. This resistance, r_B, typically assumes values between 0.1 Ω (for high-power diodes) and 2 Ω (for low-power units).

A graphical analysis of diode conduction helps one to visualize the influence of the nonlinear characteristic. Refer to Figure 6.9. If $i = 0$, the full value of the input voltage appears across the diode, $iR = 0$, and we have the intercept point on the x-axis (v_{in}). When i is so large that the entire drop appears across R, $v_D = 0$, and we have the intercept point on the y-axis. The straight line between these two points is the *load line,* whose slope is equal to the negative reciprocal of R $[(v_{in}/R)/v_{in} = -\text{slope} = -1/R]$. The load line intercepts the static diode curve at the equilibrium point termed the **Q-point** (*Q* standing for quiescent).

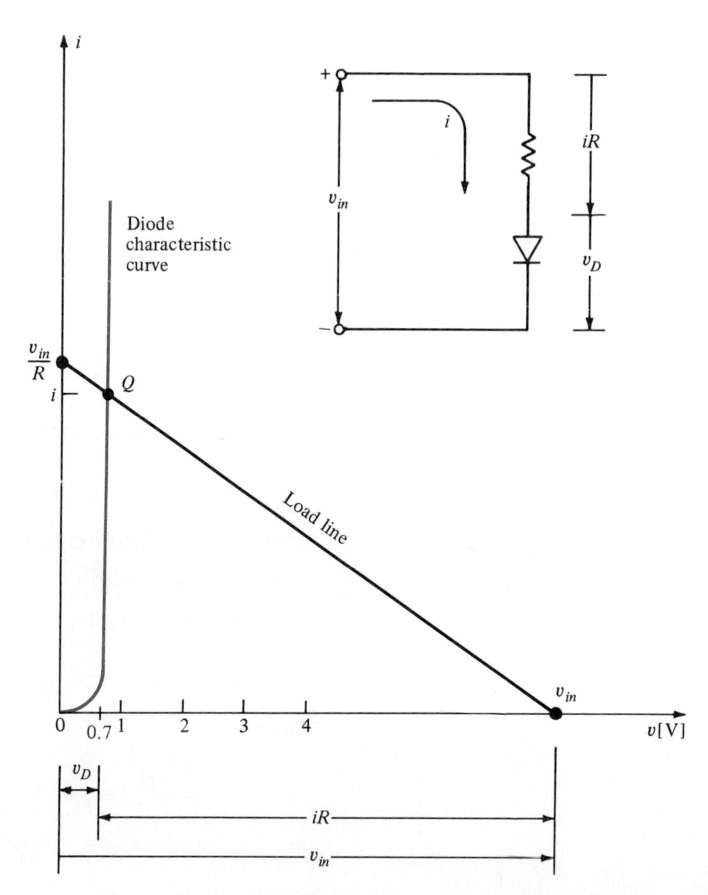

FIGURE 6.9. Intersection Between Diode Characteristic Curve and Load Line, Establishing the *Q*-Point, Which Indicates Static Current (*i*) and Diode Voltage Drop (∼0.7 V)

Typically, the voltage across the diode is about 0.7 V, and the value across R is iR. If v_{in} changes, the load line moves parallel to itself (since the value of R remains fixed), and the resulting change in the intercept with the diode characteristic indicates the instantaneous value of i and the instantaneous value of the voltage across the diode. The latter stays fairly constant at 0.7 V as long as v_{in} is more than a few volts.

If v_{in} is an ac signal, the x-intercept moves along the x-axis from zero, to V_{peak}, back down to zero, and then in the negative x-direction. There will be an output current only during the positive half-cycle. There is some distortion of the current's half-wave sinusoid when the instantaneous value of the ac voltage is < 1 V because of the curvature of the diode's response curve. When dealing with large peak voltages, this is of little consequence, but if the diode is being used with small signals (for example, in radio detection), this curvature can constitute considerable distortion. Subsequently, in our discussion of operational amplifiers, we will see how such distortion may be minimized.

Diodes can broadly be divided into two categories according to use:

1. *Rectifiers* are diodes intended to be used in conjunction with power conversion from ac into dc. Their current rating is generally 1 A or more.
2. *Signal diodes* usually have smaller current ratings and are used as signal detectors, wave shapers, voltage limiters, and so forth.

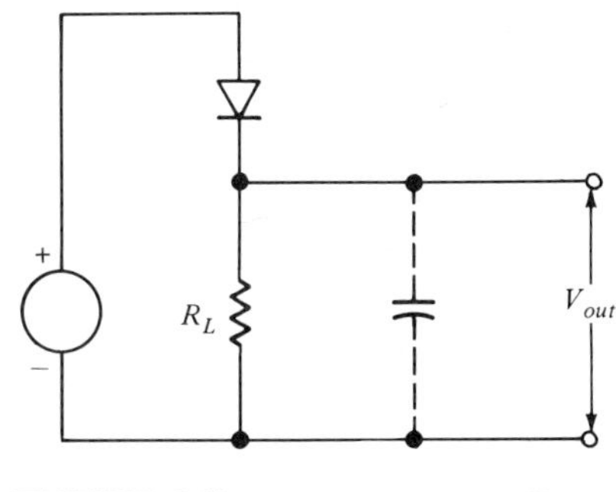

FIGURE 6.10. **Half-Wave Rectifier with Capacitive Filter**

6.5.2 Rectifiers

If we now attach an ac source to the diode, through a series resistor R_L, the resulting current will take the form of pulsating dc. The current will only be present when the applied ac constitutes a forward bias. Having placed a capacitor across R_L (Figure 6.10), we find that during the first positive swing of the ac the current through R_L will rise in unison, and the voltage across the capacitor will reach the peak value of the applied waveform. As soon as the applied ac passes its peak value and starts to drop, the diode stops conducting because the anode is less positive than the cathode (whose potential is being held near the peak value by virtue of the charge on the capacitor).

The charged capacitor now starts to discharge through R_L, thereby maintaining the current through the load despite the fact that the diode is nonconducting. But as the capacitor discharges, the load voltage will diminish, the rapidity of the drop being determined by the values of R_L and C.

The diode will not start to conduct again until the applied ac becomes more positive than the charge remaining on the capacitor (point C in Figure 6.11). It will again stop at D, and the sequence will repeat. We have thereby diminished the magnitude of the pulsation across the load. Rather than an amplitude A to B, it is now C to D. If we wish to diminish the fluctuations to a still greater degree, we may insert a larger capacitance across the load. (The fluctuation will then be C' to D.) There is a precaution in this regard, however. Note that as the capacitance is increased, the duration of the conducting pulse is diminished. During this interval of time, the system must replace all the power used by the load during the remainder of the cycle. Thus, with a very large capacitor, the magnitude of the current pulse may exceed the current-handling capacity of the diode.

One may also approach the smoothing process from the Fourier viewpoint. The applied ac consists of a single frequency, 60 Hz, but once rectified, it is no longer sinusoidal, and hence there are additional frequencies present. We have already seen (Section 5.3) that it takes the following form:

$$v = V_m\left[\frac{1}{\pi} + \frac{1}{2}\sin\omega t - \left(\frac{2}{\pi}\right)\left(\frac{1}{3}\right)\cos 2\omega t - \left(\frac{2}{\pi}\right)\left(\frac{1}{15}\right)\cos 4\omega t - \cdots\right] \tag{6.17}$$

The capacitor allows the ac components to be bypassed around the load resistor.

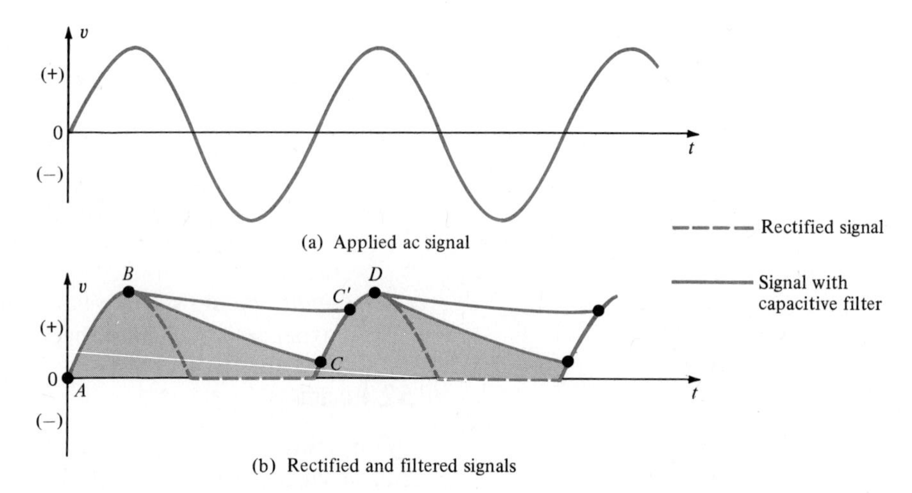

FIGURE 6.11. **Half-Wave Rectification and Capacitive Filtering**

Rectifiers that allow only the alternate half-cycles to be passed are termed **half-wave rectifiers.** A greater efficiency of rectification can be achieved by employing two diodes in the configuration shown in Figure 6.12. Note the center-tapped transformer, as well as the instantaneous polarities of the ac. As shown, the lower diode would conduct but not the upper one. During the next half-cycle, the upper diode would conduct, but the direction of the current through the load would *not* change. We would then have the temporal sequence shown in Figure 6.13.

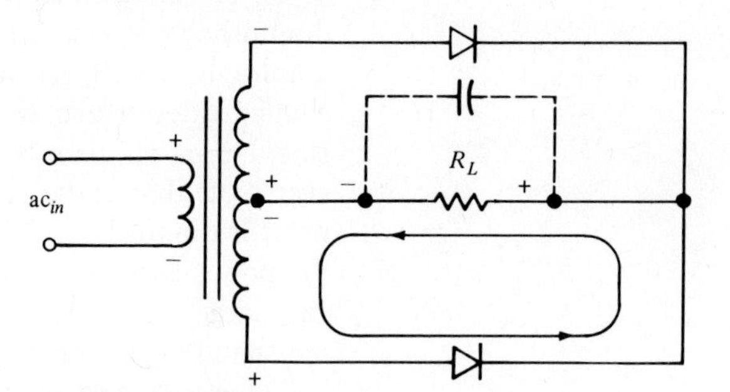

FIGURE 6.12. **Full-Wave Rectifier** During the next half-cycle, the upper diode will conduct.

The Fourier components for this **full-wave rectifier**, as it is called, are

$$v = V_m\left(\frac{2}{\pi} - \frac{4}{3\pi}\cos 2\omega t - \frac{4}{15\pi}\cos 4\omega t - \ldots\right) \tag{6.18}$$

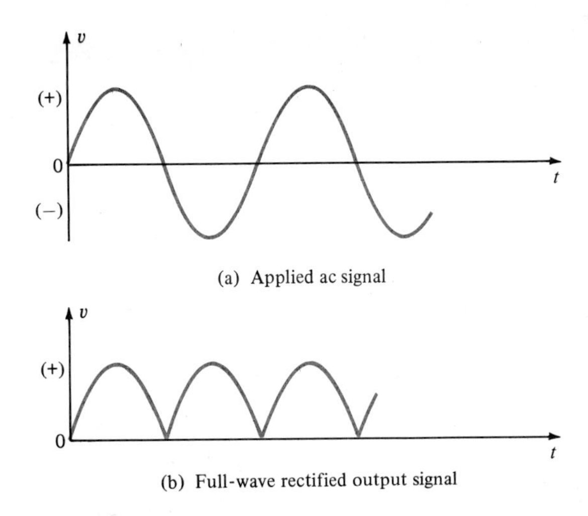

FIGURE 6.13. **Full-Wave Rectification**

Note that the dc level here is twice what it was in the case of the half-wave rectifier. Also, with a 60 Hz power source, in the half-wave case the lowest *ripple* component was 60 Hz; here it is 120 Hz—that is, 2ω. The nipple that arises with full-wave rectification makes it easier to filter since only half the capacitance is required for the same degree of filtering.

One disadvantage of the full-wave rectifier stems from the need for a center-tapped transformer, which is somewhat more expensive than a comparable one without the tap. Also, at a given time, only half of the transformer is being utilized, a rather inefficient usage. Since semiconductor diodes are relatively inexpensive, it is advantageous to use the bridge rectifier shown in Figure 6.14. Here again, on alternate half-cycles the current will pass through the load in the same direction. A capacitor across the load can again be used to reduce the ripple.

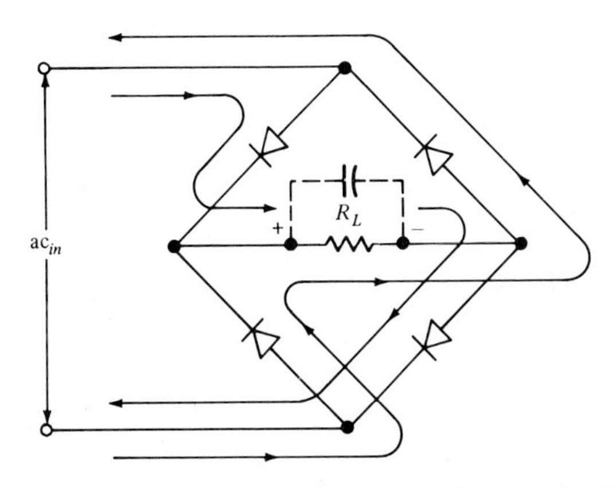

FIGURE 6.14. **Bridge Rectifier Showing Alternate Half-Cycle Conduction Paths**

FOCUS ON PROBLEM SOLVING: Estimating Ripple-Reducing Capacitance

If V_{dc} is the voltage delivered by a rectified power source, ΔV is the maximum desired ripple voltage, I_L is the load current demand, R_L is the load resistance, and f is the power frequency, we have

$$Q = CV$$

$$I = \frac{dQ}{dt} = C\left(\frac{dV}{dt}\right) \simeq C\left(\frac{\Delta V}{\Delta t}\right)$$

For a half-wave rectifier, $T \sim 1/f \equiv \Delta t$, and

$$C = \frac{I_L \Delta t}{\Delta V} \simeq \frac{I_L}{f\Delta V} = \frac{V_L/\Delta V}{fR_L}$$

since $I_L = V_L/R_L$.

For a full-wave rectifier, $t \sim 1/2f$, and

$$C = \frac{I_L}{2f\Delta V} = \frac{V_L/\Delta V}{2fR_L}$$

EXAMPLE 6.6 An amplifier requiring a dc voltage of 15 V and representing a load current of 2.0 mA is to be subjected to a maximum ripple of 1% peak to peak. What is the necessary value of the smoothing capacitor used in conjunction with full-wave rectification?

Solution:

$$15 \text{ V} \times 0.01 = 0.15 \text{ V} = \Delta V$$

$$C = \frac{I_L}{2f\Delta V} = \frac{2 \times 10^{-3}}{2(60)(0.15)} = 1.1 \times 10^{-4} = 110\ \mu\text{F}$$

A standard 150 μF capacitor or a 200 μF capacitor will suffice.

6.5.3 Signal Diodes

When a diode that has been passing current in the forward direction is suddenly reverse-biased, the diode current will not immediately fall to its steady-state reverse-voltage value. There can be a significant reverse current for a period of time following such a reversal, as can be seen from Figure 6.15. This current arises because, with forward bias, minority carriers from each side have drifted across the junction where they now constitute majority carriers. Thus both the p-side and the n-side have carrier concentrations in excess of their equilibrium value. Upon reversing the applied voltage, the concentrations attempt to return to values appropriate to the reverse bias.

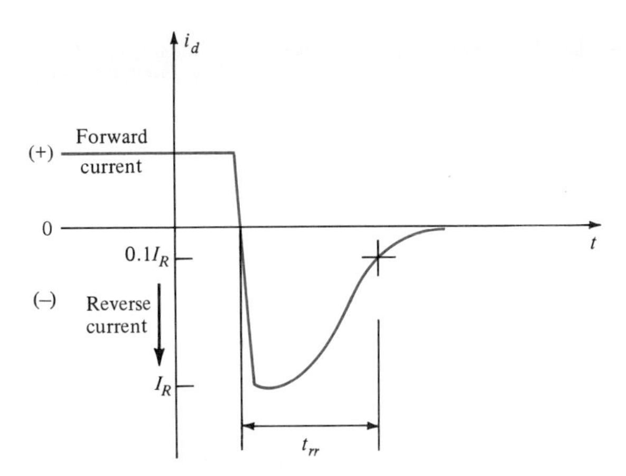

FIGURE 6.15. **Reverse Recovery Time (t_{rr}) of a Diode** The reverse recovery time is defined as the interval between zero crossing and the time to reach 0.1 of the peak reverse current.

There then arises a transient reverse current while the equilibrium values of charges are being reestablished.

While the initiation of current reversal arises at a rather specific time, termination does not. The termination time may be taken to be that time at which the reverse current diminishes to a value equal to 0.1 of its peak (Figure 6.15). The time between the current crossing the zero axis and reaching the 0.1 value constitutes the **reverse recovery time** (t_{rr}) of the diode. Small signal diodes have a $t_{rr} \sim 5$ ns. For rectifiers, the values of $t_{rr} \sim 3000$–4000 ns. There is also a type of fast recovery power rectifier, designed to be used at frequencies as high as 250 kHz, whose $t_{rr} \sim 100$–200 ns.

Clippers and Clampers. Consider the application of a 30 V peak ac signal to the circuit shown in Figure 6.16. The diode will not conduct unless its anode is more positive than its cathode (by about 0.6 V for silicon diodes). Therefore, conduction will not start until the input signal exceeds about 10.6 V, and the output signal will consist only of the shaded peaks. At any lower voltage there is no drop across the resistor, and the output voltage is zero. We have just described the basic operation of a *clipping circuit.* The clipping

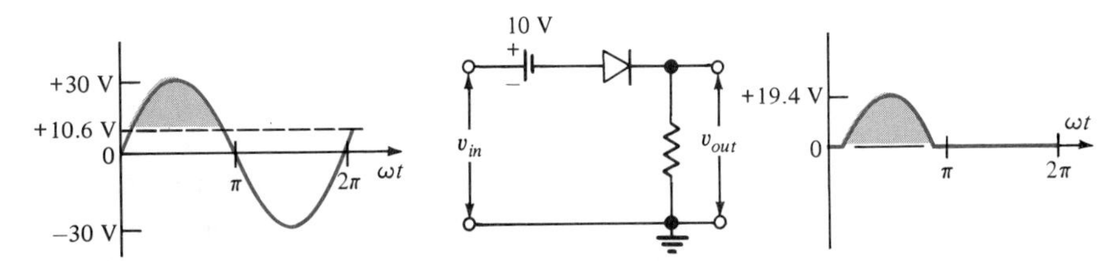

FIGURE 6.16. **Clipping Circuit** Without diode conduction, the output voltage is zero. The diode conducts when the applied voltage exceeds the bias (+10 V) plus the diode threshold (0.6 V). Output then follows input.

level can be altered by changing the value of V. If the battery polarity is reversed, the entire positive half-cycle will emerge—plus a portion of the negative one, depending on the value of V. The portions of the ac waveform that are clipped away may be altered by reversing the diode with, again, two alternative battery polarities.

Alternatively, one could place the diode in parallel with the output. Also, parallel arrangements of oppositely directed diodes allow simultaneous clipping of positive and negative half-cycles, thereby creating a square wave from a sinusoidal one (Figure 6.17).

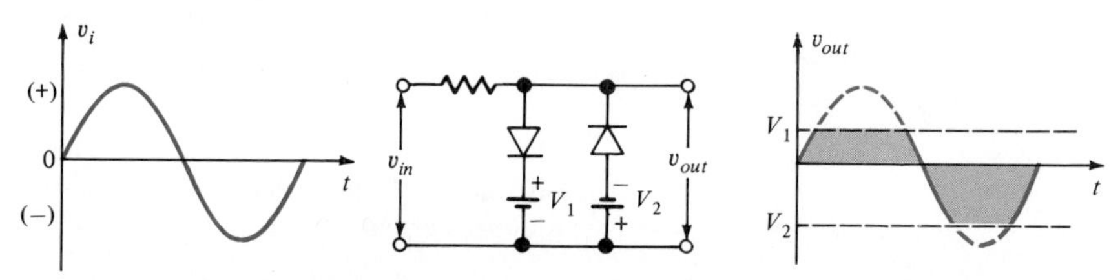

FIGURE 6.17. Clipping Circuit The output voltage follows the input only when neither diode is conducting. Otherwise the output voltage equals the bias voltage associated with the conducting diode.

EXAMPLE 6.7

The sine wave $v = 25 \sin \omega t$ [V] is applied to the clamping circuit in Figure 6.18. Indicate the nature of the output if the voltage drop across the diodes is disregarded.

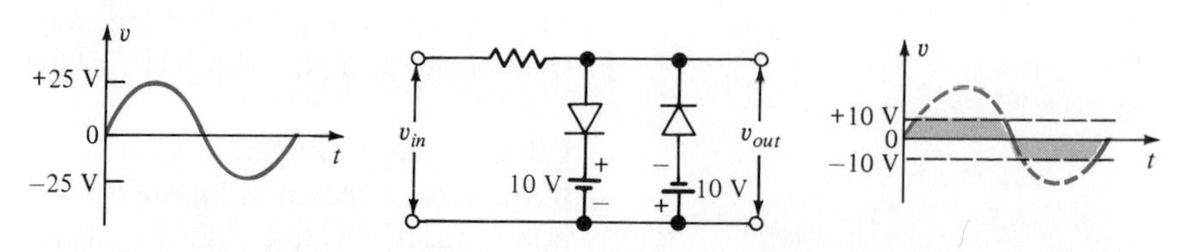

FIGURE 6.18. Output Limiter That Prevents Output Signal from Exceeding Values +10 V and −10 V.

Solution: It will approximate a square wave with a 10 V peak value.

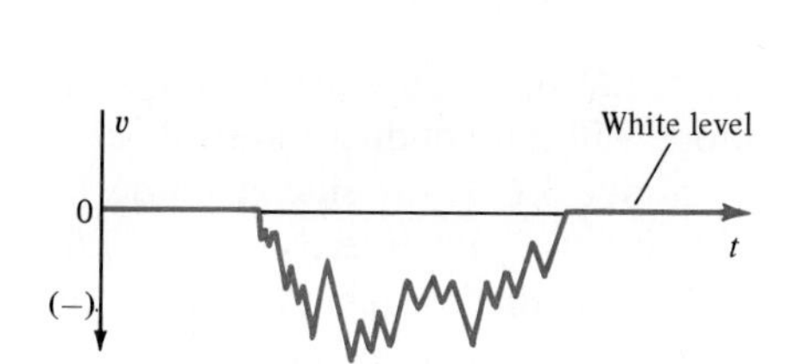

(a) Magnitude of negative voltage determines degree of darkness

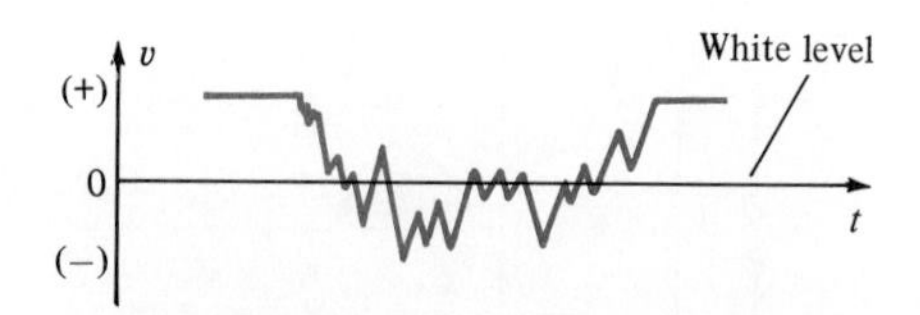

(b) When this signal is passed through a capacitor, the dc reference is removed, and the degree of darkness (magnitude of negative voltage variations) is reduced, leading to diminished picture contrast

FIGURE 6.19. Effect of Capacitor on the Voltage Level

Figure 3.58, associated with Problem 3.18, depicted an ac signal atop a +10 V pedestal. When such a signal is applied to a high-pass filter, such as is employed in coupling between amplifiers, we "lose" the input dc pedestal—as expected, since a capacitor represents an open circuit to dc. At the output of such a coupling circuit, the ac signal will establish itself about an average that represents the dc level at the output. For the case at hand (Figure 3.58), the output dc level is ground. Thus, the ac output signal has a zero reference level.

Figure 6.19 depicts what might represent a signal corresponding to a single scan on a TV picture tube. If we try to amplify this signal by passing it through the usual RC coupling network—that is, a high-pass filter between amplifier stages—the capacitor will adjust the signal in such a manner that the integrated signal above the average will equal that below. Such a signal would present a "washed out" appearance on the TV screen. The situation may be corrected by using a **clamping** circuit, such as shown in Figure 6.20.

When the input signal first goes positive, the diode conducts, and the capacitor charges to the maximum positive value, V_{max}, with the indicated polarity. With the anode at a negative potential the diode cannot conduct. If the time constant of the circuit (τ = RC) is large compared to the period of the signal, little charge leaves the capacitor and the anode is held at the negative potential. Using KVL, $v_{in} = V_{max} + v_{out}$, or, $v_{out} = v_{in} - V_{max}$. From the latter equation it can be seen that the entire input signal is shifted downward by a magnitude equal to V_{max}, resulting in an output signal that is clamped to the zero value. A rough rule of thumb is to have the time constant at least five times greater than the signal period. (Often, for example, in television sets, the circuit is termed a dc restorer.)

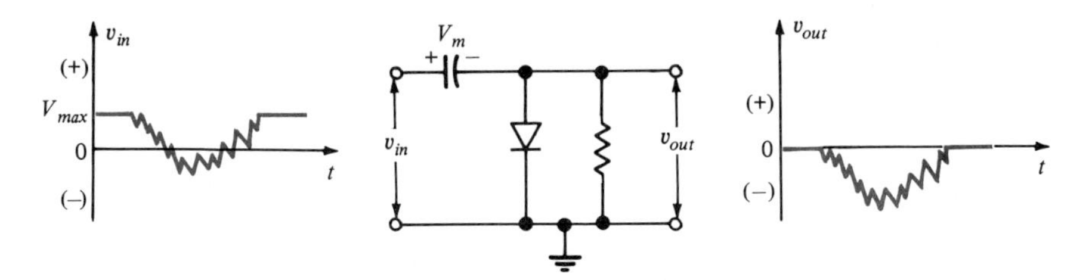

FIGURE 6.20. Clamping Circuit to Reestablish Zero Signal Reference

EXAMPLE 6.8

a. A square wave with a 1000 Hz repetition rate is applied to a clamping circuit with $C = 0.01\ \mu F$. (See Figure 6.21.) Compute the necessary value of R.

b. What would be the appearance of the output waveform if $\tau = T$?

c. The results depicted in Figure 6.21 are idealized. With $\tau = 5T$, what is the magnitude of the output just before the input returns to +5 V? What is it if $\tau = 10T$?

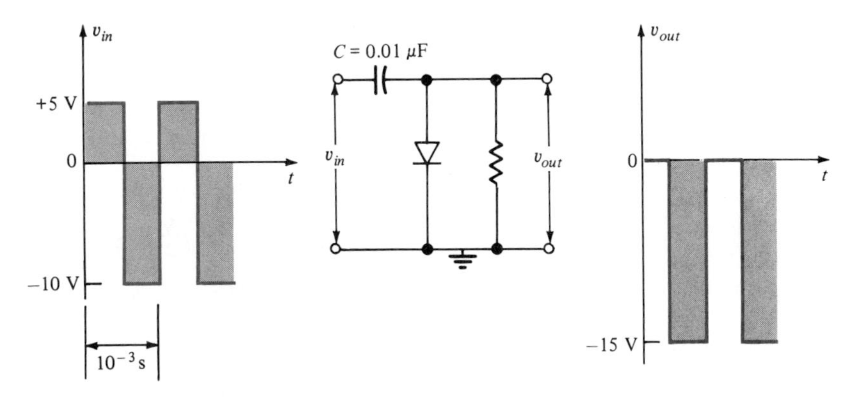

FIGURE 6.21. Clamping Circuit to Establish Zero Reference This circuit is also called dc restorer. (The output waveform is idealized.)

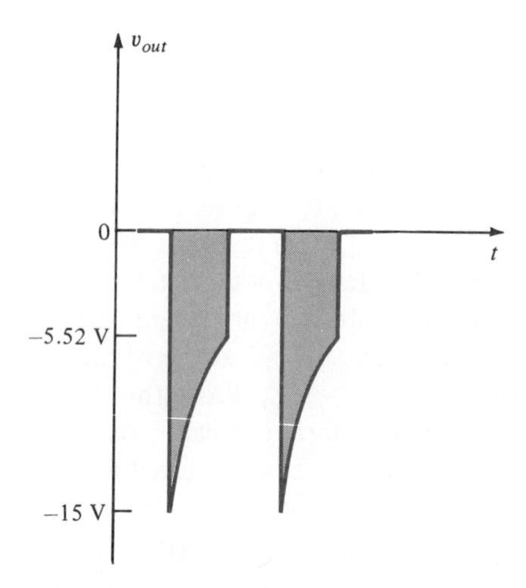

FIGURE 6.22. Clamping Circuit with a Small Time Constant

Solution:

a. $T = 1/f = 1/10^3 = 10^{-3}$ s

Let $\tau = RC = 5T = 5 \times 10^{-3}$ s.

Then we have

$$R = \frac{5 \times 10^{-3}}{10^{-8}} = 5 \text{ k}\Omega$$

b. During the interval when the input is at +5 V, the output remains at 0 V. When the square wave input is at −10 V, there is a discharge current through R that diminishes exponentially (Figure 6.22). We have

$$V_{out} = -15e^{-\tau/T} = -15e^{-1} = -5.52 \text{ V}$$

c. When $\tau = 5T$,

$$v_{out} = -15e^{-1/5} = -12.3 \text{ V}$$

and when $\tau = 10T$,

$$v_{out} = -15e^{-1/10} = -13.6 \text{ V}$$

Amplitude Modulation (AM) Application. We turn now to a consideration of diodes used to extract information transmitted by radio waves (as well as other forms of communication). In terms of transmitting radio waves, we may say that the higher the frequency utilized, the easier and more efficient is the radiation process. Suppose we desire to send speech or music. A microphone would convert the sound into electrical pulsations, and these pulsations (after amplification) would be applied to an antenna and radiated. To achieve efficient electromagnetic radiation, ideally we would require an antenna whose dimensions are of the order of either a quarter- or half-wavelength of the electrical signals. Wavelength and frequency are related through the equation

$$\lambda = \frac{c}{f} \tag{6.19}$$

where c = velocity of radio wave propagation ($3 \times 10^8 \mu$ m/s)

f = frequency in hertz

λ = wavelength in meters

By taking 1000 Hz as a typical sound frequency, the corresponding electromagnetic wavelength is

$$\lambda = \frac{3 \times 10^8}{10^3} = 3 \times 10^5 \text{ m} = 186 \text{ (mi)}$$

For efficient radiation we would need an antenna that was about 93 mi long,[3] which is a rather impractical requirement. Therefore, we utilize a much higher frequency (like 1000 kHz) that may be radiated efficiently using a more compact and reasonably sized antenna and then impress the desired lower frequency onto the higher one.

Thus, the audio signal is carried by the radio wave in "piggyback" fashion. Figure 6.23 illustrates the process.

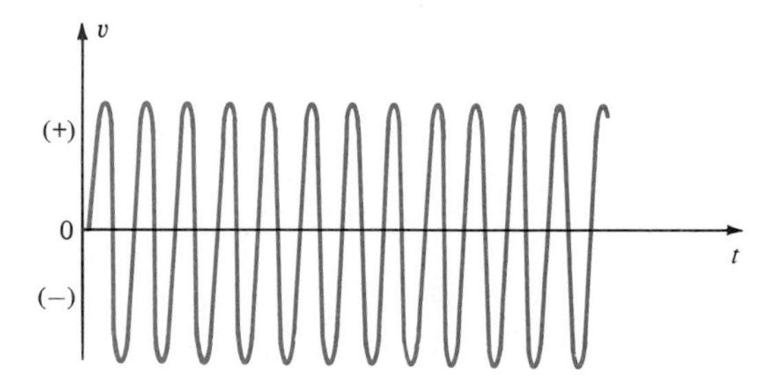

(a) Carrier wave: high-frequency radio wave (~ 1000 kHz)

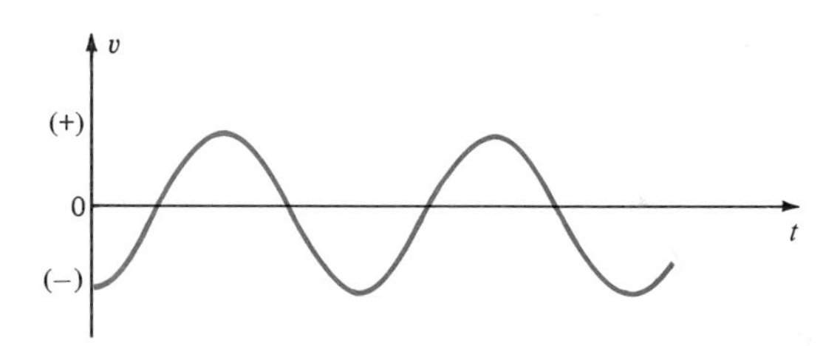

(b) Modulating signal: audio signal to be transmitted (~ 1000 Hz)

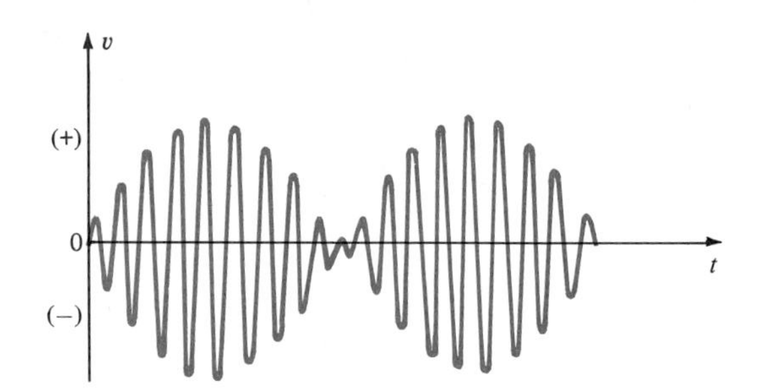

(c) Modulated carrier: high-frequency radio wave with audio signal superimposed

FIGURE 6.23. **Amplitude-Modulated Carrier Wave**

The amplitude of the high frequency varies in response to the low one, thus representing **amplitude modulation** (AM).

At the receiver we wish to extract the audio signal. Figure 6.24 illustrates how diode rectification is employed in this process, termed **demodulation.**

[3]Smaller fractions of a wavelength may also be used, with diminished efficiency.

Again the diode converts the ac into pulsating dc. If we now insert a capacitor across the load resistor, we can "fill in" between pulsations, or, looking at it from the Fourier viewpoint, the radio frequency is bypassed around the load resistance. Of course, the capacitor cannot be too large, because the audio signal would also be bypassed.

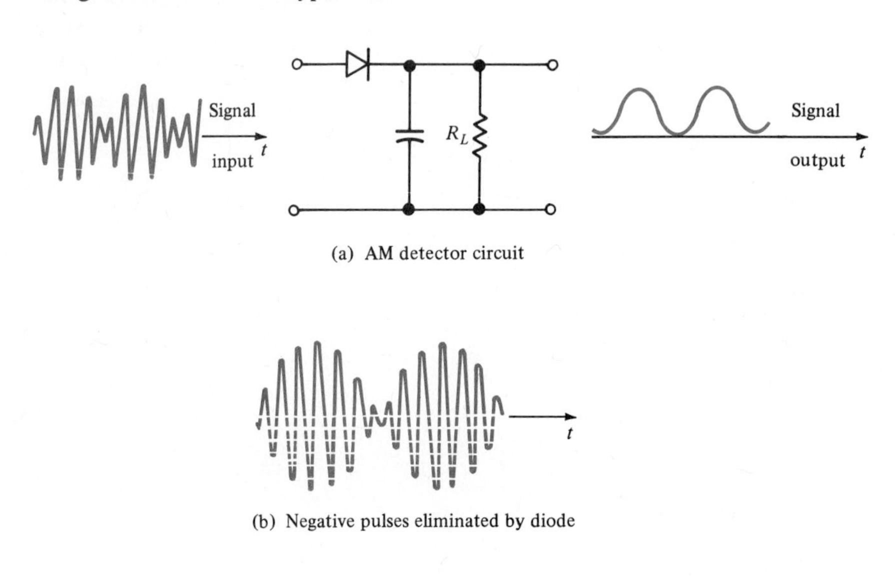

(a) AM detector circuit

(b) Negative pulses eliminated by diode

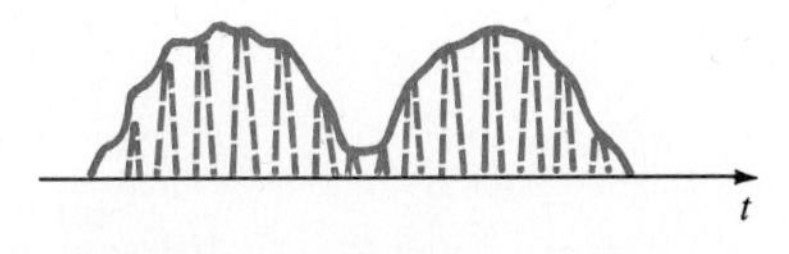

(c) Original audio signal recovered as capacitor "fills in" between pulses

FIGURE 6.24. Demodulation Extraction of an audio signal from an amplitude-modulated carrier is depicted.

The load resistor in this case would be a pair of headphones that would reproduce the original speech or music. Alternatively, additional amplification could follow the demodulator, and the resultant audio signal could be made to power a loudspeaker.

Since from a Fourier viewpoint only a sinusoid of constant amplitude constitutes a single frequency, an amplitude-modulated (AM) radio frequency wave includes additional adjoining frequencies. The radio frequency is termed the carrier frequency (1000 kHz in our numerical example), and the adjoining frequencies (termed sidebands) have values equal to the carrier frequency plus and minus the modulation frequencies. (See Figure 6.35.) Thus, when a 1000 Hz audio signal is transmitted on a carrier frequency of 1000 kHz, the radio wave consists of three frequencies: 999, 1000, and 1001 kHz. The higher the audio frequency to be transmitted, the greater the channel width necessary, and the fewer the number of stations that can be accommodated in the broadcast band. By regulation, the highest allowed audio frequency in the AM band is about 4000 Hz, and so the channel occupied is about 8 kHz. (Actually, about 10 kHz per channel is allowed.)

EXAMPLE 6.9

An AM signal consists of a 1000 kHz carrier modulated by a 1000 Hz audio signal. If the load resistance is 2 kΩ, what should be the value of C if the radio frequency but not the audio frquency is to be bypassed around R_L? What percentage of the audio current passes through R_L? (See Figure 6.24.)

Solution: A rule of thumb in such problems is to make the alternate path for the higher frequency—that is, the path through the capacitor—have a reactance that is one-tenth that of the resistance R_L. Thus, at 1000 kHz, set $X_C \leq \frac{1}{10} \times 2\ \text{k}\Omega = 200\ \Omega$. Then we have

$$C = \frac{1}{2\pi f X_C} = \frac{1}{2\pi(10^6)(200)} \simeq 800\ \text{pF}$$

At 1000 Hz this capacitance represents a reactance of

$$X_C = \frac{1}{2\pi(10^3)(8 \times 10^{-10})} \simeq 200\ \text{k}\Omega$$

By using the current division rule, the current through R_L is X_C/Z times that current entering the parallel combination, where Z must now be used in the denominator in place of the R_{total}, which was used for strictly resistive circuits. We have

$$\left|\frac{X_C}{Z}\right| = \frac{2 \times 10^5}{\sqrt{(2 \times 10^5)^2 + (2 \times 10^3)^2}} > 0.99$$

Therefore, more than 99% of the audio current passes through R_L.

Frequency Modulation (FM) Application. One major shortcoming of AM radio is its susceptibility to static interference. The electrical discharge accompanying a lightning flash, for example, superimposes sharp amplitude pulses atop the smoothly varying AM wave that, when demodulated, lead to the ear-jarring noise so familiar on summer evenings.

In the 1930s considerable effort was expended in developing "noise-free" radio, and this work culminated in the development of frequency modulation (FM). In this type of broadcasting, the information to be transmitted is not contained within amplitude variations, but rather the radio frequency itself is varied.

Presume again that we wish to transmit an audio frequency of 1000 Hz, using a radio frequency of 1000 kHz. In this case, however, we will vary the radio frequency continually between, say, 1000 and 1100 kHz, with the variation between the two extremes taking place at a *rate* of 1000 Hz, which is the audio frequency we wish to transmit. We may schematically depict this process as in Figure 6.25. The time taken to go between the two extremes (and return) is 10^{-3} s (the reciprocal of the audio frequency to be transmitted).

How do we convert such variations back into the audio signal? The method we will consider, although not used in practice, does work and most

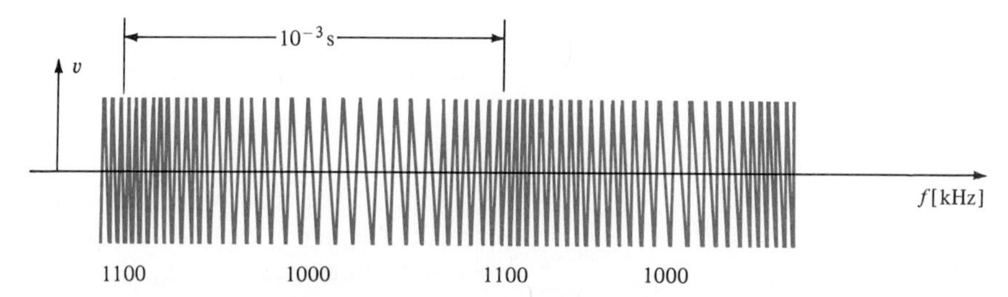

FIGURE 6.25. **Frequency Variation Between 1000 and 1100 kHz at a 1 kHz Rate** $f = 1/T = 1/10^{-3}\,\text{s} = 10^3\,\text{Hz}$.

easily demonstrates the basic principle without the necessity of getting bogged down in involved circuitry. In Figure 6.26 is shown the impedance variation of a parallel *LC* resonant circuit that we presume to be tuned to a resonant frequency considerably higher than 1000 kHz (for the specific numerical example we are using). As the applied frequency is varied between 1000 and 1100 kHz, the signal voltage developed across the tuned circuit will vary at a frequency that depends on the rate at which the signal "rides" up and down the skirt of the response curve.

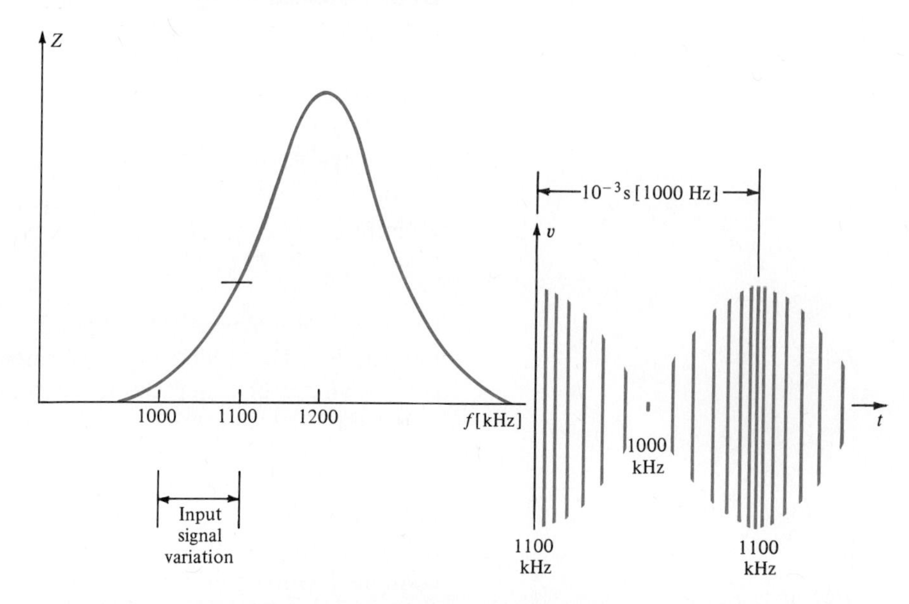

FIGURE 6.26. **Rudimentary Demodulation of FM Signal** Impedance variation of a parallel resonant circuit tuned to 1200 kHz is shown. The signal from Figure 6.25 leads to a rate of output amplitude variation equal to the rate of input frequency variation. FM has been converted into AM, which may be demodulated to regain the audio signal.

If the transmitter variation rate between 1000 and 1100 kHz is increased, the frequency of the output voltage from the receiver will likewise increase, corresponding to the higher audio frequency being transmitted. If we wish to increase the amplitude of the audio signal, we leave the rate at which it rides up and down the curve the same, but increase the frequency deviation from 1000 kHz to, let us say, 1200 kHz.

Since the frequency is continually changing, not being a single frequency, from a Fourier viewpoint we might suspect that there will be numerous harmonics present. Indeed there are, and their number depends on the frequency deviation. Additionally, a large frequency deviation leads to a large audio amplitude variation, which, in turn, leads to a large dynamic range that is second only to frequency bandwidth in terms of quality sound reproduction.

In the case of frequency modulation, the radio frequency itself is varied by the modulating frequency. Thus,

$$\begin{aligned} v &= A \cos(\omega_c t + m \sin \omega_m t) \\ &= A \cos \omega_c t \cos(m \sin \omega_m t) - A \sin \omega_c t \sin(m \sin \omega_m t) \end{aligned} \tag{6.20}$$

where the **modulation index** (m) in the case of FM is defined as the ratio of frequency deviation to modulation frequency.

Using the following expansions (to simplify matters, we retain only the first few terms), we may compute the frequency content:

$$\begin{aligned} \sin(m \sin \omega_m t) &= 2[J_1(m) \sin \omega_m t + \cdots] \\ \cos(m \sin \omega_m t) &= J_0(m) + 2[J_2(m) \cos 2\omega_m t + \cdots] \end{aligned} \tag{6.21}$$

These J_n coefficients are termed Bessel functions of the nth order. Their respective values as a function of the modulation index are shown in Figure 6.27. Note that when m is zero (no modulation) only J_0 is finite; all the power appears in the center frequency. As the modulation amplitude increases, m increases, and the center frequency may actually disappear—that is, J_0 goes to zero. The total radiated power stays the same, but power from the carrier is transferred into the sidebands. We would then expect that FM is a much more efficient process than AM.

Unlike AM in which a single modulation frequency (ω_m) results in two sidebands (at $\omega_c \pm \omega_m$), in FM a single modulation frequency can result in a whole series of sidebands (at $\omega_c \pm \omega_m$, $\omega_c \pm 2\omega_m$, $\omega_c \pm 3\omega_m$, and so on). The number of these "extra" sidebands depends on the modulation index.

The ratio of the *maximum* carrier frequency deviation to the *highest* modulation frequency is called the **deviation ratio**. In FM broadcasting the maximum carrier frequency deviation is ± 75 kHz, and the audio bandwidth extends from 50 Hz to 15 kHz. When the transmitted signal occupies a bandwidth equal to that for which the receiver was designed, 100% modulation is reached. If the frequency deviation is made greater, the received signal will be distorted.

Since the entire AM broadcast band is only a little over 1000 kHz wide (535–1605 kHz), operation with an FM deviation of 75 kHz would severely restrict the number of channels available. But by operating with a center frequency of about 100 MHz (100,000 kHz), such bandwidths represent but a small fraction of the available band (88–108 MHz), which is why FM broadcasting is confined to such high radio frequencies.

To minimize any static interference that, as we mentioned, represents an amplitude variation, the FM signal may be passed through a clipping circuit

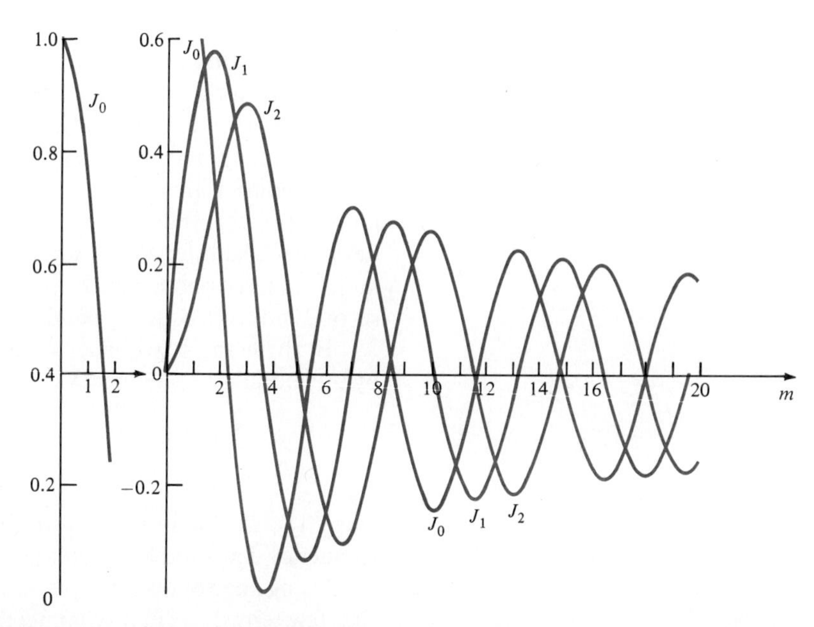

FIGURE 6.27. **Bessel Functions of Zero, One, and Two Orders vs. Modulation Index**

similar to that depicted in Figure 6.16. Figure 6.28 shows an FM signal that includes some noise pulses. The biasing levels of the two diodes in the clipping circuit are set so as to remove all amplitude variations. The output signal will have a uniform amplitude, but the frequency variations containing the information will still be there, ready to be demodulated.

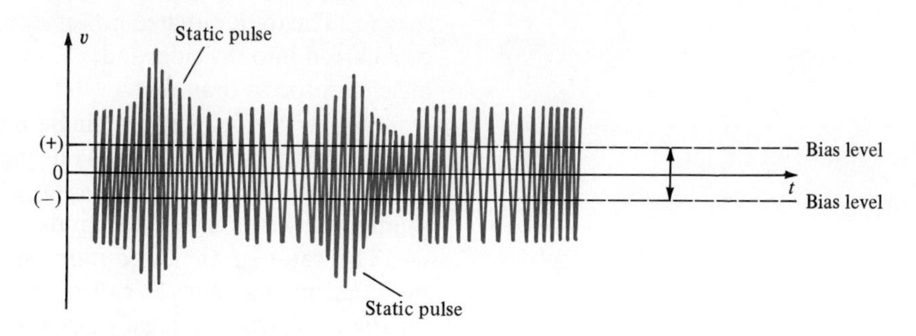

FIGURE 6.28. **FM Signal with Static (Noise) Superimposed**

6.5.4 Diode Specifications

The following constitute the most pertinent diode specifications:

- Maximum static forward voltage (V_F), at a specified static foward current (I_F)
- Maximum reverse voltage that can be applied before breakdown, variously designated as either PIV (peak inverse voltage), PRV (peak reverse voltage), V_{BR} (breakdown voltage), or $V_{R(max)}$

—Peak power dissipation
—Reverse recovery time (t_{rr})

One of the earlier signal diodes was the 1N914, which is rated as having V_{BR} = 100 V and 1 V (V_F) at 10 mA (I_F) under static conditions. There are two other versions of this diode, the 1N914A (1 V at 20 mA) and the 1N914B (1 V at 100 mA). Their reverse recovery time is 4 ns. This characteristic is important in pulse and high-frequency operation.

By way of comparison, we can cite a couple of diodes intended for power rectification, the 1N4002 with t_{rr} = 3500 ns and the 1N4007 at 5000 ns. These two diodes are members of a family that provides rectifiers suitable for a wide range of operating conditions with a 1 A rating.

Designation	V_{BR} [V]	I_F [A]
1N4001	50	1
1N4002	100	1
1N4003	200	1
1N4004	400	1
1N4005	600	1
1N4006	800	1
1N4007	1000	1

With their long recovery times, these diodes would not be suitable for signal processing.

Very fast recovery times are available from **Schottky diodes**. Rather than being constructed from the juncture between two semiconductors, they are made by joining a metal and a semiconductor. The forward voltage drop for such diodes is about 0.3 V, half that of a silicon diode, and the reverse recovery time is much smaller than a nanosecond.

SUMMARY

6.1 Introduction

- Resistance to electrical conduction in a vacuum arises from the collision of charged carriers with residual gas atoms.

6.2 The Physics of Electrical Resistance

- Resistance to electrical conduction in crystalline solids arises from the collision of charged carriers with impurity atoms, lattice defects, and the displacement of crystalline atoms from their equilibrium position due to thermal vibration.

6.3 Metallic Conduction

- Most metals exhibit positive temperature coefficients of resistance, the resistance increasing with a rise in temperature. This increased resistance is largely due to the thermal motion of the lattice atoms.

6.4 Semiconductors

- In a semiconductor, as the temperature rises, additional chemical bonds are broken, leading to the release of charged carriers, which become available for conduction. Therefore, semiconductors generally have negative temperature coefficients, their conductivity increasing with increasing temperature.

- A semiconductor having an equality of positive and negative carriers constitutes an intrinsic material.
- The purposeful introduction of excessive negative carriers (electrons) or positive carriers (holes) represents a doped semiconductor and determines that the material is, respectively, *n*- or *p*-type.

6.5 Semiconductor Diodes

- A semiconductor diode basically consists of a junction between *p*-type and *n*-type semiconductors. The *p*-type represents the anode and the *n*-type, the cathode.
- Forward conduction, with relatively small resistance, arises when the anode is positive relative to the cathode. A high resistance results when these polarities are reversed.
- The most common use of a diode is to convert ac into pulsating dc. These pulsations may be minimized by placing a capacitor across the load resistance, which is in series with the diode.
- With 60 Hz applied to a diode, the most significant Fourier component from a half-wave rectifier is 60 Hz. With a full-wave rectifier, it is 120 Hz.
- Significant forward conduction through a silicon diode commences at about 0.6 V. With typical currents, it amounts to about 0.7 V. Under saturated conditions, with large currents, it amounts to about 0.8 V.
- Diodes are broadly divided into signal diodes and rectifiers. The demarcation is generally decided by the forward current, which is 1 A or greater for rectifiers.
- The dc resistance of a diode may be significantly different from the resistance it presents to ac, termed the dynamic resistance.
- Signal amplitudes may be restricted by the use of clipping circuits. The simplest clipping circuit requires a resistor, a diode, and a biasing voltage.
- Clamping circuits are used to impose a desired dc level on an ac signal. In addition to a resistor and a diode, a clamping circuit requires a capacitor.
- The introduction of information onto a sinusoidal carrier signal constitutes modulation. If it is the amplitude of the carrier that is varied, it is amplitude modulation (AM). If it is the frequency itself that is varied, the result represents frequency modulation (FM).

KEY TERMS

bias: the application of a fixed voltage intended to establish a desired current
  forward: the application of a voltage polarity that results in easy conduction
  reverse: the application of a voltage polarity that retards conduction

clamper: a circuit used to establish a desired dc level for an ac signal

conductance (G): reciprocal of resistance [S]

conductivity (σ): a measure of the ease with which a material conducts $[\Omega \cdot m]^{-1}$

demodulation: the extraction of signal information from a composite signal

depletion region: the junction of a diode containing "uncovered" ionic charges; also termed the space-charge region

deviation ratio: in FM, the ratio of maximum carrier deviation to the highest modulation frequency

diffusion: charge movement because of concentration gradients

diode: device formed by the junction of *n*- and *p*-type semiconductors
  dc resistance: dc voltage divided by dc current
  dynamic resistance: ac resistance (r), approximated as $r = 26/I$ [mA], where I is the dc current

drift velocity: charge movement under the influence of an electric field

frequency deviation: in FM, the range of frequency variation due to modulation

majority carriers: the dominant species of the two types of charges present in a semiconductor

mean free path: travel distance that, as a result of collisions, leads to a diminution in the number of directed particles by a factor $1/e$

minority carriers: of the two types of charges present in a semiconductor, those present in the smaller concentration

mobility (μ): ratio of drift velocity to electric field intensity $[m^2/V \cdot s]$

modulation: the imposition of information on a waveform

amplitude: information varies with signal amplitude

frequency: information varies with signal frequency

modulation index: in FM, the ratio of frequency deviation to the modulation frequency; in AM, the ratio of modulating signal amplitude to carrier amplitude

Q-point: current and voltage under equilibrium conditions

rectifier: a device for converting ac into dc

full-wave: each half-cycle is rectified

half-wave: only alternate half-cycles are rectified

resistance (R): opposition to current [Ω]

resistivity (ρ): reciprocal of conductivity [Ω·m]

reverse recovery time (t_{rr}): upon terminating forward conduction, the time necessary for the ensuing reverse current to decay

reverse saturation current: the small diode current that flows under reverse bias conditions

Schottky diode: a metal-semiconductor diode

semiconductor: a material whose conductivity is intermediate between that of a conductor and an insulator (Other criteria also need to be satisfied for a material to be classed as a semiconductor.)

acceptor: doping atoms that provide conduction holes

donor: doping atoms that provide conduction electrons

doped: created by the purposeful introduction of impurity atoms that furnish charged carriers

extrinsic: a doped semiconductor

intrinsic: containing an equality of positive and negative charges

sensistor: a positive temperature coefficient semiconductor

temperature coefficient of resistance (α): rate of resistance change with temperature

negative: decrease of resistance with temperature

positive: increase of resistance with temperature

thermistor: negative temperature coefficient material, often used as a temperature sensor

SUGGESTED READINGS

Smith, R. A., "Physics and the New Electronics," J. SCI. INST., vol. 34, pp. 377–382, 1957. *The basis for our treatment of solid-state conduction.*

FENWAL THERMISTOR MANUAL EMC-6A (1974). Fenwal Electronics, P.O. Box 585, Framingham, Mass. 01701. *A thermistor reference.*

Fitzgerald, A. E., David E. Higginbotham, and Arvin Grabel, BASIC ELECTRICAL ENGINEERING, 5th ed., pp. 338–350. New York: McGraw-Hill, 1981. *Graphical analysis of diode circuits.*

Martin, Thomas L., Jr., ELECTRONIC CIRCUITS, pp. 467–479. Englewood Cliffs, N.J.: Prentice-Hall, 1955. *A readable treatment of the mathematics of amplitude and frequency modulation; also a comparison between FM and AM, including a consideration of interference and noise.*

Horowitz, Paul, and Winfield Hill, THE ART OF ELECTRONICS, pp. 41–43. New York: Cambridge University Press, 1980. *Diode clamps and limiters.*

EXERCISES

1. What is the estimated mean free path in a vacuum corresponding to a pressure of 10^{-8} mm Hg and a temperature of 150°C?

2. Semiconductor resistors that have positive temperature coefficients are called ______. Those with negative coefficients are ______.

3. Semiconductors that depend primarily on the free carriers released due to thermal vibration for their conduction are called ______ semiconductors. In those cases where additional atoms (either three- or four-valence type) have been added to increase the conductivity, they constitute ______ semiconductors.

4. Depending on the polarity of the major charge carriers, semiconductors are classed as either ______ type for plus (+) carriers or ______ for minus (−) carriers.

5. To achieve the same current-carrying capacity with aluminum as with copper, by what factor should the aluminum conductor be increased with regard to cross-sectional area?

6. While the mobility of electrons in semiconductors is greater than in copper, their lower conductivity is due to ______.

7. Conductivity involves the product of ______, ______, and ______.

8. Two common semiconductor materials used as transistors are ______ and ______. They both consist of ______ valence atoms.

9. Conductivity is the reciprocal of ______.

10. ______ is a resistive material used where very low temperature coefficients of resistance are desired.

11. ______ is a material used as electrical heater elements.

12. Voltage divided by distance constitutes ______.

13. The ratio of drift velocity to electric field intensity is termed ______.

14. What is the process of converting ac to dc called?

15. If a current is unidirectional but changing in magnitude, what is the name applied to the process?

16. What is the polarity applied to the anode and cathode of a forward-biased diode?

17. For a semiconductor diode, what types of semiconductors are used for the cathode and anode, respectively?

18. What is the function of the capacitor placed across the load resistor in a rectifier system? What precaution should be observed concerning the value of the capacitance?

19. For a half-wave rectifier operating at 60 Hz, what is the necessary relationship between the load resistance and the capacitor's reactance? How is this modified in the case of a full-wave rectifier?

20. What are the advantages of full-wave over half-wave rectification?

21. For silicon diodes significant conduction commences with a forward bias of about 0.6 V. What is this voltage called?

22. What are two of the resistances associated with a conducting diode?

23. What is the quantity $T/11{,}600$ called? What is its approximate value at room temperature?

24. What is the approximate expression for the dynamical resistance of a diode?

25. What designation is given to the intersection between the diode characteristic curve and the load line?

PROBLEMS

6.1
The SI unit of energy is the *joule* (J). In some problems this unit is too large, and the *erg* is used instead (10^7 ergs = 1 J). But in many electronic devices even the erg is too large a unit, and one uses the *electron volt* (eV). One electron, whose charge is 1.60×10^{-19}, "falling" through a potential difference of 1 V equals 1 eV. Express 1 eV in joules and in ergs.

6.2
Compute the mean free path of an electron in a vacuum whose pressure is 10^{-8} mm Hg at 20°C.

6.3
The following thought-provoking problem is taken from *The Art of Electronics* by P. Horowitz and Winfield Hill (New York: Cambridge University Press, 1980), p. 7:

> New York City requires 10^{10} watts of electrical power, at 110 volts (this is plausible: 10 million people averaging 1 kilowatt each). A heavy power cable might be an inch in diameter. Let's calculate what will happen if we try to supply the power through a cable 1 foot in diameter made of pure copper. Its resistance is 0.05 μΩ (5×10^{-8} ohms) per foot. Calculate (a) the power lost per foot from "I^2R losses," (b) the length of cable over which you will lose all 10^{10} watts [assuming dc], and (c) how hot the cable will get, if you know the physics involved ($\sigma = 6 \times 10^{-12}$ W/K^4 cm^2).

(This σ should not be confused with conductivity. It is, instead, Stefan's constant, which appears in the equation for power radiated from a surface at temperature T K: $P = \sigma T^4$.)

6.4

a. Determine the units of μ dimensionally.

b. With 10 mV applied to a copper conductor whose length is 1 cm, what is the drift velocity of the free electrons in the metal if the mobility of copper is 40 cm^2/V·s?

c. The same 10 mV is applied to 1 cm of intrinsic silicon whose electron mobility is 1300 cm^2/V·s. What is the electron drift velocity?

d. Why would the computation of drift velocity upon application of 1 V to a 1 cm length of copper represent an unrealistic problem?

e. Would 1 V applied to a 1 cm length of intrinsic silicon probably represent a realistic situation if the cross-sectional area of the silicon were comparable to that of 0000-gauge wire? (Take the resistivity to be 2.3×10^5 Ω·cm.)

6.5
The flowing of electrons with drift velocity u constitutes a current. If the concentration of charged carriers is n [per m^3], the current density J [A/m^2] is $J = neu$, where e is the charge per carrier (1.6×10^{-19} C). If the resistivity of copper at 20°C is 1.76×10^{-8} Ω·m, what is the average drift velocity in copper

carrying 1 A through an area of 10^{-2} cm^2? (Take the mobility to be 40 cm^2/V·s.)

6.6

a. At 20°C, compute the conductivity of intrinsic silicon.

b. At the same temperature, compute the conductivity of copper.

c. Compare the mobilities and conductivities of copper and intrinsic silicon. Why, with its much greater mobility, is the silicon called a *semi*conductor and copper, with a much smaller mobility, a conductor?

d. Using the results of part a, determine the resistivity of intrinsic silicon. Why does this result differ somewhat from that of Example 6.3?

6.7

Taking β (in the thermistor equation) to have a value of 2500 K, calculate the temperature coefficient α at 25°C and compare it with that for copper (0.0039 Ω/Ω·°C). What can you conclude concerning the use of a thermistor as a thermal sensor? (*Hint*: Solve the resistance equation for α, and express it in terms of a differential resistance variation.)

6.8 

With current passing through a thermistor, its equilibrium temperature will be determined by the rate at which the dissipated heat is carried away. Using a number of thermistors, each in conjunction with an indicator light, design a simple liquid-level indicator.

6.9

a. A particular thermistor has a resistance of 3100 Ω at 25°C. Its value of β is 3440 K. Plot its resistance between 0°C and 60°C, in 10°C steps, but include 25°C.

b. Compute the resistance at the same temperature points if the thermistor is paralleled by a 2040 Ω shunt resistance. Plot the results and compare with part a.

6.10

A relay using copper wire has a resistance of 5 kΩ at 25°C. It is meant to "pull in" at 1 mA of current.

a. Considering the temperature coefficient of resistance for copper (α = 0.0039/°C), compute the coil resistance between 0°C and 60°C at 10°C intervals (but also include T = 25°C).

b. Compute the "pull-in" voltage at the same temperature values ($V_{pull\ in} = R_{total} \times 1$ mA).

c. The relay is to be placed in series with the thermistor-resistor parallel combination of Problem 6.9b in the manner shown in Figure 6.29. Compute the total resistance at the same temperature points as in parts a and b.

d. Compute the percent change in resistance with and without the compensating network and plot the results.

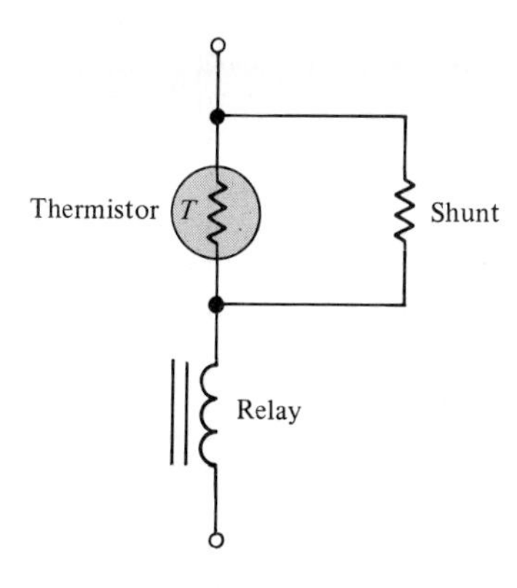

FIGURE 6.29

e. What is the required voltage for pull-in with the compensating network in place?

6.11

a. If a 120 V rms sinusoidal voltage is passed through a half-wave rectifier, what is the average value of the resulting voltage?

b. Answer the question in part a for a full-wave rectifier.

6.12

a. Draw the circuit for a bridge rectifier. Assume a polarity for the input terminals and draw the direction of the current.

b. Redraw the circuit, reverse the polarity of the input, and again draw the direction of the current.

6.13

a. Units of the 1N4153 high-speed switching diode, used for computer and general-purpose applications, are listed in the specification sheet as having (at 25°C) a reverse saturation current of 50 nA. With a forward current of 10 mA, the same specification sheet indicates a minimum static forward voltage drop of 0.7 V and a maximum of 0.81 V. In each of these two cases, using Eq. 6.14, compute the reverse saturation current I_0.

b. In each of these two cases, you will end up with a reverse saturation current that is miniscule compared to the expected value. The difficulty arises from the fact that at low values of current in silicon diodes the diode equation should take the form

$$I = I_0(e^{V/\eta V_T} - 1)$$

where η = 2 for silicon diodes and η = 1 for germanium diodes. Recalculate parts a and b using this modified equation. You should now obtain results that at least agree with the order of magnitude expected from the specification sheet data.

6.14 

Two silicon diodes are connected in series-opposition across a

5 V source. Find the voltage across each diode at room temperature (300 K). (Calculate using Eq. 6.14 and recalculate using the modified equation given in Problem 6.13. Which would you expect to be more nearly correct?)

6.15
Semiconductor diode current is rather sensitive to temperature variations. The reverse saturation current increases by about 7%/°C. Determine the change in temperature necessary to approximately double the diode current. (*Hint*: Use the compound interest law: $S = P(1 + i)^n$, where P is the principal, i is the interest per period, n is the number of interest periods, and S is the sum accumulated.)

6.16
a. What is the PIV to which a half-wave rectifier is subjected if the transformer's secondary voltage is 12 V rms?

b. What is the PIV with a capacitor across the load in part a?

c. What is the PIV to which a full-wave diode is subjected if the secondary transformer voltage is 12 V rms on each side of the center tap?

d. What is the PIV for a bridge-rectified diode with a 12 V rms secondary transformer voltage?

6.17 △
A measure of rectifier effectiveness is provided by the *ripple factor* (r), defined as

$$r = \frac{V_{ac}}{V_{dc}} = \frac{\text{rms value of ac components}}{\text{dc component}}$$

Obviously, the smaller the ripple factor, the more effective the rectification. Since

$$I_{rms}^2 R_L = I_{dc}^2 R_L + I_{ac}^2 R_L$$

$$I_{ac}^2 = I_{rms}^2 - I_{dc}^2$$

we have

$$r = \frac{\sqrt{I_{rms}^2 - I_{dc}^2}}{I_{dc}} = \sqrt{\left(\frac{I_{rms}}{I_{dc}}\right)^2 - 1}$$

Determine the ripple factor for half-wave and full-wave rectifiers.

6.18 △
With a capacitor filter, if the capacitor discharge rate between the peaks is small, it can be presumed to decay in a linear fashion. The peak-to-peak ripple can then be approximated as follows:

$$Q = CV$$

$$I = \frac{dQ}{dt} = C\left(\frac{dV}{dt}\right)$$

$$\Delta V = \left(\frac{I}{C}\right)\Delta t$$

where I is the assumed constant load current. For a full-wave rectifier, $t \simeq 1/(2f)$, and $t \simeq 1/f$ for a half-wave rectifier.

a. Calculate the necessary value of C for a bridge-rectified 10 V dc supply, with $I_L = 10$ mA and with a ripple of 0.1 V or less. (Disregard the diode drops.)

b. If each diode represents a 0.7 V drop, what should be the rms voltage rating of the necessary transformer?

6.19
If an AM radio station utilizes a carrier frequency of 960 kHz, what frequencies are being radiated by the station's antenna upon transmitting a pure audio tone of 3000 Hz?

6.20
Since an optimum length for an antenna is approximately half a wavelength, what should be the length of an antenna to be used for FM reception? (The FM band extends from 88 to 108 MHz.)

6.21
A signal voltage of $v_1 = 20 \sin \omega t$ [volts] is applied to the circuit in Figure 6.30. Sketch the form of the output voltage (v_2), including an indication of the magnitudes. (Disregard the diode voltage drops.)

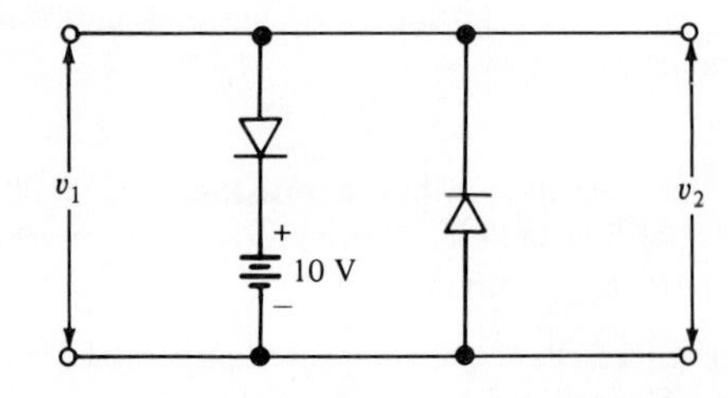

FIGURE 6.30

6.22
After passing through a capacitor, a signal assumes the form of a symmetric square wave, as shown in Figure 6.31.

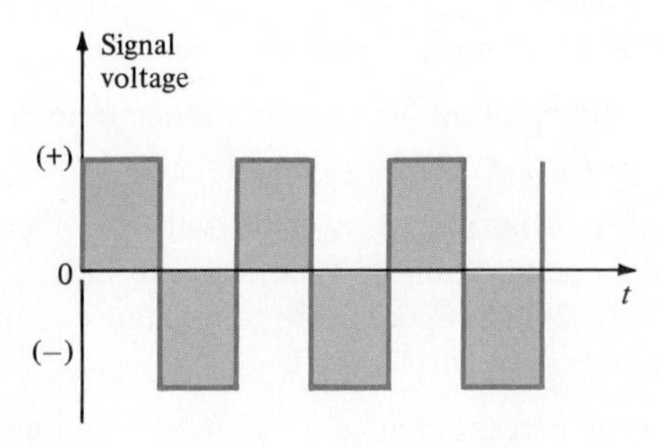

FIGURE 6.31

a. Sketch the circuit you would use to obtain the output shown in Figure 6.32.

b. Sketch the circuit you would use to obtain the output shown in Figure 6.33.

6.23

The *carrier* wave, which can be made to carry audio information, is represented as a constant amplitude sine wave whose frequency (f_c) in the AM broadcast band is, typically, 1000 kHz. Mathematically, as a voltage, the carrier wave can be represented as

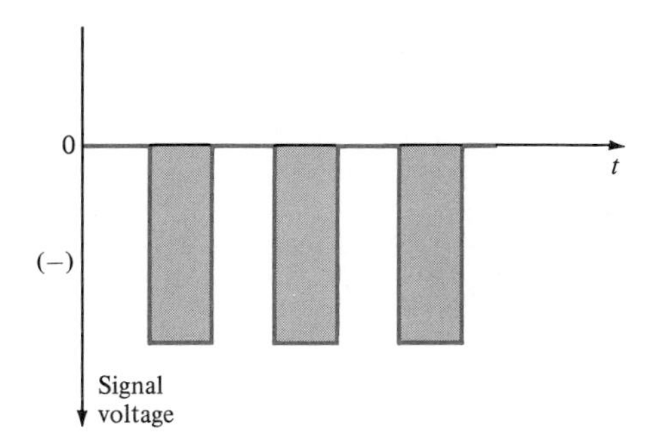

FIGURE 6.32

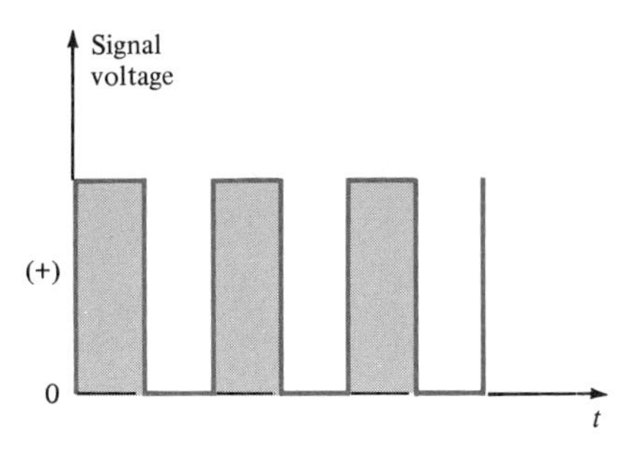

FIGURE 6.33

$$v_c = A \cos \omega_c t$$

The amplitude variation is given by the expression

$$a_m = 1 + m \cos \omega_m t$$

where m, called the *modulation index,* represents the magnitude of the amplitude vibration. (See Figure 6.34.) The modulated wave is

$$v = a_m v_c = A(1 + m \cos \omega_m t) \cos \omega_c t$$

where Am is the peak value of audio signal and $2Am$ the peak-to-peak value. With $m < 1$, using the trigonometric relationship, we have

$$\cos x \cos y = \tfrac{1}{2} \cos(x + y) + \tfrac{1}{2} \cos(x - y)$$

a. Compute the frequency content of the amplitude-modulated wave.

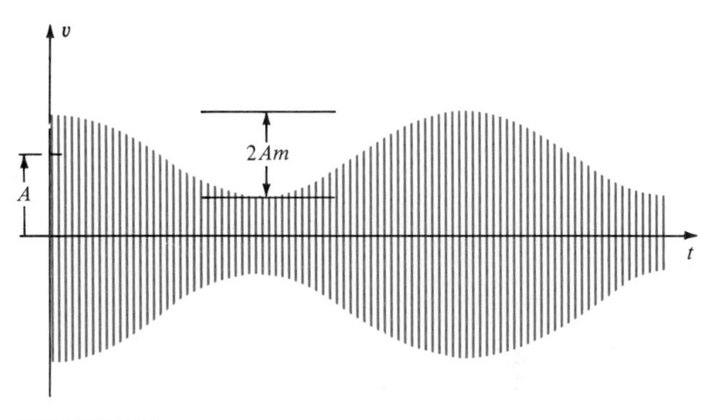

FIGURE 6.34

b. If the carrier frequency is 1000 kHz and the audio signal to be transmitted is 1 kHz, what frequencies are present in the transmitted signal?

c. The Federal Communications Commission limits the transmitted AM audio frequency to a maximum of 5 kHz. What bandwidth must be allowed for each channel?

6.24

a. The power content of an AM wave depends on the mean value of the voltage squared ($\overline{v^2}$). Compute its expression in terms of the carrier amplitude (A) as well as the modulation index (m).

b. For 100% modulation ($m = 1$), what are the relative portions of the power in the carrier and in the sidebands? What do you conclude about the efficiency of an AM wave? (See Figure 6.35.)

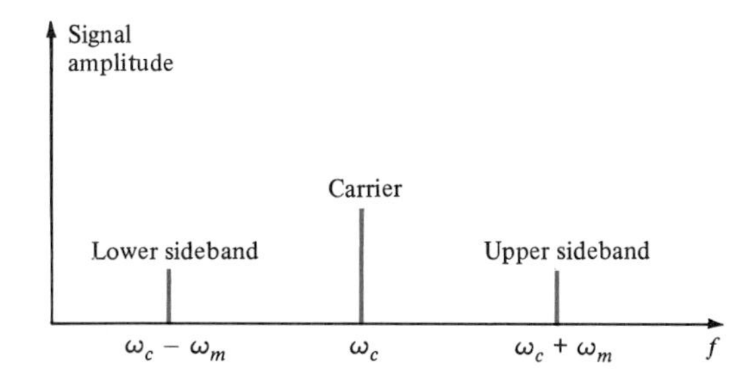

FIGURE 6.35

6.25

If the modulation index is greater than unity, the AM waveform is cut off during an interval of time. From the Fourier viewpoint, how might you expect this to affect the frequency content of the AM wave? (See Figure 6.36.)

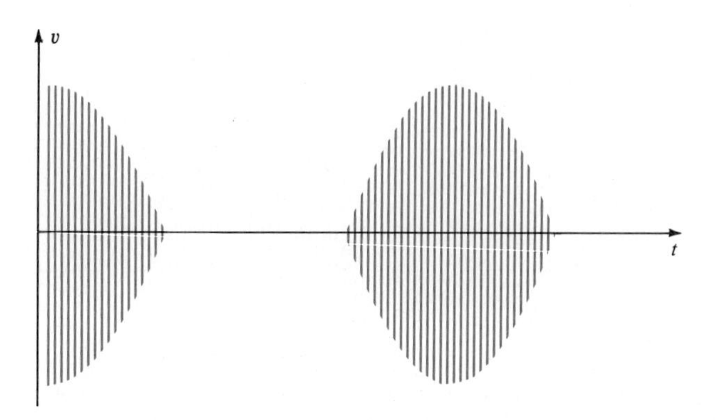

FIGURE 6.36

6.26

Using Eq. 6.20, and the following identities in conjunction with Eq. 6.21, compute the frequency content of v and sketch the results in bar graph form:

$$\cos a \cos b = \tfrac{1}{2}[\cos(a - b) + \cos(a + b)]$$
$$\sin a \sin b = \tfrac{1}{2}[\cos(a - b) - \cos(a + b)]$$

6.27

a. What is the modulation index of the signal whose deviation is 3 kHz with a modulation frequency of 1 kHz? 3000 Hz with the same deviation? 100 Hz with the same deviation?

b. What is the deviation ratio used in FM broadcasting?

6.28

With a sinusoidal voltage of peak amplitude greater than V_R impressed at the inputs of the circuits in Figure 6.37, qualitatively sketch the nature of the outputs.

6.29

Show the validity of the current division rule as applied to a parallel combination of R and C as in Example 6.9.

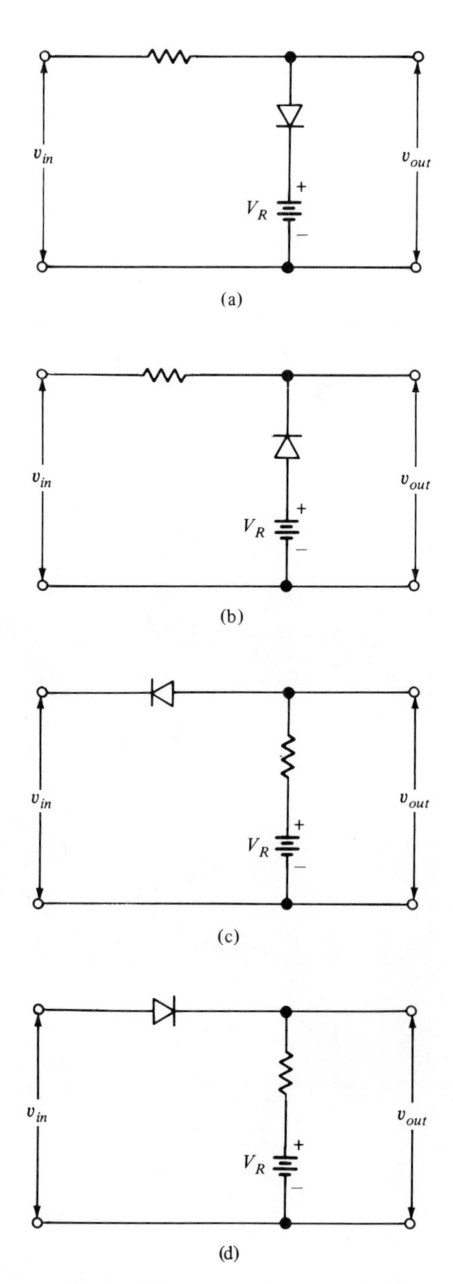

FIGURE 6.37

7 BIPOLAR JUNCTION TRANSISTORS

OVERVIEW

We have now arrived at one of the culmination points of this text, the transistor amplifier. More accurately, this objective should be expressed as "one form of transistor amplifier," for a variety of transistor categories exists: the bipolar junction transistor (BJT), which we consider in this chapter; the field-effect transistor (FET), which we consider in its diverse forms in Chapter 9; and, finally putting together the different categories of transistors (together with the necessary capacitors and resistors that accompany them when made into amplifiers), the integrated circuit, which we discuss in Chapter 11.

To arrive at this point, we have first had to consider the three basic elements of electrical circuits—resistor, capacitor, inductor—together with the two basic forms of electrical power—dc and ac—and finally the single most important foundation of our intellectual construct—the nature of electrical conduction through crystalline solids.

OUTLINE

7.1 ▪ INTRODUCTION

We return now to the consideration of an automotive ignition system. Consider the two displays shown in Figure 7.1. The first prevails at an engine speed of 500 rpm and the second, at an engine speed of 2000 rpm. One of the most notable differences between these two graphs is the shorter duration of the arc line in the second case. The reason for this behavior is easily explained: At the higher engine speed, there is less time for the current to build up the magnetic field of the "coil," and consequently less energy is available to be delivered in the ignition process.

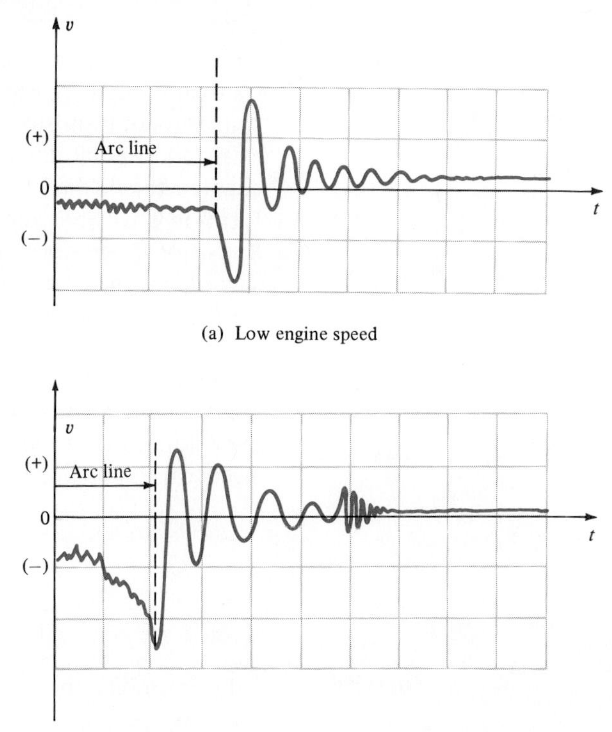

FIGURE 7.1. Ignition Arc Line Terminations Higher engine speed results in an earlier termination than lower speed due to the longer dwell time in the latter case.

With less time available, the proper corrective action would call for either a larger current flow during the time that the contacts are closed or a longer dwell time. But in either case this would probably lead to a greater degree of heating at the points and, consequently, a more rapid erosion of the contact points. The development of the transistor ignition system represents a significant advance in alleviating this difficulty. Figure 7.2 is meant to be representative of the improved performance that is available.

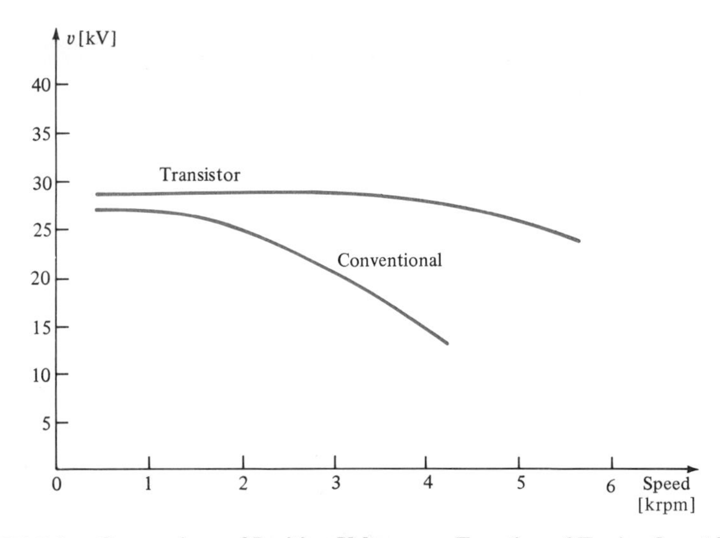

FIGURE 7.2. **Comparison of Ignition Voltage as a Function of Engine Speed Between Conventional and Transistorized Systems**

7.2 ▪ THE BIPOLAR JUNCTION TRANSISTOR: AN INITIAL CONSIDERATION

If another section of *p*-material is added to an existing *pn* junction of a diode (Figure 7.3), with the lower *pn* junction forward-biased—that is, with significant current flow—a considerable current will also flow in the upper *p*-section. The lower section is termed the *emitter,* the middle section is termed the *base,* and the upper section is termed the *collector.* The polarity of the battery between the base and emitter (V_{BB}) is such that holes are injected into the emitter and move toward the base. But because the base region is very thin, the majority of these carriers pass through the base region and find themselves in the collector, where they are subject to the attractive negative potential due to V_{CC}.

Some of the emitter current (I_E) does end up constituting a current in the base-emitter circuit (I_B). (Some of this current is also made up of electrons passing from the *n*-type base into the *p*-type emitter, but as we've previously indicated, this is equivalent to classical current passing in the other direction, so we consider I_B to be made up of both components.)

Denote I_E as the emitter current and α as the fraction that goes to the collector. Since α typically has a value between 0.95 and 0.99, the collector current is

$$I_C = I_E - I_B \tag{7.1}$$

$$\alpha I_E = I_E - I_B \tag{7.2}$$

$$I_B = (1 - \alpha)I_E \tag{7.3}$$

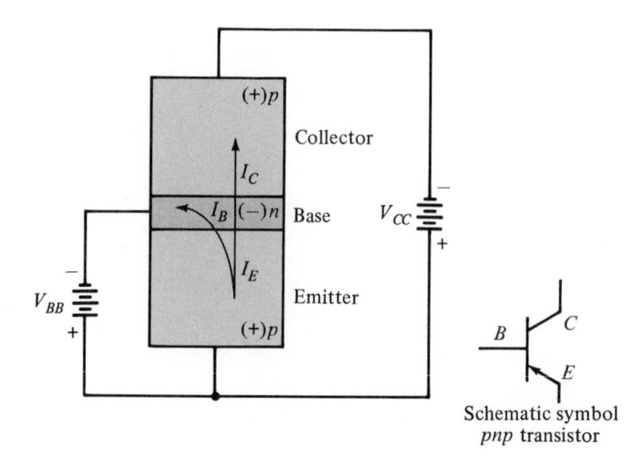

FIGURE 7.3. **Emitter Current Division Between Base and Collector of a *pnp* Transistor**

$$I_C = \alpha I_E \tag{7.4}$$

$$I_C = \frac{\alpha}{1 - \alpha} I_B \tag{7.5}$$

$$\Delta I_C = \frac{\alpha}{1 - \alpha} \Delta I_B \tag{7.6}$$

Eq. 7.6 indicates that a change in the base current (ΔI_B) leads to a collector change of ΔI_C. Since $\alpha \simeq 0.99$, we have

$$\Delta I_C = \frac{0.99}{1 - 0.99} \Delta I_B \simeq 100\, \Delta I_B$$

We see, therefore, that a small change in the base current can lead to a collector current change that is, typically, 100 times greater. Here, then, is the basis of amplification with a transistor. The quantity $\alpha/(1 - \alpha)$ constitutes the transistor's **beta** (β) and is a measure of the current-amplifying ability of the transistor.

Alternatively, the transistor's β is often designated as h_{fe}.[1] Since we will use β to designate something entirely different in a subsequent chapter, we will henceforth use h_{fe} to indicate transistor current gain. (The reader should keep in mind, however, that this transistor parameter is frequently called its beta value.) We have

$$\frac{\alpha}{1 - \alpha} \equiv \beta \equiv h_{fe} \tag{7.7}$$

Rather than an *npn* combination, one could alternatively sandwich a *p*-section between two *n* sections. Such an *npn* transistor operates much in

[1]Such designations as h_{fe} arise from an approximation (termed the *h*-parameter model) used to describe small-signal BJT operation and apply to ac signals. h_{FE} applies to dc conditions—that is, I_C/I_B. Generally, $h_{fe} \simeq h_{FE}$.

the same manner except that all dc polarities must be reversed since one is dealing primarily with electron flow. However, in keeping with current convention, we continue to treat the circuit as if we were dealing with positive carriers. (See Figure 7.4.) Additionally, it is not necessary to employ two separate batteries (or other power sources) in conjunction with BJT circuits. We have done so initially merely for the sake of simplicity. In the next section we will show how this is accomplished with but a single power source and voltage dividers.

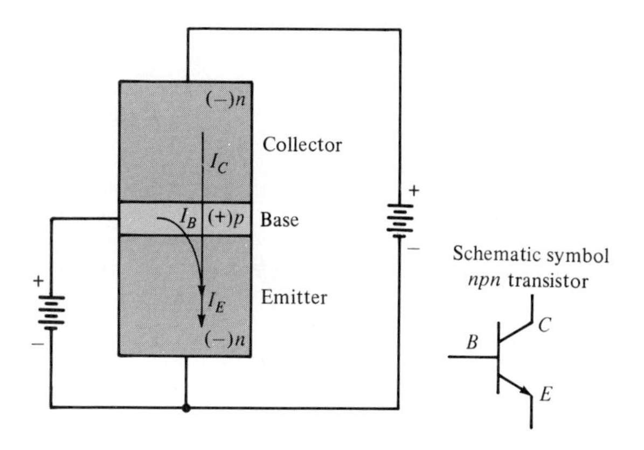

FIGURE 7.4. **Convergence of Collector and Base Currents to Form Emitter Current of an *npn* Transistor**

An important design feature of transistor amplifiers stems from the fact that the values of h_{fe}, even for transistors with the same designation, can vary by as much as 100%. Thus, upon replacing a transistor, you cannot depend on the value of h_{fe} being the same as that for the initial unit. Also, h_{fe} varies considerably with temperature. A properly designed circuit must minimize the influence that such variations have on its operation.

7.3 ■ dc OPERATING CONDITIONS

The base of a BJT acts in a fashion comparable to that of a valve in a hydraulic system—that is, it controls the current between emitter and collector. For proper circuit operation the base current must be adjusted to a specific initial value I_B.[2] We turn first to a consideration of how this adjustment is made, after which we will turn to treating the amplification process—how a signal is applied and how an amplified version is extracted.

Consider Figure 7.5. Since the base and emitter constitute a forward-biased diode with a typical voltage drop of 0.7 V, if I_B represents the desired operating base current, the value of R_B may be determined by the equation

[2]A summary of notation usage is contained in "A Closer Look" at the end of this chapter.

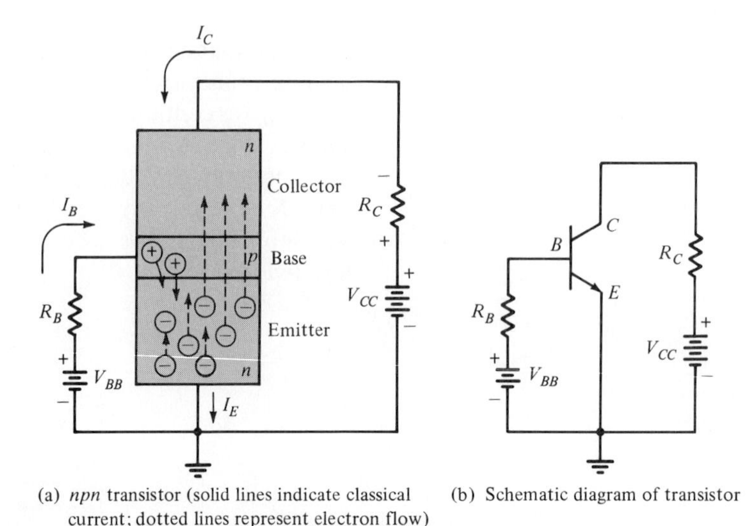

(a) *npn* transistor (solid lines indicate classical current; dotted lines represent electron flow)

(b) Schematic diagram of transistor

FIGURE 7.5. Bipolar Junction Transistor (BJT)

$$R_B = \frac{V_{BB} - 0.7}{I_B} \tag{7.8}$$

Figure 7.6, however, shows the promised method of achieving the same condition with a single power source. The same desired value of I_B may now be obtained using the equation

$$R_B = \frac{V_{CC} - 0.7}{I_B} \tag{7.9}$$

The circuit arrangement in Figure 7.6, however, tends to be unstable, and the biasing arrangement in Figure 7.7 is preferred. The base current value is adjusted by means of the voltage divider (R_1, R_2) in conjunction with R_E (called the emitter resistor). The amplified base current appears in the resistive load R_C (called the collector resistor).

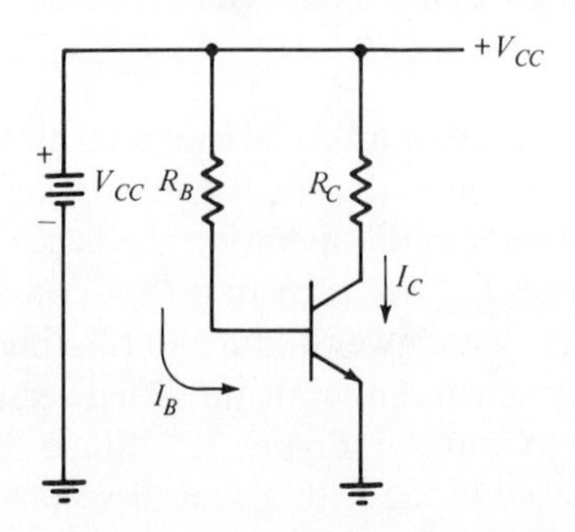

FIGURE 7.6. Fixed Bias *npn* Transistor with Grounded Emitter

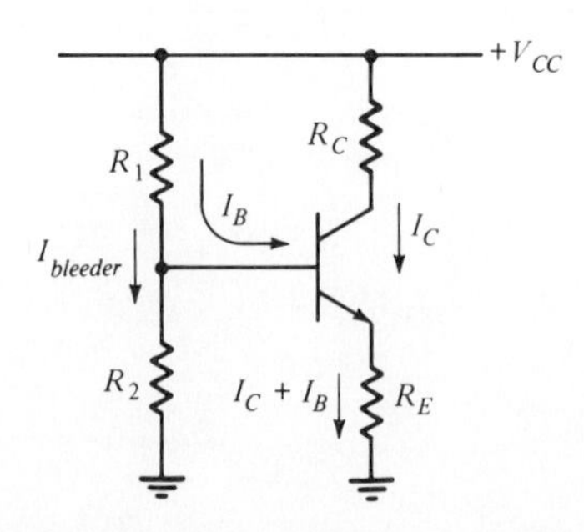

FIGURE 7.7. Self-Biased *npn* Transistor Via Emitter Resistor and Voltage Divider

To assure proper operation, for a given transistor there will be a specific value of dc collector current (I_C) that, in conjunction with the voltage drops across the external loads, will lead to a specific value of voltage between collector and emitter (V_{CE}). This combination of dc current and voltage constitutes the operating quiescent point of the transistor, termed its ***Q*-point**.

It is important to remember that the basic process of electronic amplification is accomplished by converting the dc of the power source into current (or voltage) variations that represent magnified versions of the signals applied for amplification.

FOCUS ON PROBLEM SOLVING: Base Current Determination

1. If a single resistor (R_B) is employed (as in Figure 7.6), with the voltage drop between the base and emitter being approximately 0.7 V (V_{BE}), we have

$$R_B = \frac{V_{CC} - 0.7}{I_B}$$

where I_B is the desired dc base current.

2. If a voltage divider is employed (such as R_1 and R_2 in Figure 7.7), it can be replaced by its Thevenin equivalent.

$$V_T = \frac{R_2}{R_1 + R_2} V_{CC} \quad \text{and} \quad R_T = \frac{R_1 R_2}{R_1 + R_2}$$

The latter represents R_1 and R_2 in parallel, which arises when the voltage source (V_{CC}) is shorted as part of the Thevenin evaluation. The base-emitter KVL equivalent then takes the form of Figure 7.10—that is,

$$V_T - I_B R_T - 0.7 - (I_C + I_B)R_E = 0$$

But $I_C = h_{FE} I_B$. Therefore,

$$V_T - I_B R_T - 0.7 - h_{FE} I_B R_E - I_B R_E = 0$$
$$V_T - I_B R_T - 0.7 - I_B R_E(h_{FE} + 1) = 0$$

With factors V_T, R_T, R_E, and h_{FE} specified, one may solve for I_B. If R_E is absent—that is, the emitter is grounded—the solution reduces to a consideration of only the first three terms.

EXAMPLE 7.1

In the circuit in Figure 7.8, determine the value of R_B that will lead to a value $I_C = 1$ mA. Compute the percentage change in the collector current when h_{FE} changes from the initial value of 100 to a value of 200.

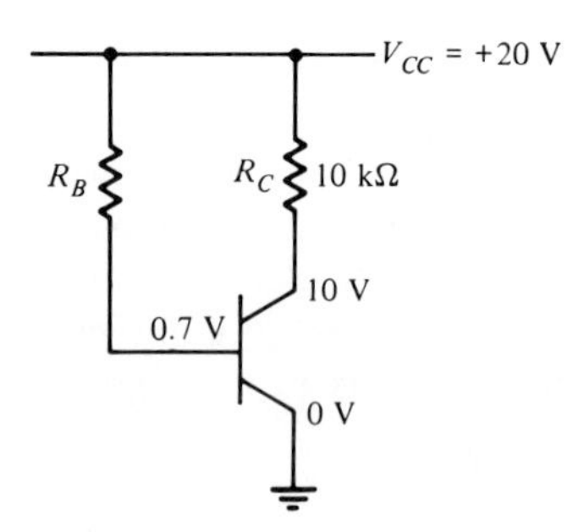

FIGURE 7.8

Solution:

$$I_B = \frac{I_C}{h_{FE}} = \frac{1\text{ mA}}{100} = 0.01\text{ mA}$$

$$R_B = \frac{V_{CC} - V_{BE}}{I_B} = \frac{20 - 0.7}{10^{-5}} = 1.93 \times 10^6 = 1.93\text{ M}\Omega$$

If $h_{FE} = 200$, we have

$$I_C = h_{FE} I_B = (200)(0.01)\text{ mA}) = 2\text{ mA}$$

$$\%\text{ variation} = \frac{I_{C2} - I_{C1}}{I_{C1}} \times 100 = \frac{2\text{ mA} - 1\text{ mA}}{1\text{ mA}} \times 100 = 100\%$$

EXAMPLE 7.2 Compute the percentage change in collector current when the h_{FE} of a transistor in the circuit in Figure 7.9 changes from 100 to 200. Take $I_C = 1.0$ mA when $h_{FE} = 100$. (*Note:* 1.71 is *not* the open-circuit voltage of the divider.)

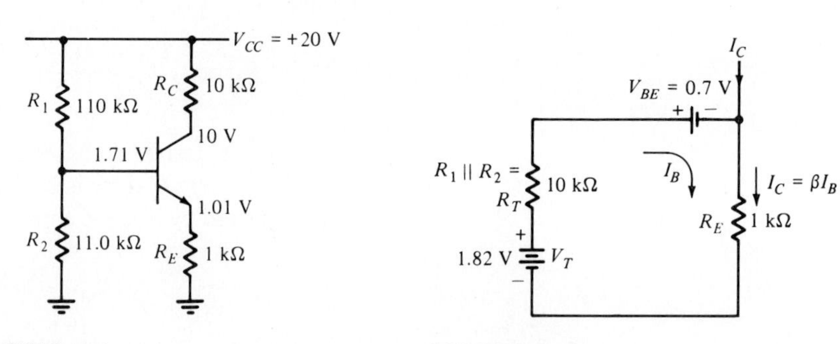

FIGURE 7.9

FIGURE 7.10

Solution:

$$V_T = \frac{R_2}{R_1 + R_2} V_{CC} = \frac{11 \times 10^3}{(110 + 11) \times 10^3} 20 = 1.82\text{ V}$$

$$R_T = \frac{R_1 R_2}{R_1 + R_2} = \frac{(110 \times 11) \times 10^6}{(110 + 11) \times 10^3} = 10 \times 10^3 = 10\text{ k}\Omega$$

The Thevenin equivalent circuit then looks as shown in Figure 7.10, and we have

$$V_T - R_T I_B - V_{BE} - (h_{FE} + 1) I_B R_E = 0$$

$$1.82 - 10^4 I_B - 0.7 - (100 + 1) I_B \times 10^3 = 0$$

$$I_B = \frac{1.82 - 0.7}{10^4 + 101(10^3)} = \frac{1.12}{1.11 \times 10^5} = 1.01 \times 10^{-5} = 0.0101\text{ mA}$$

$$I_C = h_{FE} I_B = 100 I_B = 100 \times 1.01 \times 10^{-5} = 1.01\text{ mA}$$

For $h_{FE} = 200$, we have

$$1.82 - 10^4 I_B - 0.7 - (200 + 1) I_B \times 10^3 = 0$$

$$I_B = \frac{1.12}{2.11 \times 10^5} = 5.3 \times 10^{-6} = 0.0053 \text{ mA}$$

$$I_C = 200 I_B = 1.06 \text{ mA}$$

$$\% \text{ change in } I_C = \frac{1.06 - 1.01}{1.01} \times 100 \simeq 5\%$$

7.4 ■ TRANSISTORIZED IGNITION

Figure 7.11 shows a simplified version of a transistorized ignition system. There is considerable current passing through the collector to the primary of the transformer, but unlike the mechanical ignition system, this high current does not pass through the contacts of S_1 but rather through the transistor. A transistor capable of handling the necessary power is utilized. Typically, the collector current might be 8 A, while the base current is 0.3 A. This is to be compared to the 3–4 A flowing through the contacts of a mechanical system.

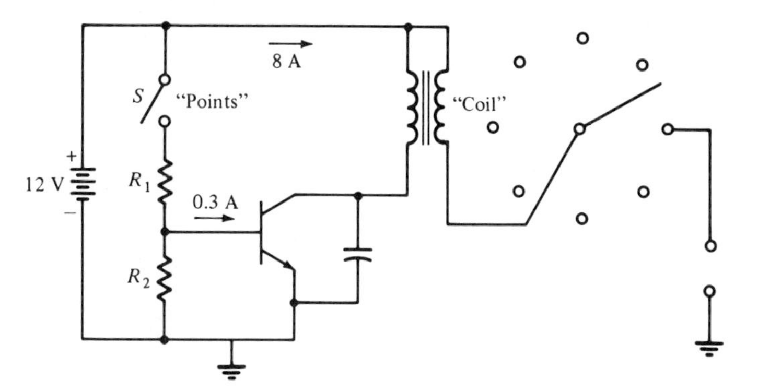

FIGURE 7.11. **Simplified Transistor Ignition** The opening of "points" removes forward bias, terminating collector current and initiating induction of a high-voltage secondary pulse.

When the points are opened, the current from the resistive branch is removed, thereby removing the forward-biasing current in the emitter-base region. With this current disrupted, there is a sharp drop in the collector current, and the magnetic field collapses, leading to the sequence of events as discussed in Section 3.4. Since the contacts do not carry the full ignition current, the dwell time and coil current both may be increased without concern about increased contact erosion.

7.5 ■ VOLTAGE AMPLIFICATION

A frequent demand put upon a transistor is to provide an amplified version of an input signal *voltage*. Figure 7.12 illustrates the successive steps. A small signal voltage variation V_{be} is applied between the base and emitter. Since these electrodes make up a diode, a relatively small voltage change can lead to a large change in base current. (Capital letters with lowercase subscripts signify rms values. See "A Closer Look" at the end of this chapter.) Diode current in the forward direction varies exponentially with voltage (Eq. 6.14), providing that a threshold of about 0.6 V (for silicon) is exceeded. The base current variation (I_b) is magnified by a factor h_{fe}, leading to the collector current variation I_c. If I_c is passed through a collector resistance (R_C), one obtains an output signal voltage V_{out} that exceeds that of the input, V_{be}.

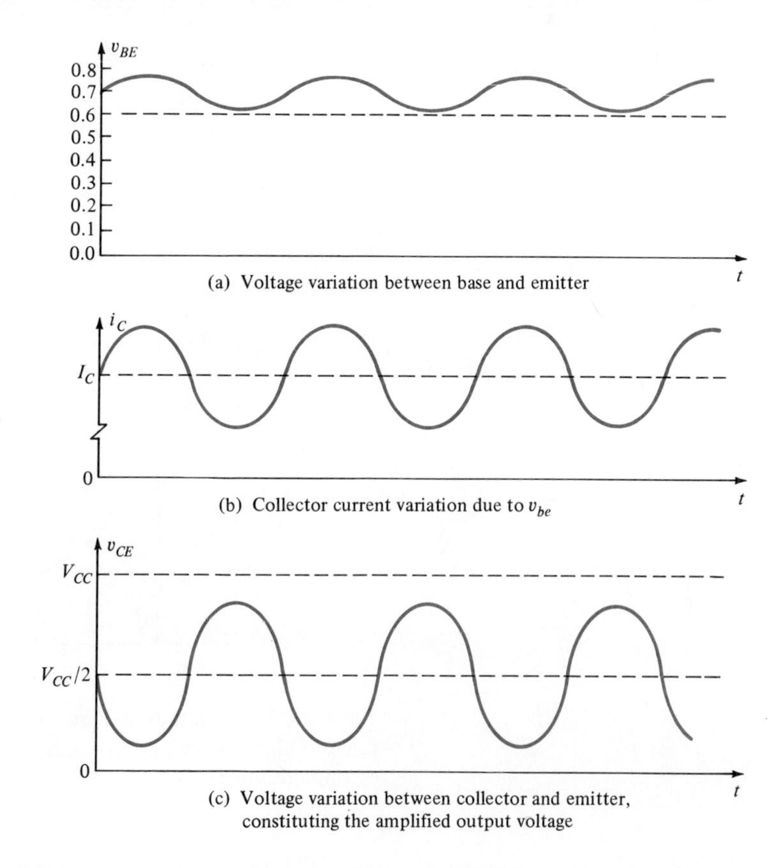

FIGURE 7.12. Transistor Voltage and Current Variation Note the 180° phase reversal between parts a and c.

The output voltage is 180° out of phase with the input (Figure 7.12c). When the input signal assumes its greatest positive value, the collector voltage drops to its lowest value and vice versa ($v_{ce} = V_{CC} - i_c R_C$).

It can be seen that the collector excursions should be limited to values between 0 and $2I_C$ if the waveform is not to be grossly distorted. The transistor is said to be in *saturation* upon reaching its upper current limit, at which time the output voltage reaches a value of approximately zero. On the other hand, when the collector current is made to assume a value of zero, the transistor is said to have reached its *cutoff* condition, and the output voltage is at its maximum value, V_{CC}. The variation in v_{be} that leads to this range of i_c variation is only about ± 20 mV, justifying the common assumption that the base-emitter voltage remains at approximately its central value of 0.7 V.

Recall again that the base-emitter junction represents a forward-biased diode, while the equivalent base-collector diode is normally reverse-biased. (For example, the collector is more positive than the base for an *npn* transistor.) By considering the base to remain at $+0.7$ V, should the collector current through R_C be such as to drop the collector voltage to 0.7 V, the base-collector junction bias is then zero. A slightly greater collector current, resulting in a drop of the collector voltage to $+0.2$ V, gives rise to a significant forward bias between them (0.7 V $-$ 0.2 V $=$ 0.5 V). This forward bias represents the onset of transistor saturation, with the collector current becoming essentially independent of the base current and resulting in a significant drop in the transistor's h_{FE} value.

EXAMPLE 7.3

The h_{FE} of the transistor in the circuit in Figure 7.13 is 100. Compute the collector current and the collector-to-emitter voltage when the switch is closed.

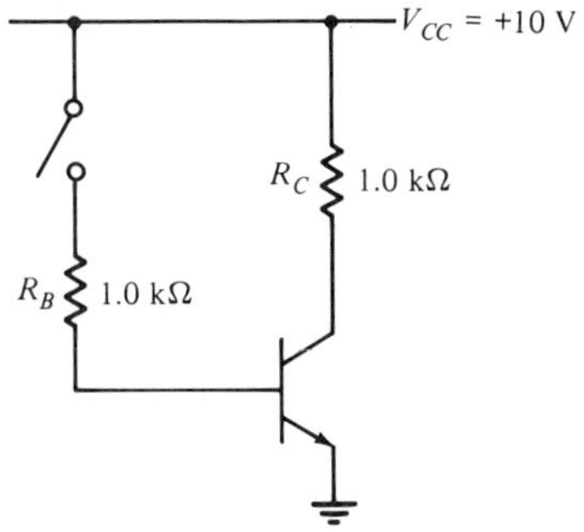

FIGURE 7.13

Solution:

$$I_B = \frac{V_{CC} - V_{BE}}{R_B} = \frac{10 - 0.7}{10^3} = 9.3 \times 10^{-3}\text{ A} = 9.3\text{ mA}$$

$$I_C = h_{FE} I_B = 100(9.3 \times 10^{-3}) = 9.3 \times 10^{-1}\text{ A}$$

$$V_{CE} = V_{CC} - I_C R_C = 10 - (9.3 \times 10^{-1})(10^3)$$

$$= 10 - 930 = -920\text{ V!!}$$

This solution, of course, is an impossible one. Depending on the collector current, the collector voltage can only assume values between $+10$ V (if $I_C = 0$) and 0 V (if $I_C = 10$ mA).

Actually, for a typical transistor, saturation is reached when the collector-to-emitter voltage is reduced to about 0.2 V. Saturation means that the base current variation is no longer effective in controlling the collector current. Pulse circuits are operated in this fashion, but linear circuits intended to faithfully amplify the input signal are not.

Therefore, the *collector* current with the switch closed is

$$I_C = \frac{V_{CC} - V_{CE\ saturated}}{R_C} = \frac{10 - 0.2}{10^3} = 9.8 \times 10^{-3}\text{ A} = 9.8\text{ mA}$$

What this example is also meant to show is that h_{FE} is not a constant independent of the operating conditions. In this case it has been reduced to approximately unity.

From Figure 7.12 one can note that the optimum average dc voltage between the collector and emitter (V_{CE}) is about $V_{CC}/2$. The value of V_{CE}, together with the average value of collector current (I_C), constitutes the operating Q-point of the transistor.

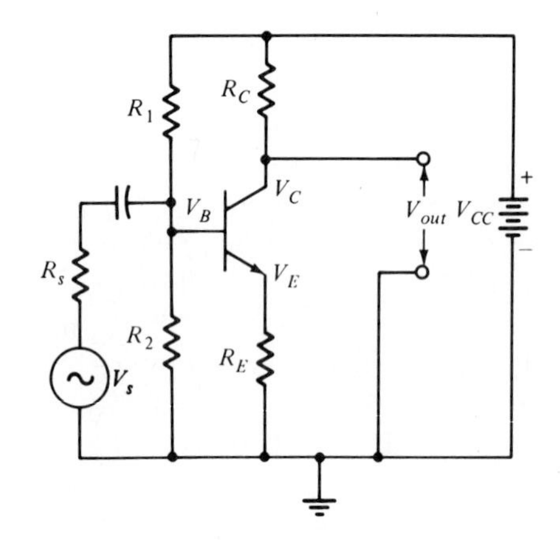

FIGURE 7.14. **Self-Biased Transistor with Signal Output Between Collector and Ground, with Input Signal Applied Via Blocking Capacitor *C***

While it is the base-emitter voltage that determines the collector current, we do not "set" the Q-point by such means; it is much too sensitive. A 60 mV variation of V_{BE} would have I_C change by a factor of 10. Instead it is the much less sensitive base current that is adjusted to yield the desired Q-point I_C. Since the collector current variation due to the signal varies by less than a factor of 2, the v_{be} variation due to the signal is very small and provides justification for assuming that V_{BE} remains at ~0.7 V.

An actual amplifier circuit, with signal source V_s, source resistance R_s, bias network (R_1 and R_2) in place, emitter resistor (R_E), and collector resistance (R_C), is shown in Figure 7.14. The amplified current passes through R_C, and the resulting voltage variation between the collector and ground constitutes the output signal voltage. Such a configuration, where the input is applied between the base and emitter and the output is withdrawn from between the collector and emitter, constitutes the most frequently encountered transistor configuration and is known as a **common emitter** (CE) **amplifier.**

Since V_{BE} remains fairly constant at 0.7 V, when the base voltage changes in response to a signal voltage, the emitter voltage follows that change. Thus, $\Delta v_B \simeq \Delta v_E$. The change in emitter current is approximately equal to the change in collector current since h_{fe} is assumed to be large:

$$\Delta i_E = \frac{\Delta v_E}{R_E} \simeq \frac{\Delta v_B}{R_E} = \Delta i_C \tag{7.10}$$

Or, in terms of rms values, we have

$$\frac{V_b}{R_E} = I_c \tag{7.11}$$

$$V_c = -I_c R_C = -V_b \frac{R_C}{R_E} \tag{7.12}$$

$$A_V \equiv \text{amplifier voltage gain} = \frac{V_{out}}{V_{in}} = \frac{V_c}{V_b} = -\frac{R_C}{R_E} \tag{7.13}$$

where the negative sign indicates a 180° phase reversal between input and output signal voltages (Figure 7.12).

Suppose $R_E = 0$. Do we get infinite gain? No, because the base-emitter diode itself has a small ac resistance equal to

$$r_e \simeq \frac{26\ [\text{mV}]}{I_C\ [\text{mA}]} \tag{7.14}$$

Thus, for the case when $I_C = 1$ mA, a typical value, we have

$$A_V \simeq -\frac{R_C}{26} \tag{7.15}$$

Can we increase the voltage gain by increasing R_C? We can within limits. As R_C increases in value, if V_{CE} gets too small, the allowable collector current swing is limited, and the value of h_{fe} diminishes.

Since the collector signal current passes through both R_C and R_E, the total amplified signal appears in part across each of these resistors. But only that across R_C represents usable output.

R_E is necessary for proper dc operation, but if no signal loss in R_E is to occur, we must place a capacitor in parallel with R_E. Such an emitter bypass capacitor shorts the signal currents around R_E. Thus, no signal voltage loss occurs at R_E, but since the capacitor represents an open circuit for dc, it does not disturb the dc operating conditions. Since the dynamic resistance r_e appears in series with R_E, the usual design condition requires that the capacitive reactance of C_E be small compared with the value of r_e at the lowest frequency of interest.

One rule of thumb is to make $X_C \leq 0.1R_E$ at the lowest frequency. But since, in reality, r_e appears in series with R_E and generally represents a smaller resistance, it would appear desirable to make $X_C \leq 0.1r_e$.

EXAMPLE 7.4

For the circuit in Figure 7.15, what should be the value of the bypass capacitor placed in parallel with R_E if the amplifier is to be effective down to 20 Hz?

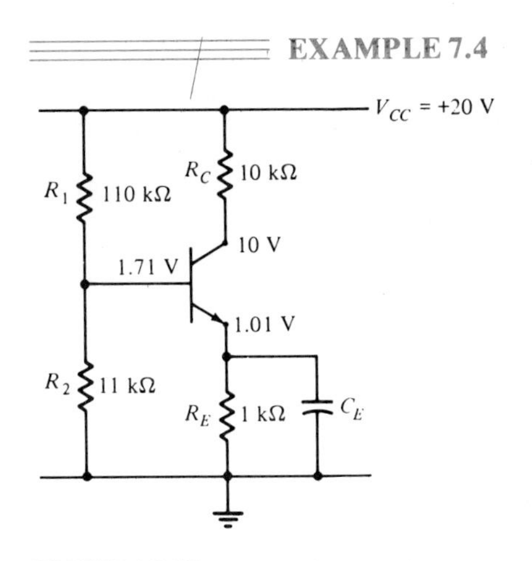

FIGURE 7.15

Solution: Across R_E we need to place a capacitor whose reactance at 20 Hz is much less than the resistance r_e. A rule of thumb is to have $X_C \leq 0.1r_e$. Therefore, with the collector current at 1 mA (since the voltage drop across R_C = 10 V and 10 V/10 kΩ = 1 mA), we have

$$r_e \simeq \frac{26\ [\text{mV}]}{I_C\ [\text{mA}]} = \frac{26}{1} = 26\ \Omega$$

Set $X_C = 2.6\ \Omega$, and we have

$$C = \frac{1}{2\pi f X_C} = \frac{1}{2\pi(20)(2.6)} = 3000\ \mu\text{F}$$

7.6 ■ CASCADED AMPLIFIERS

Only very infrequently will a single stage of voltage amplification prove sufficient. We turn now to a consideration of how signal gain may be increased by cascading amplifier stages, as in Figure 7.16.

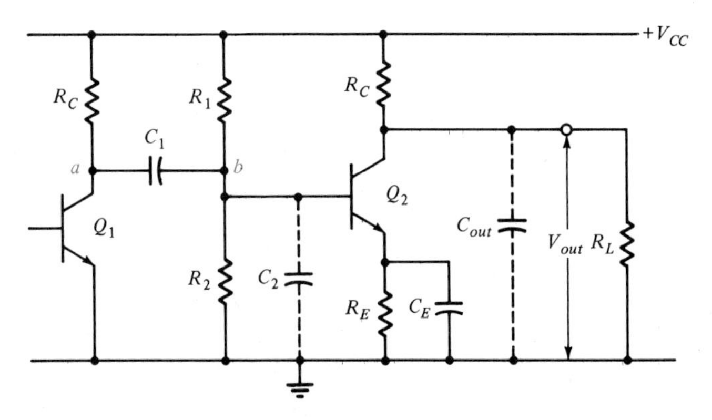

FIGURE 7.16. **Interstage Coupling Between Amplifiers Q_1 and Q_2** The low-frequency limit is influenced by C_1; the high-frequency limit is affected by C_2 and C_{out}.

An amplified signal voltage is developed across R_C of the first stage. We wish to pass this on for further amplification in the second stage. But a direct connection may upset the biasing conditions. (We alluded to this situation in Section 3.7.) Point *a* is at one potential, and point *b* often needs to be kept at another.

How do we transfer the ac signal without affecting the dc potentials? We do so by means of capacitor C_1. In order not to significantly reduce the signal amplitude, we must have X_{C_1} much less than the resistive load at the input to the second stage. If we presume a constant value of input signal amplitude, a plot of output amplitude versus frequency will appear as in Figure 7.17. At the frequency designated as f_1, the capacitive reactance (X_{C_1}) will equal the equivalent resistive input at the second stage. (This frequency represents the point at which the gain is down by 3 dB.)

Will the amplifier continue to have a constant gain at all the higher frequencies? No. The connecting leads between stages, although small, constitute a capacitance, and the transistor input itself represents a considerable capacitance as well. Cumulatively, we represent these contributions as C_2. Therefore, at some high frequency which we designate as f_2 (the upper 3 dB point) this shunting capacitance starts to pass some of the signal around the resistances, and the gain diminishes (Figure 7.18).

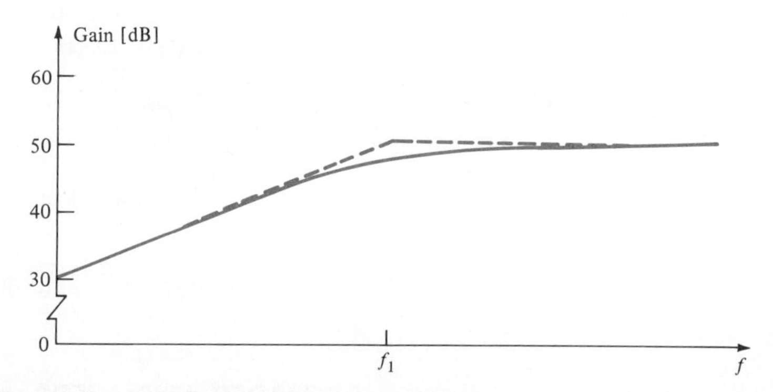

FIGURE 7.17. Diminishing Low-Frequency Gain of an Amplifier Due to Interstage Coupling Capacitor The dashed curve shows the idealized response with the drop-off at 20 dB/decade. The actual drop-off (solid curve) shows a 3 dB loss at frequency f_1.

The signal drop-off at both low and high frequencies takes place so gradually that it would be difficult to determine precisely at what frequencies it starts to take place. Therefore, by convention, the upper and lower band limits of such amplifiers are specified as being those frequencies at which the signal *power* drops to half of the power at midband frequencies, or, in terms of signal voltage, those frequencies at which the signal voltage has dropped to 0.707 of the midband value. Since these statements are equivalent, f_1 and f_2 are correspondingly the same whether dealing with power or voltage amplification.

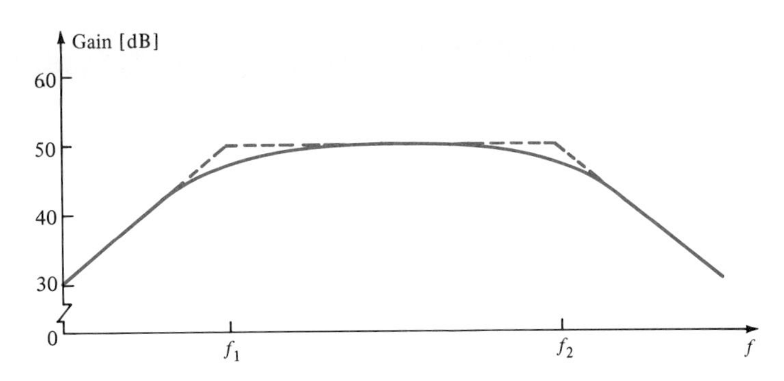

FIGURE 7.18. **Overall Gain vs. Bandwidth Curve for an Amplifier** The 3 dB drop-off points are at f_1 and f_2. The dashed curve represents idealized response, with 20 dB/decade drop-off at both high and low frequencies.

Since the 0.707 points represent a 3 dB drop, the upper and lower limits of an amplifier are those frequencies at which the signal has dropped by 3 dB from the midband value, and the bandwidth of the amplifier is the frequency interval between f_1 and f_2. For an audio amplifier the bandwidth might be 25,000 Hz; for a TV set, 5 MHz; and for an oscilloscope amplifier, as much as 100 MHz.

It is possible to design amplifiers such that the collector potential of one stage is equal to the base potential of the succeeding stage. Most integrated circuits are of this form. One may then eliminate the coupling capacitor and the drop in gain at low frequencies. The value of the bandwidth thus becomes identical with the value of the upper band limit.

EXAMPLE 7.5

What is the lower band limit of an amplifier with $C = 1.0\ \mu\text{F}$ and an equivalent input resistance (R) of 2000 Ω to the succeeding stage? (See Figure 7.19.)

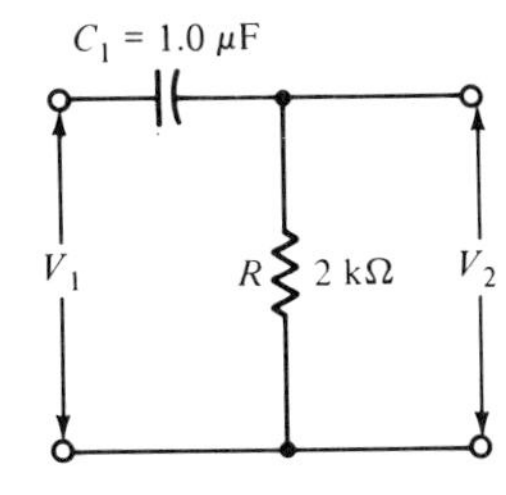

FIGURE 7.19

Solution:

$$V_2 = \frac{R}{R - (j/\omega C_1)} V_1$$

$$= \frac{1}{1 - (j/\omega C_1 R)} V_1$$

$$= 0.707 V_1 = \frac{1}{\sqrt{2}} V_1 \qquad \text{when } \frac{1}{\omega C_1 R} \equiv 1$$

Calling this frequency ω_1, we have

$$f_1 = \frac{1}{2\pi R C_1} = \frac{1}{2\pi(10^{-6})(2 \times 10^3)} = 79.6 \text{ Hz}$$

7.7 ■ INPUT AND OUTPUT IMPEDANCE

To determine the band limits of an amplifier, we need to know the input and output impedances of the transistor stages. Once again, we can profit by considering the Thevenin equivalent of the base circuit. (We are now considering the ac equivalent, *not* the dc.)

Refer to Figure 7.20 in conjunction with Figure 7.16. V_s and R_s may represent the Thevenin equivalent output of transistor Q_1. An *unbypassed* emitter resistor (R_E) would appear to the transistor input to have a much larger value than the actual value. This apparent increase in the value of R_E arises because the source is inserting a signal current I_b but experiencing a disproportionately larger drop because of the simultaneous presence of I_c in the resistance. The dynamic resistance (r_e) also appears larger because of the presence of the collector current ($r_e \simeq 26/I_c$; $h_{fe}r_e \simeq I_c r_e/I_b = 26/I_b$).

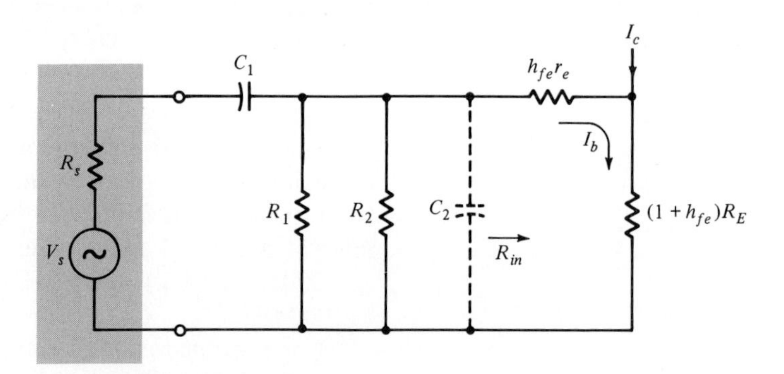

FIGURE 7.20. ac Equivalent Transistor Input Circuit

The signal voltage drop across R_E is $(I_b + I_c)\,R_E = (1 + h_{fe})I_b R_E$. Hence, we have

$$R_{in} = \frac{(1 + h_{fe})R_E I_b}{I_b} + h_{fe}r_e = (1 + h_{fe})R_E + h_{fe}r_e \qquad (7.16)$$

But it was noted in Section 7.5 that the portion of the output signal that appears across R_E represents an output loss that may be effectively reduced by placing a large bypass capacitor across R_E. While conserving the output signal, this capacitor considerably lowers the input resistance of the amplifier by effectively eliminating the first term in Eq. 7.16. The lowered input resistance, in turn, raises the lower band limit (f_1), as will be shown in a moment. If one is willing to forego some output signal gain in exchange for a reduction in the lower band limit, capacitor C_E may be deleted from the circuit.

Returning again to Figure 7.20, note that the biasing network appears across the input terminals as R_1 in parallel with R_2. The reason for this, again, lies in the fact that the voltage source (V_{CC}) presents a negligible resistance to ground. Therefore, from a signal viewpoint the top of R_1 is at ground potential, and the total input resistance of the amplifier becomes R_1 in parallel with R_2 and R_{in}.

In any well-designed amplifier, the bias network (R_1, R_2) is so much larger than R_{in} that its contribution to the parallel resistance can be neglected, and the equivalent circuit at midband frequencies takes the form shown in Figure 7.21a. The series effect of C_1 has been disregarded since X_{C_1} is negligible. C_2, on the other hand, is so small that X_{C_2} is large, and its shorting effect on the input can also be disregarded. We have

$$\text{voltage gain} \equiv A_V = \frac{V_{out}}{V_{in}} \tag{7.17}$$

where V_{out} is the amplifier's output signal voltage and V_{in}, the input. One should distinguish this voltage gain from the *overall* voltage gain that takes into account the source resistance R_s:

$$\text{overall voltage gain} \equiv A_{V_s} = \left(\frac{V_{out}}{V_{in}}\right)\left(\frac{V_{in}}{V_s}\right) = A_V\left(\frac{R_{in}}{R_{in} + R_s}\right) \tag{7.18}$$

It can be seen that if R_s is significant compared to R_{in}, then $A_{Vs} < A_V$.

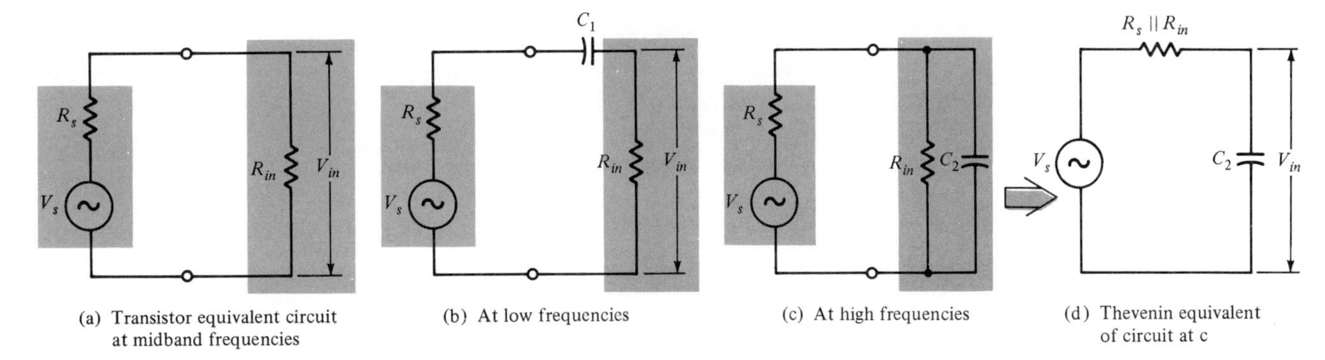

(a) Transistor equivalent circuit at midband frequencies

(b) At low frequencies

(c) At high frequencies

(d) Thevenin equivalent of circuit at c

FIGURE 7.21. Equivalent Amplifier Circuits as a Function of Frequency

By definition the lower band limit (f_1) arises when the amplifier's input is "down" by 3 dB from its midband value. It is down by this amount when $X_{C_1} = R_s + R_{in}$—that is, when

$$f_1 = \frac{1}{2\pi C_1(R_s + R_{in})} \tag{7.19}$$

This expression differs from the result of Example 7.5 because here we have included the effect of R_s, the source resistance. (See Figure 7.21b.)

The upper band limit also arises when the gain drops by 3 dB from its midband value. But trying to express the upper band limit (f_2) in terms of circuit parameters is much more involved than is the case at the lower limit (f_1). Therefore, what follows takes on a more qualitative coloring than has been the case heretofore.

At high frequencies C_1 can again be disregarded, but C_2 now begins to exert a significant shorting effect on the input (Figure 7.21c). The Thevenin equivalent of Figure 7.21c is shown as Figure 7.21d. Additionally, in the output circuit at high frequencies, C_{out} also begins to make its presence known (Figure 7.16) and contributes to a diminishing gain. There is a 3 dB frequency associated with C_2 that arises when $X_{C_2} = R_s \,||\, R_{in}$. There is another associated with C_{out} that arises when $X_{C_{out}} = R_C \,||\, R_L$. The upper band limit of the amplifier is taken to be the lower of these two frequencies. A dependence of R_{in} on R_L constitutes a major complication.

With a cascade of amplifier stages, when the input impedance of one stage is low enough to act as a shunt on the output of the preceding stage, it is not possible to consider each as an isolated stage, and under such circumstances individual 3 dB frequencies cannot be defined.

FOCUS ON PROBLEM SOLVING: Estimating an Amplifier's Band Limits

1. The transistor's input resistance (R_{in}) is obtained by using Eq. 7.16 if there is no bypass capacitor (C_E) across R_E. If there is such a capacitor and its reactance at low frequencies is much less than the value of R_E, R_{in} essentially consists of only the second term in Eq. 7.16.
2. Generally, the equivalent parallel resistance of the biasing network is much larger than R_{in}, and its influence on the signal may be disregarded. In the absence of C_E, however, R_{in} will be much larger, and the effective input may have to be taken to be $R_{in} \,||\, R_1 \,||\, R_2$.
3. The lower band limit (f_1) is obtained using Eq. 7.19.
4. Estimates of the upper band limits may be obtained by using

$$f_2' = \frac{1}{2\pi C_2(R_s \,||\, R_{in})}$$

where R_{in} is subject to the conditions stated in item 2, and

$$f_2'' = \frac{1}{2\pi C_{out}(R_C \,||\, R_L)}$$

The smaller of these two values represents an estimate of the upper band limit.

Again the reader is cautioned about placing too much reliance on the computed value for the upper band limit. While C_2 stays fairly constant, R_{in} varies with both frequency and with the value of R_L.

Insofar as the output resistance is concerned, since the collector current shows little dependence on V_{CE} in many cases, the transistor behaves like a current source. Ideally, it has an infinite output resistance. In many instances the resistance is large but finite, 100 kΩ at $I_C = 1$ mA being typical. (It is usually expressed in terms of output conductance—that is, in this case 10^{-5} S — and designated as h_{oe}.) Since the collector resistor (R_C) is generally significantly smaller (perhaps a few kilohms), in terms of the Norton equivalent, R_C appears in parallel with the transistor's output resistance, and the amplifier's output resistance is essentially that of R_C. The reader is cautioned, however, about the wide range of values that the transistor's output resistance can assume—sometimes as small as 100 Ω in the case of power amplifiers. (See Section 7.8.)

EXAMPLE 7.6 Compute the input resistance "seen" by a signal voltage source with a 500 Ω source resistance applied through a 1 μF capacitor (C_1) between base and ground of the transistor in Figure 7.9. With $h_{fe} = 100$, what is the lower band limit? If the capacitance between base and ground is 300 pF, disregarding any output circuit effect, what is the upper band limit? How do these values change if a 3000 μF capacitor (C_E) is placed across R_E?

Solution:

$$R_{in} = (1 + h_{fe})R_E + h_{fe}r_e = (101)(10^3) + (100)(26) = 103.6 \text{ k}\Omega$$

$$R_1 \,||\, R_2 = \frac{R_1 R_2}{R_1 + R_2} = \frac{(110)(11) \times 10^6}{121 \times 10^3} = 10^4 = 10 \text{ k}\Omega$$

$$R_1 \,||\, R_2 \,||\, R_{in} = 9.1 \text{ k}\Omega$$

$$X_{C_1} \equiv 9.6 \text{ k}\Omega$$

(That is, the lower 3 dB point arises when the input reactance equals the effective input resistance plus R_s.) We have

$$f_1 = \frac{1}{2\pi C_1(9.6 \times 10^3)} = \frac{1}{2\pi(10^{-6})(9.6 \times 10^3)} = 16.6 \text{ Hz}$$

The upper 3 dB point arises when the parallel combination of R_s and the effective value of R_{in} equals X_{C_2}. We have

$$R_s \,||\, 9.1 \text{ k}\Omega = 474 \,\Omega$$

$$f_2 = \frac{1}{2\pi C_2(474)} = \frac{1}{2\pi(3 \times 10^{-10})(474)} = 1.12 \text{ MHz}$$

In the presence of C_E, we have

$$R_{in} \simeq h_{fe}r_e = 100(26) = 2600 \,\Omega$$

$$R_s + (R_1 \,||\, R_2 \,||\, R_{in}) = 0.500 \text{ k}\Omega + 2.06 \text{ k}\Omega = 2.56 \text{ k}\Omega$$

$$f_1 = \frac{1}{2\pi(10^{-6})(2.56 \times 10^3)} = 62.2 \text{ Hz}$$

The presence of C_E provides a greater midband gain, but the lower band limit is more restrictive. At high frequencies, we have

$$R_s \,||\, R_1 \,||\, R_2 \,||\, R_{in} = 402 \,\Omega$$

$$f_2 = \frac{1}{2\pi(3 \times 10^{-10})(402)} = 1.3 \text{ MHz}$$

This result would seem to indicate a greater bandwidth with C_E present. However, while C_2 remains about the same, in actuality R_{in} shows a frequency dependence as well as a dependence on the output load. Hence, the result should not be taken too literally. In fact the absence of C_E will lead to a wider bandwidth, as we will see when discussing negative feedback in Chapter 8.

7.8 ■ POWER AMPLIFIERS

7.8.1 Voltage Amplifiers, Power Amplifiers, and Sound Amplification

Why speak of sound reproduction in a book intended to introduce electronics to engineering students? First, students generally have had some experience with such systems, and this experience allows us to discuss various aspects of electronics in conjunction with familiar equipment. Second, the distinction between voltage and power amplifies is nicely illustrated. Third, in passing through the system, the variety of impedance levels the signal encounters also serves as a nice illustration of some practical aspects of electronics.

The bandwidth of present-day stereo systems extends from about 20 to 20,000 Hz. The signal sources represent a rather wide range of impedance and voltage levels:

—350 μV for a low-impedance microphone
—1.5 mV from a tape pickup
—8 mV from a phono pickup
—250 mV from an FM tuner

The function of a preamplifier is to accept these various sources and yield a common output signal level of about 1 V into a load impedance of about 10,000 Ω. Thus, the major function of the preamp is to boost the signal *voltage* level, and hence it constitutes a *voltage amplifier.*

The 10,000 Ω load into which the preamp operates is the input to the *power amplifier.* The 1 V input to the power amplifier comes out as not much greater than (perhaps) 8 V, representing a voltage gain of only 8. The output power, however, might be of the order of tens or hundreds of watts. This power is usually delivered into a loudspeaker whose impedance is 8 Ω. The 1 V input signal, being delivered into a 10,000 Ω load, represents a power input of 100 μW ($P = V^2/R = 1^2/10{,}000 = 10^{-4}$ W). Thus, the *power* gain of the amplifier is about 10^5.

7.8.2 Amplifier Configurations

The transistor amplifiers that we have thus far considered have been of the *common emitter* variety (Figure 7.22a), called so because the input signal is applied between the base and emitter, and the output is taken from between the collector and emitter. It is most easily recognized when the emitter is grounded. In the CI configuration there may be an emitter resistor between the emitter and ground, but as long as there is a larger one between the collector and ground, it remains a CE configuration.

One might suspect then that rather than grounding the emitter one could, alternatively, ground either the base or the collector. A grounded collector constitutes a unity-gain voltage amplifier or voltage follower, which we will consider in some detail in Chapter 8 (See Figure 7.22b). (It must be stressed that term *grounded* in these cases refers to the ac signal. Obviously, in the voltage follower the collector must be maintained at a dc potential, but

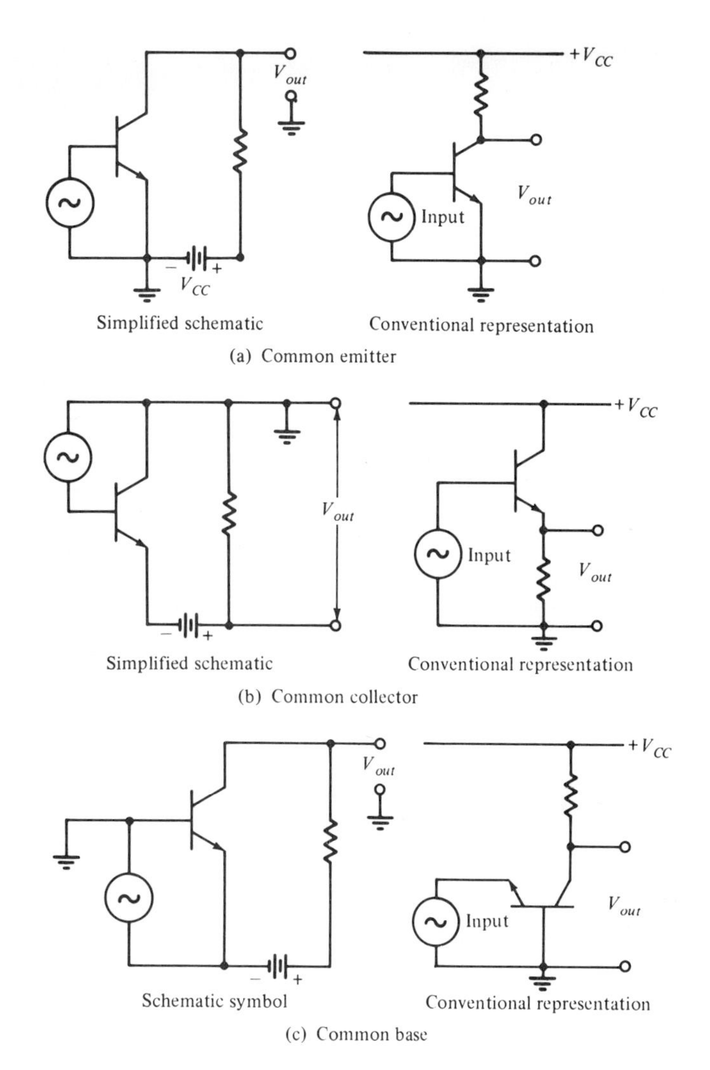

FIGURE 7.22. **Three Possible Transistor Configurations** In each instance the biasing networks and source resistances have been deleted for the sake of clarity.

from the signal point of view it is at ground.) This configuration is also known as the *common collector* (the signal is applied between the base and collector and removed from between the emitter and collector), abbreviated CC. We will see that the main function of such a circuit is impedance matching; it leads to a very large input impedance and a very low output impedance.

The *common base* circuit is shown in Figure 7.22c. It is characterized by a low input impedance (some tens of ohms) and a high output impedance (about 1 MΩ). While the CC circuit gives unity voltage gain, the CB circuit gives approximately unity current gain.

The power gain of a stage is the product of its voltage gain and its current gain. While the CC may have a unity voltage gain, its current gain is significant, as is its power gain. Likewise, while the CB circuit has roughly unity current gain, its voltage (and power) gain is significant.

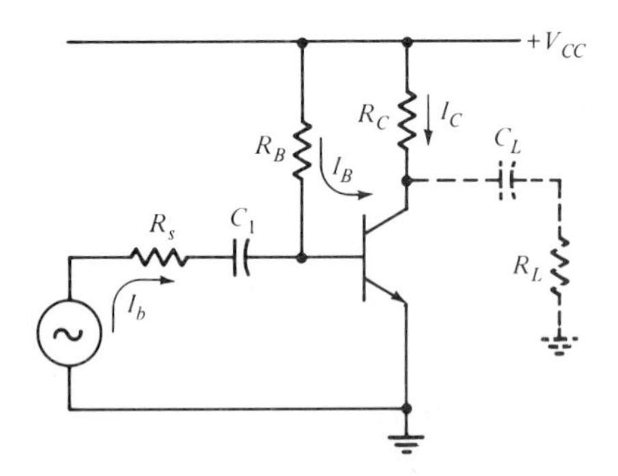

FIGURE 7.23. Common Emitter (CE) Configuration

7.8.3 Power Amplifier Classification

Class A Power Amplifiers. Since the dc power drawn by a voltage amplifier is small, we are not usually concerned with its efficiency.[3] But we usually are concerned with the efficiency of power amplifiers. If the input signal (and hence the output signal) is momentarily zero, we do not want the amplifier drawing any considerable amount of dc power.

The configuration of the amplifier shown in Figure 7.23 is the same as that for the CE amplifiers we have thus far considered. It would be a voltage amplifier if the output resistance of the amplifier were significantly smaller than R_L. It would be a power amplifier if R_{out} were of the order of magnitude of R_L. To a large degree the distinction depends on the characteristics of the transistor employed.

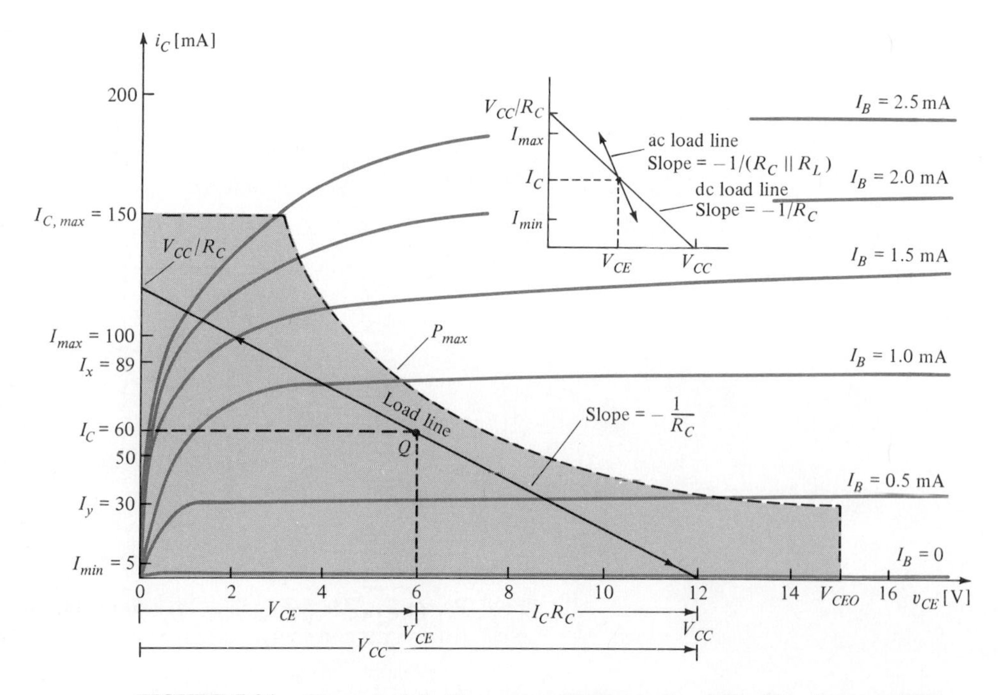

FIGURE 7.24. Characteristic Curves for CE Power Amplifier Showing Load Line, *Q*-Point, and Presumed Input Base Current Swing, Up to 1.5 mA and Down to 0.1 mA Other values of collector current are used in distortion computation. The insert shows the distinction between ac and dc load lines.

A graphical solution will pose the problem more clearly and likewise show the direction to go in resolving the problem of efficient power amplification. Consider the transistor characteristic curves shown in Figure 7.24. For a given value of base current I_B, we plot the increase in collector current (I_C) as a function of the voltage between the collector and emitter (V_{CE}). Starting with a base current of zero, Figure 7.24 shows the succession of curves obtained upon progressively increasing I_B. On this same plot is shown

[3]This might, however, be a factor in battery-operated systems if battery lifetime is a consideration.

V_{CC}, the dc supply voltage. As the input signal is varied, the voltage drop across R_C can vary between zero and V_{CC}. This variation constitutes the output signal. The maximum signal amplitude will prevail if the average value of the collector voltage is about $V_{CC}/2$, designated as V_{CC}. Additionally, probably the least distortion will arise when the base current is about halfway between the allowable extremes. In our graphical example this base current is about 0.8 mA.

Having selected V_{CE} and I_B, the average collector current is thus fixed at the value I_C. Of course, V_{CE} and I_C constitute the Q-point, being the quiescent values, the values when nothing is happening in terms of amplification.

Now draw the straight line that connects the Q-point and the point V_{CC}, extending this line to where it intercepts the y-axis. When the collector current is so large that the full dc supply voltage (V_{CC}) is dropped across the resistance, the voltage between the collector and emitter (V_{CE}) is zero. This value of current is V_{CC}/R_C.

What is the slope of this line? It is

$$\text{slope} = -\frac{V_{CC}/R_C}{V_{CC}} = -\frac{1}{R_C}$$

Thus, the slope is the negative reciprocal of the collector resistance, just as it was in the case of the diode. But in the diode case the load line moved parallel to itself as the input signal varied. Here the load line remains fixed.

The graphical design procedure thus proceeds as follows: Having selected V_{CC} and the Q-point, we then have a value for the slope of the load line and hence can determine the value of R_C that will yield the proper dc operating conditions. With the application of the input signal, the base current might extend upward to 1.5 mA and down to 0.1 mA—that is, centered on 0.8 mA—in our example.

What output signal power does this variation represent? With this variation in base current, the collector current reaches a positive peak at about 100 mA, and on the negative peak of the input it reaches a value of i_C equal to 5 mA. The peak output signal current is then

$$\frac{100 \text{ mA} - 5 \text{ mA}}{2} = 47.5 \text{ mA}$$

If V_{CC} is taken to be 12 V, $R_C = 100\ \Omega$—that is, 12/0.120—the power output is about 0.113 W ($0.0336^2 \times 100$, since $I_{rms} \simeq 33.6$ mA).

The dc power expended is $12 \text{ V} \times 0.060 \text{ A} = 0.72$ W (or, alternatively, $V_{CC}^2/2R_C = 12^2/200 = 0.72$ W). Thus, the efficiency of the amplifier is

$$\frac{P_{ac\,out}}{P_{dc\,in}} = \frac{0.113 \text{ W}}{0.72 \text{ W}} = 0.157 \qquad \text{or} \qquad 15.7\%$$

Summarizing, we have

$$P_{dc} = \text{dc power input} = 0.72 \text{ W}$$

$$P_{ac} = \text{ac power output in } R_C = 0.113 \text{ W}$$

$$P_{R_C} = \text{dc power dissipated in } R_C = 0.060^2 \times 100 = 0.36 \text{ W}$$

The difference must be the power dissipated in the transistor itself, P_t:

$$P_t = P_{dc} - P_{ac} - P_{R_C} = 0.72 - 0.113 - 0.36 = 0.247 \text{ W}$$

It would appear that a larger signal output current might be obtained by resorting to either a smaller value of R_C or a larger value of V_{CC}, thus moving the Q-point upward and to the right. Such alteration of the operating conditions, however, is limited by the maximum power dissipation rating of the transistor. In the hypothetical case depicted in Figure 7.24, this dissipation limit is 0.45 W, and safe operation is limited to a region to the left of the P_{max} demarcation line. Additional power transistor specifications include I_{Cmax} and V_{CEO}, the former being self-explanatory and the latter representing the maximum safe V_{CE}. Since V_{CE} represents a reverse bias between the collector and emitter, as with diodes there is a limit to the reverse voltage that may be applied before electrical breakdown. (V_{CEO} means voltage between C and E with the third element—the base—open-circuited.)

In Figure 7.23 the collector resistor R_C has also served as the load resistor in our computations. In many instances the collector resistor is distinct from the load resistor, wihch is connected to the collector terminal through a large capacitor (C_L) (dashed components in Figure 7.23). In such cases the load line, which was used in Figure 7.24 for both ac and dc computations, now becomes the *dc load line.* The ac load consists of R_C in parallel with R_L and would be represented by an *ac load line* that would pass through the selected Q-point but with a larger slope [equal to $-1/(R_C \sqrt{\surd} R_L)$]. The signal power should now be computed using the ac load line. (See the insert in Figure 7.24.)

In our specific example the transistor is called upon to dissipate a fraction of a watt, but in larger power amplifiers this dissipation could be tens or even hundreds of watts, and the transistor would have to be provided with a heat sink to prevent damage.

The selection of a Q-point near the center of the load line results in a power amplifier that operates in a rather inefficient manner. Such a location of the Q-point constitutes a *class A amplifier.*

In the case of power amplifiers, the selection of the load line is generally dictated by that slope that yields a maximum of power output with minimal distortion. But that value of load will usually differ from the actual load into which the amplified power is to be introduced. One may then use transformer coupling to match the two impedances (Figure 7.25).

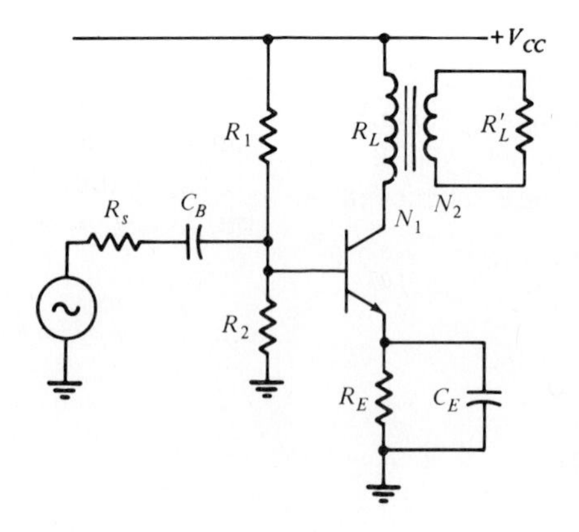

FIGURE 7.25. Transformer-Coupled Class A Amplifier

By assuming the dc resistance of the transformer primary to be very small, the operating Q-point of the transistor will now be V_{CC} and I_C, which is 12 V and 0.060 A in the example shown in Figure 7.26. Whatever the value of R'_L, the turns ratio is adjusted so that R_L equals the previous (optimum) load line. (See Eq. 1.23.) The ac signal "sees" the ac load line. The dc load line represents the transformer resistance plus R_E, and this total, being rather small, leads to a large negative slope.

How can the value v_{CE} exceed V_{CC}, the applied dc voltage? There is a counter-emf developed across the primary inductance that, during part of the cycle, adds to V_{CC}. The power balance equation now assumes the following form:

$$P_t = P_{dc} - P_{ac}$$

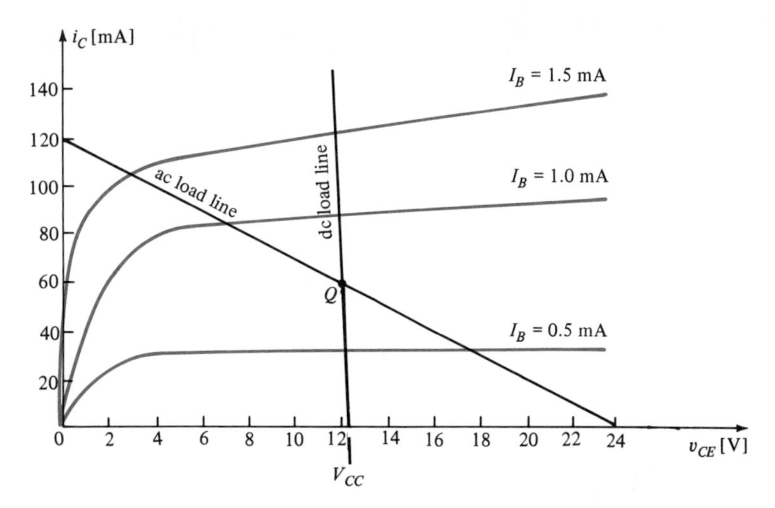

FIGURE 7.26. **Characteristic Curves for Transformer-Coupled Class A Amplifier Showing ac and dc Load Lines**

EXAMPLE 7.7

a. The dc power input to an amplifier of the form shown in Figure 7.23 is 10 W, the ac power output is 1.25 W, and the dc power dissipated in the resistive load is 5 W. What is the transistor power dissipation (P_t)?

b. For a class A amplifier, the maximum efficiency is 25%. What would be the transistor dissipation in the numerical example under consideration under such conditions?

c. What is the power dissipation of an ideal transformer-coupled class A power amplifier with a dc input of 10 W?

Solution:

a. $P_t = P_{dc} - P_{out} - P_R = 10\text{ W} - 1.25\text{ W} - 5\text{ W} = 3.75\text{ W}$

b. $P_t = 10\text{ W} - 2.5\text{ W} - 5\text{ W} = 2.5\text{ W}$

c. $P_t = 10\text{ Q} - 5\text{ W} = 5\text{ W}$

EXAMPLE 7.8

A transformer-coupled power transistor having the characteristics shown in Figure 7.27 is to be made to drive a loudspeaker whose impedance is 8 Ω.

a. What should be the turns ratio?

b. With an operating Q-point of $I_C = 175$ mA and $V_{CE} = V_{CC} = 10$ V, determine the power output, the transistor dissipation, and the distortion, taking the peak driving current to be 7 mA. In computing the distortion, use the values $I_{max} = 320$ mA, $I_x = 260$ mA, $I_y = 87$ mA, and $I_{min} = 0$ mA in conjunction with Eq. 5.8.

Solution:

a. $$\frac{1}{R_L} = \frac{0.175\text{ A}}{10\text{ V}} = 0.0175\text{ S}$$

$$R_L = 57.1\ \Omega$$

$$\frac{N_1}{N_2} = \frac{\sqrt{R_{in}}}{R_{out}} = \frac{\sqrt{57.1}}{8} = 2.67$$

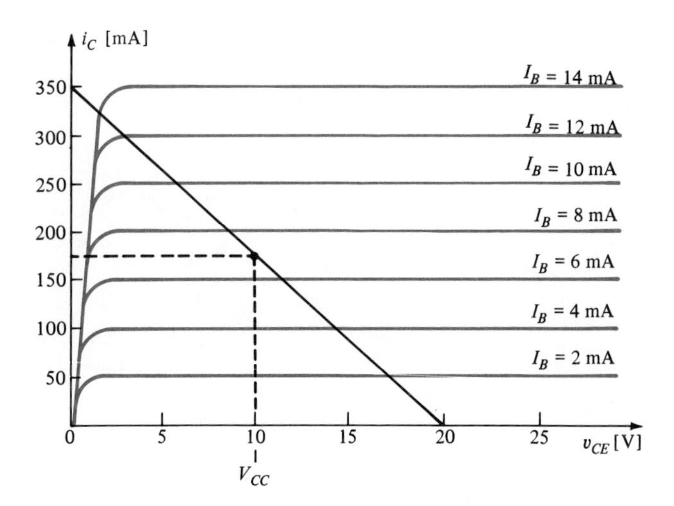

FIGURE 7.27

b. With distortion present, the ac power should not be computed by utilizing the rms value obtained from I_{max}. Instead, the current represented by A_1 must be used. Also, in computing the dc input, I_C must be used in conjunction with A_0 since the two half-cycles may not be symmetric. We have

$$A_0 = \frac{1}{6}(320) + \frac{1}{3}(260 + 87) - 175 = -6$$

$$A_1 = \frac{1}{3}(320) + \frac{1}{3}(173) = 164$$

$$A_2 = \frac{1}{4}(320) - \frac{175}{2} = 7.5$$

$$A_3 = \frac{1}{6}(320) - \frac{1}{3}(173) = 4.33$$

$$A_4 = \frac{1}{12}(320) - \frac{1}{3}(347) + \frac{1}{2}(175) = 1.5$$

$$D_2 = \frac{7.5}{164} = 0.046 \quad \text{or} \quad 4.6\%$$

$$D_3 = \frac{4.33}{164} = 0.026 \quad \text{or} \quad 2.6\%$$

$$D_4 = \frac{1.5}{164} = 0.0091 \quad \text{or} \quad 0.91\%$$

$$D = \frac{\sqrt{7.5^2 + 4.33^2 + 1.5^2}}{164} = 0.054 \qquad 5.4\%$$

$$P_{ac} = I_{ac}^2 R_L = \left(\frac{A_1}{\sqrt{2}}\right)^2 R_L = \frac{0.164^2}{2}(57.1) = 0.768 \text{ W}$$

$$P_{dc} = (I_C + A_0)V_{CC} = (0.175 - 0.006)(10)$$
$$= 1.69 \text{ W}$$

$$P_t = P_{dc} - P_{ac} = 1.69 - 0.768 = 0.922 \text{ W}$$

$$\eta = \frac{P_{ac}}{P_{dc}} = \frac{0.768}{1.69} = 0.454 \quad \text{or} \quad 45.4\%$$

It might be thought that with power amplifiers one again strives to have the output resistance equal to the load resistance in order to achieve maximum power transfer. But instead the prime consideration usually is minimal distortion, achieved by having an equality of spacing of the I_B line intercepts along the load line.

An appreciation of a transistor's output resistance may be obtained by again referring to characteristic curves such as shown in Figure 7.24. The area where the I_B curves are fairly flat constitutes the transistor's *active region* of operation. If these curves were perfectly flat, i_C would be independent of v_{CE}, and $\Delta v_{CE}/\Delta i_C$ (which would be the transistor's ac output resistance) would be infinity. In other words, the transistor would behave like an ideal current source. By considering the $I_B = 1.0$ mA curve in Figure 7.24 as typical, for $\Delta v_{CE} \simeq 9$ V (13 V $-$ 4 V), the corresponding Δi_C is 0.005 A (0.085 A $-$ 0.080 A), and the transistor's output resistance is 1800 Ω—that is, 9/0.005—and represents a positive slope of 1/1800. For a matching load, the load line should have a *negative* slope of 1/1800, corresponding to a resistance of 1800 Ω. To achieve such a condition, V_{CC} would have to be moved to the right by a considerable amount, much beyond the allowed value of V_{CEO}.

As we move along the load line, notice that the slope of the I_B intercepts keeps changing. Such a change means that the resistance presented by the transistor keeps changing—that is, it is a nonlinear resistance, which accounts for the creation of distortion.

Class B Power Amplifiers. One could considerably improve upon the efficiency if the Q-point were moved down the load line so that the dc current drain would be very low. But if we did that, the result would be a very distorted output signal since the negative portions of the input sinusoidal current variation would be cut off.

How can we achieve a small Q-point current and yet not be subject to the accompanying distortion? We can do so by using a push-pull arrangement, such as shown in Figure 7.28.

In the push-pull arrangement, the base is biased at the positive threshold (for *npn* transistors) so that, with no signal applied to the transistor, it is cut off—that is, zero collector current. Upon applying the input signal, Q_1 will amplify the positive pulses. When the polarity of the input signal reverses, the negative portions of the input will appear as positive pulses to transistor Q_2, because of the phase reversal through the input transformer. Therefore, Q_2 will amplify the negative portions of the cycle. The pulsed outputs from

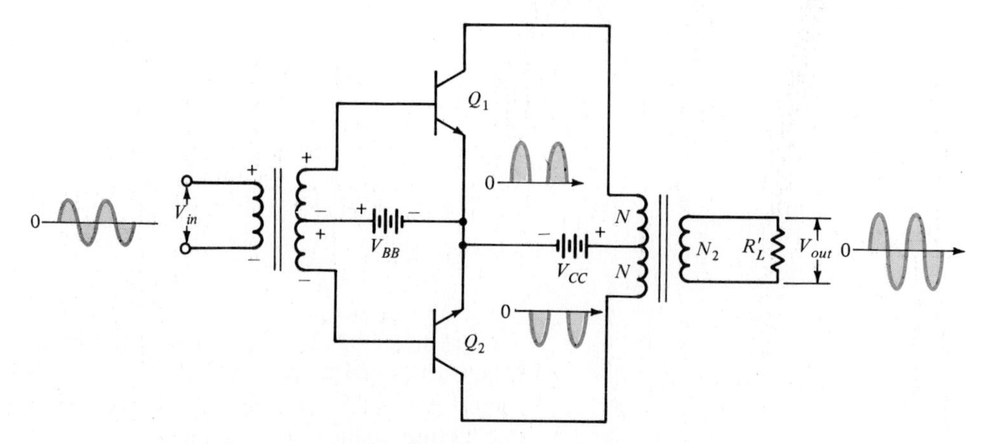

FIGURE 7.28. **Push-Pull Amplifier Configuration**

Q_1 and Q_2 are then combined in the output transformer and appear as an amplified sinusoidal output across the secondary load R'_L. Graphically, this situation is as depicted in Figure 7.29.

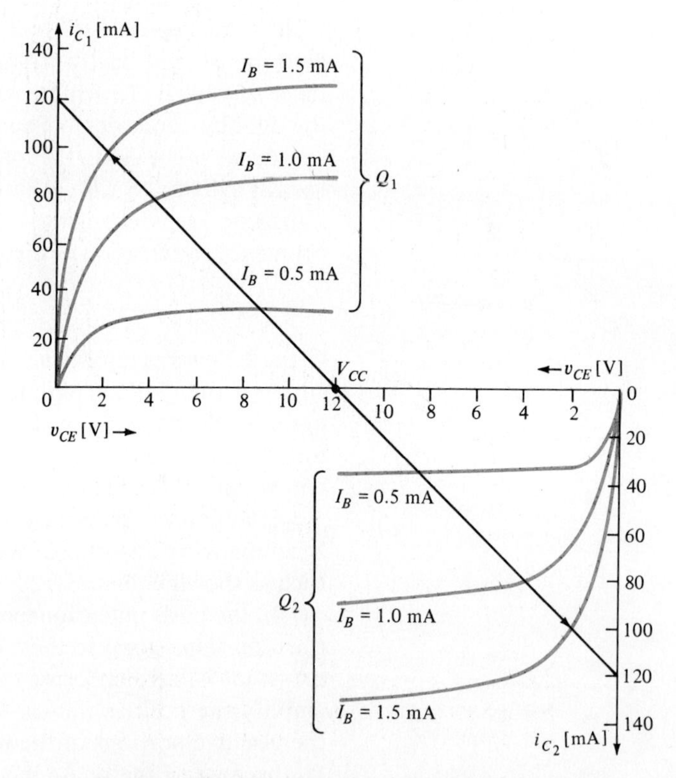

FIGURE 7.29. **Characteristic Curves for Push-Pull, Class B Amplifier, with Presumed Peak Base Current Swing of 1.5 mA**

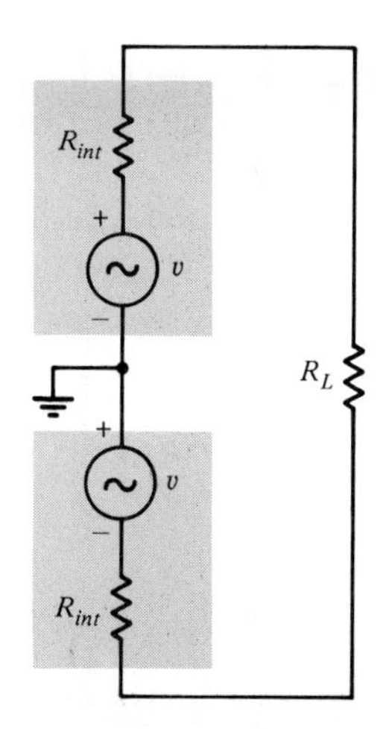

FIGURE 7.30. **Equivalent Output Circuit for Push-Pull Amplifier**

The output from the power amplifier can be depicted as two equivalent generators in series (Figure 7.30). R_L differs from R_L' because of the impedance transformation in the transformer. We have

$$R_L = \left(\frac{2N}{N_2}\right)^2 R_L' \tag{7.20}$$

$$R_L = 4\left(\frac{N}{N_2}\right)^2 R_L' \tag{7.21}$$

$$I_C = \frac{2v}{2R_{int} + (2N/N_2)^2 R_L'} = \frac{v}{R_{int} + 2(N/N_2)^2 R_L'} \tag{7.22}$$

$$P_{out} = \left[\frac{v}{R_{int} + 2(N/N_2)^2 R_L'}\right]^2 \left(\frac{2N}{N_2}\right)^2 R_L'$$
$$= \left[\frac{v}{(R_{int}/2) + (N/N_2)^2 R_L'}\right]^2 \left(\frac{N}{N_2}\right)^2 R_L' = I_{equiv}^2 R_{equiv} \tag{7.23}$$

Therefore, we have

$$I_{equiv} = \frac{v}{(R_{int}/2) + (N/N_2)^2 R_L'} \quad \text{and} \quad R_{equiv} = \left(\frac{N}{N_2}\right)^2 R_L' \tag{7.24}$$

From the preceding it can be seen that for a push-pull arrangement the optimum load (for maximum power transfer) is equal to one half of the internal output impedance of one of the transistors.

Each of the transistors behaves like a half-wave rectifier. Therefore, the I_{dc} per transistor is I_{max}/π. The I_{eff} of the output signal current is $I_{max}/\sqrt{2}$. The ac output power is

$$P_{out} = \frac{I_{max}^2 R_L}{2} \tag{7.25}$$

The total dc power is

$$P_{dc} = \frac{2I_{max}}{\pi} V_{CC} \tag{7.26}$$

The efficiency is

$$\eta = 100\left(\frac{I_{max}{}^2 R_L}{2}\right)\left(\frac{1}{2I_{max}V_{CC}/\pi}\right) = \left(\frac{\pi}{4}\right)\left(\frac{I_{max}R_L}{V_{CC}}\right)100 = 78.5\% \tag{7.27}$$

Since from the slope of the load line, $I_{max}R_L = V_{CC}$.

This type of push-pull operation, which has the transistors biased at cutoff, is termed class B. While class B transformer-coupled push-pull amplifiers lead to a high efficiency, there are many disadvantages that accompany transformer coupling. Transformers are expensive, and the output signals (being dependent on the variation of magnetic flux) can be decidedly nonlinear, thereby leading to distortion, particularly at high power levels. There is another aspect of transformers that also poses problems: Thus far we have looked upon the windings of transformers as inductors, but each turn of the transformer, being a conductor separated from an adjoining turn, constitutes a capacitance (Figure 7.31). Together with the large transformer inductance, such capacitance constitutes a resonant system. A properly designed audio transformer will place the resulting resonant peak near the high-frequency limit of the amplifier and thereby extend the amplifier's bandwidth (Figure 7.32). But this peaking (unless its magnitude is flattened by proper design) constitutes frequency distortion.

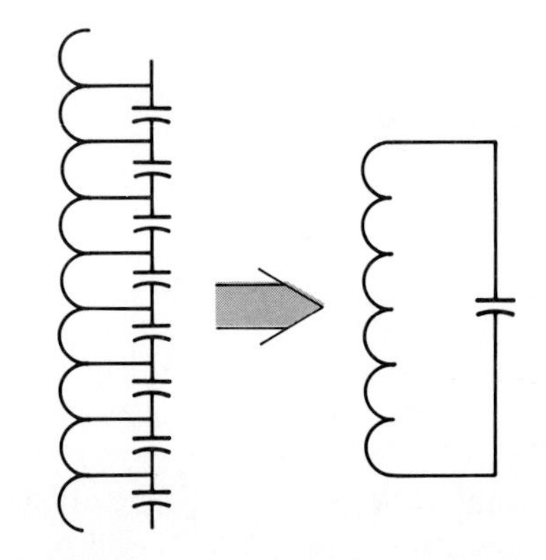

FIGURE 7.31. Transformers Exhibit Small Capacitances Between Turns, Leading to an Equivalent Resonant Circuit

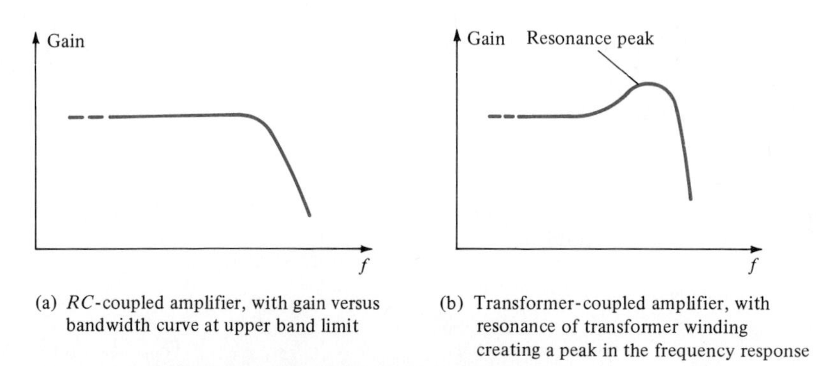

(a) *RC*-coupled amplifier, with gain versus bandwidth curve at upper band limit

(b) Transformer-coupled amplifier, with resonance of transformer winding creating a peak in the frequency response

FIGURE 7.32. Transformer-Induced Frequency Distortion

The existence of low output impedance transistor power amplifiers has made it possible to eliminate the need for transformer coupling with its attendant difficulties. Loudspeakers may now be directly driven.

Distortion tends to diminish with the order of the harmonic. The symmetric distortion that arises with class B push-pull operation constitutes half-wave symmetry, with only odd harmonics being present. The absence of the second harmonic, usually the most significant, represents a major attractive feature of such operation.

There is no one speaker that will satisfactorily cover the entire audio range. To cover the upper spectral frequency, one speaker, termed the *tweeter*, is generally used; a second, the *woofer*, covers the lower frequency range. Often a third, the *squawker*, can be used (in conjunction with the other two) to cover the midfrequency range.

The use of a multispeaker output requires the use of a crossover network. Without such a network, the woofer in a tweeter-woofer combination will reproduce some of the same frequencies at its high end that the tweeter will reproduce at its low end. This duplication will lead to an overemphasis of the midfrequency range. Also, the tweeter cannot withstand the high power provided to the woofer, which also militates against merely connecting them in parallel.

The simplest crossover network consists of the insertion of a capacitor in series with the tweeter, the two speakers being connected in parallel (Figure 7.33). If the tweeter has an impedance of 8 Ω, a 4 μF capacitor will equal the speaker impedance at about 5 kHz, which will constitute the crossover frequency. (If the capacitor is an electrolytic, it is necessary that it be of the nonpolarized type.)

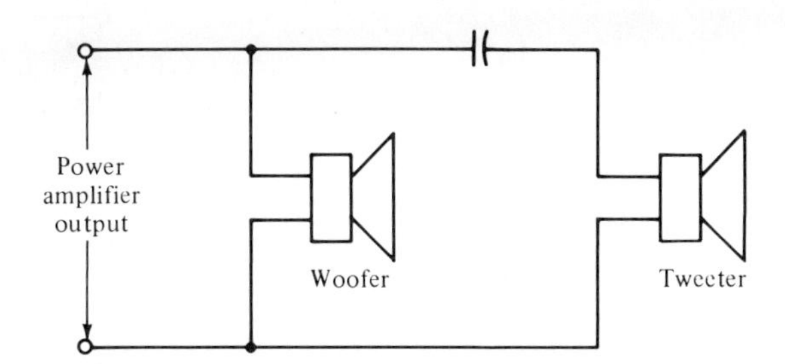

FIGURE 7.33. **Capacitor Restricts Tweeter to Reproducing Only High Audio Frequencies**

This type of an arrangement prevents the tweeter from reproducing frequencies at the high end of the woofer range, but it may not prevent the woofer from reproducing frequencies at the low end of the tweeter range. This would lead to an overemphasis of the intermediate frequency range. This defect may be remedied by connecting a coil in series with the woofer. For a 5 kHz crossover, L should have a value of about 0.25 mH if the woofer, again, is equal to 8 Ω (Figure 7.34). Presuming the voice coil impedances to be resistive and recognizing that while X_L is proportional to the frequency while X_C bears an inverse relationship to frequency, the load seen by the amplifier remains substantially 8 Ω over the entire frequency range.

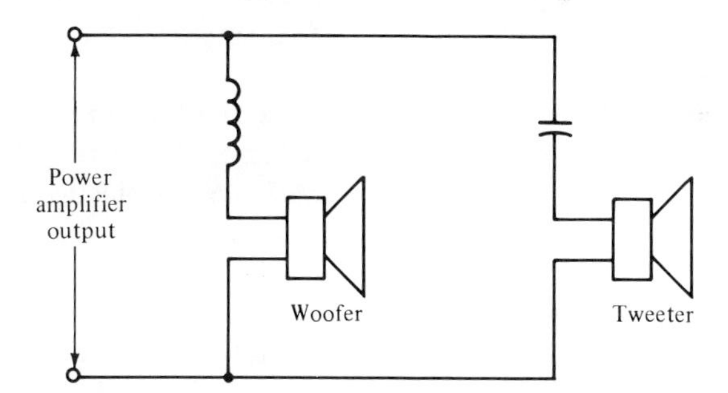

FIGURE 7.34. **Low Audio Frequencies Channeled into Woofer by Inductor *L* and High Frequencies Channeled into Tweeter by Capacitor *C***

But a voice coil does have some inductance, which might lead to some disparity from the calculated values. Also, since the tweeter tends to be more efficient than the woofer, it is generally necessary to attenuate the tweeter drive. This may be achieved by the insertion of a rheostat in series with the tweeter (Figure 7.35).

Other Classes of Power Amplifiers. By definition class A operation arises when there is collector current during the entire cycle. In class B operation, on the other hand, the transistor is biased at cutoff so that its output current persists for about 180° of the cycle. This type of operation requires push-pull amplification.

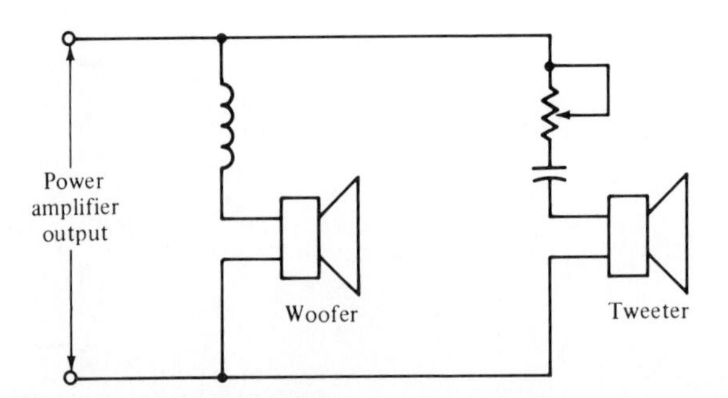

FIGURE 7.35. Somewhat Elaborate Crossover Network for Tweeter and Woofer

In class B operation the junctures between the positive and negative half-cycles do not quite coincide and lead to a condition called crossover distortion. This distortion may be cured by biasing the push-pull transistors into slight conduction with no signal present. This constitutes class AB operation and leads to efficiencies intermediate between classes A and B.

In class C operation the transistor is biased considerably beyond cutoff and conducts during a very small fraction of the cycle. By itself this type of operation would lead to a very distorted output, but such amplifiers are used in conjunction with resonant *LC* circuits. Recall that during each cycle the circulating current of a resonant circuit needs only to have its losses for that cycle replenished by a pulse of short duration. Such is the function of class C amplifiers. They are used to amplify signals at radio and television frequencies.

Heat Dissipation. A particular application calls for a power transistor capable of dissipating 25 W. A perusal of power transistor listings shows an MJE3055, a plastic-encased silicon power transistor, to have a rated dissipation of 90 W at 25°C. That would appear to satisfy our requirement with a considerable safety factor. Let us see if it does.

Power transistors come encased within a variety of packages that may broadly be divided into two categories: all-metal casings with a maximum temperature rating of 200°C and a cheaper plastic package with a metal backing for thermal contact with a maximum rating of 150°C. The substantial heat generated within a power transistor must be dissipated if the transistor is not to be damaged. In most cases the collector is internally connected to the metallic case.

If P is the power being dissipated, between the junction temperature T_J and the ambient temperature T_A, the heat flow will encounter various **thermal resistances** (θ). Thermal resistance is defined as the temperature rise per unit power transferred. Generally, there are three such thermal resistances encountered:

1. θ_{JC} between the collector junction and the case,
2. θ_{CS} between the case and the heat sink,
3. θ_{SA} between the heat sink and the ambient air.

Thus,

$$T_J = T_A + (\theta_{JC} + \theta_{CS} + \theta_{SA})P \quad (7.28)$$

Since the transistor is generally located in a rather confined space, an assumed ambient temperature of 50°C is not uncommon.

The values of θ_{JC} depend on the transistor packaging and take on values from 1°C/W to 7°C/W. Some power transistors with a threaded stud firmly attached to the case can reduce this specification to 0.5°C/W.

To assure good thermal contact between the package and the heat sink, a heat-conducting silicon grease can be used, leading to $\theta_{CS} \simeq 0.25$°C/W. Otherwise, even with a well-mated clean surface the value will be about 0.5°C/W.

The value of θ_{SA} depends greatly on the nature of the heat sink employed. For a 5 × 5 in., $\frac{1}{8}$ in. thick aluminum plate, it might be 5°C/W. Commercial units can vary from 1°C/W to 5°C/W. Heat sinks designed to be used with blower motors may have values as small as from 0.05°C/W to 0.2°C/W.

If the collector must be *electrically* insulated from the heat sink, a mica, Teflon, or beryllia washer (with both sides coated with heat-conducting grease) may be interposed. The first two each have $\theta_{CS} = 0.5$°C/W. For the beryllia, which is more expensive, $\theta_{CS} = 0.14$°C/W.

Let us return now to the problem that commenced this discussion. $\theta_{JC} = 1.4$°C/W for the MJE3055, while a greased mica washer represents $\theta_{CS} = 0.5$°C/W and a good sink with radiating fins yields $\theta_{SA} \simeq 2.5$°C/W. The total rise in the transistor temperature will be

$$(1.4 + 0.5 + 2.5)\text{°C/W} \times 25\text{ W} = 110\text{°C}$$

By taking the ambient temperature to be 50°C, the resultant transistor temperature will be 160°C, which exceeds the rating for this plastic transistor.

EXAMPLE 7.9 A metal-case silicon transistor equivalent to the MJE3055 is the 2N3055 with a 115 W rating at 25°C and with $\theta_{JC} = 1.5$°C/W. What will be its operating temperature when dissipating 25 W using an ambient temperature of 50°C?

Solution: With $\theta_{CS} = 0.5$°C/W and $\theta_{SA} = 2.5$°C/W, we have

$$(\theta_{JC} + \theta_{CS} + \theta_{SA})P = \Delta T$$
$$(1.5 + 0.5 + 2.5)\text{°C/W} \times 25\text{ W} = 112.5\text{°C}$$
$$\Delta T + T_{ambient} = T_{junction}$$
$$112.5\text{°C} = 50\text{°C} = 162.5\text{°C}$$

which is within the capabilities of this transistor.

7.9 ▪ SOME BJT SPECIFICATIONS

Bipolar junction transistors are generally distinguished by a 2N prefix. (Recall that 1N was used for diodes.) They fall into four main categories: general-purpose, power, radio frequency, and switching.

The general-purpose devices are low-power units with ratings up to hundreds of milliwatts. There are three categories of current rating—low (1mA), medium (10 mA), and high (100 mA)—and three grades of voltage ratings—low (30 V), medium (60 V), and high (100 V).

In addition to h_{fe} (the current gain) and the current and voltage ratings, V_{CEO} is an important parameter, being the maximum allowable voltage between the collector and emitter with the base circuit open.

The type numbers of general-purpose BJTs number in the thousands. We will start the reader off with an easy one to remember to illustrate the category, 2N2222. It is rated as a high-current unit (I_C = 150 mA, I_{Cmax} = 600 mA) and has h_{fe} = 150 and V_{CEO} = 30–60 V.

Two widely used power transistors will illustrate the category: 2N3054 (I_C = 4 A, V_{CEO} = 55 V, with a 25 W power rating) and 2N3055 (I_C = 15 A, V_{CEO} = 60 V, with a 115 W rating).

Radio frequency transistors are divided into four categories, starting from high-frequency units (20 kHz–30 MHz) and ending with microwave units (above 1000 MHz).

Switching transistors are broken down into five categories according to switching speed, from the slowest at 1 μs to those acting at the ultra high speed of 10 ns.

SUMMARY

7.2 The Bipolar Junction Transistor

- The ratio of collector current to base current constitutes the transistor's beta, also termed its h_{fe}. The dc current ratio is designated as h_{FE} and the ac current ratio, as h_{fe}. Usually, $h_{fe} \simeq h_{FE}$.
- Typical transistor betas range between 20 and 250, although super beta transistors may have values of the order of 1000.

7.3 dc Operating Conditions

- The presence of an emitter resistor tends to stabilize the Q-point against various changes, such as those due to temperature, and the wide tolerance values that characterize transistor betas.

7.5 Voltage Amplification

- The most frequently encountered transistor circuit employs a common emitter (CE). It presents modest input and output impedances.
- In the absence of a bypass capacitor C_E, the voltage gain of a common emitter configuration is approximately $-R_C/R_E$, being the collector resistance divided by the emitter resistance.
- When a dependent variable, in response to a variation of an independent variable, is pressed to attempt to assume values in excess of an allowed maximum, it is said to be in saturation. At the other extreme, when pressed by a variation of the independent variable to attempt to assume a value less than zero, it is said to be in cutoff.

7.6 Cascaded Amplifiers

- Capacitor coupling between cascaded amplifier stages reduces the gain at low frequencies. Transistor input capacitance and stray wiring capacitance are responsible for the high-frequency drop-off in gain.
- The bandwidth is defined as the frequency interval between a low frequency (f_1) and a higher frequency (f_2) at which the voltage gain has dropped to 0.707 of its midband value. At these same frequencies the power gain will have dropped to one half of its midband value. In both instances, as a voltage ratio and as a power ratio, the diminished gain relative to midband represents a 3 dB change.

7.8 Power Amplifiers

- Amplifiers are broadly divided into voltage and power types.
- To actuate loudspeakers, electromagnetic shakers, and other devices requiring signal power, power amplifiers are needed. A power amplifier's output resistance is of the same order of magnitude as the load resistance.
- Power amplifiers usually require a minimum signal voltage of the order of 1 V. To raise the small signal

voltages available from most sources (recording tapes, FM receivers, etc.) to such a voltage level, voltage amplifiers are needed. Their output resistance generally should be much lower than the resistive load into which they operate.

- The common collector (CC) amplifier has a high input impedance and a low output impedance.
- The common base (CB) amplifier has a low input impedance and a high output impedance.
- Power amplifier transistors are large in physical size and are often mounted on heat sinks in order to dissipate the considerable power that is internally generated.
- Power amplifier efficiency represents the ratio of signal output power to dc input power. This efficiency may be improved by using transformer coupling and still further improved by using push-pull arrangements.
- Power amplifiers with transistor conduction taking place during 100% of the signal cycle constitute class A operation. Conduction during 50% of the signal cycle constitutes class B operation. Conduction for less than 50% represents class C. Conduction for slightly more than 50% represents class AB.
- Class A amplifiers with a resistive load have an ideal efficiency of 25%; transformer coupled class A, 50%; and push-pull class B, 78.5%.
- An amplifier's distortion diminishes as the order of the harmonic increases. Push-pull operation is generally devoid of second harmonic distortion.

KEY TERMS

ac load line: a line on the characteristic curves that traces out the collector current and collector-emitter voltage in response to the signal input (The slope is the negative reciprocal of the ac load resistance.)

beta: BJT current amplification factor; alternatively designated as either β or h_{fe}

bipolar junction transistor (BJT): a combination of three segments of *p*- and *n*-type semiconductors

cascaded amplifier: a succession of amplifier stages

class A amplifier: transistor conduction throughout duration of signal cycle

class AB amplifier: biasing intermediate between classes A and B

class B amplifier: transistor conduction during 180° of signal cycle

class C amplifier: used in conjunction with resonant *LC* circuit to amplify radio and TV power transmission during a small fraction of signal cycle

common base amplifier: common base terminal between emitter input and collector output; constitutes low input and high output impedances

common collector amplifier: common collector terminal between base input and emitter output; constitutes high input and low output impedances; also variously known as voltage follower, emitter follower, unity-gain amplifier

common emitter amplifier: common emitter terminal between base input and collector output; constitutes modest input and output impedances

cutoff: as applied to a transistor, a base voltage insufficient to give rise to conduction

dc load line: a line on the characteristic curves that connects V_{CC} on the v_{CE}-axis with the operating *Q*-point (Its slope is equal to the negative reciprocal of the dc load resistance.)

distortion, amplitude: nonlinear response of the output signal relative to the input, due to unequal spacing between the base current characteristic curves

overall voltage amplification: output signal voltage divided by signal source voltage

power amplification: ratio of output signal power to input signal power

push-pull amplifier: an arrangement allowing two transistors to conduct on alternate half-cycles and yet produce a rather accurate version of the input signal

***Q*-point:** a combination of collector-emitter voltage and collector current that determines the operating point of the BJT

saturation: an attempt to force a dependent variable (for example, output signal voltage amplitude) beyond an established limit

squawker: a loudspeaker that reproduces intermediate audio frequencies; also termed the midrange speaker

thermal resistance: opposition to heat conduction, measured in terms of °C/W

tweeter: a high-frequency loudspeaker

voltage amplification: output signal voltage divided by transistor input voltage

woofer: a bass loudspeaker

SUGGESTED READINGS

Fitzgerald, A. E., D. E. Higginbotham, and A. Grabel, BASIC ELECTRICAL ENGINEERING, 5th ed., pp. 355–359. New York: McGraw-Hill, 1981. *Graphical analysis of BJT circuits.*

Millman, J., MICROELECTRONICS, pp. 375–390. New York: McGraw-Hill, 1979. *h-parameter model of BJT (if you want further information on the nature of h_{fe}, h_{ie}, and so on).*

Horowitz, P., and Winfield Hill, THE ART OF ELECTRONICS. New York: Cambridge University Press, 1980. Pp. 65–66, *the input voltage of a BJT (rather than input current) in reality controls the output current*; p. 88, *a selected listing of small-signal transistors (with major parameter values) according to category: general-purpose, high current, and so on*; and p. 178, *a selected listing of power transistors (with major parameter values).*

Tepper, Marvin, ELECTRONIC IGNITION SYSTEMS. Rochelle Park, N.J.: Hayden, 1977. *Ignition systems.*

Ribbens, Wm. B., and N. P. Mansour, UNDERSTANDING AUTOMOTIVE ELECTRONICS. Ft. Worth: Radio Shack, 1982. *Automotive electronics.*

Boylestad, R., and L. Nashelsky, ELECTRONIC DEVICES AND CIRCUIT THEORY, 3rd ed., pp. 393–398. Englewood Cliffs, N.J.: Prentice-Hall, 1982. *Distortion calculations.*

A CLOSER LOOK: Notation Summary for Combined Circuits

Quiescent value:

V_B, V_C, I_B, I_C

Instantaneous total value:

v_B, v_C, i_B, i_C

Instantaneous value of varying component:

v_b, v_c, i_b, i_c

Effective value of varying component:

V_b, V_c, I_b, I_c

Source voltages:

V_{BB}, V_{CC}

The sinusoidal waveform in Figure 1 is used to illustrate the notational conventions employed in conjunction with constant and varying electrical quantities, as well as their designation when they rise in combination with one another.

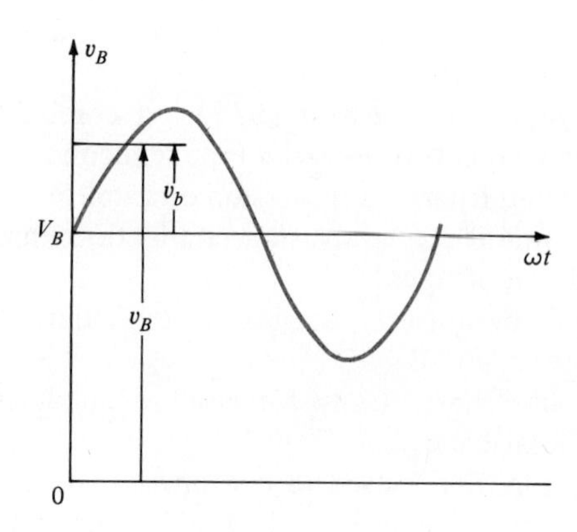

FIGURE 1

EXERCISES

1. The three segments of a bipolar transistor are called the ______, the ______, and the ______.

2. The fraction of transistor emitter current that is present in the collector is termed the ______ of a transistor.

3. The ratio of collector current variation to base current variation constitutes the ______ of a transistor. In terms of α, the β is equal to ______.

4. The collector of a *pnp* transistor should be operated at a ______ potential relative to the emitter. The collector of an *npn* transistor should be operated at a ______ potential relative to the emitter.

5. What is the meaning of the term *Q-point* as applied to transistor operation?

6. Sketch a transistor amplifier of the common emitter (CE) type.

7. For an *RC*-coupled amplifier, what determines the low-frequency 3 dB point?

8. What determines the upper 3 dB point on the gain vs. frequency curve of a transistor amplifier?

9. What defines the bandwidth of an amplifier?

10. a. For a transistor with emitter resistance R_E, what is the dc input resistance?

b. Why does this resistance appear much larger to the input circuit?

11. a. Using voltage divider biasing in a transistor amplifier, with $R_1 = 100\ \text{k}\Omega$, $R_2 = 50\ \text{k}\Omega$, $R_E = 500\ \Omega$, $h_{fe} = 100$, and $I_C = 1$ mA, what is the ac input resistance to the transistor?

b. If the emitter is grounded, what would be the ac input resistance under like conditions?

c. What advantage does the arrangement in part a offer over that in part b?

12. a. What is the order of magnitude of the output resistance of a small-signal transistor? Express this in terms of conductance.

b. What is the order of magnitude of the output resistance of a transistor suitable for power amplification?

13. a. What is the voltage gain for an amplifier with $R_C = 10\ \text{k}\Omega$ and $R_E = 500\ \Omega$?

b. What is the voltage gain for an amplifier with $R_C = 10\ \text{k}\Omega$, $R_E = 0$, and $I_C = 1$ mA?

c. What is the major disadvantage in the second case?

d. What precaution must be observed when using large values for R_C in order to obtain large gains?

14. If two identical stages with individual band limits $f_1 = 100$ Hz and $f_2 = 50$ kHz are operated in cascade, how much lower, in dB, is the gain at these frequencies compared to mid-band? What can you say about the overall bandwidth under such circumstances?

15. What are some typical bandwidths for stereo systems? A TV set? An oscilloscope amplifier?

16. What is the function of a preamplifier in an audio system?

17. In terms of amplifier output resistance and the load resistance, distinguish between a power amplifier and a voltage amplifier.

18. What typical input signal voltage is required by a power amplifier?

19. In terms of voltage and current gains, what is the power gain of an amplifier?

20. Compare the input and output resistances of the three amplifier configurations: CE, CB, and CC.

21. A class A power amplifier, employing $V_{CC} = 12$ V and $I_C = 100$ mA, results in a sine output of 80 mA peak value, operating into a resistive load of 60 Ω. What is the efficiency?

22. By what means can one obtain an operating Q-point involving a small value of I_C and yet obtain large power outputs with relatively low distortion?

23. What are the disadvantages of class B transformer coupling?

24. Why, in an improperly designed transformer, will there be a frequency distortion in the form of a response peak at high frequencies?

25. What is the function of a crossover network used in conjunction with an audio system? What is the simplest form of crossover network? What is its disadvantage?

26. Because of unequal spacing between the base current characteristic curves (see Figure 7.24), single-transistor amplifiers will be subject to amplitude distortion having unequal positive and negative half-cycles. Push-pull arrangements, however, will have symmetrical half-cycles, but they will be distorted sinusoids. (See Figure 7.29.) What will be the major distortion harmonic in each case? (Recall Chapter 5, devoted to Fourier analysis.)

27. a. If the y-intercept of a transformer-coupled load line on the I_C-V_{CE} curves of a transistor is at 1 mA and the value of V_{CC} is 15 V, what is the value of the load resistance?

b. In using a power transistor, if the y-intercept is 250 mA, with a 15 V source voltage, what is the value of the load resistance?

28. In terms of power dissipation, why is transformer coupling superior to a resistor load in the form of R_C?

29. Equation 7.26 indicates the maximum (idealized) efficiency of class B push-pull amplifiers to be 78.5%. Compute the corresponding maximum efficiency of a class A transformer-coupled amplifier with peak signal current I_m and with the Q-point at I_C and V_{CC}.

30. Distinguish among class A, class B, and class C amplification.

PROBLEMS

7.1

A transistor has $h_{fe} = 150$. What is the relative fraction of the emitter current that reaches the collector?

7.2

a. Sketch the schematic symbol for an *npn* transistor.
b. Do the same for a *pnp* transistor.

7.3

a. By means of two batteries, with one in the base-emitter circuit and the other between the collector and emitter, indicate the appropriate polarities for operating an *npn* transistor.

b. Do the same for a *pnp* transistor.

7.4

For the characteristic curves shown in Figure 7.26, what is the value of the ac load resistance? Estimate the value of the dc load.

7.5

With regard to the circuit in Figure 7.36, the instantaneous output voltage (v_{out}) is also the instantaneous voltage between the collector and emitter (v_{ce}). Using Kirchhoff's voltage law (KVL) as applied to the output circuit, with V_{CC} the dc voltage, we have

$$v_{ce} = V_{CC} - i_c R_C$$

a. Justify that the output signal voltage is, in general, 180° out of phase with the input signal voltage.

b. What is the relationship between the output signal *current* and the input signal *voltage*?

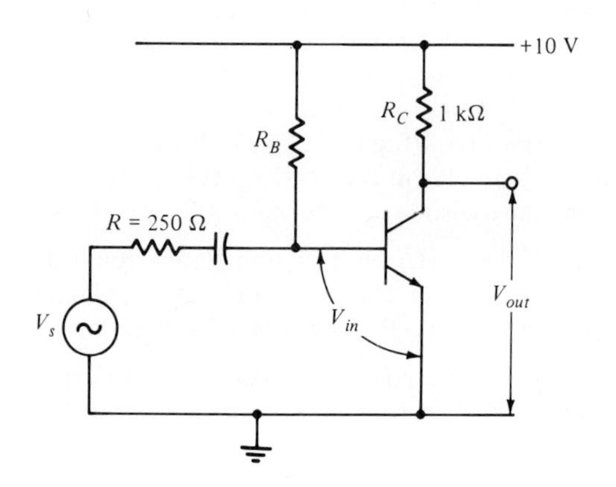

FIGURE 7.36

7.6

If $V_{CC} = 15$ V, $V_{CE} = 9$ V, and $R_C = 1.5$ kΩ, what is the value of the quiescent current I_C?

7.7

It was noted that a V_{BE} variation of 60 mV leads to a collector current change by a factor of 10. A maximum collector signal current variation might have the value 2. What is the corresponding change in V_{BE}?

7.8

The output admittance of a transistor is given as 25 μS. What is the corresponding output resistance?

7.9

Since the value of the load resistance can be determined from the slope of the load line drawn on the characteristic i_c vs. v_{ce} curves, how can the transistor's output resistance be determined from these same curves? What qualitative justification do these curves provide for considering that the transistor acts as a current (rather than a voltage) source? Under what conditions may the transistor act as a voltage source?

7.10

Assume the base current of a transistor to be furnished by a constant current source, with $I_B = 10\ \mu$A. For a grounded emitter circuit with $h_{FE} = 100$ and $V_{CC} = +10$ V,

a. What is V_{CE} when $R_C = 1$ kΩ? What is the value of V_{BC}?

b. When $R_C = 9.3$ kΩ, what is the value of V_{CE}? The value of V_{BC}?

c. What happens when $R_C = 9.7$ kΩ?

7.11

For the circuit in Figure 7.36, $h_{fe} = 75$ and $V_{BE} = 0.7$ V. Find R_B if $I_C = 5$ mA. Estimate the voltage gain

$$A_{Vs} = \frac{V_{out}}{V_s}$$

as well as

$$A_V = \frac{V_{out}}{V_{in}}$$

Assume X_C to be negligible at the signal frequency. (Note the distinction between A_V and A_{Vs}.)

7.12

With $V_{CC} = 15$ V, $V_{CE} = 7.5$ V, $I_C = 5$ mA, and $I_B = 0.1$ mA, what is the necessary value of R_C for a grounded emitter CE amplifier? If the applied signal current is

$$i_i = 0.05 \sin \omega t \qquad \text{[mA]}$$

what is the approximate current gain of the amplifier? What is the value of the output signal voltage?

7.13

The gain of an *RC*-coupled amplifier drops off at low frequencies because of the divider action of the coupling network (Figure 7.37):

$$\mathbf{V}_{out} = \frac{R}{R - (j/\omega C)} V_{in} = \frac{1}{1 - (j/\omega CR)} V_{in}$$
$$= \frac{1}{1 - j(f_1/f)} V_{in}$$

where $f_1 = 1/(2\pi RC)$, with this frequency representing the 3 dB point. The magnitude of the transfer function is

$$\left|\frac{V_{out}}{V_{in}}\right| = \frac{1}{\sqrt{1 + (f_1/f)^2}}$$

Show that the rate of drop-off of gain with frequency at $\frac{1}{10}f_1$ is at the rate of 20 dB/decade.

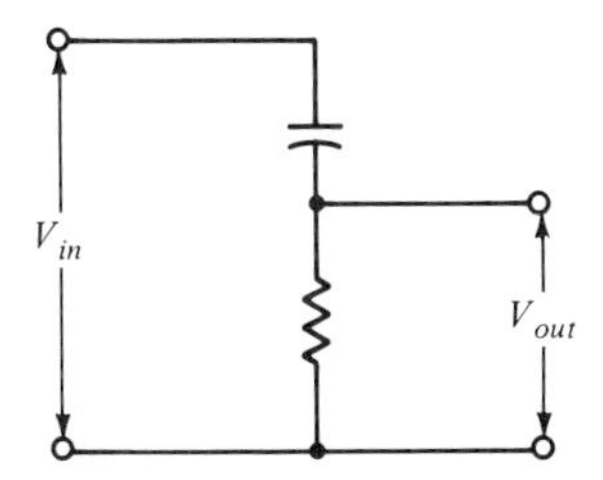

FIGURE 7.37

7.14
If the 3 dB corner frequency of a high-pass filter is f_1, what is the attenuation and phase shift at $0.1f_1$ and at $10.0f_1$?

7.15
See Figure 7.38. Assuming that $R >> 1/\omega C$, show that $V_{out} = 0.707V_{in}$ at the frequency satisfying the equation $R_s = 1/\omega C$. This, of course, is the situation that arises at the upper band limit of an amplifier. (*Note*: Since $R >> 1/\omega C$, $\omega^2C^2R^2 >> 1$.)

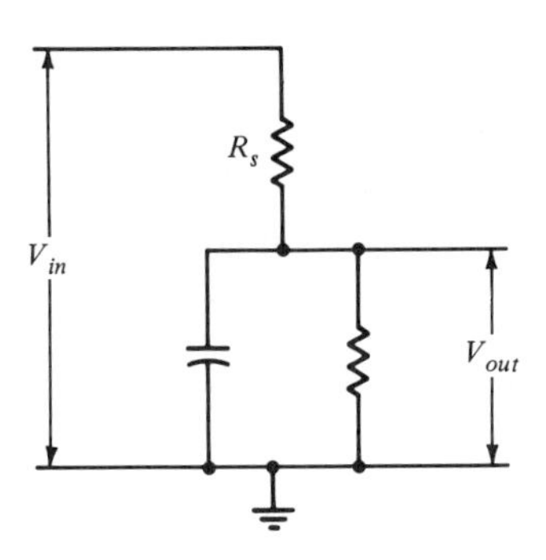

FIGURE 7.38

7.16
We have seen that for good regulation the bleeder current of a voltage divider should be considerably greater than the load current. This has been reflected in the design of the biasing network (R_1 and R_2) in Example 7.2, the parallel combination being made to equal approximately one tenth of the value of $(1 + h_{fe})R_E$. Such a design, however, reduces the effective input resistance to the amplifier as seen by the source, and one might be tempted to make $R_1 \parallel R_2 \sim 10$ times R_{in}. Calculate R_1 and R_2 under such conditions using the parameters of Example 7.2. Determine the change in I_C when h_{fe} changes from 100 to 200, and compare this change with that of the design in Example 7.2.

7.17
Design the bias network for Example 7.2 so that $R_1 \parallel R_2$ equals $(1 + h_{fe})R_E$ for $h_{fe} = 100$. Calculate the percent change in the collector current if h_{fe} becomes 200.

7.18
To illustrate the distortion arising from the use of a grounded emitter amplifier—that is, no R_E resistor—consider a triangular input waveform that should lead to the output waveform shown in Figure 7.39.

a. With $V_{CE} = 10$ V and $R_C = 10$ kΩ, compute the voltage gain for $I_C = 1$ mA.

b. Using the voltage gain calculated in part a, what is the corresponding input voltage change for a 2 V change in the output voltage?

c.
1. If the gain remained constant at the value calculated in part a, compute the values of V_{CE} for I_C in the range 0.2–2.0 mA, in 0.2 mA increments.
2. Compute the actual gain at each value of I_C in part 1 and the average gain for each current interval.
3. Since $\Delta v_{out} = \Delta v_{in} \times$ average interval gain, using the same Δv_{in} as in part b, compute the value of V_{CE} for each value of I_C, starting with $I_C = 1.0$ mA and progressing to both higher and lower values.
4. Plot V_{CE} vs. time, showing the resulting distortion of the triangular waveform depicted in Figure 7.39.

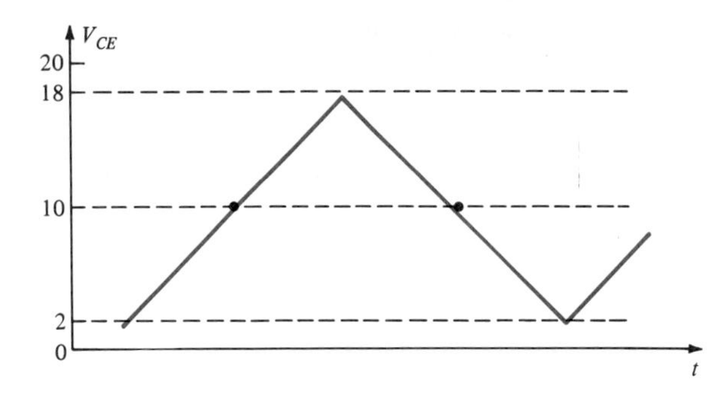

FIGURE 7.39

7.19
Compare the stored magnetic energy of a transistor coil to that of a conventional ignition coil. For the conventional coil, take the current to be 3 A and the inductance to be 5 mH. For the transistor coil, take the current to be 8 A and the primary inductance to be 1.25 mH. (Refer to Eq. 2.45 and the accompanying footnote.)

7.20
A typical conventional ignition coil has 200 primary turns of

#20 wire and 14,000 secondary turns of #38 wire, giving a turns ratio of 70:1. The inductance of the primary coil lies between 4 and 6 mH. A transistor coil might have 100 turns of #18 wire as the primary and 25,000 turns of #41 as the secondary. The primary inductance would have values in the range 1–1.5 mH. Taking #18 wire to have a resistance of 0.0198 Ω/m and #20 wire, 0.0316 Ω/m,

a. Compare the two time constants.

b. What does the larger number of turns on the primary do to the back emf?

c. Why can fewer turns of larger wire be used in the transistor case?

7.21

If a spark plug (for whatever reason) fails to fire, the resulting oscillatory behavior of the primary circuit can lead to a back emf of the order of hundreds of volts. This reverse voltage would constitute a large forward bias between the collector and emitter and could lead to permanent damage of the transistor. What means of protection can be employed against such an eventuality and yet not disturb the proper circuit operation?

7.22

Utilizing a transformer-coupled class A amplifier, one desires to deliver the maximum signal power output into a load R_L = 5 Ω, with the following Q-point: I_C = 200 mA, $V_{CE} = V_{CC}$ = 50 V. Using the idealized load line in Figure 7.40,

a. What should be the turns ratio (N_1/N_2)?

b. What is the maximum signal output power?

c. What is the rms signal voltage across the 5 Ω load?

d. How can V_{CE} assume values larger than V_{CC}?

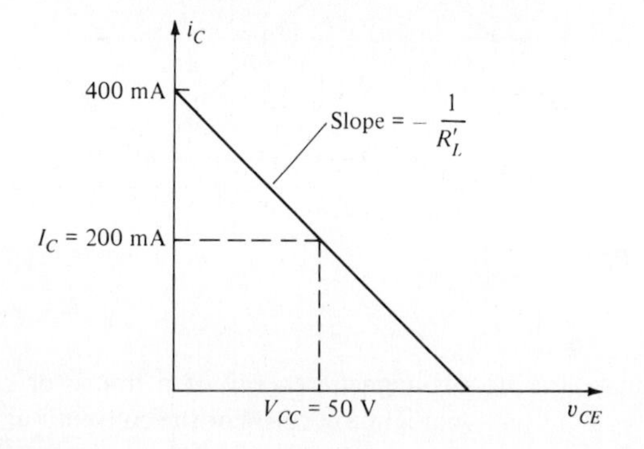

FIGURE 7.40

7.23

a. Using the graphical data given as Figure 7.24, what are the following dc values: I_C, I_B, V_{CE}?

b. With a base current of 1.5 mA, from the graphical data, determine the value of V_{CE} and the voltage drop across the load resistor.

c. Assuming a dc supply voltage of 12 V, what is the value of the load resistance in this design?

d. What value should R_B have, presuming again that there is a 0.7 V drop between the base and emitter? What is the closest 10% tolerance standard resistance value?

7.24

If two transistors with the same characteristics as the one used in Problem 7.22 are used in a push-pull arrangement,

a. What is the value of the load line—that is, the resistance?

b. Compute the maximum signal power available.

c. What is the dc power input, and how does it compare with that of a single transistor?

d. What are the efficiencies in the two cases?

7.25

Design a biasing network for the circuit in Figure 7.25 utilizing the characteristics given in Example 7.8. Design for a lower band limit of 100 Hz. (One often selects $V_E = \frac{1}{10} V_{CE}$.) Let R_s = 1 Ω.

7.26

Using the circuit in Figure 7.41 as the equivalent of the crossover network for a woofer and a tweeter, compute the impedance in the vicinity of 5 kHz. What values of C would give precisely a value of 8 Ω at 5 kHz?

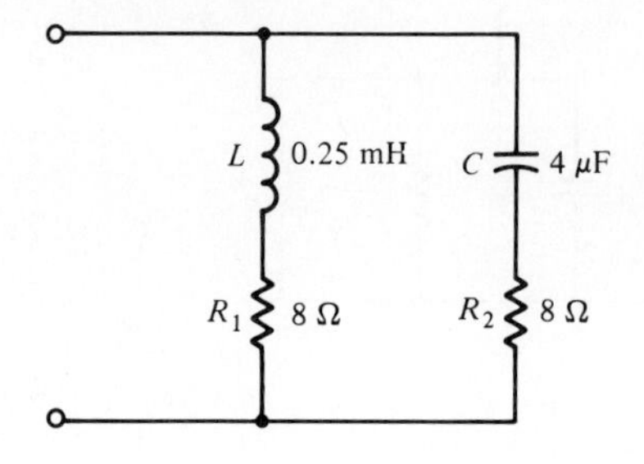

FIGURE 7.41

7.27

With a sinusoidal input, sketch the output signals at 1 and 2 for the circuit shown in Figure 7.42, termed a phase splitter.

7.28

Refer to Figure 7.24. If the operating Q-point is moved to a value corresponding to a significant reverse bias, I_C will be limited to the very small reverse saturation current between the collector and emitter due to minority carrier flow.

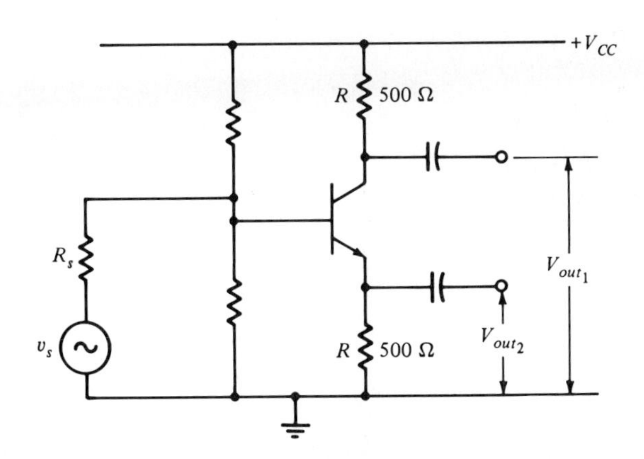

FIGURE 7.42

a. If a sinusoidal signal is applied that is sufficient to drive the base into forward bias at the very tips of the positive half-cycles, sketch the appearance of the transistor output signal.

b. If this output is applied to a high-Q LC resonant circuit, how can the deficiency of part a be remedied?

7.29

For the transistor cited in Example 7.9, what should be the maximum allowed power dissipation in the transistor, assuming, again, a 50°C ambient temperature? For a 25°C ambient temperature?

8 FEEDBACK

OVERVIEW

Because of their nonlinear characteristics, amplifiers subject the impressed signal to some degree of distortion. In Chapter 7 we saw why the distortion arises and how its influence may be measured. Now we turn to a consideration of how its influence may be minimized.

We cannot state "how distortion is *corrected*," for that implies that we can eliminate it. What we *can* do is to minimize it to any desirable degree we wish. The procedure used to minimize various forms of distortion is termed *negative feedback*. There is also a form called *positive feedback*, which has the opposite effect on a signal—it distorts it. Since there are circuits that employ positive feedback (and we also consider such circuits in this chapter), one might legitimately raise the question as to why one might purposely wish to distort a signal.

The answer to that question involves an acknowledged engineering principle: Anytime one achieves an improvement in one engineering parameter, it generally is at the expense of diminishing performance with regard to some other parameter. In the case of negative feedback, the decrease in distortion is accompanied by a diminution in the amplifier's gain. By implication, while positive feedback increases the amount of distortion, it can be used to greatly increase the amount of gain. We will consider some of the circumstances in which this might prove to represent a desirable trade-off.

OUTLINE

8.1 ■ INTRODUCTION

By an appropriate selection of R and C between stages, we can extend an amplifier's bandwidth to some practical limit. But, eventually, upon approaching the band limits, the gain will begin to drop. This constitutes **frequency distortion**: Not all frequencies are being subjected to the same degree of amplification.

Another problem, frequently encountered with electronic measuring equipment, arises from the difficulty in obtaining repeatable outputs for successive inputs of equal magnitude. Fixed gain may prove difficult to maintain over extended periods of time since performance changes with component aging, temperature, and so forth. Also, transistors and integrated circuits show considerable variability even among units of the same type designation. Such is the problem of **amplifier stability.**

Finally, in the process of producing an amplified version of an input signal, an electronic amplifier may (and usually does) introduce various types of **amplitude distortion.** For example, a pure sinusoidal input may have a magnified version emerging with other than a sinusoidal variation.

It may prove surprising that all of these enumerated shortcomings may be improved by one and the same process, **negative feedback.**

8.2 ■ NEGATIVE FEEDBACK

8.2.1 A Mechanical Analog

Consider Figure 8.1, which shows a succession of amplifier stages. Note that as the input signal is amplified, there is a reversal of phase at each stage of amplification. (We are assuming that the amplifiers are of the common emitter variety.) To understand what follows, it must be realized that, in general, there is no time delay involved in the appearance of these signal voltages at the respective output points. Thus, the points labeled a in Figure 8.1 appear simultaneously at the outputs of A_1, A_2, and A_3. And since a was chosen in a general manner, other portions of the amplified input waveform also appear simultaneously at their respective equivalent locations in the circuit.

An analogy might prove useful at this point: Consider the lever shown in Figure 8.2. Application of force F_1 will raise a weight W that is much larger

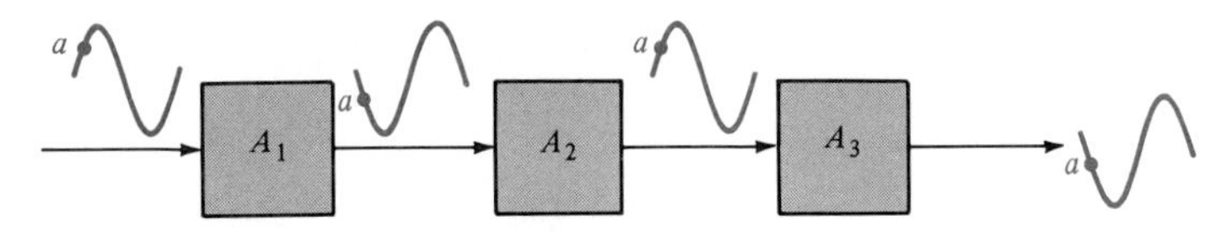

FIGURE 8.1. **Three Successive CE Amplifier Stages** The signal is inverted as it passes through each stage. The instantaneous corresponding value of the voltage at each stage is indicated by point a, implying that in general there is no delay between the appearance of the signal at each output point.

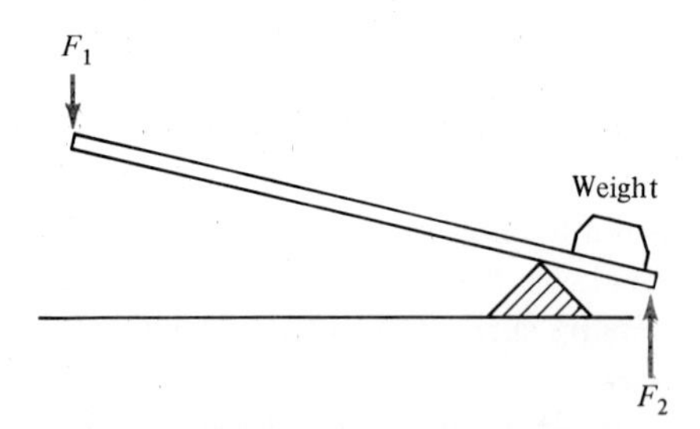

FIGURE 8.2. **Mechanical Advantage of Simple Lever** $F_2 > F_1$, and the two are oppositely directed—that is, 180° out of phase for oscillatory forces.

than the applied force. This, of course, comes about because of the mechanical advantage of the lever.

With two levers in succession (Figure 8.3), the magnified force F_3 is now in the same direction as F_1. Again note that there is no delay between the application of F_1 and the appearance of F_3. The addition of another lever (Figure 8.4) leads to a still greater force (F_4) with output and input now out of phase. The mechanical advantage of these levers corresponds to gain in electronic amplifiers.

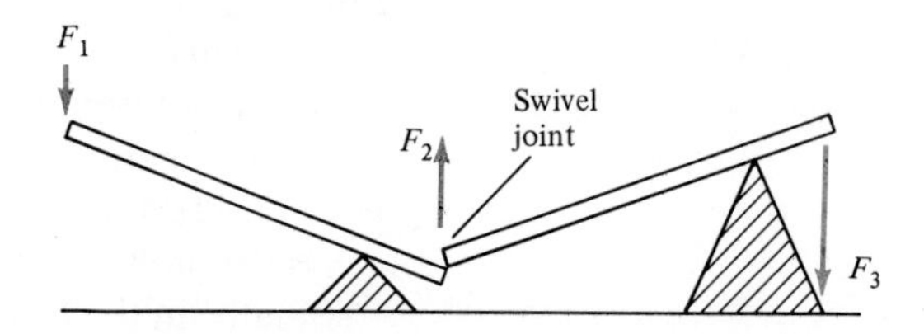

FIGURE 8.3. **Two Levers in Series** $F_3 > F_2 > F_1$, and F_3 is in phase with F_1 for oscillatory forces.

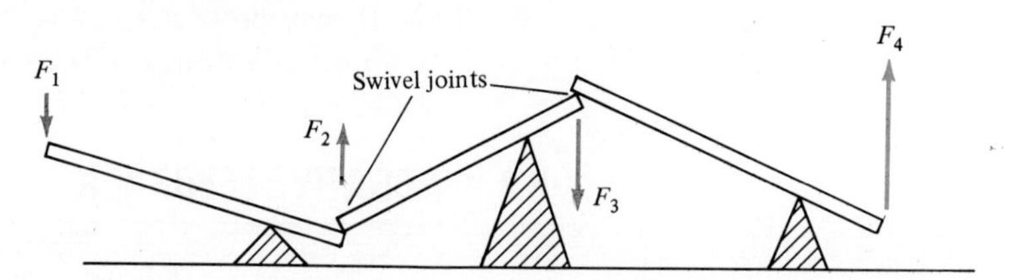

FIGURE 8.4. **Three Levers in Series** $F_4 > F_3 > F_2 > F_1$, with F_4 180° out of phase with F_1, the input.

If F_{in} is the input force and F_{out} the output, we have

$$\frac{F_{out}}{F_{in}} = A \tag{8.1}$$

If we now take some fraction of the output force, say β,[1] and couple it back to the input, because of phase opposition, this procedure will require the application of a greater input force if the output is to be maintained. The required input will now be

$$F'_{in} = F_{in} + \beta F_{out} \tag{8.2}$$

Designating the gain with feedback to be A', we have the following equation[2]:

[1]This β should not be confused with the current gain of a transistor. That is why we chose to represent the β of a transistor by its alternative designation, h_{fe}.

[2]The general equation for gain with feedback is written in the form $A' = A/(1 - \beta A)$ and is applicable to both negative and positive feedback. For negative feedback, β is taken to be negative, and we have $A' = A/(1 + \beta A)$. For positive feedback, β is taken to be positive, and we have $A' = A/(1 - \beta A)$. Positive feedback arises when the feedback signal is in phase with the input and negative feedback, when they are out of phase.

$$A' = \frac{F_{out}}{F'_{in}} = \frac{F_{out}}{F_{in} + \beta F_{out}} = \frac{F_{out}}{(F_{out}/A) + \beta F_{out}} = \frac{A}{1 + \beta A} \tag{8.3}$$

Let's put in some numbers. Assume $A = 50$ and $\beta = \frac{1}{10}$; then $A' = 8.33$. We have considerably reduced the gain as a result of employing negative feedback.

What we have done in the mechanical case may seem like something silly, but let's see what happens when we apply these concepts to electronic amplifiers.

8.2.2 Amplifier Stability

We consider three stages of CE amplification. Thus, the output is 180° out of phase with the input.

The rms signal voltage applied to the amplifier is V_{in}. The amplifier's gain being A, the rms output voltage (V_{out}) is

$$V_{out} = AV_{in} \tag{8.4}$$

If a fraction (β) of the output voltage is combined with the input signal, since the two are in phase opposition, one must increase the signal applied from the external source (V'_{in}) if the output is to be maintained. Thus, we have

$$V'_{in} = V_{in} + \beta V_{out} \tag{8.5}$$

or, alternatively,

$$V_{in} = V'_{in} - \beta V_{out} \tag{8.6}$$

There are two different gains with which we deal, that of the amplifier,

$$A = \frac{V_{out}}{V_{in}} \tag{8.7}$$

and that of the signal source,

$$A' = \frac{V_{out}}{V'_{in}} \tag{8.8}$$

It is the latter one that is of prime interest to us. If there were no feedback, V_{in} would equal V'_{in}, and the two gains would be identical.

What is the relationship between gain with and without feedback? It takes the same form as in the mechanical case (Eq. 8.3):

$$A_{of} = \frac{A_o}{1 + \beta A_o} \tag{8.9}$$

where A_{of} is the midband gain with feedback, termed the midband **closed-loop gain**, and A_o is the midband gain without feedback, being the midband **open-loop gain.** Feedback has considerably reduced the overall midband gain.

If the gain of the amplifier is very large, $\beta A_o >> 1$, and

$$A_{of} \simeq \frac{A_o}{\beta A_o} = \frac{1}{\beta} \tag{8.10}$$

Thus, we see that the gain, although significantly reduced, is now independent of the original gain (A_o), which may have shown considerable variability due to age, temperature, replacement, and so on. The amplifier's gain now depends only on the value of β.

How stable is β? By way of an answer, look at a segment of an actual circuit employing negative feedback. In this case the output load resistor has a variable tap, and the value of β depends on the location of this tap (Figure 8.5). (If a fixed value of feedback is desired, one may use two fixed resistors, properly proportioned.) It will be zero when the tap is at the bottom of the resistor and unity, when at the top. Thus, the stability of the overall system depends only on resistor stability, and resistors can be made to be rather immune to variations due to temperature, aging, and so forth.

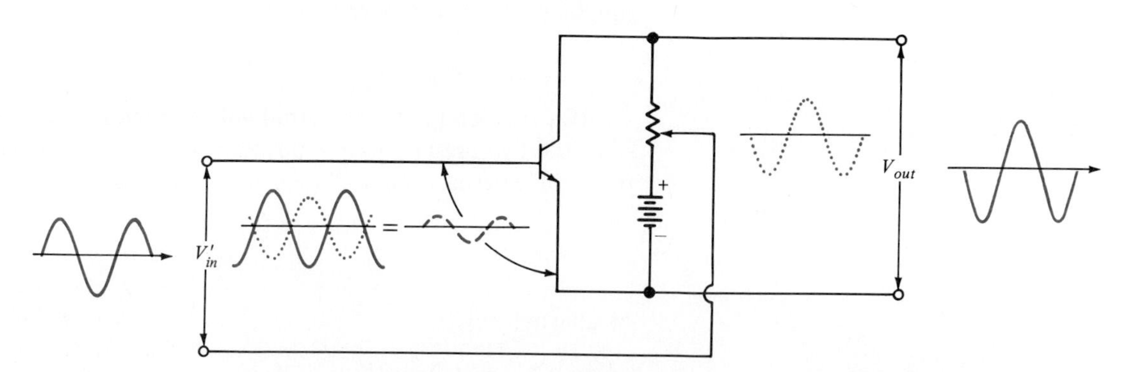

FIGURE 8.5. Basic Negative Feedback Circuit A fraction of the output βV_{out} (dotted waveform) is added to the input signal V'_{in} (solid waveform at left), producing signal V_{in} (dashed waveform), which is what is actually amplified. The degree of feedback is varied by pot adjustment at the output.

EXAMPLE 8.1 If $A_o = 500$ and $\beta = 0.1$, what is the midband gain with feedback?

Solution:

$$A_{of} = \frac{A_o}{1 + \beta A_o} = \frac{500}{1 + (0.1)(500)} \simeq 10$$

Or, in general, to a good approximation, we have

$$A_{of} \simeq \frac{A_o}{\beta A_o} = \frac{1}{\beta} = \frac{1}{0.1} = 10$$

Rather than an approximation to the gain, it is often desired to know how stable the gain will be. What enters such considerations, and many others involving negative feedback, is the quantity $1 + \beta A_o$, known as the **desensitivity factor**. In this case the gain variability is reduced by the factor $1/(1 + \beta A_o)$.

EXAMPLE 8.2 By using negative feedback, it is desired that an amplifier have a stable gain (A_{of}) of 100. What value of negative feedback should be used if the gain without feedback is 1000? If the gain without feedback is specified as 1000 ± 100 —that is, ± 10%—what is the accuracy of the gain with feedback?

Solution:

$$A_{of} = \frac{A_o}{1 + \beta A_o}$$

$$100 = \frac{1000}{1 + \beta(1000)}$$

$$\beta = \frac{10^3 - 10^2}{10^5} = \frac{900}{10^5} = 9 \times 10^{-3} \qquad \text{or} \qquad 0.9\%$$

$$\frac{10\%}{1 + \beta A_o} = \frac{10\%}{10} = 1\%$$

8.2.3 Frequency Distortion

The gain of an *RC*-coupled amplifier drops off at low and high frequencies. What happens with negative feedback? As the gain drops off, so does the magnitude of the feedback voltage βV_{out}. Therefore, there is less available to be subtracted from V'_{in}. The smaller the signal subtracted from V'_{in}, the larger the fraction of the input signal available to be amplified. Even though the gain of the amplifier has dropped, there is an automatic increase in the effective input voltage, and the output voltage magnitude is maintained out to a much higher (and lower) frequency. We have thereby achieved a wider bandwidth but at the expense of a much smaller midband frequency gain. Amplifiers are characterized by a **gain-bandwidth product**: In return for reducing the midband gain, we may extend the bandwidth. For a transistor this parameter is specified as f_T, being the bandwidth for unity gain. Any increase in gain is obtained at the expense of this bandwidth.

At high and low frequencies, the gain of an *RC*-coupled amplifier suffers reduction because of capacitances—at low frequencies because of the coupling capacitors and at high frequencies because of the input capacitances. Such capacitive effects will introduce a phase shift between the input and output waveforms. In negative feedback we have depended on the feedback voltage being 180° out of phase with the input signal voltage. With an added phase shift, the phase difference between them becomes less than 180°—and progressively more so as the frequency increases (and decreases). At some high (and/or low) frequency, if this additional total phase shift becomes 180°, the feedback signal will now be in phase with the input, and rather than negative feedback, we will now have **positive feedback**. We will consider positive feedback in Section 8-4, but for now let it simply be noted that *in amplifiers* such conditions are generally to be avoided since they can lead to unstable operation.

EXAMPLE 8.3 The high-frequency gain of a CE amplifier drops off as a function of frequency in the following manner:

$$A(f) = \frac{A_o}{1 + j(f/f_2)} \tag{8.11}$$

with $f_2 = 1/(2\pi R_2 C_2)$; R_2 and C_2 are the series input resistance and parallel input capacitance, respectively. At f_2, $R_2 = 1/j\omega C_2$.

a. If $f = f_2$, what is the gain of the amplifier?
b. What is the phase shift between the input and output at this frequency?
c. What is the maximum phase shift that can arise with a single CE stage (that is, in addition to the usual 180° reversal)?
d. What is the minimum number of *RC*-coupled CE stages that can lead to instability using negative feedback?

Solution:

a. At $f = f_2$, we have

$$A(f_2) = \frac{A_o}{1 + j1} = \frac{A_o}{\sqrt{2}} \angle -45°$$

b. The phase shift is $+180° - 45° = +135°$, where $+180°$ is the normal phase shift through the CE stage.
c. When $f \gg f_2$, we have

$$j\frac{f}{f_2} \gg 1 \qquad \text{and} \qquad \tan^{-1}(-\infty) = -90°$$

which leads to a $+90°$ phase shift between input and output.
d. Since a total phase shift of $-180°$ is necessary (to cancel out the original $+180°$ shift in negative feedback), three stages of *RC*-coupled CE amplifiers are necessary. (Two stages can give rise to $-180°$, but their combined gain would be zero when they do so.)

With positive feedback the sign in the denominator of Eq. 8.9 becomes negative, and under such circumstances represents one condition that may lead to unstable amplifier operation. The other condition to be satisfied is that the value of βA_o (termed the **loop gain**) becomes equal to unity. The denominator then assumes a value of zero, with the mathematical consequences that that entails.

EXAMPLE 8.4 An amplifier has an open-loop gain of 38 dB and an upper band limit of 10 kHz. (See Figure 8.6.) If the drop-off in gain is uniform at 20 dB/decade, what is the frequency at which the gain becomes unity?

Solution:

$$\frac{38 \text{ dB}}{20 \text{ dB/decade}} = 1.9 \text{ decades}$$

$$10 \text{ kHz to } 100 \text{ kHz} = 1 \text{ decade}$$

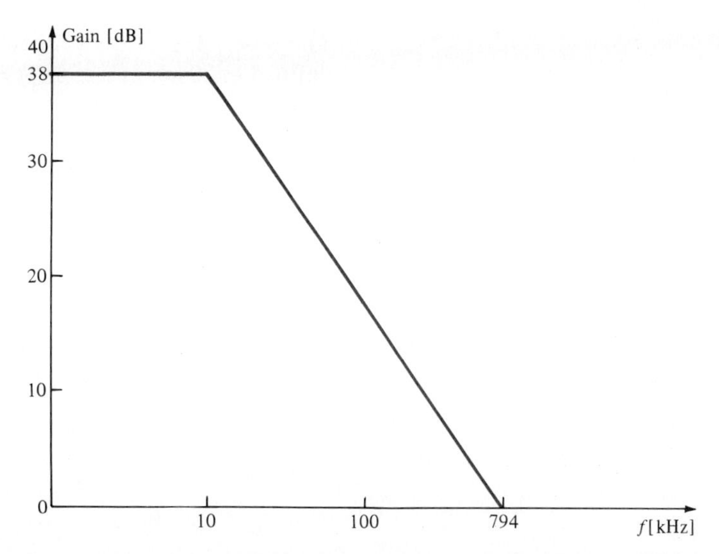

FIGURE 8.6. Idealized Bode Plot Showing 20 dB/Decade in Gain Beyond 10 kHz Upper Band Limit Gain at 10 kHz is actually down by 3 dB (from 38 to 35 dB).

Note that 0.9 decade = $10^{0.9} = 7.94$. Therefore, 100 kHz to 794 kHz = 0.9 decade. The frequency would be 794 kHz. (Remember that 0 dB = 1.)

EXAMPLE 8.5

In Section 7.6 we noted that, by proper design, coupling capacitors between amplifier stages could be eliminated, giving rise to gain drop-off only at high frequencies. Consider such an amplifier with a 1 MHz gain-bandwidth product and a midband gain of 38 dB.

- **a.** What is its bandwidth?
- **b.** What value of feedback will yield a midband gain of 25 dB?
- **c.** What will be the bandwidth with feedback?
- **d.** Plot the results.
- **e.** What is the value of the feedback factor β?

Solution:

a.

$$38 \text{ dB} = 20 \log_{10}(A_o)$$

$$\text{antilog}\left(\tfrac{38}{20}\right) = A_o = 79.4$$

$$\frac{1 \text{ MHz}}{79.4} = 12.6 \text{ kHz}$$

b.

$$A_{of} = \frac{A_o}{1 + \beta A_o}$$

$$20 \log_{10}(A_{of}) = 20 \log_{10}(A_o) - 20 \log_{10}(1 + \beta A_o)$$

$$20 \log_{10}(A_o) - 20 \log_{10}(A_{of}) = 20 \log_{10}(1 + \beta A_o)$$

$$38 \text{ dB} - 25 \text{ dB} = 13 \text{ dB}$$

Therefore, 13 dB of feedback is needed, where the *feedback in decibels* is defined as $20 \log_{10}(1 + \beta A_o) \simeq 20 \log_{10}(\beta A_o)$ if $\beta A_o \gg 1$.

c. $25 \text{ dB} = 20 \log_{10}(A_{of})$

$\text{antilog}(\frac{25}{20}) = A_{of} = 17.8$

$$\frac{1 \text{ MHz}}{17.8} = 56.2 \text{ kHz}$$

d. At the corner frequencies (12.6 and 56.2 kHz), the gain is actually 3 dB down, but the idealized plots shown in Figure 8.7 (called idealized **Bode plots**) simplify matters. At 0 dB—that is, unity gain—the plots intersect the axis at 1 MHz, the gain-bandwidth product.

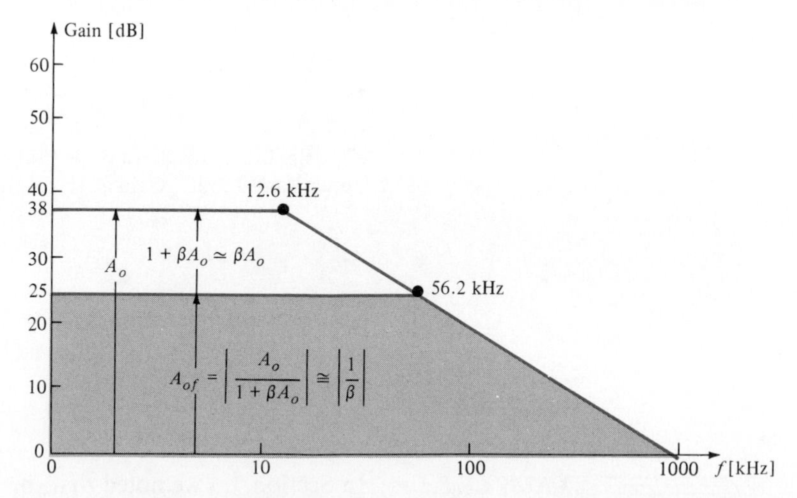

FIGURE 8.7. Idealized Bode Plot Shown are uniform band gain without feedback, A_o = 38 dB; midband gain with feedback, A_{of} = 25 dB; and also feedback, $1 + \beta A_o$ = 13 dB. The band limit without feedback is 12.6 kHz; with 13 dB of feedback, it is 56.2 kHz.

e. $13 \text{ dB} = 20 \log_{10}(1 + \beta A_o)$

$\text{antilog}(\frac{13}{20}) = 1 + \beta A_o = 4.47$

$\beta A_o = 3.47$

$$\beta = \frac{3.47}{79.4} = 0.044 \quad \text{or} \quad 4.4\%$$

Bode plots actually consist of two parts: the gain vs. frequency (as shown, for example, in Figure 8.7) as well as the phase angle versus frequency. The latter may more appropriately be discussed in conjunction with Chapter 10; see "A Closer Look" at the end of that chapter.

8.2.4 Amplitude Distortion

We turn now to a consideration of how negative feedback can improve upon the amplitude distortion that the amplifier introduces. With no distortion we have the sequence of waveforms identified in Figure 8.8a. In Figure 8.8b we start with a sinusoidal input signal, and because of amplitude distortion, we obtain a waveform whose negative excursion we presume to be smaller than the positive one. The negative feedback signal is similarly distorted, and when combined with the original sinusoidal input, it produces an effective input signal that has a positive excursion that is greater than its negative swing. When this effective signal is passed through the amplifier, it emerges as a fairly decent amplified sinusoidal signal. In other words, the negative feedback has predistorted the input signal so that it will compensate for the amplifier's distortion.

Well, if the output signal has thereby been corrected, how can there be a predistorted signal to be fed back? The point to be remembered is that the distortion cannot be completely eliminated—it can only be reduced (again by the value of the desensitivity factor). A numerical example might help in this regard.

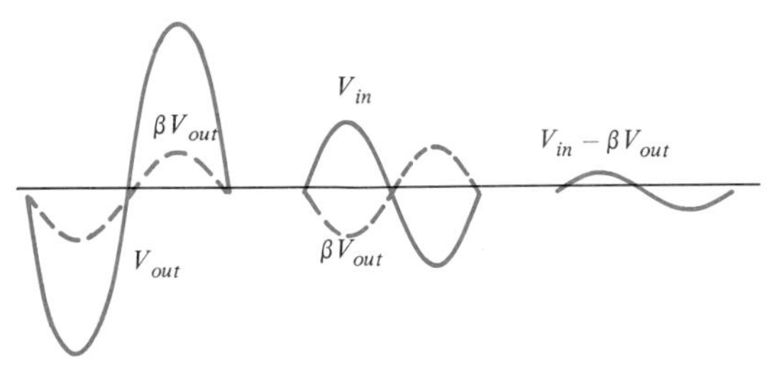

(a) Signal sequence for an ideal amplifier

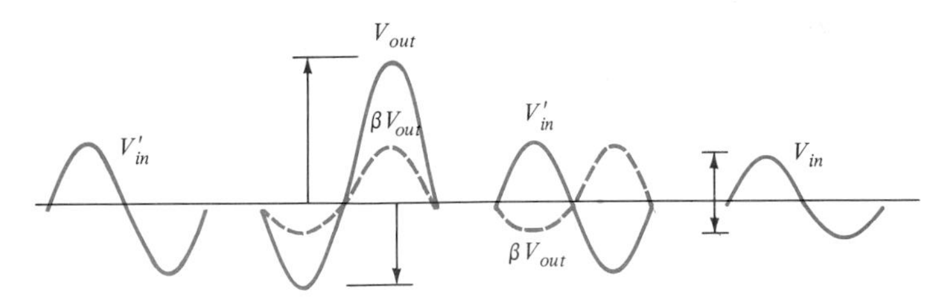

(b) Signal sequence for a distorting amplifier

FIGURE 8.8. **Negative Feedback**

An amplifier, presumed to yield a 10 V rms signal at 1 kHz, is said to contain 20% second harmonic distortion—that is, the second harmonic signal at 2 kHz has an rms value of 2 V. This distortion is introduced by the amplifier. The actual situation might be as depicted in Figure 8.9a.

While the distortion actually arises in the amplifier, we can instead consider the amplifier to be ideal—that is, distortionless—and introduce the distortion as a signal V_D (= 2 V) as in Figure 8.9b. Or we can instead move the distortion to the input side of the amplifier and presume that there it has a magnitude $V_D/100$ (equal to 0.2 V rms; see Figure 8.9c). All three representations will yield the same output: a 10 V rms signal at 1 kHz and a 2 V rms signal at 2 kHz.

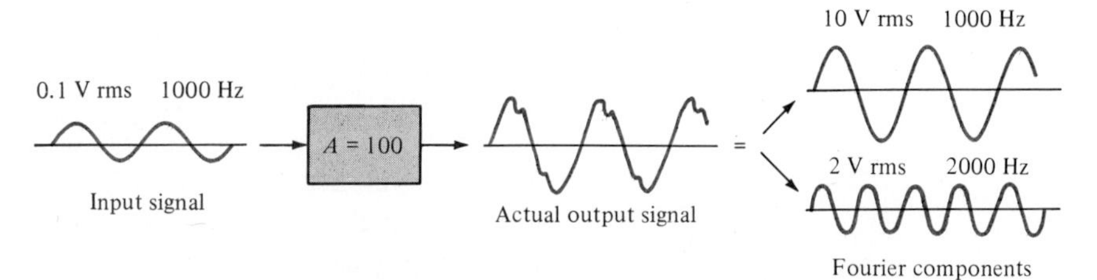

(a) Distorting amplifier with Fourier components

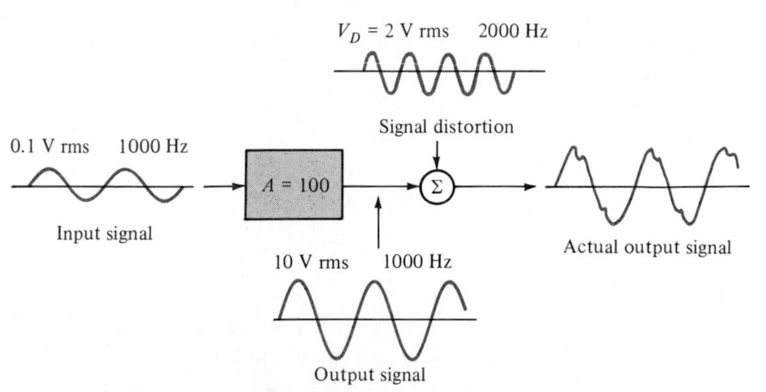

(b) Ideal amplifier, with externally introduced distortion at the output

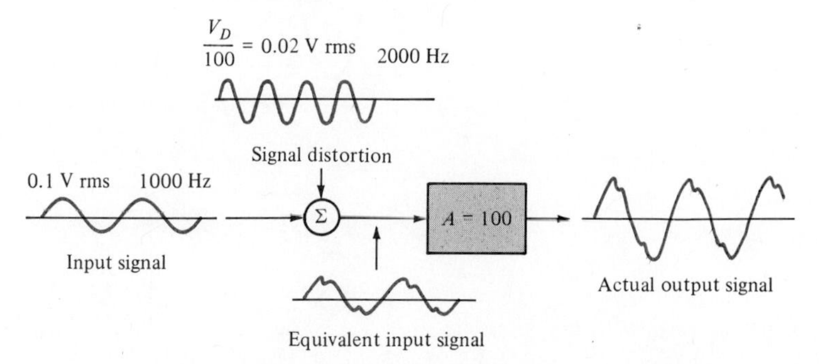

(c) Ideal amplifier, with externally (reduced) distortion at the input

FIGURE 8.9. Alternative Distortion Analysis

Let us now apply 19% negative feedback as in Figure 8.10a. We can consider separately what happens with negative feedback at 1 kHz (Figure 8.10b) and at 2 kHz (Figure 8.10c). In Figure 8.10b there is a 2 V input signal at 1 kHz that is being applied externally. In Figure 8.10c there is no 2 kHz signal external to the system since the distortion is generated within the amplifier itself.

With 19% negative feedback we still have a 10 V output at 1 kHz (although, of course, we had to increase the input signal to 2 V from the original value of 0.1 V without feedback), and the distorted output has been reduced from 2 to 0.1 V.

Rather than resorting to a graphical analysis, one may instead determine the degree of distortion reduction by a simple analytical procedure: The distortion voltage with feedback is equal to the original distortion magnitude

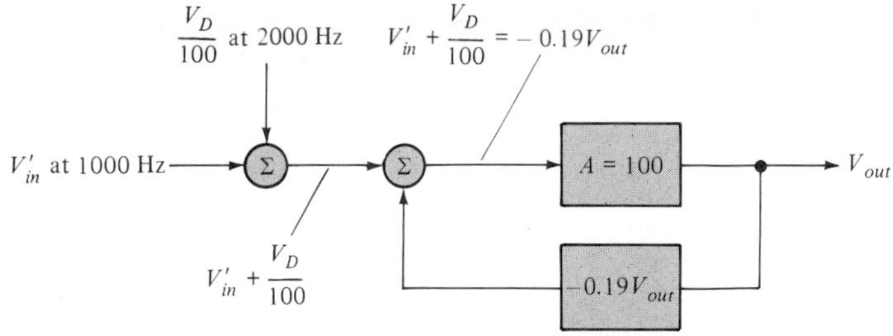

(a) Ideal amplifier with externally introduced second harmonic distortion at input, employing 19% negative feedback

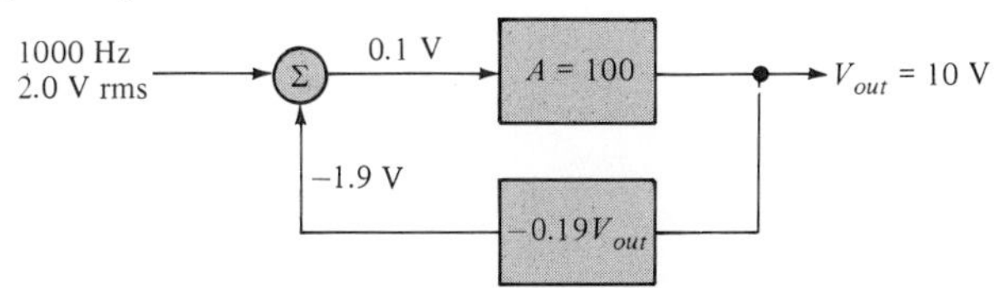

(b) Effect of feedback on fundamental signal

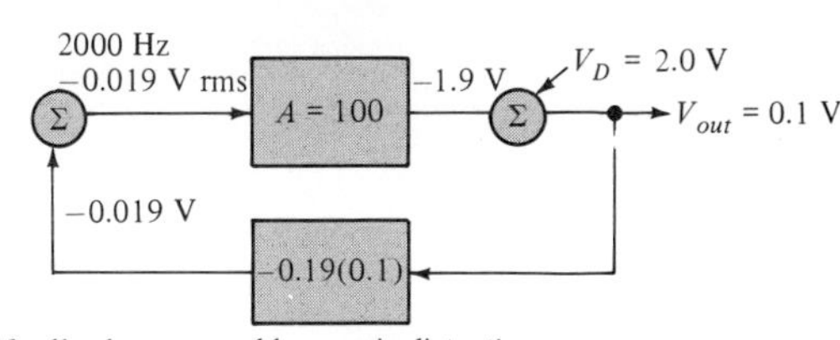

(c) Effect of feedback on second harmonic distortion

FIGURE 8.10. Reduction of Distortion by Negative Feedback

divided by the desensitivity factor. Thus, in the preceding case, with a 2 V distortion, $\beta = 0.19$, and $A = 100$, we have

$$\frac{2\text{ V}}{1 + \beta A} = \frac{2}{1 + (0.19)10^2} = 0.1\text{ V}$$

8.2.5 Emitter Follower

Among the more frequently encountered circuits is one employing 100% negative feedback. Refer to Figure 8.4. This degree of feedback would arise with the resistor tap "all the way up." Circuit operation will not be altered if the resistor and battery are interchanged, and if the resistor tap is at the top, we can redraw the circuit as in Figure 8.11.

Using Eq. 8.9, with β (the feedback ratio) equal to unity and presuming again that A_o (the midband gain without feedback) is large, we have $A_{of} \simeq A_o/\beta A_o \simeq 1$. Thus, the voltage gain of this circuit is close to unity—and hence the name *unity-gain amplifier* (also called **emitter follower** because input and output signal voltages are about the same). This amplifier also constitutes the common collector (CC) classification; input is applied between the base and collector—that is, signal ground—and the output appears between the emitter and collector (ground).

If the gain is unity, the reader may well ask, "Why bother?" Well, while the voltage gain is unity, there is a considerable current gain. And while the

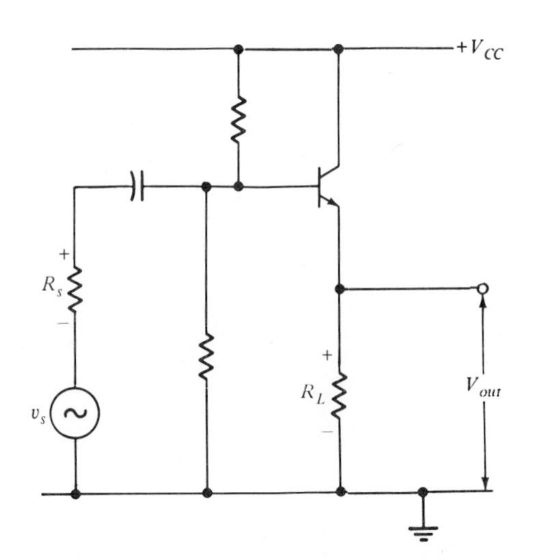

FIGURE 8.11. **Emitter Follower** The load resistor appears between the emitter and ground.

input signal current is I_b, the current through the load resistor is $(1 + h_{fe})I_b$, where h_{fe} is the current gain of the transistor, a number generally between 50 and 250. (See Figure 8.12.)

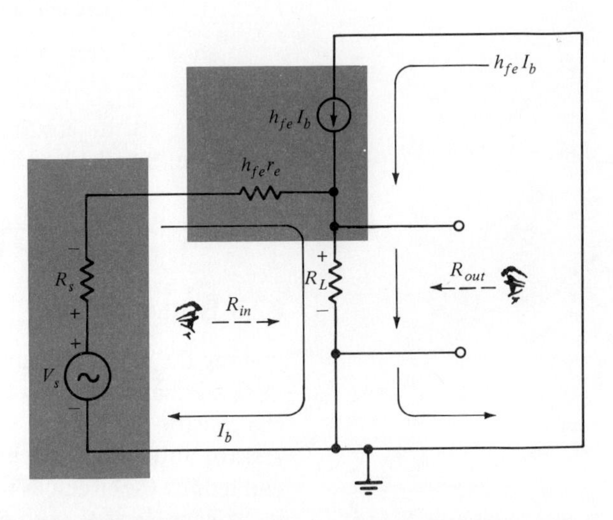

FIGURE 8.12. **Emitter Follower Equivalent Circuit, Indicating Nature of Input and Output Resistances** (The bias network is not included.)

But there is also another very important function served by such circuits. The input signal voltage V_{in} applied to the transistor is

$$
\begin{aligned}
V_{in} &= I_bR_L + I_cR_L + I_bh_{fe}r_e \\
&= (1 + h_{fe})I_bR_L + h_{fe}r_eI_b
\end{aligned}
\tag{8.12}
$$

The input resistance that the amplifier presents to the source is

$$R_{in} = \frac{V_{in}}{I_b} = (1 + h_{fe})R_L + h_{fe}r_e \tag{8.13}$$

The output resistance may be found by using Thevenin's theorem. The open-circuit output is V_s (that is, unity gain). The short-circuit output current ($R_L = 0$) is $I_b + h_{fe}I_b$. With R_L shorted, $I_b = V_s/(R_s + h_{fe}r_e)$. We have

$$I_{SC} = (1 + h_{fe})I_b = \frac{(1 + h_{fe})V_s}{R_s + h_{fe}r_e} \tag{8.14}$$

$$R_{out} = \frac{V_{OC}}{I_{SC}} = \frac{V_s(R_s + h_{fe}r_e)}{(1 + h_{fe})V_s} = \frac{R_s + h_{fe}r_e}{1 + h_{fe}} \tag{8.15}$$

A numerical example might make clear the significance of Eq. 8.15. The coaxial cable—that is, a central conductor surrounded by a shielding braid—used as a connecting link between various pieces of electronic apparatus constitutes a rather small impedance, typically 52 Ω. The typical output impedance of a CE small-signal transistor amplifier might be 2500 Ω (primarily due to R_C). Suppose we wish to feed this signal into a 52 Ω coaxial cable. A direct connection would result in a considerable mismatch. But if between the CE amplifier and the coax we insert an emitter follower, with $I_C \simeq 1$ mA and $h_{fe} = 90$, the output of the emitter follower would present a fairly satisfactory match:

$$\frac{2500 + (90)(26)}{91} = 53\ \Omega$$

Thus, the emitter follower is used as an impedance matching device. While the output impedance has been lowered, the input impedance has been made desirably high—the emitter follower loading on the preceding stage has been significantly decreased. (See "A Closer Look" at the end of this chapter for additional discussion of input impedance.)

EXAMPLE 8.6

With regard to the emitter follower in Figure 8.13, with $h_{fe} = 100$ and $I_C = 1$ mA, what is the output impedance? What is the input impedance R_{in}? What is the input impedance (including the biasing network) R'_{in}?

Solution:

$$R_1 \,||\, R_2 = 60\ \text{k}\Omega$$

$$r_e = \frac{26}{I_C\ [\text{mA}]} = \frac{26}{1} = 26\ \Omega$$

$$R_{out} = \frac{R_s + h_{fe}r_e}{1 + h_{fe}} = \frac{250 + 2600}{101} = 28.2\ \Omega$$

This resistance appears in parallel with 6 kΩ for R_L. Therefore, we have

$$R'_{out} \simeq 28.1\ \Omega$$

$$R_{in} = (1 + h_{fe})R_L + h_{fe}r_e = 101(6000) + 100(26) = 608.6\ \text{k}\Omega$$

$$= R_1 \,||\, R_2 \,||\, R_i = 54.6\ \text{k}\Omega$$

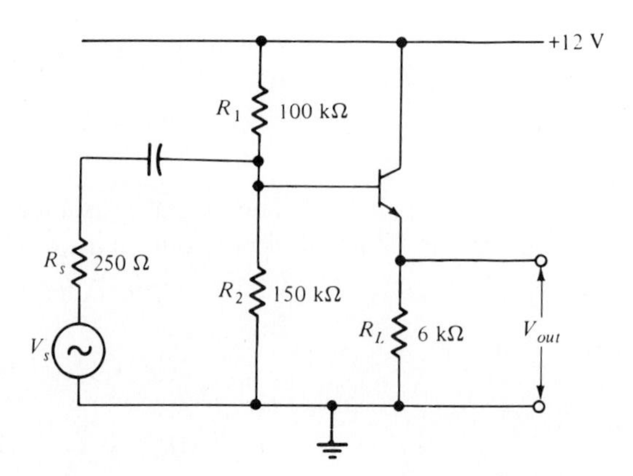

FIGURE 8.13

In the general case of the type of negative feedback that we have been considering, if the input resistance without feedback is R_{in}, with feedback it will be $R_{if} = (1 + \beta A_o)R_{in}$, while the output resistance without feedback, R_{out}, becomes $R_{of} = R_{out}/(1 + \beta A_o)$.

As to why negative feedback has increased the input resistance of an emitter follower (and will do likewise for other *similar* negative feedback circuits), one can gain an understanding by again considering the mechanical analog with which we commenced this chapter (see Figure 8.4). Without any feedback, we apply a certain force and experience an opposition due to the load imposed at the output of the lever system. If a fraction of that oppositely directed force is applied at the input in opposition to F_1, we must exert a larger external force to actuate the system. Therefore, the input resistance has been increased in proportion to the degree of feedback employed.

Caution: The reader should be aware, however, that we have considered but one form of negative feedback called *voltage sampling*, *series mixing*. There are other types that alter some of our results. Some types, for example, decrease the input resistance and increase the output resistance. We will not be considering such modifications at this time.

8.2.6 Darlington Circuit

Since an emitter resistor R_E may be made to appear at the input circuit of an emitter follower as a resistance of much larger value $(1 + h_{fe})R_E$, there would appear to be no limit as to how high one might make the input resistance. The difficulty arises when one considers the dc biasing function of R_E. If, for example, we want a 5 MΩ input resistance with $h_{fe} = 100$, we would require $R_E = 49.5$ kΩ and with a modest $I_C = 1$ mA the result would be a dc drop of 49.5 V, a rather unrealistic situation, made even more so if one is dealing with a power amplifier with amperes of collector current.

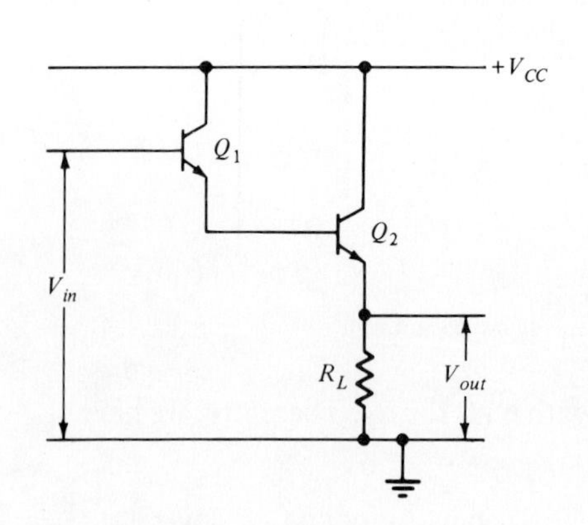

FIGURE 8.14. Darlington Circuit
The biasing network is not shown.

What we need in the emitter circuit is a large *ac* resistance but a relatively small *dc* value. This condition may be realized by one emitter follower as the emitter load for another emitter follower, as shown in Figure 8.14. Such a configuration is known as a *Darlington circuit*.

As a numerical example, if $R_L = 490$ Ω, the input resistance to Q_2 should be about 49.5 kΩ [490(101) = 49.5 kΩ]. In turn, with this resistance as an emitter load for Q_1, the input resistance to the latter should be 5 MΩ [49.5(101) = 5 MΩ]. But this result is moderated by the output resistance of Q_1.

Recall that the output conductance of a transistor (h_{oe}) might typically be 10^{-5} S. This value of conductance corresponds to an output resistance of 100 kΩ. Since this resistance is much larger than that of the collector resistor, R_C, the output resistance of a transistor amplifier is considered to be equal to R_C. But in the case of a Darlington circuit, the input resistance of Q_2 is of the same order of magnitude as the output resistance of Q_1. Therefore, the input impedance of Q_2 in the preceding numerical example must be taken to be about 33.1 kΩ rather than 49.5 kΩ (49.5 || 100 = 33.1). In turn, the input impedance of Q_1 is about 3.3 MΩ rather than 5.5 MΩ [33.1(101) kΩ = 3.3 MΩ]. The biasing circuit for Q_1 may additionally lower this value. The Darlington circuit may also be used to advantage under some circumstances to provide a lower output impedance than may be available from an emitter follower.

Let us return to a consideration of the push-pull amplifier of Figure 7.28. The disadvantages of an output transformer have been noted. Similar objections of cost, distortion, physical space, and so forth apply to the input transformer. Figure 8.15 shows how a complementary arrangement of transistors (that is, *npn* and *pnp*) in an emitter follower push-pull arrangement, with its attendant low-impedance output, may be used to drive a loudspeaker directly. The *npn* conducts on the positive half-cycles of the input, the *pnp* conducts on the negative half-cycles, and the need for an input transformer is thereby eliminated. Figure 8.16 shows a push-pull arrangement using Darlington stages, providing an even lower output impedance.

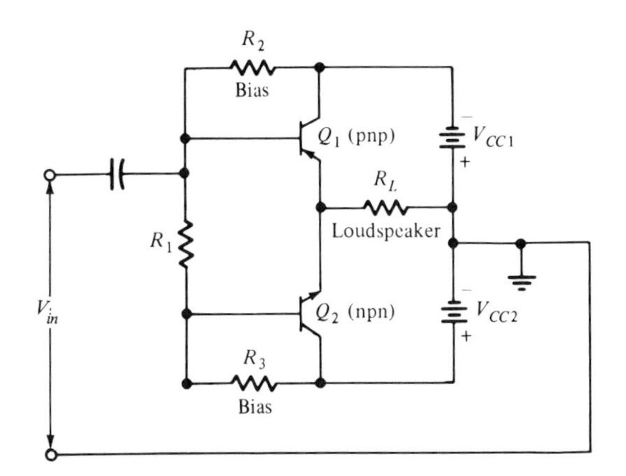

FIGURE 8.15. Complementary Arrangement of Emitter Followers in a Push-Pull Circuit

The nominal value of h_{fe} for a Darlington circuit is equal to the product of the individual h_{fe} values. Thus, h_{fe} = 80 might be expected to lead to a value of about 6400. Darlington transistors are available as single packages.

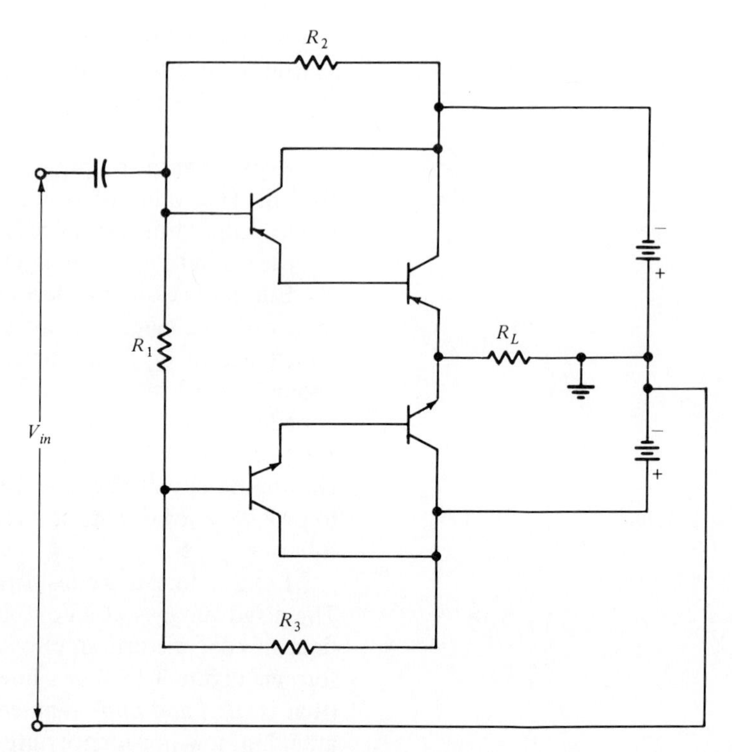

FIGURE 8.16. **Complementary Arrangement of Darlington Circuits in a Push-Pull Circuit**

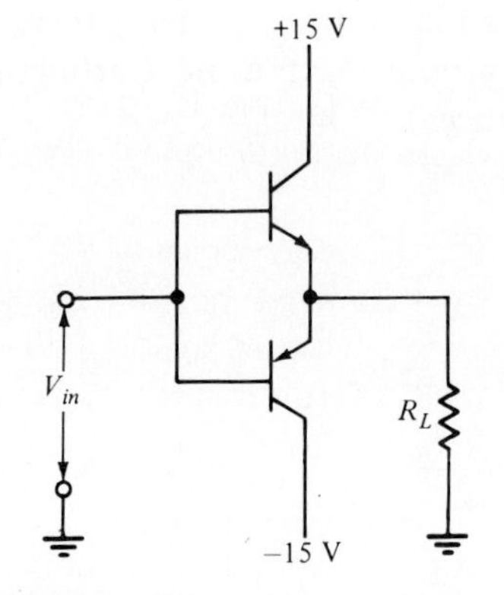

FIGURE 8.17. **Complementary Push-Pull Amplifier**

Thus, a 2N6285 Darlington has an $h_{fe} = 4000$ and a collector current rating of 10 A.

The Darlington units should not be confused with so-called superbeta transistors. The latter derive their large values of h_{fe} via the manufacturing process employed. 2N5963 is a typical superbeta with $h_{fe} = 900$ at a collector current of but a few milliamperes.

There is a major difficulty with the complementary transistor push-pull amplifier schematically sketched in Figure 8.17. Since a transistor does not start to pass significant current until V_{BE} exceeds about 0.6 V, the output voltage across the load continually lags by about 0.6 V below the impressed input. Consider a sine wave input; Figure 8.18 shows the corresponding output.

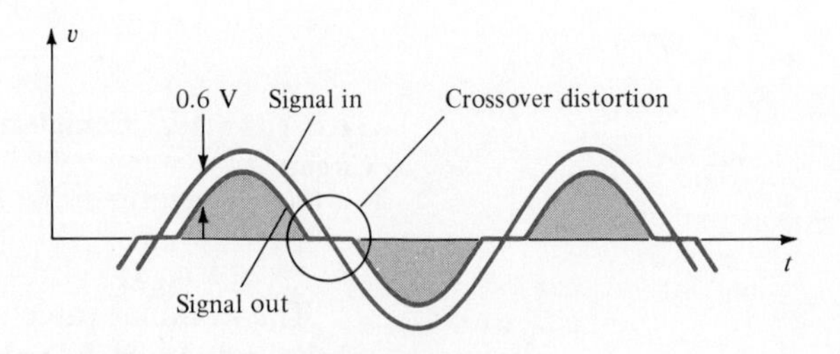

FIGURE 8.18. **Illustration of Crossover Distortion**

The major problem is the crossover distortion. One cure for this defect is to use negative feedback; another is to bias the transistors into slight conduction, as in Figure 8.19.

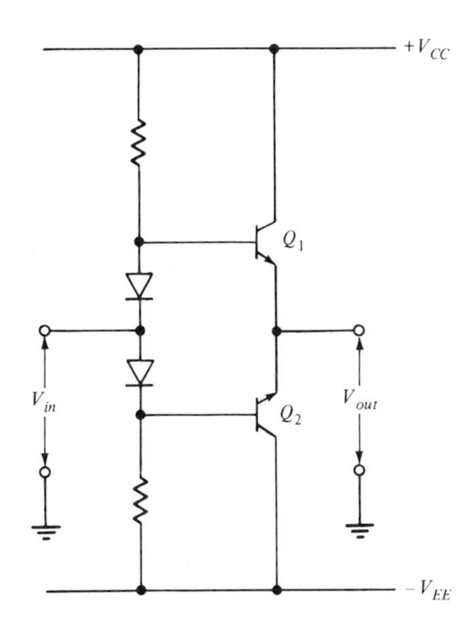

FIGURE 8.19. Correction for Crossover Distortion Provided by Series Diodes

The forward-conducting diodes hold the base of Q_1 one V_{BE} above the input signal and Q_2 one V_{BE} below. By this means one of the output transistors is always on.

Modifications of this basic arrangement are generally necessary if the system is to be thermally stable. As the transistor temperature rises, V_{BE} diminishes, resulting in the flow of quiescent current with no signal applied. This flow raises the temperature still further and can result in a thermal runaway.

8.3 ▪ TRANSISTOR STABILITY

Thus far we have considered the transistor base bias voltage as being obtained either from a separate battery or from a voltage divider, each of which in turn may be represented by its Thevenin equivalent. Since semiconductors are rather temperature-sensitive, the collector current (I_C) may show a significant increase with temperature. Since $I_C = h_{fe}I_B$, this current increase amounts to a variation in the transistor's beta value. We have seen that the insertion of an emitter resistance may be used to minimize the effect of such variations. Typically, the value of R_E selected is such as to have $R_E I_C$ equal to a few volts.

We turn now to a consideration of the degree to which transistor stability is aided by such a measure. We have

$$V_B = I_B R_B + (I_C + I_B)R_E$$

$$= \frac{I_C}{h_{fe}} R_B + I_C\left(1 + \frac{1}{h_{fe}}\right)R_E \tag{8.16}$$

$$= \frac{I_C}{h_{fe}} [R_B + (1 + h_{fe})R_E] \tag{8.17}$$

If I_{C1} corresponds to h_{fe1} and I_{C2} corresponds to h_{fe2}, since V_B is assumed not to change, we have

$$\frac{I_{C2}}{I_{C1}} - 1 = \frac{I_{C2} - I_{C1}}{I_{C1}} = \frac{\Delta I_C}{I_{C1}} = \frac{(h_{fe2} - h_{fe1})(R_B + R_E)}{[R_B + R_E(1 + h_{fe2})]h_{fe1}} \tag{8.18}$$

$$\frac{\Delta I_C}{I_{C1}} = \frac{(\Delta h_{fe}/h_{fe1})[1 + (R_B/R_E)]}{(R_B/R_E) + 1 + h_{fe2}}$$

$$= \frac{(\Delta h_{fe}/h_{fe1})[1 + (R_B/R_E)]}{[(R_B/h_{fe2}R_E) + (1/h_{fe2}) + 1]h_{fe2}}$$

$$\simeq \frac{[1 + (R_B/R_E)]\ \Delta h_{fe}/h_{fe1}}{[1 + (R_B/h_{fe2}R_E)]h_{fe2}} \tag{8.19}$$

For a given fractional change in h_{fe}, it can be seen that the effect on the collector current is reduced as the value of R_E is increased. (It is primarily the R_E in the numerator that is effective in this reduction as can be seen from Example 8.7.)

EXAMPLE 8.7 The computation in Example 7.2, in conjunction with the circuit in Figure 7.9, resulted in a 5% change in I_C when $R_E = 1\ \text{k}\Omega$, $R_C = 10\ \text{k}\Omega$, and h_{fe} changed from 100 to 200. Using a bias divider circuit with $R_1 = 110\ \text{k}\Omega$ and $R_2 = 11.0\ \text{k}\Omega$, show that Eq. 8.19 leads to a similar result.

Solution:

$$\frac{\Delta I_C}{I_C} = \frac{[1 + (R_B/R_E)]\ \Delta h_{fe}/h_{fe1}}{[1 + (R_B/h_{fe2}R_E)]h_{fe2}} = \frac{[1 + (10/1)](100/100)}{\{1 + [10/200(1)]\}200}$$

$$= \frac{11}{(1.05)200} = 0.0523 \simeq 5.2\%$$

In addition to changes in h_{fe}, the stability of I_C is also influenced by the temperature dependence of V_{BE} and the reverse saturation current I_{CO}. I_{CO} doubles approximately for each 10°C rise in temperature. V_{BE} decreases at approximately 2.5 mV/°C.

The reverse saturation current (I_{CO}) arises from the minority carriers; for them the collector acts like an emitter and vice versa. Therefore, as the emitter current for majority carriers is $(1 + h_{fe})I_B$, the reverse saturation current is $(1 + h_{fe})I_{CO}$. We have

$$I_C = h_{fe}I_B + (1 + h_{fe})I_{CO} \tag{8.20}$$

Combining this with Kirchhoff's law for the base circuit—that is,

$$V_B = I_B R_B + V_{BE} + (I_B + I_C)R_E \tag{8.21}$$

we obtain the following:

$$\Delta I_C \simeq \frac{R_B + R_E}{(R_B/h_{fe}) + R_E} \Delta I_{CO} \tag{8.22}$$

$$\Delta I_C \simeq -\frac{h_{fe}}{R_B + h_{fe}R_E} \Delta V_{BE} \tag{8.23}$$

Here again it can be seen that large values of R_E stabilize transistor operation but reduce gain since part of the signal appears across R_E. If such a loss is to be avoided, one may place across R_E a capacitor whose reactance at the low band limit is significantly smaller than r_e, with a rule of thumb being to have the capacitor reactance at that frequency equal to, or less than, $\frac{1}{10} r_e$.

Finally, a condition that used to be quite bothersome might be mentioned: While I_{CO} (the reverse saturation current) may be small, it gets multiplied by $1 + h_{fe}$. I_C thereby increases, in turn increasing the temperature, which increases I_{CO}, resulting in a still further increase in I_C. Under some circumstances a thermal runaway condition can develop, leading to a destruction of the transistor in a badly designed circuit.

8.4 ■ POSITIVE FEEDBACK

8.4.1 Oscillation

When the sign in the denominator of Eq. 8.9 becomes negative, the gain will be greater than it was without feedback. This condition represents *positive feedback*, and so positive feedback increases gain (but narrows the bandwidth).

If $\beta A = 1$, the gain with positive feedback becomes infinite. Of course, this cannot happen, but the gain does become very large, and the amplifier becomes an **oscillator**, emitting a frequency independent of any input. It's as if we attempted to talk while whistling. Once oscillation starts, the operating conditions of the amplifier may change so drastically that a completely different analysis must be employed.

For an amplifier to function as an oscillator, it must satisfy two conditions:

1. There must be some type of frequency selector that determines the frequency of oscillation; otherwise, the oscillation frequency would be determined strictly by chance.
2. At the oscillation frequency there must be zero phase shift for the signal in passing around the loop from input, through the amplifier, and then back to the input again—that is, positive feedback—and its magnitude must be such that the product of feedback factor times amplifier gain—that is, loop gain—must equal unity. We have

$$\beta A = 1 \tag{8.24}$$

$$\beta = \frac{1}{A} \tag{8.25}$$

Thus, when A is the amplifier gain, the feedback factor must be $1/A$.

8.4.2 *RC* Oscillator

In the audio range probably the most common generator of sinusoidal waveforms is the Wien bridge oscillator (Figure 8.20). By again presuming a 180° phase shift through each stage of amplification, to obtain positive feedback, two amplifier stages are used.

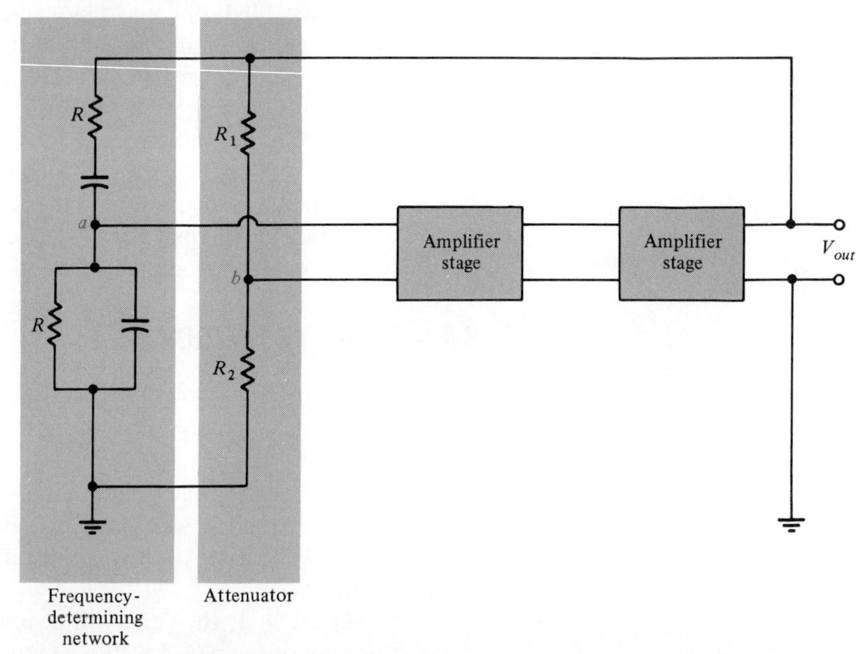

FIGURE 8.20. Wien Bridge Oscillator The unbalanced bridge leads to positive feedback between points *a* and *b*, causing a sinusoidal signal to be generated at a frequency determined by the values of *R* and *C*.

We have previously (in Section 3.9) considered the Wien bridge filter. In Chapter 3 it was shown that, with respect to ground, the potential at point *a* at resonance equals one third of the voltage applied to the bridge (being, in this case, the output signal). When we set, in the attenuator branch, $R_2 = \frac{1}{2} R_1$, the potential at point *b* also was equal to one third of the applied voltage, and at the resonant frequency the output voltage of the bridge (between points *a* and *b*) was equal to zero. (Remember that the circuit was a rejection filter.) How, then, can there be any feedback signal? When used as part of an oscillator, the Wien bridge is not operated in a balanced condition. By changing the ratio R_1/R_2, we can cause the system to behave as either a negative or a positive feedback system.

Consider Figure 8.21. When $R_1/R_2 = 2$, signals at *a* and *b*, measured with respect to ground, are equal and in phase with the amplifier's output voltage. While the potentials at *a* and *b* vary, there is no potential difference at any time—hence the bridge output is zero. If the value of R_2 is reduced so that $R_1/R_2 > 2$, while the magnitude of the signal at a stays the same (all other things being equal), the magnitude of the signal at b is diminished. By taking the difference between the signals at a and b on a point-by-point basis, the resulting waveform can be seen to constitute positive feedback. For $R_1/R_2 < 2$, the signal amplitude at b is greater than that at a, and this amounts to negative feedback. Thus, we see that the condition for oscillation with a Wien bridge requires $R_1/R_2 > 2$.

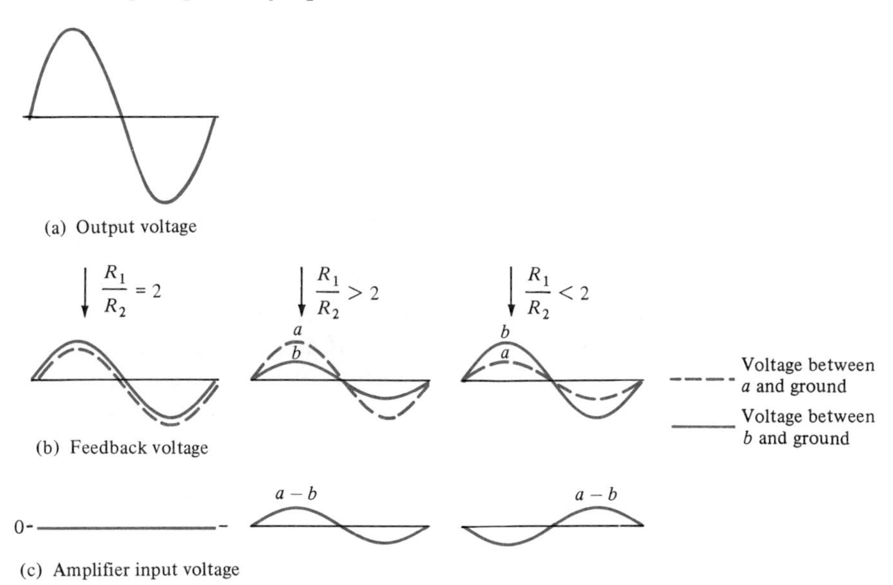

FIGURE 8.21. Wien Bridge Oscillator Waveforms Left: The potentials at a and b are identical, and the potential difference between them is zero. Center: The magnitude at a remains larger than at b, resulting in output and input being in phase. Right: The magnitude of b remains larger than at a, résulting in input and output being out of phase.

The frequency generated by a Wien bridge oscillator is confined to relatively low frequencies, sonic and supersonic. The reason for this limitation may be realized by looking at the expression for the resonant frequency (having taken both Rs and Cs in the frequency-determining section to be the same):

$$f_o = \frac{1}{2\pi RC} \tag{8.26}$$

The smallest values of R and C will be set by the input R and C of the amplifier and result in the mentioned limitation.

8.4.3 *LC* Oscillator

The most basic frequency-determining circuits generally utilize L and C. Due to the large resistance which accompanies the large values of L needed at low frequencies, one obtains low-Q resonant circuits with correspondingly low selectivity. It was for this reason that we resorted to the use of R and C as in the

Wien bridge. At higher frequencies this limitation no longer holds since the requisite small L leads to a small R, and we can obtain large values of Q.

For an oscillator we need positive feedback with $\beta A = 1$. Therefore, we have

$$\beta = \frac{1}{A} = \frac{1}{V_{out}/V_{in}} = \frac{V_{in}}{V_{out}}$$

That is, the feedback factor must be $1/A$. Thus, if we start with an LC resonant circuit being supplied with a signal V_{out}, we must have $V_{in} = V_{out}/A$.

We commence with a properly biased transistor amplifier, using a parallel LC resonant circuit tuned to the desired frequency as the amplifier's collector load (in place of the load resistor we've thus far generally used).

If the circuit in Figure 8.22 were meant to be an amplifier, the signal source (V_s) might be inserted across R_2. With instantaneous polarity as indicated, the output polarity would show the 180° phase reversal characteristic of a CE amplifier. But, being an oscillator, a portion of the output signal must be returned to the input. This feedback voltage may be obtained by tapping off a proportional amount from across the inductance. Looking at the instantaneous polarities, we would conclude that the top of the inductance should be connected to the transistor base and the tap connected to the emitter. But we cannot do this for two reasons:

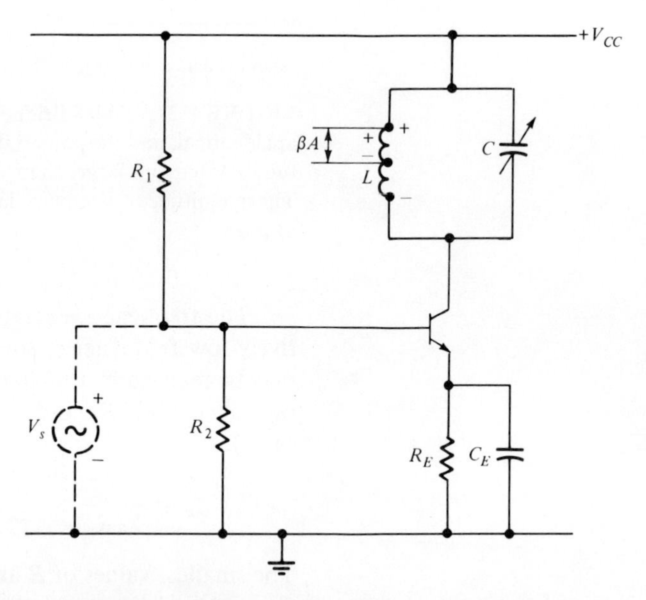

FIGURE 8.22. Transistor Amplifier with LC Resonant Circuit Showing Loop Gain (βA) Needed for Positive Feedback

1. Such direct connections would upset the dc polarities; we could easily remedy this difficulty through the use of capacitors.
2. There would probably be an impedance mismatch.

A much simpler way of accomplishing this result is to make the inductor into a transformer, as in Figure 8.23, making sure that the secondary winding is connected into the base circuit in such a manner that it constitutes positive feedback.

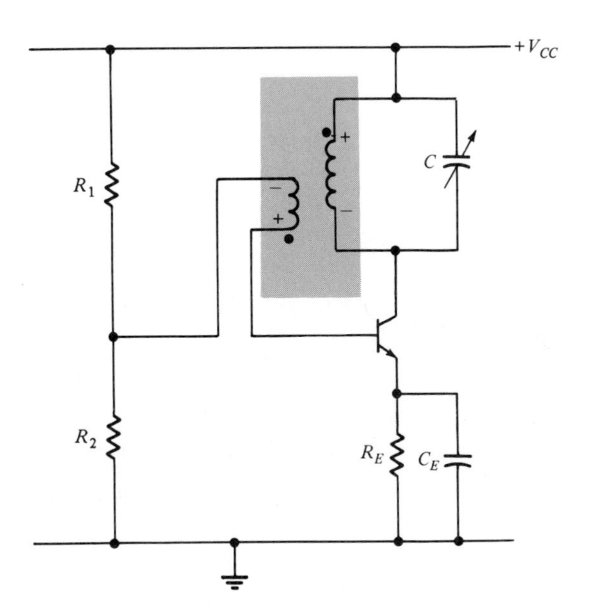

FIGURE 8.23. ***LC* Oscillator** Because of pulsed (class C) operation, C_E is charged through the low resistance of the transistor but discharges (during nonconduction) through R_E. If the time constant R_EC_E is too large, after a few cycles loop gain drops below unity, terminating oscillation. It does not restart until C_E has discharged sufficiently.

For oscillation we have indicated that we need $\beta = 1/A$. What if $A >> 1/\beta$, as will probably be the case with the circuit in Figure 8.23? In this case the amplitude of oscillation across LC will be large and will probably cause the base (on its negative swing) to drop below the point needed for conduction. As a result the transistor, rather than conducting in a continuous fashion, will conduct in pulses, one pulse per cycle. This constitutes class C operation.

The LC circuit, however, will continue to be subjected to a sinusoidal oscillation since it acts like a mechanical vibrator with considerable inertia (high Q). The periodic pulses from the transistor merely act to replenish the losses that occurred during the previous cycle. However, in extreme cases C_E might acquire a charge sufficient to completely cut off transistor conduction for a number of cycles while it discharges through R_E. This will result in *squegging*—that is, the oscillator circuit will emit a train of damped pulses. This situation may be prevented by lowering the gain through the use of some negative feedback, in this case by increasing the portion of the dc bias provided by the unbypassed portion of the resistance in the emitter circuit, R_f.

How do we get a usable output from the oscillator? We get it by placing yet another secondary winding on the transformer, as shown in Figure 8.24.

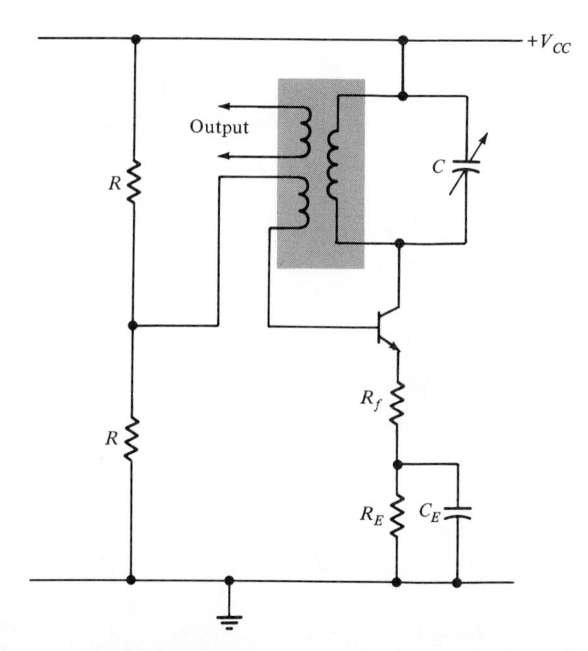

FIGURE 8.24. **Preventing Disruption of Oscillation by Reducing Amplifier Gain with Negative Feedback Resistor R_F**

How does the oscillator get started? There is always some electrical noise present in any circuit, and from a Fourier viewpoint noise consists of all frequencies. With a frequency-selective circuit like *LC*, one Fourier component will be selected and then preferentially amplified until the system "breaks" out into oscillation at the desired frequency.

When a high degree of stability is desired, the *LC* resonant circuit may be found to be deficient. Temperature and humidity conditions may change the physical characteristics and lead to a variation in the oscillator's frequency. Under such conditions a crystal, which behaves like an *LC* resonant circuit, may be used instead. The frequency in this case is not readily variable, being determined by the thickness of the crystal. A **crystal oscillator** is widely used where a fixed frequency of high stability and accuracy is needed. Such, for example, is the case with a color TV set: A crystal oscillator is used to accurately reproduce a frequency generated in the TV station, 3.579545 MHz.

SUMMARY

8.1 Introduction

- Negative feedback may be used to correct a number of amplifier deficiencies: amplitude distortion, frequency distortion, instability.

8.2 Negative Feedback

- $1 + \beta A$, where β is the feedback factor and A is the amplifier's gain, constitutes the feedback circuit's desensitivity factor. In the type of feedback we have considered, distortion, midband gain, output resistance, and lower band limit are all diminished by the reciprocal of this factor; upper band limit and input resistance are increased by this factor.
- The emitter follower, also called the unity-gain amplifier, constitutes a common collector (CC) configuration and represents a very high input impedance and a very low output impedance.

8.4 Positive Feedback

- Positive feedback results in narrower bandwidth and increased gain. In the limit the amplifier becomes an oscillator, generating rather than amplifying signal frequencies.
- The Wien bridge oscillator is the most common generator of electrical signals at audio frequencies.
- The high frequencies are usually generated by means of an *LC* resonant circuit in conjunction with a transistor.
- The highest stability of frequency generation is achieved with a crystal oscillator.

KEY TERMS

amplifier stability: the maintenance of constant amplifier characteristics, generally gain, in the face of temperature variations, voltage fluctations, and so forth

Bode plot: a plot of dB gain and phase shift vs. log of the frequency (where approximated by straight lines, these plots represent *idealized* Bode plots, and the intersection of the straight-line gain characteristics defines the corner frequencies.)

closed-loop gain: gain with feedback

corner frequency: 3 dB point on a gain vs. frequency curve

crystal oscillator: the equivalent of a very stable *LC* oscillator but one whose frequency is not generally adjustable

Darlington circuit: use of an emitter follower as a dynamic load for a second emitter follower (Leads to a very high input impedance and a very low output impedance.)

decade: the frequency interval between a frequency f and 10 times that frequency, $10f$—for example, 100–1000 Hz and 5000–50,000 Hz

desensitivity factor $(1 + \beta A)$: the factor by which impedances, gain, distortion, and so forth are altered, in conjunction with negative feedback processes

emitter follower: also called unity-gain amplifier, common collector amplifier, voltage follower (Output is obtained from across the emitter resistor, with the collector connected directly to V_{CC}.)

feedback: the return to the input of a portion of a device's output signal

frequency distortion: the unequal amplification of the various frequencies

gain-bandwidth product: a transistor characteristic indicating the relationship between an amplifier's gain and its bandwidth (It is designated by the parameter f_T, being the bandwidth at unity gain.)

***LC* oscillator:** use of inductance and capacitance to control the generated signal frequency (Frequency is inversely proportional to $\sqrt{LC}$, and therefore the range of frequency variability is more restricted than is the case with *RC* oscillators.)

loop gain: the product of feedback fraction times amplifier gain—that is, βA

negative feedback: feedback that is generally employed as a means of furnishing some sort of corrective action to the response of a device

octave: the frequency interval between a frequency f and twice that frequency, $2f$—for example, between 20–40 Hz and 5000–10,000 Hz

open-loop gain: gain without feedback

oscillation: excessive positive feedback that results in the creation of a signal generator

positive feedback: feedback used to increase amplifier gain; in extreme cases, leads to oscillation

***RC* oscillator:** use of resistance and capacitance to control the generated signal frequency (Frequency is inversely proportional to *RC*.)

SUGGESTED READINGS

Millman, Jacob, MICROELECTRONICS, pp. 409–442. New York: McGraw-Hill, 1979. *Details concerning various forms of feedback.*

Malmstadt, H.V., C. G. Enke, and S. R. Crouch, ELECTRONICS AND INSTRUMENTATION FOR SCIENTISTS, pp. 436–455. Menlo Park, Calif.: Benjamin/Cummings, 1981. *Use of feedback in control systems; qualitative treatment.*

Fitzgerald, A. E., D. E. Higginbotham, and A. Grabel, BASIC ELECTRICAL ENGINEERING, 5th ed., pp. 278–286. New York: McGraw-Hill, 1981. *Bode plots, showing gain and phase angle variation with frequency.*

Horowitz, Paul, and Winfield Hill, THE ART OF ELECTRONICS, pp. 53–64, 67–70. New York: Cambridge University Press, 1980. *Emitter followers; additional discussion and design details.*

Tucker, D. G., "The History of Positive Feedback," RA-

DIO AND ELECTRONIC ENGINEER, vol. 42, pp. 69–80, 1982. *Positive feedback.*

Morris, Noel, ADVANCED INDUSTRIAL ELECTRONICS, pp. 122–149. New York: McGraw-Hill, 1974. *Detailed discussion of various oscillator types.*

Black, Harold S., "Inventing the Negative Feedback Amplifier," IEEE SPECTRUM, vol. 14, no. 12, pp. 55–60, Dec. 1977. *How negative feedback was invented on the old Lackawanna ferry.*

A CLOSER LOOK: Input Impedance

The performance of electronic circuits shows considerable dependence on their input impedance. This is not just simply the value of the resistor that might exist between the input terminals. Suppose we have the situation depicted in Figure 1. We are "looking into" the input terminals, and we have an ac ammeter in series with one of the input leads. Ohm's law, as applied to a dc circuit, defined the resistance as the applied voltage divided by the resulting current—that is, $R = V/I$. Instead of a constant voltage, we can apply a voltage *variation* Δv to a circuit and observe the resultant current variation Δi and thus define a *dynamic* resistance r. We have

$$r = \frac{dv}{di} \simeq \frac{\Delta v}{\Delta i} \text{ [ohms]}$$

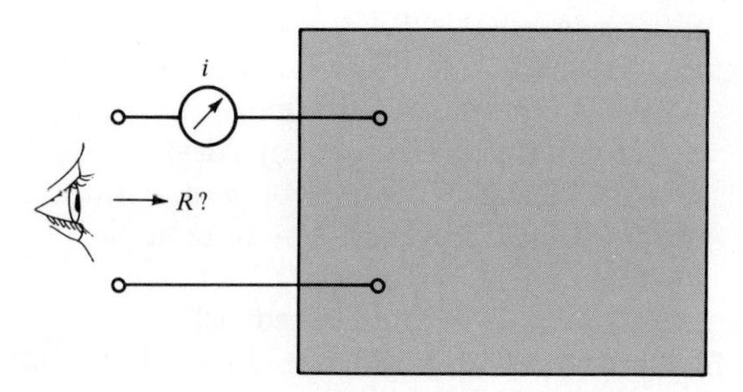

FIGURE 1

If we apply Δv to the "black box" in Figure 1 and observe a current variation Δi that is quite small, we interpret this as a large dynamic resistance.

Let us now look into the box: Suppose that things are arranged in such a fashion that as we increase the input voltage there is a little demon inside who increases the opposition voltage proportionally. Outside the box we find that in order to get a significant current we must furnish a much larger Δv for a given Δi. This condition we interpret as representing a high input resistance (impedance). (See Figure 2.)

Of course there is no such demon, but in utilizing series negative feedback, the circuit behaves in precisely this manner. As the magnitude of the input signal increases, there is a proportional increase in the opposition due to negative feedback, and to achieve a certain input current variation, the input signal must be greatly increased over what would be necessary without the feedback. This behavior accounts for the increased input impedance experienced with amplifiers employing series negative feedback.

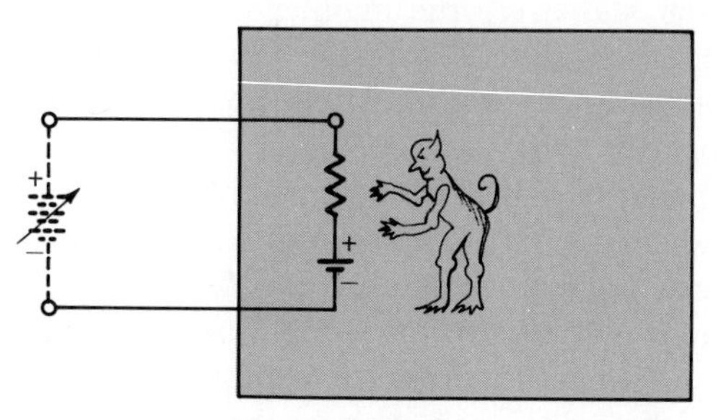

FIGURE 2

In Figure 3 is shown the current versus voltage curve for a typical diode. Application of V_{dc} leads to an I_{dc}, and the dc resistance is V_{dc}/I_{dc}. But if an ac voltage whose amplitude is Δv is superimposed on the dc voltage, the corresponding variation of Δi depends on the slope, which depends on V_{dc}. The smaller the slope at the bias point, the greater the dynamic resistance. This dynamic resistance is given by the expression

$$r_e = \frac{26}{I_C \text{ [mA]}}$$

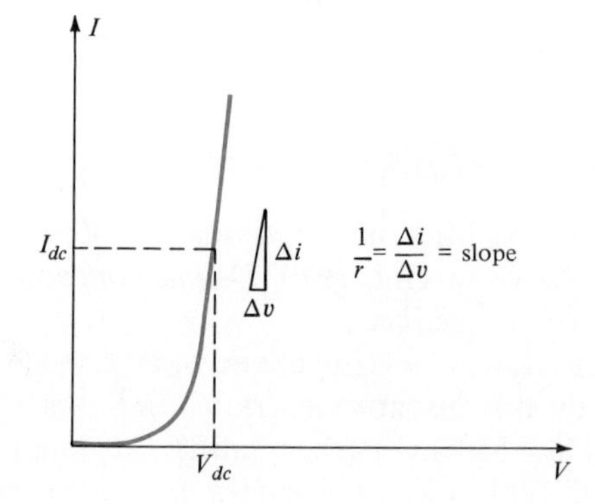

FIGURE 3

EXERCISES

1. Name three amplifier deficiencies that may be improved by negative feedback.

2. **a.** If an amplifier, without feedback, presents a gain of 10,000, what is the gain with 20% negative feedback?

   b. To achieve the same output voltage as prevails without negative feedback, how much larger must the input be in the feedback case?

3. Repeat Exercise 2 with feedback equal to 1%.

4. An amplifier with a gain of 1000 provides a 20 V output signal with 10% distortion. What is the magnitude of the output distortion? What amount of negative feedback will reduce the distortion to 1%?

5. What two conditions must be satisfied for an amplifier to function as an oscillator?

6. **a.** What is the most popular version of an audio oscillator called?

   b. By what means can its feedback be varied between being positive and negative?

7. Why are *LC* resonant circuits feasible at high frequencies but of rather limited use at low frequencies?

8. Why is the range of frequency variability greater in the case of *RC* oscillators than for *LC* oscillators?

9. How much feedback is employed in an emitter follower circuit? What is its voltage gain?

10. In terms of relative magnitude, evaluate the input and output impedances of an emitter follower? How do these impedances compare with those of a common emitter amplifier?

11. In terms of the base current, what is the magnitude of the current passing through the emitter resistance of an emitter follower?

12. In what manner is the stability of the collector current as a function of h_{fe} variability determined by R_E, the emitter resistor? What penalty might we have to pay for the resultant increase in stability? How may this penalty be minimized?

13. **a.** If an amplifier's high-frequency gain, as a function of frequency, varies as

$$A(f) = \frac{A_o}{1 + j(f/f_2)}$$

where A_o is the midband gain and f_2 is the upper 3 dB frequency, what is the phase shift between the signal at frequency f_2 and that at midband?

   b. If the midband gain is $-A_o$, what is the phase shift between midband and that at f_2?

   c. For $f >> f_2$, with a midband gain of $-A_o$, what is the nature of the phase shift between midband and that at frequency f?

   d. What possible result, generally to be avoided, can arise in a multistage negative feedback amplifier having finite gain at $f >> f_2$?

14. By means of a sketch, indicate the distinction between the dc and dynamic resistances of a diode.

15. What is the approximate gain of a negative feedback amplifier with a feedback factor of β?

16. What is the name of the circuit that uses an emitter follower as a dynamic load for another emitter follower?

17. If f_T = 5 MHz, what will be an amplifier's bandwidth if the gain is 100?

18. What term is applied to the intersection of the straight-line gain curves on an idealilzed Bode plot?

19. Draw the circuit for an emitter follower.

PROBLEMS

8.1
An amplifier with a gain of 1000—that is, A_o = 1000—exhibits a 10% reduction when a faulty resistor is replaced.

a. With a 1% negative feedback, what is the overall percentage change in gain?

b. Why might an amplifier gain change so significantly—that is, by 10% in the preceding case—if a faulty resistor is replaced by one of supposedly like value?

8.2
The high-frequency gain drop-off of an amplifier without feedback may be represented by Eq. 8.11. With negative feedback the gain, as a function of frequency, may be represented by

$$A_f(f) = \frac{A}{1 + \beta A} = \frac{A_o/[1 + j(f/f_2)]}{1 + \{\beta A_o/[1 + j(f/f_2)]\}}$$

If A_o = 1000, β = 0.009, and f_2 = 10,000 Hz, what is the upper band limit with feedback?

8.3
a. At low frequencies the gain drop-off of an amplifier without feedback may be represented by the relation

$$A(f) = \frac{A_o}{1 - j(f_1/f)}$$

where f = frequency of interest
f_1 = lower band limit

If an amplifier has a midband gain of 100 but only 55.5 at a frequency of 10 Hz, what negative feedback factor should be used to obtain a 10 Hz gain that is not less than 95% of the new midband value?

SUGGESTION: Find the corner frequency—that is, the lower band limit—for the original amplifier. Then using the expression for gain with feedback in terms of the gain without feedback, determine the new corner frequency.

b. What will be the new midband gain?

8.4
In observing the results of Problems 8.2 and 8.3, by what factor should an amplifier's lower band limit be multiplied to find the amplifier's new limit with feedback? By what factor should the upper band limit be multiplied?

8.5
a. With a 1 mV signal input and an amplifier with a gain of 1000, what will be the output signal at midband using 10% negative feedback?

b. What value of input signal must now be provided to yield an output equal to that before feedback was employed?

c. With a 1000 Hz input, when referred to the input, the original second harmonic distortion was 0.2 V/1000 = 0.2 mV. What will be the magnitude of the second harmonic distortion with negative feedback?

d. Draw, in block diagram form, the circuitry as it exists at 1000 Hz and at 2000 Hz.

8.6
An amplifier with 45 dB open-loop gain has an upper band limit of 10 kHz. With a 20 dB/decade drop-off, what is the frequency at which the gain becomes unity?

8.7
An idealized Bode plot of an amplifier's frequency response is shown in Figure 8.25, with and without feedback. The high-frequency variation of gain without feedback is given by Eq. 8.11.

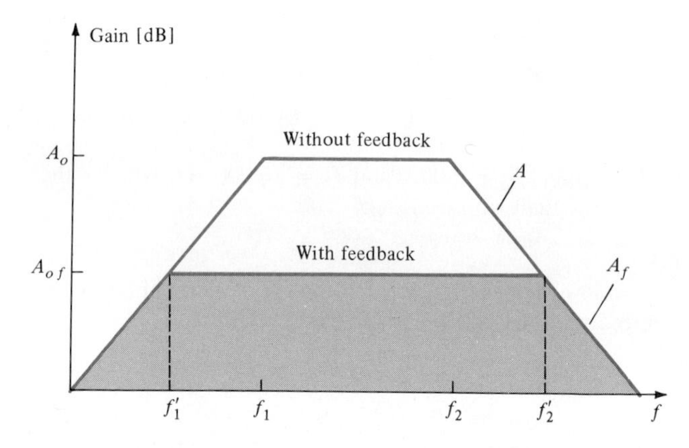

FIGURE 8.25

a. What is the magnitude of the actual gain (relative to A_o) at $f = f_2$, expressed in decibels?

b. What is the phase angle of A at $f = f_2$?

c. What is the phase angle for $f >> f_2$?

8.8
Refer to problem 8.7. With negative feedback the gain variation at high frequencies is

$$A_f(f) = \frac{A_{of}}{1 + j(f/f_2')}$$

where

$$A_{of} = \frac{A_o}{1 + \beta A_o}$$

The gain without feedback (A) and the gain with feedback (A_f) intersect at the upper band limit of the feedback amplifier, that is, at f_2'. Determine the relationship between f_2 and f_2'.

8.9
Three identical (noninteracting) amplifier stages are operated in cascade. Thus, we have

$$A = \frac{A_o}{1 + j(f/f_2)} \times \frac{A_o}{1 + j(f/f_2)} \times \frac{A_o}{1 + j(f/f_2)} = \left(\frac{A_o}{1 + j(f/f_2)}\right)^3$$

a. Expressed in decibels, for a single stage, what is the gain at $f = f_2$ relative to A_o?

b. Expressed in decibels, for three stages, what is the gain at $f = f_2$ relative to A_o?

c. What is the total phase shift at $f = f_2$?

d. If f_2 = 1 MHz, what is the total phase shift at $f = \sqrt{3}$ MHz—that is, at 1.732 MHz?

e. Expressed in decibels, what is the gain at $f = \sqrt{3}$ MHz relative to A_o?

8.10
a. At some high frequency, if each stage of a three-stage negative feedback amplifier contributes a $-60°$ phase shift (for a total of $-180°$), the feedback signal at that frequency will be *in phase* with the input—that is, the feedback has become positive. If, additionally, $|\beta A| = 1$, the system becomes unstable. Why?

b. Using the three-stage amplifier of part a with f_2 = 1 MHz, if the midband gain is -10^3, what is the maximum value of β that can be used to assure stable operation?

8.11
Rather than using identical stages of amplification, if the 3 dB frequencies of three amplifier stages are made progressively

higher to a sufficient degree, much more negative feedback may be employed before the possibility of instability sets in. This procedure, of course, allows greater consequent bandwidth. For a three-stage amplifier with A_o = 60 dB and with 3 dB frequencies of 1, 10, and 50 MHz, respectively, the phase shift will become $-180°$ at 22 MHz. At this frequency the amplifier gain is 26 dB. Since the critical value for stability is represented by $|\beta A| = 1$, what is the maximum value of β that can be used with this amplifier? (Express the result in terms of decibels and as a percentage of the midband gain.)

8.12
If $A = 1000$, $R_{in} = 1.5$ kΩ, and $R_{out} = 100$ kΩ, what would be the input and output resistances of this amplifier employing voltage-sampling, series-mixing feedback with $\beta = \frac{1}{10}$?

8.13
a. Assume that a transistor amplifier whose output impedance is 2500 Ω is meant to drive a like transistor connected as an emitter follower. With transistor h_{fe}'s of 100 and I_C = 1 mA, the emitter follower operates into a 50 Ω load. What is the input impedance presented to the transistor amplifier (disregarding the loading effects of the biasing network)?

b. Why is it desirable to have the Thevenin equivalent generator with an R_T = 2500 Ω "working" into a load that is considerably greater, rather than have it operating into a matched load?

8.14
If the output conductance (h_{oe}) of a transistor is 20×10^{-6} S, what is the input impedance of a Darlington circuit using a transistor with h_{fe} = 100 and R_L = 1 kΩ, disregarding any input loading due to the biasing network?

8.15
The output resistance of an emitter follower is given by Eq. 8.15. Compute the output resistance if h_{fe} = 100, r_e = 25 Ω, and R_s = 1 kΩ. Using two identical transistors, compute the output impedance of a Darlington circuit for which the output impedance of Q_1 becomes the source resistance for Q_2. Compare the result to that obtained for the emitter follower. If R_s is negligible, how do the two compare?

8.16
The objective of this problem is to compare the stability of two biasing arrangements (fixed and self-bias) when h_{fe} changes from 100 to 200. The desired operating Q-point for the emitter follower is V_{CE} = 6 V, and I_C = 1 mA.

a. See Figure 8.26. Compute the value of the base current for h_{fe} = 200 and the collector current, and note the percentage change in the latter from the value when h_{fe} = 100.

b. Do likewise for the circuit shown in Figure 8.27.

8.17 △

See Figure 8.28. The usual transistor design procedure suggests the selection of an R_C such that for the stated I_C, $V_C = \frac{1}{2}$ V_{CC}. Furthermore, the value of R_E is selected so that $V_E \simeq \frac{1}{10}$ V_C. The purpose of this problem is to show how a value for R_E that is too small leads to a considerable change in the collector current as the value of V_{BE} changes because of temperature variation.

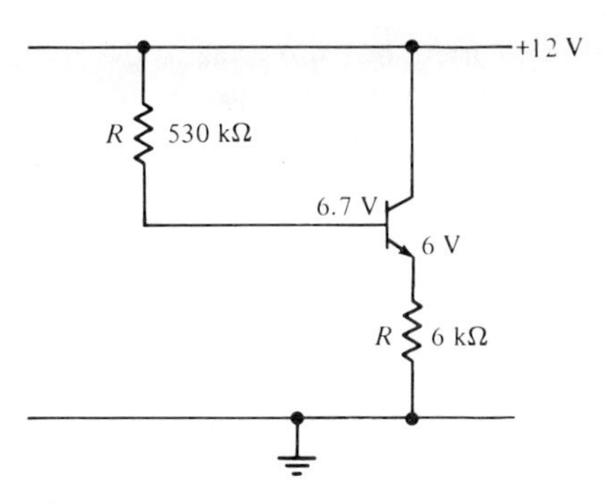

FIGURE 8.26

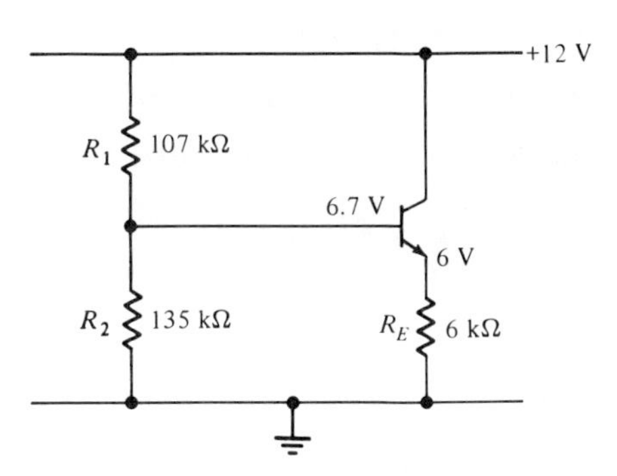

FIGURE 8.27

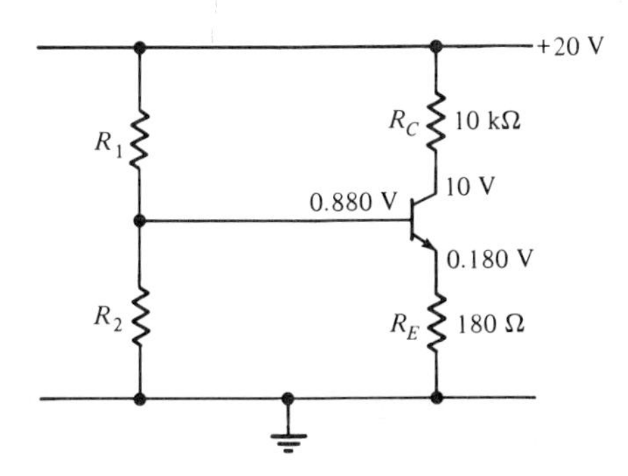

FIGURE 8.28

a. V_{BE} changes at the rate of -2.5 mV/°C. The biasing network regulation is presumed to be excellent and maintains the base at $+0.880$ V as V_{BE} changes. A 20°C change causes the emitter, therefore, to assume a potential of 0.230 V. Presuming an initial collector current of 1 mA, what is the new collector current? (Take $I_E \simeq I_C$.) Compute the percentage change.

b. Compute the change under like circumstances if R_E is 1 kΩ, making the emitter potential 1 V, and the biasing network is designed so that the base remains at 1.7 V.

8.18

Ralph J. Smith in *Circuits, Devices, and Systems* (4th ed., Wiley, New York, 1976, p. 543) derives the stability equation with regard to changes in h_{fe} in the following manner:

$$I_C = h_{fe}I_B$$

$$\Delta I_C = \Delta h_{fe}I_B + h_{fe}\Delta I_B \tag{1}$$

From Thevenin's theorem applied to the base circuit, we have

$$V_{BB} = I_BR_B + V_{BE} + (I_C + I_B)R_E$$

Taking V_{BB} and V_{BE} to remain constant, we have

$$\Delta I_B(R_B + R_E) - \Delta I_CR_E = 0 \tag{2}$$

$$\Delta I_B = -\frac{\Delta I_CR_E}{R_E + R_B} \simeq \Delta I_C\left(\frac{R_E}{R_B}\right) \tag{3}$$

since $R_E \ll R_B$. Using Eqs. 1 and 2, we have

$$\frac{\Delta I_C}{I_C} = \frac{\Delta h_{fe}I_B}{I_C} + h_{fe}\left(\frac{\Delta I_B}{I_C}\right)$$

$$= \frac{\Delta h_{fe}}{h_{fe}} - \left[\left(\frac{h_{fe}}{I_C}\right)\left(\Delta I_C\right)\left(\frac{R_E}{R_B}\right)\right]$$

$$\frac{\Delta I_C}{I_C}\left(1 + \frac{\Delta h_{fe}R_E}{R_B}\right) = \frac{\Delta h_{fe}}{h_{fe}}$$

$$\frac{\Delta I_C}{I_C} = \frac{\Delta h_{fe}/h_{fe}}{1 + h_{fe}(R_E/R_B)} \tag{4}$$

Eq. 4 is termed the *stability factor* and very nicely shows the advantage of using large h_{fe} transistors.

a. However, comparing this result with Eq. 8.19 of this text, there would seem to be a functional discrepancy between the two equations. Can you reconcile them? (*Hint*: In Eq. 8.19, start with the form before the approximation sign.)

b. See Figure 8.29. Smith compares Eq. 4 to a feedback problem. Note the similarity of the denominator to the 1 + βA encountered in feedback theory in the expression for distortion reduction. Show how the flow diagram in Figure 8.29 applies to his equations.

c. State the respective applicability of the two equations that have been discussed.

8.19

To compare the relative importance of change in h_{fe}, V_{BE}, and I_{CO}, consider the following: silicon transistor amplifier with

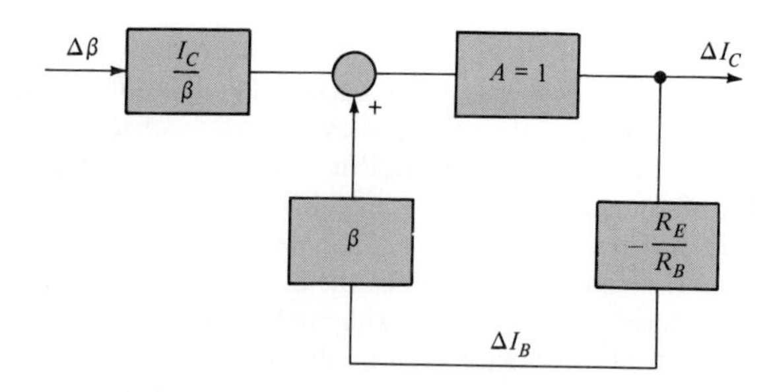

FIGURE 8.29

R_E = 1 kΩ, R_C = 10 kΩ, R_B = 10 kΩ, h_{fe} at 25°C = 100, h_{fe} at 175°C = 200, and I_C = 1 mA.

a. Compute the percent change in I_C due to Δh_{fe}, ΔI_{CO}, and ΔV_{BE}, recognizing that V_{BE} changes at −2.5 mV/°C and that I_{CO} changes at 7%/°C. Take I_{CO} = 1 mA at 25°C.

b. Repeat if R_E is increased to 2.5 kΩ.

8.20

What form do the stability equations take for emitter followers?

8.21

In Section 8.2.5 the reader was cautioned that there are other forms of feedback that alter some of our prior conclusions. Figure 8.30 constitutes one such example. The feedback resistor is connected between the output terminal and the input. Because of the phase reversal between the input and output of the transistor, the instantaneous current divides as indicated. Ordinarily, R could be connected to $+V_{CC}$ and adjusted to provide the requisite bias. As connected, R furnishes bias but, being connected between output and input, also constitutes a form of negative feedback. With instantaneous polarity as shown, some of the signal current (I_s) is diverted from entering the base terminal and hence diminishes the gain to which I_s is subjected—that is, we have negative feedback. What do you conclude concerning the input resistance that the transistor presents to the source with this form of negative feedback (termed *voltage-sampling*, *shunt-mixing*) in contrast to the input resistance that would prevail in the absence of negative feedback? How does this type of feedback influence the output resistance? We will meet this type of feedback again in Chapter 10 when discussing inverting operational amplifiers.

8.22

a. With a unit step voltage impressed on the RC circuit in Figure 8.31, express the output voltage (v_C) in terms of the time constant τ. (See "A Closer Look" at the end of Chapter 3.) Sketch the time variation of the output and label the point representing the time constant τ.

b. The circuit in Figure 8.31, of course, constitutes a low-pass filter. Sketch the amplitude vs. ω curve when subjected to

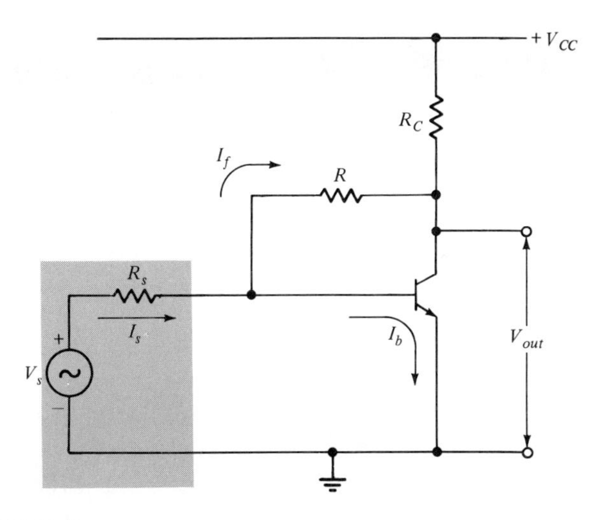

FIGURE 8.30

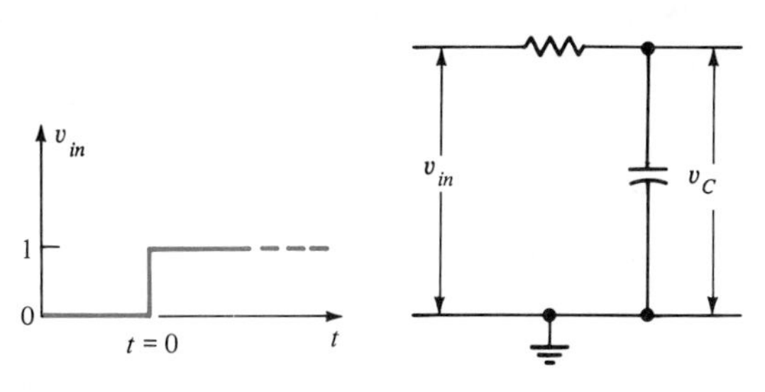

FIGURE 8.31

the same step input voltage and label the point ω_H representing the band limit.

c. What is the relationship between the time constant (τ) and the band limit (ω_H)?

d. The high-frequency drop-off of an amplifier was attributed to the presence of a low-pass filter in the circuit. To clearly demonstrate the role of negative feedback in improving circuit performance, we will follow the procedure used in Figure 8.9—that is, we will separate out the offending item and declare the amplifier to be ideal. (See Figure 8.32.) Derive the expressions for the transfer functions $\mathbf{V}_5/\mathbf{V}_4$ and $\mathbf{V}_5/\mathbf{V}_3$. What is the expression for the output voltage (v_C) in terms of the time constant?

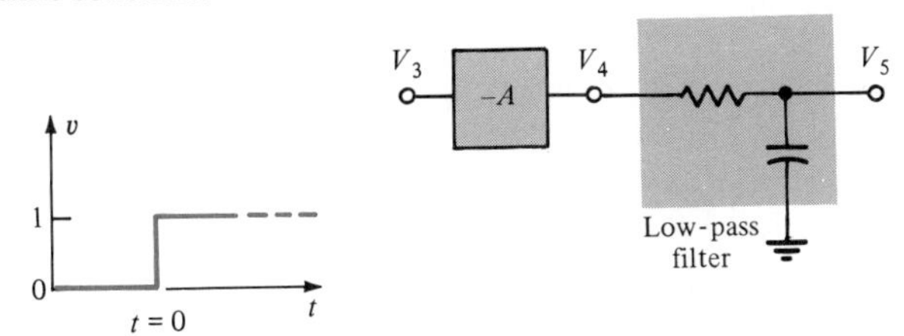

FIGURE 8.32

e. Add the negative feedback system as shown in Figure 8.33. State the expression for the transfer function $\mathbf{V}_5/\mathbf{V}_2$ considering a unit step function to be applied at terminal 2.

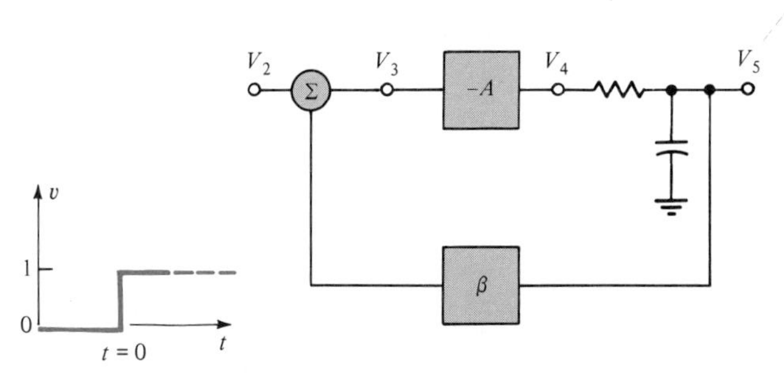

FIGURE 8.33

f. Add an ideal amplifier with a negative gain of magnitude $-(1 + \beta A)/A$ as per Figure 8.34 and derive the transfer function $\mathbf{V}_5/\mathbf{V}_1$ for a step function input at terminal 1. Noting the new band limit in this transfer function, what is the expression for v_C?

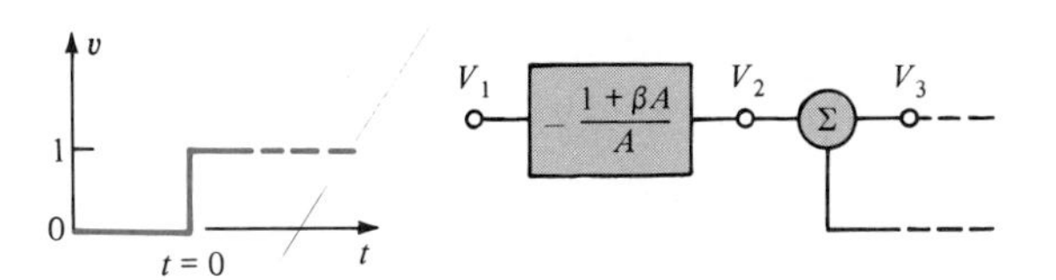

FIGURE 8.34

g. As a function of time, sketch the response to a unit step function of the *RC* network alone and with the completed circuit of part f. Do the same as a function of frequency.

COMMENT: If the low-pass filter is replaced by its mechanical equivalent, one can see that negative feedback may also be used to reduce the time constant of mechanical systems.

8.23 ⚠

The phase shift oscillator (Figure 8.35a) may be constructed from a single transistor and an *RC* network that provides a 180° phase shift to compensate for that of the amplifier.

a. Assuming R_C to be negligible, the network in Figure 8.35b may be solved by means of loop equations as follows:

$$V_{out} = I_1(R - jX) - I_2R$$
$$0 = -I_1R + I_2(2R) - I_2jX - I_3R$$
$$0 = -I_2R + I_3(2R) - I_3jX$$

Defining $\alpha \equiv X/R = 1/\omega RC$, we have

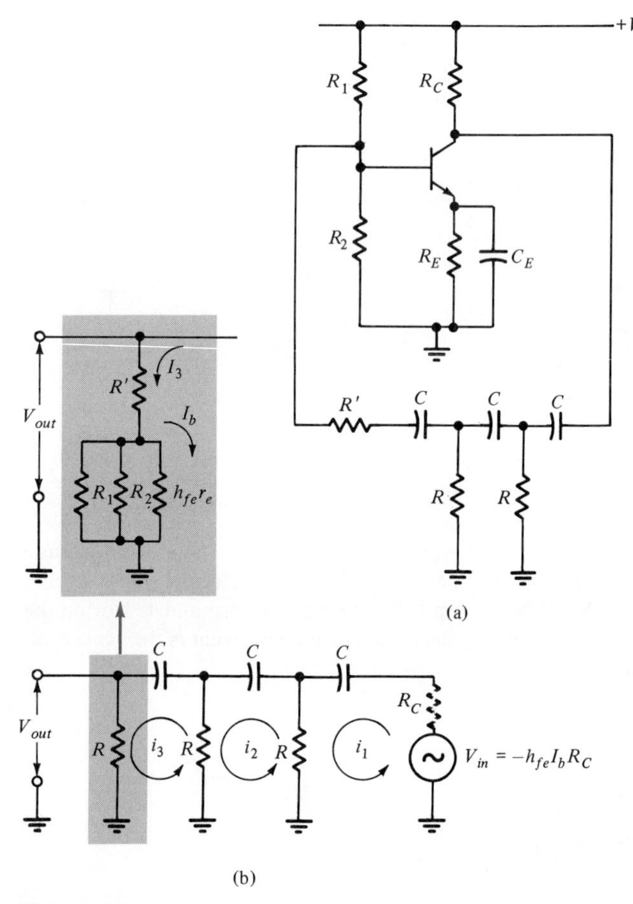

FIGURE 8.35

$$\Delta = \begin{vmatrix} R - jX & -R & 0 \\ -R & 2R - jX & -R \\ 0 & -R & 2R - jX \end{vmatrix}$$

$$= R^3 \begin{vmatrix} 1 - j\alpha & -1 & 0 \\ -1 & 2 - j\alpha & -1 \\ 0 & -1 & 2 - j\alpha \end{vmatrix}$$

$$\Delta_3 = \begin{vmatrix} R - jX & -R & V_{in} \\ -R & 2R - jX & 0 \\ 0 & -R & 0 \end{vmatrix}$$

$$I_3 = \frac{\Delta_3}{\Delta}$$

The voltage transfer function for the network is

$$\frac{V_{out}}{V_{in}} = -\frac{I_3 R}{V_{in}}$$

Determine the expression for the oscillation frequency. (This condition arises when the imaginary part of the transfer function equals zero. Why?)

b. If the collector resistance is considered, the only modification to the solution arises from a change in the upper left element of the Δ determinant. Thus, we have

$$\Delta = \begin{vmatrix} R + R_C - jX & -R & 0 \\ -R & 2R - jX & -R \\ 0 & -R & 2R - jX \end{vmatrix}$$

Determine the expression for the oscillation frequency in this case.

c. The last R in the phase shift network is actually made up of R' in series with the parallel combination $R_1 \,||\, R_2 \,||\, h_{fe}r_e$, where $h_{fe}r_e$ is the transistor's input resistance (designated as $h_{ie} = h_{fe}r_e$). From the requirement that $I_3 \geq I_b$, in terms of R_C/R, determine the minimum h_{fe} needed for oscillation to occur.

d. From the expression derived in part c, determine the value of R_C/R that leads to a minimum value of h_{fe}.

e. What is the minimum value of h_{fe} needed for oscillation?

8.24 △

The equivalent circuit for an oscillator crystal is shown in Figure 8.36. This equivalent can be used in a positive feedback circuit. There are both a series resonance and a parallel resonance, at different frequencies. The former constitutes the resonant frequency (f_r), while the latter represents the antiresonant frequency (f_a). (See Section 9.9 for a discussion of crystal oscillators used in electronic watches.) The admittance of the equivalent circuit is

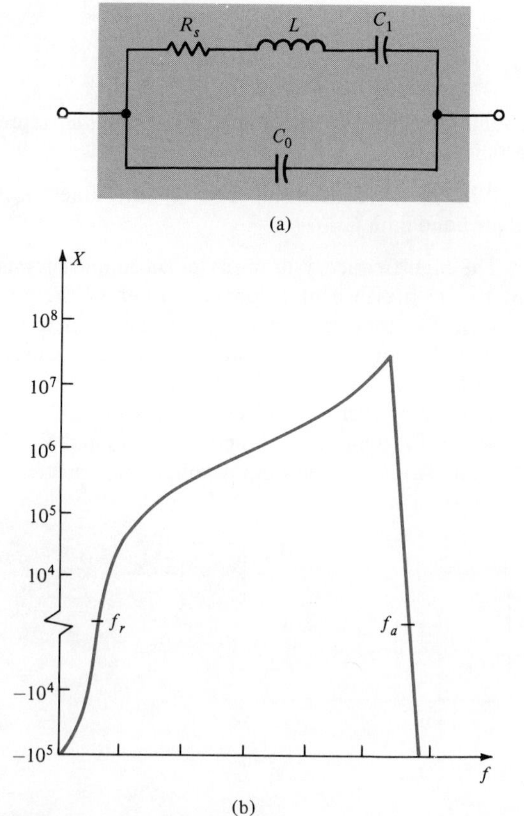

FIGURE 8.36

$$Y = j\omega C_0 + \frac{1}{j\omega L + R_S + (1/j\omega C_1)}$$

whose solution consists of an equation the sum of whose imaginary terms, on being set equal to zero, may be used to determine the values of the resonant and antiresonant frequencies. This solution takes the form

$$\omega^4 + \omega^2\left(\frac{R_S^2}{L^2} - \frac{2}{LC_1} - \frac{1}{LC_0}\right) + \frac{C_0 + C_1}{L^2C_1^2C_0} = 0$$

Typical values for a 32 kHz crystal such as used in electronic watches are

$$C_0 = 2.85 \text{ pF} \qquad R_S = 40 \times 10^3\ \Omega$$
$$C_1 = 0.00491 \text{ pF} \qquad L = 4800 \text{ H}$$

a. Determine the resonant and antiresonant frequencies. Between f_a and f_r the crystal exhibits an inductive reactance. (See Figure 8.36b.) An external loading capacitance in parallel with the crystal can be used to trim the resonance between the values of f_a and f_r.

b. The large value for R_S would seem to *imply* a low value of Q. What is the value of Q? Explain the result.

9 THE FIELD-EFFECT TRANSISTOR

OVERVIEW

We have been forewarned that there are transistors other than those involving the bipolar junction. The other major category is the field-effect transistor (FET). But it in turn can be divided into a number of subcategories, and that is what makes it more involved than the BJT. The major shortcoming of the BJT—its basically low input impedance and the accompanying power demand it places upon the signal source—is corrected in the FET by a very high input impedance. This high impedance constitutes a major advantage.

Comparable to the *npn* and *pnp* choices one experiences with BJTs, with FETs one has a choice between an *n*-channel device and a *p*-channel device. The former is usually selected because of the greater mobility of the electrons. However, a combination of *n*-channel and *p*-channel units leads to a device that in a quiescent state draws very little electrical power. Such devices are particularly useful in the digital circuits used in computers since they spend considerable time in just such a condition.

The integrated circuits that we will be considering in Chapter 10 will make significant use of the FET. We will also find FETs used in the digital circuits of Chapter 11. In Chapter 13 we will see that under certain conditions their inherent noise is smaller than that of BJTs, and finally in Chapter 17 they will enter into the consideration of some measurement techniques.

OUTLINE

9.1 ■ INTRODUCTION

In most instances bipolar junction transistors (BJTs) are operated with forward bias input circuits that lead to significant signal power demands being placed on the driving circuit. The input of a **field-effect transistor** (FET), on the other hand, can operate under reverse bias conditions, leading to very high input resistances and requiring miniscule amounts of driving signal power.

While the very high input resistance of FETs proves desirable under many circumstances, these devices, in discrete form, do have some disadvantages. They have relatively small gain-bandwidth products, and they are rather susceptible to damage from simply being handled. Another difficulty comes about because of the rather large tolerances that accompany their specifications. In this respect FETs are even worse than BJTs.

When employed as elements in an integrated circuit, however, FETs occupy very little space and hence exhibit a very high packing density that makes them attractive in such cases. Probably the most involved aspect of FET circuitry is the recognition of their various schematic symbols, for there is a variety of FET devices. Our main goal in this chapter will be a disentanglement of these distinctions.

9.2 ■ THE JUNCTION FIELD-EFFECT TRANSISTOR

Figure 9.1 depicts the construction of one type of field-effect transistor: Between the metal electrodes, which constitute the source (*S*) and drain (*D*), there may be a continuous "bridge" of *n*-type material. The gate electrode makes contact with a segment of the *p*-type semiconductor, and as this gate is progressively made more negative (relative to the drain), electrons are pulled out of the adjoining *n*-channel, thereby creating a depletion region.

Figure 9.2 will aid in understanding why depletion layers are formed.[1] Assume 4 V is to be applied between source and drain. By using a voltmeter with one terminal placed at the source (and therefore at the gate since $V_{GS} = 0$), as the other terminal is moved along the channel, the potential difference—that is, the voltage—progressively increases, reaching a value of +4 V at the drain. Since the region between *p*- and *n*-materials constitutes a reverse-biased diode [the *n*-material (cathode) being positive relative to the *p*-material (anode)] whose magnitude of reverse bias increases as we approach the drain, the thickness of the depletion region also increases. Increasing the value of V_{DS} will bring the depletion regions closer together, and at some given value of V_{DS} they will (almost) meet. This voltage is called the **pinch-off voltage** (V_p). (The regions cannot actually meet since that would prevent conduction, and conduction is necessary to create the equipotential lines.)

[1]This structure is a simplified version of what is actually used and is intended primarily to illustrate the basic principles involved.

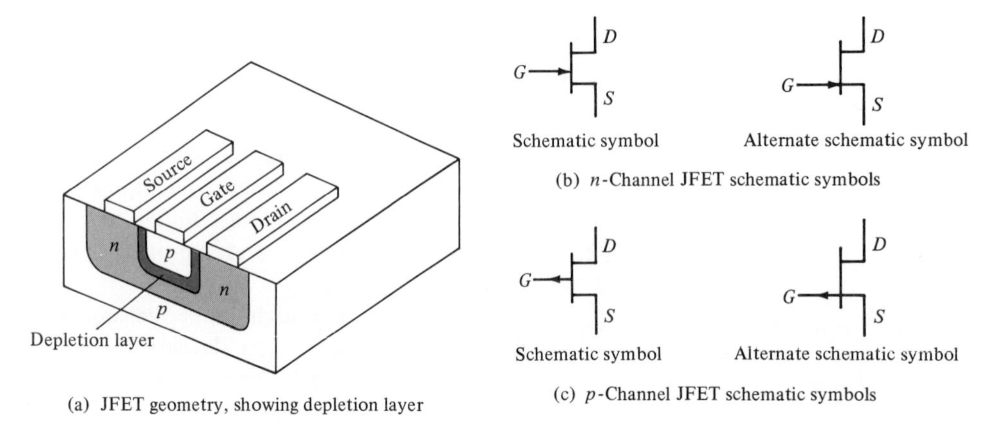

(a) JFET geometry, showing depletion layer

FIGURE 9.1. JFET

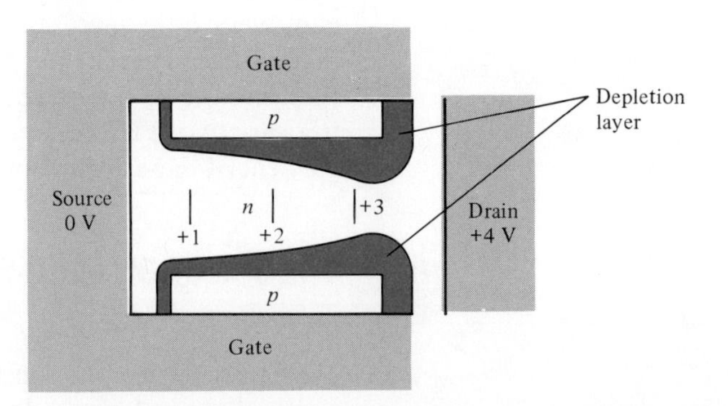

FIGURE 9.2. Growth of Depletion Region Thickness Due to Increasing Reverse Bias Between *p* and *n*-Regions as Drain Is Approached

The *n*-channel constitutes a resistance $R = L/\sigma A$, where A is the channel area, L is the length of the conductor, and σ is the material conductivity. We have

$$I_{DS} = \frac{V_{DS}}{R} = \frac{\sigma A V_{DS}}{L} \tag{9.1}$$

V_{DS}/L constitutes the electric field intensity (E) measured in volts per meter. The current density, on the other hand, is given by the expression

$$J = \frac{I_{DS}}{A} = \frac{\sigma V_{DS}}{L} = \sigma \mathrm{E} \tag{9.2}$$

J increases with increasing V_{DS} as long as σ remains constant; it does so for $\mathrm{E} < 10^5$ V/m. For E in the range of 10^5–10^6 V/m, $\sigma \propto 1/\sqrt{\mathrm{E}}$, and for still higher fields $\sigma \propto 1/\mathrm{E}$, thereby leading to a constant value of J. Therefore, as V_{DS} progressively increases, the current first rises linearly, slows down as the pinch voltage is approached, and thereafter flattens out, as shown in Figure 9.3. At large values of V_{DS}, there is a sudden increase in current, termed the breakdown region.

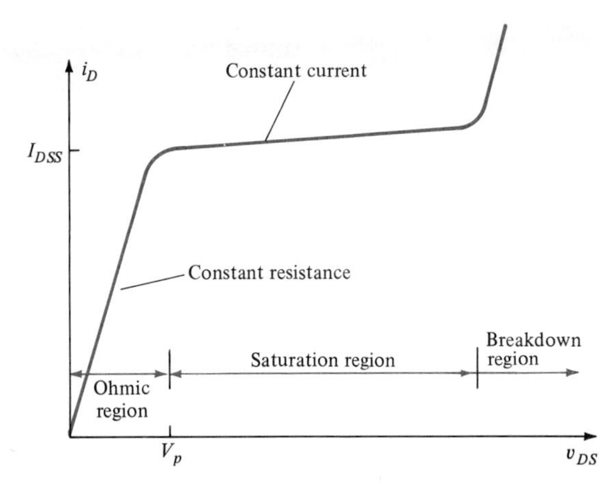

FIGURE 9.3. Drain Current Variation vs. Drain-Source Voltage V_p represents pinch voltage.

Upon applying a negative potential difference between gate and source, successive increases in V_{DS} will lead to similar characteristic curves, but the pinch-off points will now arise sooner since this biasing voltage aids in the creation of depletion layers. We may, in such fashion, obtain the characteristic curves shown in Figure 9.4.

Figure 9.4 shows a load line and an operating Q-point. Since a large

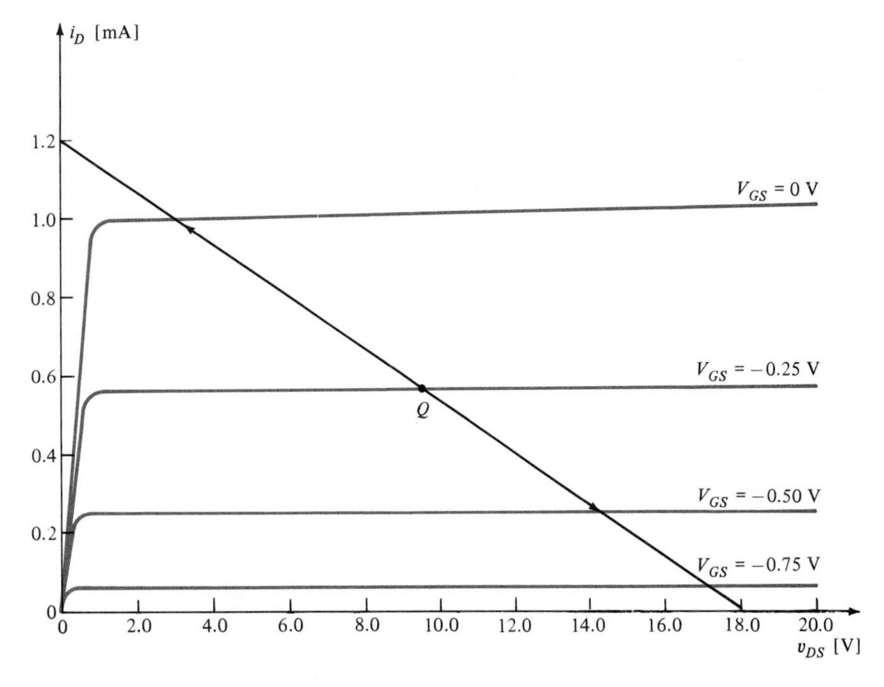

FIGURE 9.4. i_D vs. v_{DS} Characteristic Curves for JFET Shown are a load line and operating Q-point assuming V_{DD} = 18 V and an impressed signal voltage of 0.25 V, peak amplitude.

source-drain voltage *change* can be produced by a relatively small gate voltage variation, the system acts as an amplifier. In this type of field-effect transistor (called the **junction field-effect transistor**, JFET) the gate is never forward-biased since that would divert most of the current to the gate. (Actually, it may be driven up to about 0.5 V positive since at that voltage a silicon *pn* junction is still not forward-biased.)

The resistance between the gate and the remainder of the device is thus high since the gate is always reverse-biased (just as the reverse bias resistance of a diode is large). Therefore, the JFET has a high input resistance, of the order of tens or hundreds of megohms. But there is also a type of field-effect transistor that has an input resistance of the order of millions of ohms, the MOSFET (metal-oxide-semiconductor FET), which we will consider later in this chapter.

9.3 ■ JFET CHARACTERISTIC CURVES

Refer again to Figure 9.4. To the left of the pinch-off points, the drain current (I_D) varies linearly with drain-to-source voltage (V_{DS}). This constitutes the **ohmic region**; $V_{DS} \ll V_p$. For $V_{DS} > V_p$ the characteristic curves are flat; the current is essentially independent of the applied V_{DS}. This constitutes the **saturation region**.[2] When voltage amplification is desired, it is the saturation region that is utilized.

In the saturation region the drain current can be obtained from the equation

$$I_D = I_{DSS}\left(1 - \frac{V_{GS}}{V_p}\right)^2 \tag{9.3}$$

where V_{GS} = gate voltage

V_p = pinch voltage

V_{GS} and V_p both have the same sign (negative for an *n*-channel device). I_{DSS} is the drain current when $V_{GS} = 0$. [The subscripts refer to current between the drain (*D*) and source (*S*), with the third subscript indicating the status of the third electrode; in this case that the gate is shorted (*S*)—that is, $V_{GS} = 0$.] By assigning V_{GS} a succession of values, one obtains the relatively flat curves shown in Figure 9.4. The specifications of a JFET include values of I_{DSS} and V_p.

For operation in the linear region, one utilizes the following equation:

$$I_D = I_{DSS}\left[2\left(1 - \frac{V_{GS}}{V_p}\right)\left(\frac{V_{DS}}{-V_p}\right) - \left(\frac{V_{DS}}{V_p}\right)^2\right] \tag{9.4}$$

[2]When dealing with BJT amplifiers, the region where the output characteristics have the collector current virtually independent of the collector-to-emitter voltage represents the *active* region, and the region where these curves run together near the zero voltage axis (a region to be avoided for linear amplifiers) is the *saturation* region. For FETs, the region where the curves are virtually independent of the drain-to-source voltage represents the *saturation* region (used for linear amplification), and the region where they run together near the zero voltage (and are to be avoided in linear amplifiers) is the *ohmic* region.

EXAMPLE 9.1

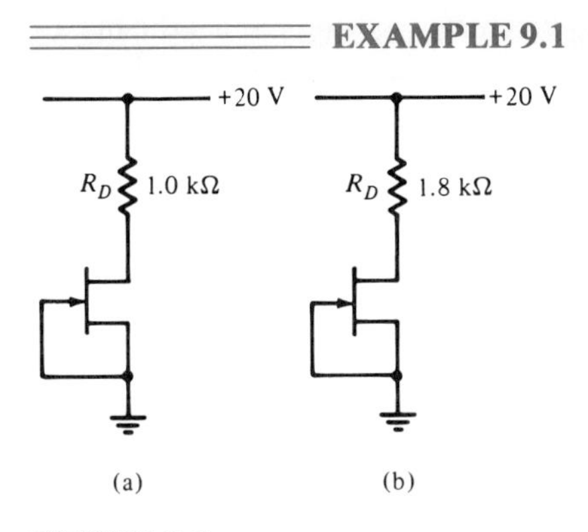

FIGURE 9.5

Determine for parts a and b of Figure 9.5 whether the JFET is operating in the ohmic or in the saturated region. In each case I_{DSS} = 10 mA, and V_p = −4 V. Since in both parts a and b the gate is at 0 V, if $V_D > |V_p|$, the operation is in the saturated region and if $V_D < |V_p|$, in the ohmic region.

Solution:

a. $V_D = V_{DD} - R_D I_D = 20 - 10^3(10^{-2}) = 10\ \text{V}$

Therefore, operation is in the saturated region.

b. $V_D = 20 - 1.8 \times 10^3(10^{-2}) = 2\ \text{V}$

Since this is less than the pinch voltage, operation is in the ohmic region.

EXAMPLE 9.2

Using the characteristic curves and load line for the JFET shown in Figure 9.4, what will be the voltage gain resulting from an impressed sine wave of 0.25 V peak amplitude. The Q-point is V_{DS} = 9.5 V, I_D = 0.56 mA.

Solution:

$$\text{load resistance} = \frac{18\ \text{V} - 0\ \text{V}}{(1.2 - 0) \times 10^{-3}\ \text{A}} = 15\ \text{k}\Omega$$

$$\text{peak-to-peak output current} = (1.0 - 0.25)\ \text{mA} = 0.75\ \text{mA}$$

$$\text{peak-to-peak output voltage} = 7.5 \times 10^{-4} \times 1.5 \times 10^4$$

$$= 11.25\ \text{V}$$

$$\text{voltage gain} = \frac{11.25\ \text{V}}{2(0.25)\ \text{V}} = 22.5$$

The voltage variations one introduces at the input gate are used to produce varying currents in the drain circuit that, when passed through the drain resistance (R_D), produce an amplified voltage output. For a fixed value of V_{DS}, the change in the drain current produced by a given gate voltage change constitutes the **transconductance** g_m. Typical values for a JFET are of the order of 1 mS—that is, 1 mA/V.

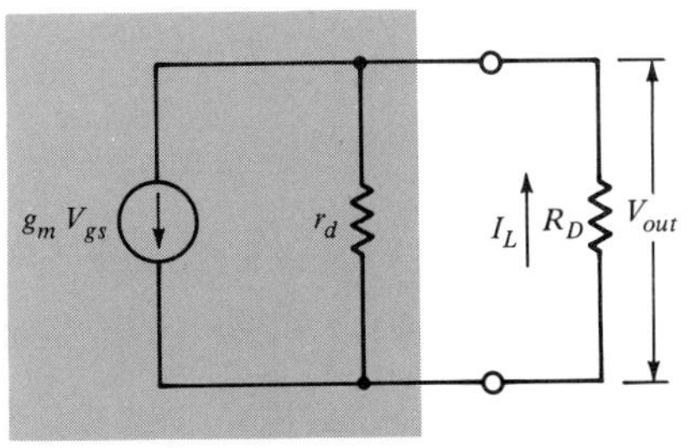

The flat saturation curves signify that the drain current to a large degree is independent of the voltage applied between the drain and source. In other words, such a transistor is a current source, having a high internal drain resistance (r_d). The transistor output circuit can be represented as a signal current generator of magnitude $g_m V_{gs}$ (Figure 9.6), where V_{gs} is the rms value of the input signal voltage, in parallel with the drain resistance r_d. From this result we see the advantage of making R_D large. But we have to be careful, and for saturated operation we must make sure that V_{DS} exceeds V_p. It would be nice if we could have an R_D that looks large to the ac signal but small to dc.[3] Another transistor may be made to provide just such a distinction, as we will see subsequently.

FIGURE 9.6. Equivalent Output Circuit for JFET JFET output resistance (r_d) is typically of the order of a megohm, while the drain resistance (R_D) is generally much smaller if V_{DS} is to be maintained in the saturated operating region. This relationship causes V_{out}/V_{gs} to be much smaller than the amplification factor μ, where $\mu = g_m r_d$.

[3]We have previously seen that a transformer-coupled load may be made to behave in this manner.

Refer to Figure 9.6. The ratio V_{out}/V_{gs} when $R_D = \infty$ represents the maximum theoretical voltage gain available from the device and is termed the **amplification factor** μ. We have

$$V_{out} = g_m V_{gs} r_d \tag{9.5}$$

$$\mu \equiv \left(\frac{V_{out}}{V_{gs}}\right)_{R_D=\infty} = g_m r_d \tag{9.6}$$

EXAMPLE 9.3

Using the characteristic curves and parameters from Example 9.2, estimate the values of g_m and r_d. What is the value of μ, the amplification factor? Why is the voltage gain so much smaller than the amplification factor?

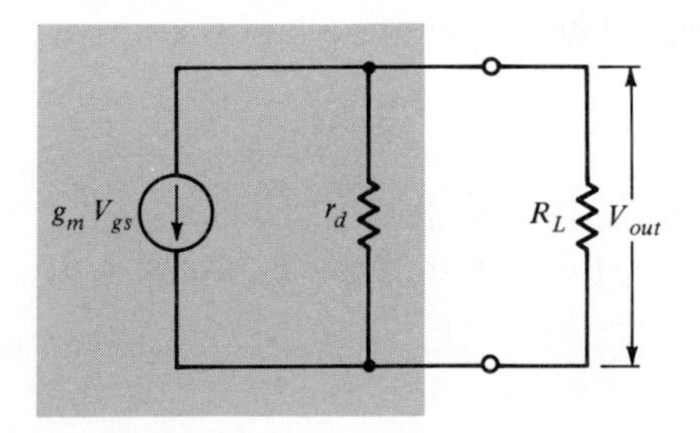

FIGURE 9.7

Solution: At $V_{DS} = 9.5$ V, corresponding to $V_{GS} = 0$, is a drain current of 1.02 mA. At $V_{GS} = -0.50$ V, the drain current is 0.25 mA. We have

$$g_m = \left.\frac{\Delta I_D}{\Delta V_{GS}}\right|_{V_{DS}=\text{constant}} = \frac{(1.02 - 0.25) \times 10^{-3}}{0.50 - 0.0}$$
$$= 1.54 \times 10^{-3}\text{ S} = 1.54\text{ mS}$$

At $V_{DS} = 20.0$ V, $I_D = 0.57$ mA. At $V_{DS} = 1.0$ V, $I_D = 0.56$ mA. We have

$$r_d = \left.\frac{\Delta V_{DS}}{\Delta I_D}\right|_{V_{GS}=constant} = \frac{20 - 1}{(0.57 - 0.56) \times 10^{-3}} = 1.9\text{ M}\Omega$$

$$\mu = g_m r_d = 1.54 \times 10^{-3} \times 1.9 \times 10^6 \simeq 2900$$

The equivalent circuit is shown in Figure 9.7.

$$A_V = \frac{V_{out}}{V_{gs}} = -\left(\frac{r_d}{r_d + R_L}\right)\left(\frac{g_m V_{gs} R_L}{V_{gs}}\right)$$
$$= -\left(\frac{1900}{1900 + 15}\right)(1.54 \times 10^{-3})(15 \times 10^3) = -22.9$$

To keep the operation in the saturation region, R_L is much smaller than r_d, resulting in a relatively small output voltage.

EXAMPLE 9.4

See Figure 9.8.

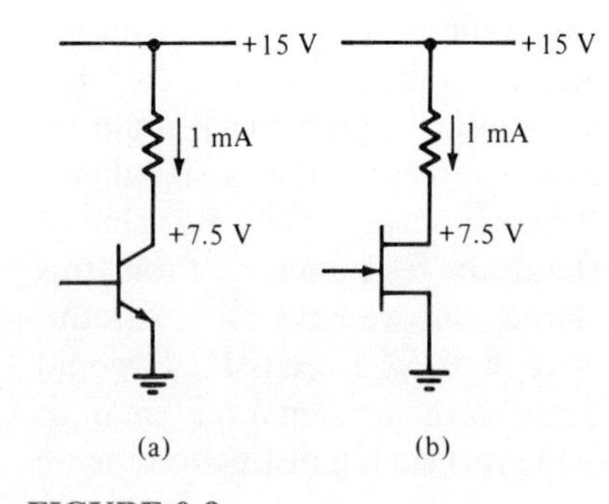

FIGURE 9.8

a. What will be the voltage gain of an amplifier utilizing a Q-point of 1 mA at 7.5 V (with a supply voltage of 15 V) if g_m is 40 mS (typical of a BJT)?
b. What will the voltage gain be if g_m is 1000 μS (typical of an FET)?
c. What conclusion can you draw from these results?

Solution:

$$R = \frac{V_{CC} - V_{CE}}{I_C} = \frac{15 - 7.5}{10^{-3}} = 7.5 \times 10^{-3}\ \Omega$$

$$g_m = \frac{i_{out}}{v_{in}} = -\frac{v_{out}}{v_{in}}\frac{1}{R} \qquad \text{since } v_{out} = -i_{out}R$$

$$A_V = -\frac{v_{out}}{v_{in}} = -g_m R$$

a. $A_V = -40 \times 10^{-3} \times 7.5 \times 10^3 = -300$
b. $A_V = -10^{-3} \times 7.5 \times 10^3 = -7.5$
c. Because of their low transconductance, one seldom utilizes FETs as simple amplifier stages unless one wishes to take advantage of their unique input properties, particularly the high impedance. Their gain may be improved by using another transistor as a load in place of a resistor.

9.4 ■ CURRENT SOURCES

Since the drain current remains essentially constant as long as V_{DS} is more than a few volts, a JFET can be used as a **current source** (Figure 9.9) providing a rather constant current through the load. While the current remains rather constant, its value is subject to some uncertainty due to the spread in values of the transistor parameters. For the same unit the I_{DSS} can vary by as much as a factor of 5.

One can, of course, adjust the current on an individual basis by using a source resistor to adjust the bias, as in Figure 9.10. Commercially, a series of FETs is available to provide currents from 0.22 to 4.7 mA, with 10% tolerances.

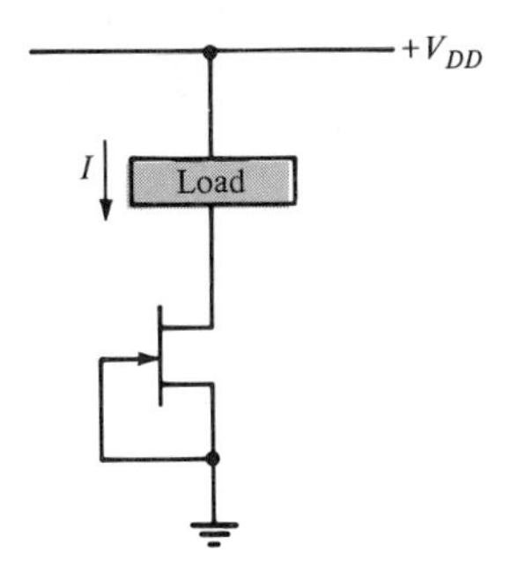

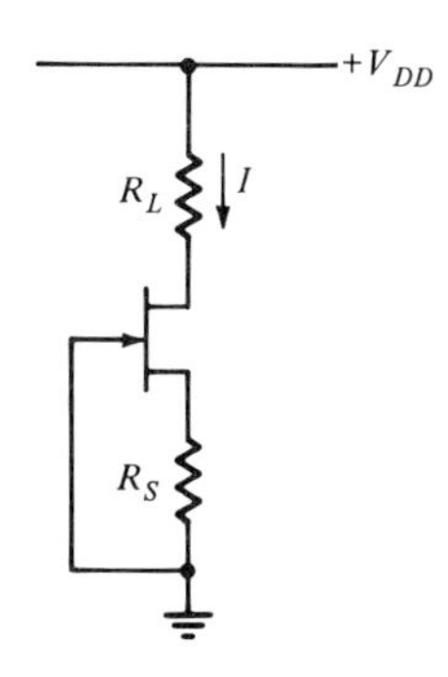

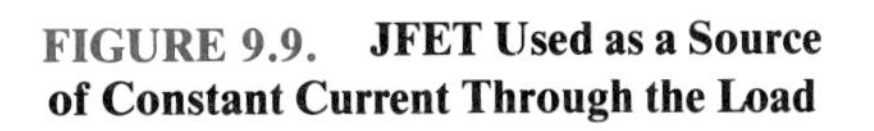
FIGURE 9.9. **JFET Used as a Source of Constant Current Through the Load**

FIGURE 9.10. **Source Resistor R_S Used to Adjust Constant Current Value Through R_L**

9.5 ■ SOURCE FOLLOWERS

A **source follower** (Figure 9.11) represents a rather common usage for FETs, serving a like purpose to that of the emitter follower but with a higher input resistance. Because of this characteristic, it is commonly found as the input

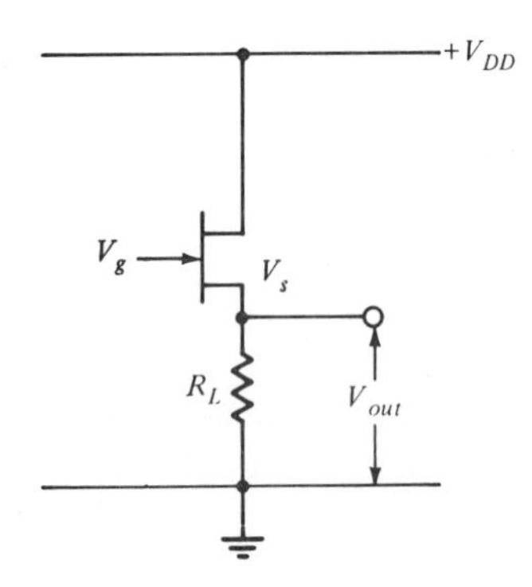

FIGURE 9.11. **Source Follower Employing JFET**

stage for measuring equipment where negligible loading effects are important. We have

$$V_s = R_L I_d$$
$$I_d = g_m V_{gs} = g_m(V_g - V_s)$$
$$V_s = \frac{g_m R_L}{1 + g_m R_L} V_g$$

For $R_L \gg 1/g_m$, $V_s \simeq V_g$—that is, we have approximately unity gain.

With respect to the output impedance, where a low value is usually desirable, the JFET is somewhat inferior to the BJT. Consider the following:

$$I_d = g_m V_{gs} \simeq g_m V_{out}$$
$$\frac{V_{out}}{I_d} = \frac{1}{g_m} \tag{9.7}$$

The transconductance of a BJT, typically, is about 40,000 μS—that is, since $g_m = 1/r_e \simeq I_C\ [\text{mA}]/26 \simeq 10^{-3}/26 = 38{,}500\ \mu$S. Thus, where the output resistance of a BJT is about 25 Ω, that for a JFET is about 1 kΩ. This difference has several consequences, including the following:

1. The output signal magnitude may be significantly smaller than the input since, even with a large value for R_L, there is a voltage divider action at the output.
2. Since g_m changes with drain current, the output resistance will vary over the signal waveform, producing distortion. FETs with large g_m help, but the use of a BJT as a dynamic load in place of a source resistance is even better.

For the circuit shown in Figure 9.12, the source current remains approximately constant at V_{BE}/R_B, yet the output resistance is the much lower value characterized by a BJT. Figure 9.13 represents another solution to the problem. Q_1 and Q_2 are matched pairs available on a single silicon chip. The cur-

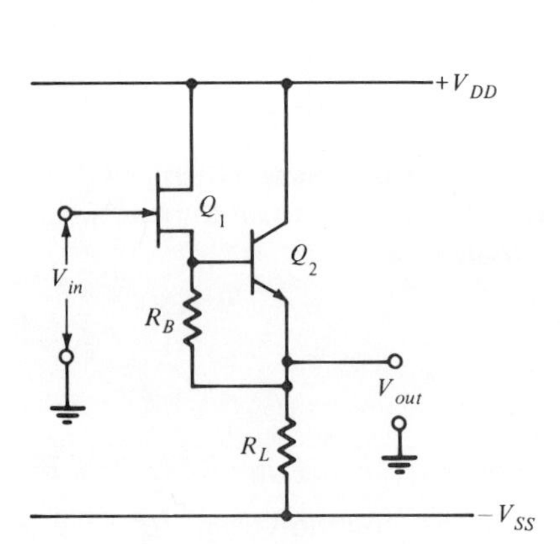

FIGURE 9.12. **BJT Used as a Dynamic Load for a Source Follower in Order to Decrease Output Resistance**

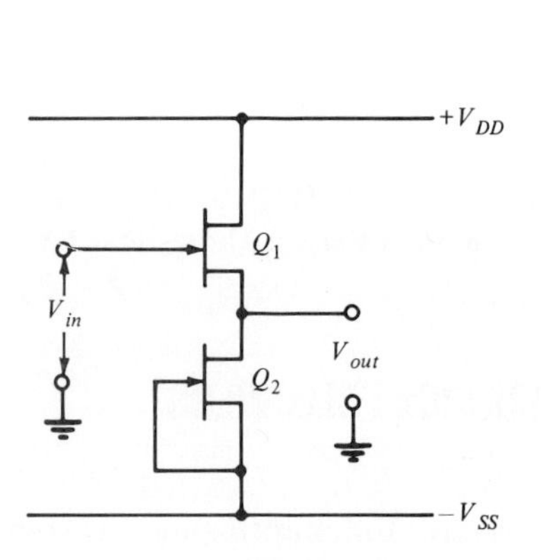

FIGURE 9.13. **JFET Used as a Dynamic Resistance for Source Follower Q_1**

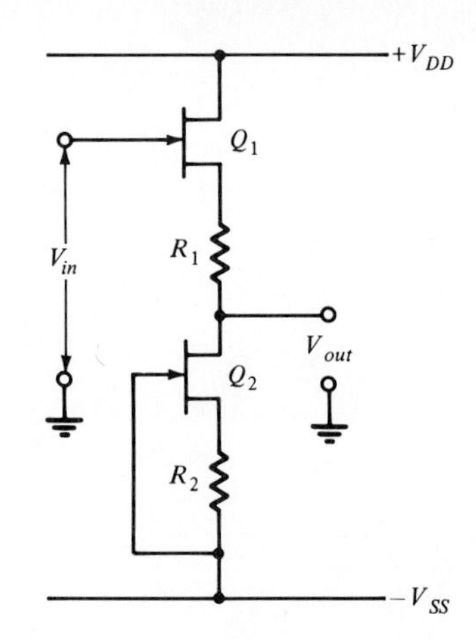

FIGURE 9.14. **Source Follower (Q_1) Using Q_2 as a Dynamic Load, with R_1 and R_2 Used to Adjust Bias**

rent through Q_2 is that appropriate to $V_{GS} = 0$, maintaining this same current through Q_1. This current through Q_1, in turn, establishes the same value for V_{GS} in Q_1 as in Q_2.

Improved linearity results from biasing Q_2 with R_2, allowing one to select a current other than I_{DSS}. (See Figure 9.14.) With $R_1 = R_2$ and symmetric power sources—that is, $V_{DD} = |V_{SS}|$—the output signal is centered at ground potential.

9.6 ■ FET AMPLIFIERS

The **common-source (CS) FET** (Figure 9.15) is analogous to the CE amplifier. The gate biasing voltage necessary to establish the Q-point may be obtained by virtue of the voltage drop across R_S. Since there is essentially no dc current passing through R_1, the dc voltage between the gate and the source V_{GS} is equal to that across R_S.

C_S is again necessary if negative feedback across R_S is to be prevented from diminishing the gain. The value of C_S is dictated by making its reactance small compared to r_s, the small signal impedance looking into the source terminal of the FET and equal to $1/g_m$. But in the case of a CS amplifier this may be comparable in magnitude to R_S, and one should then make the reactance small compared to r_s in parallel with R_S.

The choice of R_1's magnitude is primarily dictated by the output resistance of the signal source, the former being made much larger than the latter. In many instances, however, rather than simply a single resistor between gate and ground, the considerable spread in FET specifications causes one to use a more elaborate procedure.

Figure 9.16 depicts the maximum and minimum I_D versus V_{GS} curves that might characterize a given n-channel JFET designation. The desired Q-point is presumed to be $I_D = 2$ mA, $V_{GS} = -3$ V. If obtained from a self-biasing arrangement using the voltage drop across R_S', as depicted the value would be 1.5 kΩ (3 V/2 mA). But with the wide variation in the values of V_p that might be encountered, I_D could be anywhere between 0.75 and 3.5 mA. These values represent the current values where the R_S' bias line intercepts the $V_{p(min)}$ and $V_{p(max)}$ curves.

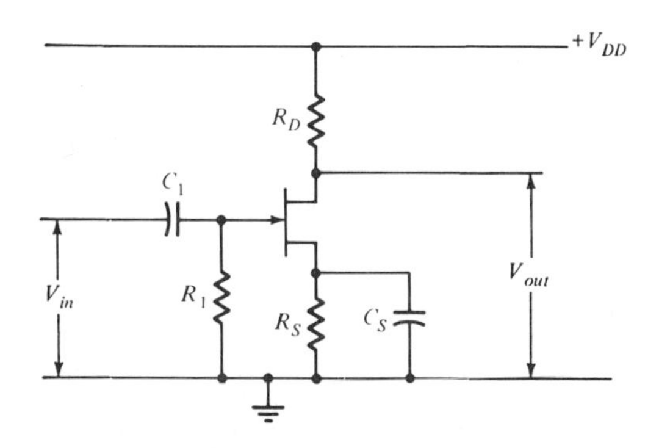

FIGURE 9.15. **Common-Source Amplifier**

If one imposes stricter limits on the allowable deviation of the drain current (say points A and B, representing 1.5 and 2.5 mA, respectively, in Figure 9.16), a larger source resistance, depicted as the R_S'' bias line, is necessary. The intercept of this line with the V_{GS}-axis, however, arises at V_{GG}. This fixed bias, in addition to the self-bias provided by R_S'', may be obtained from a voltage divider such as was used in the BJT case. (See Figure 9.17.)

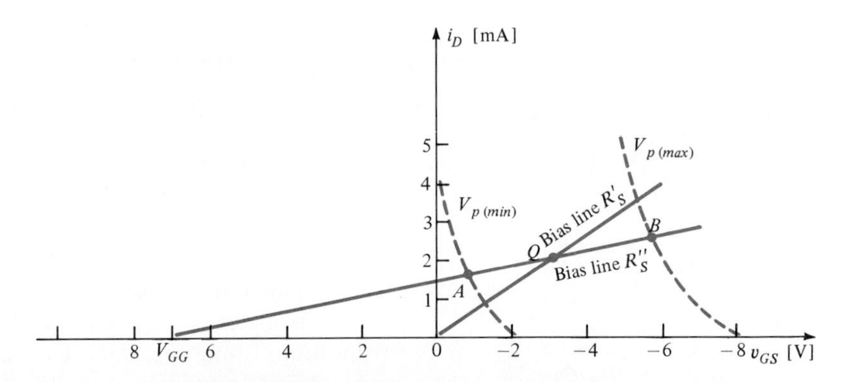

FIGURE 9.16. Extreme Transfer Characteristics ($V_{p(min)}$ and $V_{p(max)}$) for a JFET Note the nominal Q-point and self-bias line (R'_S) as well as bias line combining self-bias and fixed bias (R''_S). V_{GG} is the requisite fixed bias magnitude.

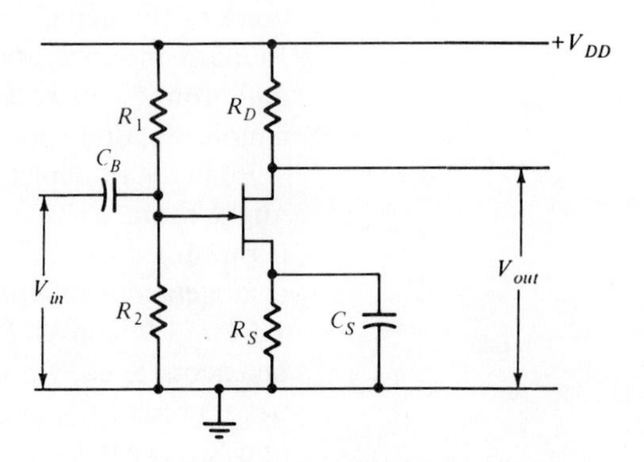

FIGURE 9.17. JFET Amplifier Showing Self-Bias Provided by R_S and Fixed Bias by R_1 and R_2

EXAMPLE 9.5

An MPF105 n-channel JFET has a recommended Q-point consisting of I_D = 0.4 mA, V_{DS} = 15 V, and V_{GS} = −4.5 V. I_{DSS} maximum and minimum values are 16 and 4 mA, corresponding, respectively, to V_p = −8 V and −2 V. V_{DD} = 30 V. (See Figure 9.18.)

a. What value of R_S will provide the necessary bias?
b. What will be the possible range of I_D values that may be encountered?
c. What value of R_S will limit the possible I_D variation to within ±20%?

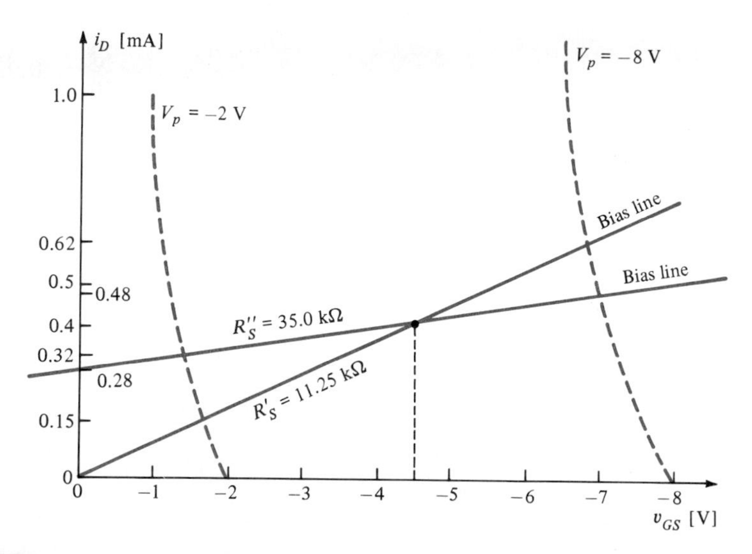

FIGURE 9.18

d. What will then be the necessary value of V_{GG}?

e. Design a divider circuit that will provide the necessary V_{GG} if the blocking capacitor (C_B) is equal to 1.0 μF and the lower 3 dB point is to be 100 Hz.

f. What should be the value of the source bypass capacitor C_S? Take g_m = 4500 μS.

Solution:

a. $$R'_S = \frac{V_{GS}}{I_D} = \frac{4.5\text{ V}}{4 \times 10^{-4}\text{ A}} = 11.25\text{ k}\Omega$$

b. From the $V_p = -2$ V intercept, $I_D = 0.15$ mA. From the $V_p = -8$ V intercept, $I_D = 0.62$ mA. These values represent a variation greater than ±50%.

c. A ±20% variation leads to extreme values of $I_D = 0.32$ mA (with $V_{GS} = 1.4$ V) and $I_D = 0.48$ mA (with $V_{GS} = 7.0$ V). We have

$$R''_S = \frac{(7 - 1.4)\text{ V}}{(0.48 - 0.32) \times 10^{-3}\text{ A}} = 35.0\text{ k}\Omega$$

d. $$\text{slope of } R''_S \text{ bias line} = \frac{1}{35 \times 10^3} = \frac{I_{D\text{ intercept}}}{V_{GG}} = \frac{0.28 \times 10^{-3}\text{ A}}{V_{GG}}$$

$$V_{GG} = 0.28 \times 10^{-3} \times 35 \times 10^3 = +9.8\text{ V}$$

e. Letting $X_{C_B} = \frac{1}{10}(R_1 \parallel R_2)$ at 100 Hz, we have

$$X_{C_B} = \frac{1}{2\pi(100)(10^{-6})} = 1.59 \times 10^3\ \Omega$$

$$R_1 \parallel R_2 = 10X_{C_B} = 10(1.59 \times 10^3) = 1.59 \times 10^4\ \Omega$$

$$\frac{R_2}{R_1 + R_2} V_{DD} = \frac{R_2}{R_1 + R_2}(30) = V_{GG} = 9.8$$

$$\frac{R_2}{R_1 + R_2} = \frac{9.8}{30} = 0.327$$

$$\frac{R_1 R_2}{R_1 + R_2} = 0.327R_1 = 1.59 \times 10^4\ \Omega$$

$$R_1 = \frac{1.59 \times 10^4}{0.327} = 4.86 \times 10^4\ \Omega$$

48.7 kΩ is the closest 1% standard resistance.

$$\frac{R_1 + R_2}{R_2} = \frac{1}{0.327} = 1 + \frac{R_1}{R_2}$$

$$R_2 = \frac{R_1}{(1/0327) - 1} = \frac{4.86 \times 10^4}{2.06} = 2.36 \times 10^4\ \Omega$$

23.7 kΩ is the closest standard 1% resistor.

f. With r_s ($= 1/g_m$) the source resistance and R_S the source resistor, bypass capacitor C_S should have a reactance $\leq \frac{1}{10}(r_s \,||\, R_S)$ at the low-frequency 3 dB point. We have

$$\frac{1}{g_m} = \frac{1}{4.5 \times 10^{-3}\ \text{S}} = r_s = 222\ \Omega$$

$R_S \gg r_s$, and therefore, $X_{C_S} \equiv \frac{1}{10} r_s = 22\ \Omega$, and we have

$$C_S = \frac{1}{2\pi(100)(22)} = 7.2 \times 10^{-5}\ \text{F} = 72\ \mu\text{F}$$

9.7 ■ VOLTAGE-CONTROLLED RESISTOR

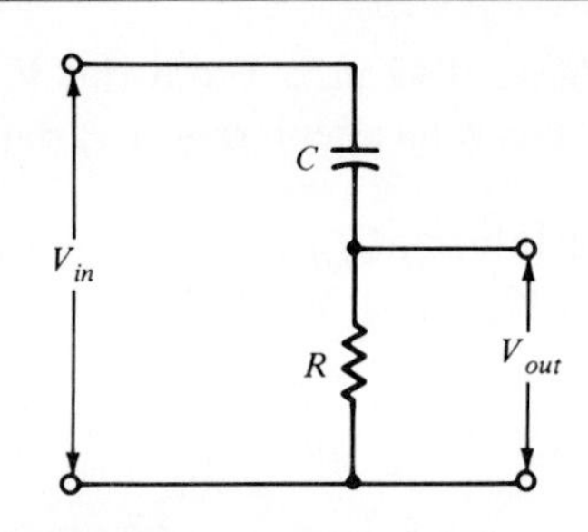

FIGURE 9.19. **High-Pass Filter**

We turn now to an important use made of the JFET in modern sound amplifier circuitry. It is sometimes desirable to have a high-pass filter whose corner frequency—that is, the 3 dB point—can be made to vary in response to a dc voltage. The *RC* circuitry shown in Figure 9.19 is a high-pass filter whose corner frequency we recognize as depending on the product *RC*.

Now, in place of the fixed resistor, we may use an FET as in Figure 9.20. This will still operate like a high-pass filter but one whose corner frequency may be varied. By changing the gate voltage on the FET, we alter the source-to-drain resistance and thereby change the corner frequency. In such applications one uses a completely different region of the FET operating curves than that encompassed by the load line in Figure 9.4.

Looking at Figure 9.21, note that the dashed line connects the pinch-off points for the various gate voltages. Thus, with $V_{GS} = 0$, the drain-to-source voltage is V_p, and the corresponding drain current is I_{DSS}. As the gate voltage is made progressively more negative (for an *n*-channel device), the pinch-off sets in earlier. At each point where this curve crosses a gate characteristic, the slope between them is the same as that for the tangent to that gate curve at the origin. Also, the horizontal line extending from any pinch voltage point to the corresponding drain current will be intersected by the appropriate tangent line at a value of V_{DS} equal to one half of that pinch voltage.

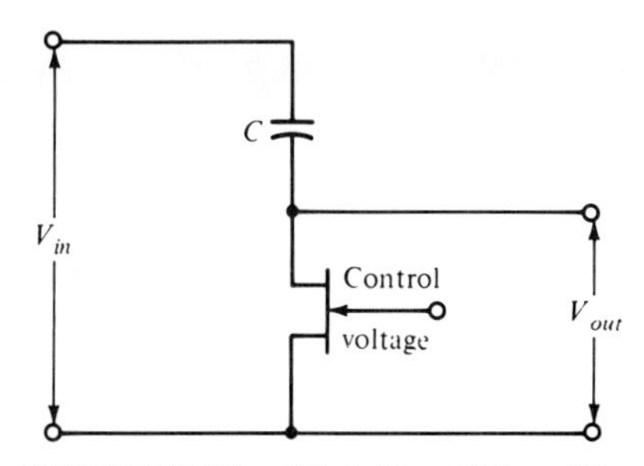

FIGURE 9.20. **High-Pass Filter with Variable Corner Frequency**

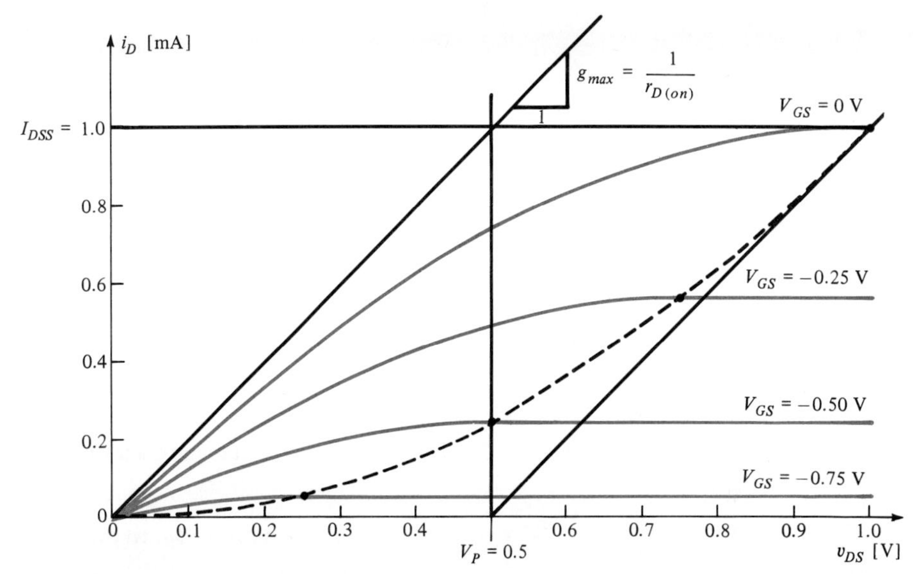

FIGURE 9.21. **JFET's Ohmic Region** Note the small values of V_{DS}. The dotted line connects pinch points at various values of V_{GS}. For a given value of V_{GS}, the slope at a pinch point equals the slope for that V_{GS} *at the origin* (illustrated for $V_{GS} = 0$). The reciprocal of the slope equals the static resistance.

The slope of each tangent line is equal to the conductance, and its reciprocal represents the zero-current channel resistance. But because of the slope equality we have noted, we can determine the slope at the pinch points and use those expressions as being applicable for the zero-current channel resistance.

The slope of the curve connecting the pinch-off points for the various values of V_{GS} is equal to the corresponding i_D divided by $V_p/2$. We have

$$i_D = I_{DSS}\left(1 - \frac{V_{GS}}{V_p}\right)^2 \tag{9.8}$$

$$g_D = \frac{di_D}{dV_{GS}} = -\frac{2I_{DSS}}{V_p}\left(1 - \frac{V_{GS}}{V_p}\right) \tag{9.9}$$

For $V_{GS} = 0$, we have

$$g_D = -\frac{2I_{DSS}}{V_p} \equiv g_{max} = \frac{1}{r_{D(ON)}} \tag{9.10}$$

A modification of Figure 9.20 (shown as Figure 9.22a) can be used for automatic gain control. When the FET is pinched off—that is, when $V_{AGC} > V_p$—the transmission depends only on the resistance R_S and the output loading. If the load can be neglected—that is, if it is a very high impedance—we have

$$\frac{V_{out}}{V_{in}} = \frac{r_d}{R_S + r_d} = \frac{1}{1 + g_d R_S} \tag{9.11}$$

where r_d is the dynamic drain resistance.

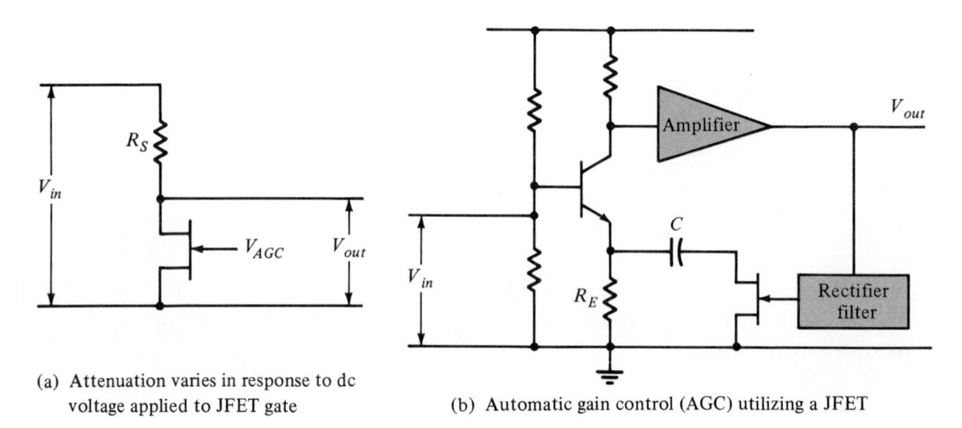

(a) Attenuation varies in response to dc voltage applied to JFET gate

(b) Automatic gain control (AGC) utilizing a JFET

FIGURE 9.22. Automatic Gain Control

Since the generalized zero drain-voltage channel resistance is

$$r_d = \frac{1}{(2I_{DSS}/V_p)[(V_{GS}/V_p) - 1]} \tag{9.12}$$

we have

$$\frac{V_{out}}{V_{in}} = \frac{1}{1 + (2I_{DSS}/V_p)[(V_{AGC}/V_p) - 1]R_S} \tag{9.13}$$

The *smallest* output will arise with $V_{AGC} = 0$, for g_d will then be g_{max}, and the resistance r_d will have its smallest value. The *maximum* value of transmission will be unity, and this condition will arise when $V_{AGC} \geq V_p$.

Figure 9.22b shows how the circuit in Figure 9.22a may be used to serve as an automatic gain control (AGC). In this case the value of C can be considered large enough to present negligible reactance. The gain of the BJT can be diminished by utilizing more negative feedback, provided by increasing the effective emitter resistance. This change can be accomplished by causing r_d of the JFET (in parallel with R_E) to increase. It will increase as the dc gate voltage increases.

With an assumed increase in the input signal (V_{in}), the dc voltage applied to the gate from the rectifier-filter increases, thereby increasing the negative feedback in Q_1 and lowering the output signal, compensating for the increase in V_{in}.

It should be noted that the output of the circuit in Figure 9.22a increases with increasing (negative) gate voltages, while that in Figure 9.22b decreases.

EXAMPLE 9.6 Using the curves shown in Figure 9.18, what are the drain resistance values corresponding to the following gate voltages: $V_{GS} = 0$ V, -0.25 V, -0.50 V, and -0.75 V?

Solution: In general, we have

$$g_d = \frac{I_D \text{ (at pinch voltage)}}{\frac{1}{2}V_p} = \frac{1}{r_d}$$

At $V_{GS} = 0$, we have

$$g_{max} = \frac{(1-0)\text{ mA}}{0.5\text{ V}} \qquad r_d = \frac{0.5}{10^{-3}} = 500\ \Omega$$

At $V_{GS} = -0.25$ V, we have

$$g_d = \frac{0.565\text{ mA}}{0.75/2} \qquad r_d = \frac{0.75}{2(0.565)\times 10^{-3}} = 664\ \Omega$$

At $V_{GS} = -0.50$ V, we have

$$r_d = \frac{0.50\text{ mA}}{2(0.25)\times 10^{-3}} = 1000\ \Omega$$

At $V_{GS} = -0.75$ V, we have

$$r_d = \frac{0.75}{2(0.06)\times 10^{-3}} = 6250\ \Omega$$

Note again in this application that no dc voltage is applied between the source and drain—only the ac signal. There is no gain in such systems. As with all FET applications, the applied signal amplitude should be small; otherwise, large amounts of distortion are introduced. For the case in point, with the operating point at the origin of the characteristic curves and a bias voltage on the gate, the conduction through the channel is not symmetric, and harmonic distortion will arise. This distortion may be considerably reduced by picking off some of the input signal and feeding it to the gate (superimposed on the dc control voltage). Why this works can be seen by looking at Eq. 9.4; it is the square term that leads to the difficulty. The addition of $\frac{1}{2}V_{DS}$ to the gate will compensate for this nonlinearity.

9.8 ■ THE MOSFET

It is also possible to construct FETs without any permanent conducting channel. Such procedures create **enhancement-mode** FETs, and one example of their construction is shown in Figure 9.23a. The metallic gate electrode is insulated from the n-type **substrate** (as the bulk material is called) by an oxide layer. Because of this insulation, the device is now termed a MOSFET (**metal-oxide-semiconductor** FET). In the schematic symbol (Figure 9.23b), the lack of a permanent conducting channel is indicated by the three segments on the right. A negative gate voltage attracts holes from the adjoining p-regions and forms a p-conduction channel—hence the expression enhancement-mode. The inversion layer arises only after a certain threshold value of V_{GS} has been exceeded, termed $V_{threshold}$ (V_T). It is common practice to connect the substrate to the source terminal. This procedure reverse-biases the junction between the channel and the substrate, which is generally desirable.

In our example, the enhanced region being of the p-type, in the schematic symbol (Figure 9.23b) the arrow points outward. (The arrow on the

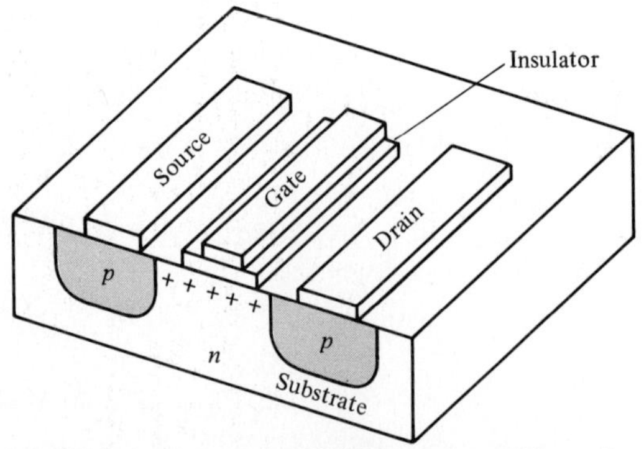

(a) p-Channel enhancement MOSFET geometry, with negative gate voltage creating positive carrier bridge between source and drain

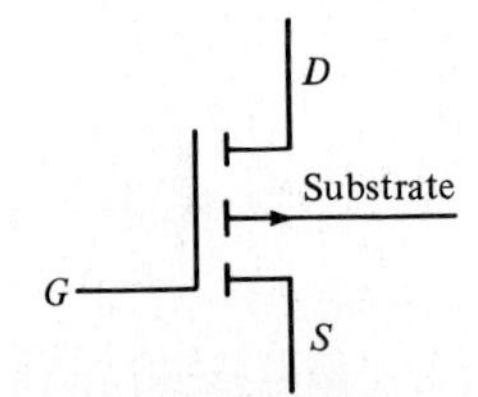

(b) p-Channel enhancement MOSFET schematic symbol

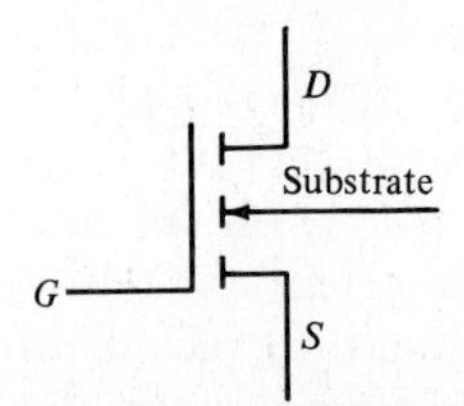

(c) n-Channel enhancement MOSFET schematic symbol

FIGURE 9.23. **Enhancement MOSFET** *P*-channel enhancement MOSFET geometry with negative gate voltage creating positive carrier bridge between source and drain.

substrate line points in the forward direction of the substrate-to-channel *pn* junction.) One could again invert everything and create an enhanced *n*-channel MOSFET, which would have the arrow pointing inward, the substrate being of *p*-type material (Figure 9.23c).

It is also possible to combine the two types of FETs. A narrow permanent conducting channel is provided, and it can be widened (leading to enhancement) by applying one polarity to the gate or narrowed (leading to depletion) by means of the opposite polarity (Figure 9.24). (We do not have to worry now about forward bias between the gate and drain since the insulated gate effectively prevents dc from flowing between them.) This device is called a **depletion-mode** MOSFET. The advantage of such a device is that under quiescent conditions the gate voltage is zero and can be driven in either the positive or the negative direction—that is, the required bias is zero.

As noted, while the high input impedance of FETs is generally desirable, these transistors do have some disadvantages. For a given device type, they

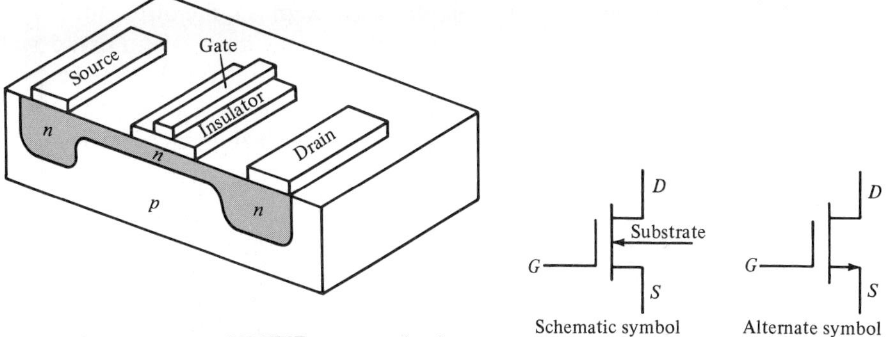

(a) n-Channel depletion MOSFET geometry, showing permanent n-channel

(b) n-Channel depletion MOSFET schematic symbols

FIGURE 9.24. **Depletion MOSFET** Positive gate voltage will attract additional negative carriers into the channel from the heavily doped drain and source (enhancement mode). Negative gate voltage will attract positive carriers from the substrate, diminishing the channel width (depletion mode).

show a wider spread of values than even that experienced with BJTs, they provide lower gain, and they have poorer high-frequency performance. Additionally, in the case of MOSFETs, they are very susceptible to electrical damage, which can arise from simply handling them. This damage derives from a puncturing of the insulation beneath the gate. The buildup of small charges on the gate (as little as 10^{-10} C) with a gate-substrate capacitance of 1 pF amounts to 100 V ($V = Q/C = 10^{-10}/10^{-12} = 100$ V). Many units have built-in diodes for protection. Also, to prevent such ruptures, the MOSFETs may be shipped with flexible conductors intertwined between the leads, thus shorting the elements to one another and thereby preventing a buildup of static charges. After the device is wired into the circuit, the flexible shorts can safely be removed.

The protection diodes may themselves create problems. Diode currents should be generally limited to less than 10 mA. The diode may also adversely affect some circuits, particularly multivibrators and other pulse circuits.

While Eq. 9.3 is applicable to both JFETs and to depletion MOSFETs, with I_{DSS} specified at $V_{GS} = 0$, with enhancement MOSFETs current does not commence until $V_{GS} > V_T$. The corresponding current equation in this case is

$$I_D = k(V_{GS} - V_T)^2 \tag{9.14}$$

For a given E-MOSFET, specifications generally state a value of I_D (called $I_{D(ON)}$) with a stipulated value of gate voltage. These specifications allow one to calculate k for the MOSFET, which can be used to evaluate I_D at other values of V_{GS}. The constant k has dimensions of milliamperes per square volt, with I_D in milliamperes. The most recent E-MOSFET development has been the creation of high-power units suitable for use as power amplifiers with dissipation values approaching 100 W.

While the biasing arrangement utilized with an E-MOSFET may often appear to be identical to that employed with other transistors—that is, a combination of self-bias (via R_S) and fixed-bias (via R_1 and R_2; see Figure

9.39)—the design is quite different. E-MOSFETs require forward bias; hence the divider voltage must exceed that developed across the source resistance by at least the threshold value (V_T). The only function of R_S is to stabilize the operating point through negative feedback; in prior cases it served both as a bias source and a stabilizing element.

FOCUS ON PROBLEM SOLVING: FET Bias Determination

Self-biasing is accomplished through the use of a source resistor, $R_S = V_{GS}/I_D$, where V_{GS} is the desired gate-to-source bias and I_D is the desired drain current (Figure 9.32). The gate-to-ground resistance (R_G) has little bearing on the bias value since it is essentially devoid of any dc current. The value of this resistor is primarily determined by the loading to be presented to the signal source. To prevent a loss of gain due to negative feedback, R_S must be bypassed by a capacitor C_S.

Self-biasing, which is limited to providing reverse bias only, is sufficient for JFET and depletion MOSFET operation. Enhancement MOSFETs require forward bias, which is generally obtained from V_{DD} by means of either

a. A fixed resistor between the drain and gate (Figure 9.37),
b. The fixed bias from a voltage divider between the drain and ground (Figure 9.38),
c. A combination of fixed divider bias and self-bias (Figure 9.39).

For case a, since there is no dc current through R_G, V_{GS} equals V_{DS}, which in turn is determined by the value of R_D. V_{DS} must exceed the E-MOSFET threshold V_T. The value of R_G must be large enough to minimize loading down the output circuit.

For case b, the bias voltage (V_{GS}) is the open-circuit voltage of the divider (using V_{DS} as a dc source). Values of R_1 and R_2 must be large enough to minimize loading both the input and output circuits.

For case c, the bias voltage (V_{GS}) is the open-circuit voltage of the divider (using V_{DD} as a source) minus the dc voltage developed across R_S. The function of the latter is to stabilize the operating point. Obviously, forward bias from the divider must exceed the reverse bias from the source resistor. R_1 and R_2 are selected to minimize input loading.

FOCUS ON PROBLEM SOLVING: Source Bypass Capacitor

Between the source terminal and ground, one can visualize a Thevenin signal generator in series with a Thevenin resistance equal to $1/g_m$, where g_m is the FET transconductance. To minimize gain loss due to negative feedback, let $X_{C_S} \leq 0.1(1/g_m)$ at the lowest frequency of interest (f_L). Generally, this is the low-frequency 3 dB point. Then we have

$$C_S \geq \frac{1}{2\pi f_L(0.1/g_m)}$$

9.9 ▪ COMPLEMENTARY MOSFETS

It has previously been mentioned that large single-stage FET voltage gain may be obtained by using an active load in place of a resistor. Figure 9.25 shows the use of **complementary** enhancement-mode **MOSFETs** (CMOSs) in such an arrangement—that is, complementary since the lower one is an n-channel device and the upper one, a p-channel unit. The upper one is an active load for the lower one and vice versa. Also note the source-drain inversion between the two units. This arrangement is similar to connecting two diodes back to back with the consequent small flow of current. This small current flow is one of the attractive features of CMOS circuits: They draw very little power.

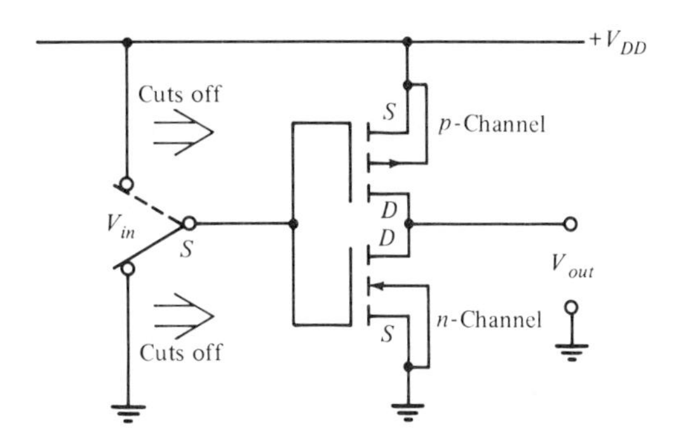

FIGURE 9.25. **CMOS Circuit** Positive input grounds the output; grounded input connects the output to $+V_{DD}$.

If the combined gates are driven in a positive direction, the n-channel unit connects the output to ground through a high-conductance path, but there is very little current flow through the series MOSFETs because the upper one has very low conductance. (We assume the output load draws negligible current.) When the gates are driven negative, the MOSFET roles reverse, and the output is connected to $+V_{DD}$, again resulting in little current flow. Thus, the output will alternately be connecting to $+V_{DD}$ or ground with significant current flow only during the transition periods.

CMOS circuits are primarily used as switches or, more accurately, in pulse circuitry where their low current drain makes them particularly attractive. They may also be used as high-gain amplifiers, and it is that function that is of prime interest to us in this section.

Refer to Figure 9.26. As the input rises from zero in a positive direction, the output voltage drops slowly at first and then precipitously and finally slowly approaches zero. It is the region of the precipitous drop that represents very high gain for small signals. Such amplifiers do have some disadvantages, particularly poor linearity and unpredictable gain. But they are simple, inexpensive, and in integrated circuit form occupy very little space. We will consider one example of their use, as an amplifier in conjunction with a crystal oscillator used to generate the rather precise frequencies needed in electronic watches and clocks.

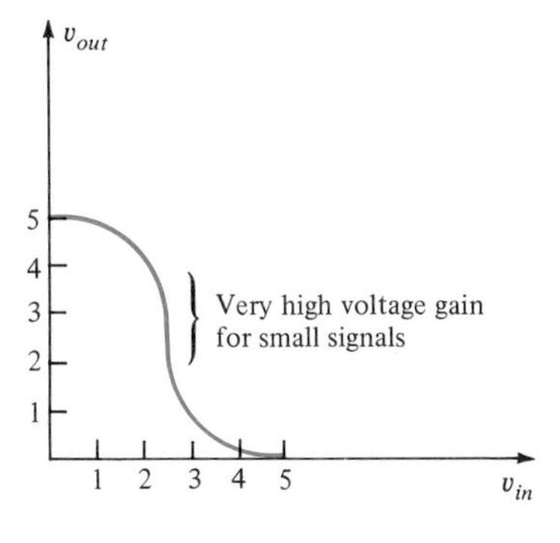

FIGURE 9.26. **V_{out} vs. V_{in} for CMOS Amplifier**

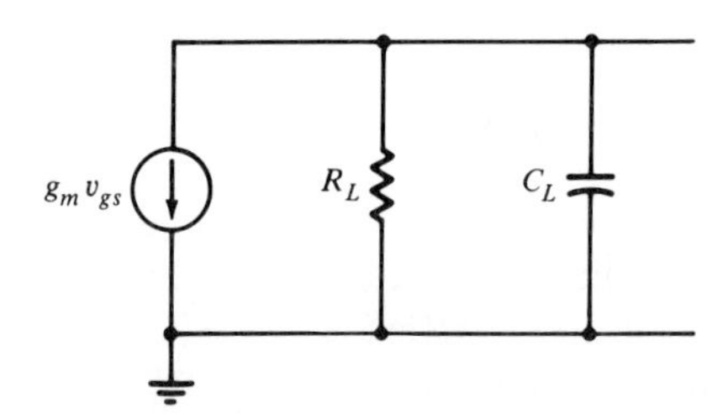

FIGURE 9.27. **Equivalent Circuit for CMOS Output**

We have previously mentioned crystal oscillators (Section 8.4.3). They are used where high-frequency accuracy is needed. In the present instance we consider their use in digital timers. A common frequency used in such applications is 32.768 kHz. This frequency is passed through divider circuits needed to provide the 1 s intervals; since $2^{15} = 32{,}768$, the division process is simplified by this selection.

As so often is the case in engineering practice, the selected frequency represents a compromise between opposing factors, in this case stability and power consumption. At high frequencies the gain of an amplifier diminishes because of the shunting capacitance. By representing the output of the CMOS amplifier as a current generator (Figure 9.27), in parallel with a load resistance (R_L) and the shunting capacitance (C_L), at frequencies approaching $f \simeq 1/C_L R_L$, the capacitance introduces a phase shift accompanied by a drop in output signal. [See the discussion of phase-magnitude (Bode) plots in "A Closer Look" at the end of Chapter 10.] Reducing the value of R_L allows one to extend the *uniform* gain to higher frequencies but at a sacrifice in the value of the gain. Compensation for the gain reduction, in turn, may be achieved by increasing the transconductance (g_m) of the transistor. For a BJT, $g_m = 1/r_e \simeq (I_C\text{ [mA]})/25$. Thus, an increase in transconductance can be brought about by an increase in collector current. For an FET, $g_m = 2\sqrt{kI_D}$ since $I_D = k(V_{GS} - V_T)^2$ and $g_m = \Delta I_D/\Delta V_{GS} = 2k(V_{GS} - V_T) = 2\sqrt{kI_D}$, and in the FET case also an increase in g_m is achieved by increasing the current. Thus, in the case of digital timers, higher operating frequencies mean higher power consumption.

The oscillator consists of a feedback circuit with attenuation factor β and an amplifier whose gain is $1/\beta$. The pi network of Figure 9.28 provides the 180° phase shift needed to compensate for the −180° due to the amplifier and thus leads to positive feedback. In the pi network the crystal serves as a high-Q resonant circuit with a Q of the order of 50,000. Slight adjustments of the oscillator frequency are made by varying one of the capacitors in the pi network.

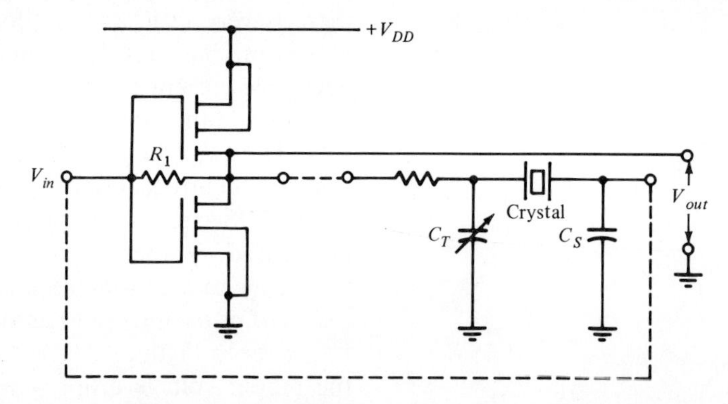

FIGURE 9.28. **CMOS Crystal Oscillator** The amplifier is at the left and the pi network, at the right.

The role of a high-Q resonant circuit in oscillator stability may be realized by observing the results depicted in Figure 9.29. Figure 9.29a shows the narrowing response curves as the Q increases. Figure 9.29b shows the corresponding variation in phase. (Remember that at the ± 0.707 points on the amplitude curve the phase angles are $\mp 45°$, and hence the narrower the curve, the more rapid the phase variation with frequency.) If at the operating frequency the amplifier exhibits a phase change due to a temperature variation, a power source voltage change, etc., to provide a compensating phase change, a low-Q resonant circuit would have to shift the operating frequency by a considerable amount. A high-Q circuit, on the other hand, would need to change the frequency only slightly to achieve the same degree of phase compensation. Thus, we see the influence of resonant Q on the oscillator stability.

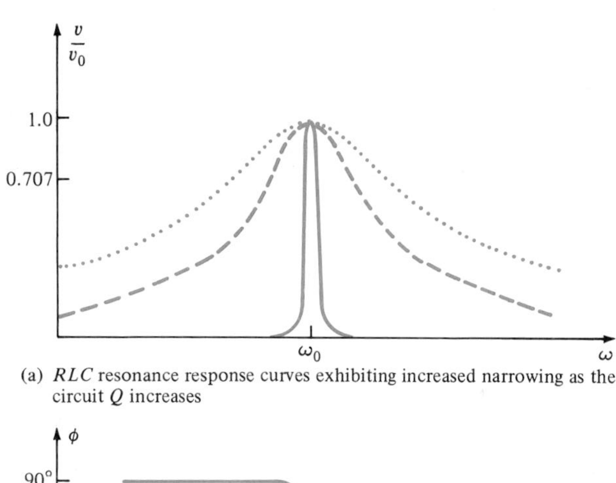

(a) RLC resonance response curves exhibiting increased narrowing as the circuit Q increases

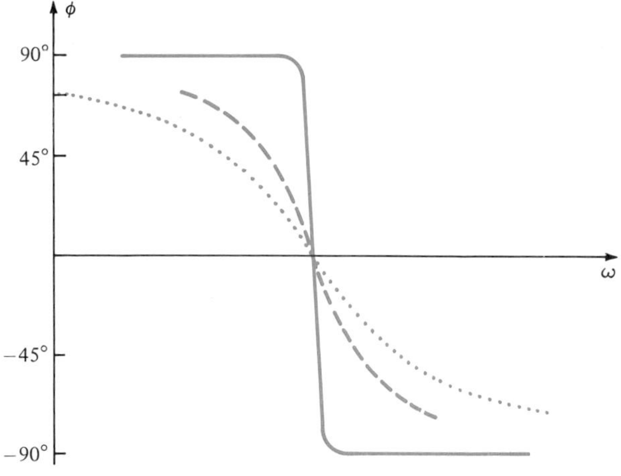

(b) Accompanying phase shifts as a function of frequency

FIGURE 9.29. ***RLC* Resonant Circuit**

The phase variation due to the crystal in Figure 9.28, however, is somewhat more complicated than that of the simple resonant circuits depicted in Figure 9.29. The low-pass filter RC_T leads to a phase change that can approach 90°. In practice it is somewhat less, perhaps 74°. At the operating frequency the crystal behaves as an inductor (see Problem 8.24). Potentially, in combination with C_S, it can lead to a total shift of 180°. In practice it is adjusted for about 105°. While the amplifier is said to exhibit a phase shift of −180°, some slight capacitive effects as it responds to the higher-frequency region might make it −181°. The summation of all three shifts provides the 0° phase shift needed for oscillation.

An interesting facet of operation arises in conjunction with the attenuation of the pi network. If the attenuation is small, the amplifier's input signal amplitude is large, and this large signal *decreases* the power consumption because with large voltage swings the current is low due to the resistance of both *n*- and *p*-channel transistors being high during most of the cycle. The nature of the pi network also helps in the realization of this low attenuation; it exhibits a high input impedance and a low output value, both favorable in this regard.

The physical dimensions of a crystal are inversely proportional to the operating frequency. Thus, the higher the operating frequency, the smaller the crystal size. (The dimensions of a 32 kHz crystal are about 0.6 × 0.2 × 0.11 in.) Additionally, the cost of a crystal decreases with frequency, up to about 1 MHz. The Q also tends to increase with increasing frequency, resulting in greater stability.

In summary, lower-frequency crystals mean increased size, but higher frequencies mean higher power consumption. Higher frequencies, however, mean greater stability.

The amplifier must be provided with dc feedback if it is to operate in the active region. This is provided by resistor R_1, leading to $V_{out} = V_{in} \simeq V_{DD}/2$ (Figure 9.28). This resistor has to be large enough to prevent loading down the feedback network yet low compared to the amplifier's input resistance. 15 MΩ is the value generally used.

The capacitors in the pi network play a major role in determining the power consumption as well. Their values are selected as a compromise between power consumption and stability.

Temperature-compensated crystal oscillators can deliver stabilities of 0.1 ppm over a temperature range of 0°–50°C. Crystals within a temperature-controlled oven can furnish stability of 1 part in 10^{11}. The Bureau of Standards maintains stabilities approaching 1 part in 10^{14} (using masers).

9.10 ■ NMOS, PMOS, AND CMOS

Rather than CMOS technology, large-scale integrated (LSI) circuits often employ MOSFETS of one polarity only, usually in the enhancement mode. This procedure simplifies the manufacturing process and also allows higher packing densities. Such transistors may be used to provide essentially all circuit functions, serving as resistors, diodes, and transistors.

The *n*-channel devices of this sort are called NMOS devices and those employing *p*-channels, PMOS devices. Fabrication of PMOS transistors was the first to be developed; however, the slower mobility of their charged carriers in contrast to those in NMOS transistors has led to the dominance of the latter and their popularity in fast pulse circuit operation.

In contrast to CMOS techniques, NMOS pulse circuits dissipate more power during switching operations. In very large circuit integration (VLSI), power dissipation may impose a limit on NMOS packing density.

PMOS transistors are made to be about twice the size of NMOS types in order to achieve equal driving currents and thus compensate for the smaller charge mobility in PMOS. But the larger sizes cause input capacitances of PMOS transistors to be greater than those of NMOS transistors, and when both are utilized in CMOS units, there results an inequality in the switching times between on and off. It is also the large PMOS size that contributes to the inferior packing density of CMOS circuitry, although some circuit advances have diminished the number of PMOS units now needed in CMOS circuits.

9.11 ■ SOME FET SPECIFICATIONS

A well-known FET is the Motorola MPF105, now designated as 2N5459. This is an *n*-channel JFET. The spread in its I_{DSS} value is 4–16 mA, with a pinch voltage spread of 2–8 V. The breakdown of gate-to-source voltage BV_{GSS} is 25 V. There is an equivalent *p*-channel unit, the 2N5462.

With regard to MOSFETs, 2N4351 is a popular enhancement unit, an *n*-channel device. While JFET and depletion units are rated by their I_{DSS} (at $V_{GS} = 0$), enhancement units are rated by specifying $I_{DS(ON)}$ at a specific V_{DS}. For the 2N4351 it is 3 mA at 10 V. Rather than a pinch voltage, one deals with a threshold voltage; in this case we again have a wide spread, 1.5–5 V. BV_{DS} is 25 V. There is no diode protection on this unit. Thc 2N4352 is a comparable *p*-channel unit.

Power MOSFETs are available with maximum drain current ratings as high as 28 A and a maximum V_{DS} of 100 V. This unit is the IRF150, manufactured by International Rectifier Co. and one of their HEXFET's line. Some lower-valued power units are the 2N6656-8 series with $I_{D(max)} = 2$ A and power dissipation of 15 W, with progressive BV_{DS} values of 35, 60, and 90 V. The 2N6659-61 is a comparable series rated at 5 W.

SUMMARY

9.1 Introduction

- FET tolerances have an even wider spread than those that characterize BJTs.

9.2 The Junction Field-Effect Transistor

- The junction field-effect transistor (JFET) is a voltage-amplifying device, characterized by a high input impedance of the order of tens of megohms.

9.3 JFET Characteristic Curves

- A linear relation between drain current and drain-to-source voltage of an FET constitutes its ohmic region. The region where the current is essentially independent of the drain voltage represents the saturation region. This contrasts with a BJT where the linear relation is called the saturation region, and the region where the collector current is essentially independent of the collector voltage is the active region.
- A change in output current as a result of an input voltage change constitutes the transconductance (g_m) of a device.
- The maximum voltage gain available from an FET (obtained with $R_D = \infty$) is termed the amplification factor (μ). In practice the relatively small value of

R_D needed for proper dc operation results in a voltage gain that is significantly smaller.

9.7 Voltage-Controlled Resistor

- A JFET operated in its linear (ohmic) region constitutes a voltage-controlled resistance (VCR).

9.8 The MOSFET

- A metal-oxide-semiconductor FET (MOSFET) has an insulated gate that leads to an input impedance of the order of millions of megohms.
- MOSFETs are available in either enhancement or depletion types.

KEY TERMS

amplification factor: change in output voltage in response to input voltage variation
common-source (CS) amplifier: equivalent to common-emitter BJT amplifier
current source: current essentially independent of load resistance
depletion mode: diminished channel width in response to gate voltage
depletion-mode FET: in practice an FET that operates with zero bias, combining depletion and enhancement modes
enhancement mode: conduction channel in an FET created by induced charges due to gate voltage
field-effect transistor (FET): a transistor whose conduction path utilizes either an *n*-type or *p*-type semiconductor exclusively
 complementary MOSFET (CMOS): circuitry combining *p*-channel and *n*-channel FETs
 junction field-effect transistor (JFET): an FET whose gate electrode is in direct contact with the semiconductor
 metal-oxide-semiconductor field-effect transistor (MOSFET): an FET with an insulator interposed between the gate electrode and the semiconductor
 NMOS and PMOS: MOSFETs utilizing only one polarity of semiconductor (either *n*-channel or *p*-channel, respectively) in fabricating an integrated circuit
ohmic region: linear relationship between gate voltage and drain current
pinch-off voltage: V_{GS} demarcation point between ohmic and saturation regions of FET operation
saturation region: drain current essentially independent of drain-source voltage
source follower: unity-gain amplifier employing an FET
substrate: the semiconductor chip on which a transistor is fabricated
threshold voltage: minimum forward bias needed for enhancement MOSFET conduction
transconductance: current change in the output in response to input voltage variation
transfer characteristics: relation between output and input parameters, for example, between drain current and gate-to-source voltage for an FET, or collector current and base-to-emitter voltage of a BJT, or output and input of a filter
voltage-controlled resistance (VCR): FET utilizing ohmic region

SUGGESTED READINGS

Colclaser, Roy C., D. A. Neamen, and C. F. Hawkins, ELECTRONIC CIRCUIT ANALYSIS, Chap. 5, pp. 114–137. New York: Wiley, 1984. *For more elaboration on FETs.*

Lancaster, Don, CMOS COOKBOOK. Indianapolis: Howard W. Sams Publ., 1977. *CMOS circuit details, mainly dealing with digital circuits.*

Horowitz, Paul, and Winfield Hill, THE ART OF ELECTRONICS. New York: Cambridge University Press, 1980. Pp. 230–231, *a listing of small-signal JFETs and MOSFETs, together with major parameter values,* and pp. 258–259, *a listing of some power MOSFETs and their major parameter values.*

Millman, Jacob, MICROELECTRONICS, pp. 689–691. New York: McGraw-Hill, 1979. *Power field-effect transistors; construction details.*

Eaton, S. S., "Timekeeping Revolution Through COS/MOS Technology," RCA COS/MOS TECHNOLOGY, pp. 33–41. 1973. *Further details on MOSFET crystal oscillators.*

Gnädinger, A. P., "Electronic Watches and Clocks," in L. Marton and C. Marton (eds.), ADVANCES IN ELECTRONICS AND ELECTRON PHYSICS, Vol. 51, pp. 183–263. New York: Academic Press, 1980. *Perhaps a more accessible reference for further details on MOSFET crystal oscillators.*

EXERCISES

1. What is the essential distinction between a BJT and an FET?

2. What is the limiting forward bias that may be applied to a silicon JFET?

3. What is the order of magnitude of the input impedance of an FET? Of a BJT?

4. What constitutes the saturation region of a JFET? How does it differ from the saturation region of a BJT?

5. What is the typical value for the transconductance of a JFET?

6. What region of the operating curves is used when a JFET is employed as a VCR?

7. Why is it necessary to employ small signals in conjunction with an FET amplifier?

8. How does a MOSFET differ from a JFET?

9. What are the two types of MOSFETs?

10. What are some of the advantages and disadvantages of MOSFETs?

11. What is the meaning of I_{DSS}? Why is it not applicable to enhancement MOSFETs? What is the corresponding specification for enhancement MOSFETs?

12. For FET I_D versus V_{DS} characteristic curves, what parameter does the slope of the curves in the saturation region measure?

13. With regard to their respective output resistances, how does a CS differ from a CE amplifier?

14. Why does the gate resistor not play any significant role in setting the bias of a JFET?

15. What value of ac signal applied to the gate of an automatic gain control reduces the distortion?

PROBLEMS

9.1
Show that the transconductance of a JFET or a D-MOSFET is related to I_{DS} by the equation

$$g_m = \frac{2}{|V_p|} \sqrt{I_{DSS} I_{DS}}$$

9.2
Determine the expression for the transconductance of an E-MOSFET.

9.3
a. Sketch a series of I_C versus V_{CE} curves for a BJT with varying I_B and label the active and saturation regions.

b. With varying V_{GS}, sketch a like series of curves for I_D versus V_{DS} for a JFET and label the ohmic and saturation regions.

9.4
Determine v_{DS} for the circuit in Figure 9.30 utilizing a 2N4869 n-channel JFET whose specifications include $I_{DSS} = 7.5$ mA and $V_p = -5$ V.

9.5
Determine v_{GS} for the circuit in Figure 9.31 for an n-channel JFET with the same specifications as in Problem 9.4.

9.6
Determine v_{GS} for the circuit shown in Figure 9.32 with transistor specifications as in Problem 9.5.

9.7
Using the circuit and parameters of Problem 9.5, upon increasing the value of R_D, what value would first place the operation in the ohmic range?

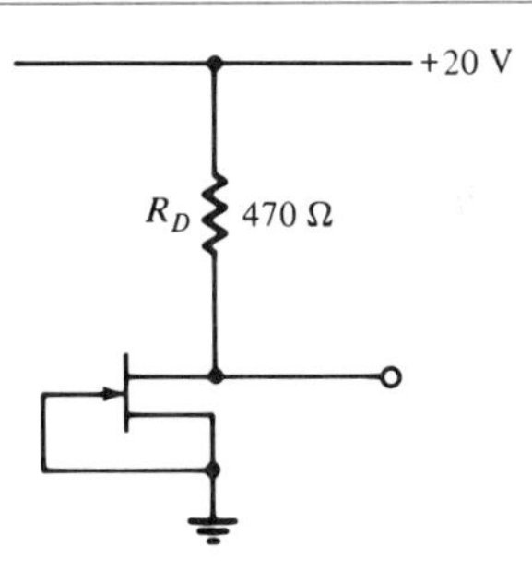

FIGURE 9.30

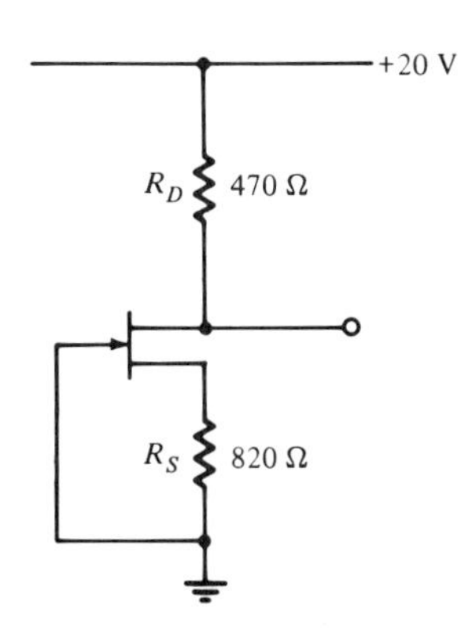

FIGURE 9.31

9.8
Using the circuit and parameters of Problem 9.6, upon increasing the value of R_D, what value would first place the operation in the ohmic range?

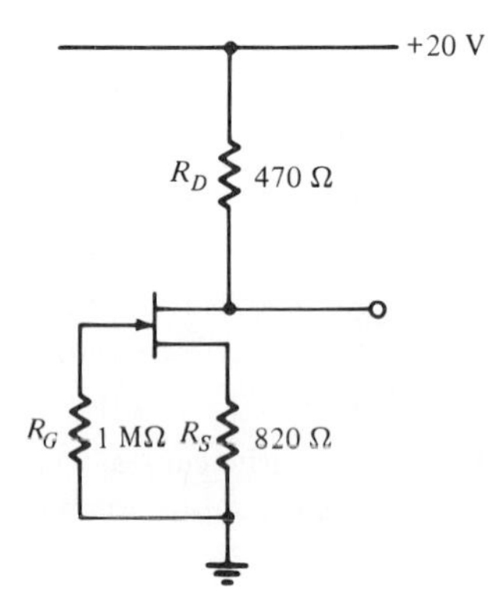

FIGURE 9.32

9.9
For the depletion NMOS shown in Figure 9.33, with I_{DSS} = 10 mA and V_p = −2 V, what value of R_D places the transistor at the boundary between the ohmic and saturation regions?

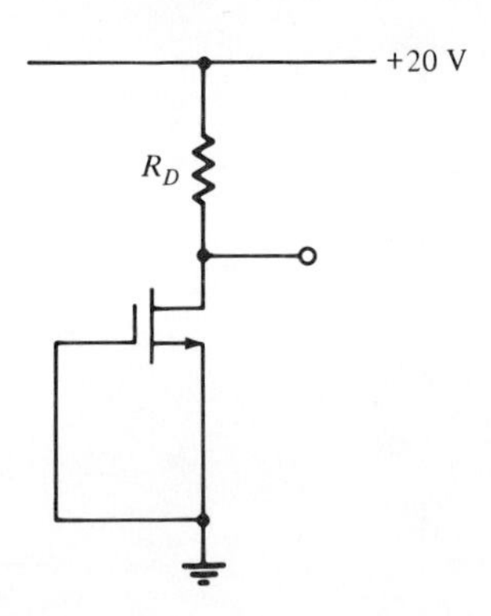

FIGURE 9.33

9.10
For the circuit in Figure 9.33 and the specifications of Problem 9.9, using a fixed bias of V_{GS} = +1 V, what value of R_D places the transistor at the boundary between the ohmic and saturation regions?

9.11
For the circuit in Figure 9.33 and the specifications of Problem 9.9, using a fixed bias of V_{GS} = −1 V, what value of R_D places the transistor at the boundary between the ohmic and saturation regions?

9.12
If, rather than a fixed bias, one decides to employ a source resistor in Problem 9.10, what can you say about the onset of ohmic operation as R_D is varied?

9.13
Consider Problem 9.11 if V_{GS} = −1 V is to be provided by a source resistance. What is the value of R_S? What is the value of R_D that places the transistor at the boundary between ohmic and saturated operation?

9.14
The operating Q-point for the JFET circuit in Figure 9.34 is specified as I_D = 4.0 mA, V_{DS} = 8 V, with V_{GS} = −2 V.

a. If V_p = −5.45 V and I_{DSS} = 10 mA, compute the necessary values for R_S and R_D.

b. What determines the value of R_G?

c. If the lowest frequency to be amplified is 20 Hz and R_G = 5 MΩ, what should be the value of C_B?

d. What should be the value of C_S?

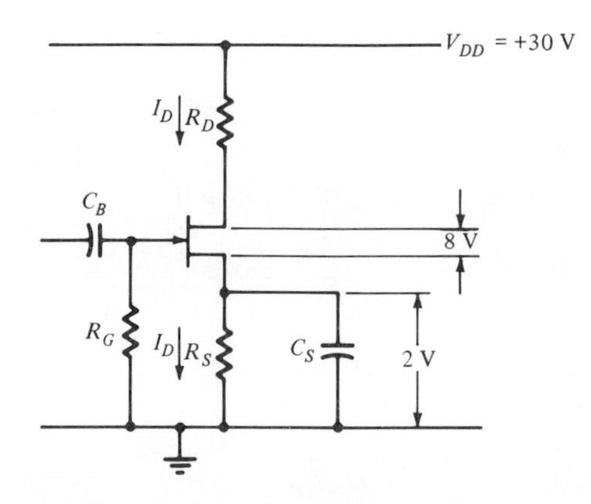

FIGURE 9.34

9.15
The circuit in Figure 9.35 has I_{DSS} = 5 mA and V_p = −4.0 V.

a. If V_{in} = 0, find V_{out}.

b. If V_{in} = 10 V, find V_{out}.

c. If V_{out} = 0, find V_{in}.

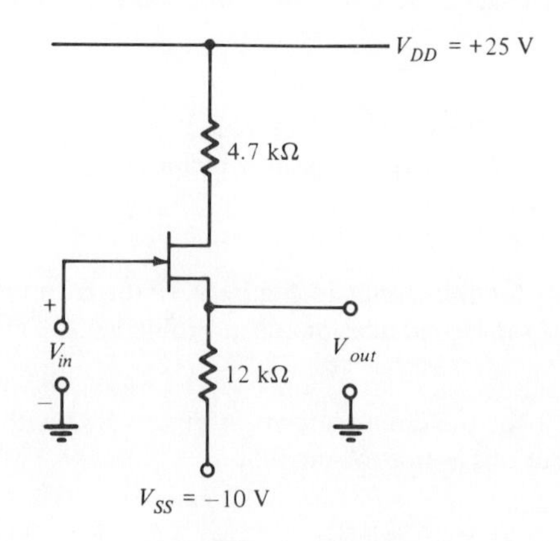

FIGURE 9.35

9.16
The gate leakage current of a JFET (Figure 9.36) is of the order of nanoamperes. Estimate the percent alteration in the bias voltage that arises due to the leakage current passing through R_G.

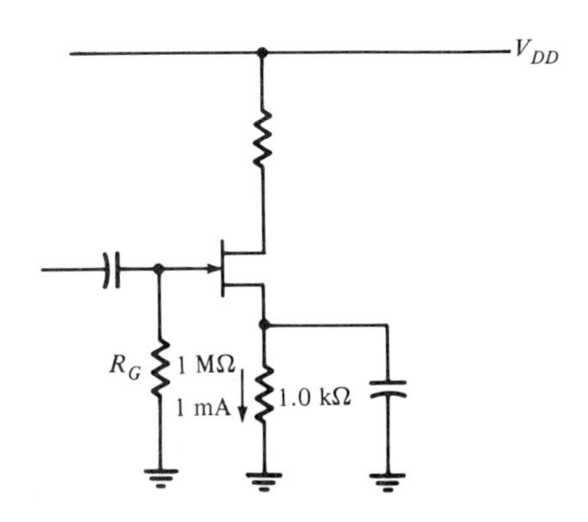

FIGURE 9.36

9.17
Since an enhancement MOSFET must be forward-biased—that is, the polarity of the gate must be the same as that on the drain—one cannot resort to the exclusive use of a source resistor alone for bias purposes. One uses either (1) a gate-to-drain connection via a resistor (Figure 9.37), (2) a voltage divider connected to the drain (Figure 9.38), or (3) a voltage divider in conjunction with a source resistor (this combination leads to the greatest stability) (Figure 9.39).

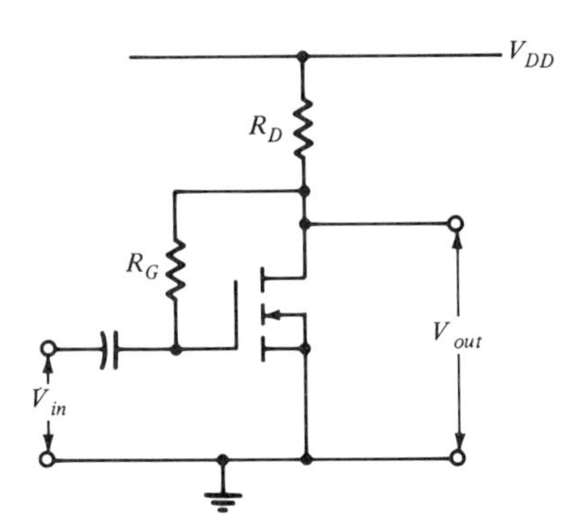

FIGURE 9.37

a. For a given n-channel E-MOSFET, the manufacturer specifies $V_T = 4$ V and $I_{DS} = 7.2$ mA at $V_{GS} = 10$ V. Taking $V_{DD} = 24$ V and $R_G = 100$ MΩ, what value of R_D is necessary to achieve an operating Q-point of $V_{DS} = 8$ V?

b. If either for the sake of linearity or for the sake of obtaining the maximum output signal it is desired to have $V_{GS} \neq V_{DS}$, one may use the circuit of Figure 9.38. Calculate I_D, V_{GS}, and V_{DS}. (You will note there are two solutions; choose the one that is physically real.)

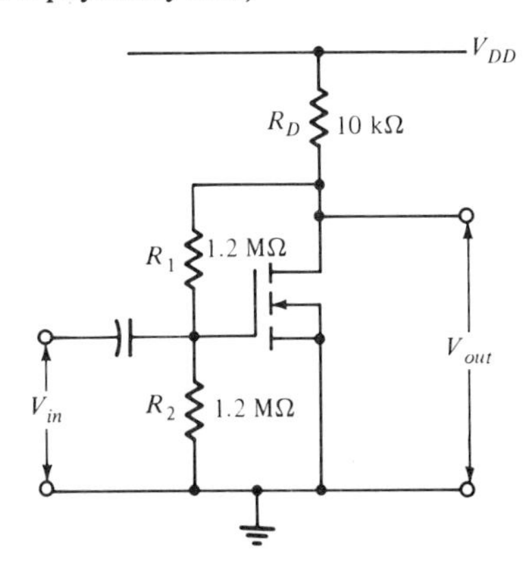

FIGURE 9.38

c. What must be the value of R_S if the circuit in Figure 9.39 is to have the same Q-point as in part b? What will be the necessary value for C_S for a 10 Hz lower band limit, taking $g_m = 2\sqrt{kI_D}$?

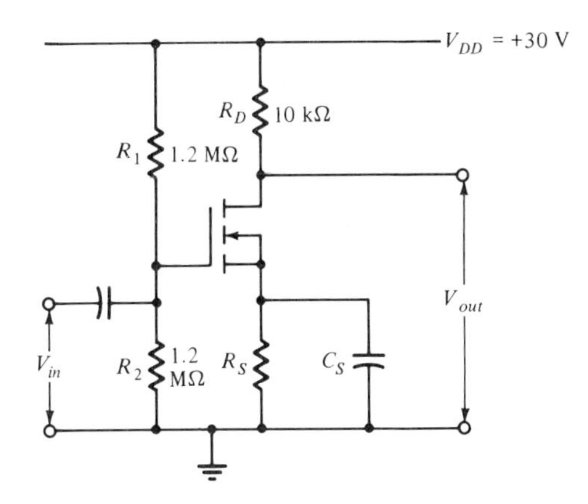

FIGURE 9.39

d. Noting where R_1 is connected in the circuit in part c in contrast to that in part b, can you surmise what the main effect would be in the operation of the two circuits?

9.18
See Figure 9.40. A 3N163 enhancement p-channel MOSFET has $I_D = 30$ mA with $V_{GS} = -10$ V and $V_T = -5$ V. What value of resistance places the operation on the threshold between the ohmic and saturation regions?

9.19
The forward bias for an enhancement MOSFET may be obtained by connecting the gate to the drain through a large resistor (Figure 9.37). Since there is no voltage drop across R_G,

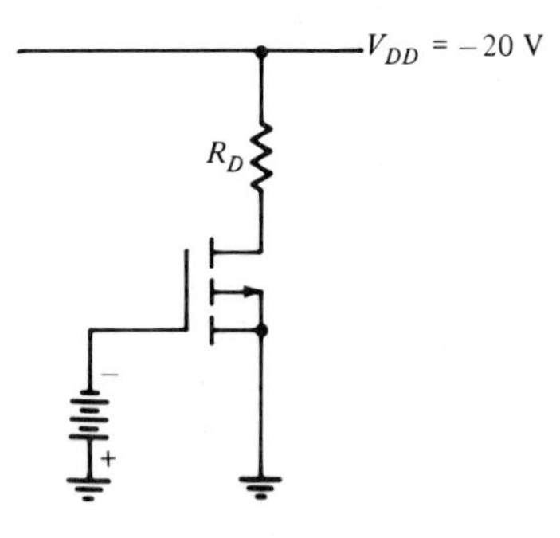

FIGURE 9.40

$v_{GS} = v_{DS}$. If $k = 1.2 \times 10^{-3}$ A, $R_G = 1$ MΩ, $R_D = 470$ Ω, and $V_T = 5$ V, find v_{GS} and i_D.

9.20

Utilizing the circuit in Figure 9.38 with $V_{DD} = 30$ V and $k = 0.3$ mA/V^2, compute the drain currents for values of $R_D = 1$, 2, 5, 10, 15, and 20 kΩ. Plot the results. What value of R_D would you select for optimum output voltage swing?

9.21

Sketch I_D versus V_{DS} using Eq. 9.3 (for a JFET and a D-MOSFET) and Eq. 9.14 (for an E-MOSFET). Indicate why one of each of the solutions is physically unrealistic.

9.22

a. Using Eq. 9.3, with $V_p = -2$ V and $I_{DSS} = 1$ mA, draw the JFET transfer curve representing I_D as a function of V_{GS}.

b. The desired Q-point is $I_D = 0.35$ mA, to be provided by a source resistor R_S. On the plot from part a, draw the load line from the coordinate origin through the desired Q-point, and from the resultant slope determine the value of R_S.

c. Check your result analytically.

d. Through the same Q-point, draw a load line corresponding to a source resistance of 4 kΩ. The x-axis intercept constitutes the additional fixed bias that must be provided by a voltage divider between V_{DD} and ground. What is the value of this fixed bias?

9.23

a. A 2N4869 n-channel JFET has maximum and minimum values of I_{DSS} given as 2.5 and 7.5 mA, respectively, with corresponding pinch voltages of -1.8 and -5 V. Using Eq. 9.3, draw the two transfer curves showing I_D versus V_{GS}.

b. It is desired to bias the circuit so that $I_{D(min)} = 1.0$ mA and $I_{D(max)} = 2.75$ mA. If the bias value to be provided by a voltage divider (V_{GG}) is 4 V, what is the range of values of R_S that will satisfy the collector current requirement. (*Hint*: Insert the extreme values of the Q-points, and with an x-intercept at $+4$ V, determine the slope of the load lines that satisfy the requirement.) What is the minimum value of R_S that can be used and still satisfy the specifications? What would then be the value of V_{GG}?

9.24

a. In terms of signal current and voltage, the drain current of a depletion MOSFET may be written as

$$i_D = I_{DSS}\left(1 - \frac{v_{GS}}{V_p}\right)^2$$

If $I_{DSS} = 2$ mA and $V_p = -4$ V, with an applied signal

$$v_{GS} = V_m \sin \omega t$$

identify the various terms appearing in the output voltage present across the resistor $R_L = 2$ kΩ. What is the signal gain?

b. In part a the dc gate bias was assumed to be zero. Compute the magnitude of the signal gain and the distortion with the dc gate bias equal to -1 V and equal to $+1$ V.

9.25

For enhancement power MOSFETs not only may the drain currents be large (ca. amperes), but also the characteristic curves in the saturation region are much flatter than those for low-power enhancement MOSFETs.

a. What is the parameter affected by the latter observation, and what would you conclude concerning the comparison between the two types of MOSFETs.

b. The spacing between the V_{GS} curves for these power transistors also shows greater uniformity than is the case with the low-power units. What parameter is affected, and what do you conclude concerning the comparison?

9.26

What would be the magnitude of distortion experienced in conjunction with the signal applied to the amplifier characteristic curves depicted in Figure 9.4?

9.27

a. The output resistance of either an emitter follower or a source follower is $1/g_m$. The respective equivalent circuits are shown in Figure 9.41. With g_m of a JFET $= 1$ mS and with that of a BJT equal to 40 mS, what are the respective voltage gains with $R_L = 1$ kΩ? What conclusion can be drawn from these results?

b. One way to remedy this difficulty is to utilize a BJT as an active source resistance as in Figure 9.42. (Alternatively, one could use another JFET in place of a BJT.) If I_D is equal to 1 mA, what should be the value of R_B? What will be the approximate output resistance?

9.28

a. To establish a channel within an E-MOSFET, we must have $v_{GS} \geq V_T$. Because of symmetry, for pinch-off $v_{GD} \leq V_T$.

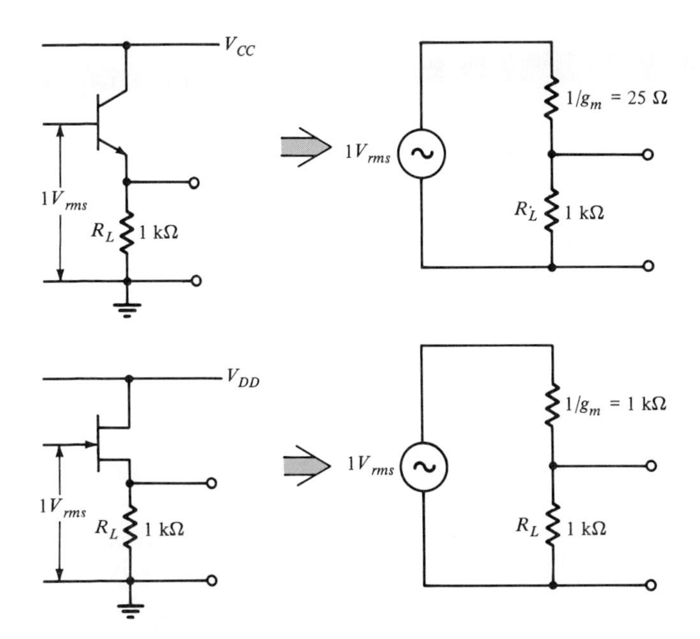

FIGURE 9.41

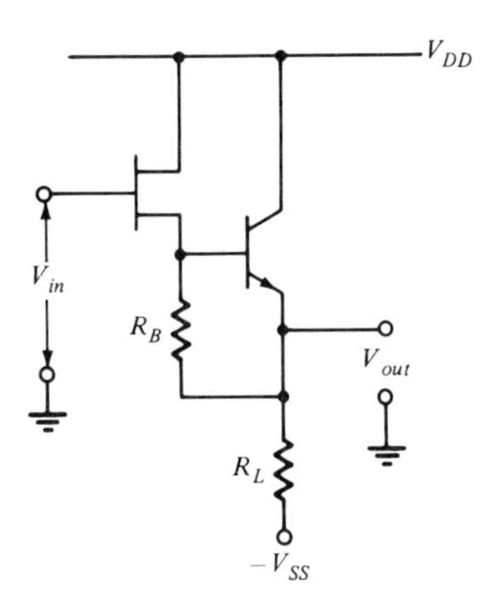

FIGURE 9.42

Since $v_{DS} = v_{DG} + v_{GS} = -v_{GD} + v_{GS}$,

$$v_{GD} = v_{GS} - v_{DS}$$

and pinch-off arises when $v_{GS} - v_{DS} \leq V_T$. This inequality implies that the gate-to-source voltage must not exceed the drain-to-source voltage by more than the threshold voltage. What can you surmise concerning the operation of the connection in Figure 9.43?

b. If you are unable to answer part a, do the following: For an E-MOSFET with $V_T = 2$ V and $k = 1.875 \times 10^{-4}$ A/V^2, we wish to ascertain the boundary between the ohmic and saturated regions. This boundary is the plot of the pinch-off points and for i_D versus v_{DS} is represented by $v_{DS} = v_{GS} - V_T$. Draw the curve, having computed i_D for v_{DS} = 2, 6, and 10 V. Also plot i_D versus v_{DS} when $v_{DS} = v_{GS}$ (which represents the circuit in Figure 9.43). *Now* can you answer the question of part a?

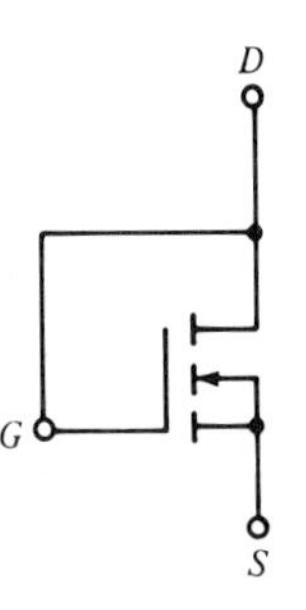

FIGURE 9.43

9.29

a. The *n*-channel E-MOSFET of Problem 9.27 is to be used as an active load for a like E-MOSFET driver (Figure 9.44). We wish to determine the *load curve* for Q_2. We have

$$v_{DS2} = V_{DD} - v_{DS1} = 12 - v_{DS1}$$

The drain currents must be equal—that is, $v_{DS1} = v_{DS2}$. Plot i_D versus v_{DS2} to obtain the load curve.

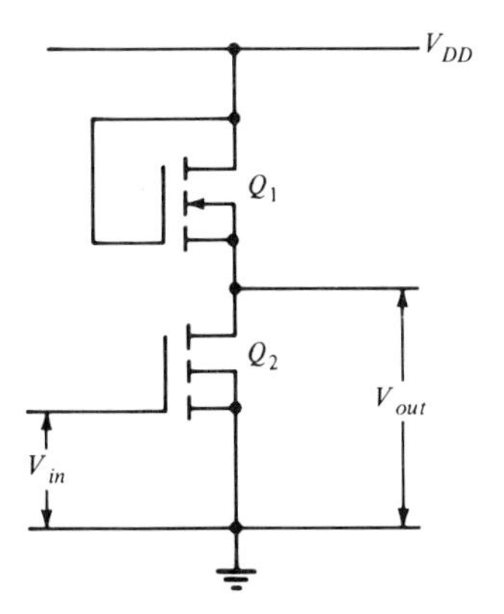

FIGURE 9.44

b. Plot the output voltage—that is, v_{DS2}—versus the input voltage (v_{GS2}). (Suggested solutions would be for currents of 0, 3×10^{-3}, and 12×10^{-3} A.) From the result, justify the characterization of this circuit as that of an *inverter*.

10 OPERATIONAL AMPLIFIERS

OVERVIEW

A typical three-stage amplifier using discrete components—transistors and resistors—connected together by individual wires in what is called a hard-wired circuit might typically contain 7 transistors and 11 resistors. A single transistor might carry a price tag of about $1, certainly a very modest amount. Yet for that single dollar one can purchase a complete three-stage amplifier containing 20 transistors and 11 resistors, already wired! In the latter case all the connections, resistors, and transistors are fabricated simultaneously on a small sliver of silicon $\frac{1}{16} \times \frac{1}{16}$ in. and constitute an integrated circuit, often simply called an IC amplifier.

In addition to linear ICs, devoted to the amplification of analog signals, there are ICs designed to process digital signals. Such units may contain in excess of 1000 components (representing LSI, large-scale integration) and even more than 10,000 (representing VLSI, very large-scale integration).

We will consider a few simple digital ICs in Chapter 11 and illustrate their use in the form of a counting circuit employing digital readout. In Chapter 14 we will encounter ICs that can be made to form a complete computer, in Chapter 15 we will undertake consideration of ICs designed for automotive engine control, in Chapter 16 we will consider ICs used as voltage regulators, and in Chapter 17 we will find some special versions useful in measurement technology.

OUTLINE

10.1 ■ INTRODUCTION

Just as the transistor replaced the vacuum tube, the **integrated circuit** (IC) is rapidly replacing the transistor. The linear IC consists of a multiplicity of permanently connected transistor amplifier stages[1] that, to a large degree, may literally be looked upon as the proverbial black box to which one applies a signal and (usually) obtains an amplified version of it at the output. To make the most intelligent use of such devices, the details of what lies within is necessary, but we can go a long way without pausing to consider such details, and that is how we will proceed.

10.2 ■ BASIC PRINCIPLES

To start, it should be pointed out that one usually measures voltages relative to ground.[2] In the case of many ICs that constitute what are known as **operational amplifiers** (OPAMPs), there are two input signal ports and one exit signal port, schematically represented as in Figure 10.1. There are also two terminals in which the dc power source is connected; they are often omitted in the schematics, it being understood that a dc power source is needed to activate the amplifier. Typically, one power terminal night be labeled +15 V (V_{CC}) and the other, −15 V (V_{EE}). In Figure 10.1 are shown two batteries that provide this power. In practice they may be dc power supplies, including rectifiers and filters, and may be referred to as dual supplies.

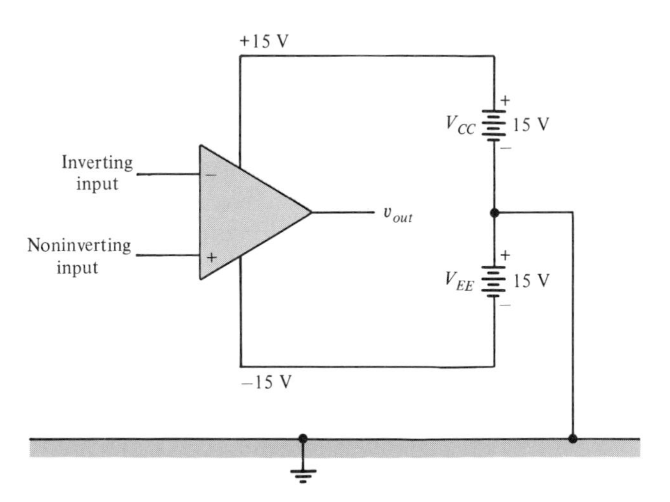

FIGURE 10.1. **Operational Amplifier Symbol with Inverting (−) and Noninverting (+) Inputs and v_{out} Outputs** Dual supplies, V_{CC} and V_{EE}, are generally employed.

[1]For example, the Fairchild μA741A contains 20 transistors, 11 resistors, and 1 capacitor, all in a single unit not much larger than a transistor of yesteryear. Transistors occupy the least space, followed by resistors and capacitors, in that order. Thus, few capacitors appear in IC circuits. Inductors are difficult to fabricate and, when required, appear outside the IC.

[2]Perhaps the term *common terminal* might represent a better description. Often OPAMP grounds are isolated from the ground of other circuits with which they might be used (such as an oscilloscope).

The minus (−) input is called the **inverting input**, and any voltage applied to this terminal will cause the output voltage to move toward the opposite polarity. The **noninverting input** is the plus (+) input terminal, and any voltage applied to this terminal will cause the output voltage to move toward the same polarity from a (presumed) initial zero value.

In the circuit on the left in Figure 10.2a, both terminals have a positive voltage (relative to ground) applied to them. If $V_1 > V_2$, the inverting terminal dominates, and the output will be negative. If $V_2 > V_1$, the noninverting terminal dominates, and the output will be positive.

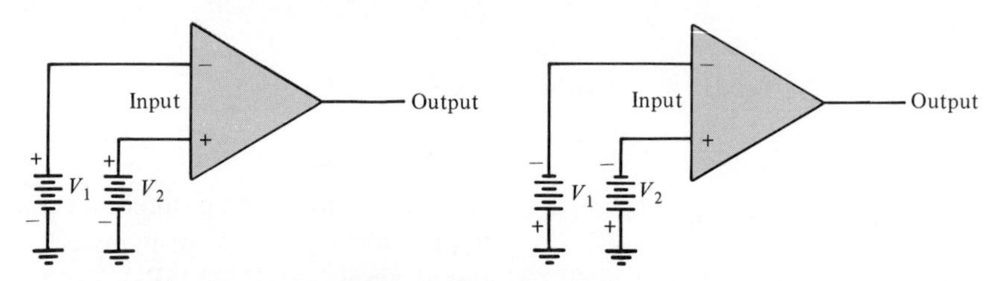

(a) Output polarity of OPAMP determined by dominant voltage and whether it appears at the inverting or noninverting terminal

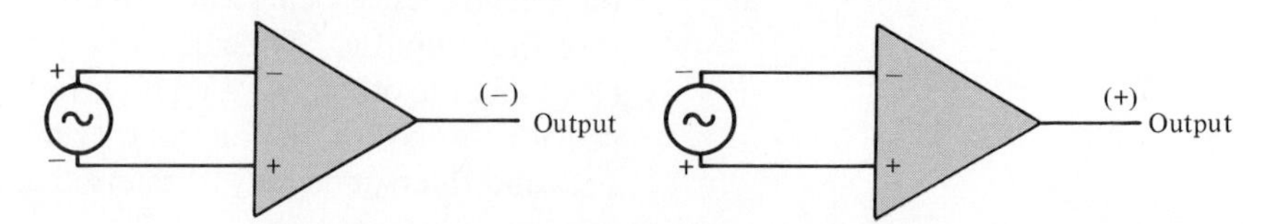

(b) Applied ac between input terminals leads to amplified ac at the output but with a 180° phase reversal

FIGURE 10.2. **Input-Output Phase Relationship**

Consider the circuit on the right in Figure 10.2a, in which each input terminal has a negative voltage (relative to ground) applied to it. If $V_1 > V_2$, the inverting terminal again dominates, but now the output voltage will be positive, and if $V_2 > V_1$, the noninverting terminal dominates, causing the output to be negative.

What is the degree to which the output voltage can move in either the positive or negative direction? This voltage variation is determined by the values of the dc voltage used to power the amplifier, $+V_{CC}$ and $-V_{EE}$. Typically, those voltages would be ±15 V.

With regard to the input, what is the allowable maximum difference $(V_2 - V_1)$? Since operational amplifiers have a very high voltage gain, typically 200,000, and the output is typically limited to perhaps a 30 V peak-to-peak value [+15 − (−15) = 30 V], the maximum input signal should be

$$\frac{30\text{ V}}{2 \times 10^5} = 15 \times 10^{-5} = 150\ \mu\text{V} = 0.00015\text{ V (peak to peak)}$$
$$= 53\ \mu\text{V} = 0.000053\text{ V (rms)}$$

What would happen if an ac voltage were applied between the two input terminals? As shown in Figure 10.2b,with the instantaneous polarities at the two inputs, the output voltage would represent an amplified version of the applied ac signal but with a 180° phase reversal centered on zero (Figure 10.3). However, if distortion is to be minimized, the maximum output voltage should be a few volts less than the dc power source voltages, although some types of ICs allow swings all the way to both dc voltages (or to the **dc rails**, as they are called). Thus, with ±15 V being used, the output voltage might be limited to ±13 V—that is, 26 V peak to peak, demanding an input voltage whose maximum should not exceed 130 μV peak to peak or 46 μV rms.

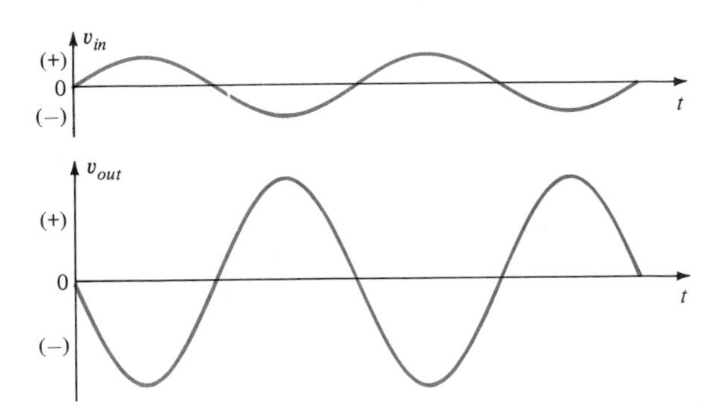

FIGURE 10.3. **Phase Reversal Between Output and Inverting Input of OPAMP.** With symmetric dual dc supplies, the output is centered on 0 V.

If a single power source is used to energize the amplifier (30 V in our hypothetical case) and the negative side of the power supply is grounded, the output signal will be centered at the +15 V level (Figure 10.4). This is generally of little consequence, for if the signal is passed on to additional stages for further amplification, it may be passed through capacitors, and the dc level may be altered as desired.

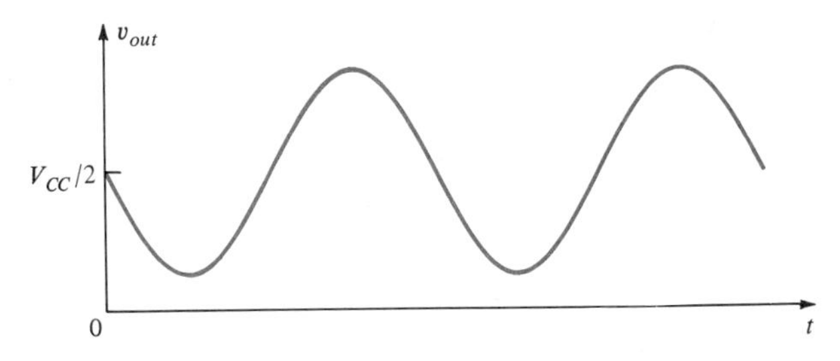

FIGURE 10.4. **OPAMP Utilizing a Single dc Power Supply** The output signal is centered on $V_{CC}/2$.

In many instances one deals with input signals that are much larger than 46 μV. In such cases the signal cannot be applied directly to the two inputs in the manner shown in Figures 10.2a and b.

A plot of the bandwidth characteristics of such a straight-through amplifier will also show another rather serious limitation (Figure 10.5). Such amplifiers have a bandwidth of about 5 Hz!

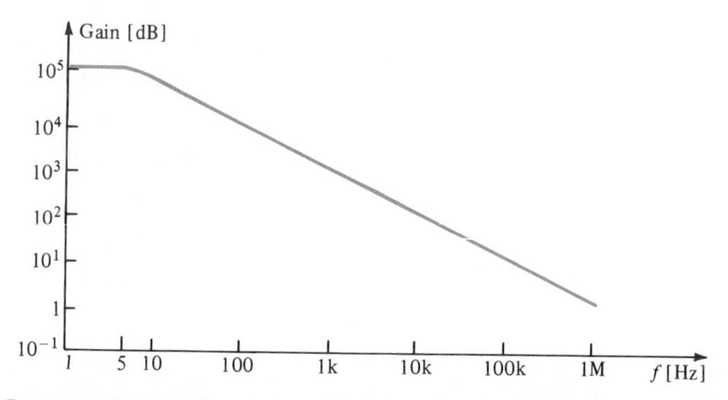

FIGURE 10.5. Open-Loop Gain vs. Frequency of a Typical OPAMP There is no drop-off at low frequencies because direct coupling is used—that is, no blocking capacitors. Drop-off at high frequencies occurs because of the effects of shunting capacitances.

The latter observation should give us a clue as to what must be done. To provide sufficient bandwidth, we must utilize negative feedback. Feedback will also lower the gain significantly so that we may impress more reasonably sized input signals without overloading the amplifier.

One basic feedback arrangement is shown in Figure 10.6. To understand the feedback operations used in such amplifiers, it is imperative to understand the concept of *virtual equality.* The input impedance of these operational amplifiers is very large, and hence it can reasonably be assumed that the current drawn by the amplifier's input terminals is negligible. If no current passes through a resistance, there is no potential drop, and with no potential drop, whatever the potential is at one end of the resistance, that same potential exists at the other end—that is, there is no potential difference. In the circuit shown in Figure 10.6 the (+) terminal is grounded. This means that the (−) terminal is also essentially at ground potential, and the junction between the feedback resistor R_F and R_1 is also essentially at ground potential. We may refer to this junction as a **virtual ground.**

Note that we have said "essentially at ground potential." There must be some small signal voltage between the input terminals; otherwise, there would be nothing to be amplified. Call this input signal v_{in}. With $Z_{in} = \infty$, whatever current flows through R_1 must also be the current flowing through R_F.[3] Therefore, we have

[3]Students are often puzzled by the indicated current in Figure 10.6. If instantaneously v_1 is positive, a small positive potential appears at the inverting terminal and creates a large negative potential at the output terminal. Therefore, there is a large potential difference across $R_1 + R_F$, and the current flows as indicated. When the input signal reverses polarity, the input is negative, a large positive potential appears at the output, and i_1 and i_F reverse direction.

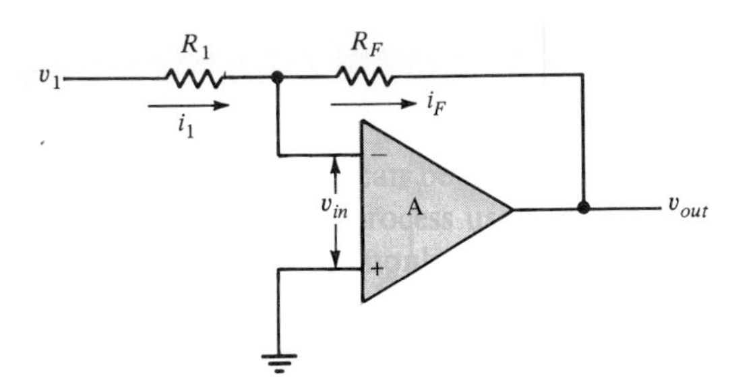

FIGURE 10.6. **OPAMP Inverting Input with Feedback Network**

$$\frac{v_1 - v_{in}}{R_1} = \frac{v_{in} - v_{out}}{R_F} \tag{10.1}$$

$$v_{out} = Av_{in} \tag{10.2}$$

$$\frac{v_1}{R_1} - \frac{v_{out}}{AR_1} + \frac{v_{out}}{R_F} - \frac{v_{out}}{R_F A} = 0 \tag{10.3}$$

$$\frac{v_1}{R_1} = v_{out}\left(\frac{1}{AR_1} + \frac{1}{AR_F} - \frac{1}{R_F}\right) \tag{10.4}$$

$$\left(\frac{R_F}{R_1}\right)v_1 = v_{out}\left(\frac{R_F}{AR_1} + \frac{1}{A} - 1\right) \tag{10.5}$$

$$\left(\frac{R_F}{R_1}\right)v_1 = v_{out}\left[\frac{1}{A}\left(\frac{R_F}{R_1} + 1\right) - 1\right] \tag{10.6}$$

$$v_{out} = -\frac{v_1(R_F/R_1)}{1 - (1/A)\,[1 + (R_F/R_1)]} \tag{10.7}$$

Since the open-loop gain is very large—that is, $A >> 1$, we have

$$v_{out} \simeq -\frac{R_F}{R_1}v_1 \tag{10.8}$$

The negative sign means that there is a 180° phase reversal between v_{out} and v_1. This result is to be expected since the signal is being fed into the (−) terminal, the inverting input.

In Figure 10.7 we apply the signal to the (+) terminal, the noninverting input. With essentially no current being drawn through the input impedance of the amplifier, the (−) and (+) terminals are again virtually at the same potential, v_1 in this case. R_1 and R_F act as a voltage divider for the output signal v_{out}. Therefore, the signal voltage across R_1 is $[R_1/(R_1 + R_F)]v_{out}$. But this signal is being applied to the inverting terminal. Therefore, it subtracts from the signal being applied to the (+) input. We then have

$$v_{in} = v_1 - \frac{R_1}{R_1 + R_F}v_{out} \tag{10.9}$$

$$v_{out} = Av_{in} \tag{10.10}$$

$$\frac{v_{out}}{A} = v_1 - \frac{R_1}{R_1 + R_F}v_{out} \tag{10.11}$$

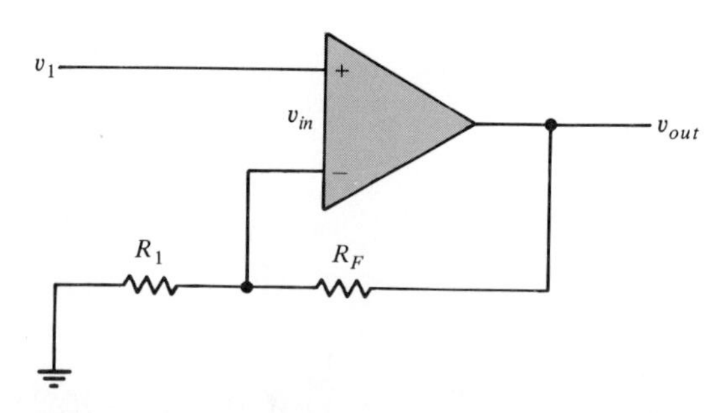

FIGURE 10.7 OPAMP Noninverting Input with Feedback Network

$$v_1 = v_{out}\left(\frac{1}{A} + \frac{R_1}{R_1 + R_F}\right) \tag{10.12}$$

Again, with $A \gg 1$, we have

$$v_1 \simeq v_{out}\left(\frac{R_1}{R_1 + R_F}\right) \tag{10.13}$$

$$v_{out} = \frac{R_1 + R_F}{R_1} v_1 \tag{10.14}$$

Note that here, as in the previous case, the gain is independent of the amplifier characteristics and depends only on the values of the stable resistors. In each instance the gain is adjustable through the values of R_1 and R_F that are employed.

EXAMPLE 10.1 Design an OPAMP with a

a. Voltage gain of $-100 \pm 1\%$
b. Voltage gain of $+100 \pm 1\%$

Solution: The feedback resistor must be large enough so that it does not load down the output, yet it should not be so large that the input bias current produces sizable offsets. With a 741, values between 2 and 100 kΩ are typical. Let $R_F = 49.9\text{ k}\Omega \pm 1\%$.

a. $$\frac{v_{out}}{v_1} = -\frac{R_F}{R_1} = -100$$

$$R_1 = \frac{R_F}{100} = \frac{49.9\text{ k}\Omega}{100} = 0.499\text{ k}\Omega = 499\ \Omega \pm 1\%$$

b. $$\frac{v_{out}}{v_1} = 1 + \frac{R_F}{R_1} = 100$$

$$\frac{R_F}{R_1} = 99$$

$$R_1 = \frac{49.9\text{ k}\Omega}{99} = 0.504\text{ k}\Omega = 504\ \Omega$$

The closest 1% standard value is 499 Ω, which should prove satisfactory.

FOCUS ON PROBLEM SOLVING: OPAMP Gain

For an inverting OPAMP the gain is

$$A = -\frac{R_F}{R_1}$$

For a noninverting OPAMP the gain is

$$A = \frac{R_1 + R_F}{R_1}$$

In what way is the bandwidth of these amplifiers altered when used with negative feedback? An important parameter in amplifier design is the gain-bandwidth product. Typically, for OPAMPS this product might be 1 MHz. Thus, if we have adjusted R_1 and R_F to provide a gain of 10, the bandwidth will be 10^5 Hz. If the gain is increased to 100, the bandwidth will be reduced to 10^4 Hz.[4]

There is an interesting variant of the noninverting circuit. If $R_F = 0$ and $R_1 = \infty$, we obtain the circuit shown in Figure 10.8. From Eq. 10.14 we note that the gain will be unity. This circuit is another example of a voltage follower. We encountered this in single-stage form in Section 8.2.5; now we see how easy it is to implement this performance using ICs.

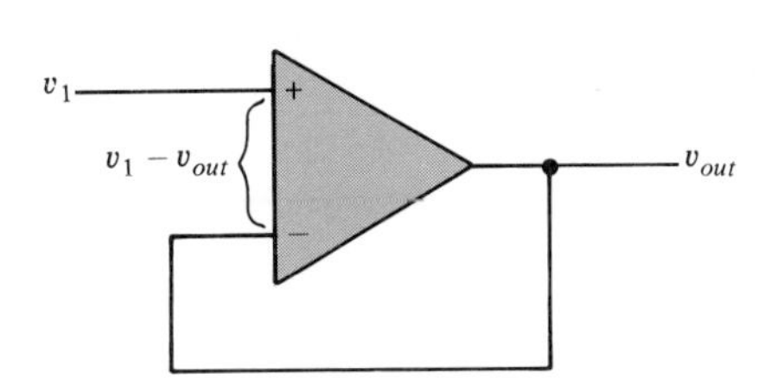

FIGURE 10.8. **OPAMP Unity-Gain (Voltage Follower) Amplifier**

We may draw the equivalent circuit for this amplifier as in Figure 10.9. Between the output terminals is an equivalent Thevenin generator in series with the output impedance of the voltage follower (R_{out}). Recognizing that $v_+ = v_1$ and $v_- = v_{out}$, this may in turn be redrawn as in Figure 10.10. We then have

$$v_1 - R_{in}i_1 - R_{out}(i_1 - i_2) - A(v_1 - v_{out}) = 0 \tag{10.15}$$

$$v_1 - R_{in}i_1 - R_L i_2 = 0 \tag{10.16}$$

$$v_{out} = R_L i_2 \tag{10.17}$$

Since $i_2 = (v_1 - R_{in}i_1)/R_L$, substitution into Eq. 10.15 may be used to determine the input impedance with feedback. We have

$$v_1 - R_{in}i_1 - R_{out}i_1 + R_{out}\frac{v_1 - R_{in}i_1}{R_L} - Av_1 + AR_L\left(\frac{v_1 - R_{in}i_1}{R_L}\right) = 0$$

$$v_1\left(1 + \frac{R_{out}}{R_L}\right) = i_1\left(R_{in} + R_{out} + \frac{R_{out}R_{in}}{R_L} + AR_{in}\right)$$

[4] These amplifiers are designed with potentials such that a direct connection is made between stages. Thus, in the absence of blocking capacitors, the bandwidth extends down to dc (hence their name, dc amplifiers). The stated bandwidth truly represents the upper 3 dB point.

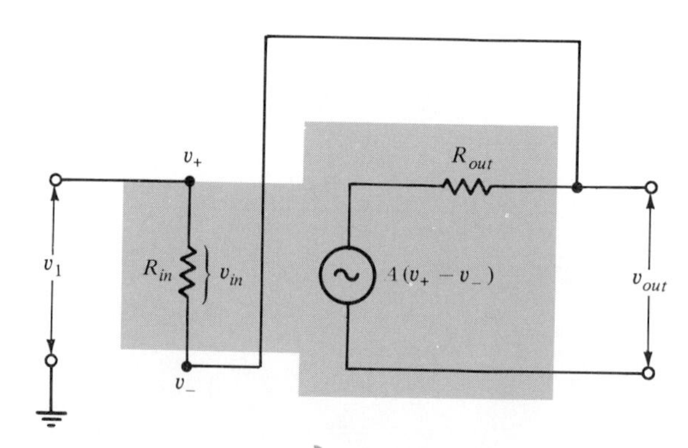

FIGURE 10.9. **Equivalent Circuit for Unity-Gain Amplifier**

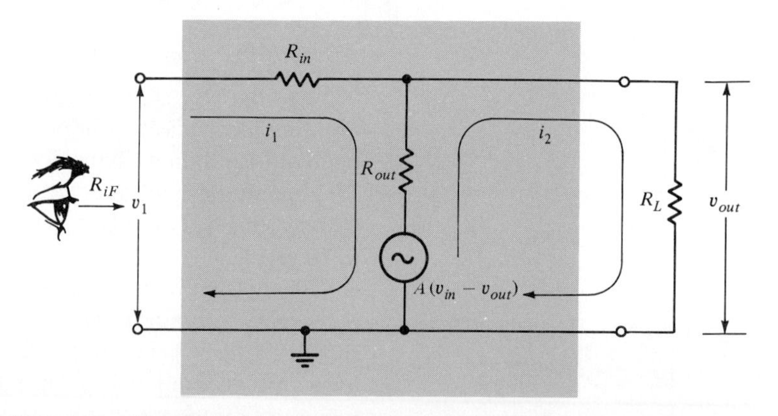

FIGURE 10.10. **Equivalent Circuit for Unity-Gain Amplifier with Load R_L**

Designating the input impedance with feedback as R_{iF} and recognizing that A is very large and that $R_L \gg R_{out}$, we have

$$R_{iF} = \frac{v_1}{i_1} \simeq \left(1 + \frac{R_{out}}{R_L} + A\right)R_{in} \simeq AR_{in} \tag{10.18}$$

Since A is very large (about 10^5), the amplifier's input impedance (R_{in}), which is already very large (about 10^5 Ω), is now made even larger ($\sim 10^{10}$ Ω). This result means that the loading effect of such an amplifier on the preceding circuit is very small. Since its output impedance is simultaneously made very small ($\sim R_{out}/A$), to minimize loading effects, such voltage followers (or unity-gain amplifiers) are used as buffers between amplifiers.

EXAMPLE 10.2

In the case of the voltage follower, the signal applied to the noninverting input is used in conjunction with 100% negative feedback, and the feedback factor (β) is equal to 1 (Figure 10.8). What is the feedback factor (β) for unity gain if the signal is applied to the inverting input?

Solution:

$$A = -\frac{R_F}{R_1} \equiv -1$$

$$R_F = R_1$$

$$\beta = \frac{R_1}{R_1 + R_F} = \frac{R_1}{2R_1} = \frac{1}{2}$$

$$\beta = \frac{1}{2}$$

10.3 ■ DIFFERENTIAL AMPLIFIERS

High-input impedance amplifiers are desirable in terms of the minimal loading that they impose on the driving circuit, but such high input impedance increases an amplifier's susceptibility to power line interference. Between the amplifier's input terminals and nearby power lines, there exists a small capacitance through which some of the 60 Hz interference is introduced into the amplifier (Figure 10.11). This capacitance and the input impedance of the amplifier act as a voltage divider, and so it can be seen that the larger R_{in}, the greater will be the degree of interference.

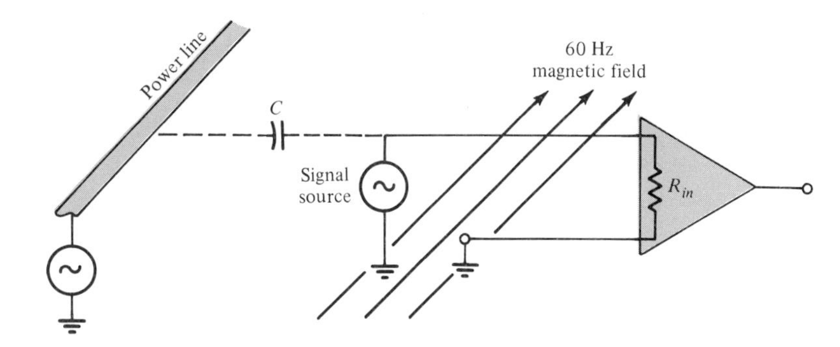

FIGURE 10.11. **Capacitively Coupled Common-Mode Interference from Power Lines and Induced Voltage Interference Due to Changing Magnetic Field**

Additional power line interference arises if magnetic fields from nearby power equipment thread through the area formed by the input leads to the amplifier. This fluctuating magnetic field creates an induced emf that is amplified along with the signal. This contribution to the interference may be minimized by making the enclosed area small. Parallel wires for the input leads will minimize the field to some extent—better yet is to twist the leads. In the latter case the induced emf in adjoining loops will effectively cancel. Still better results might come about from using shielded leads. Elimination of the capacitively coupled component of interference may be accomplished to a large degree through the use of a **differential amplifier** (Figure 10.12).

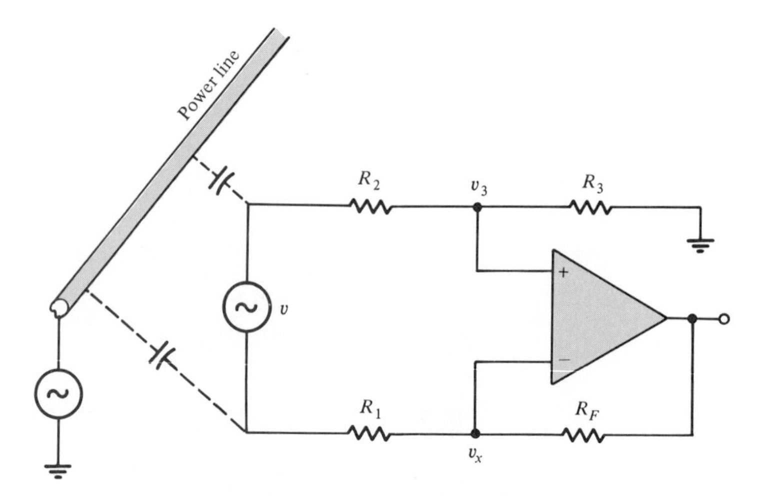

FIGURE 10.12. **OPAMP Differential Amplifier with Feedback Network**

Consider the application of a differential signal to this amplifier (Figure 10.13). If one end of the source is at a positive potential at a given moment and the other end at a negative potential, then somewhere in the source is a point at zero potential. We may simplify the treatment of the problem by splitting up the source into two equivalent sources (Figure 10.14), noting that $(v/2) + (v/2) = v$. This results in the application of a positive voltage (measured relative to ground) to the noninverting input and a negative voltage (also relative to ground) to the inverting input. Both, therefore, make a like (and equal) contribution to the output.

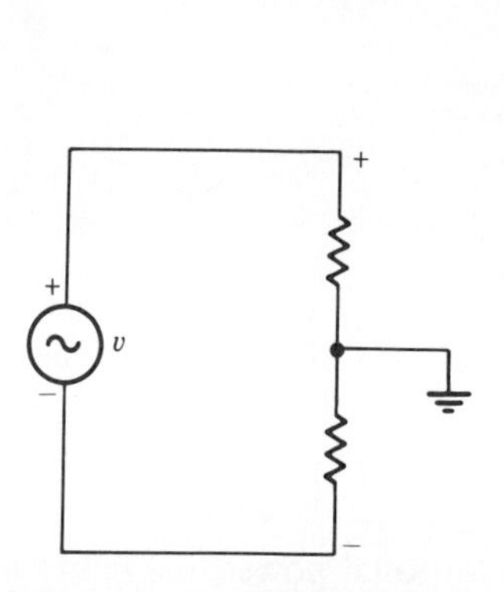

FIGURE 10.13. **Differential Input Signal Applied to OPAMP Input Terminals (+) and (−), Showing Input Resistances of OPAMP**

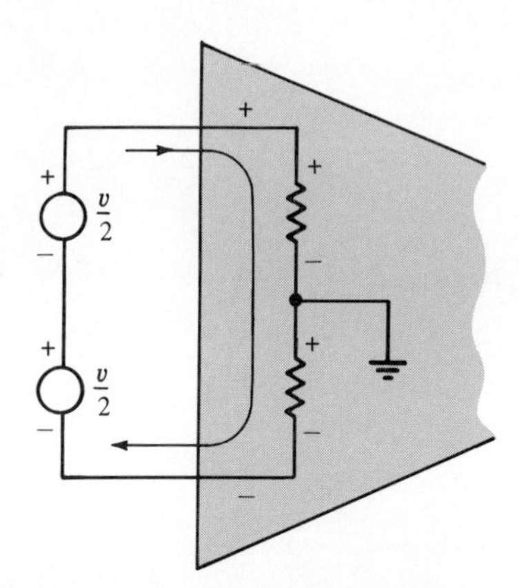

FIGURE 10.14. **Differential Input Signal Source Split into Two Equivalent Sources with Each One Being Half of the Value of the Original**

Turning now to the nature of the interfering signal, recall from our initial consideration of OPAMPs (Figures 10.2a and b) that an output voltage arose only as a result of an inequality between the magnitudes of V_1 and V_2. By implication, if $V_1 = V_2$, the output will be zero (if a dual dc supply is used).

While the power line interfering voltage induced via capacitive coupling into the two amplifier inputs is not constant, the two magnitudes at each moment may be presumed to be the same. Since the interfering signal, called the **common-mode signal** (v_{cm}), simultaneously applies the same polarity to both inverting and noninverting inputs (Figure 10.15), the two contributions to the output should effectively cancel, providing there is perfect balance between the two inputs.

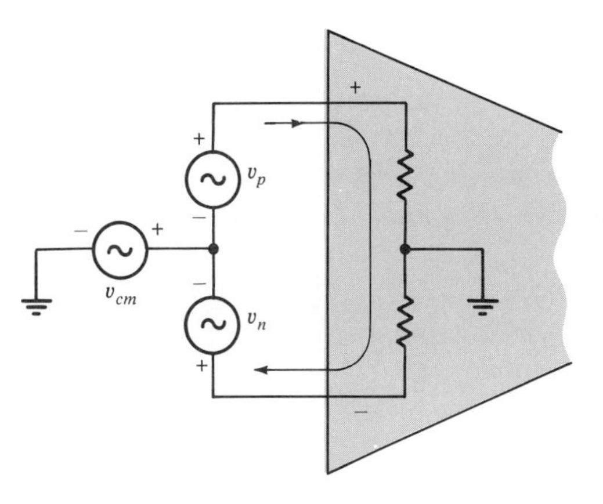

FIGURE 10.15. **Differential Input ($v_p - v_n$) Equaling Original Input Signal** The common-mode signal acts in phase at both inputs.

Turning to Figure 10.16, note that for calculation purposes we have inverted the polarity of v_n. This alteration simply means that the total differential input will now be $v_p - v_n$ rather than their sum. Because of the high input impedance of the amplifier, again there is a virtual equality of the potentials at the two inputs, making the potential at x equal to that at the juncture between R_2 and R_3. We have

$$v_3 = (v_{cm} + v_p)\left(\frac{R_3}{R_2 + R_3}\right) = v_x \tag{10.19}$$

Since i_{in} is taken to be zero, $i_1 = i_F$. We have

$$i_1 = \frac{v_{cm} + v_n - v_x}{R_1} = i_F = \frac{v_x - v_{out}}{R_F} \tag{10.20}$$

$$\frac{v_{cm}}{R_1} + \frac{v_n}{R_1} - \frac{(v_{cm} + v_p)[R_3/(R_2 + R_3)]}{R_1} = \frac{(v_{cm} + v_p)[R_3/(R_2 + R_3)]}{R_F} - \frac{v_{out}}{R_F} \tag{10.21}$$

$$v_{out} = v_{cm}\left(-\frac{R_F}{R_1} + \frac{R_F}{R_1}\frac{R_3}{R_2 + R_3} + \frac{R_3}{R_2 + R_3}\right) + v_p\left(\frac{R_3}{R_2 + R_3} + \frac{R_F}{R_1}\frac{R_3}{R_2 + R_3}\right) - v_n\left(\frac{R_F}{R_1}\right) \tag{10.22}$$

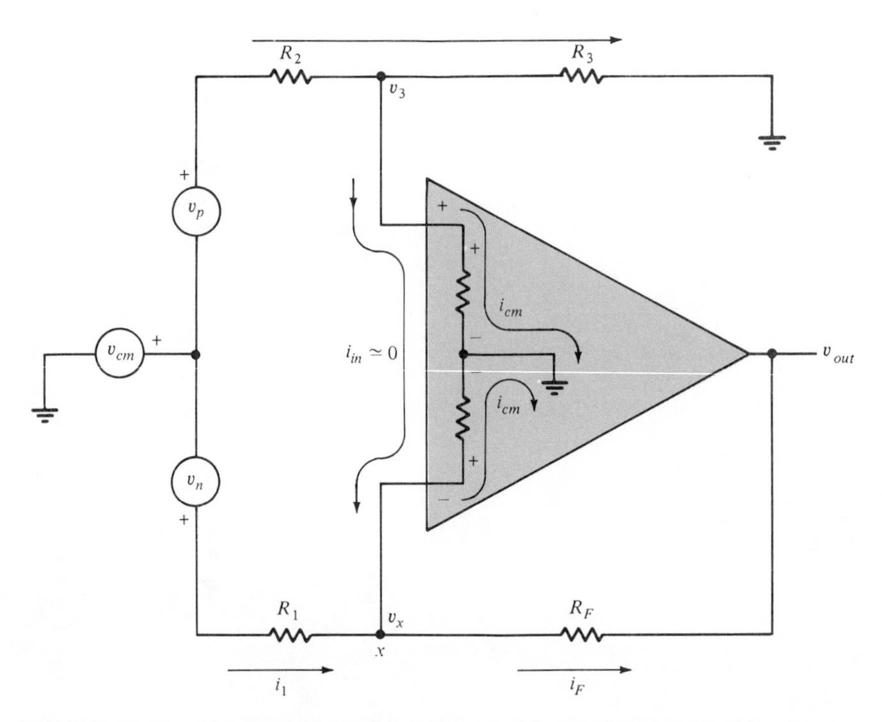

FIGURE 10.16. Common-Mode Signal Currents Flowing in Opposite Directions to Ground Through the Amplifier Input Resistances and Mutually Cancelling The differential signal $(v_p - v_n)$ leads to a potential difference at the two input terminals of the amplifier and is amplified.

Since we wish to be rid of the common-mode signal, set its coefficient equal to zero:

$$\frac{R_F}{R_1} = \left(1 + \frac{R_F}{R_1}\right)\left(\frac{R_3}{R_2 + R_3}\right) = \left(1 + \frac{R_F}{R_1}\right)\left[\frac{R_3/R_2}{1 + (R_3/R_2)}\right] \tag{10.23}$$

The condition for cancellation will be met if $R_F/R_1 = R_3/R_2$. Then we have

$$\begin{aligned} v_{out} &= v_p\left(\frac{R_3}{R_2 + R_3} + \frac{R_F}{R_1}\frac{R_3}{R_2 + R_3}\right) - v_n\left(\frac{R_F}{R_1}\right) \\ &= v_p\left(\frac{R_3}{R_2 + R_3}\right)\left(1 + \frac{R_F}{R_1}\right) - v_n\left(\frac{R_F}{R_1}\right) \\ &= v_p\left[\frac{R_3/R_2}{1 + (R_3/R_2)}\right]\left(1 + \frac{R_F}{R_1}\right) - v_n\left(\frac{R_F}{R_1}\right) \\ &= v_p\left(\frac{R_3}{R_2}\right) - v_n\left(\frac{R_F}{R_1}\right) \end{aligned} \tag{10.24}$$

$$v_{out} = -\frac{R_F}{R_1}(v_n - v_p)$$

$$\text{differential gain} = \frac{v_{out}}{v_n - v_p} \equiv A = -\frac{R_F}{R_1} \tag{10.25}$$

A voltage applied to the (+) terminal is subjected to a gain A_+ and that to the (−) terminal, A_-. Ideally, both A_+ and A_- are the same. If not, tying the two terminals together and applying a common-mode signal leads to an output of

$$v_{out} = A_+ v_{cm} - A_- v_{cm}$$

$$\text{common-mode voltage gain} = |A_+ - A_-| \equiv A_{cm} \tag{10.26}$$

Since $A_+ \simeq A_-$, we have

$$\text{common-mode rejection ratio (CMRR)} = \frac{A}{A_{cm}} \tag{10.27}$$

This quantity is often expressed in decibels, in which case we have

$$\text{CMRR [dB]} = 20 \log_{10}\left(\frac{A}{A_{cm}}\right)$$

Typically, CMRR might be 100 dB. Such a value means that if the signal to be amplified is 1 μV, the interfering signal can be as large as 0.1 V before its output magnitude is equal to that of the desired signal.

EXAMPLE 10.3

A typical common-mode rejection ratio for the 741 series of OPAMP is 90 dB. If the differential signal ($v_p - v_n$) is equal to 10 μV, how large should the common mode input be to yield an output signal contribution that is equal to that of the amplified differential signal? Take $A = 2 \times 10^5$.

Solution:

$$\text{CMRR} = \frac{A}{A_{cm}}$$

$$20 \log_{10} \text{CMRR} = 90 \text{ dB}$$

$$\log_{10} \text{CMRR} = \tfrac{90}{20} = 4.5$$

$$\text{CMRR} = 10^{4.5}$$

$$A_{cm} = \frac{A}{\text{CMRR}} = \frac{2 \times 10^5}{10^{4.5}} = 2 \times 10^{0.5} = 6.32$$

$$v_{diff} = 10^{-5} \times 2 \times 10^5 = 2 \text{ V}$$

$$\frac{2 \text{ V}}{6.32} = 0.316 \text{ V} = 316{,}000 \ \mu\text{V}$$

10.4 ■ DIFFERENTIATORS AND INTEGRATORS

Rather than the basic feedback OPAMP employing a single resistive feedback element R_F, we now turn to a consideration of two modifications of this basic circuit. Suppose the feedback element to be a capacitor instead of a resistor, as in Figure 10.17. By again using the concept of virtual ground, the negative

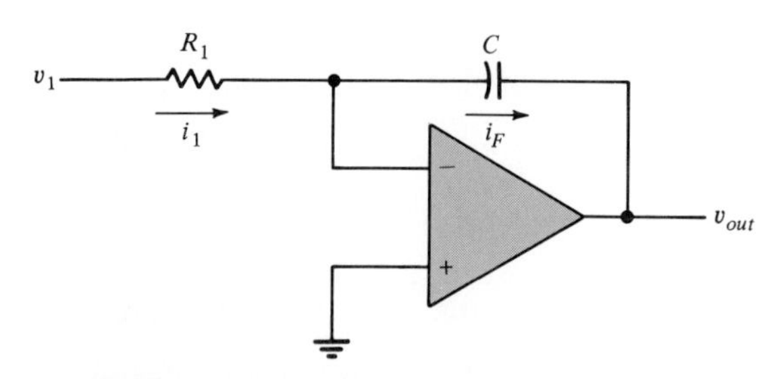

FIGURE 10.17. **OPAMP Integrator**

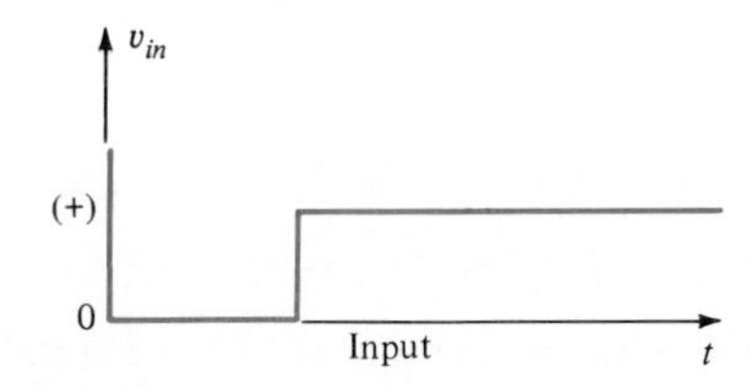

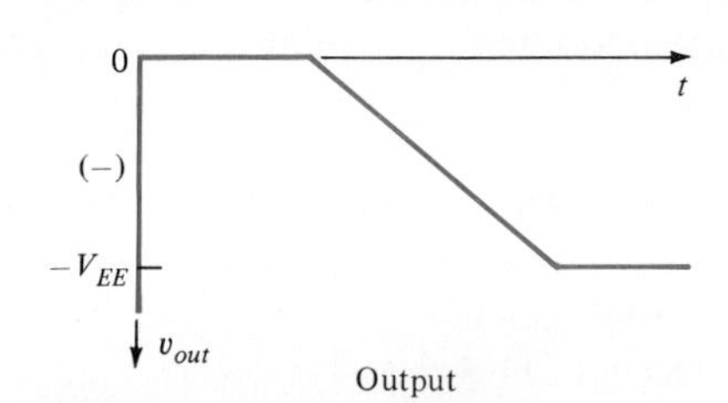

FIGURE 10.18. **Integrator Waveforms** The output voltage is proportional to the duration of the input step function and, therefore, is proportional to the integral of the input. The output voltage is limited to $-V_{EE}$.

input terminal is effectively at ground potential, and the current through R_1 must equal that through the feedback capacitor. With the direction of i_F as shown, v_{out} is negative. We have

$$q = C(-v_{out}) \qquad i_F \frac{dq}{dt} = -C\frac{dv_{out}}{dt} \qquad i_1 = \frac{v_1}{R_1}$$

Setting the currents equal to one another, we have

$$\frac{v_1}{R_1} = -C\frac{dv_{out}}{dt} \tag{10.28}$$

$$v_{out} = -\frac{1}{R_1C}\int_0^t v_1\, dt \tag{10.29}$$

The output represents an integral of the input signal. At any time, the output voltage is proportional to the integral of the input. Refer to the input signal shown in Figure 10.18. The output signal will continue to drop until it reaches the saturation limit of the amplifier, which is $-V_{EE}$, as depicted.

By interchanging the capacitor and the resistor, the circuit becomes a **differentiator** (Figure 10.19). Again equating currents, we have

$$C\frac{dv_1}{dt} = -\frac{v_{out}}{R} \tag{10.30}$$

$$v_{out} = -RC\frac{dv_1}{dt} \tag{10.31}$$

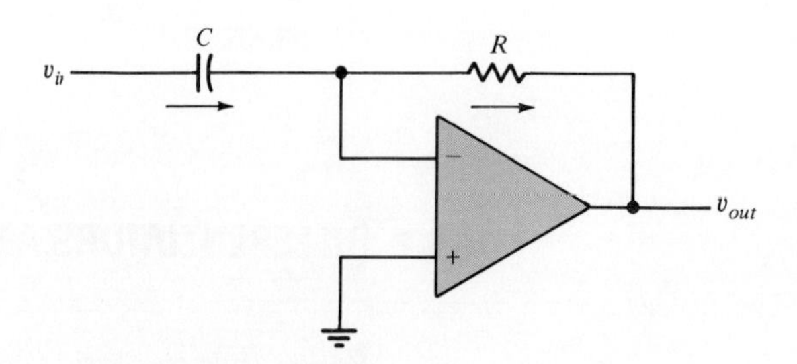

FIGURE 10.19. **OPAMP Differentiator**

By way of an example, Figure 10.20 shows a square wave signal with the differentiated output.

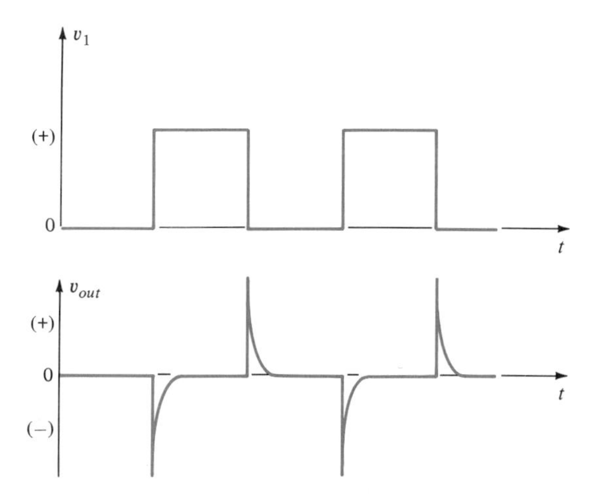

FIGURE 10.20. **Input and Output Differentiator Waveforms** The output approximates derivative of the input except for sign reversal.

In actual use these basic integrator and differentiator circuits must generally be modified. The integrator must be bias-stabilized by providing a dc feedback path. The differentiator, to account for the phase shift instabilities at high frequencies, must be prevented from going into oscillation. In the most fundamental manner, these difficulties are rectified by placing an appropriate resistor across the integrating capacitor and a capacitor across the differentiating resistor.

One may also approximate integrator and differentiator operation in a more rudimentary fashion by means of a low-pass filter (Figure 3.26) and a high-pass filter (Figure 3.25), respectively. By presuming the input to be a square wave of period T, in the integrator csse the time constant ($\tau = R_1C_1$) should be much larger than T and in the differentiator case ($\tau = R_2C_2$), much smaller than T. The voltage rise across the capacitor will approximate a linear increase as a function of time for the integrator and lead to alternate narrow positive and negative pulses in synchronism with the positive-going and negative-going edges of the square wave in the case of the differentiation output. (There is no sign reversal as there was with the OPAMP differentiator.)

10.5 ■ SOME PRACTICAL ASPECTS OF OPAMPS

10.5.1 741 Description

The most widely used OPAMP is the Fairchild μA741C, most simply called the 741 and sketched in Figure 10.21. It is also produced by a variety of other manufacturers. The packaging illustrated is called *mini-DIP* (mini-dual-in-line package) and consists of eight pins to which connections are made (actually one pin is a blank). The amplifier itself occupies a piece of silicon that is $\frac{1}{16} \times \frac{1}{16}$ in.

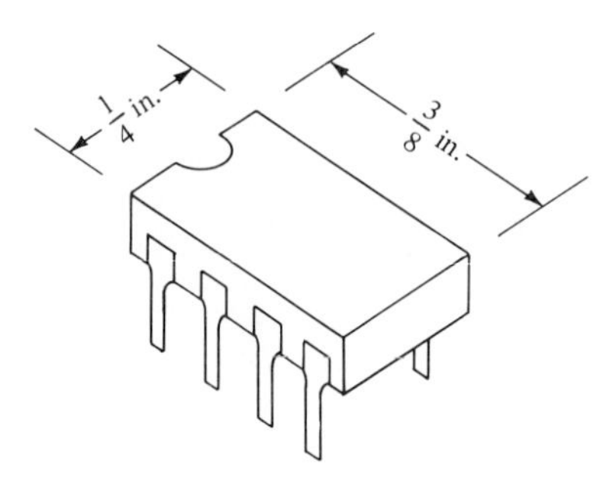

(a) Mini-DIP encasement used for 741 OPAMP

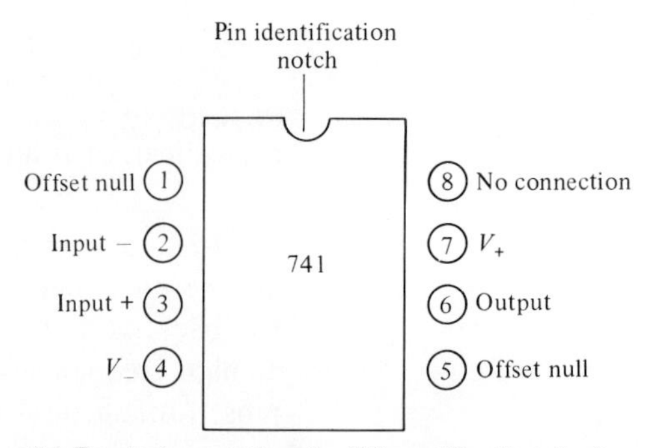

(b) Terminal connections for 741 amplifier (top view)

FIGURE 10.21. **741 Operational Amplifier**

The 741 is also available in other packaging forms including a 14-lead DIP and an 8-lead circular can. The 14-lead DIP actually contains two 741 units and is called a 747.

The 741 also appears in various disguises, produced by manufacturers other than Fairchild Semiconductor, the original source. There is, for example, a fast version (741S), an even faster version (NE530), a low-power version (4132; its dc input current is 35 μA rather than the customary 2.8 mA), a low-noise version (MC741N), and a precision version (OP-02). The 741 has an operating temperature range from $-55°$C to $+125°$C, thereby meeting military specifications. The 741A has a similar temperature range but tighter characteristics. The 741C is the commercial version with a temperature range of 0° to $+70°$C. The 748 is a 741 without internal compensation. (The function of either uncompensated or decompensated OPAMPs is to allow for wider bandwidths if large loop gains are not contemplated.)

10.5.2 Input Offset Voltage

If the two inputs are shorted together, the output voltage, rather than being zero (presuming a dual supply is being used), will probably assume some finite (but unpredictable) value. The reason for this is an inevitable asymmetry in the construction. The situation can be depicted as in Figure 10.22, where v_{os} represents a fictitious voltage generator, used in conjunction with an ideal amplifier, that leads to the observed output. v_{os} is called the **input offset voltage**.

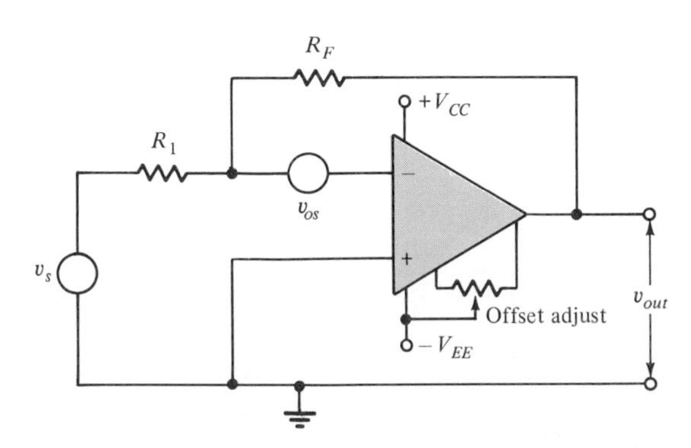

FIGURE 10.22. **OPAMP Showing Amplifier Offset Voltage Generator (v_{os}) Referred to Input and Offset Null Adjustment**

If R_F = 100 kΩ and R_1 = 100 Ω, with v_s shorted, the input offset voltage (expressed in millivolts) is equal to the output voltage (expressed in volts).

With the 741 amplifier this offset voltage can be reduced to zero by utilizing the offset null terminals. A 100 kΩ pot, connected between terminals 1 and 5 and with the wiper arm going to $-V_{EE}$, can be used to cancel the offset. However, the offset voltage will drift with temperature and time.

A typical offset value for the 741 is 2 mV (with 6 mV maximum); the drift rate is unspecified. For the OP-02E (a precision version of the 741), however, the offset is 0.3 mV with a typical drift rate of 2 μV/°C. On the other hand, there are some precision OPAMPs with a 30 μV offset, a temperature coefficient of 0.2 μV/°C, and a long-term drift of 0.2 μV/month! (This is the OP-07.)

10.5.3 Input Bias Current and Offset Current

Contrary to the idealized situation, there is a slight dc current through the inputs of an OPAMP. These currents can be thought of as originating from fictitious current generators connected from each input to ground (Figure 10.23). Distinct from the offset voltage, these bias currents create an additional voltage offset in the following manner: If the noninverting lead is grounded, I_{ib+} makes no contribution to this offset. But current I_{ib-} flows through R_1 in parallel with R_F. (The output impedance to ground is taken to be very small.) At the inverting terminal this current creates an offset voltage of magnitude $I_{ib-}(R_1 \,||\, R_F)$. To find the magnitude of the output voltage offset, we need to multiply this offset voltage by the gain. But the signal gain depends on whether the inverting or the noninverting input is being utilized. Since we could just as well have considered a noninverting amplifier in our example, we might have decided to use either $-R_F/R_1$ or $[1 + (R_F/R_1)]$—whichever was appropriate

But we have seen that the gain in feedback amplifiers is $1/\beta$, in this case called the *noise gain*. The advantage of this form is that it is equally applicable to both inverting and noninverting inputs.[5]

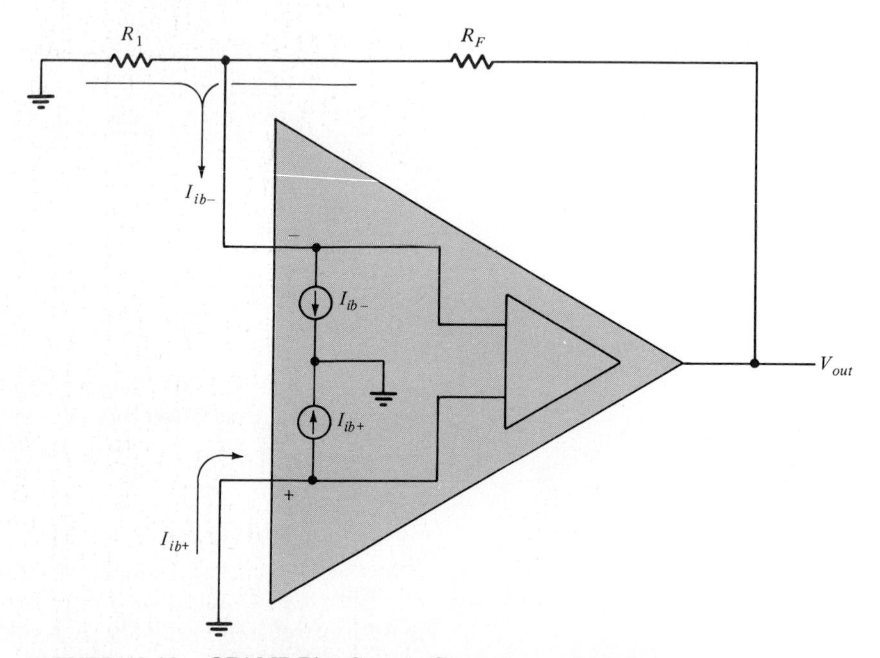

FIGURE 10.23. OPAMP Bias Current Generators at the Input

We then have

$$V_{out} = I_{ib-}(R_1 \,||\, R_F)\left(\frac{R_1 + R_F}{R_1}\right) \tag{10.32}$$

Since the output is near ground potential, we see that R_1 can be taken to be in parallel with R_F.

We may minimize the offset by having the other input terminal see the same impedance. In the case of Figure 10.24, this diminished offset may be

[5]Since $\beta = R_1/(R_1 + R_F)$, using the feedback equation $A_o/(1 + \beta A_o)$, for the inverting input we have

$$\frac{-R_F/R_1}{1 + \{[R_1/(R_1 + R_F)](-R_F/R_1)\}} \simeq \frac{-R_F/R_1}{-[R_F/(R_1 + R_F)]} = \frac{R_1 + R_F}{R_1} = \frac{1}{\beta}$$

For the noninverting input we have

$$\frac{(R_1 + R_F)/R_1}{1 + \{[R_1/(R_1 + R_F)][(R_1 + R_F)/R_1]\}} \simeq \frac{R_1 + R_F}{R_1} = \frac{1}{\beta}$$

accomplished by placing a resistance [$R_{equiv} = (R_1R_F)/(R_1 + R_F)$] in series with the (+) input lead to ground.

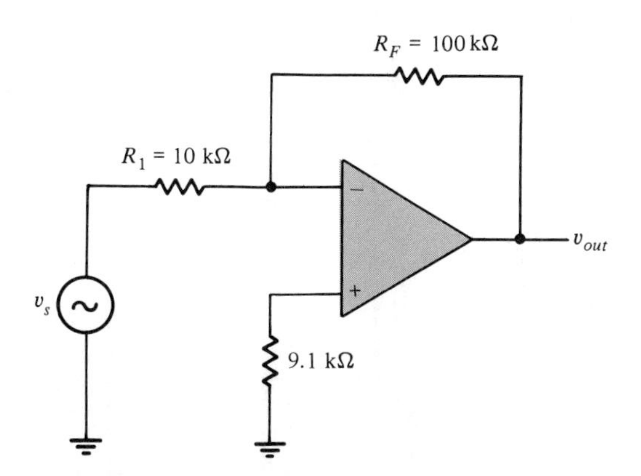

FIGURE 10.24. **OPAMP with Balanced Input Loading to Minimize Effects of Bias Currents**

We then have

$$V_{out} = [I_{ib-}(R_1 \,||\, R_F) - I_{ib+}(R_{equiv})]\frac{R_1 + R_F}{R_1}$$
$$= (I_{ib-} - I_{ib+})\frac{R_1R_F}{R_1 + R_F}\frac{R_1 + R_F}{R_1}$$
$$= (I_{ib-} - I_{ib+})R_F \tag{10.33}$$

If the two lead currents were the same, we could get complete cancellation, but, again, due to manufacturing limitations, there is generally a difference between them, termed the **input offset current** (I_{os}). To find the magnitude of the output voltage due to I_{os}, use the equation

$$V_{out} = I_{os}R_F \tag{10.34}$$

If one is not interested in amplification down to dc, a simple way of eliminating all these difficulties with offsets is to place a capacitor in series with either the signal lead or in the feedback loop to ground.

For the 741 a typical bias current is 80 nA, with a typical offset current of 20 nA. In the OP-02E version these currents are reduced, respectively, to 18 and 0.5 nA. In this regard one can see the advantage of FET inputs. For the LF13741, which is a 741 equivalent with such inputs, the respective values are 0.05 and 0.01 nA.

EXAMPLE 10.4

With dc amplification not necessary, to minimize the effect of drift and offset voltages, the signal source for the noninverting amplifier in Figure 10.25 is applied through a capacitor.

a. What is the lower 3 dB point for this circuit? (Note the need for a resistor between the (+) input and ground in order to carry the bias current.)

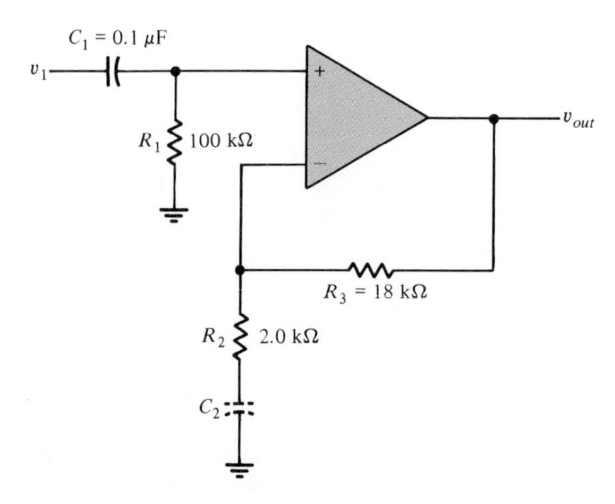

FIGURE 10.25

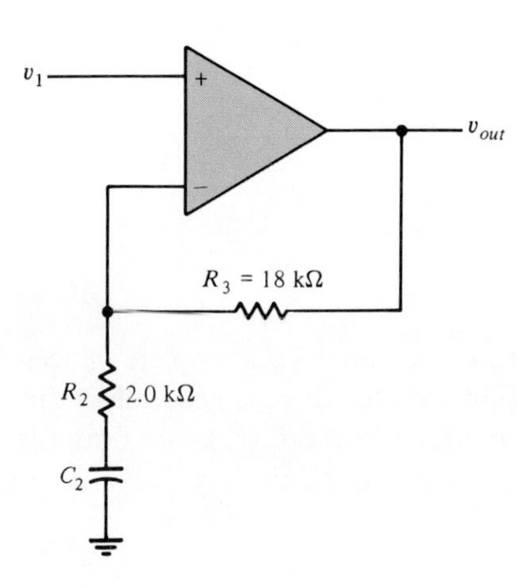

FIGURE 10.26

b. If, instead of C_1, a capacitor (C_2) were inserted into the feedback circuit, what should be its value to achieve the same 3 dB point? (See Figure 10.26.)

One must always provide a dc path for the feedback; therefore, the capacitor cannot be inserted in series with the feedback resistor.

Solution:

a. $$f_1 = \frac{1}{2\pi R_1 C_1} = \frac{1}{2\pi(10^5)(10^{-7})} = 16 \text{ Hz}$$

b. $$f_1 = \frac{1}{2\pi R_2 C_2}$$

$$C_2 = \frac{1}{2\pi(2 \times 10^3)(16)} = 5\ \mu\text{F}$$

10.5.4 Slew Rate

The application of a sudden step input to an OPAMP will not necessarily lead to a correspondingly identical output swing. It takes time for the internal capacitor and biasing networks to adjust. The limiting speed of the output variation is specified in terms of volts per microseconds, or **slew rate**. For the 741 it is 0.5 V/μs.

When, in fact, in Section 10.5.1 we mentioned *fast* versions of the 741 it was to a large degree the slew rate that we were talking about. Thus, the 741S has a slew rate of 12 V/μs, while the NE530 has a value of 35 V/μs.

There are other OPAMPs that have larger values—the LM318 at 70 V/μs and some even in the hundreds of volts per microsecond. The ultimate, however, is probably the LH0063C at 6000 V/μs.

It might be noted that the slew rate is also an important parameter in stereo systems. We have spoken of frequency bandwidth, output power, and so forth. There is also a power bandwidth f_p that comes into play when the reproduced sound is suddenly required to go from a small output voltage up to a peak value. Calling V_{op} the peak output voltage, we have

$$f_p = \frac{SR}{2\pi V_{op}} \tag{10.35}$$

where SR is the slew rate.

Because of slew rate limitations, above a specific frequency the maximum sine wave output swing will drop. The limiting amplitude is given by

$$\text{peak-to-peak voltage} \leq \frac{SR}{\pi f_p} \tag{10.36}$$

The results are sketched in Figure 10.27. At low frequencies the amplitude is flat, being due to power supply limits imposed on the output voltage swing.

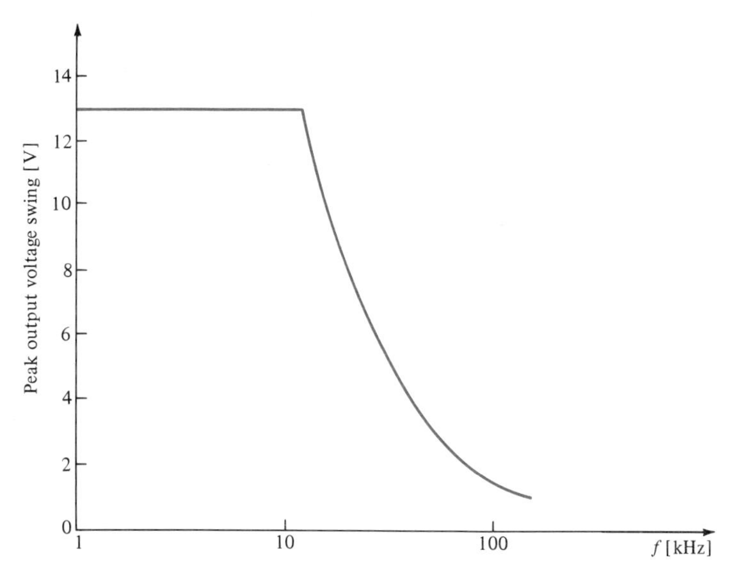

FIGURE 10.27. Reduction of OPAMP Output Amplitude Due to Slew Rate Limitations

10.5.5 Input Range Limits

There are limits placed on the maximum input signals that can be applied to OPAMPs if damage to them is to be avoided. Consider, for example, a voltage follower driven by a fast-rising input. During the slewing time the amplifier is not operating under normal closed-loop conditions where the differential input is close to zero. This condition causes large voltage spikes to appear at the input. Combined with a low source resistance, they can lead to excessive currents. For the 741 the *maximum differential input* is rated at 30 V, but it must not exceed the total supply voltage if that is less. High-voltage

OPAMPs are available if high differential inputs are to be expected. Thus, the 3583 has a rating of 300 V when used with a 300 V supply.

There are also limits placed on the *common-mode input*. The common-mode input range represents the values of common-mode signal that should not be exceeded if linear operation is to be maintained in the amplifier. For the 741 these limits are approximately −12.6 and +14.4 V.

10.6 ■ CURRENT AND VOLTAGE CONVERTERS

We have concentrated our attention on voltage amplification. There arise instances when we wish to have a voltage source give rise to a current of some substantial value, such as to drive a relay coil, for example. This may be done with the circuit shown in Figure 10.28.

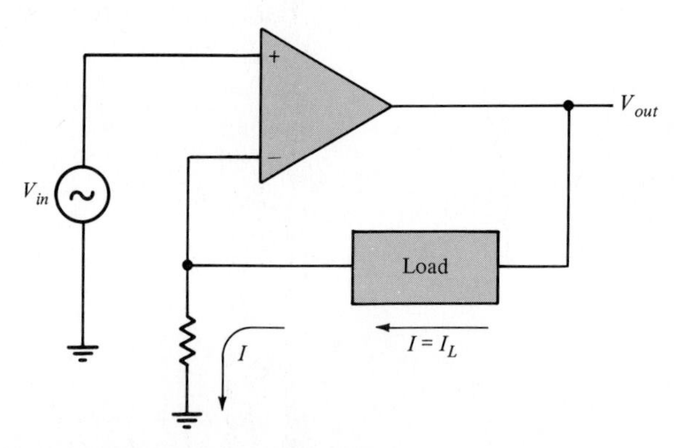

FIGURE 10.28. Voltage-to-Current Converter

V_{in} produces a current $I = I_L = V_{in}/R$ since both input terminals are at virtual ground. The current therefore is essentially independent of the load.

One may also need to convert current into a voltage. Such conversions may most simply be accomplished by passing the current through a resistance, but an operational amplifier offers an opportunity to do so without introducing an undesirable resistance into the circuit being measured (Figure 10.29). Since $I = I_F$ and the inverting input is at virtual ground, we have

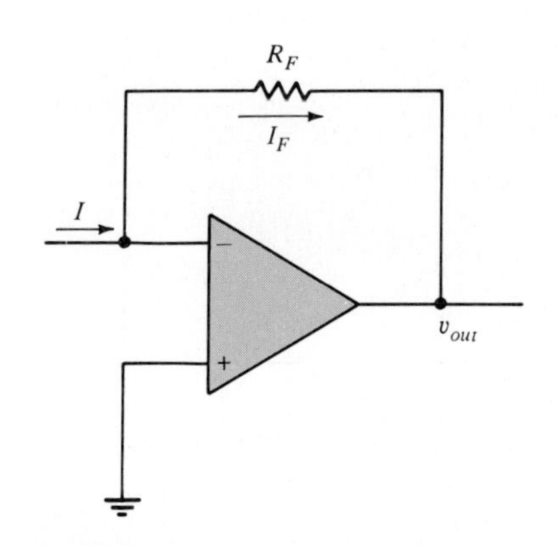

FIGURE 10.29. Current-to-Voltage Converter

$$V_{out} = -I_F R_F = -IR_F \tag{10.37}$$

EXAMPLE 10.5

In a photodiode the junction region is exposed to light, and the resulting electron-ion pairs generated lead to a diode current that may be used to measure the light intensity.

A typical photodiode generates 0.25 μA/μW of light illumination. In principle the resulting current could be passed through a resistance and the voltage measured, thereby providing a measure of the radiant power. But such diodes are operated with a reverse bias, and the resulting current would diminish the bias voltage.

Show how an OPAMP may be advantageously used in this application.

Solution: With a measurement across a series resistance, the circuit would look as shown in Figure 10.30. The OPAMP circuit for a current-to-voltage converter looks as shown in Figure 10.31. Since $I_D \simeq I_F$, we have

$$V_{out} = -I_F R_F = -I_D R_F$$

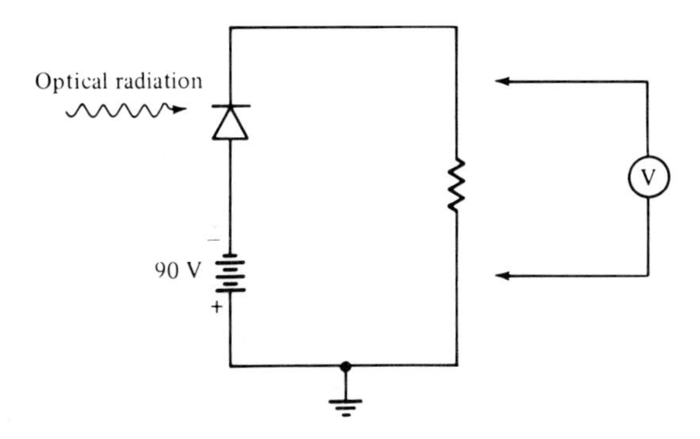

FIGURE 10.30

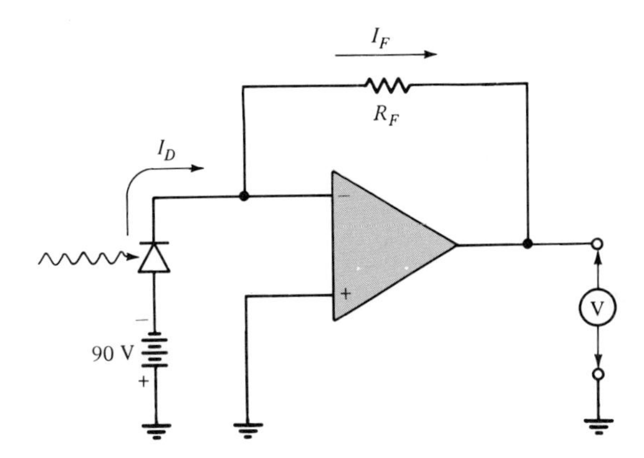

FIGURE 10.31

To achieve full-scale reading with 1 mW of light power, we have

$$\frac{0.25 \times 10^{-6}\ \text{A}}{\mu\text{W}} \times 10^3\ \mu\text{W} = I_D = 0.25 \times 10^{-3}\ \text{A}$$

$$V_{out} = 0.25 \times 10^{-3} \times R_F$$

For a voltmeter having a 1 V full-scale reading, we have

$$R_F = \frac{1\ \text{V}}{0.25 \times 10^{-3}\ \text{A}} = 4\ \text{k}\Omega$$

The sensitivity is

$$\frac{1\ \text{mW} \times 10^3\ [\mu\text{W/mW}]}{1000\ \text{mV}} = \frac{1\ \mu\text{W}}{\text{mV}}$$

10.7 ■ PROGRAMMABLE OPERATIONAL AMPLIFIERS

Battery-operated amplifiers, when minimal power consumption is desirable, often utilize **programmable OPAMPs**. That designation derives from their internal operating currents being set by an externally controlled current I_{set}. The internal currents of the various amplifier stages, in turn, are related to I_{set} via current mirrors. The first such amplifier was the 4250, which is still widely used. The 776 is a somewhat improved version.

The pin arrangements of the 4250 and 776 are similar to that of a 741 OPAMP. But that pin 8 that we noted was blank on the 741 DIP (Figure 10.21) is used as a single master bias terminal. An external resistor connected from this terminal to $-V_{EE}$ (or to ground in some cases) establishes the value of I_{set}. Its value determines amplifier parameters such as power consumption, gain-bandwidth product, and slew rate. When programmed for low power consumption, such OPAMPs are called micropower amplifiers. I_{set} can be varied from 100 μA down to 100 nA.

As an example of 4250 operation, utilizing two 1.5 V flashlight batteries as a dc power source, using a 2.5 MΩ resistor for R_{set} establishes $I_{set} = 1\ \mu$A. That value of resistance sets the quiescent input current at about 7 μA, corresponding to an input power of 21 μW plus approximately 2.5 μW drawn in $R_{set}[(10^{-6})^2(2.5 \times 10^6)]$. At this setting the open-loop gain is 10^5, the gain-bandwidth product is about 60 kHz, and the slew rate is slightly more than 0.01 V/μs.

Near the other extreme, with $I_{set} \simeq 80\ \mu$A, the gain is 2×10^5 (the gain actually peaks at 4.5×10^5 with $I_{set} = 10\ \mu$A), the gain-bandwidth product is 300 kHz, and the slew rate is about 1 V/μs.

10.8 ■ OPAMP DESIGNATIONS

Since our main objective in this chapter has been the introduction of operational amplifier concepts, most of the discussion has centered upon the 741 amplifier. In this section the objective is to improve upon the reader's IC vocabulary by introducing a few more representative OPAMPs and clarifying some of the designations.

We mentioned the 4250, a programmable amplifier, and a somewhat improved version, the 776. More accurately, these designations should be LM4250 and μA776, where the prefixes are generally indicative of the original manufacturer of the unit, although many of them have subsequently become available from other sources.

The following represents a partial listing of prefixes identifying various manufacturers:

– LM, LH, NH, DM: National Semiconductor, Santa Clara, CA.
– μA: Fairchild Semiconductor, Mountain View, CA.
– CA: RCA, Somerville, NJ.
– SN: Texas Instruments, Dallas, TX.

—ULN: Sprague Electric, Worcester, MA.
—MC, ML, M: Motorola Semiconductors, Phoenix, AZ.
—N, NE, S, SE: Signetics, Sunnyvale, CA.

While the 741, employing BJT transistors, has established itself as an industry standard, one might inquire as to whether there is a comparable industry standard utilizing FET transistors. The answer is no, because again the gain available from FETs is so much smaller, their output impedance tends to be high, and there is a much wider spread in their parameter values.

What has been rather widely developed is the **BiFET**, a combination of FET input stages (leading to high input resistances) followed by succeeding stages of BJTs. The 355, 356, and 357 series are representative examples. Their inputs are *p*-channel JFETs (*p*-channels are easier to fabricate in conjunction with BJTs than are *n*-channel devices). These OPAMPs lead to bias currents of about 0.03 nA and offset currents of 0.003 nA (compared to the 741's 80 and 20 nA, respectively).

SUMMARY

10.2 Basic Principles

- The two inputs of an operational amplifier (OPAMP) represent, respectively, a noninverting (+) input and an inverting (−) input.
- OPAMPs operating in an open-loop mode—that is, without negative feedback—have a very high signal voltage gain but a very small bandwidth, typically 100,000 and 10 Hz, respectively.
- In open-loop mode OPAMPs are easily driven into saturation—that is, the output voltage is equal to the dc voltage used to power the amplifier.
- With negative feedback the gain of an inverting OPAMP is $-R_F/R_1$, where R_F is the feedback resistance and R_1 the series input resistance. The gain of a noninverting OPAMP with negative feedback is $1 + (R_F/R_1)$.
- Unity gain OPAMPs are often used as buffers between stages of voltage amplification.

10.3 Differential Amplifiers

- Differential amplifiers provide a high degree of discrimination against common-mode (interfering) signals, such as hum.
- A/A_{CM}, where A is the differential gain and A_{CM} the common-mode gain, constitutes the common-mode rejection ratio (CMRR) of an amplifier.

10.4 Differentiators and Integrators

- OPAMPs may be employed to perform the mathematical equivalents of differentiation and integration.

KEY TERMS

BiFET: IC combining FETs and BJTs (the 355-356-357 series is representative)

buffer amplifier: most usually a voltage follower inserted to minimize loading effects of one amplifier stage on the previous amplifying stage

closed-loop mode: amplifier operation with feedback

common-mode gain: gain to which a differential amplifier subjects a common-mode signal (generally much smaller than the differential gain)

common-mode rejection ratio (CMRR): ratio of differential gain to common-mode gain, usually expressed in terms of decibels

common-mode signal: like voltages simultaneously applied to the two inputs of a differential amplifier

dc rail: the dc limits set by the power supply voltages (The expression stems from the method of drawing common dc terminals in schematic diagrams as parallel horizontal lines.)

differential amplifier: an amplifier that responds primarily to voltage differences between its two input terminals but discriminates against simultaneously applied common voltages (measured relative to ground) such as typified by hum pickup

differential gain: gain to which a differential signal is subjected

differential mode signal: the voltage difference applied to the two inputs of a differential amplifier

differentiator: a circuit whose output represents the mathematical equivalent of differentiating the input signal

dual-in-line package (DIP): the rectangular encasement of an integrated circuit with connecting pins emerging symmetrically from the two longer sides of the package (Common pin totals are 8, 16, 24, 28, and 40; an 8-pin total is called a mini-DIP.)

feedback impedance: impedance connecting input and output terminals of an OPAMP

input bias current: the small current that enters (or leaves) the input terminals of an OPAMP

integrated circuit (IC): a complete operational electronic system consisting of numerous transistors and associated components

digital IC: an integrated circuit intended to process digital (pulse) signals

linear IC: an integrated circuit intended to process analog signals

integrator: a circuit whose output represents an integral of the input signal

inverting input: condition when output is 180° out of phase with signal applied to input

noninverting input: condition when input results in an in-phase output signal

offset current: the unbalance between the two input bias currents of an OPAMP

offset voltage: an asymmetry between the two inputs of an OPAMP, resulting in a finite output with inputs shorted

open-loop mode: straight-through amplifier employing no feedback

operational amplifier (OPAMP): a linear IC circuit

programmable OPAMP: OPAMP whose power consumption, gain-bandwidth product, and slew rate are adjustable

saturation: as applied to OPAMPs, the output voltage is equal to one or the other of the dc rail voltages

slew rate: time rate of change of output voltage in response to a sudden input step function

virtual ground: a circuit point that can be considered to be at ground potential (Generally it is connected to actual ground through a very large resistance through which little current passes, making the voltage drop (potential difference) negligible relative to ground.)

voltage follower: an OPAMP with 100% negative feedback, resulting in approximately unity gain (Its high input impedance and low output impedance make it useful as a coupling element between voltage amplifiers.)

SUGGESTED READINGS

Jung, Walter G., IC OP-AMP COOKBOOK. Indianapolis: Howard W. Sams, Publ., 1974. Chap. 2, "History of OPAMPs," and pp. 403–440, *programmable OPAMPs.*

Horowitz, Paul, and Winfield Hill, THE ART OF ELECTRONICS. New York: Cambridge University Press, 1980. Page 115, *Brief history of the 741 OPAMP,* and pp. 108–113, *an extensive list of various OPAMPs, with major parameter listings.*

Wooley, Bruce A., Sing-Yui J. Wong, and D. O. Pederson, "A Computer-Aided Evaluation of the 741 Amplifier," IEEE JOURNAL OF SOLID STATE CIRCUITS, vol. SC-6, no. 6, pp. 357–366, Dec. 1971. *A detailed analysis of the 741 circuit.*

van Kessel, Th. J., and R. J. van de Plassche, "Integrated Linear Basic Circuits," PHILIPS TECHNOLOGICAL REVIEW, vol. 32, no. 1, pp. 1–12, 1971. *Current sources used to bias OPAMPs.*

A CLOSER LOOK: Bode Plots

The high-frequency drop-off in amplifier gain is due to the presence of shunt capacitance. (There can, of course, also be a drop-off at low frequencies due to coupling capacitors, but since the treatment in that case is a similar matter, we will confine our attention to the high-frequency drop-off. Also, internally, OPAMPs are direct-coupled and devoid of such coupling capacitors.) The presence of a capacitance means an associated phase shift, and thus, as a function of frequency, an amplitude change is always accompanied by a phase angle change. This hand-in-hand relationship is known under a variety of names, to some extent dependent on the physical phenomenon involved. In optics it is known as the Kramers-Kronig relationship, being the link between absorption and dispersion. In electronics it is the name of H. W. Bode that is associated with the relationship. (It has been suggested that the first words spoken by the best theoretical physicists while still in their cribs are "Kramers-Kronig" rather than "Ma" and "Da.")

Figure 1a is meant to illustrate the gain vs. frequency response of a single-stage amplifier. At the corner frequency (shown as 1.0 on the frequency scale), the actual response is down by 3 dB, and the phase shift is $-45°$. This phase shift begins to manifest itself about a decade earlier, at 0.1. Being due to a single capacitance, it reaches an ultimate maximum value of $-90°$ a decade later, at 10.0.

$-180°$ due to the capacitances. However, the actual approach to $-180°$ is asymptotic, and the zero overall phase shift does not arise until infinite frequency. Therefore, a two-stage negative feedback amplifier should be inherently stable (providing the feedback loop itself does not introduce some additional reactance).

Incidentally, with two stages of CE amplification, the output is normally in phase with the input, and the reader might question how one arrives at negative feedback in that instance. One can achieve negative feedback with two stages by having a CE amplifier followed by a stage of CC amplification, for example.

In Figure 1c we consider a three-stage amplifier with identical corner frequencies. In this case even at the common corner the cumulative phase shift is already $-135°$, so that progressing not too far up the frequency scale we can arrive at a situation where the phase shift is $-180°$ and unity loop gain, leading to an unstable situation.

Rather than having identical corner frequencies for each

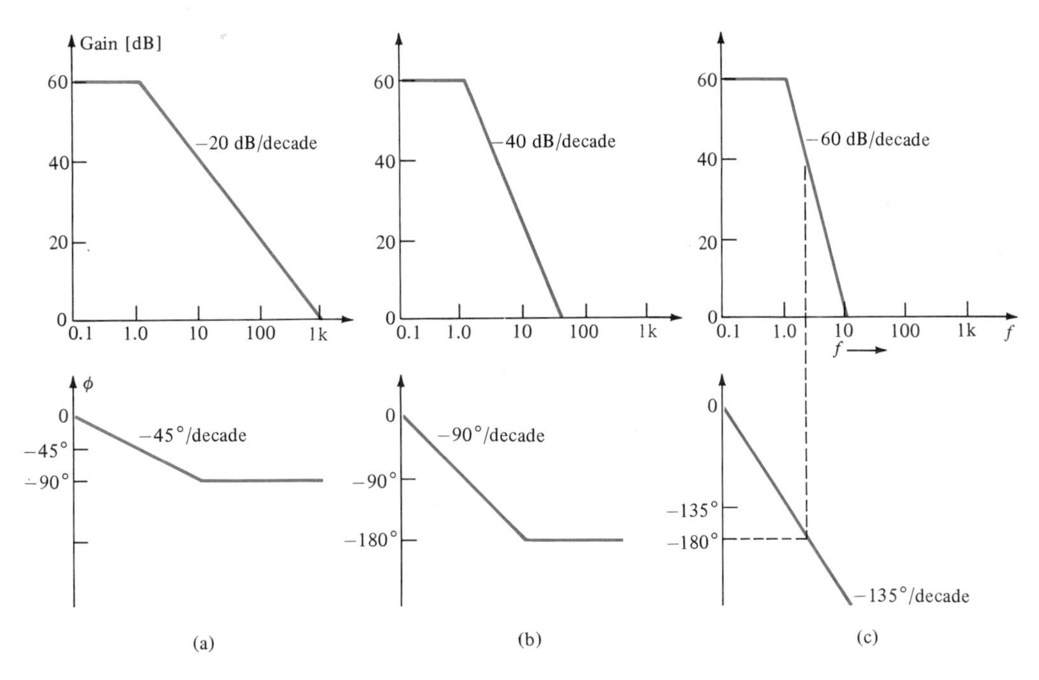

FIGURE 1 **Bode Plots for Negative Feedback Amplifiers**

In Figure 1b we presume we are dealing with a two-stage amplifier with identical corner frequencies. Two shunt capacitances are involved. At the turnover frequency, each contributes a shift of $-45°$, for a total of $-90°$, and an ultimate value of $-180°$ one decade later, at frequency 10.0. It might seem in this instance that it is possible for such an amplifier with negative feedback to develop an instability, with the 180° of negative feedback being cancelled by the

stage (in mathematical parlance they are called *poles*), it is common practice to place each of them at successively higher frequencies. Such disposition leads to a wider effective bandwidth. It should be noted that with identical poles, as each additional stage is added the overall 3 dB point moves to a lower frequency, thereby diminishing the bandwidth.

Figure 2a shows the effect of poles at 1.0, 10, and 100.

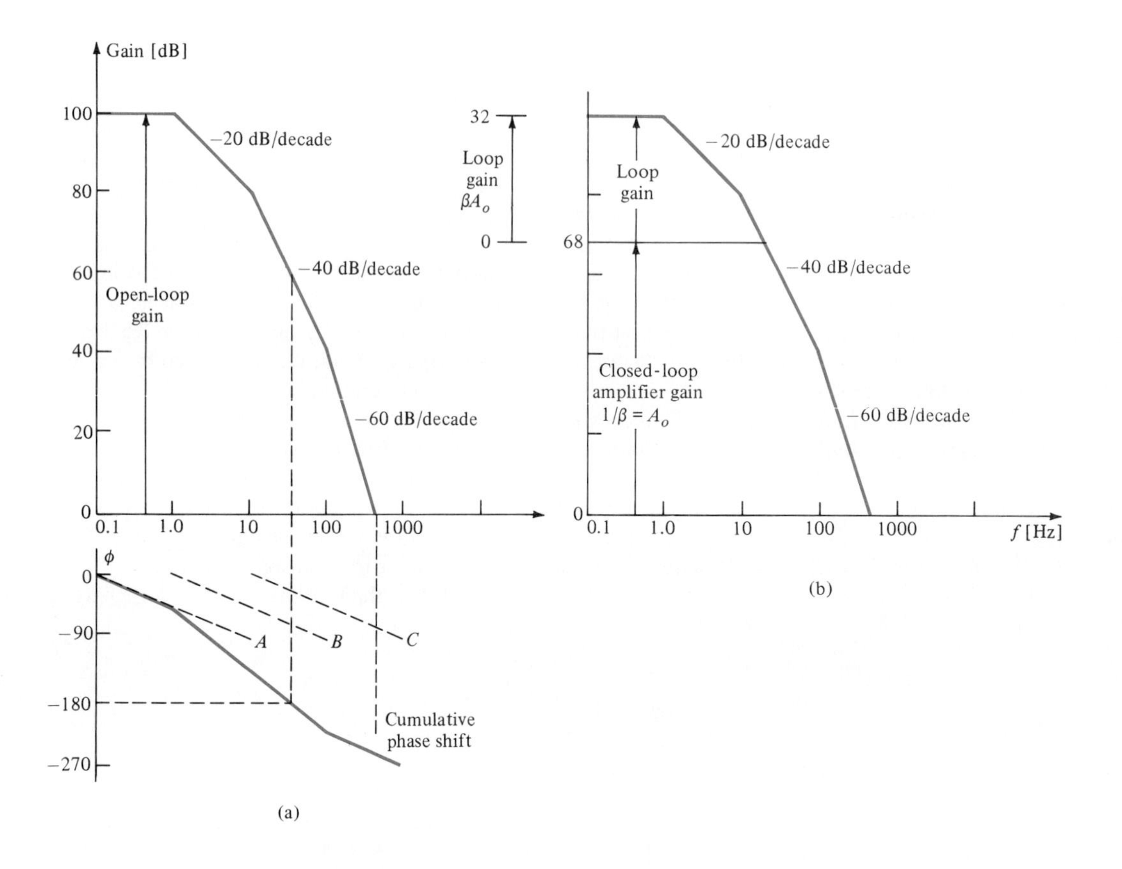

FIGURE 2

The individual phase shift variations arc shown as A, B, and C. In each case their influence extends one decade to each side of the pole, and the cumulative phase shift is shown by the solid line. The critical frequency arises at the $-180°$ mark, corresponding in this example to *frequency* of about 30.

In Figure 2b we assume feedback to be in play, having diminished the midband gain from 100 to 68 dB. The loop gain is 32 at midband (100 − 68 = 32 db). As the amount of negative feedback increases—that is, by increasing β—the loop gain increases.

The result of increasing the loop gain by increasing the amount of negative feedback is shown in Figure 3. In the initial case we assume 20 dB of feedback (loop gain), and we see that the loop gain goes to unity (0 dB) at a frequency equal to 10. By having increased the feedback to 30 dB, the unity-gain frequency is somewhat higher. Finally, an increase of feedback for a loop gain of 40 dB leads to a unity-gain intercept at the critical frequency in agreement with Figure 2a. To be on the safe side with this amplifier, the feedback used should be something less than 40 dB since there may be some stray capacitances or temperature effects that could push the amplifier into an unstable situation. Such safety

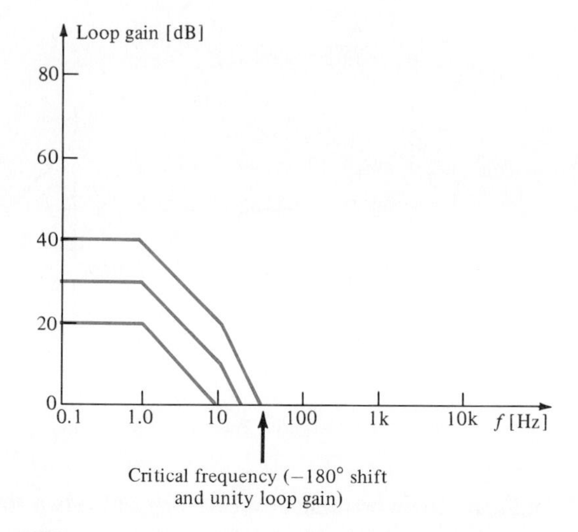

FIGURE 3

factors are termed either *phase margins* or *gain margins.* Under no circumstances should this amplifier be used as a voltage follower without some alteration in its design.

To see how a correction could be effected, consider Figure 4. The judicious use of a compensating network (including a capacitor) has caused the pole at 1 MHz to be removed and a new one to be introduced at a much lower frequency. As a result the phase shift does not reach $-180°$ until after the gain has dropped below 0 dB. For the case illustrated there is a phase margin of 45° (at 0 dB the phase shift is $-135°$; $180 - 135 = 45°$), and the gain margin is -10 dB (that is, 180° is not reached until gain $= -10$ dB).

Most OPAMPs have such a corrective network within the circuit. Compensation, however, diminishes the amplifier bandwidth. For a 741 the corner frequency is at 10 Hz, and unity gain arises at 1 MHz.

Direct coupling is generally used in OPAMP circuits, and capacitors are studiously avoided. The one exception is the capacitor that is used in the compensating network. (In footnote 1 to this chapter, it was mentioned that a 741 contained 20 transistors, 11 resistors, and 1 capacitor. That's the one!)

Many OPAMPs are available in both compensated and uncompensated forms (witness the 741 and the 748). If an amplifier is not going to be used with large amounts of loop gain (for example, it will not be used as a unity-gain amplifier), the omission of the compensation leads to a wider bandwidth and, in some cases, to a higher slew rate. There are also some amplifiers classed as being "decompensated," meaning that their degree of compensation limits their use to some minimum value of gain. Examples in the latter category are the 349 and the 357, having minimum gain ratings of 5 and slew rates of 2 and 50 V/μs, respectively.

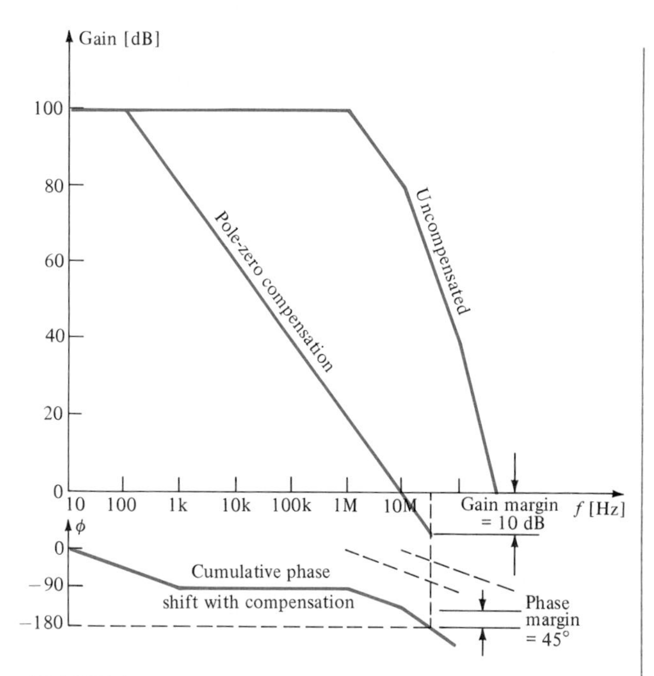

FIGURE 4

EXERCISES

1. What is the nature of the two input terminals of an OPAMP?

2. **a.** If an OPAMP operates from a dual symmetrical supply, what ideally is the output voltage if the inputs are shorted?

   b. When used with a single power source, what is the output voltage of an OPAMP with the inputs shorted?

3. What is a typical value for the voltage gain of an OPAMP?

4. What is meant by virtual ground?

5. With the (+) terminal grounded and the signal applied to the (−) terminal through resistor R_1, using a feedback resistor R_F, what is the expression for the gain of an OPAMP?

6. For a noninverting OPAMP, in terms of R_1 and R_F, what is the expression for gain?

7. For an OPAMP with gain-bandwidth product 5×10^5, what is the bandwidth if the gain with negative feedback is 100?

8. What is the nature of an OPAMP with noninverting input and 100% negative feedback?

9. How does the voltage gain of an inverting OPAMP with negative feedback of the type we have considered affect the input and output resistances?

10. If an OPAMP has 60 dB of CMRR and the differential signal is 1 μV, how large can the magnitude of the common-mode signal become before there is equality between them in the output?

11. Sketch the circuit for an OPAMP differentiator.

12. Sketch the circuit for an OPAMP integrator.

13. An OPAMP employing negative feedback in the inverting input has $R_F = 100$ kΩ and $R_1 = 100$ Ω. With input terminals shorted, the output voltage is 3 V. What is the magnitude of the offset voltage?

14. Some precision OPAMPs can have offset voltage as small as 30 μV. With such an amplifier, what would be the output voltage if $R_F = 100$ kΩ and $R_1 = 100$ Ω?

15. An OPAMP employs 10% negative feedback. What is the noise gain?

16. With an OPAMP employing the inverting input and $R_F = 100\ k\Omega$ and $R_1 = 10\ k\Omega$, to measure the input bias current, what value of resistance should be placed between the noninverting terminal and ground?

17. With $R_1 = 10\ k\Omega$ and $R_F = 100\ k\Omega$, using balanced input impedances, what will be the value of the output voltage due to an input offset current of 10 nA?

18. If ac amplification is contemplated, what is an easy way of eliminating the offsets in an OPAMP?

19. For a 741 OPAMP using $V_{CC} = V_{EE} = 15$ V, with a slew rate of 0.5 V/μs, what is the limiting peak voltage for output at 100 Hz? At 20 kHz?

PROBLEMS

10.1

For an operational amplifier (Figure 10.32) with the characteristics $R_s = 1\ k\Omega$, $R_L = 10\ k\Omega$, $A = 10^5$, $R_{in} = 100\ k\Omega$, $R_{out} = 100\ \Omega$, and $v_{out} = 10$ V (rms), calculate v_s and v_s/v_{out} and estimate the circuit's input resistance.

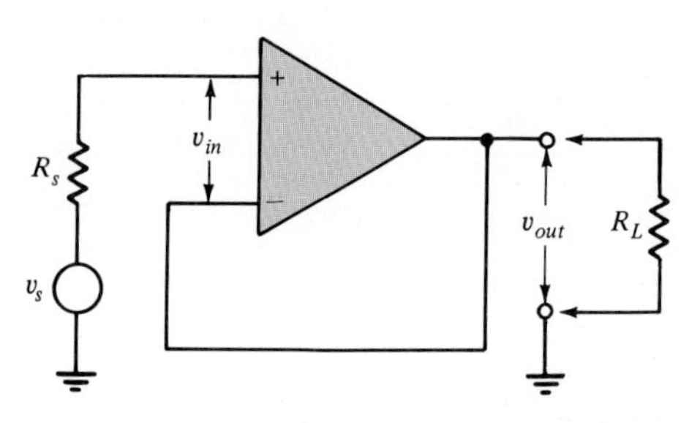

FIGURE 10.32

10.2

Consider the operational amplifier in Figure 10.33.

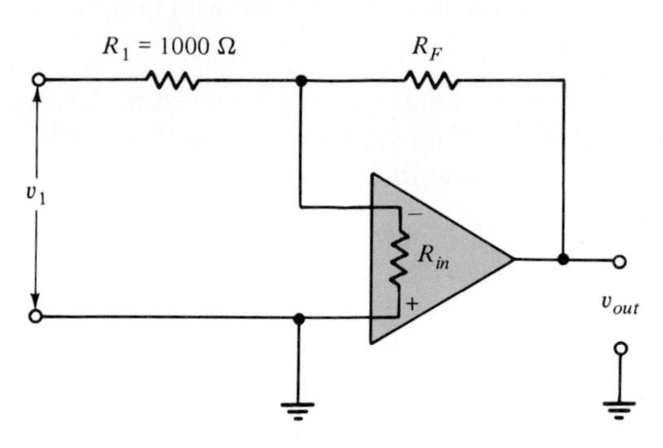

FIGURE 10.33

a. What should be the value of R_F presuming it is desired to have a gain of -100?

b. If the amplifier uses a dual ± 15 V power source, about what should be the maximum rms value for the input signal v_1 if linear amplification is to be maintained?

c. Sketch the operational amplifier circuit that will yield a gain of $+100$. What is the value of the feedback resistor in this case?

d. What are the values of input resistance seen by the generator in parts a and c?

10.3

A transducer can be represented by a Thevenin generator of 10 V (V_T) in series with a Thevenin resistance (R_T) of 5 kΩ. It is made to operate a 1 mA meter—that is, 1000 Ω/V. If the expected full-scale reading is 10 V, what will be the voltage and power delivered to the indicator *with and without* a unity-gain buffer?

10.4

In a differential amplifier the dc input signals are $v_p = +1.01$ V and $v_n = +0.99$ V. Predict the output signal (v_{out}) if

a. $R_1 = R_2 = 10\ k\Omega$ and $R_F = R_3 = 100\ k\Omega$.

b. $R_1 = 10\ k\Omega$, $R_2 = 9\ k\Omega$, and $R_F = R_3 = 100\ k\Omega$.

10.5

A differential amplifier has a common-mode rejection ratio (CMRR) of 100 dB. If the signal to be amplified is 1 μV, how large will the interfering signal input be if its output is to equal that of the desired signal?

10.6

An OPAMP with $A = 10^5$ and 100 dB of CMRR is subjected to a common-mode signal of 12 V amplitude. What is the magnitude of the output common-mode signal?

10.7 ⚠

That 0.7 V voltage drop that we employed between the base and emitter of a transistor arises because the junction between them is in reality a forward-biased diode. (See Figure 10.34.) When used as a rectifier, it has been assumed that the amplitude of the applied signal was considerably greater than this "offset" voltage. Should we try to rectify a signal that is as small as a few tenths of a volt, this would be insufficient to overcome this built-in potential drop. An OPAMP may be used to allow rectification of signals as small as some tens of microvolts. Such a low-level rectifier assumes the form in Figure 10.35. If v_1 is positive by at least V_γ/A_o, then $v' > V_\gamma$, and the diode will conduct. Since the $(-)$ input is virtually equal to v_1, $v_{out} \simeq v_1$. Thus, the system behaves as a voltage follower for positive signals in excess of approximately $0.6/A_o$ V. When v_1 goes negative, the diode is cut off, and practically no current is delivered to the load. Sketch the response to a sinusoidal input with the diode as shown in Figure 10.35 and with the diode reversed, presuming $A_o = 10^5$.

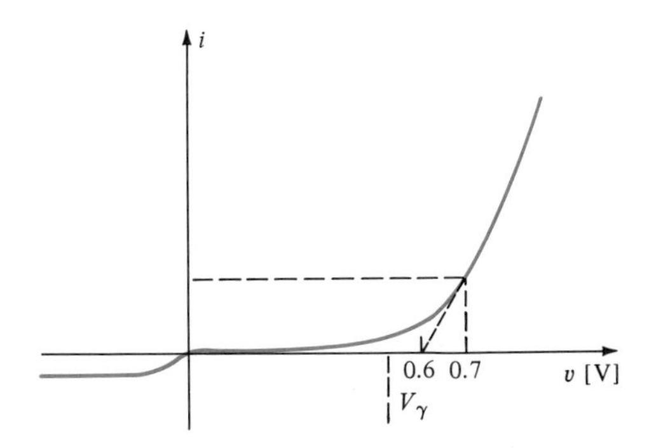

FIGURE 10.34

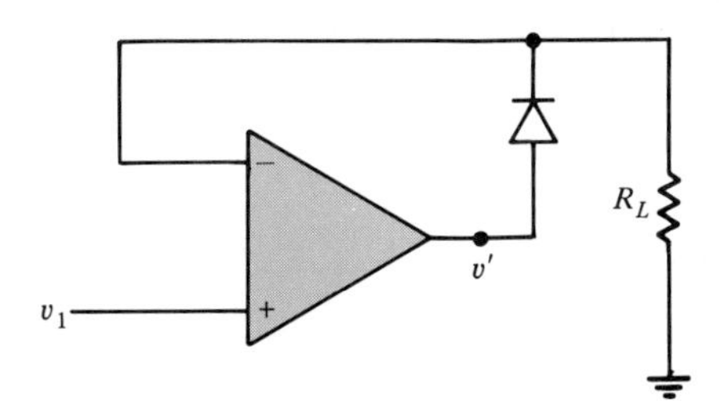

FIGURE 10.35

10.8

In light of Problem 10.7, explain the manner in which the circuit in Figure 10.36 operates. What is its function? (V_R is a reference voltage.)

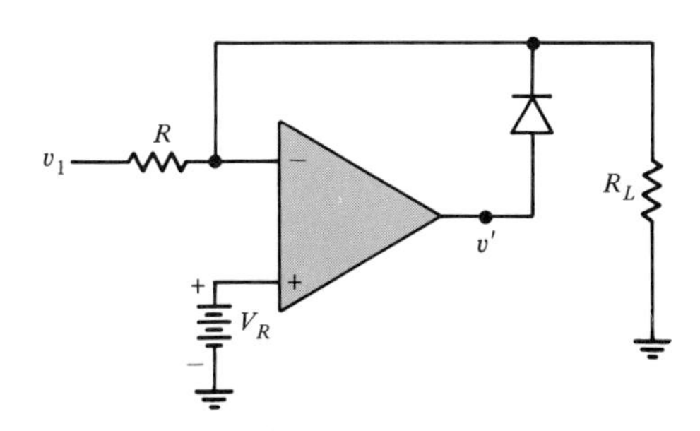

FIGURE 10.36

10.9

A voltage follower operates with 100% negative feedback. But it has been indicated that it is possible with excessive feedback for a circuit to go into oscillation at some high frequency at which the amplifier contributes a 180° phase shift, above the usual 180° due to negative feedback. One of the earliest of IC OPAMPs was the 702, which has a midband gain of 68 dB. At a frequency of 12.5 MHz, there arose a −180° phase shift, and the gain at this frequency was 36 dB. What is the maximum feedback (in decibels) that could be employed with this amplifier? What would be the numerical voltage gain with this feedback? What would be the maximum actual value of β?

10.10

As a further example of the noise gain leading to identical results in the noninverting and inverting cases, use it to compute the necessary *amplifier* voltage gains in order to achieve unity overall gain in both cases.

10.11

For the 741 the bias current can be as large as 200 nA. Refer to Figure 10.24. If the noninverting input were grounded, what would be the output voltage due to the bias current? For the same gain of 10, if R_1 = 100 kΩ and R_F = 1 MΩ, what would be the output voltage due to the bias current? What would be the output voltage due to I_{os} = 10 nA in the two cases?

10.12

Using four-cycle semilog paper, with corner frequencies at 1, 5, and 50 MHz and an open-loop gain of 80 dB, draw Bode diagrams for amplitude and phase. In terms of decibels, what is the maximum allowable amount of negative feedback that may be used with this amplifier? (See "A Closer Look" in this chapter for a discussion of Bode plots.)

10.13

An OPAMP has corner frequencies (that is, poles) at 1, 10, and 100 MHz.

a. What is the rate of phase change in the range 1 to 100 MHz?

b. What is the phase shift at 1.0 MHz?

c. What is the frequency at which the phase shift becomes −180°?

d. What is the rate of gain drop-off in the range 10 to 100 MHz?

e. What is the maximum allowable loop gain if the midband gain is 100 dB?

f. For stable operation (with no phase margin), what is the maximum allowable percentage of negative feedback?

10.14

Using the idealized Bode plots, it is assumed that at $f/f_p = 0.1$ and $f/f_p = 10$ the phase angle for a single-stage amplifier is 0° and −90°, respectively (f_p is the corner frequency). What are the actual values of the phase shifts at these frequencies? Ideally, at $f/f_p = 0.1$ the actual gain A is taken to be the midband gain A_o. Thus, ideally, $|A/A_o| = 0$ dB. What is the actual value?

10.15
With an output swing of 10 V (peak) and full power response at 18 kHz, what is the required slew rate?

10.16

a. The device in Figure 10.37 is known as a *summing* amplifier. Since the noninverting input is at virtual ground, the current through each of the input resistances can be taken to be independent of the others. Using the superposition principle, express the total output voltage in terms of the inputs and the resistances.

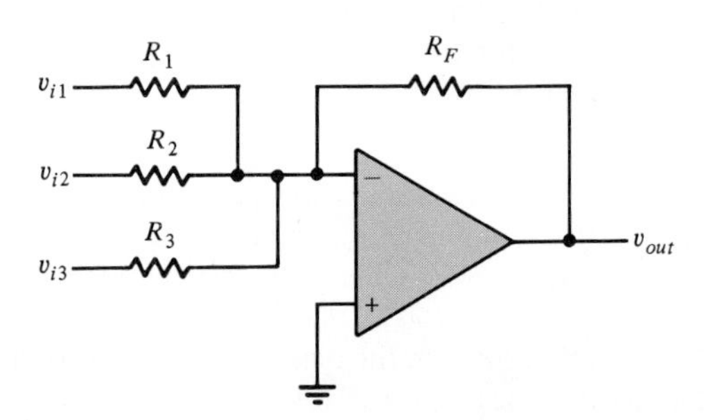

FIGURE 10.37

b. If $R_1 = R_2 = R_3 = 3R_F$, what function is performed by this circuit?

10.17

a. To take into account the finite input current and resistance of an inverting OPAMP, Eq. 10.1 takes the form

$$\frac{v_1 - v_{in}}{R_1} - \frac{v_{in} - v_{out}}{R_F} = i_{in} = \frac{v_{in}}{R_{in}}$$

Derive the exact expression for the gain (A_V) in terms of the open-loop gain A_o.

b. Using values applicable to the 741,—that is, $R_{in} = 2\ \text{M}\Omega$ and $A_o = -2 \times 10^5$—compute the percentage error incurred by using the approximate gain equation ($A_V = -R_F/R_1$) for gains of 10^4, 10^3, and 10^2. Use $R_F = 50\ \text{k}\Omega$.

10.18
See Figure 10.38.

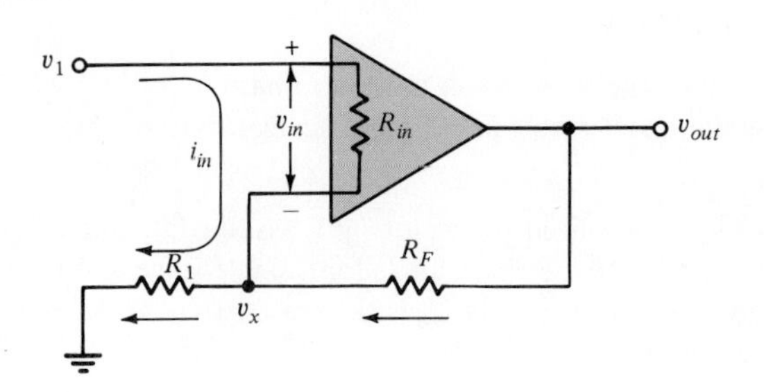

FIGURE 10.38

a. To take into account the finite input current and resistance of a noninverting OPAMP, we obtain the equation

$$\frac{v_1 - v_x}{R_{in}} + \frac{v_{out} - v_x}{R_F} = \frac{v_x}{R_1}$$

while $A(v_1 - v_x) = v_{out}$. Derive the exact expression for the gain (A_V) in terms of the open loop gain A_o.

b. Using values applicable to the 741—that is, $R_{in} = 2\ \text{M}\Omega$ and $A_o = 2 \times 10^5$—compute the percentage error incurred by using the approximate gain equation—$A_V \simeq (R_F + R_1)/R_1$—for gains of 10^4, 10^3, and 10^2. Use $R_F = 50\ \text{k}\Omega$.

10.19
In Problem 10.18, if one resorts to a solution with all the currents reversed from that indicated in the figure, in what manner must the open-loop gain (A_o) be altered?

10.20
Justify that the circuit in Figure 10.39 represents a constant current source for R_1 under varying load conditions—that is, variation of R_1. What is a function of R_2? What is the disadvantage of the circuit?

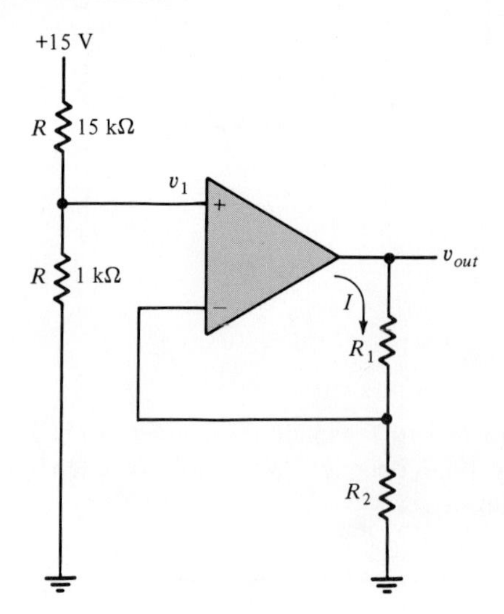

FIGURE 10.39

10.21

a. With regard to Problem 10.7, if the OPAMP is a 741, what sets a limit as to the frequency of the input signal? (*Hint*: When the signal swings negative, the feedback path is disengaged, and the output saturates at $-V_{EE}$.)

b. This deficiency may be improved upon by using the half-wave rectifier circuit in Figure 10.40. Determine its mode of operation by successively applying a positive and a negative input voltage and noting the nature of the conduction (or nonconduction) of the diodes. (*Hint*: It is the negative half-cycles that are rectified. If positive rectification is desired, the output may be passed on to a unity-gain inverter.)

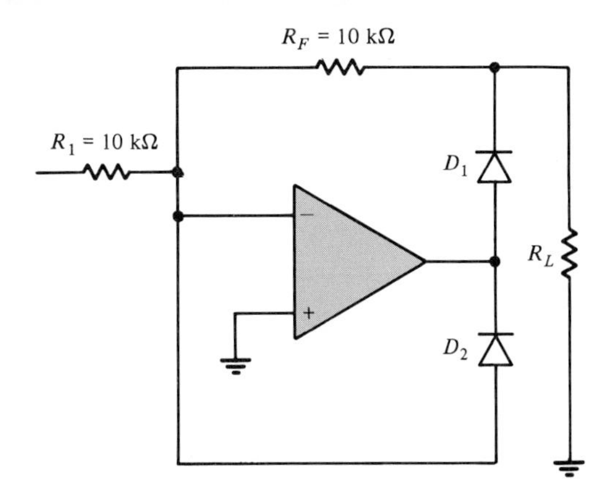

FIGURE 10.40

10.22

This problem might appropriately be entitled "making a mountain out of a molehill." To achieve large gains and a substantial input resistance, the ratio of feedback to input resistance of an inverting amplifier often demands a value for the feedback resistance that is too large to conveniently be fabricated by integrated circuit techniques. The T-network, employed as shown in Figure 10.41, eliminates this difficulty.

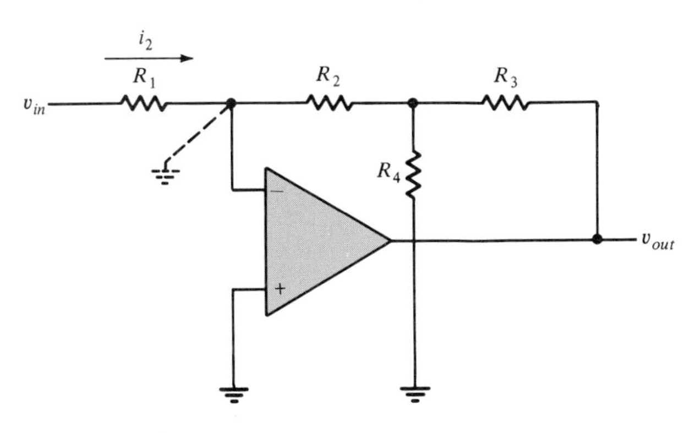

FIGURE 10.41

Since the inverting input is at virtual ground, the network can be represented as in Figure 10.42.

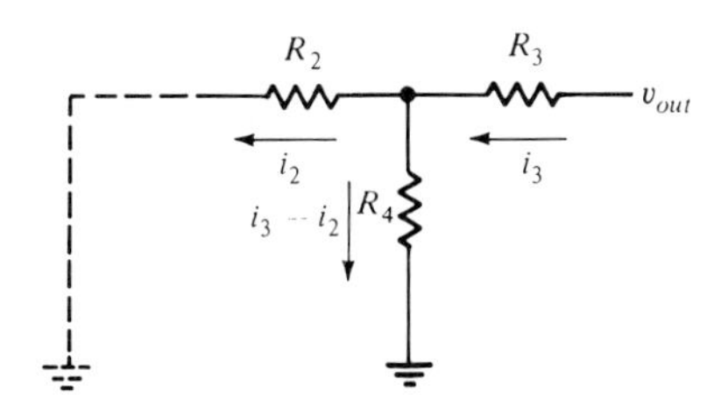

FIGURE 10.42

a. Derive the expression for v_{out}/i_2 utilizing Figure 10.42.

b. With the result from part a, determine the amplifier's gain, that is, v_{out}/v_{in}.

c. If $R_1 = R_2 = R_3 = 100$ kΩ, what is the effective value of the feedback network if $R_4 = 1$ kΩ?

10.23

The same procedure as used in Problem 10.22 may be employed with differential amplifiers (Figure 10.43). In this case the voltage across R_4 is twice that of the single-sided circuit, and it behaves like a resistance that has half the actual value. In this case, what is the expression for the gain? If $R_1 = 25$ kΩ, $R_2 = 250$ kΩ, $R_3 = 99$ kΩ, and $R_4 = 2$ kΩ, what is the gain? What is the *effective* value of each of the feedback networks?

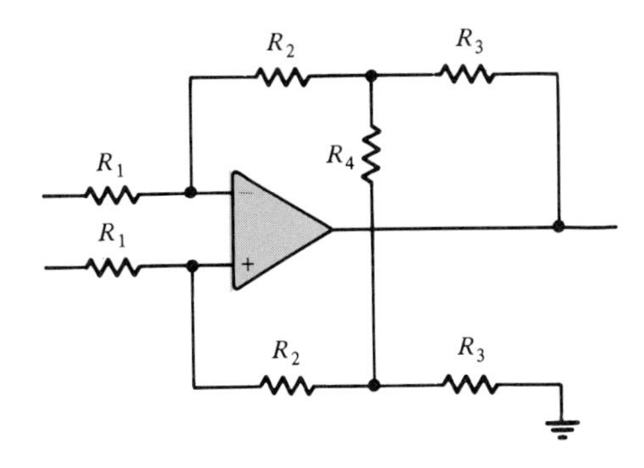

FIGURE 10.43

PART

ELECTRONIC CONCEPTS: Pulse Signals

11 DIGITAL ELECTRONICS

OVERVIEW

An analog signal is one that is continuously variable both in terms of magnitude and time. In Parts I and II we have primarily dealt with analog signals. Commencing with this chapter, we begin a consideration of digital signals. A transistor, in such cases, is either in an on or off state, and a major advantage of such operation is its diminished susceptibility to noise.

The generation of digital signals is most conveniently accomplished through the use of logic circuits, which are available in integrated circuit form. However, our discussion commences with their generation using discrete-form electronics—that is, through the use of individual transistors, resistors, and capacitors—which provides a better insight as to what transpires. We then review digital pulses obtained through the use of IC operational amplifiers; one can consider such usage to represent a transition mode between the analog and digital worlds. We conclude with a consideration of the aforementioned logic circuits.

The basic circuits introduced in this chapter will be utilized in Chapter 12 (oscilloscopes), Chapter 14 (microcomputers), Chapter 15 (transistorized automotive controls), and Chapter 17 (measurements).

OUTLINE

11.1 ■ INTRODUCTION

Despite the apparently high degree of complexity represented by systems such as a digital computer, to a large degree they consist of repetitions of a few basic circuits. Rather than having an output that varies in proportion to the input signal, electronic amplifiers used in conjunction with such circuits assume one of two extreme output conditions—either they are conducting or they are nonconducting. To represent such conditions, one employs a type of mathematics termed **Boolean algebra**, consisting of but two numbers, 0 and 1. A fundamental circuit used in conjunction with such a system is the *flip-flop*. In response to an incoming sequence of pulses, its output voltage will alternate between being "full on" and being "full off." Such conditions may, respectively, be equated to 1 and 0.

With analog circuitry we found it necessary to be able to generate sinusoidal waveforms with controllable frequencies. The comparable procedure with pulse circuitry is the generation of square waves with controllable periods. Such signal generation is the function of an **astable multivibrator** (often simply called an MV).

Another name for the flip-flop is **bistable multivibrator**. It has two stable states between which it switches in response to input signals, in contrast to the astable MV, which continually switches between two states without the need for an external stimulus.

There is also a **one-shot MV** that, in response to an input pulse, changes state, remains there for a fixed controllable length of time, and then reverts back to the original state to await another input pulse. It also has another name, **monostable multivibrator**.

We will first consider the discrete versions of the flip-flop and the astable MV, that is, circuitry employing two separate transistors. Such analyses will provide the reader with an appreciation of the processes involved. We will then turn to a consideration of how these and other digital functions may be implemented utilizing integrated circuits, first in terms that represent modifications of analog IC circuits and then in terms of logic IC circuits.

By way of preparation for the discussion of microcomputers in a subsequent chapter, we will compare decimal and binary numbering systems, and we will conclude by putting together a number of the circuits discussed in this chapter into a system that may be used for frequency measurement.

11.2 ■ DISCRETE COMPONENT DIGITAL CIRCUITS

11.2.1 The Flip-flop

We turn now to a series of fundamental circuits that have widespread applications. The first, alternatively called the **Eccles-Jordan circuit**, the bistable multivibrator, or (more flippantly) the **flip-flop**, finds application in a wide variety of devices: in electronic counters, in oscilloscope displays, and in computers, to name but a few of its uses.

We start by considering two resistive branches (Figure 11.1). By simple Ohm's law one may compute the voltage drop across each resistor. As we have indicated, it is customary to measure voltages relative to ground. Let us, therefore, ground the negative side of the power source and use this as a reference (Figure 11.2).

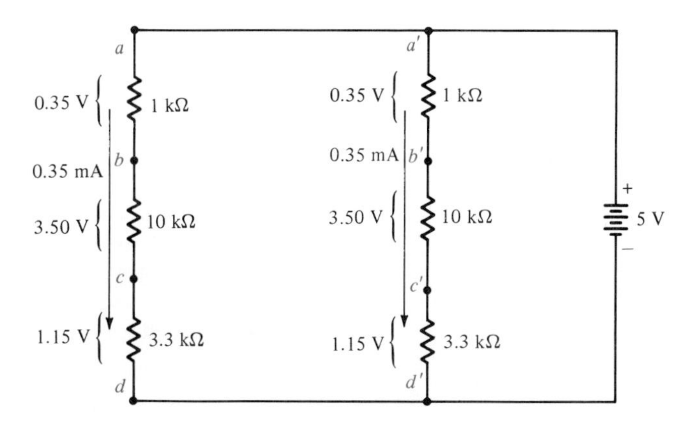

FIGURE 11.1. Identical Resistive Branches That Form the Basis of a Flip-flop Circuit The voltage drops across respective resistors are indicated.

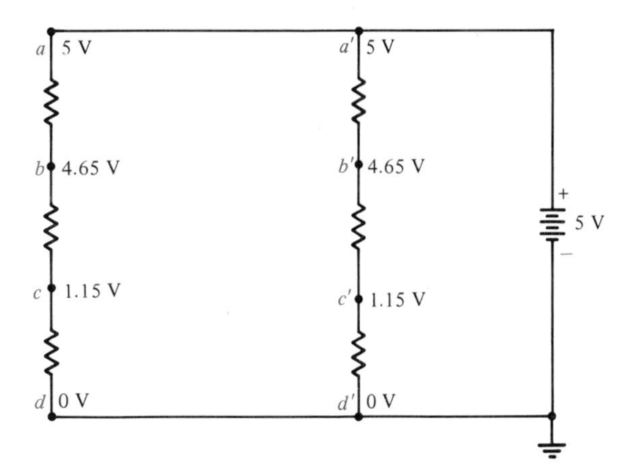

FIGURE 11.2. Voltage Dividers With the negative side of the power source grounded, the (positive) potential (relative to ground) at each junction is as indicated.

Refer to Figure 11.3. At point b we connect the collector of an *npn* transistor. Its base is tied to point c', and its emitter is grounded. Since point c' is initially at $+1.15$ V, the transistor will be driven into saturation (see Example 7.3), with point c' assuming a value of about $+0.8$ V, due to the base current drawn through R_1' and R_2'. In turn, the collector current through R_1 causes b to assume a value of about $+0.2$ V, being V_{CEsat}.

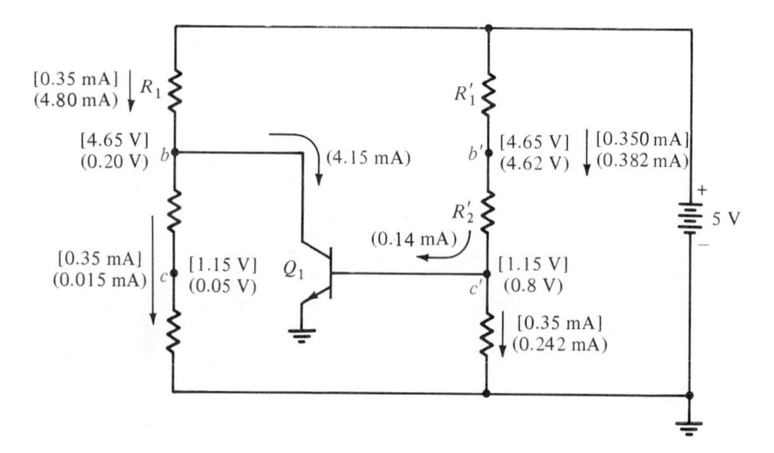

FIGURE 11.3. Divider Modification by a Saturated Transistor The initially substantial positive potential at the base of the *npn* transistor (Q_1) drives it into saturation, resulting in a base potential of about 0.8 V, and a collector potential of 0.2 V. Initial currents and voltages are indicated by brackets and values after connecting Q_1, by parentheses.

Let us now connect a second transistor (Q_2) in the manner shown in Figure 11.4. With Q_1 conducting, the base potential of Q_2 is at +0.05 V relative to its emitter, and this voltage is insufficient to bring about conduction; Q_2 remains cut off. The collector potential of Q_2 is +4.62 V, down from +4.65 V because point c' is now at +0.8 V.

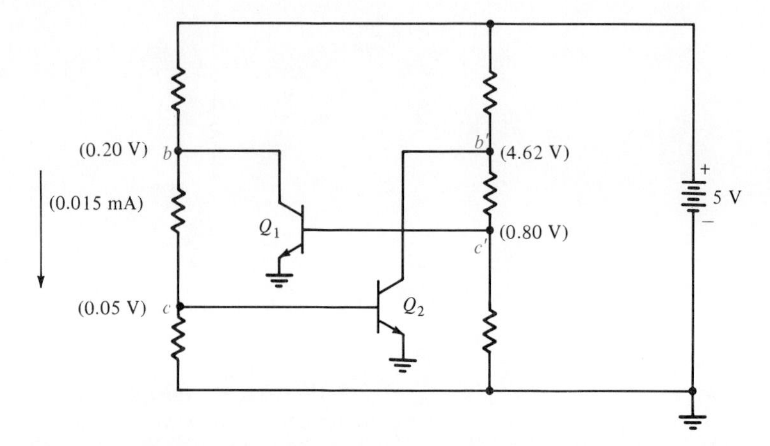

FIGURE 11.4. Stable State of a Flip-flop As a result of the saturated condition of Q_1, the base of Q_2 is insufficiently positive, and Q_2 remains in a cutoff condition.

The system will stay in this condition indefinitely. But, if we apply a negative pulse to the base of Q_1, this will cut it off, and the roles of Q_1 and Q_2 will be interchanged, with Q_2 conducting. Let us follow the sequence.

The negative pulse applied to the base of Q_1 stops conduction in Q_1.

With the diminished current through R_1, the potential at point b moves in the positive direction toward 4.65 V (as shown originally in Figure 11.3). Simultaneously, point c tends toward 1.15 V. With this positive voltage at the base of Q_2, the latter conducts, and (referring to Figure 11.4) all the voltages at the left resistive branch are now applicable to the right branch and vice versa. Now, indefinitely, Q_2 will remain conducting, and Q_1 will be cut off. We may have the circuit revert to its original state, with Q_1 on and Q_2 off, by applying a negative pulse to the base of Q_2.

The alternate placement of negative pulses at points c and c' may be most easily accomplished by applying such pulses through capacitors, as shown in Figure 11.5. Assume Q_1 to be conducting. The base of Q_2 is at 0.05 V and the base of Q_1, at 0.8 V. A negative trigger pulse applied simultaneously to each capacitor means that point c momentarily goes negative. As far as the base of Q_2 is concerned, this change in voltage does nothing. It was cut off at +0.05 V, and it remains cut off at any negative voltage. Point c', however, is diminished from +0.8 V to a negative voltage that cuts off the collector current in Q_1. With Q_1 cut off, the diminished current through R_1 (now consisting only of the bleeder flow) causes point b to move in a positive direction; point c will do likewise. This resultant voltage at c will cause Q_2 to conduct, and the large collector current through R_1' causes point c' to assume a voltage below cutoff, assuring that Q_1 remains nonconducting. (The applied negative pulse can be presumed to have been removed in the meantime, and the situation remains stable.)

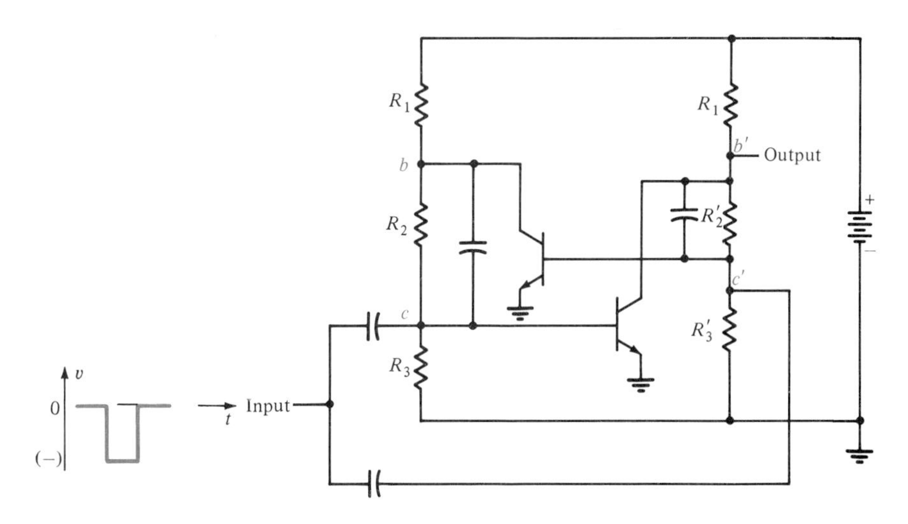

FIGURE 11.5. **Initiating Transitions in a Flip-flop** Successive negative pulses applied to the transistor bases through capacitors will cause Q_1 and Q_2 to conduct in alternate fashion.

The next negative pulse will initially do nothing to the base of Q_1 (already cut off), but the sequence of events will be reversed, and Q_1 will eventually go into conduction, with Q_2 cut off. Thus, we see that each applied pulse will cause the system to change its state.

There is a cross-connection between Q_1 and Q_2. Upon being triggered, the altered collector potential of one transistor influences the base of the opposite one and vice versa. Such coupling takes place through R_2 and R_2'. Without going into the detailed explanation of the physics involved, let it simply be noted that if capacitors are placed in parallel with R_2 and R_2', the coupling is made more intimate. They act as short circuits across the resistors and assure fast transition between states, but under static conditions the circuit is in no way affected by their presence.

The useful output from this circuit may be obtained, for example, from point b'. For every two input pulses, the output will give rise to a single pulse (4.62 V, down to 0.8 V, and back up to 4.62 V). Thus, this circuit can be used as an arithmetic divider. But it also has many other applications as well, particularly in its integrated circuit forms. We have presented the operation in terms of discrete transistors because this approach will help the reader to understand the evolution of the various IC forms of this circuit.

One point remains to be considered: Generally, this circuit reposes within the "bowels" of some more extensive circuitry. Sometimes we wish to know whether it is Q_1 or Q_2 that is conducting at any given time. If we place a small bulb across, say, R_1', when Q_1 is conducting, the drop across R_1' is very small; when Q_2 is conducting, the drop is substantial, sufficient to "fire" the bulb. Whenever the bulb is lit, we know that Q_2 is conducting. The bulb, of course, would be located at a point where it can be observed.

11.2.2 The Astable Multivibrator

The astable multivibrator is shown in Figure 11.6. We will see that Q_1 and Q_2 conduct alternately without any need for external pulsing and may be used to produce a train of square waves.

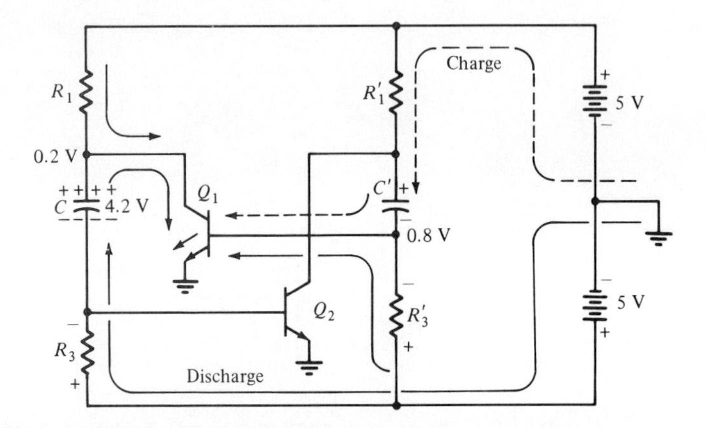

FIGURE 11.6. **Astable Multivibrator** Assuming Q_1 is conducting, the solid lines indicate the discharge path of C and the broken lines, the simultaneous charging path of C'. Associated base polarities are such as to assure conduction in Q_1 and the maintenance of Q_2 in a cutoff condition.

Assume Q_1 has just started to conduct. While it was off, C charged to 4.2 V. With Q_1 conducting heavily, its base and collector are at their saturation values (0.8 and 0.2 V, respectively). The capacitor C will then discharge, and the large initial current through R_3 puts a negative bias on the base of Q_2, thereby cutting it off. This condition can be observed in Figure 11.7 where the voltage drop across R_3 has been designed to be greater than the positive potential provided by the battery.

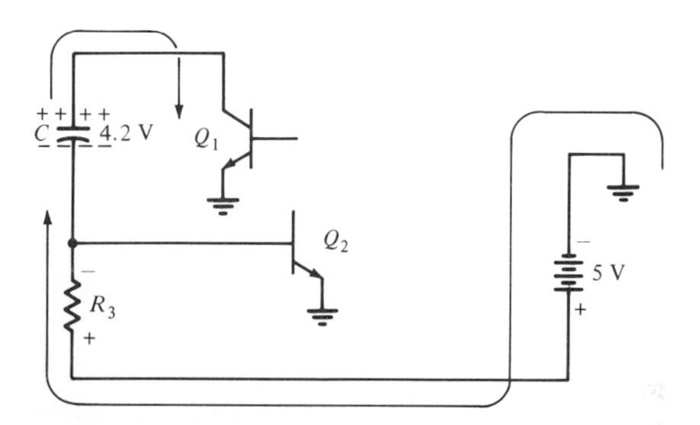

FIGURE 11.7. Current as C Commences Discharging The source voltage is the battery voltage *plus* the voltage across C. The base voltage of Q_2 is the difference between the battery voltage and the voltage drop across R_3, with the latter initially prevailing and keeping Q_2 in a cutoff condition. As the capacitor discharges, the voltage drop across R_3 diminishes due to reduced discharge current, and ultimately the positive battery potential prevails, causing Q_2 to start conduction.

In the meantime C' is connected across the upper 5 V battery and quickly is charged to 4.2 V [quickly because R_1 $(R_1') \ll R_3$ (R_3')]. The current through R_3' at this time is less than that through R_3 because only the battery is the source potential in the former case, while in the case of R_3 it is the battery *plus* the charged capacitor. Hence, the voltage drop across R_3' is insufficient to place a negative bias on Q_1 (Figure 11.8).

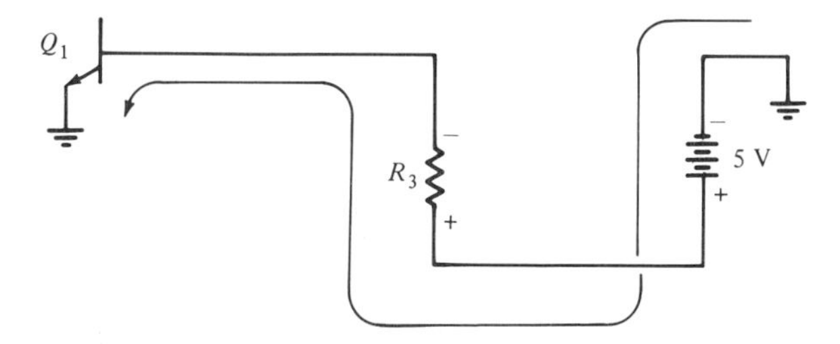

FIGURE 11.8. Base Current During Capacitor Discharge During the discharge of C, the small current through R_3' is insufficient to override the +5 V from the battery, and hence Q_1 continues to conduct.

Eventually the discharge current from C (which diminishes in an exponential fashion) leads to a voltage across R_3 that is insufficient to override the positive bias of the lower battery, and Q_2 will start to conduct. With the onset of such conduction, the roles of Q_1 and Q_2 will reverse. Q_1 will stop conducting, C will start to recharge, and C' will start to discharge.

In reality only one battery (or power supply) is necessary. The "bottoms" of R_3 and R_3' could just as well have been connected to the positive terminal of the upper battery. This procedure is followed in Figure 11.9.

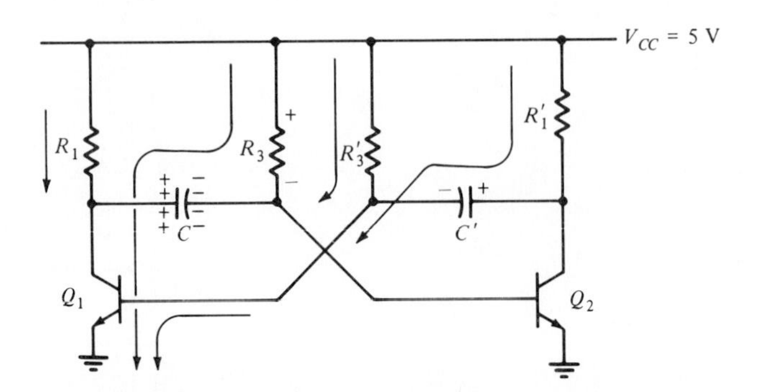

FIGURE 11.9. **Astable Multivibrator—Single Power Source Equivalent to Circuit Shown as Figure 11.6** Q_1 is presumed to be conducting, Q_2 is cut off, C is discharging, and C' is charging.

Consider Figures 11.9 and 11.10. Assume Q_1 has just started to conduct. v_{CE1} suddenly drops from $+V_{CC}$ to $+0.2$ V (the saturated value of v_{CE}). This negative swing causes C to discharge via R_3, leading to a negative bias at the base of Q_2 and cutting off the latter. In the meantime C' is charging via R_1', and v_{CE2} reaches a value $+V_{CC}$ when C' becomes fully charged.

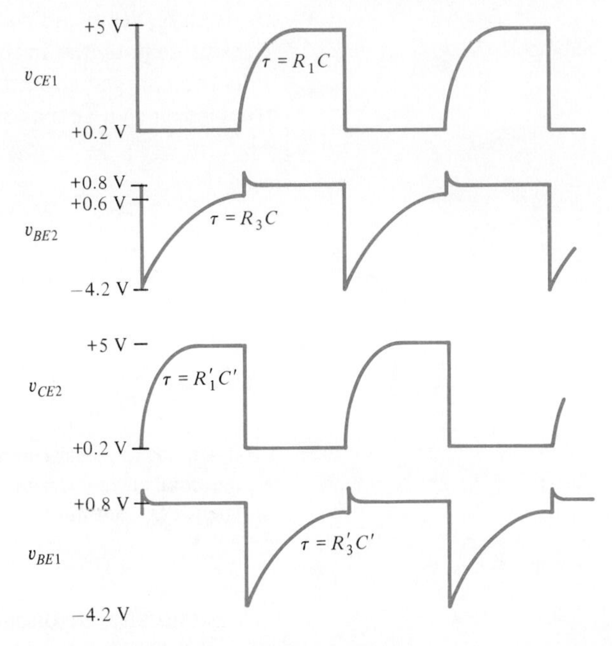

FIGURE 11.10. **Base and Collector Voltage Variations of an Astable Multivibrator** Base positive voltage peaks are due to positive-going collector voltage transfer through the coupling capacitors.

C continues to discharge through R_3 exponentially. As this discharge current diminishes to the extent that v_{R3} drops below about 4.4 V, v_{BE2} becomes sufficiently positive to cause Q_2 to start conducting.

Voltage v_{CE2} suddenly drops from $+V_{CC}$ to $+0.2$, and this negative-going step is communicated via C' to the base of Q_1. Conduction stops in Q_1 as C' discharges through R_3', giving rise to a negative potential at the base of Q_1. With Q_1 off, its collector-emitter voltage v_{CE1} should increase to $+V_{CC}$; however, this increase takes place at an exponential rate as C charges through R_1. When C is fully charged, v_{CE1} has reached $+V_{CC}$.

As the discharge of C' through R_3' diminishes, the base of Q_1 will again go sufficiently positive to start conduction. v_{CE1} goes negative, as does the base of Q_2, thereby cutting it off. With Q_1 conducting, C' charges through R_1'.

If at the collectors a square wave output is wanted, a small time constant R_1C ($R_1'C'$) is desired. But small Cs would minimize the base-switching signals, and small R_1s would limit the size of the output pulses. A solution is shown in Figure 11.11. The charge currents, rather than flowing through R_1s, are made to flow through the R_Cs. For example, if Q_2 suddenly cuts off, a positive voltage appears at the cathode of diode D_2. The charging current flow to C' must then come through R_{C1}' rather than R_1'. Without this charging flow through R_1', v_{CE2} rapidly goes to $+V_{CC}$.

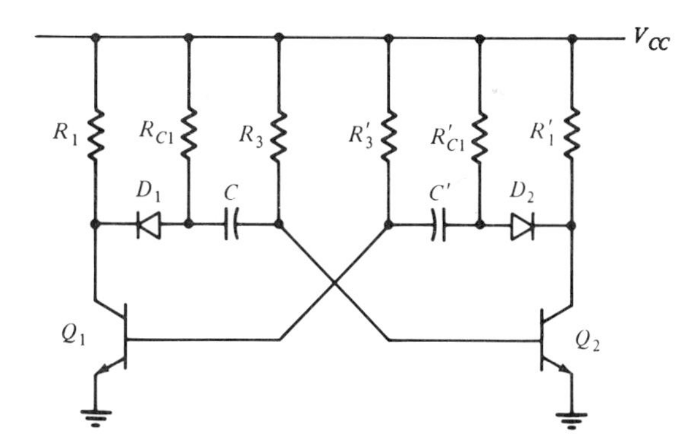

FIGURE 11.11. Astable Multivibrator Circuit That Minimizes Influence of Capacitor Charging on Output Waveform

If the two branches of an MV are identical, the two halves of the square wave will be identical and characterized as having a 50% duty cycle. The **duty cycle** is defined as the pulse "on" duration time divided by the period of the square wave.

How does the MV get started in the first place? There is always some unbalance between the two transistors, and this condition will lead to one going into conduction and the other being shut off. If one wishes to avoid the unlikely event that they are equally balanced, there are slight circuit modifications that can be employed to provide for this eventuality. (See the Malmstadt and Enke reference in Suggested Readings at the end of this chapter.)

11.3 ■ DIGITAL INTEGRATED CIRCUITS

11.3.1 Introduction

Flip-flops, multivibrators, and other forms of digital circuits may be conveniently constructed through the use of comparators. A **comparator** is an electronic device that may be used to detect whether an input signal is larger or smaller than some chosen reference. While an OPAMP such as the 741 can be made to serve as a comparator, other IC circuits are specifically designed for such operation, which they perform in a much more effective manner. We start our discussion with the OPAMP version, however, since its shortcomings in this regard more clearly delineate the unique features of the IC circuits specifically designed to serve as comparators.

11.3.2 Comparators

Refer to Figure 11.12. In Figure 11.12a is the comparator and in Figure 11.12b, its idealized response. As long as $v_{in} < V_{ref}$, the output is at the saturated value $+V_{out}$. When $v_{in} > V_{ref}$, the output is at the other saturated value $-V_{out}$. Depending on the magnitude of the gain, there is a small range of values (termed the *window*) that has the amplifier operating in a linear fashion (see Figure 11.12c). Reversing the reference will give rise to the noninverting characteristics also shown in Figure 11.12c. Because of the high voltage gain of the OPAMP, a difference of only a fraction of a millivolt between the two inputs is sufficient to cause the output to assume a saturated value.

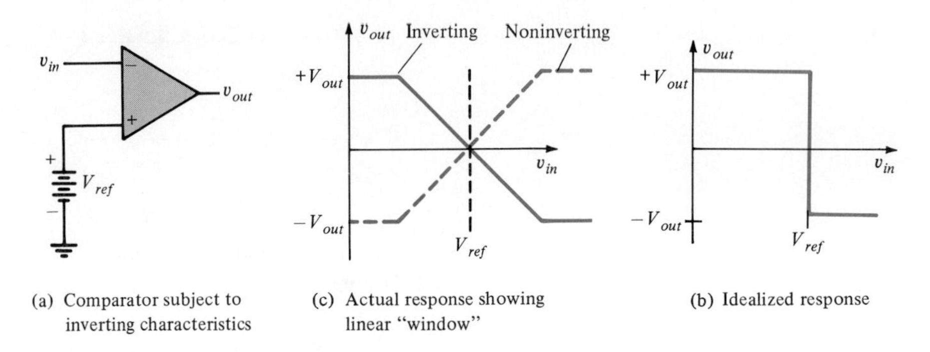

(a) Comparator subject to inverting characteristics (c) Actual response showing linear "window" (b) Idealized response

FIGURE 11.12. Comparator Operation

However, for digital operation, as the two stable states, one prefers 0 and $+V_{out}$ rather than the $-V_{out}$ and $+V_{out}$ outputs. Also, for high-speed operation a high slew rate is desirable but is inhibited by the stabilizing network incorporated into most OPAMPs. Finally, one often needs to place the outputs of two or more comparators in parallel; such might be the case if it were necessary to decide when two or more comparators were simultaneously above the reference value. The push-pull output of OPAMPs is not suited for such operation.

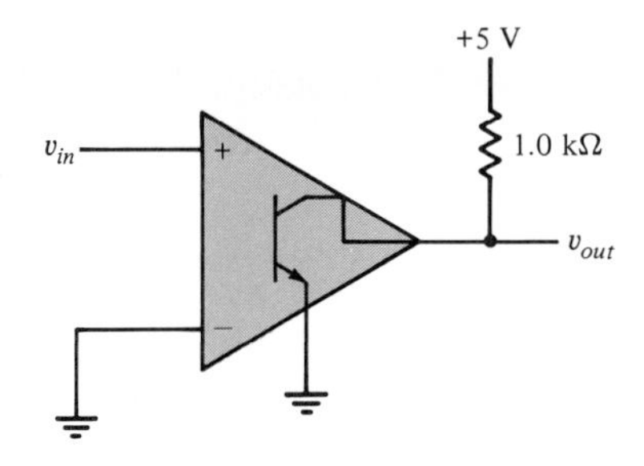

FIGURE 11.13. **IC Comparator Depicting an Open-Collector Output Requiring an External Pull-up Resistor**

ICs specifically designed to be used as comparators have no compensating networks. Such deletions lead to high values of slew rate (as much as several thousand volts per microsecond) but prevent such amplifiers from being used with negative feedback. Also, rather than the push-pull output of OPAMPs such as the 741, they have their output consisting of a grounded emitter and an "open collector." Such an output requires an external "pull-up" resistor, which is connected to a voltage source of one's own choice (Figure 11.13). The comparator thus operates between 0 V and the chosen voltage. The open collector also allows a number of comparator outputs to be operated in parallel. There is a wide choice available in the value of the collector resistance, a few hundred to a few thousand ohms being typical. The lower values provide faster switching at the expense of greater power consumption.

Rather than slew rate, in conjunction with comparators one uses the concept of propagation delay vs. input overdrive. With regard to the dc input voltage necessary to carry the amplifier into saturation (typically 100 μV), the overdrive is the ac voltage excess. The greater the overdrive, the shorter the time constant. The diminished time constant comes about because comparators, not being used with negative feedback, need no compensation. Thus, while they have a wider bandwidth than an OPAMP, which decreases their response time, the bandwidth does eventually fall off at the higher frequencies. The overdrive tends to compensate for this drop and thereby reduces the response time to an even greater degree.

The response time is the time necessary for the circuit to respond after the arrival of the signal. It is specified for both positive and negative input steps. A popular comparator, the 311, has a typical response time of 200 ns (obtained with a 100 mV step plus a 5 mV overdrive) and a typical gain of 200,000.

One other use of comparators might be mentioned at this time. If $V_{ref} = 0$, the output of a comparator will change state every time the input signal passes through zero. Such a circuit constitutes a **zero-crossing detector**. If the input to a zero-crossing detector is a sine wave, the output will be a square wave of like frequency. When applied successively to a differentiator and a clipper, one obtains a set of pulses with spacing equal to that of the sine wave period. These pulses may be used for timing markers (Figure 11.14).

11.3.3 IC Version of the Astable MV

In Figure 11.15 one should recognize the simultaneous presence of both negative and positive feedback. In the negative feedback circuit, however, rather than a strictly resistive voltage divider, one of the elements is a capacitor. The output will consist of a sequence of square waves whose period may be altered by changing the values of R_t and C_t as well as R_1 and R_2.

The circuit acts just like a comparator, comparing the voltage at the two junctions and changing the output accordingly. Assume the output voltage to be at $+V_{out}$, having just switched from $-V_{out}$. Capacitor C_t will start to discharge through R_t, rising toward the $+V_{out}$ value. As the voltage rises and exceeds the value βV_{out}, the $(-)$ terminal will now be more positive than the $(+)$ terminal, and this condition will cause the output to suddenly go to

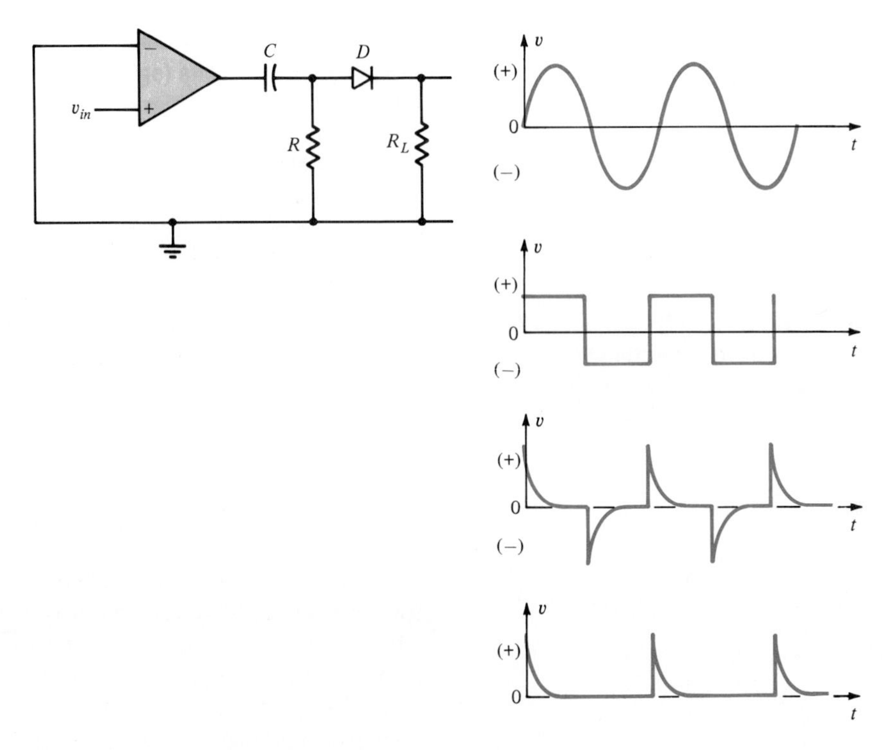

FIGURE 11.14. **Timing Pulse Generation** With a sine wave input to the comparator, followed by a differentiator (RC) and a clipper (DR_L), one may obtain timing pulses whose period is equal to the period of the incoming sine wave.

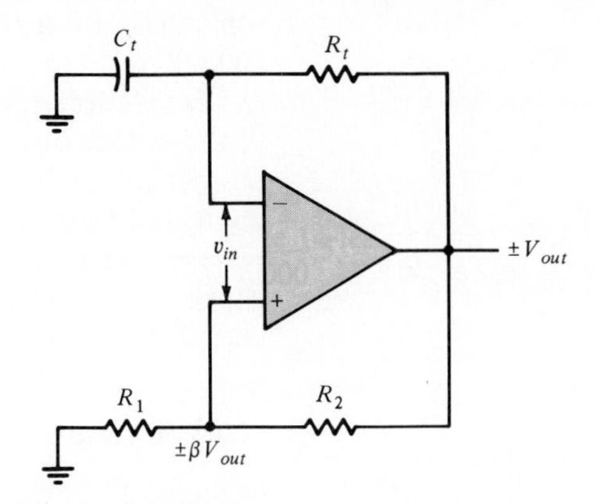

FIGURE 11.15. **OPAMP Version of an Astable Multivibrator**

$-V_{out}$. With v_{out} now at $-V_{out}$, the capacitor starts to charge toward $-V_{out}$. When it reaches a value more negative than $-\beta V_{out}$ [which is on the (+) terminal], the amplifier will again switch to $+V_{out}$ (Figure 11.16).

If t_1 and t_2 are the half-periods, $T = t_1 + t_2$, and

$$T = 2R_t C_t \ln\left(\frac{1+\beta}{1-\beta}\right) \tag{11.1}$$

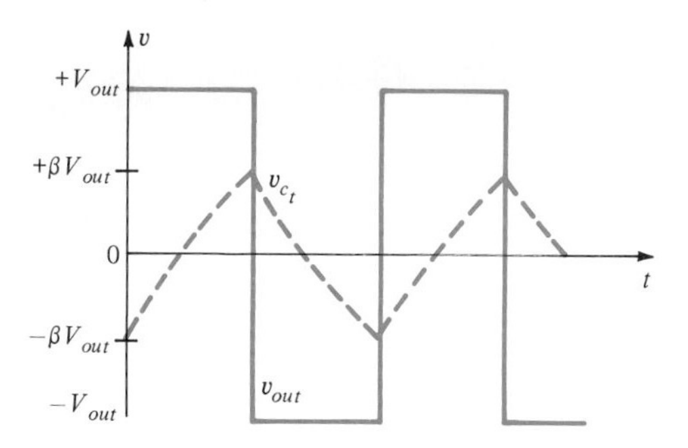

FIGURE 11.16. **Output Voltage Swing of an Astable Multivibrator (v_{out}) and the Voltage across the Capacitor v_{c_t}**

where β represents the degree of positive feedback. If the logarithmic factor is made equal to unity, the period will depend only on R_tC_t, usually a desirable condition.

11.3.4 The Monostable MV

The multivibrator that emits a succession of square waves is termed an astable MV. A useful variant of this circuit emits only a single pulse in response to a triggering signal. Such a circuit is termed a *monostable MV* as well as a *one-shot MV*. The generated pulse constitutes a gate pulse, leading to yet another name, a **gating circuit** (Figure 11.17).

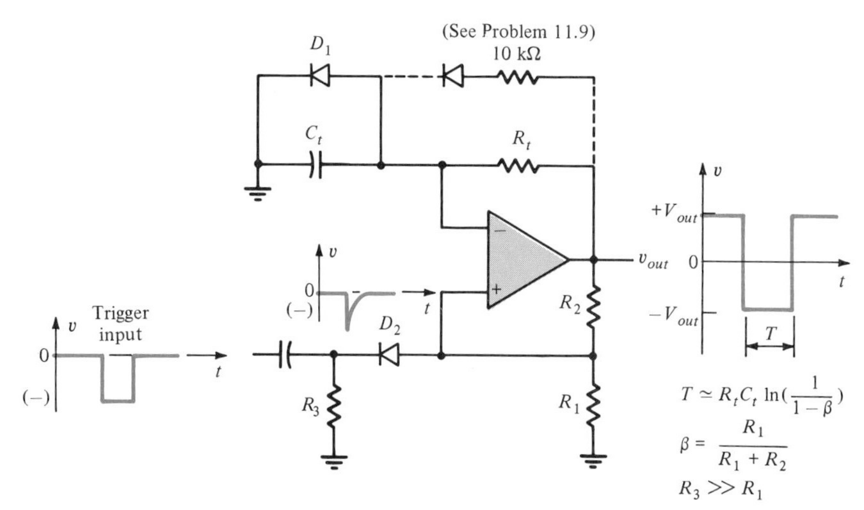

FIGURE 11.17. **Monostable Multivibrator** Each input trigger pulse will generate one output pulse whose duration (T) is determined by the circuit parameters.

Assume the output to be $+V_{out}$. Current drawn through the diode D_1 leads to a voltage (V_d) of about 0.7 V across the capacitor. The positive feedback, provided by R_1 and R_2, is proportioned so that $\beta V_{out} > V_d$, and so the (+) input retains control (Figure 11.18).

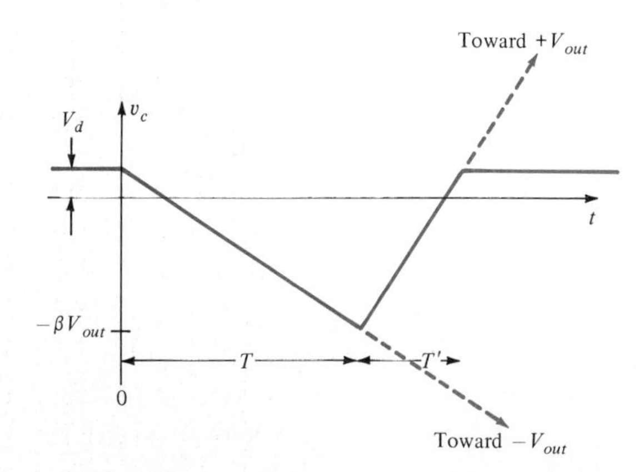

FIGURE 11.18. Time Variation of the Capacitor Voltage of a Monostable Multivibrator

The application of a trigger amplitude greater than $\beta V_{out} - V_d$ will cause the comparator to switch to $v_{out} = -V_{out}$. The capacitor will charge at an exponential rate toward $-V_{out}$, with time constant $\tau = R_t C_t$ because the diode is now reverse-biased. When v_c becomes more negative than $-\beta V_{out}$, the comparator switches back to $+V_{out}$. The capacitor now charges toward $+V_{out}$ through R_t until v_c reaches V_d, at which point everything stops. The system will remain in that equilibrium state, awaiting another triggering pulse.

The pulse width (T) is given by

$$T = R_t C_t \ln\left[\frac{1 + (V_d/V_{out})}{1 - \beta}\right] \tag{11.2}$$

If $V_{out} >> V_d$ and $R_2 = R_1$ (that is, $\beta = \frac{1}{2}$) $T = 0.69R_t C_t$. The triggering pulse must be of much smaller duration than the generated pulse, but there are minimum conditions that must also be met, depending on the comparator used.

Diode D_2 is not absolutely necessary but prevents malfunction in case there are any positive pulses on the trigger line. The time interval $(T' - T)$ represents the *recovery time*, and the next trigger pulse must be delayed accordingly.

There are commercial units of the monostable MV available for timing purposes, of which the Signetic NE/SE555 timer is the most notable. Using external *R*s and *C*s provides delays from microseconds to tens of seconds. Their use with long time delays is not recommended, being subject to numerous difficulties, including the need for expensive low-loss capacitors.

EXAMPLE 11.1 Derive the expression for the pulse width of a monostable MV in terms of the time constant, diode voltage, feedback ratio, and saturated voltage (Eq. 11.2).

Solution: Refer to Figure 11.18. It can be seen that the capacitor voltage is a composite of a constant value and an exponentially varying value—that is,

$$v_c = A + Be^{-t/R_tC_t}$$

At $t = \infty$, $v_c \to -V_{out}$; therefore, $A = -V_{out}$. We have

$$v_c = -V_{out} + Be^{-t/R_tC_t}$$

At $t = 0$, $v_c = V_d$; therefore, $B = V_d + V_{out}$. We have

$$v_c = -V_{out} + (V_{out} + V_d)e^{-t/R_tC_t}$$

At $t = \mathrm{T}$, we have

$$v_c = -\beta V_{out} = -V_{out} + (V_{out} + V_d)e^{-T/R_tC_t}$$

$$T = R_tC_t \ln\left(\frac{V_{out} + V_d}{V_{out} - \beta V_{out}}\right) = R_tC_t \ln\left[\frac{1 + (V_d/V_{out})}{1 - \beta}\right]$$

If $V_{out} >> V_d$, we have

$$T \simeq R_tC_t \ln\left(\frac{1}{1 - \beta}\right)$$

11.3.5 Schmitt Trigger

The basic performance expected of a **Schmitt trigger** circuit is illustrated in Figure 11.19. In response to a sequence of random pulses, we wish to count only those above a certain (selectable) amplitude level. Ideally, the output voltage should be raised when the input penetrates the trigger threshold and diminish when it drops below the threshold. But more satisfactory performance is achieved when there is a slight disparity between trigger on and trigger off levels for reasons that will become evident as we proceed with our discussion.

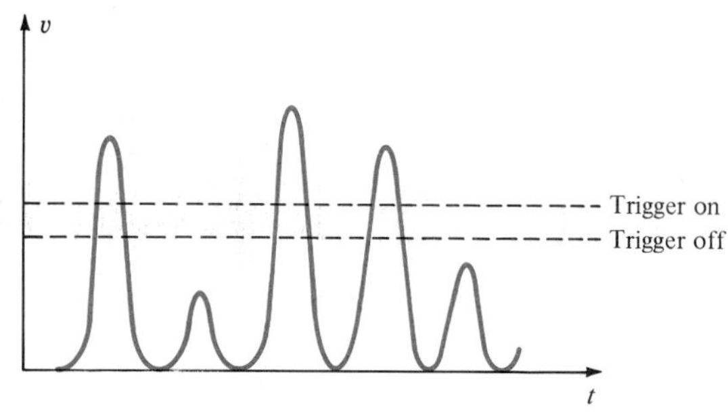

(a) Time sequence of input pulses to a Schmitt trigger circuit

(b) Output pulses resulting from input pulses whose amplitude exceeds the (adjustable) "trigger on" level

FIGURE 11.19. **Schmitt Trigger Signals**

We could again have started our consideration of Schmitt trigger operation with a circuit employing discrete transistors as we did when discussing the flip-flop and the MV, but instead we pass on immediately to the IC version employing a comparator (Figure 11.20).

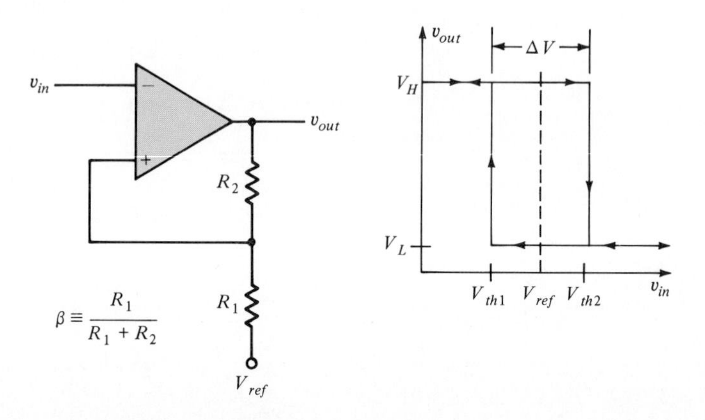

FIGURE 11.20. Schmitt Trigger Circuit, Together with v_{out} vs. v_{in} Characteristic Showing Hysteresis ΔV

The Schmitt trigger combines both the inverting and noninverting curves of the comparator (Figure 11.12c). We will call that value of v_{in} that leads to the switchover V_{th2}, the upper threshold voltage. It is called the *upper* threshold because when v_{in} is diminished, the switchback takes place at a lower level, namely V_{th1}. This difference $(V_{th2} - V_{th1})$ is termed the **hysteresis** of the circuit. It might be thought that zero hysteresis might be desirable, but such a condition represents an unstable equilibrium, with the response being carried in a random direction by the ever-present electrical noise.

With $v_{out} = V_H$ and v_{in} increasing, at $v_{in} = V_{th2}$, we have

$$V_{th2} = V_{ref} + \beta(V_H - V_{ref}) \tag{11.3}$$

Designating the voltages at the $(-)$ and $(+)$ terminals as v_- and v_+, respectively, we have

$$v_+ - v_- = v_+ - V_{th2} = \frac{V_H}{A} \tag{11.4}$$

$$V_{th2} = -\frac{V_H}{A} + v_+ = -\frac{V_H}{A} + V_{ref} + \beta(V_H - V_{ref}) \tag{11.5}$$

In comparable fashion the transition from the lower state at V_{th1} is

$$V_{th1} = -\frac{V_L}{A} + V_{ref} + \beta(V_L - V_{ref}) \tag{11.6}$$

The value of the hysteresis is then

$$\Delta V = V_{th2} - V_{th1} = \left(\frac{\beta A - 1}{A}\right)(V_H - V_L) \tag{11.7}$$

For $V_{ref} = 0$ the hysteresis is symmetrical—but not for finite values of V_{ref}. Also, if the respective values of saturation voltages are not symmetrical, neither will be the hysteresis.

EXAMPLE 11.2

A 4 V peak-to-peak 100 Hz sine wave (Figure 11.21) is applied to a Schmitt trigger whose $V_{th2} = 0$ V and that has a hysteresis of 0.2 V. Calculate the duration of the negative and positive peaks of the output as a result of applying a sine wave to the inverting input.

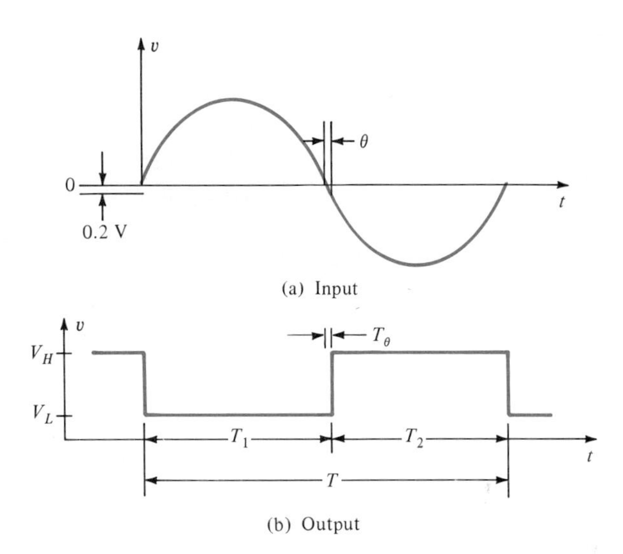

FIGURE 11.21. Sine Wave Input Applied to a Schmitt Trigger Exhibiting a 0.2 V Hysteresis, Together with the Resulting Output Square Wave

Solution:

$$2 \sin \theta = 0.2$$

$$\theta = \sin^{-1} 0.1 = 0.1 \text{ rad}$$

$$\frac{1}{f} = T = \frac{1}{100} = 10^{-2} \text{ s} = 10 \text{ ms}$$

$$\omega T_\theta = 2\pi(100)T_\theta = 0.1 \text{ rad}$$

$$T_\theta = \frac{0.1}{2\pi(100)} = 1.59 \times 10^{-4} \text{ s}$$

$$T_1 = \frac{T}{2} + T_\theta = 0.5 \times 10^{-2} + 1.59 \times 10^{-4} = 5.16 \times 10^{-3} \text{ s}$$

$$T_2 = \frac{T}{2} - T_\theta = 0.5 \times 10^{-2} - 1.59 \times 10^{-4} = 4.84 \times 10^{-3} \text{ s}$$

The Schmitt trigger employs positive feedback, which is responsible for the sudden switch to the opposite state as the threshold is approached, rather than the gradual one characteristic of a comparator window.

11.4 ■ TRUTH TABLES AND BOOLEAN ALGEBRA

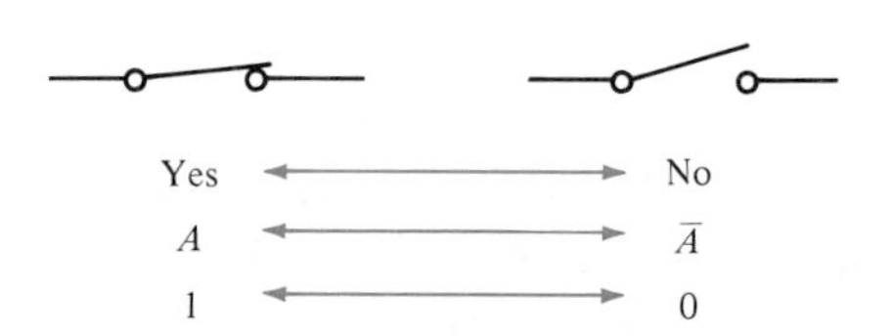

FIGURE 11.22. **Logical Alternatives—Yes, No; $A, \overline{A}$; 1, 0—Represented by a Single Switch**

Thus far we have considered some devices that may be made to interface between the analog and digital worlds (the comparator and Schmitt trigger) as well as various means for the generation and manipulation of digital pulses (multivibrators). We now temporarily depart from a consideration of digital circuit development and consider the mathematics of digital pulse manipulation.

The simplest logical statement, either a yes or a no, may be physically represented by means of a switch that can be either closed or open (Figure 11.22). We can represent a closed switch by means of a symbol such as A and an open switch by $\overline{A}$ (which is to be read as "not A"). A can be interpreted as a true statement and $\overline{A}$, as a false statement.

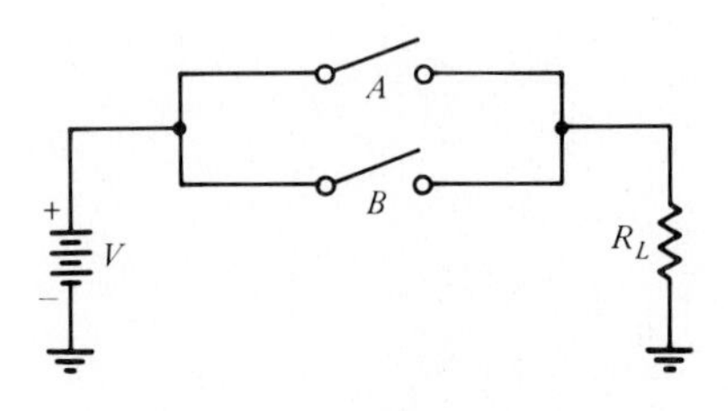

FIGURE 11.23. **Logical OR Represented by Parallel Switches**

Consider a battery supplying a load resistance through two parallel switches either of which can be independently closed or open (Figure 11.23). If X is taken to represent the voltage across the load, we may write the logical statement

$$A + B = X \tag{11.8}$$

In a logical statement + is read as OR. Thus, the preceding statement reads "A OR B equals X," meaning that with either A or B closed, the voltage appears across the load. With regard to this circuit, we may construct the **truth table** shown as Table 11.1. In the second instance 0 represents a "no" and 1 represents a "yes." We can see that the voltage does not appear (that is, $\overline{X}$) if both switches are open (that is, $\overline{A}$ OR $\overline{B}$).

TABLE 11.1

	A	B	X		A	B	X
If not A, OR not B, then not X	$\overline{A}$	$\overline{B}$	$\overline{X}$		0	0	0
If not A, OR B, then X	$\overline{A}$	B	X	or	0	1	1
If A, OR not B, then X	A	$\overline{B}$	X	alternatively	1	0	1
If A OR B, then X	A	B	X		1	1	1

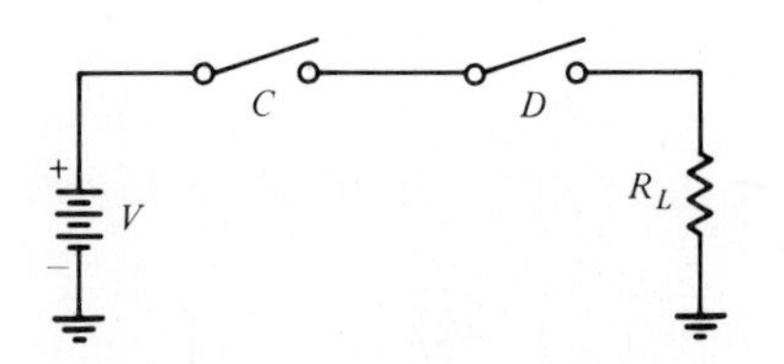

FIGURE 11.24. **Logical AND Represented by Series Switches**

Now consider two switches C and D in series (Figure 11.24). This configuration may be expressed by means of the logical equation

$$C \cdot D = Y \tag{11.9}$$

which is read as "If C AND D, then Y" and sometimes simply written as $CD = Y$. In this case the voltage appears across the load (that is, Y) if both C and D are closed. The truth table now reads as shown in Table 11.2

TABLE 11.2

C	D	Y
0	0	0
0	1	0
1	0	0
1	1	1

Figure 11.25 illustrates the symbolic representation of the logical AND and OR operations. With an AND gate the output exists only if both *A and B* are present. With an OR gate the output exists if *either A* or *B* is present. These operations can be extended to a multiplicity of inputs: The AND gate will give an output W if *A and B and C, . . . , and N* are all present, otherwise $\overline{W}$. The multiple OR gate will give an output if *either A, or B, or C, . . . , or N* is present. In the case of the EXCLUSIVE OR, an output will appear only if the two inputs are different

The type of mathematics utilized in logical manipulation is termed **Boolean algebra**. There are a number of useful theorems that we will use. At this point let's consider a few of them. See Table 11.3.

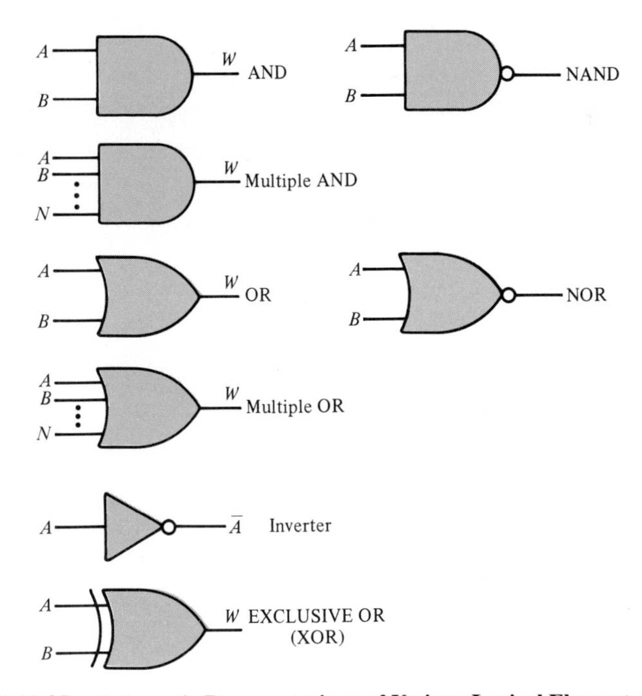

FIGURE 11.25. **Schematic Representations of Various Logical Elements**

Refer to Figure 11.26 where two A switches are *ganged*—that is, they are operated together. We may write this operation logically as

$$A + A = A \tag{11.10}$$

meaning "If A OR A, then A." In other words, one A is redundant. The operation in Figure 11.27 is equivalent to that of Figure 11.26.

Two particularly useful theorems are known as **De Morgan's theorems:**

$$\overline{A + B} = \overline{A} \cdot \overline{B} \tag{11.11}$$

$$\overline{A \cdot B} = \overline{A} + \overline{B} \tag{11.12}$$

TABLE 11.3 Some Basic Boolean Theorems

Boole's Theorems in One Variable

$A + 0 = A$	Proof: $0 + 0 = 0$ $1 + 0 = 1$ $\uparrow \quad \uparrow$ $A \quad A$
$A + 1 = 1$	Proof: $0 + 1 = 1$ $1 + 1 = 1$ $\uparrow \quad \uparrow$ $A \quad A$

Boolean Theorems in More than One Variable

Commutative:	$A + B = B + A$ $A \cdot B = B \cdot A$
Associative:	$A + (B + C) = (A + B) + C$ $A \cdot (B \cdot C) = (A \cdot B) \cdot C$
Distributive:	$A \cdot (B + C) = (A \cdot B) + (A \cdot C)$ $A + (B \cdot C) = (A + B) \cdot (A + C)$ ← not allowed in ordinary algebra
Absorption:	$A + (A \cdot B) = A$ $A \cdot (A + B) = A$

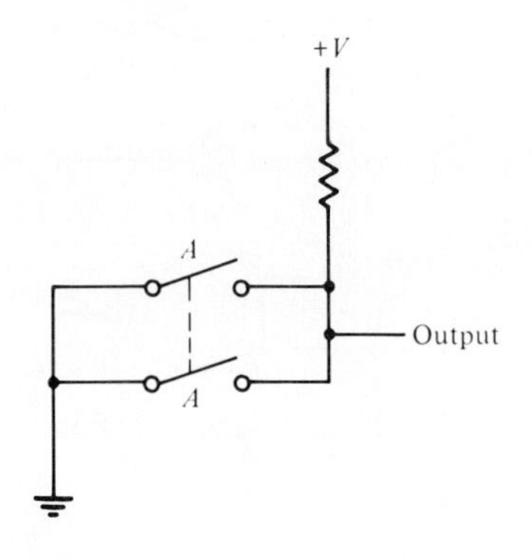

FIGURE 11.26. Physical Representation of Logical Statement $A + A = A$ The dashed connection indicates simultaneous operation of switches. (In electrical parlance, they are said to be *ganged*.)

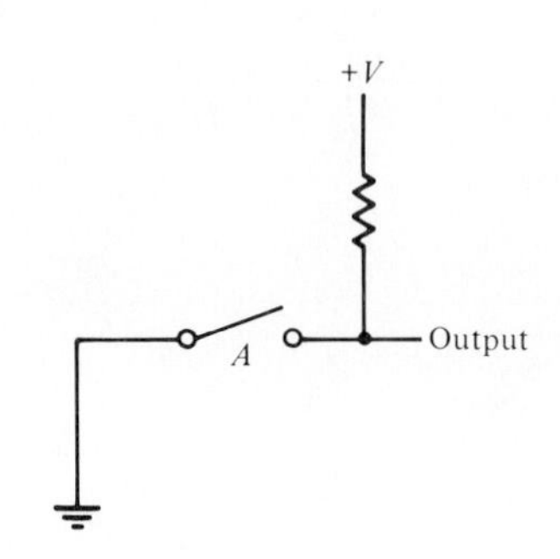

FIGURE 11.27. Redundancy Simplification Removal of redundancy in Figure 11.26 leads to this result.

EXAMPLE 11.3

A room has a light switch alongside each of its two entrances. We wish to have each switch capable of turning the lights either on or off. If A is one switch and B the other and we let L stand for the light being on, we may construct Table 11.4.

TABLE 11.4

A	B	L
0	0	0
0	1	1
1	0	1
1	1	0

Indicate what the physical realization of this truth table might be in terms of the problem posed.

Solution: See Figure 11.28.

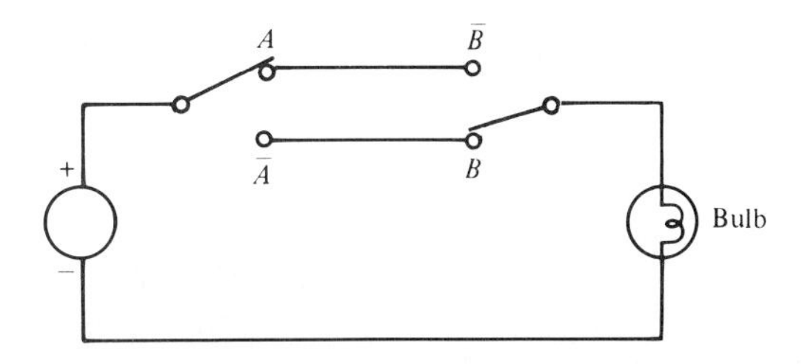

FIGURE 11.28. **Arrangement of Light Switches That Allows Control from Either of Two Room Entrances**

The electronic circuits that perform various logical functions may be quite simple. Consider the switching transistor shown in Figure 11.29. In dealing with positive logic—that is, a positive pulse represents a 1 and the absence of a pulse, 0—if the input to the transistor is 0 V, we presume that it does not conduct and that the output level is at a high positive potential since there is no drop across R. Upon applying a 1 (in this case a positive pulse), the transistor is driven into saturation, and the output voltage drops to essentially zero. Thus, there is an inversion of the pulse polarity. Such a circuit constitutes an **inverter**, and its logical symbol is as shown in Figure 11.25. If an AND gate or an OR gate is followed by an inversion process, this procedure is indicated by a small circle following the logical symbol. The AND and OR gates then, respectively, become NAND and NOR gates (again see Figure 11.25). The inverting gates—that is, NOR and NAND—are the more popular since they are more versatile. One can combine two inverting gates to make a noninverting one, but there is no way to make a noninverting gate into an inverting one.

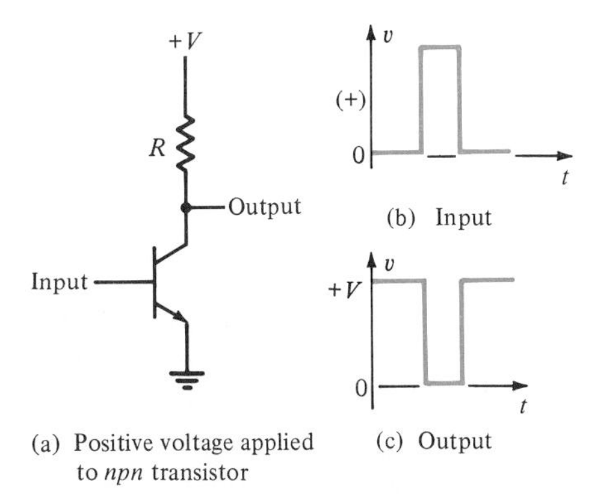

FIGURE 11.29. **Simple Inverter Circuit** Application of a positive voltage to an *npn* transistor leads to saturation and (essentially) zero output; (essentially) zero input leads to cutoff and a positive output voltage.

Given two voltage levels, if the more positive level represents a 1 and the other level a 0, the combination constitutes **positive logic**. On the other hand, if the 1 state is represented by the more negative of the levels and 0 the more positive, that combination constitutes a **negative logic** system. At times negative logic is found to be useful. Negative logic, however, does not mean that the pulse must be negative. In this text we will deal exclusively with positive logic.

In referring to the OR truth table, we see that we also get a 1 output if both inputs are 1. In many instances we wish to have an output *only* if one or the other input is 1 but not both. Such was the case when we considered turning on a room's lights from two different locations (Example 11.3). Such a circuit constitutes an **EXCLUSIVE OR** (XOR) and is indicated by the symbol $\oplus$. We have shown how the EXCLUSIVE OR may be implemented with switches. We now show how it may be done with logical circuits (Figure 11.30). The output of an EXCLUSIVE OR is $(A + B) \cdot (\overline{A \cdot B})$, meaning "$A$ or B, but not (A and B)." The reader should be able to verify that the two circuits in Figure 11.30 equate to this result.

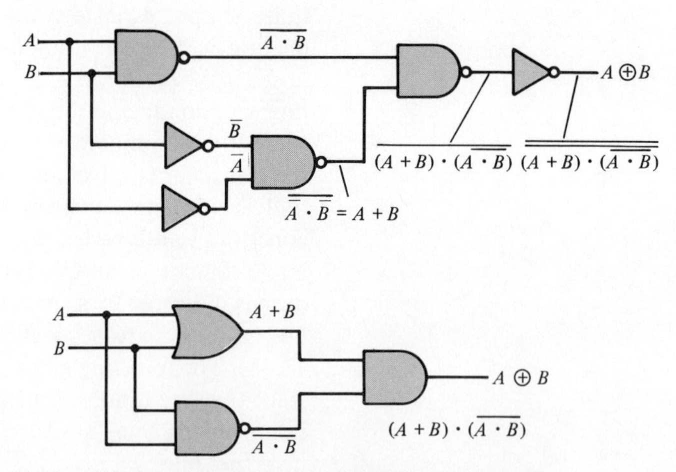

FIGURE 11.30. **Two Manifestations of EXCLUSIVE OR Logic**

11.5 ▪ LOGIC CIRCUITS

11.5.1 The Digital Flip-flop

In prior sections we discussed the operation of a flip-flop (FF) employing discrete components (transistors, resistors, capacitors) as well as versions employing IC comparators. The FF can also be constructed by a combination of logical elements, with the complete unit being available in IC form. Figure 11.31 shows how two NOR gates may be combined to form an ***RS* flip-flop**, the R and S standing, respectively, for RESET and SET. A 1 applied to the SET input yields a 1 output at Q. (The output at $\overline{Q}$ will always be the complement of Q—in this case, 0.) A 1 applied to the RESET input yields a 0 output at Q (and a 1 at $\overline{Q}$). With both inputs at 0, the FF retains its existing state. A major difficulty with the *RS* flip-flop arises if both *R* and *S* inputs are simultaneously 1. This combination is a disallowed state. Both Q and $\overline{Q}$ will temporarily go to zero. Upon termination of the input pulses, the last one terminated will determine the final status.

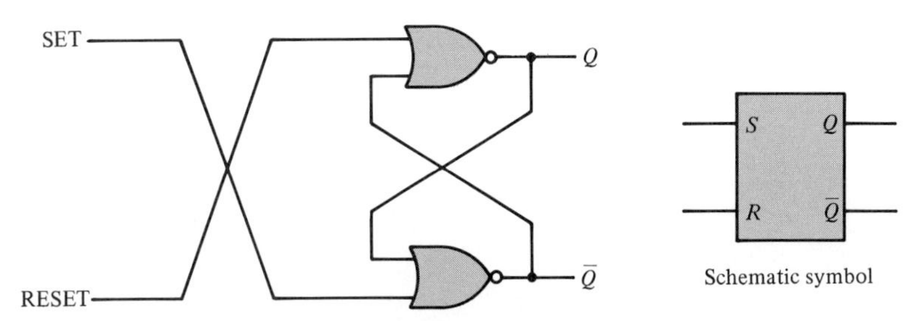

FIGURE 11.31. **NOR Gates Used to Form *RS* Flip-flop**

Figure 11.32 shows a clocked *RS* FF. Inputs at the *R* and *S* terminals will have no effect unless the clock pulse is present. Depending on what we want the flip-flop to do, the inputs are set accordingly, but this operation will not be carried out until the **clock pulse** is applied.

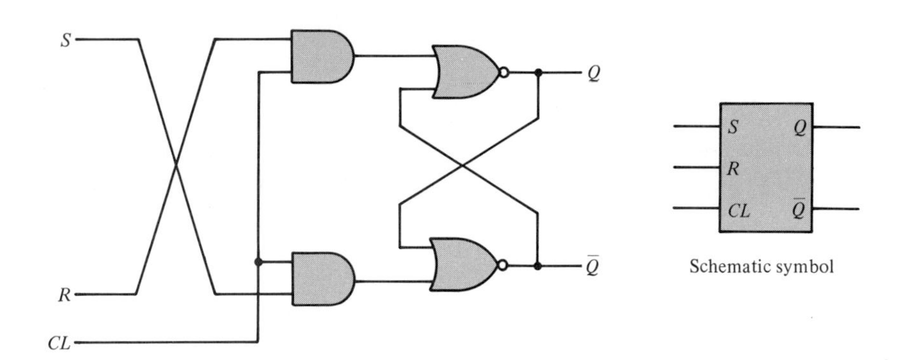

FIGURE 11.32. **Clocked *RS* Flip-flop**

In Figure 11.33 each of the NOR gates is made to have three inputs. Before reacting to the *RS* inputs, it may be desired to set the FF to 0. This clearing operation may be accomplished by applying an appropriate pulse to the clear (CLR) terminal. If instead an initial 1 is desired, this condition may be achieved through the preset (PR) terminal.

The *RS* FF could be made to divide the input pulse rate by 2 if we cross-couple the outputs back to the inputs, Q to R and $\overline{Q}$ to S (Figure 11.34). (This operation, of course, is quite reminiscent of what we did with the discrete element FF.) Each time a clock pulse is applied, the FF changes state. Thus, for every two clock pulses, we get one output pulse (from Q).

But there is a potential difficulty with this circuit. If the clock pulse stays

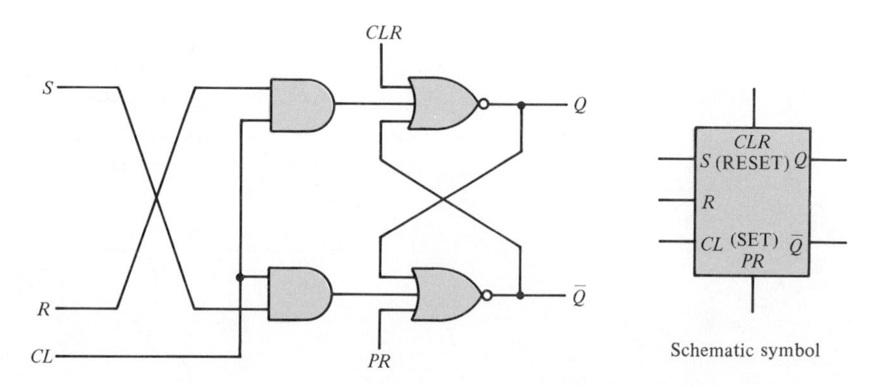

FIGURE 11.33. **Clocked *RS* Flip-flop, Together with CLEAR and RESET Inputs**

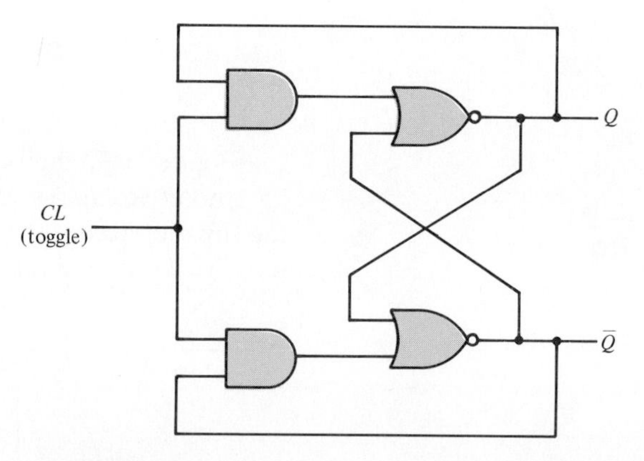

FIGURE 11.34. **Toggle Flip-flop** *T* (toggle) flip-flop changes state in response to each successive input pulse.

on, the new output may change the input (because it is coupled back), which again changes the input, etc. Thus, we could end up with a gated oscillator, one that puts out a sequence of square waves as long as the clock gate persists. We might try to solve this problem by applying a clock pulse of sufficiently short duration. This solution, however, is not very reliable. In fact, there is no reliable way to incorporate a single clocked FF into a counting circuit. Another major deficiency of an *RS* FF stems from the possible simultaneous presence of pulses at both inputs, an unacceptable state. Both deficiencies can be remedied by using a *JK* master-slave FF.

Figure 11.35 illustrates the *JK* **master-slave** arrangement. The truth table for the *JK* FF is given in Table 11.5. Like the *RS* FF, we again have available two input terminals, in addition to that for the clock pulse, but unlike the *RS*, there are no disallowed states with the *JK*. If both inputs are 0s, there will be no change in the output. If both inputs are 1s, the output will be reversed upon the application of a clock pulse. If $J = 1$ and $K = 0$, the output (Q) will be 1. If $J = 0$ and $K = 1$, the output (Q) will be 0.

TABLE 11.5

J	K	Q_{n+1}
0	0	Q_n
0	1	0
1	0	1
1	1	$\overline{Q}_n$

Upon application of a positive clock pulse to the *JK* master-slave, the master FF will respond to the *JK* input according to the truth table. The slave

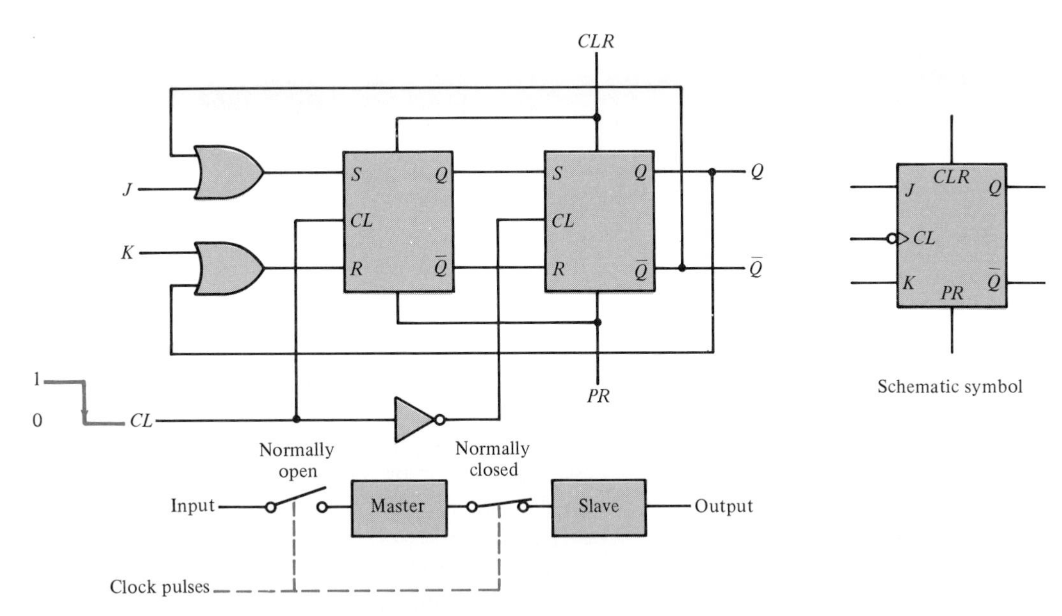

FIGURE 11.35. ***JK* Master-Slave Flip-flop Composed of Clocked *RS* Flip-flops**

remains inoperative, however, because of the negative clock pulse it has received by virtue of the inverter in the clock line. Upon termination of the clock pulse, the master is disengaged, but the slave is activated, and the output of the master is passed on to the slave as well as to the output line. Since at no time is there a completed circuit between output and input, there is no chance for regenerative pulsing.

The actual transfer of the input signals takes place on the negative-going edge of the clock pulse. On the symbolic representation of the *JK* master-slave FF, the negative aspect is indicated by the small (inverting) circle at the clock input, and that the FF is edge-triggered is indicated by a wedge alongside the circle.

Rather than utilizing gating pulses, to change the state of a system, much more effective action is obtained by the use of **edge triggering**. The data, for example, can be acquired by the master-slave at the edge of a clock pulse and the output changed to the new state on the same edge but at a later time. If the triggering takes place on the positive-going edge, a small wedge appears alongside the clock input in the symbolic representation (as in Figure 11.35, but without the circle).

It is also possible to have the master trigger on the leading edge of the pulse and have the transfer take place on the trailing edge. In the symbolic representation, such a sequence is indicated by a double wedge, together with a reversing circle (because the actual transfer takes place on the negative-going edge), and is termed a **data lockout**.

If the *JK* inputs are connected as in Figure 11.36, the circuit becomes a ***D* flip-flop**, the *D* standing for either *data* or *delay*. Data presented to the *D* input will be transferred to the *Q* output after a delay equal to one clock pulse.

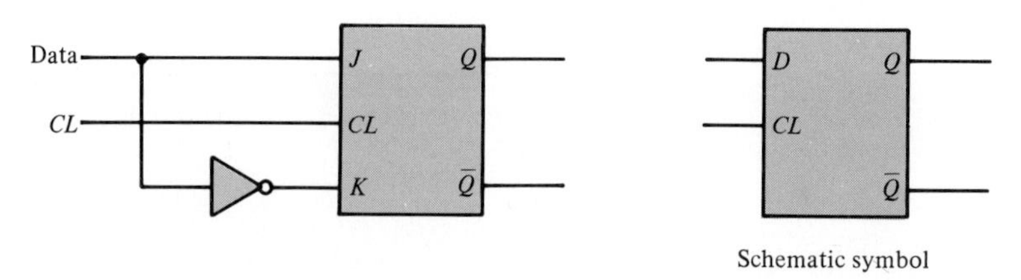

FIGURE 11.36. ***D* Flip-flop**

Figures 11.37 and 11.38 show the *D* FF and the *JK* FF being used as divider circuits. Figure 11.39 shows a *D* FF **shift register** in which the output of one FF serves as the input to the succeeding one. By such means, as the successive pulses (being any combination of 0s and 1s) enter the cascade of FFs, they are made to pass down the line. They can then be made to appear in a parallel format—that is, rather than appearing as successive pulses on a single line, the whole sequence appears simultaneously at a number of terminals. In a subsequent section we will make use of such a system when we consider the measurement of an unknown frequency and display its numerical value by a visual indicator. The parallel outputs (*A*, *B*, *C*, *D*) will represent binary integers that make up a decimal number representing the frequency.

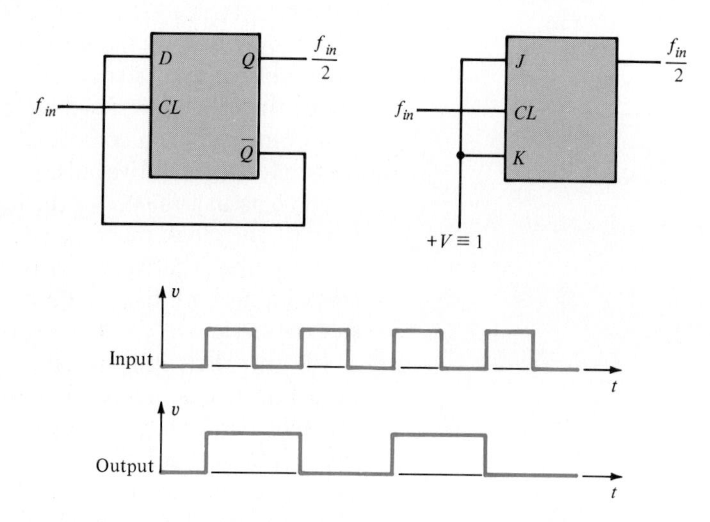

FIGURE 11.37. ***D* and *JK* Flip-flops Used in Divide-by-2 Circuits**

11.5.2 IC Logic Circuits

The best known commercially available logic circuits comprise the 7400 series.[1] Typically, the 7473 is a *JK* FF (two FFs per chip actually) in a master-slave arrangement, with a reset and with data transfer taking place on the

[1]There is also an equivalent 5400 series made to satisfy military specifications requiring operation between −55°C and +125°C. The 7400 series operates between 0°C and 70°C.

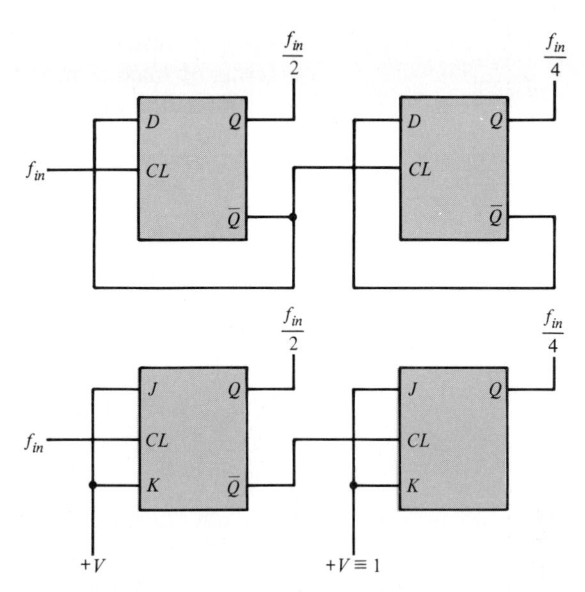

FIGURE 11.38. ***D* and *JK* Flip-flops Used in Divide-by-2 and Divide-by-4 Circuits**

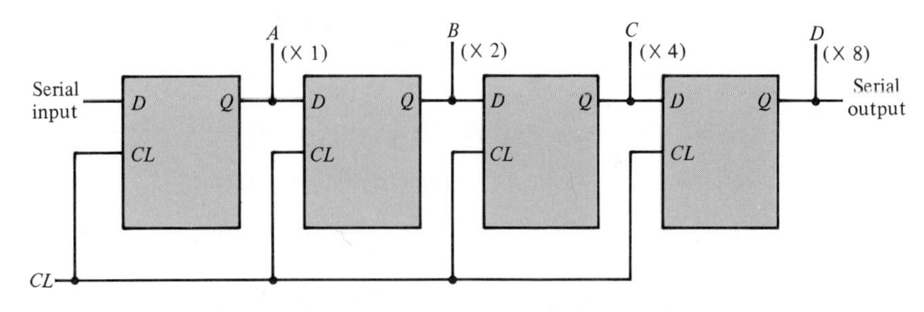

FIGURE 11.39. **Shift Register Showing Input (Serial) Pulses Converted into a Parallel Output That Forms the Basis of a Counting System**

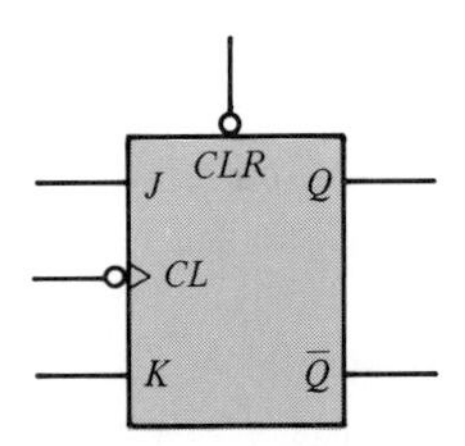

FIGURE 11.40. **Symbolic Representation of One Half of a 7473 Master-Slave *JK* Flip-flop, with Signal Transfer to the Slave Taking Place on the Negative-Going Clock Pulse**

negative edge of the clock pulse (Figure 11.40). A 7474, on the other hand (again two per chip), is a *D* FF with positive edge triggering (Figure 11.41).

These logic circuits are part of a family that has gone through a considerable evolution. Among the earliest members were the 7400 (a two-input NAND gate) and the 7410 (a three-input NAND gate). These and other members of the family are now also available in mutant forms that accomplish the same tasks but in an improved manner. For example, the 74LS00 is a lower-power and faster-acting version of the 7400. The 74S00 is an even faster version. In like manner the original 7474 dual *D* FF is available now in a variety of forms, including 74LS74 and 74S74. The LS stands for low-power Schottky series (2 mW/gate vs. 10 mW/gate for the original). (Recall that a Schottky diode is made of a junction between metal and semiconductor rather than between two semiconductors.) The LS form is slightly faster than the original (9.5 vs. 10 ns). There is also an ALS series (advanced low power) with an even greater speed (4 ns delay, with 1 mW power). The S and F series

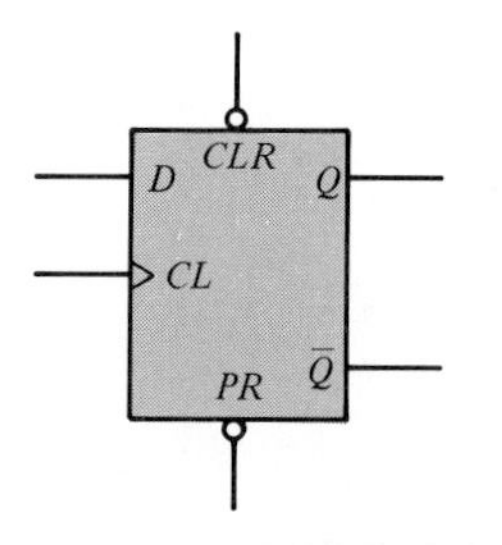

FIGURE 11.41. **Symbolic Representation of One Half of a 7474 Positive Edge-Triggered *D* Flip-flop**

are even faster (3 and 2.7 ns, but S is rated at 20 mW and F at 4 mW). In terms of maximum pulse repetition rates to which these logic elements can respond, the original 7400 series was rated at about 35 MHz, the LS runs around 45 MHz, and the S runs at 125 MHz.

11.5.3 Logic Classification

Diode Logic (DL). We commenced our discussion of logic circuits in terms of switches. Since diodes effectively act as switches, we should be able to implement logic operations by such substitutions.

Consider Figure 11.42. The diodes will conduct when their cathodes are at 0 V. With diodes conducting, most of the 5 V source voltage is dropped across the resistor, and the output voltage is very small and can be interpreted as 0.

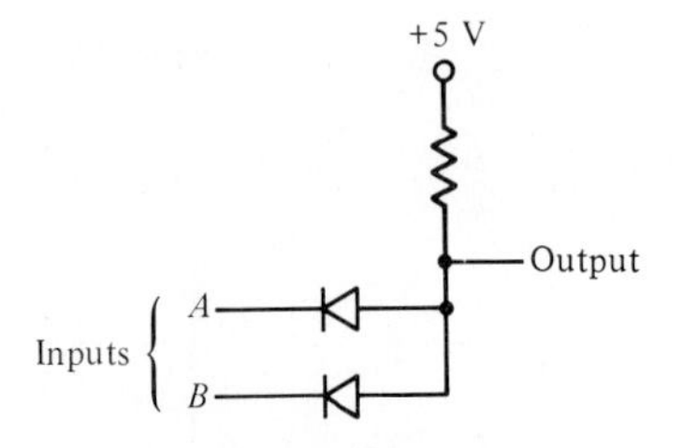

Truth table

A	*B*	Output
0	0	0
0	+5 V	0
+5 V	0	0
+5 V	+5 V	+5 V

FIGURE 11.42. **Diode Logic (DL) AND Gate**

With the diodes being in parallel, the output level is low as long as either or both are conducting. Only when both inputs are "high"—that is, at +5 V—will both be cut off, resulting in no drop across R. The output will then be at +5 V, which is interpreted as a 1. The circuit, therefore, acts as an AND gate (presuming positive logic). Other logic circuits can be implemented in a comparable manner.

A major disadvantage of the DL system is the presence of a significant voltage drop across the diodes, thereby reducing the range of voltage distinction between a 0 and a 1.

We might want to expand the number of inputs (called **fan-in**) to the AND gate as well as the number of circuits to be attached to the output (termed the **fan-out**). In both instances, for DL circuits we are limited to but three or four units because of signal degradation, a major deficiency. Although propagation delay in DL circuits is desirably small, ∼ 12 ns, they are infrequently used today.

Diode-Transistor Logic (DTL). As implied by the name, DTL circuits combine diodes with transistors. Figure 11.43 will be recognized as the diode AND gate of the prior circuit (Figure 11.42), followed by a CE amplifier

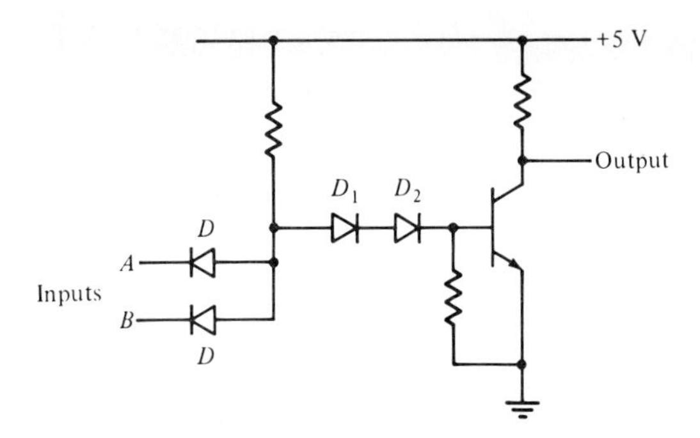

FIGURE 11.43. **Diode-Transistor Logic Used to Form a NAND Gate**

whose inverted output makes it into an overall NAND gate. Diodes D_1 and D_2 will only conduct when all inputs are 1. Such conduction causes the transistor to saturate, thereby dropping the output close to zero, resulting in a 0 output—that is, we have a NAND gate. (The series diodes D_1 and D_2 reduce the possibility that noise pulses will lead to a false output. See J. Millman, *Microelectronics*, p. 142, which is listed in the Suggested Readings at the end of this chapter, for detailed analysis.)

DTL increases fan-out capability to about 8 or 10. The propagation delay, however, is somewhat degraded compared to DL (30 vs. 12 ns) since saturated transistors require significant times to pass out of saturation. DTL, today, finds infrequent application.

Transistor-Transistor Logic (TTL or T²L). TTL is recognizable by the transistor with multiple emitters. This type of logic cannot be implemented in discrete transistor form but lends itself very nicely to IC fabrication.

To understand TTL operation, we have to digress for a moment. A BJT is a rather symmetrical device since (for example) we place a *p*-section between two *n*s. The emitter *n*-segment is generally foward-biased relative to the base, and the collector *n*-segment is reverse-biased. It is possible, therefore, by means of the biasing to make the collector and emitter reverse their roles. However, since the emitter is normally doped to a much heavier degree than the collector, the inverted-current gain is very small—that is, $h_{fei} < 1$.

The other point to be made is that when a BJT is saturated, there is a considerable buildup of minority carriers in the base region. (In a *p*-type base, the minority carriers are *n*-type, and they remain trapped by the positive potentials at the collector and emitter.) It is the reduction in the number of these charges that makes the BJT rather slow in coming out of saturation.

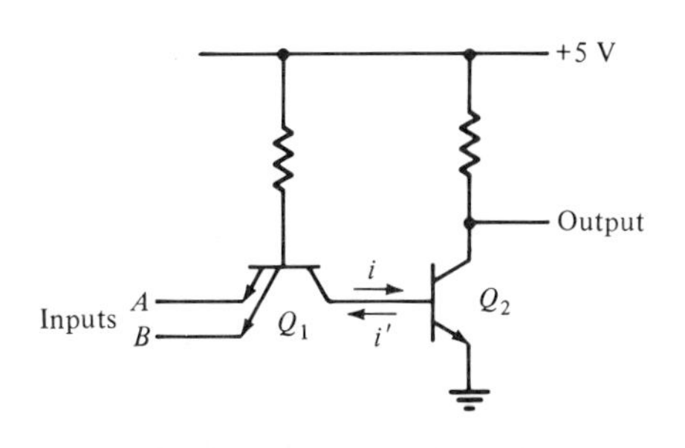

FIGURE 11.44. **Transistor-Transistor Logic (TTL) Used to Form NAND Gate**

Refer to Figure 11.44. If all the emitters are held high, the base current of $Q_1(i)$ switches over to the collector (inverted mode), which forward-biases Q_2, driving it into saturation and a "low" output. If any of the emitters of Q_1 go "low," Q_1 acts in its normal saturated mode. Initially, i' is large as it rapidly removes the saturation charge from the base of Q_2. Thereafter, the current i' simply consists of a small current characteristic of the reverse bias between the base and emitter of Q_2. The latter is cut off, its output is "high," and these results correspond to those expected of a NAND gate.

To more closely approximate the DTL circuit NAND, another transistor should be inserted between Q_1 and Q_2 of Figure 11.44, its base-emitter corresponding to D_2 in Figure 11.43.

Fan-out for TTL is around 10, and propagation delay is relatively short (10 ns compared to DTL's 30 ns). This TTL value is for standard gates; other series of TTL circuits were discussed in Section 11.5.2.

Emitter-Coupled Logic (ECL). ECL circuits basically are composed of differential amplifiers that are *not* driven into saturation as they assume extreme values of output voltage. As such their recovery time as they change states is very small, leading to ECL being the fastest of all the logic families, with propagation times as low as 0.5 ns/gate. Fan-out can be as large as 25. They do, however, exhibit higher power dissipation than other logic forms.

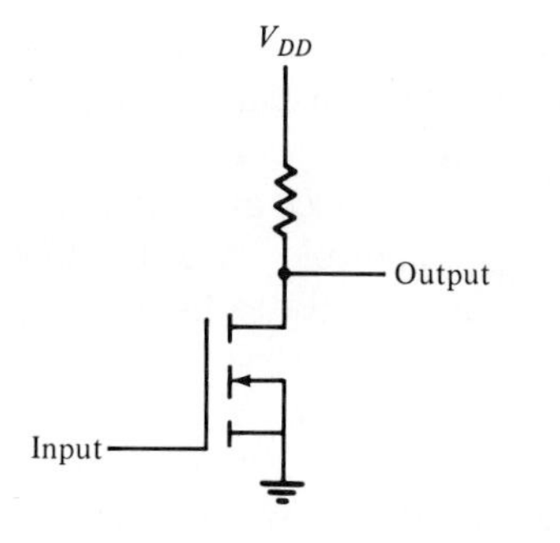

FIGURE 11.45. **MOS Logic Used to Form an Inverter**

CMOS Logic. We have already met CMOS logic in Section 9.9 and Figure 9.22. Its major advantage is very low power consumption and large fan-out ($\sim$ 50), but it is relatively slow, with propagation delays of the order of 50 ns.

MOS Logic. Enhancement MOS transistors, having insulated gates, pose very little loading and allow for large fan-out (20). They use much less chip space than BJTs, and since they can also be made to behave as resistors, they permit very large packing densities and are used in large-scale integration (LSI). An MOS-LSI may contain thousands of transistors. Pocket calculators, for example, utilize one MOS-LSI for computation, logic, and control. They are rather slow, however, with propagation delays of 100 ns. Figure 11.45 illustrates an MOS inverter.

Logic Levels. While to a large extent, considering positive logic, we have taken the low state to be either 0 or 0.2 V and the high state to be +5 V (typical of TTL logic), in each case there is a range of voltage values that the circuits will interpret as low and high signals. Manufacturers thus specify guaranteed respective ranges of values for their output voltages and a somewhat wider range of acceptable input voltages.

A well-designed circuit should function under the worst possible combination of minimum and maximum values. Such values constitute the *worst-case design*, and any reputable device, constructed from standard components—that is, without any being specially selected—should be so designed.

11.6 ■ NUMBERS

11.6.1 Binary, Decimal, and Hexadecimal

The basic operating units of electronic calculators use bistable circuits, which means that calculations must be based on a **binary system** of numbers, consisting of 0 and 1. A typical binary number might be

1	0	1	1	← binary number
×8	×4	×2	×1	

To evaluate the preceding number, multiply the ×1 value by 1; add this value to the value of ×2 times 1, which in turn is to be added to the ×4 multiplied (in this case) by 0; and to this value, finally add ×8 multiplied by 1. Thus, we have

$$\begin{aligned} 1 \times 1 &= 1 \\ 1 \times 2 &= 2 \\ 0 \times 4 &= 0 \\ 1 \times 8 &= \underline{8} \\ &\ 11 \end{aligned}$$

Thus, the number is eleven (decimal). We may express this equivalence as follows:

$$1\,0\,1\,1_2 = 11_{10}$$

where the subscripts indicate, respectively, that the numbers are binary (B) and decimal (D). Alternatively, this equivalence could be written as 1011(B) = 11(D).

Each 0 or 1 alternative represents a **bit** of information. Thus, the preceding example constitutes 4 bits of information. A 4-bit binary system can represent numbers from 0 to 15. To accommodate larger numbers, one must use more bits. Thus, an 8-bit system could handle numbers from 0 to 255:

1	1	1	1	1	1	1	1	$= 255_{10}$
×128	×64	×32	×16	×8	×4	×2	×1	

A **byte** is equal to eight bits of information; four bits equal a **nibble**. Expansion to a 2-byte system, typically, would lead to binary numbers such as

1	0	1	1	1	0	0	0	1	1	1	0	1	1	0	0
×32,768	×16,384	×8192	×4096	×2048	×1024	×512	×256	×128	×64	×32	×16	×8	×4	×2	×1

It is very easy to make a mistake when dealing with such lengthy strings of binary digits. This difficulty may be remedied by resorting to a **hexadecimal** (H) system, based on the number 16, wherein each 4-bit sequence is represented by a single equivalent symbol. Since a 4-bit sequence represents 16 different combinations, we can use 0–9 to represent the first nine numbers in increasing order and then use A to represent 10, B to represent 11, $C = 12$, $D = 13$, $E = 14$, $F = 15$. Thus, we have

$$1011_2 = 11_{10} = B_{16}$$

or

$$1011(\text{B}) = 11(\text{D}) = B(\text{H})$$

Consider some other examples:

$$1000_2 = 8_{10} = 8_{16}$$
$$1110_2 = 14_{10} = E_{16}$$
$$1100_2 = 12_{10} = C_{16}$$

Therefore, we have

$$1011 \quad 1000 \quad 1110 \quad 1100_2 = 47{,}340_{10} = B8EC_{16}$$

This system takes care of integers. How about fractional numbers? Limiting ourselves to an 8-bit example, we might have the number

1011.1011

To the left of the decimal we again progressively multiply each position by 1, by 2, by 4, and by 8. Then we add and, in the preceding case, obtain

$$\begin{aligned} 1 \times 8 &= 8 = 2^3 \\ 0 \times 4 &= 0 \\ 1 \times 2 &= 2 = 2^1 \\ 1 \times 1 &= \underline{1} = 2^0 \\ & 11 \end{aligned}$$

In a somewhat comparable manner, proceeding to the right of the decimal, multiply each successive bit by 2^{-1}, 2^{-2}, 2^{-3}, 2^{-4}, and so forth. Thus, we obtain

$$\begin{aligned} 1 \times \frac{1}{2^1} &= 0.5 \\ 0 \times \frac{1}{2^2} &= 0.00 \\ 1 \times \frac{1}{2^3} &= 0.125 \\ 1 \times \frac{1}{2^4} &= \underline{0.0625} \\ & 0.6875 \end{aligned}$$

Thus, we have

1011.1011(B) = 11.6875(D)

To go the other way—that is, to convert a decimal into a binary number—taking the integer portion first, successively divide by 2 and note whether there is any (1) remainder or not:

2 /11		.	decimal
2 /5	carry 1	↑	
2 /2	carry 1	│	
2 /1	carry 0	│	binary representation = 1011.
0	carry 1	│	

In the case of the decimal portion, multiply successively by 2 and note whether the remainder is 0 or 1 *ahead* of the decimal point. In the preceding case, we have

		. decimal
$0.6875 \times 2 = 0.375$	carry 1	
$0.375 \times 2 = 0.75$	carry 0	binary expression = 0.1011
$0.75 \times 2 = 0.50$	carry 1	
$0.50 \times 2 = 0.00$	carry 1	

In addition to the binary system, there is also the **binary-coded decimal** (BCD) system, with which it should not be confused. The simplest approach to this BCD system is to consider the binary equivalents of the decimal numbers 0–9. Four bits are necessary—for example, $1001_2 = 9_{10}$, $0101_2 = 5_{10}$, and so forth. In the BCD system one does not utilize combinations such as 1011 ($= 11_{10}$), 1100 ($= 12_{10}$), 1101 ($= 13_{10}$), and so on (in other words, no combinations representing numbers greater than 9_{10}). While this procedure makes the BCD system somewhat wasteful, it is convenient for handling numbers going into and coming out of the computer and in digital displays of the form we will consider in Chapter 12.

There are other systems of numbers used in digital computers, for example, the octal system based on the number 8. But we will forego their consideration.

EXAMPLE 11.4

Express 265_{10} in BCD notation.

Solution:

weighting factor	800	400	200	100	80	40	20	10	8	4	2	1
BCD	0	0	1	0	0	1	1	0	0	1	0	1
decimal digits		2				6				5		

11.6.2 Digital Arithmetic Operations

A digital computer must be capable of performing the four arithmetic operations. But multiplication is essentially repeated addition, and division is essentially repeated subtraction. Subtraction, in turn, can be transformed into an addition problem, reducing the demand on the computer to the performance of addition.

Consider the addition of two decimal arithmetic numbers, say the hundred's digit. The hundred's digits are added and supplemented by a carry from the ten's digit (if there is one). In like manner, in binary addition the 2 bits corresponding to the 2^n digit are added, and then to the resultant is added any carry that might have arisen from the 2^{n-1} digit. A two-input adder is called a half-adder, the complete addition being performed by two half-adders. See Table 11.6.

A half-adder has two inputs (A and B) and two outputs [one the sum (S) and one the carry bit (C)]. Table 11.7 constitutes the truth table for the half-adder. It can be implemented by the logical gates shown in Figure 11.46.

TABLE 11.6 **Binary Addition**

0 plus 0 = 0	no carry
0 plus 1 = 1	no carry
1 plus 1 = 0	carry 1

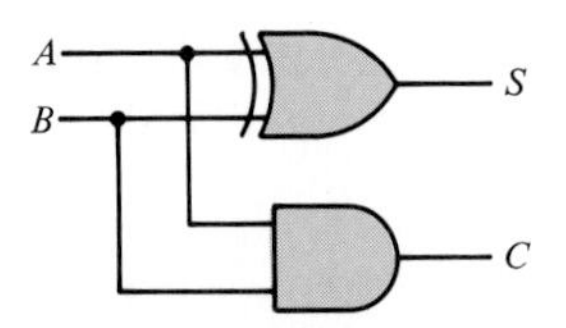

FIGURE 11.46. **Implementation of a Half-Adder Circuit with Sum (S) and Carry (C) Outputs** The OR circuit representation is that of an EXCLUSIVE OR (XOR).

TABLE 11.7 **Half-adder Truth Table**

A	B	S	C
0	0	0	0
0	1	1	0
1	0	1	0
1	1	0	1

A full-adder has three inputs A_n, B_n, and C_{n-1}, where the last represents the carry from the next lower bit. The two outputs are the sum (S_n) and the carry (C_n) (Figure 11.47). Figure 11.48 represents a 4-bit adder with subscripts 1 representing the least significant bit (LSB) and the subscripts 4 representing the most significant bits (MSB). Figure 11.49 shows how two 8-bit numbers might be added using two 4-bit units.

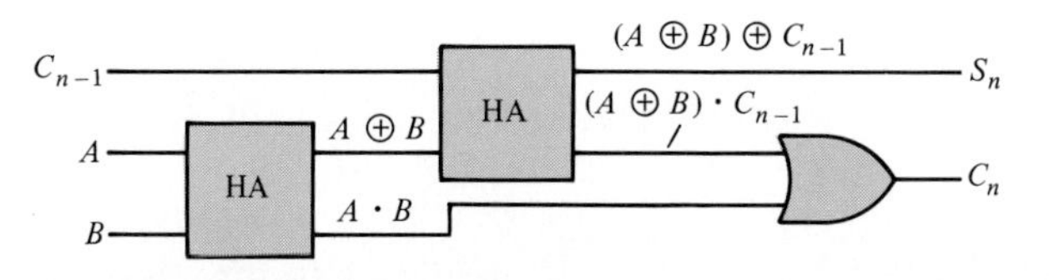

FIGURE 11.47. **Full-Adder, Composed of Two Half-Adders (HA) with Input Carry (C_{n-1}) and Output Carry (C_n), Together with the Sum (S_n)**

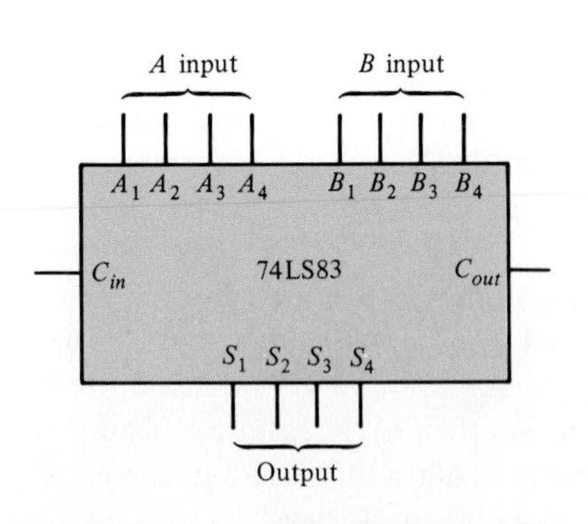

FIGURE 11.48. **Four-Bit Adder** The addition of A and B bits appears as the 4-bit sum at S outputs. C_{in} and C_{out} are input and output carries, respectively.

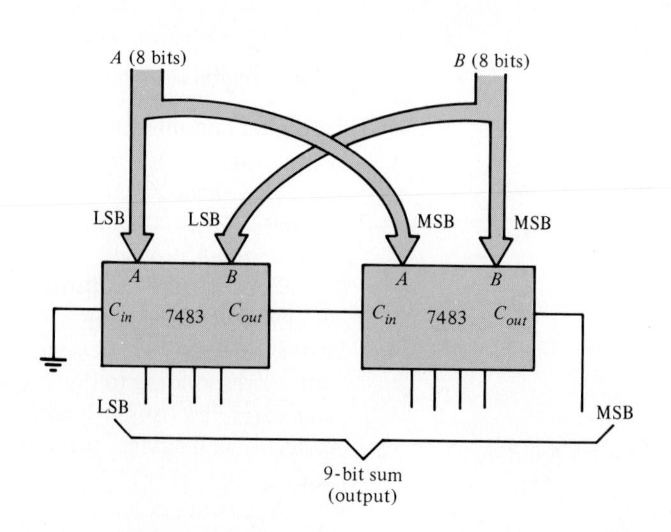

FIGURE 11.49. **Adding Two 8-Bit Numbers (A and B) Using Two 4-Bit Adders**

In digital circuits negative numbers may be handled by *sign magnitude representation*, where 1 bit (say the MSB) is made to represent the sign. Typically, 0 might be used for positive numbers and 1, for negative numbers. While this procedure might be suitable when numbers are to be displayed, it has some disadvantages, including the presence of two 0s (+0 and −0).

There are other methods of representation that have advantages under specific circumstances, but the *2's complement method*, with the MSB being reserved for the sign indication, is the most widely used for integer computation. There is only one 0, +0 (that is, 0000_2).

11.6.3 Complements: 2's and 1's

Signed binary integers utilize 0 as the MSB for positive numbers and 1 as the MSB for negative numbers. For binary words employing n bits, this system limits the expressible values to the range $+(2^{n-1} - 1)$ to $-(2^{n-1} - 1)$, with two distinct 0s (+0 and −0). For example, with $n = 4$ the expressible limits are +7 (0111) and −7 (1111) but with $+0_{10} = 0000_2$ and $-0_{10} = 1000_2$.

In the 2's complement system, the expressions for positive integers remains the same as for signed binary numbers. However, for negative integers (that is, $-N$) the 2's complement ($\equiv N^*$) employing an n-bit system is

$$N^* = 2^n - N$$

EXAMPLE 11.5 For $n = 4$, what is the 2's complement of −5?

Solution:

$$N^* = 2^n - 5 = 2^4 - 5 = 16 - 5 = 11_{10} = 1011_2$$

As an unsigned binary, this sequence equals 11_{10}. As a 2's complement, the first 1 indicates a negative number, and the remainder is the complement of the number.

There is only 0 (+0) in the 2's complement method, as can be seen from the following: $5_{10} - 5_{10} = 0_{10}$. We have

$$
\begin{aligned}
+5_{10} &= \quad 0101 \\
-5_{10} &= \quad \underline{1011} \quad \leftarrow \text{2's complement} \\
\text{discard with a 4-bit system} &\leftarrow 1 \mid 0000 = +0_{10}
\end{aligned}
$$

The 1's complement of a negative number (that is, $-N$) using an n-bit system is

$$\overline{N} = (2^n - 1) - N$$

EXAMPLE 11.6 If $n = 4$, what is the 1's complement of −5? What is the binary expression for +5? What simple method suggests itself for obtaining the 1's complement of a negative integer in binary form?

Solution:

$$\overline{N} = (2^4 - 1) - 5 = 15 - 5 = 10_{10} = 1010_2$$
$$+5_{10} = 0101_2$$

The 1's complement of a negative integer may be obtained from the binary representation of the positive integer by replacing 0s by 1s and vice versa.

Now consider the 1's complement of -8:

$$\overline{N} = (2^4 - 1) - 8 = 7 = 0111$$

which is wrong because this sequence indicates a positive integer. There can be no 1's complement representation of -8 in a 4-bit system; at least 5 bits must be used. In a 5-bit system, we have

$$+8_{10} = 0\,1000_2$$
$$-8_{10} = 1\,0111_2$$

Again, this number can be obtained by replacing 0s with 1s and vice versa. Otherwise, we have

$$\overline{N} = (2^5 - 1) - 8 = 31 - 8 = 23_{10} = 1\,0111$$

Since

$$N^* = 2^n - N = (2^n - 1) - N + 1 = \overline{N} + 1$$

the 2's complement can also be obtained by complementing N on a bit-by-bit basis—that is, 0s replace 1s and 1s replace 0s—and adding 1 to the result.

EXAMPLE 11.7 Find the 2's complement of -9 by first finding the 1's complement, to which unity is added.

Solution:

$$+9 = 0\,1001$$
$$-9 = 1\,0110 \quad \leftarrow \text{1's complement}$$
$$-9 = 1\,0111 \quad \leftarrow \text{2's complement}$$

Another way of finding the 2's complement is to start at the right of the binary representation and complement all bits to the left of the first 1. (Verify this in the preceding case.)

Finally, to find the magnitude of a negative 2's complement, since

$$N = 2^n - N^*$$

one can obtain the magnitude by taking the 2's complement of the 2's complement.

In like manner, since

$$N = (2^n - 1) - \overline{N}$$

the magnitude of a 1's complement of a negative number can be obtained by taking its 1's complement.

EXAMPLE 11.8 Perform the following addition utilizing a 4-bit binary system:

$$\begin{array}{r} 3_{10} \\ \underline{5_{10}} \\ 8_{10} \end{array}$$

Solution:

$$\begin{array}{rl} 3_{10} = & 0011 \\ 5_{10} = & \underline{0101} \\ & 1000 = 8_{10} \end{array}$$

In each column of the last three columns, 1 plus 1 equals 0, with a 1 carry.

EXAMPLE 11.9 Utilizing binary arithmetic, perform the subtraction operation 5 minus 3.

Solution: This computation is most conveniently carried out in a computer by changing the 3 to its 2's complement, which is then added to 5. (It was mentioned in the text that a subtraction can be converted into an addition problem.)

The 2's complement is obtained from the binary form by replacing 0s with 1s and vice versa. This constitutes the 1's complement. Upon adding unity, it becomes the 2's complement. We have

$$\begin{array}{rrl} 3_{10} = & 0011_2 & \\ & 1100 & \\ & \underline{1} & \\ & 1101 & \leftarrow \text{2's complement of 3} \\ 5_{10} = & 0101 & \\ & \underline{1101} & \leftarrow \text{2's complement of 3} \\ & 10010 & \end{array}$$

The carry is discarded, and we have the result

$$0010 = 2_{10}$$

EXAMPLE 11.10 Perform the following in terms of signed binary numbers:

a. $\begin{array}{r} 3_{10} \\ \underline{1_{10}} \end{array}$

b. $\begin{array}{r} 3_{10} \\ \underline{-1_{10}} \end{array}$

c.
$$\begin{array}{r} 6_{10} \\ \underline{6_{10}} \end{array}$$

d.
$$\begin{array}{r} -6_{10} \\ \underline{-6_{10}} \end{array}$$

Solution:

a.
$$\begin{array}{rcll} & & \text{indicates + sign} & \\ & & \downarrow & \\ 3 & = & 0001 & \\ 1 & = & \underline{0001} & \\ & & 0100 & = +4_{10} \end{array}$$

b.
$$\begin{array}{rcll} & & \text{signs} & \\ 3 & = & 0011 & \\ -1 & = & \underline{1111} & \leftarrow \text{2's complement} \\ & & 10010 & = +2_{10} \end{array}$$

The carry is disregarded, and what remains is a valid result. When there is a carry *into* the sign bit and *out* of the sign bit, the result is valid. When there is either a carry into the sign but not out or a carry out of the sign bit but not in, the result is not valid. When this condition arises, a computer will generally indicate that an overflow has arisen.

c.
$$\begin{array}{rcl} 6 & = & 0110 \\ 6 & = & \underline{0110} \\ & & 1100 \end{array}$$

This result is invalid because there has been a carry into the sign bit but not out. We need a system with a larger number of bits to handle the result.

d.
$$\begin{array}{rcll} -6 & = & 1010 & \leftarrow \text{2's complement} \\ -6 & = & \underline{1010} & \leftarrow \text{2's complement} \\ & & 10100 & \end{array}$$

This result is also invalid because there has been a carry out of the sign bit but not into it.

EXAMPLE 11.11

Using the following decimal multiplication as a model, perform the same operation using binary multiplication ($0 \times 0 = 0, 0 \times 1 = 0, 1 \times 1 = 1$) and adding the partial products:

$$\begin{array}{r} 13.5 \\ \underline{1.25} \\ 675 \\ 2\ 70 \\ \underline{13\ 5} \\ 16.875 \end{array}$$

Solution:

$$
\begin{array}{rr}
13.5_{10} = & 1101.1 \\
1.25_{10} = & \underline{\quad 1.01} \\
 & 11\ 011 \\
 & 000\ 00\ \\
 & \underline{1101\ 1\ \ \ } \\
 & 1\ 0000.111 = 16.875
\end{array}
$$

The computer performs a shift operation between successive partials and then sums the result.

11.7 ■ DIGITAL READOUT

11.7.1 Display Devices

Among the convenient and commonly used devices for numeric displays are seven-segment **LEDs (light-emitting diodes)**. Such diodes, when passing current in the forward direction, will have their junction region illuminated. By arranging the junctions as in Figure 11.50, upon excitation of appropriate combinations of diodes, any one of the 10 integers (0–9) may be displayed. LEDs are available with either common cathodes or common anodes. In the former case particular segments are energized upon applying to them a positive voltage with the remainder "down." In the latter case unenergized portions are "high," and those portions to be energized are "low"—that is, at 0 V.

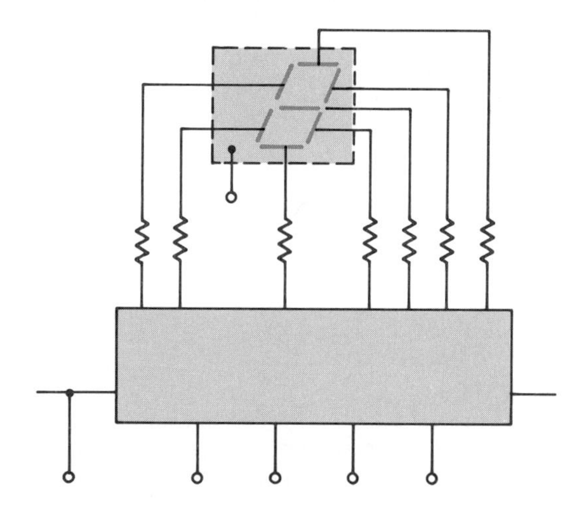

FIGURE 11.50. **LED Display** A BCD number applied to a 7447 decoder illuminates appropriate segments of the (upper) slanted lines, yielding a visual indication of the BCD number.

The numerical information to be displayed is in BCD form, and to convert this information into appropriate combinations of seven-segment excitations, a *decoder* must be used; one that is commonly used has the designation 7447. In addition to the required dc excitation (5 V), there are four input terminals (1, 2, 4, 8) for binary input and the seven output terminals. The decimal point (DP) is separately excited via a range switch.

By way of illustrating LED use, we will undertake to consider the operation of a frequency meter using an LED display. This instrument will measure an unknown frequency by converting it to an equivalent series of pulses that are counted, with the resulting value displayed numerically.

Before doing so, however, it should be mentioned that another popular form of display utilizes the **LCD (liquid-crystal display)**. In this case the numerical segments consist of transparent electrically conducting areas. The application of a voltage between a given segment and an inner electrode changes the light transparency of the fluid between them. Depending on the type of illumination employed, the numerals appear either light on a dark background or vice versa. Wristwatch displays utilize LCDs primarily because of their lower activation power (microwatts vs. milliwatts for LEDs). But LCDs have a slower response time ($\sim$ 100 ms vs. $<$100 ns), though this factor does not prove to be a detriment in such applications.

11.7.2 Cast of Characters

We will first present a short introduction to the various IC circuits that will be utilized in the frequency counter.

7400. The 7400 is characterized as a quad two-input NAND gate, consisting of four NAND gates mounted on one DIP (dual-in-line) chip (Figure 11.51). There are seven terminals on each side of the unit; their numbering is uniquely determined by observing the slot at one end when viewing the unit from the top. In many of the 74 series of logic units, the +5 V source is connected to pin 14, while pin 7 is ground. But there are some exceptions, as we will see.

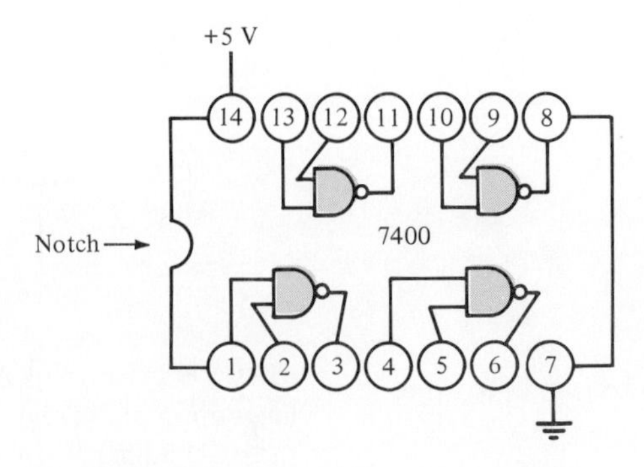

FIGURE 11.51. **7400 Quad Two-Input NAND Gates, Showing Connections (Top View)**

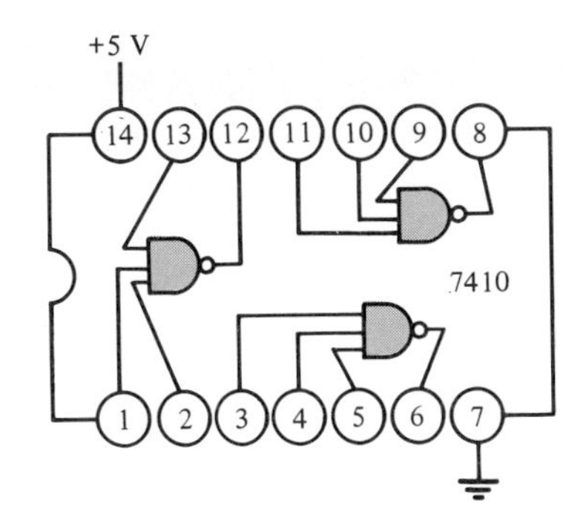

FIGURE 11.52. **7410 Triple Three-Input NAND Gates, Showing Connections (Top View)**

7410. The 7410 is a triple three-input NAND gate (Figure 11.52). The conventional power connections are observed. When all three inputs of each gate are high, the output is in a low state.

74121. The frequency counter that we will consider utilizes 60 Hz power-line sine waves to generate clock pulses. These sinusoidal signals must first be converted to 60 Hz square waves, and for this purpose we use a 74121 unit (Figure 11.53).

There are several ways to trigger this monostable MV, depending on what is done with inputs A_1, A_2, and B. The circles at the OR gate inputs indicate the usual pulse reversal. If A_1 and A_2 are both grounded, the OR gate output is high—that is, +5 V—and constitutes the reference voltage for the Schmitt trigger (signified by the input symbol, which should be recognized as that of a hysteresis loop). The ac input signal applied to B triggers the Schmitt when the 5 V threshold is exceeded.

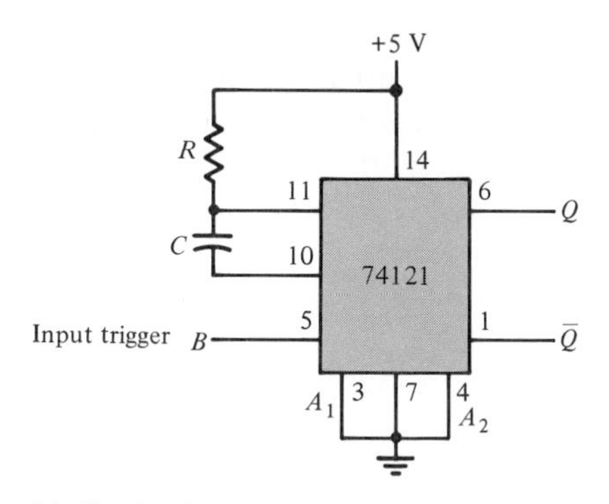

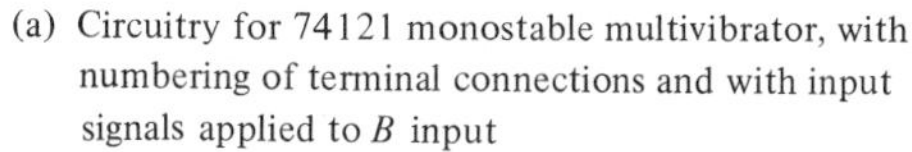

(a) Circuitry for 74121 monostable multivibrator, with numbering of terminal connections and with input signals applied to B input

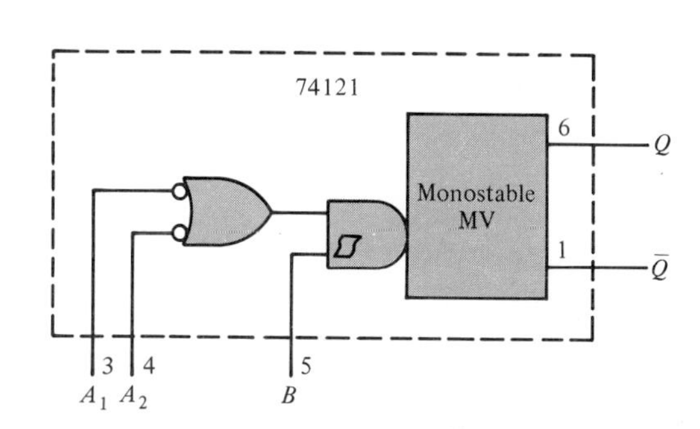

(b) Details of alternative trigger inputs; symbol alongside terminal 5 is that for a Schmitt trigger

FIGURE 11.53. **Schmitt-Triggered Monostable Multivibrator**

The output pulse persists for a time determined by the time constant RC and determines the duty cycle of the output pulses at Q. The period of the output square waves is determined by the periodicity of the input trigger pulses. Thus, if a 60 Hz input trigger is used, the output consists of 60 Hz square waves.

Many monostables can be retriggered during the input pulse duration, thereby extending the output for a time equal to the output pulse duration. The 74121, however, is nonretriggerable; it ignores any input pulses for the duration of the output pulse.

7492. The 7492 is termed a base-12 ($\div$ 12) counter (Figure 11.54). The numbers outside the rectangle indicate the pin connections. This unit consists of four flip-flops of which the last three are internally connected. The pulses at the A output will have a frequency equal to one half of that applied to the A input. Input B will yield division by 6 at output D. If the A output is connected to the B input, the D output will have divided the A input by 12; the

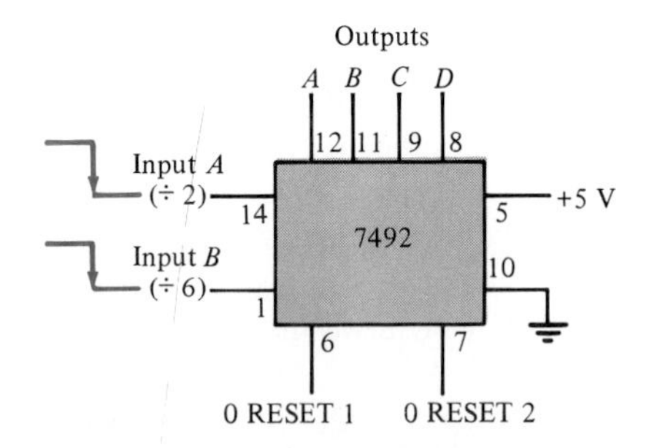

FIGURE 11.54. 7492 Divide-by-12 Counter The *A* output is required to be connected to the *B* input. The output is at *D*.

output will be low for six counts and high for six counts. The counter advances on the negative-going clock edges.

There are two reset pins. Under normal counting conditions both are grounded. If both are made to go high, the counter will reset to zero.

The 7492 is called a **ripple counter**, meaning that a given flip-flop is triggered by the changing state of the prior unit, with a consequent delay from unit to unit. Such action is in contrast to a **synchronous counter** wherein the successive stages respond in unison.

7493. Here again we have four flip-flops of which the final three are internally connected (Figure 11.55). The separate one yields a division by 2, and if output *A* is connected to input *B*, there will be an overall division by 16 (in contrast to the division by 12 from the 7492). This unit is again a ripple counter, with two resets operating the same as in the 7492.

7490. Rather than divide by 12 (7492) and divide by 16 (7493), the 7490 is a divide-by-10 ripple counter (Figure 11.56). The *A* output must be connected to the *B* input. One can initially preset—that is, reset—these units to either 0 or 9. The latter is useful when using 9's complement representation of negative numbers.[2] The operation of these presets is somewhat involved (see Figure 11.56). In Figure 11.56, zero preset arises if terminals 2 and 3 are high *and* one or the other of the 6 and 7 terminals is low. The counter is preset with nines if terminals 6 and 7 are high, irrespective of what is applied to the

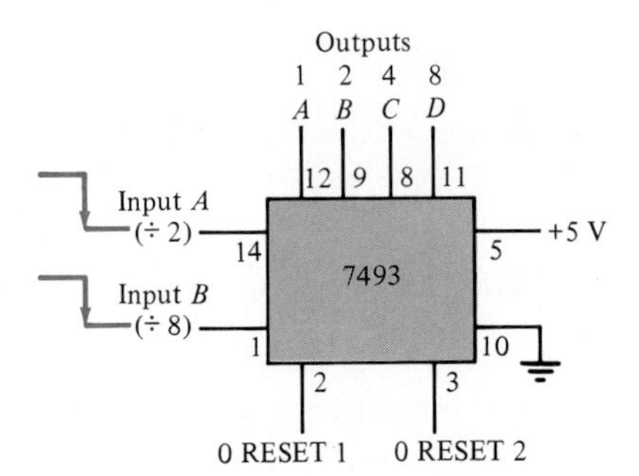

FIGURE 11.55. 7493 Divide-by-16 Counter The *A* output is required to be connected to the *B* input. The output is at *D*.

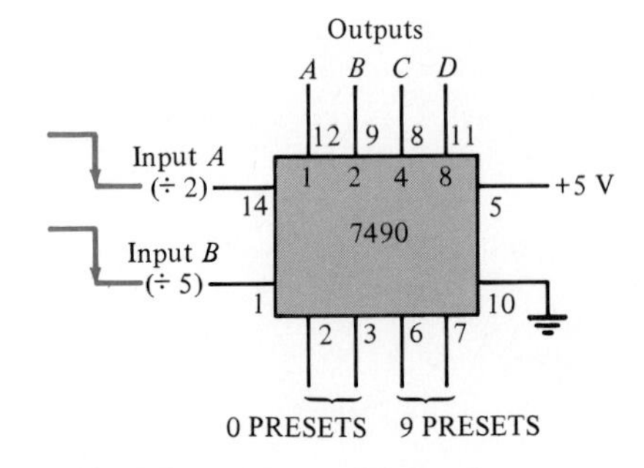

FIGURE 11.56. 7490 Decade Counter The *A* output is required to be connected to the *B* input. The output is at *D*.

[2]One obtains the 9's complement by subtracting each individual digit from 9. This procedure allows negative numbers to be distinguished from positive numbers by the initial digit:

	etc.		
+2	0002		
+1	0001		
0	0000	or	9999
−1	9998		
−2	9997		
	etc.		

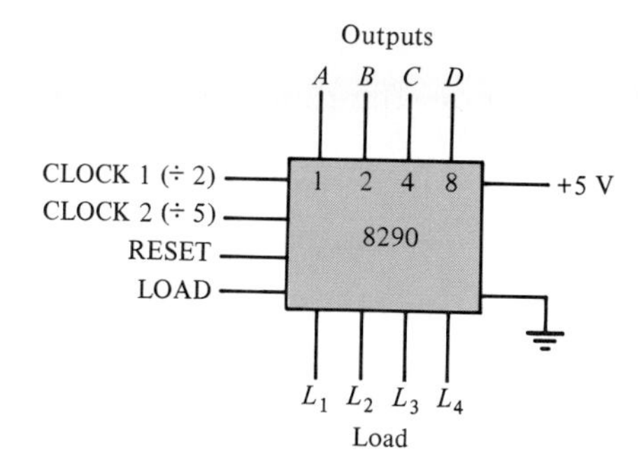

FIGURE 11.57. **8290 Decade Counter** The A output is required to be connected to the clock 2 input. The output is at D. The counter may be preloaded with the BCD number applied to L inputs, in conjunction with the load terminal.

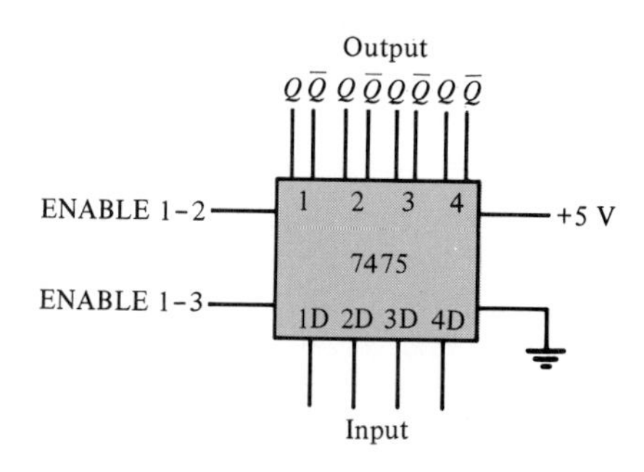

FIGURE 11.58. **7475 Latch Circuit** With the enabling terminals high, the input is transferred to the output. With the enabling terminals low, the output retains the last-transferred value from the input.

zero presets. Counting will occur whenever any paired combination of the four presets is zero, irrespective of the state of the other pair. As a counter, it will reset on the tenth pulse, and the 1248 output will yield a BCD representation of the number of input pulses.

8290. The 8290 is also a decade counter but is faster than the 7490 (Figure 11.57). In our frequency counter the 8290 will be used to perform the same counting function as the 7490s, except for being faster. It, therefore, is used as the units counter, being subject to the greatest pulse rate. It also has provision for loading any desired initial count. Such a count is applied to the L terminals, the LOAD terminal is briefly grounded, and the initial count is transferred to the output. Although we will not be making use of this feature in the counting circuit to be discussed, it is indicative of an additional versatility available in IC form.

7475. The 7475 is a QUAD LATCH, consisting of four D flip-flops, which may be used as a 4-bit memory (Figure 11.58). The operational nature of the unit is determined by the nature of two input signals applied via the **enabling lines.** When used for 4-bit storage, both enables are operated in parallel. When the enables are high, the output follows the input, and when low, the output holds the previous values.

11.7.3 Modulo-*n* Counters

The **modulo** of a counter represents the number of states through which it passes before repeating. Thus, a decade counter represents modulo-10, the 1248 binary counter is modulo-16, etc.

Figure 11.59 shows how four flip-flops may be connected to form a decade counter (modulo-10). What is necessary is a feedback to clear all the counter upon application of the tenth pulse. This clearing operation is accomplished by means of a NAND gate. Since $10_{10} = 1010_2$, that is, $Q_1 = 0$, $Q_2 = 1$, $Q_3 = 0$, and $Q_4 = 1$, if Q_1 and Q_3 are presented to the input of the NAND gate, when both are 1s, the NAND gate provides the reset—that is, clear.

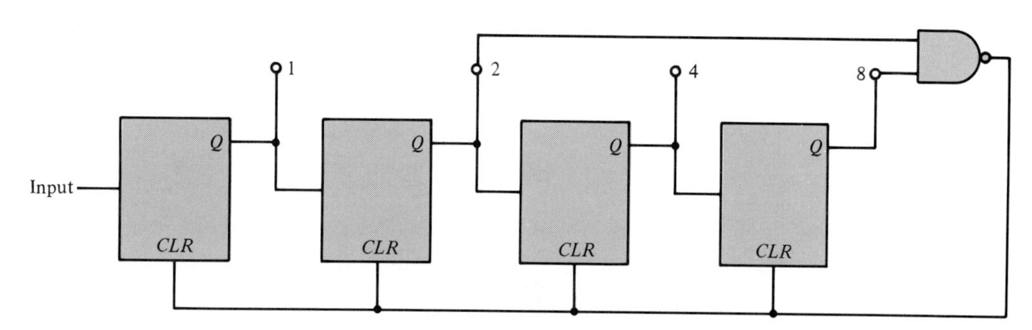

FIGURE 11.59. **Divide-by-10 (Modulo-10) Circuit** Upon entry of the tenth pulse, terminals 2 and 8 go high, and the NAND gate generates a CLEAR pulse, resetting the system to zero.

The same procedure can be used to create a divide-by-6 (modulo-6) counter (Figure 11.60). Since $6_{10} = 110_2$, only three flip-flops are necessary, and when $Q_3 = Q_2 = 1$, the counter will reset.

In a divide-by-12 circuit, it is the last two flip-flops that are brought up to the NAND gate since $12_{10} = 1100_2$ (Figure 11.61).

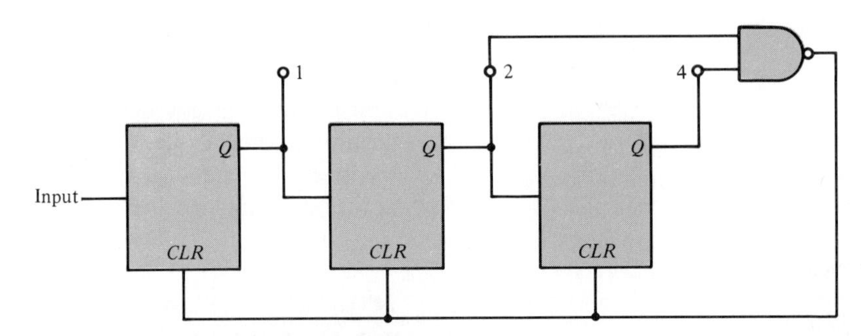

FIGURE 11.60. **Divide-by-6 (Modulo-6) Circuit** Upon entry of the sixth pulse, terminals 2 and 4 go high, and the NAND gate initiates a CLEAR pulse, resetting the system to zero.

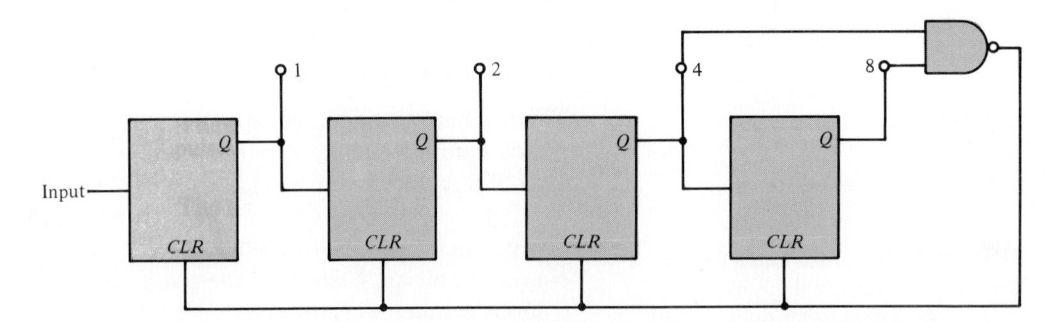

FIGURE 11.61. **Divide-by-12 (Modulo-12) Circuit** Upon entry of the twelfth pulse, terminals 4 and 8 go high, and the NAND gate initiates a CLEAR pulse, resetting the system to zero.

11.7.4 A Frequency Counter

Consider the block diagram in Figure 11.62. The frequency of the incoming signal will be converted into pulses, one for each cycle of the signal. The objective of the instrument is to count the number of pulses arriving during an accurately known time interval. One second proves to be particularly convenient since the number of cycles arriving during this interval is a direct measure of the unknown frequency.

To "measure off" an accurate 1 s interval, we need a reference frequency. We will start by taking this reference to be that of the 60 Hz power line.[3] Pass-

[3]We will subsequently consider the accuracy of such a reference signal.

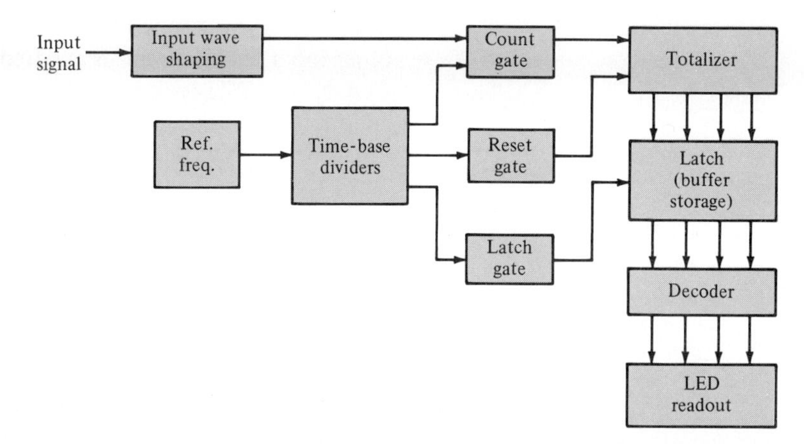

FIGURE 11.62. **Block Diagram of a Frequency Counter Employing BCD Readout**

ing the 60 Hz reference through a divide-by-6 circuit will generate square waves at the rate of 10/s (see the top line of Figure 11.63). We will then leave a time interval equal to the duration of two clock pulses (in order to perform various other operations) before allowing the gate pulse to again open up for another 1 s interval. (The start of this second 1 s interval is shown at the right of the COUNT GATE line in Figure 11.63.)

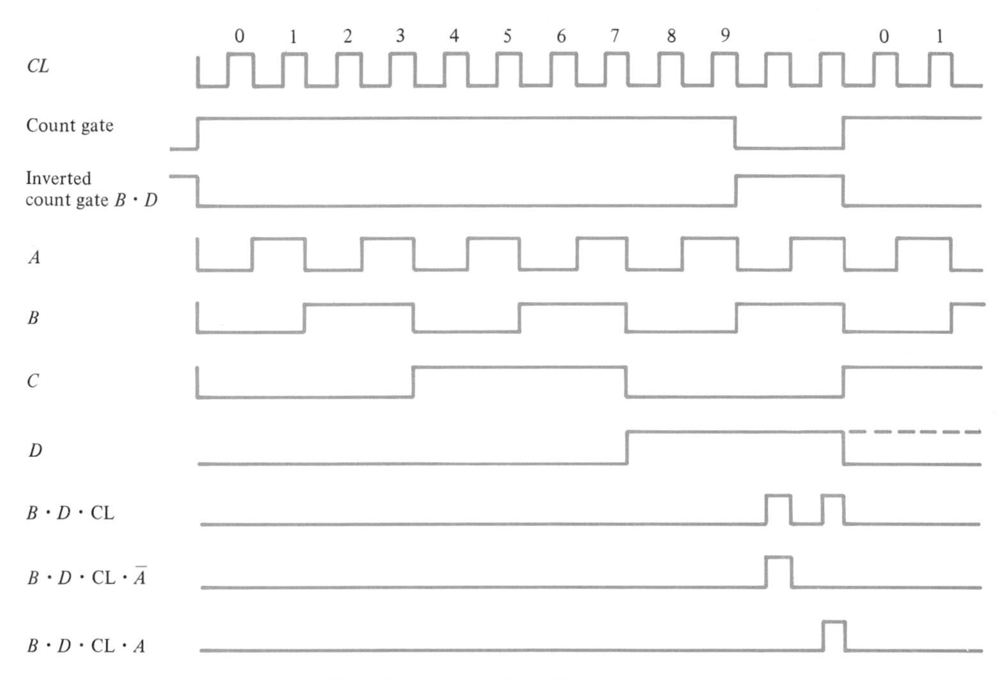

FIGURE 11.63. **Clock Pulses and Gates Required for Frequency Counter**

The 1 s gate is applied to one of the two inputs of the COUNT GATE. The other input to this circuit is the frequency to be measured. The output of the COUNT GATE is passed to the TOTALIZER, which is an adding circuit that counts the number of pulses arriving during the 1 s interval. The COUNT GATE will accept pulses only during the 1 s interval when the gating pulse is "up." Upon termination of this 1 s interval, a latch pulse will allow the total count to be passed to buffer storage, where it will remain at the same value, and displayed by the LEDs during the next counting interval. A RESET pulse follows the LATCH pulse. This reset returns the totalizer to zero, the GATE pulse then comes up, and a new count of the unknown frequency is taken for another 1 s interval. A LATCH pulse and a new readout take place, and then a RESET pulse makes everything ready for another counting interval.

We now turn to a somewhat more detailed consideration of the circuits (Figure 11.64). The direct use of a 60 Hz sinusoidal signal to trigger the digital circuits poses difficulties; the variation of the 60 Hz sine wave is not fast enough for proper triggering. We must convert the sine wave into a square wave. This conversion is done by a 74121, which is a Schmitt trigger followed by a monostable MV. The output will be a 60 Hz square wave.

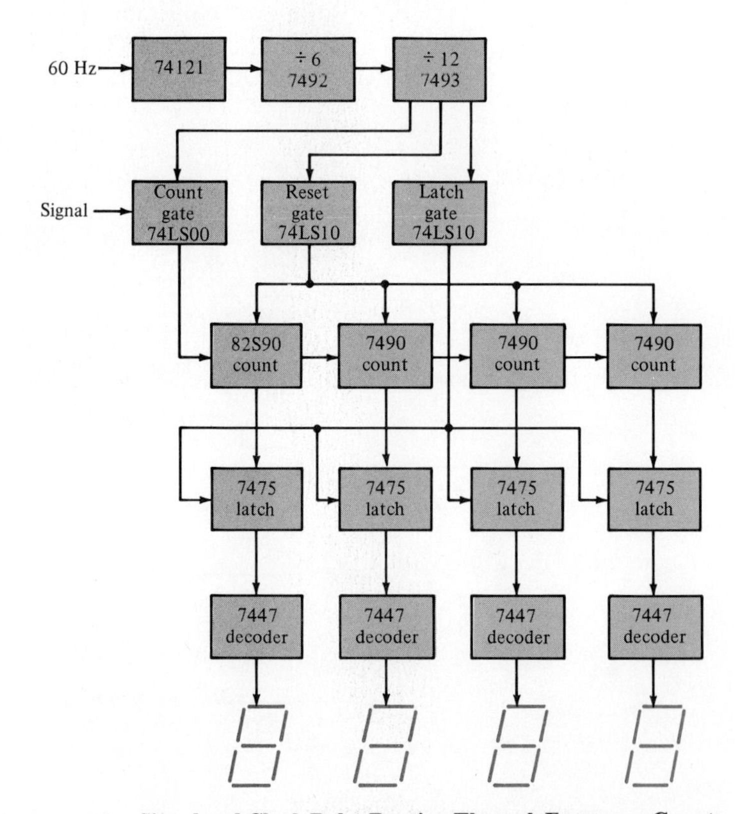

FIGURE 11.64. Signal and Clock Pulse Routing Through Frequency Counter

This output is passed on to a 7492 divider that has a divide-by-2 and a divide-by-6 input. We use only the latter and thereby obtain 10 square waves/s, which is line 1 in Figure 11.63. This 10 Hz square wave is passed on to a 7493 divider.

If the A output of the 7492 is passed on to the B input, we obtain a divide-by-16 output at D. But looking at the *desired D* output in Figure 11.63, we see that the output will be down for eight clock pulses and then be up for eight clock pulses in a divide-by-16 circuit. However, we desire to terminate the "up" state after 12 pulses—that is, after pulse 11.

The 7493 has two reset inputs that will cause the D output to return to the "down" state when they are both excited. Taking a look at the truth table for the 7493, we see that D and C outputs are 1s for pulses 12, 13, 14, and 15. Thus, if the D and C outputs are connected, respectively, to the two reset lines, after pulse 11 both D and C become 1, and the 7493 is reset to zero, not to become activated again until a RESET pulse arises.

Let's turn once again to Figure 11.63. Application of B and D to an adding circuit will yield line 3, which is an inversion of the desired gate pulse (passing it through an inverter will give us what we want). Adding B, D, and CL gives us two successive pulses that follow the 1 s base. The first of these pulses will be the LATCH pulse and the second, the RESET pulse. $B \cdot D \cdot \mathrm{CL} \cdot \overline{A}$ will isolate the first; $B \cdot D \cdot \mathrm{CL} \cdot A$ will isolate the second. The succession of logic circuits that generate these pulses is shown as Figure 11.65. It will be noted that two types of reset pulses are generated, one the negative of the other. In the totalizer two types of counters are used. One requires a 0 for reset (7490) and the other, a 1 (8290).

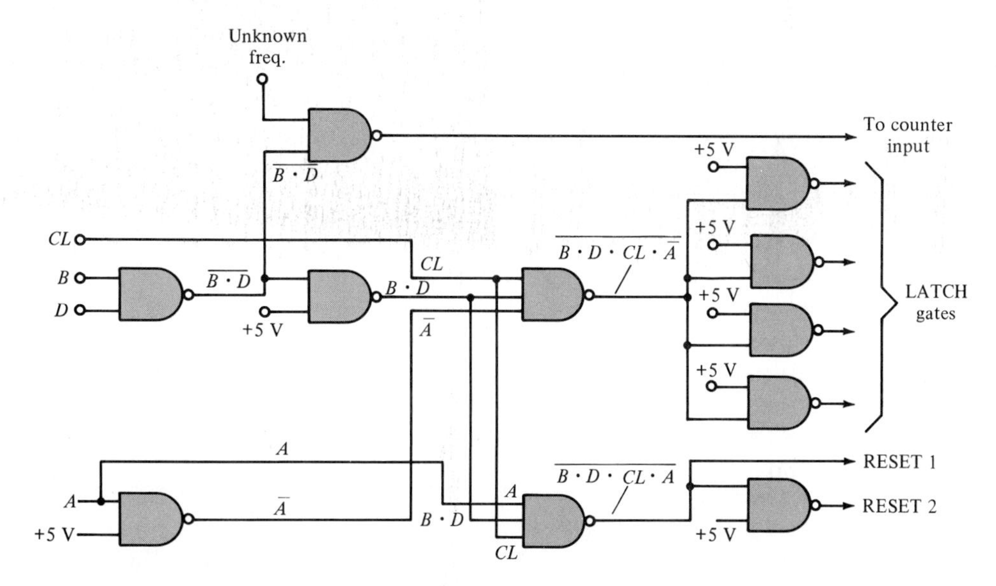

FIGURE 11.65. Logical Circuits Needed to Generate Gating Pulses for Frequency Counter

The totalizer consists of successive stages of decade counters. Each of them can retain a count in the range 0–9. Upon arrival of the tenth pulse, the first stage returns to zero and passes on a "carry" count to the next stage. Thus, if we are counting the number 3425, the first stage will count all 3425 pulses in the 1 s timing interval but will retain only the terminal 5. The second stage will have counted 342 carry pulses from the first but will retain only the terminal 2. The third stage will have counted 34 pulses, retaining 4, and the final stage will register 3 upon completion of the counting. Thus, we see that the frequency-handling capacity of the first stage is the greatest. Since the objective of the proposed frequency meter is to be able to count up to 50 MHz, the first stage must have this capacity. An 82S90 will accomplish this task. Since the remaining three stages have less stringent requirements, slower and cheaper units were used, the 7490s being capable of counting rates up to 15 MHz, which was quite adequate.

The accuracy of the 60 Hz power frequency is perhaps $\pm 0.1\%$. This accuracy represents an error of 1 kHz/MHz. At 50 MHz this value would mean a potential measuring error of 50 kHz. A typical accuracy easily obtained from a crystal oscillator is $\pm 0.001\%$. A high-frequency crystal can improve upon this performance by perhaps an order of magnitude. Thus, a 6 MHz crystal is available with $\pm 0.0005\%$ temperature accuracy. A succession of five divide-by-10 stages yields a 60 Hz output that can then be fed into the existing circuitry as we have previously considered it.

SUMMARY

11.2 Discrete Component Digital Circuits

- The flip-flop is alternatively known as the Eccles-Jordan circuit and the bistable multivibrator.
- The astable multivibrator (MV) can be termed a square wave oscillator.

11.3 Digital Integrated Circuits

- The Schmitt trigger may be made to respond only to signal pulses above an adjustable level.
- The difference between the on and off levels of a Schmitt trigger is termed the hysteresis.
- The one-shot multivibrator, also called a monostable MV, emits a single pulse for each input pulse, in contrast to a bistable MV, which emits a single pulse for each two input pulses.

11.4 Truth Tables and Boolean Algebra

- Logical mathematics utilizes Boolean algebra.
- Basic logical operations consist of AND and OR and their negations, NAND and NOR, respectively.
- De Morgan's theorems connect logical operations with their negation.

11.5 Logic Circuits

- Two basic faults of the *RS* flip-flop stem from the possibility of multiple pulse output from a single triggering and the simultaneous presence of pulses at both inputs being an unacceptable state.
- The deficiencies of an *RS* flip-flop may be corrected by using a *JK* master-slave arrangement.
- Rather than using a gating pulse, flip-flops are subject to more certain action by using edge triggering.

11.6 Numbers

- The basic unit of digital circuits is the bit, representing either a 0 or a 1. Eight bits of information constitute a byte.
- Rather than a binary system, which is subject to transcription errors when representing a large number, the hexadecimal system, using 16 as a base, constitutes a much more compacted representation.

11.7 Digital Readout

- Digital readouts employ either an LED (light-emitting diode) or an LCD (liquid-crystal display).

- For display purposes the binary-coded decimal (BCD) must be converted into a seven-segment excitation required by either an LED or an LCD. This conversion is accomplished by means of a decoder.
- A ripple counter changes state in response to a change in the state of the previous counter. A synchronous system has all counters changing state simultaneously in response to a clock pulse.
- The modulo of a counter represents the number of states through which it passes before starting to repeat the sequence. A 1248 binary system represents a modulo-16 and a BCD system, modulo-10.

KEY TERMS

binary-coded decimal (BCD): utilizes four binary digits to represent decimal numbers 0–9

binary numbers: utilize 0 and 1 exclusively

bit: the most elemental unit of information, being either a 0 or a 1

Boolean algebra: an algebra based exclusively on digits 0 and 1

byte: 8 bits of information

clock pulse: the various portions of a digital system; generally made to perform their respective functions in synchronism initiated by a common pulse termed a clock pulse

comparator: an electronic circuit that can detect whether an input signal magnitude is either less than or greater than some chosen reference

data lockout: no response to input trigger pulses until a current operation cycle is completed

decimal numbers: utilize 10 digits (0–9)

De Morgan's theorems: relationships that equate OR and AND logical operations

***D* flip-flop:** a device in which serial data are transferred after a delay of one clock pulse

duty cycle: the ratio of "on" time of a square wave to the square wave period

edge-triggered flip-flop: a device in which data are acquired and the state is changed by triggering due to the edge of the input pulse (This minimizes uncertainties due to input variations during the gating pulse of a conventional master-slave.)

enabling line: wire lead whose state of activation determines the nature of a device's output

exclusive or: in terms of positive logic, where a 1 output prevails only if two dissimilar inputs are applied and a 0 output prevails only if the two inputs are the same, either both 0 or both 1

fan-in: the number of inputs that can safely be connected to the input of a logic circuit without significant signal degeneration

fan-out: the number of circuits that can be connected to the output of a logic circuit without significant signal degeneration

flip-flop: an electronic circuit with two distinguishable stable states, adaptable to the manipulation of Boolean algebra; also called a bistable multivibrator

hexadecimal numbers: utilize 16 digits, represented by 0–9 and the letters *A–F*

hysteresis: the voltage difference between the upper and lower thresholds of a Schmitt trigger

inverter: a device that converts a logical 0 into a logical 1 and vice versa

JK flip-flop: remedies the major *RS* FF deficiency (The simultaneous presence of 1s is an acceptable input; it causes the FF to change its state.)

light-emitting diode (LED): a display device in which the segments making up the visual number display consist of diode junctions that become luminous when forward-biased

liquid-crystal display (LCD): a display device in which the symbols are applied in the form of transparent conducting areas; in conjunction with a backplate, the optical properties of the intervening liquid changes in response to a voltage, rendering the symbol visible

master-slave flip-flop: a device whose output is delayed relative to the input by one clock pulse

modulo: the number of different states that a counter can assume before repetition

multivibrator (MV): in effect, a source of square waves; also called a "multi"

- **astable MV:** an MV that, without the need for external triggering, continually switches between its two stable states, thereby generating a sequence of square waves
- **bistable MV:** an MV that, by means of trigger pulses, may be made to alternate between two stable states; also called a *flip-flop* as well as an *Eccles-Jordan circuit*
- **monostable MV:** an MV that, upon receipt of a trigger signal, is made to change its state for a fixed time interval before returning to its original state to await the arrival of another trigger

pulse; also called a *one-shot MV* as well as a *gating circuit*

negative logic: where 1 is represented by the absence of a pulse and 0, by the presence of a pulse

nibble: 4 bits of information

positive logic: where 1 is represented by a pulse and 0, by the absence of a pulse

ripple counter: where each successive counter is triggered by the previous counter

***RS* flip-flop:** a device in which an *S* (set) input yields a 1 output and an *R* (reset) input gives rise to a 0 output (The major difficulty is that simultaneous 1 inputs constitute a disallowed state.)

Schmitt trigger: a circuit that allows for the detection of signal amplitudes in excess of a (selectable) level; may also be used to generate square waves from other waveforms such as sine waves

shift register: used to convert the serial data pulses along a single line into a simultaneous sequence appearing at a number of terminals equal to the number of bits; may also be used to transfer en masse the pulse assemblage along the parallel format

synchronous counter: where all counter stages are triggered in unison by a clock pulse

***T* flip-flop:** called a toggle flip-flop; yields a 1 output for each alternate pulse applied

truth table: a tabular listing of logical operations appropriate to a particular device

zero-crossing detector:: a comparator with a reference voltage of zero

KEY DEVICES

LM311: a popular comparator (The LF311 is an FET version to be used where low input current is a necessity: 0.05 nA vs. 100 nA for the LM311.)

555: a classic timer circuit, essentially an astable MV (The 556 is a dual 555; the 7555 is a low-power CMOS version; the 7556 is a dual 7555.)

74xx: a family of logic circuits: 10 ns delay, 10 mW/gate, 0°–70°C temperature range

54xx: the military equivalent of the 74xx, with −55°C to +125°C temperature range

74LSxx: low-power Schottky: 9.5 ns delay, 2 mW/gate

74ALSxx: advanced low power: 4 ns delay, 1 mW/gate

74Sxx: high speed: 3 ns delay, 20 mW/gate

74Fxx: high speed, low power: 2.7 ns delay, 4 mW

74Lxx: obsolete

74Hxx: obsolete

7400: a quad—that is, four per chip—two-input NAND gate

7410: a triple—that is, three per chip—three-input NAND gate

7473: a *JK* flip-flop

7474: a *D* flip-flop

7447: a BCD decoder

74LS191: an IC 4-bit up-down counter (See Problems 11.27 and 11.28.)

SUGGESTED READINGS

Eccles, W. H., and F. W. Jordan, "A Trigger Relay Utilising Three-Electrode Thermionic Vacuum Tubes," ELECTRICIAN, vol. 83, p. 298, Sept. 19, 1919. *First bistable electronic trigger, used basically as a relay.*

Wynn-Williams, C. E., PROCEEDINGS OF THE ROYAL SOCIETY OF LONDON, vol. A132, p. 295, 1932. *First flip-flop used for counting. Each unit drew 7 A, and the counter was reset using a rope!*

Richards, R. K., ARITHMETIC OPERATIONS IN DIGITAL COMPUTERS. New York: Van Nostrand, 1955. *Boolean algebra; 1's, 2's, 9's, and 10's complements; binary multiplication and division; and so forth.*

Lancaster, Don, *TTL* COOKBOOK. Indianapolis: Howard W. Sams, Publ., 1974. *TTL logic, circuits, and description of some popular units.*

Lancaster, Don, CMOS COOKBOOK. Indianapolis: Howard W. Sams, Publ., 1977. *CMOS logic, circuits, and description of some popular units.*

Colclaser, Roy A., Donald A. Neamen, and Chas. F. Hawkins, ELECTRONIC CIRCUIT ANALYSIS, Chaps. 6, 7, 8. New York: Wiley, 1984. *Many numerical examples involving the analysis of electronic logic circuits, particularly TTL and Schottky TTL.*

Horowitz, Paul, and Winfield Hill, THE ART OF ELECTRONICS. New York: Cambridge University Press, 1980. P. 16, *graphical presentation of logic level tolerances for TTL, ECL, and CMOS logic families,* and pp. 351–356, *useful practical comments on the use of monostable MVs.*

Millman, J., MICROELECTRONICS, pp. 225–228. New York: McGraw-Hill, 1979. *In contrast to the ripple up-down counter we considered, for a synchronous up-down counter, see this book.*

Hall, Jerry, and Chas. Watts, LEARNING TO WORK WITH INTEGRATED CIRCUITS. Newington, Conn.: American Radio Relay League, 1977. *For construction details of the frequency counter considered in Section 11.7.4.*

Malmstadt, H. V., and C. G. Enke, DIGITAL ELECTRONICS FOR SCIENTISTS, p. 219. New York: Benjamin, 1969. *Astable MV guaranteed to be self-starting.*

EXERCISES

1. Give some alternative names for the flip-flop circuit.

2. In Figure 11.1, if the positive terminal of the 5 V battery is grounded, what are the voltages at the various resistor junctions?

3. Sketch an OPAMP version of the Schmitt trigger circuit.

4. What type of feedback is employed in a Schmitt trigger? Why?

5. Why is it not desirable to reduce the hysteresis of a Schmitt trigger to too low a value?

6. In what way does the nature of the output of an astable MV differ from that of a flip-flop?

7. In what basic way does the circuit of an astable MV differ from that of a flip-flop?

8. Sketch the respective waveforms between the base-emitter and collector-emitter of an astable MV.

9. How is the duty cycle of a square wave defined?

10. Sketch an OPAMP version of an astable MV. What is the nature of the feedback used?

11. How can the period of an OPAMP square wave generator be altered?

12. What is the difference between the output of an astable MV and that from a monostable MV?

13. For a monostable MV, what does the recovery time represent?

14. What is the required relationship between trigger and output pulses of a monostable MV?

15. What, in logical terms, is the meaning of + and ·?

16. Using De Morgan's theorems, solve the following:

$$\overline{A + B} =$$

$$\overline{A \cdot B} =$$

17. What is the function of an inverter?

18. Distinguish between positive and negative logic.

19. Distinguish between an OR circuit and an EXCLUSIVE OR circuit.

20. Using A and B, write a logic statement representing an EXCLUSIVE OR.

21. Using logical elements, show the nature of a flip-flop circuit.

22. What is an RS flip-flop?

23. What is a major difficulty with an RS flip-flop?

24. What deficiencies of an RS flip-flop are remedied by a JK master-slave flip-flop?

25. State the truth table for a JK flip-flop.

26. With regard to the symbolic representation of a flip-flop, how is the need for negative triggering indicated? How is edge triggering indicated?

27. With regard to symbolic representation, how are leading edge triggering and transfer on trailing edge indicated?

28. What does data lockout signify?

29. What does the LS designation, for example, 74LS00, applied to a logic circuit represent?

30. What is the decimal equivalent of 1101, where the least significant bit (LSB) is at the right?

31. Express the decimal number 12 in binary form.

32. What is the numerical base for the hexadecimal system?

33. How does the binary system differ from the binary-coded decimal (BCD) system?

34. What is an LED?

35. What type of circuit must be used to apply a BCD number to an LED display?

36. What is the symbolic representation for a Schmitt trigger?

37. What is the difference between a ripple counter and a synchronous counter?

38. What is the meaning of the term *modulo*?

39. What is the function of the totalizer in the frequency counter? Of the latch pulse? Of the reset pulse?

PROBLEMS

11.1

(*Note*: A *pnp* transistor is used in the circuit in Figure 11.66; all polarities are the reverse of those for an *npn* transistor.)

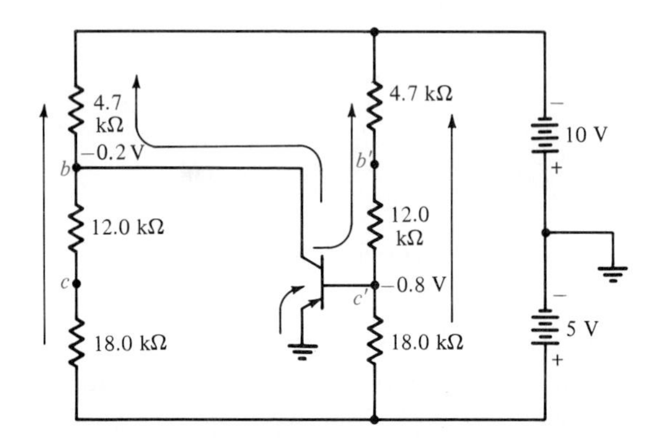

FIGURE 11.66

a. Calculate the value of the voltage drop across each resistor (Figure 11.67).

b. Measured relative to ground, what is the voltage at *a*? At *b*? At *c*? At *d*?

c. When a transistor is connected as in Figure 11.67 and draws sufficient current to bring point *c′* to essentially saturation at −0.8 V and this base current gives rise to a collector current sufficient to make v_{CE} equal to the saturation value of −0.2 V, measured relative to ground, what is the voltage at *c*? At *b′*?

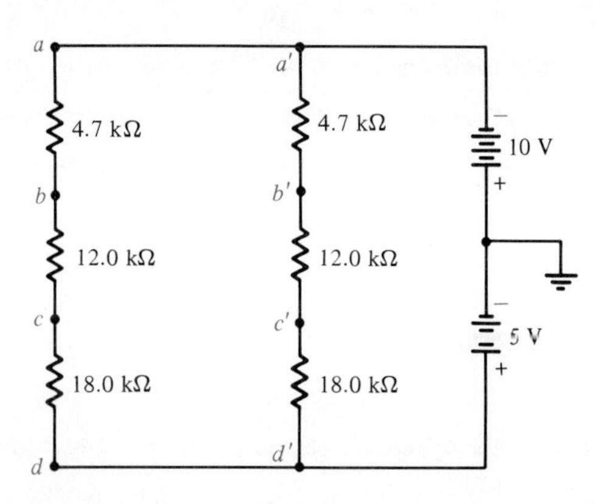

FIGURE 11.67

d. If another similar transistor has its base connected to point *c* and its emitter grounded, what can you say about its collector current? (Its collector is connected to point *b′*.)

e. What is the magnitude of the conducting transistor's base current? Of its collector current?

11.2

Refer to Figure 11.12. With saturated outputs at ±15 V and an amplifier with a voltage gain of 50,000, what is the range of input voltage that constitutes the window?

11.3

Refer to Figure 11.20. Using an OPAMP with a gain $A = 1.0 \times 10^4$, we have $V_H = +15$ V and $V_L = -15$ V.

a. With $R_1 = R_2 = 1$ kΩ and $V_{ref} = 0$ V, what is the magnitude of the hysteresis?

b. Sketch the hysteresis curve for part a.

c. Indicate how V_{ref} may be conveniently varied.

d. With $R_1 = 1$ kΩ and with all other conditions as in part a, what should be the value of R_2 for a hysteresis magnitude of 100 mV?

e. Evaluate the threshold values for the following cases:

1. $V_H = +15$ V, $V_L = -15$ V, $V_{ref} = 1$ V, $R_1 = R_2 = 1$ kΩ.
2. Change V_L to zero; all other values are as in part 1.
3. $V_H = +15$ V, $V_L = -15$ V, $V_{ref} = 1$ V, $R_1 = 1$ kΩ, $R_2 = 200$ kΩ.
4. Change V_L to zero; all other values are as in part 3.

11.4

a. With regard to the Schmitt trigger (Figure 11.20), how must V_{ref} be chosen so that V_{th1} is negative? (Take $V_H = +V_{out}$ and $V_L = -V_{out}$.)

b. What should be the value of V_{ref} if $V_{th2} = -V_{th1}$?

11.5

With regard to the equation for the period of an OPAMP multivibrator (Eq. 11.1), derive the condition that will make the period independent of β.

11.6

To trigger a *pnp* flip-flop, rather than applying a positive pulse to stop conduction, we could alternatively apply a negative pulse to start conduction. We now wish to trigger the FF *not* by a positive pulse but rather by the *leading edge* of the positive pulse. Such triggering can be accomplished by using a differentiator in conjunction with a diode as in Figure 11.68. What circuit alteration will allow the FF to be triggered by the leading edge of a negative pulse?

11.7

A square wave whose frequency is 500 Hz and whose amplitude is +5 V is impressed upon the differentiating circuit in Figure 11.69. When the +5 V is first applied, before any charge builds up on the capacitor, the full source voltage appears across *R*, and the initial current is *V*/*R* (where *V* = +5 V in this case). At any time thereafter the value of the current is

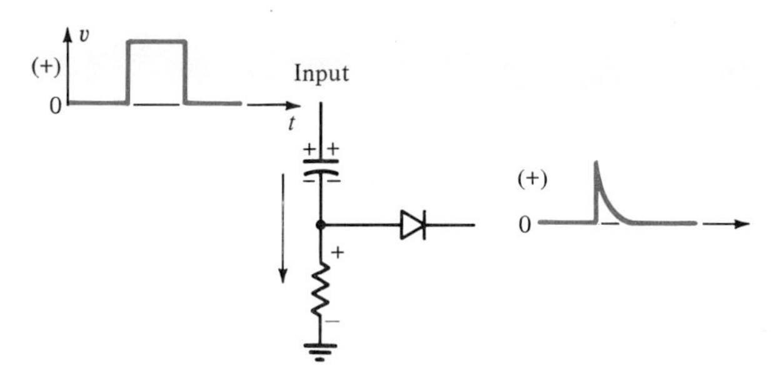

FIGURE 11.68

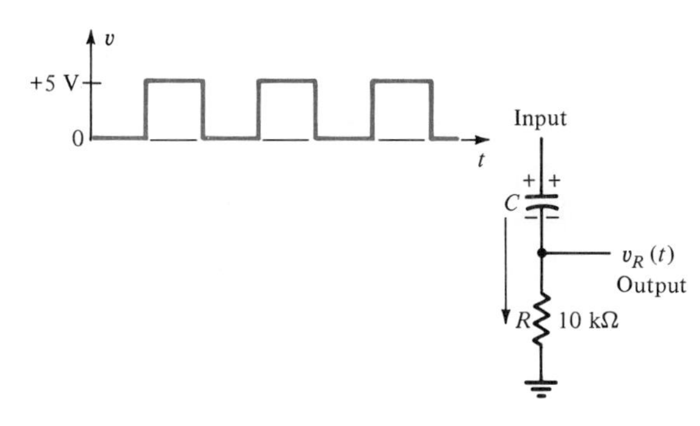

FIGURE 11.69

$$i(t) = \frac{V}{R} e^{-t/RC}$$

At corresponding times the voltage across R is

$$v_R(t) = Ri(t) = R\frac{V}{R} e^{-t/RC} = Ve^{-t/RC}$$

a. What is the period of one square wave cycle for the stated repetition rate?

b. What is the value of the capacitor if the circuit time constant is to be 0.2 ms?

c. Sketch one cycle of the square wave and immediately below it the corresponding v_R waveform—that is, the voltage across the resistance.

11.8
For the astable MV shown in Figure 11.6, if $R_1 = 2.2$ kΩ, $C = 0.01$ μF, and $R_3 = 43$ kΩ, what is the pulse repetition rate? What is the rise time of the collector voltage?

11.9
Refer to Figure 11.17. For a monostable MV, what condition makes the period independent of β? Using the following values, estimate the recovery time $(T' - T)$ (see Figure 11.18): $R_1 = 47$ kΩ, $R_2 = 27$ kΩ, $C_t = 0.01$ μF, $R_t = 100$ kΩ, $V_{out} = 15$ V. Using the optional recovery circuit (dashed connections) in Figure 11.17, what is the recovery time?

11.10
List the truth table for a two-input AND gate. Connect the two inputs together. What will be the respective outputs as a 1 and a 0 are successfully applied to the joined inputs? What will be the output from a NAND gate when the inputs are tied together and a 1 and a 0 are successively applied to the joined inputs?

11.11
Three NAND gates may be connected to act as a two-input OR gate as in Figure 11.70. Construct a truth table that justifies the result shown.

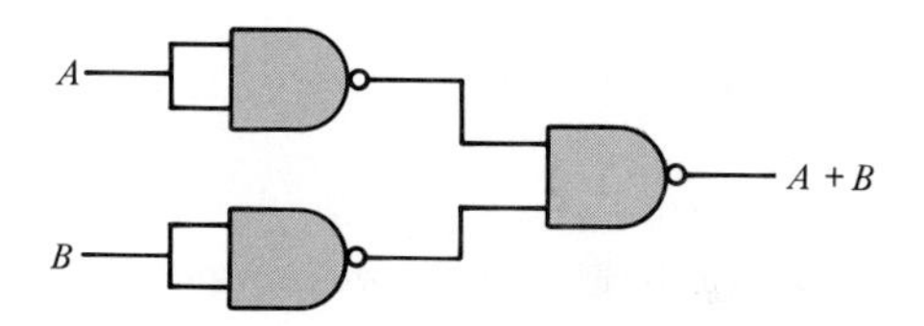

FIGURE 11.70

11.12
Using the rule that the inverse of a Boolean function can be obtained by inverting all variables and replacing ORs by ANDs and vice versa, determine

$$\overline{\overline{A} + \overline{B}} =$$

$$\overline{\overline{A} \cdot \overline{B}} =$$

11.13
Analyze the logic circuit in Figure 11.71 by constructing a truth table and demonstrate that this circuit can be replaced by a single NAND gate.

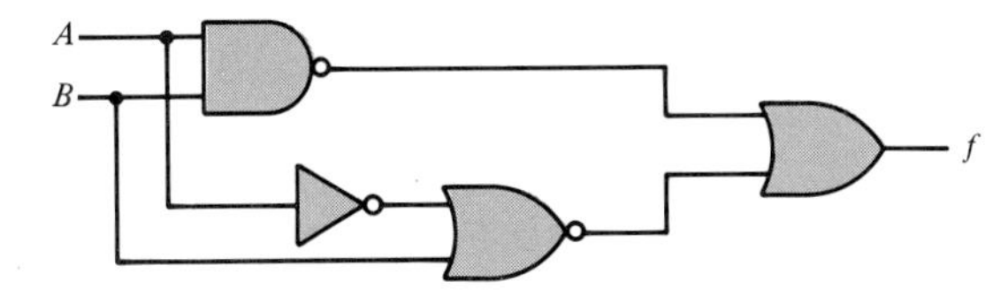

FIGURE 11.71

11.14
State De Morgan's theorems and draw the logical circuits illustrative of each side of the equations. Show the validity of the equations by means of truth tables.

11.15
Determine the truth table for the circuit in Figure 11.72 and indicate the nature of the output.

11.16
Consider the following statement of the EXCLUSIVE OR:

$$(A + B) \cdot (\overline{A \cdot B})$$

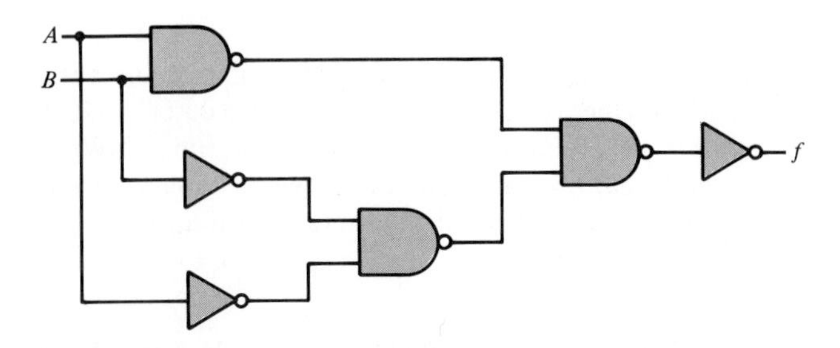

FIGURE 11.72

An alternative form of this logical statement is

$$(\overline{A} \cdot B) + (A \cdot \overline{B})$$

This statement is called an *inequality operator* because it provides a 1 output if A and B are *not* equal. The *inverse* of the inequality operator is the *equality comparator*:

$$(A \cdot B) + (\overline{A} \cdot \overline{B})$$

This statement will yield a 1 output if A and B *are* equal. Show that the inverse of an inequality comparator is an equality comparator and demonstrate that the circuit in Figure 11.73 constitutes an equality comparator.

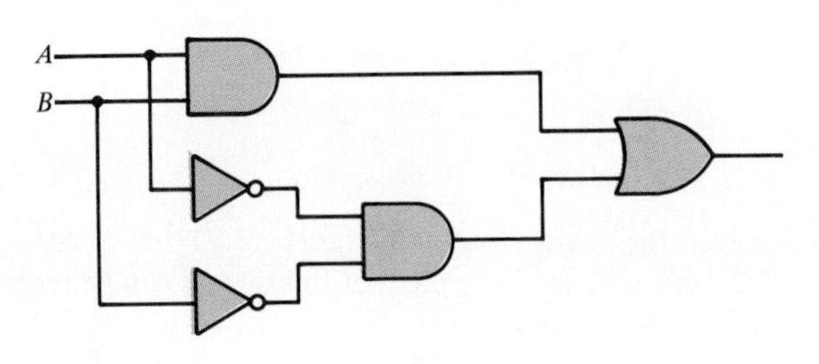

FIGURE 11.73

11.17
Identify the nature of the triggering in the circuits of Figure 11.74.

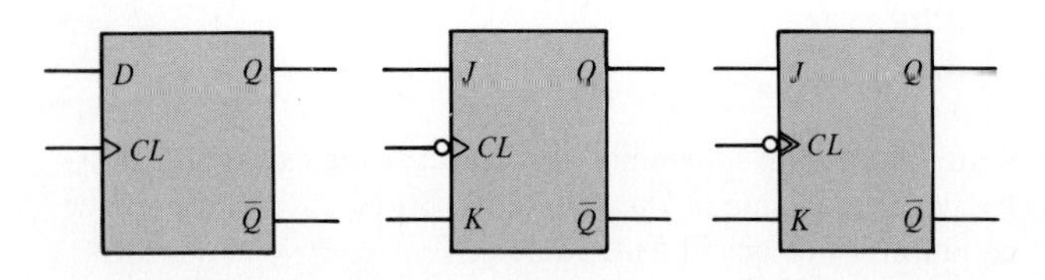

FIGURE 11.74

11.18
Write the following decimals as binary numbers:

5, 11, 16, 20, 31, 42

Write the following binary forms in terms of their decimal equivalents:

0011 0101
1010 1110
1101 1010
1110 0001
1011 1111

11.19

a. Write the following decimals as binary numbers:

4, 6, 12, 17, 30, 43

b. Write the following binary forms in their decimal equivalents:

1101 1101
1111 1111
0101 1101
0110 1111
1000 0001

c. Convert the following into their binary forms:

0.265625
0.5703125

11.20

a. Express the following decimals in BCD form:

187_{10}
887_{10}
238_{10}
112_{10}
556_{10}

b. If the following are BCD forms, express them in decimal equivalents:

0101 0101 0101
0110 0111 0111
1001 0101 0101
1100 0001 0101
0010 0111 0001

c. Express the BCD numbers of part b in binary form.

11.21
Express the following as signed 2's complement numbers, using a total of 8 bits:

+127, +5, +4, +3, +2, +1, 0, −1, −2, −4, −5, −127, −128

11.22

a. Express 1071_{10} as a BCD number.

b. What is the range of numbers represented by a four-decade BCD code?

c. How many bits does this represent?

d. What is the range of decimal numbers (in contrast to BCD) that can be represented by this same number of binary bits?

11.23

Perform the following additions in terms of signed binary forms:

a. 87_{10} / 34_{10} **b.** -87_{10} / -34_{10}

c. 87_{10} / -34_{10} **d.** -87_{10} / 34_{10}

e. -80_{10} / -52_{10} **f.** 80_{10} / 52_{10}

11.24

Consider the four-input AND gate in Figure 11.75, where A is the LSB. What 4-bit combination will yield a 1 output? To what decimal value does this result correspond? Indicate how 10 AND units may be used to convert 4-bit binary numbers into decimals.

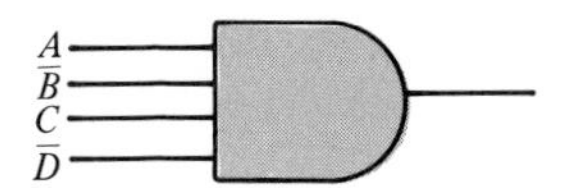

FIGURE 11.75

11.25

Monitoring the 60 Hz power-line frequency at times shows a frequency shift of as much as 950 ppm. In using the frequency counter to measure a 50 MHz frequency, how much of an error is introduced if the power line is employed as the source of clock pulses? How much of a frequency error is introduced if a 0.0005% temperature coefficient 6 MHz crystal is employed under like conditions?

11.26 △

We have discussed an up counter—that is, each successive input pulse leads to an increased count. Show how the use of the $\overline{Q}$ output of a flip-flop to trigger each successive stage may be used to create a down counter by considering a modulo-16 counter. Starting with a count of 12_{10}, consider the consequence of coupling from $\overline{Q}$ as another pulse enters.

11.27 △

Figure 11.76 represents an up-down counter—that is, one that can be made to add or subtract in response to the input pulses, depending on whether the control signal is up (1) or down (0). Explain its operation. (*Caution*: Changing the control lines of such ripple counters between input pulses may lead to spurious results. The synchronous up-down counter does not suffer from this difficulty.)

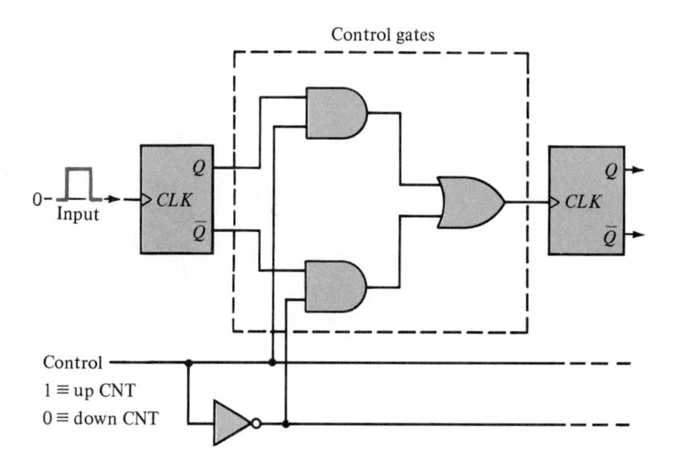

FIGURE 11.76

11.28 △

In an up-down counter, rather than using two AND and one OR gate for control between stages (Figure 11.77), this function may be handled with NAND gates exclusively. Construct the truth table in both cases and show that they lead to identical results.

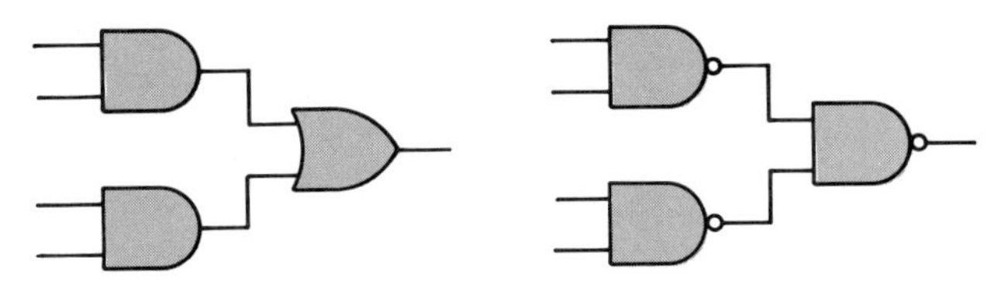

FIGURE 11.77

11.29

Sketch the switching arrangement that allows one to alternatively use a 7490 with either preset 0 or preset 9 whenever the counter resets via a positive reset pulse.

12 OSCILLOSCOPES

OVERVIEW

The visual depiction of signals has often dominated the discussions in prior chapters. As a result of our consideration of circuits in Chapter 11, we are now in a position to relate in a more informed manner how this observation is accomplished through the use of an oscilloscope. Even when used to observe strictly analog signals, the oscilloscope's circuitry utilizes digital techniques in the form of Schmitt trigger circuits, gating multivibrators, etc. This reliance on digital circuits accounts for the lengthy delay in the appearance of the oscilloscope in the table of contents.

The oscilloscope, being the most versatile tool in the realm of electronics, plays a role in perhaps every topic discussed even when not explicitly mentioned. For this reason we choose to consider its commercial aspects more fully than done so for any other device.

OUTLINE

12.1 ■ THE BASIC OSCILLOSCOPE

The cathode ray tube used to observe electrical waveforms utilizes a stream of electrons that, on striking the fluorescent screen behind a glass face on the tube, gives rise to a bright illumination spot. By means of two sets of parallel plate electrodes (Figure 12.1), this spot may be made to move either up and down or left and right.

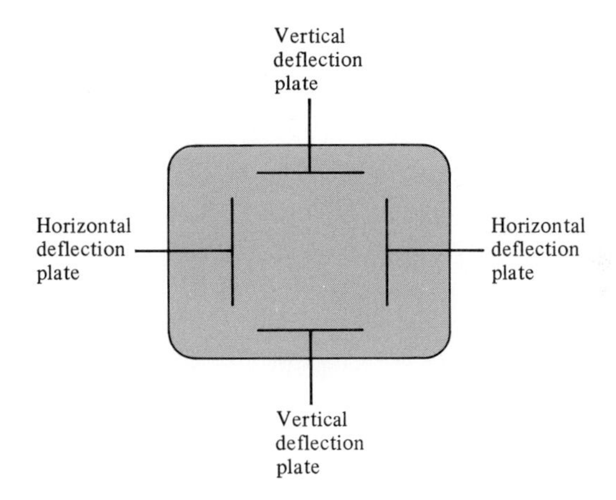

FIGURE 12.1. **Vertical and Horizontal Deflection Plates of an Oscilloscope**

Under the usual operating conditions, after suitable amplification the waveform to be observed is applied to the **vertical deflection plates.** Let us presume the signal to be observed is a sinusoid. Applying this waveform to the vertical deflection plates will give rise to a bright vertical line, the ac applied to the plates deflecting the beam up and down as the ac changes magnitude and polarity (Figure 12.2).

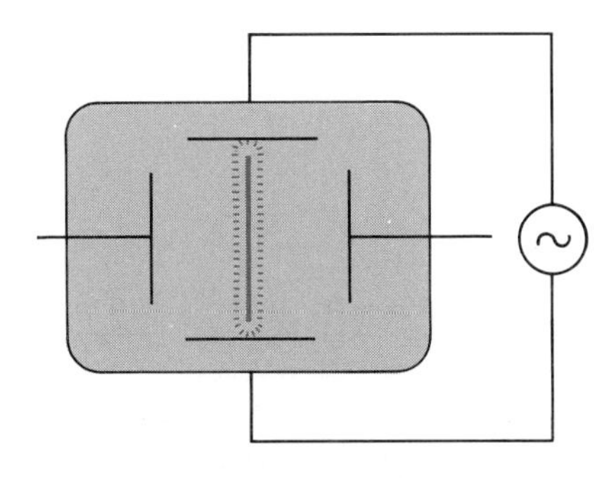

FIGURE 12.2. **Signal Voltage Applied to the Vertical Deflection Plates** Without any voltage variation across the horizontal plates, vertical beam deflection equals the peak-to-peak signal amplitude.

Presuming the left horizontal plate to be grounded, if to the other horizontal plate we apply a sawtooth wave of the form shown in Figure 12.3, when this sawtooth is at its greatest negative value, the right horizontal plate will have pushed the beam to the left of the screen. As the voltage becomes less negative, the beam approaches the center of the screen. Then as it becomes positive, it is pulled to the right. On reaching the right extremity of the screen, the beam "snaps back" to the left as the sawtooth voltage suddenly goes negative. The cycle will then be repeated.

If the period of the sawtooth is integrally related to the period of the signal to be observed, one will see an apparently stationary waveform (Figure 12.4). With the same signal voltage, if the sawtooth period is twice that of the signal, one will observe a display consisting of two cycles of the signal (Figure 12.5).

The oscilloscope has a "built-in" **sawtooth** generator that gives rise to a **time base** whose period is adjusted by front panel controls. Such controls are generally calibrated in terms of **sweep speed** *per centimeter.* Thus, with the

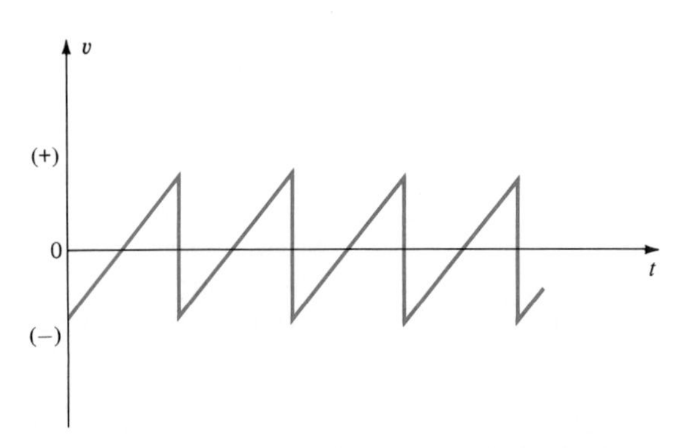

FIGURE 12.3. **Sawtooth Voltage Necessary for Linear Horizontal Sweep of Oscilloscope Beam**

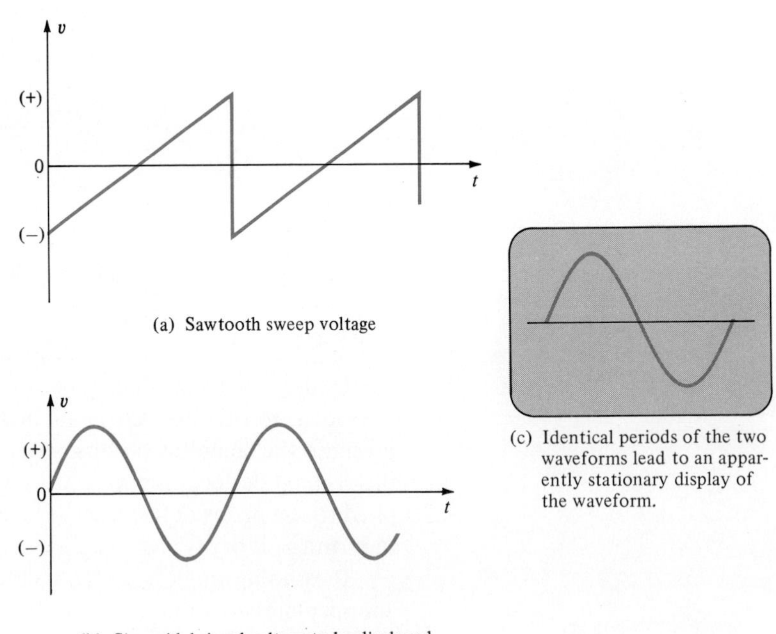

FIGURE 12.4. **Basic Oscilloscope Display**

sweep speed set at 10 ms/cm, for example, if the observed sinusoid consists of one cycle, with the usual display width of 10 cm we know the frequency of the applied signal to be 10 Hz (10 ms/cm × 10 cm/cycle = 0.1 s/cycle; frequency = 1/period = 10 Hz). If two cycles are observed with the sweep speed setting the same, the frequency of the applied signal is 20 Hz.

If the period of the sawtooth and that of the signal are not exactly matched, the display will not appear stationary and will, instead, show a continuous motion that makes observation of the display details impossible. Despite careful adjustment, even if the two are initially brought into synchro-

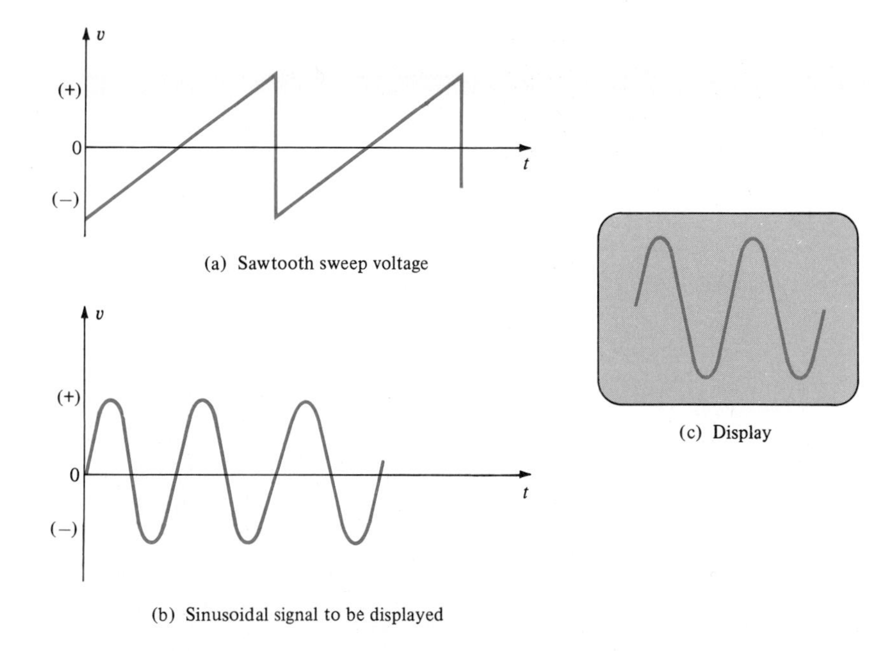

FIGURE 12.5. **Basic Oscilloscope Display**

nism, in practice it will be found that the display will soon start to move because of heating drift, source instability, etc. Some method of continuous automatic synchronization is desired. Figure 12.6 shows a series of circuit functions that can accomplish this. The input signal is simultaneously applied to two circuits:

1. The customary signal amplifier applies the signal to the vertical deflection plates after having passed through a delay line. (More about the delay line in a moment.)
2. The input signal is also applied to a Schmitt trigger. This circuit creates a corresponding square wave that is then applied to a differentiator, creating a set of negative and positive pulses. The separation between each set of pulses is thus assured of being equal to the period of the signal to be observed.

The differentiated pulses are applied to the sawtooth generator, which (in our presumed example) commences the sweep only when activated by a negative pulse from the differentiator. (A simple circuit change could alternatively make the sweep start with the positive differentiated pulses.)

Now if the period of the applied waveform starts to slowly change because of drift, the period of the trigger pulses will also change accordingly, and sawtooth and signal will remain synchronized.

That portion of the signal that precedes the triggering of the Schmitt (and hence the commencement of the sweep) would ordinarily be lost. But, by passing the signal through a delay line, it appears at the vertical deflection plates after the sweep has been initiated, and hence the complete waveform is

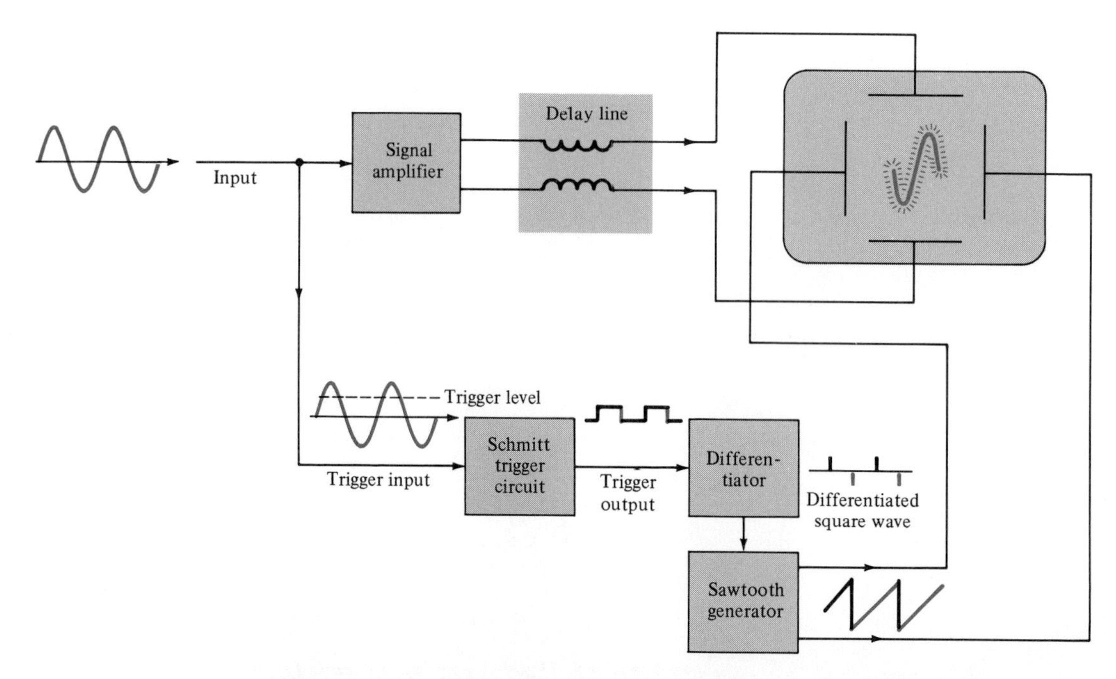

FIGURE 12.6. Basic Oscilloscope Configuration The amplified signal is applied to the vertical deflection plates after being delayed. The signal is also applied to a Schmitt trigger from which the resultant square is applied to a differentiator, and the negative pulses are used to initiate sweep voltage in synchronism with the signal to be displayed.

observed. Initially delay lines consisted of lengths of specially designed coaxial cables that often proved cumbersome. Today they can be made much more compact by using printed circuit techniques.

Suppose we wish to observe pulses of the form shown in Figure 12.7a. Since the interval between successive pulses varies in a random fashion, with a sawtooth having a constantly repeating period, each successive signal pulse would be displayed at a different position on the screen, leading to a display that would have the signal moving rapidly and randomly across the screen's face (Figure 12.7b).

Consider the same display with **synchronization**. A sawtooth arises only when there is a signal pulse to initiate a sweep, and the pulse always appears near the beginning of the sweep. The display now takes on a stationary appearance (Figure 12.7c).

To illustrate another operating facet of an oscilloscope, we might return to Figure 3.7, which showed the succession of waveforms associated with automotive ignition systems. In the display as illustrated, it was assumed that the large negative pulse associated with the "points" opening started the sweep. But suppose we wanted to observe in greater detail the structure associated with the dissipation oscillation, its nature serving to provide information about possible defects in the capacitor and the "coil." To observe greater detail, we need a faster sweep speed. But at low engine speeds when the firing line may occupy a significant fraction of the sweep, increasing the sweep speed may move the dissipation oscillation off the screen to the right. An al-

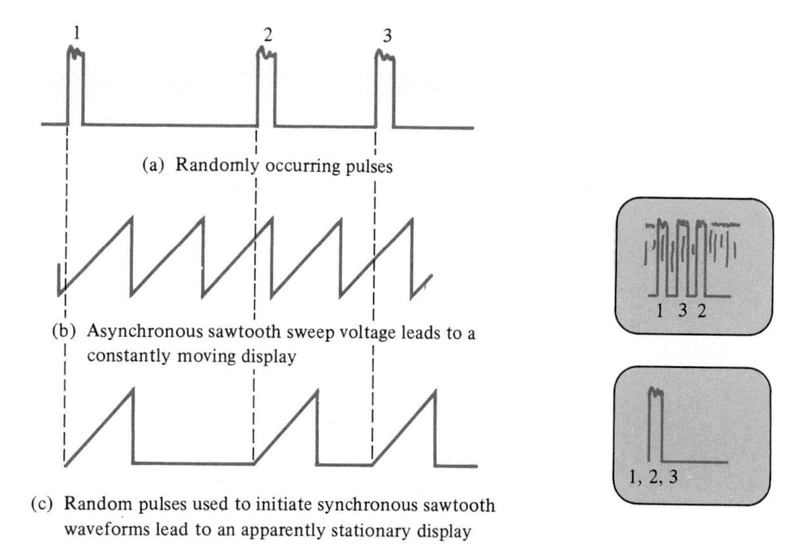

FIGURE 12.7. **Oscilloscope Display of Random Pulses**

ternative arrangement would be to make the sweep trigger on positive pulses. The sweep would then be started by the first positive excursion of the dissipation waveform thereby eliminating the ignition line from the display, and a higher sweep speed would now spread the dissipation oscillation over a considerable portion of the screen.

Not only are there front panel controls that allow us to select whether negative or positive pulses will be used to trigger the scope, but there is also a control labeled **trigger level** that determines how large the positive (or negative) signal must be to cause sweep initiation. (The reader might guess that this is accomplished by a Schmitt trigger.) (See Figure 12.8.)

Just as the sweep speed is calibrated, so is the vertical deflection, often in volts per centimeter. The vertical sensitivity is made variable in steps of 1, 2, and 5—for example, 1 mV/cm, 2 mV/cm, 5 mV/cm, 10 mV/cm, etc. The sweep speed is also usually made variable in steps of 1, 2, and 5.

12.2 ■ *X-Y* PLOTTING WITH AN OSCILLOSCOPE

The time variation between two periodic variables can be displayed on an oscilloscope upon applying the two signals to the *X*- and *Y*-amplifiers, respectively. Under such circumstances the sawtooth signal is removed from the horizontal deflection plates, and the *X*-signal is applied instead. Figure 12.9 illustrates one such application. Sine waves of equal frequency are applied to the *X*- and *Y*-channels, but they are displaced from one another (in this example) by 45°. The resultant display on the cathode ray tube (CRT) is an ellipse. The phase difference between them can be obtained from the equation

$$\theta = \sin^{-1}\left(\frac{A}{B}\right) \tag{12.1}$$

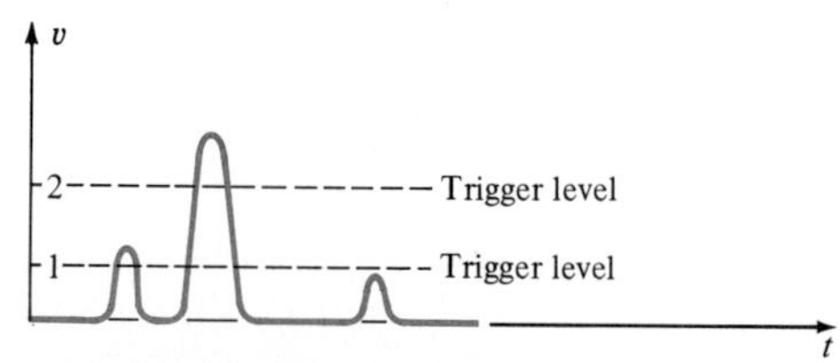

(a) Succession of pulses with two different levels of Schmitt trigger settings

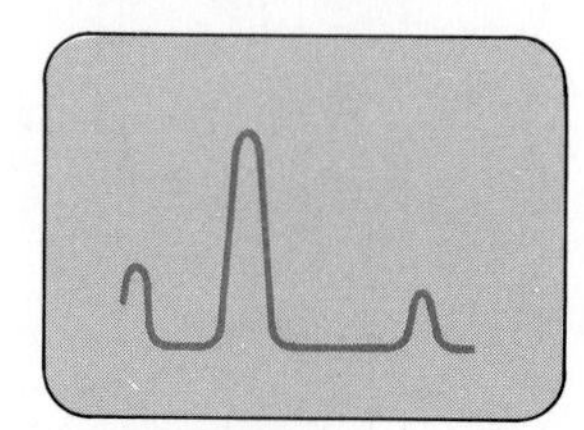

(b) Display with trigger set at level 1

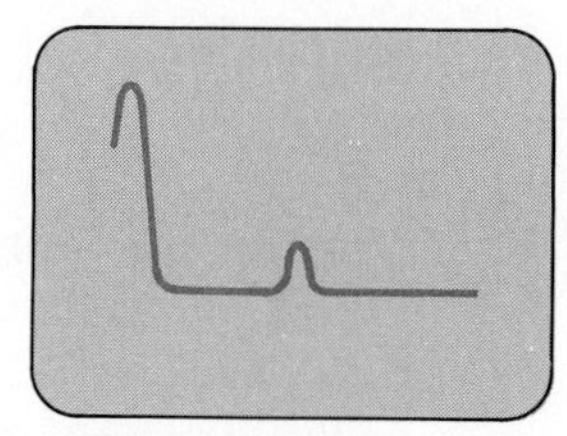

(c) Display with trigger set at level 2

FIGURE 12.8. Operation of Oscilloscope Trigger Level

where $A = +y$-axis intercept and $B =$ projection on the y-axis of the peak vertical deflection

EXAMPLE 12.1 In Figure 12.9, what is the ratio of A/B for the 45° phase shift illustrated?

Solution:

$$\sin 45^\circ = 0.707 = \frac{A}{B}$$

For angles between 0° and 90°, the ellipse has a positive slope, and for angles between 90° and 180°, the slope is negative. The slope remains negative between 180° and 270°, the ellipse resuming positive slopes between 270° and 360°. To distinguish between upper and lower quadrants, an additional known phase shift is introduced. In the range 0° to 180° an added shift causes positive slopes to increase and negative slopes to diminish. For the lower quadrants the effects are oppositely directed.

The CRT may also be used to determine an unknown frequency if it is an exact multiple of a known frequency. Respective applications to the X- and Y-channels lead to a stationary pattern of the form shown in Figure 12.10

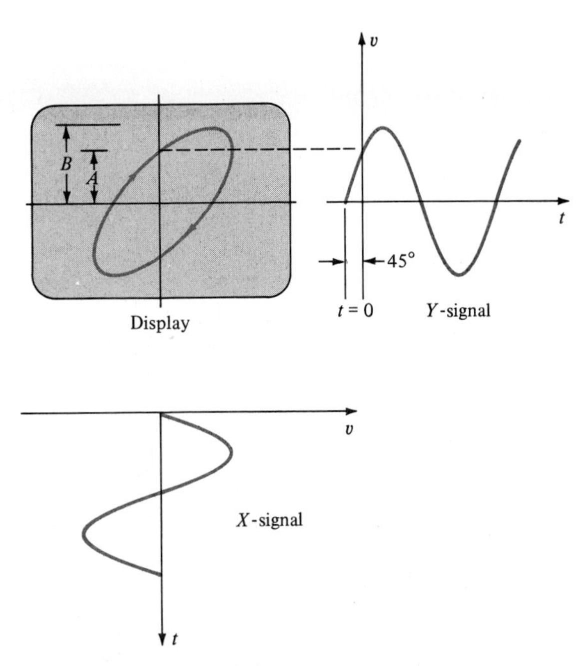

FIGURE 12.9. **Lissajous Figure Resulting from the Application of Two Sine Waves of Equal Frequency to the X- and Y-Deflection Plates But with the X-signal Leading by 45°**

and called a **Lissajous figure.** The number of tangencies in the enclosing rectangle is equal to the frequency ratio. Tangencies along the vertical axis correspond to the X-frequency. For the example shown, $f_x/f_y = \frac{2}{3}$. It can also be seen that it is not necessary that the gains of the two channels be equal. The same holds when making phase measurements, it being obvious that a phase determination (Figure 12.9) is an example of a Lissajous figure for a 1:1 frequency ratio.

X-Y inputs may also be used for other purposes: to obtain hysteresis curves, to plot crossover distortion, to plot transistor characteristics, etc. In instances such as the last, when one might desire to plot i_c vs. v_{CE}, for example, the current variation is converted into a necessary voltage variation (proportional to i_c) by passing i_c through a small resistance.

12.3 ■ DELAYED TIME BASE

Again, using the automotive ignition as an example, suppose one wishes to examine in more detail the slight oscillation that arises when the point contacts close (Figure 12.11a). (This waveform provides information about any "coil" defects.) If the oscilloscope is set to trigger with a positive pulse, the sweep will start with the commencement of the damped oscillation at the termination of the firing line, *not* at the start of the dwell time since the former event comes first. Increasing the sweep speed will not result in our objective

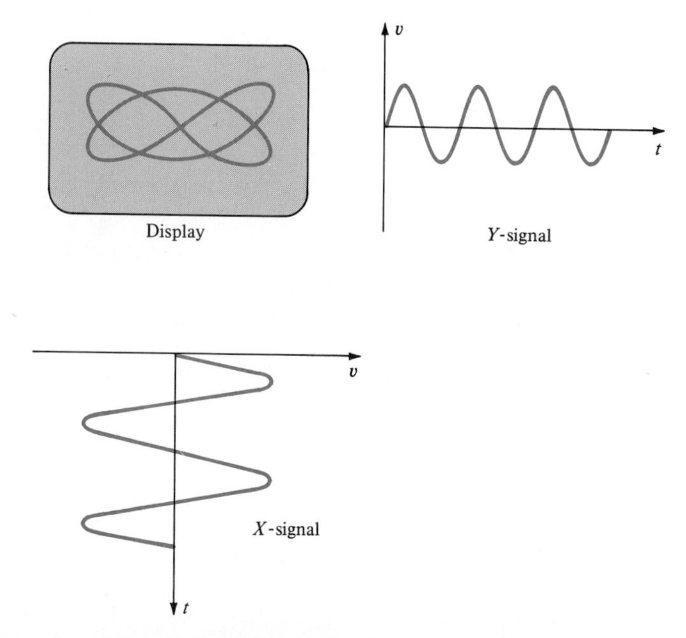

FIGURE 12.10. **Lissajous Figure Resulting from the Application of Two In-Phase Sine Waves Whose Frequencies Differ in the Ratio $X:Y = 2:3$**

since such an increase will merely expand the beginning of the trace and again probably move the desired portion off the screen to the right.

Our objective may be reached by employing an oscilloscope with a **delayed sweep.** Figure 12.11a shows the complete firing cycle and below it (Figure 12.11b) the main time base sawtooth. Additionally, the cross on the sawtooth represents an adjustable dc level. This dc level constitutes the threshold setting of a Schmitt trigger to which the sawtooth waveform is also applied. When the sawtooth reaches the Schmitt threshold, a second (faster) sawtooth (termed the **delayed time base**) commences. The sweep speeds of the main time base (MTB) and the delayed time base (DTB) are separately controlled.

The trigger from the Schmitt also initiates a gating pulse equal in duration to that of the delayed sweep. Both the MTB and the DTB produce unblanking pulses equal to their respective durations (Figure 12.11c). The sweep is visible only during an unblanking pulse. It is by such means that the return trace (at the end of each sawtooth) is made to be invisible on the CRT.

When the two unblanking pulses are superimposed (as in Figure 12.11c), that portion of the main sweep that is to be examined more closely appears brighter than the remainder of the main sweep. The width of this brightened portion can be varied (by means of the delayed time base sweep speed), as can its location along the main sweep (by altering the dc level of the Schmitt trigger). The latter is accurately adjustable by means of a 10-turn helipot. One then selects the portion of the main sweep to be observed and switches to **delayed sweep**, and now the screen will display *only* the part that was preselected, now occupying the entire screen and hence visible in expanded form.

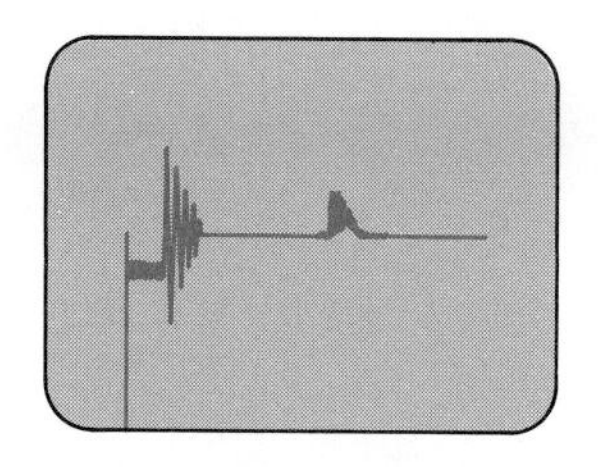

(a) Main time base display showing complete waveform associated with one ignition cycle (The brightened portion is to be expanded using delayed time base.)

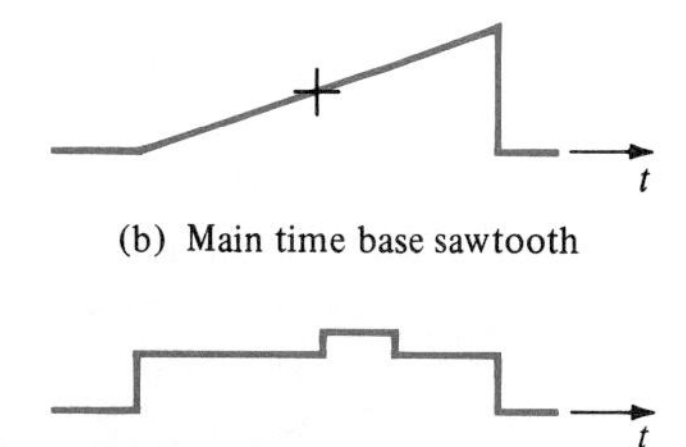

(b) Main time base sawtooth

(c) Unblanking pulses (The broad pulse is associated with the main time base and the narrower, leading to brighter trace, with delayed time base.)

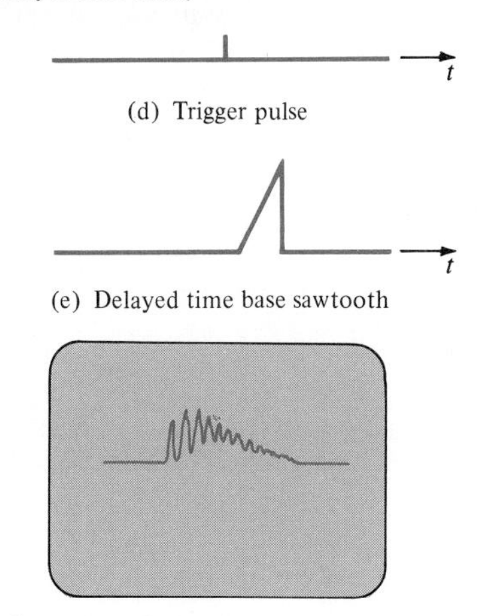

(d) Trigger pulse

(e) Delayed time base sawtooth

(f) Delayed time base display showing expanded version of brightened portion of the display in part a

FIGURE 12.11. **Oscilloscope Ignition Display**

12.4 ■ DUAL TRACE AND DUAL BEAM

Many oscilloscopes allow for the simultaneous display of two different signals, most frequently in order to allow for their comparison in terms of magnitude and time relation. In a **dual-beam** instrument, such simultaneity is accomplished by having two separate electron beams each of which is made to respond to one of the two signals. Such CRT tubes necessarily are more expensive than single-beam instruments.

In a **dual-trace** instrument, a single electron beam is alternately subjected to one signal and then the other. Switching between the two signals in dual-trace instruments can take one of two forms:

1. Chopped mode
2. Alternate mode

In the chopped mode, during each trace, the two signals are alternately applied, the switching between them generally taking place many times during a single sweep. In the alternate mode first one signal is applied for one complete trace and then the second signal, also for one complete trace. Figure 12.12 compares the three possibilities, with each pair showing two successive presentations. In Figure 12.12a, each of the two signals is available in complete form since dual beams are used. In Figure 12.12b, it can be seen that both signals in the chopped mode exhibit discontinuities. To prevent such an appearance, the chopping frequency has to be much higher than the signal frequencies. In the alternate mode first one signal is presented and then the other. If the signal frequencies (and hence also the sweep frequency) are low, the two traces can be seen to follow one another. By not having the appearance of simultaneity, one may have difficulty in comparing them. As a result, below about 10 ms/division, one generally uses the chopped mode, while at sweep speeds greater than 0.1 ms/division, the alternate mode is used. Between these limits either may be used, but the chopped mode will give somewhat less flicker. The chopping frequency usually lies between 200 kHz and 1 MHz.

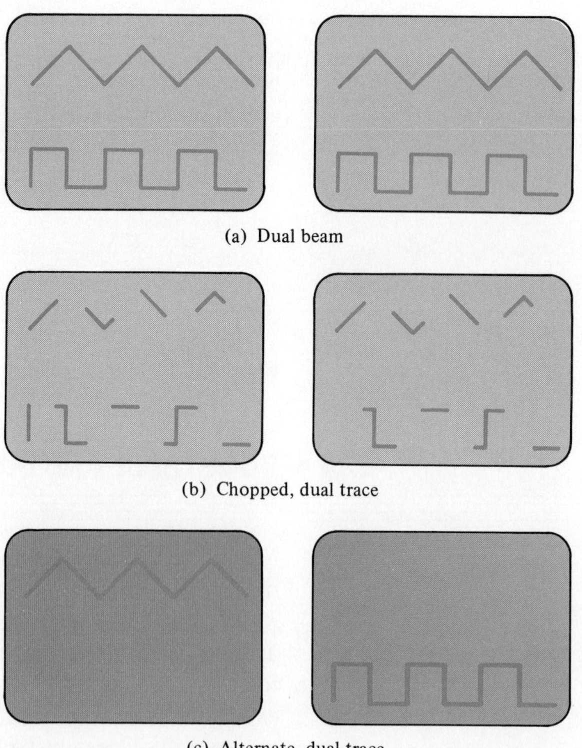

FIGURE 12.12. Two-Signal Oscilloscope Display (Successive Depictions)

12.5 ■ SIGNAL PROBES

Since the function of an oscilloscope is to amplify the usually rather small signals that are to be displayed, the newcomer to electronics is sometimes perplexed as to why one uses signal pickup probes that *attenuate* the signal before applying it to the oscilloscope. The function of such **signal probes** is twofold:

1. To disrupt the signal source to a minimal degree
2. To retain the integrity of the signal

To achieve these objectives, it usually becomes necessary to first attenuate the input signals, as we shall now proceed to show.

Assume we are dealing with an oscilloscope bandwidth of 20 MHz. The input impedance to the scope is a resistance in parallel with a capacitance. The signal is applied by means of some sort of wire leads (usually a coaxial cable) that will increase the input capacitance. The total C_{in} might now easily amount to 120 pF (Figure 12.13a). By presuming a signal source with R_s = 1 kΩ, this capacitance limits the source bandwidth to about 1.3 MHz. If we are attempting to observe a signal pulse, the large probe capacitance causes the high-frequency components of the pulse to be considerably attenuated and alters the appearance of the displayed pulse. Also, because of the low impedance, the attachment of the signal probe to the source could considerably disturb the operation of the circuit being measured.

To minimize the disturbance, we place a resistance in series with the input lead (Figure 12.13b). This resistance considerably reduces the capacitive loading on the source. The capacitance is now typically about 5 pF from the probe tip to the surrounding ground points, and the bandwidth of the source can now be as large as 31.8 MHz. But now the probe's bandwidth is small:

$$1/[2\pi(10^3)(100 \times 10^{-12})] = 1.6 \text{ MHz}$$

This may be corrected by making the series probe resistance much larger (Figure 12.13c). At low frequencies, C_{in} behaves like an open circuit, and we have a simple resistive voltage divider. This divider reduces the signal amplitude at low frequencies. But we are interested in *uniform* signal amplitudes over a wide band of frequencies rather than maximum signal amplitudes. As the frequency increases, C_{in} acts increasingly as a short. But if we place a capacitance across R_p, the "effective" value of R_p also decreases with frequency. When properly adjusted, the attenuation of the network should be constant over a wide bandwidth. This requirement is met if

$$R_pC_p = C_{in}R_{in} \tag{12.2}$$

$$C_p = \frac{R_{in}}{R_p}C_{in} \tag{12.3}$$

Figure 12.13d shows a typical result. The input resistance to the probe is 10 MΩ, the input capacitance to the probe is about 4.2 pF [(4.7 × 42.5)/(4.7 + 42.5)], and the signal is attenuated by a factor of 10.

A coaxial cable (also termed a transmission line) represents a relatively

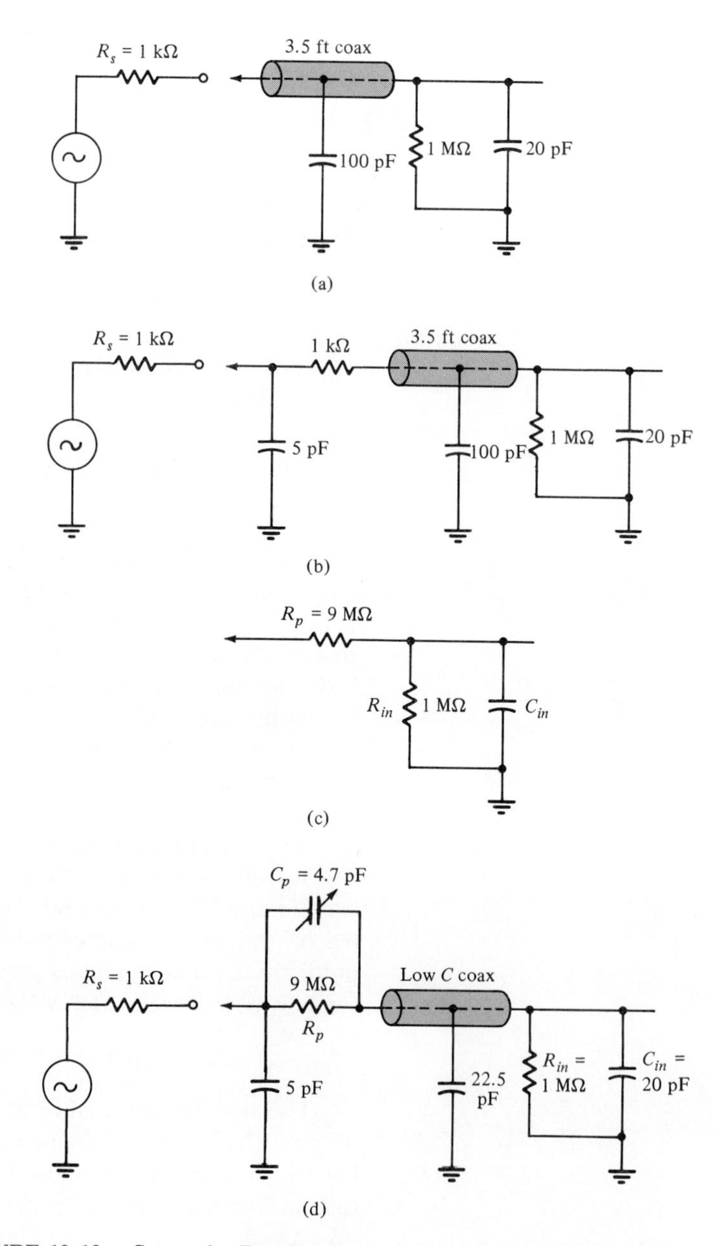

FIGURE 12.13. Successive Development of a High-Impedance Signal Probe

low impedance called the *characteristic impedance* and having a value that is typically 52 Ω, 72 Ω, etc. As the signal passes down the cable, in order not to be reflected at the end, it should be terminated by a resistance equal to its characteristic impedance (that is, a matched resistance). As a result, there arises another problem in conjunction with signal probes. At the high frequencies, when the wavelength becomes comparable to the length of the coax, we start running into cable-matching problems. It is impossible to terminate the cable in a matched fashion and simultaneously maintain high impedance. The solution is to try to absorb the reflected wave by using a

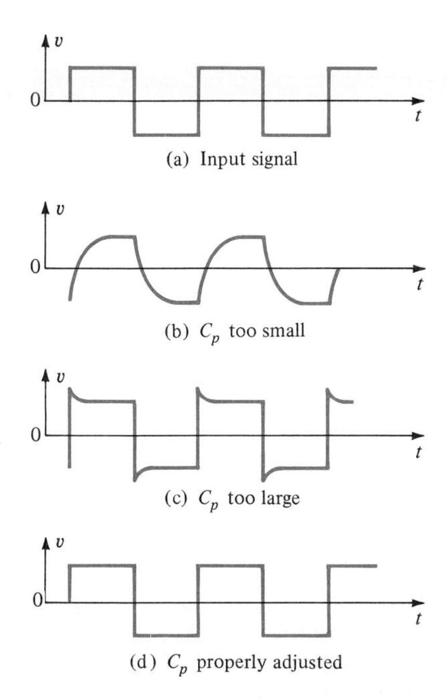

FIGURE 12.14. Oscilloscope Probe Adjustment

resistive conductor as the central wire of the coax. The probe cables are designed to provide significant resistance and low capacitance.

There is another solution to this difficulty. At high frequencies and for fast pulse circuits, 50 Ω terminations are generally used as a standard. While the use of a 50 Ω probe results in severe restrictions at low frequencies, the overall accuracy may show an improvement.

Thus, while it is true that with a 50 Ω source and a 50 Ω probe one half of the input signal is "lost" due to loading, this loss remains constant out to a much higher frequency because the low value of R and the accompanying small value of C lead to a much higher frequency limit than would be the case with a high-impedance probe.

When attenuator probes are used, the signal amplitudes, determined by observation of the calibrated scope deflection, should be multiplied by the attenuation of the probe (usually 10× or 100×) to obtain a true measure of the signal amplitude. (Some oscilloscopes display the magnitude of the attenuation on the screen to remind the user of this fact.)

The probe capacitor is generally made adjustable, and a proper setting is obtained by "picking off" an internally generated square wave. If the capacitor has too low a value, the higher frequencies are attenuated more than the lower ones, and the square wave will have rounded edges. Too large a value of C_p will accentuate the pulse edges. A proper setting of C_p yields a true square wave. (See Figure 12.14.)

EXAMPLE 12.2

The input impedance of an oscilloscope equals 1 MΩ in parallel with a 20 pF capacitance. What should be the value of a 10× probe resistance, and its parallel capacitance, if the 3 ft of coax lead represents an additional 30 pF of capacitance?

Solution:

$$\frac{1}{10} = \frac{R_{in}}{R_{in} + R_p} = \frac{1\ \text{M}\Omega}{1\ \text{M}\Omega + R_p}$$

$$1\ \text{M}\Omega + R_p = 10\ \text{M}\Omega$$

$$R_p = 9\ \text{M}\Omega$$

$$C_p = \frac{R_{in}}{R_p} C_{in} = \frac{1}{9}(20 + 30)\ \text{pF} = 5.5\ \text{pF}$$

Current waveforms may be measured either by placing a small resistance in series with the signal lead, thereby converting the current waveform into a proportional voltage, or by a current probe. The latter simply clips around the signal lead and measures the current waveform by virtue of the accompanying magnetic field variations.

Our discussion has been limited to *passive* probes. There are more elaborate signal probes that may include a high-impedance FET amplifier within the probe itself. Such *active* probes present a very high input impedance.

12.6 ■ SAMPLING OSCILLOSCOPES

As one progressively diminishes the signal rise times, or the pulse durations, one finds that to be observed on an oscilloscope the signal intensities must get correspondingly larger. For example, the Tektronix 519, which had a bandwidth of 1000 MHz, required signal amplitudes of the order of tens of volts. Such limitations, of course, are a result of the gain-bandwidth product. To observe fractional nanosecond low-level signals, sampling techniques must generally be used. This technique constitutes the electronic analog of the stroboscopic process often used to study mechanical motion.

Consider a rotating solid with one spot on its periphery. If we periodically illuminate the spot by a flash lamp whose period is equal to that of the rotation of the rotor, the latter will appear stationary. If the rotational frequency and the strobe frequency are not precisely synchronized, the rotor will appear to move either forward or backward depending on whether the strobe frequency is, respectively, slower or faster than the rotor frequency.

In like manner, *if* the fast electronic signal we wish to observe is repetitive, we may probe its successive repetitions at progressively later times and then reconstruct the signal using a relatively slow time base.

A pulse with a rise time of 10^{-9} s, to be displayed by conventional techniques, would require a rather unrealistic bandwidth in excess of 1000 MHz. (The rise time of an oscilloscope may be approximated by the expression 0.35/bandwidth.) A sampling system, with much narrower bandwidths, will reproduce the essentials of this waveform.

Prior to the beginning of the signal pulse, one must initiate a trigger pulse. At some time after the appearance of this trigger, a time that is made variable, there appears a sampling pulse that results in a momentary display of the signal amplitude at that particular time. On each succeeding repetition of the signal waveform, the sampling pulse can be moved slightly to the right (Figure 12.15). Thus, the fast waveform is reconstructed by a series of dots, each dot corresponding to one sampling process. It should again be stressed that the waveform to be observed must be *repetitive* (although not necessarily *periodic*).

How does one obtain the initiating pulse prior to the appearance of the signal? We do so in the same manner as was used to observe the initial portions of the signal in a conventional scope. The signal is applied to a Schmitt trigger, then delayed by means of a delay line, and finally subjected to the sampling pulse after time T has elapsed (Figure 12.15).

The sampling process depicted in Figure 12.15 constitutes an *open-cycle mode.* In its simplest form it utilizes a reverse-biased diode that acts as a sampling switch. The interrogate pulse forward-biases the diode for a short duration, and the capacitor starts to charge (Figure 12.16).

The basic sampling circuit is shown in Figure 12.16a, being a sampling switch, a series resistance, and a shunt capacitance to ground. With the closing of the switch, the capacitor starts to charge. Being closed for but a short time compared to the time constant RC, the capacitor will charge to only a small fraction of the actual signal amplitude; for example, $v_{sample} = 0.05v_{in}$. Figure 12.16b indicates the nature of the circuitry.

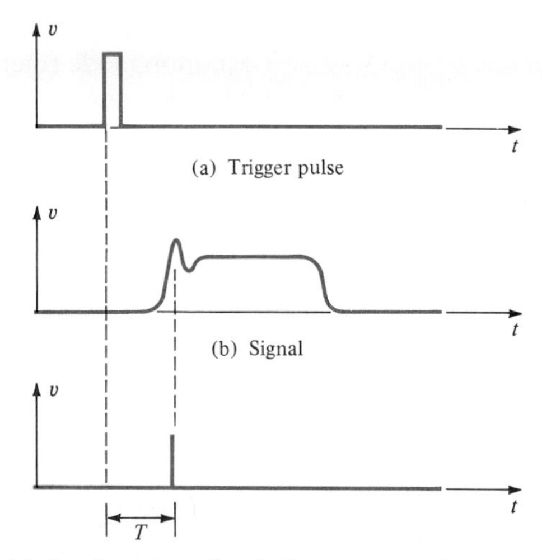

(c) Sampling pulse at time T after appearance of trigger pulse

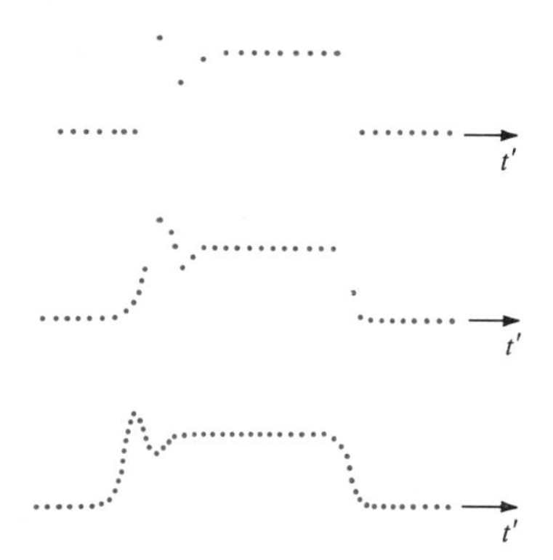

(d) Successive appearances of reconstructed signal with progressively increasing number of sample pulses

FIGURE 12.15. Sampling Oscilloscope Operation

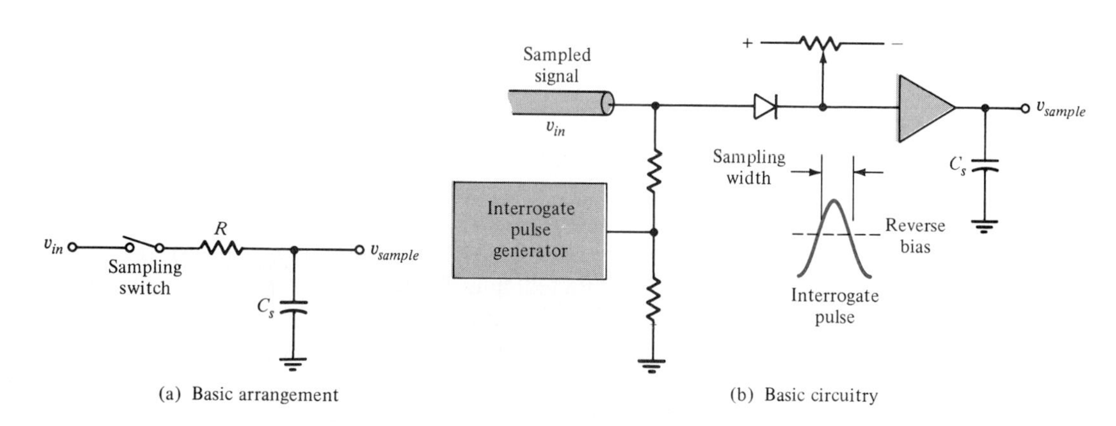

(a) Basic arrangement

(b) Basic circuitry

FIGURE 12.16. Open-Cycle Sampling

An *error-sampled feedback system* may be used to improve upon the open-loop mode (Figure 12.17). The closing of the sample switch for a short time will again charge C_{in} only to a small fraction of the sampled voltage. This fractional voltage, deposited on C_{in}, will be amplified and applied to the stretcher circuit. The stretcher switch was closed at the same time as the sampling switch but remains closed for a longer time. This time extension allows C_s to charge to the full output voltage of the amplifier, and this output voltage is fed back to C_{in}. The gain of this feedback amplifier is adjusted so that the voltage fed back to C_{in} is equal to the full sample voltage. If the next sampled level is the same, there will be no signal change, and the CRT display will remain at the same vertical level.

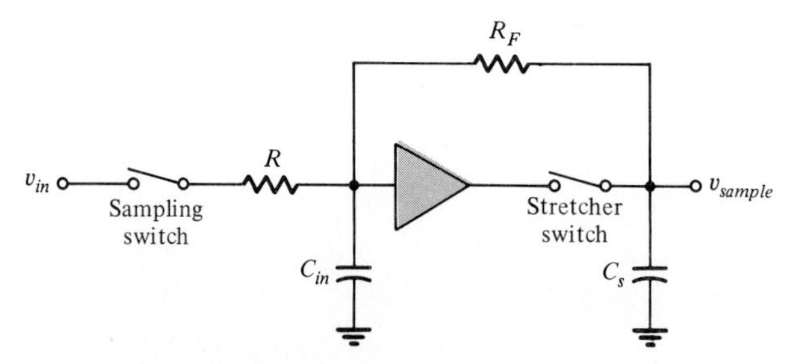

FIGURE 12.17. **Error-Sampled Feedback System**

EXAMPLE 12.3

Presume the sampled signal magnitude to be 1 V and the sampling time to be such that C_{in} charges to 5% of the input voltage. What should be the closed-loop gain of the amplifier?

Solution:

$$0.05 \times 1 \text{ V} = 0.05 \text{ V} \quad \text{on } C_{in}$$

The necessary gain is

$$\frac{1 \text{ V}}{0.05 \text{ V}} = 20$$

12.6.1 Sequential Sampling

Refer to Figure 12.18. At the desired trigger level, a Schmitt circuit gives rise to a trigger pulse that initiates a timing ramp. The signal is then applied to a delay line, resulting in time delay T.

The timing ramp is applied to a comparator whose other input is a staircase voltage. When the ramp voltage equals the step voltage, a sampling pulse is initiated, and the value of the delayed signal at that time is stored in memory as a Y-value. The pulse from the comparator also increases the staircase voltage by one step, and when the timing ramp again becomes equal to the new staircase voltage, another sample pulse is initiated, as well as another step increase in the staircase.

The staircase voltage also constitutes the sweep voltage. The sweep, therefore, progresses across the scope face in stepwise fashion, and at each

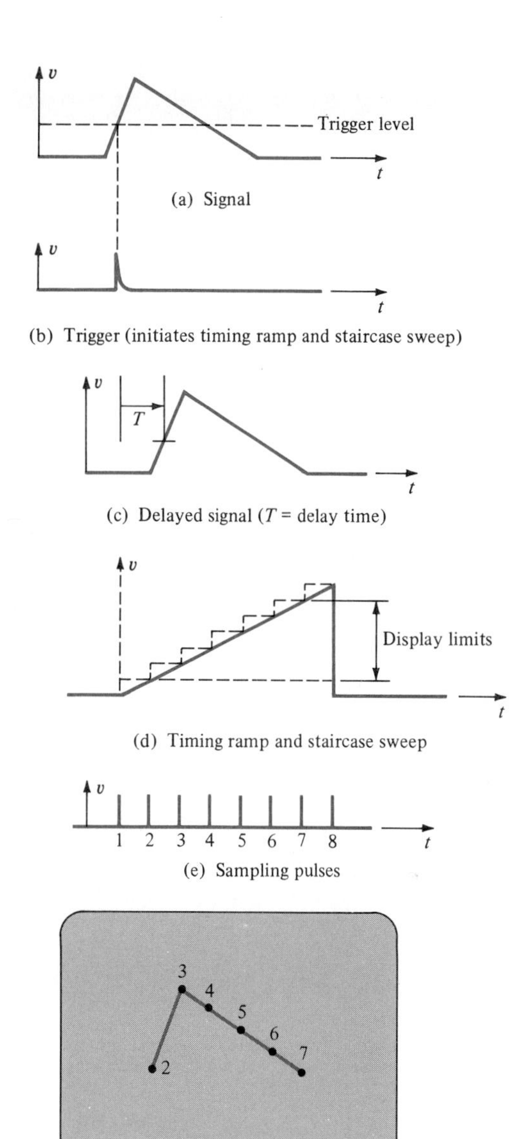

FIGURE 12.18. Sequential Sampling

step the sampled voltage is displayed. Given a sufficient number of steps of fine gradation, the depicted signal will represent a rather accurate version of the original.

The main disadvantage of sequential sampling is the need to pass the signal through a delay line whose bandwidth must be quite wide. Any deficiencies in this respect may lead to signal distortion.

The sampling rate can be much lower than the repetition rate of the signal. Thus, for example, the samples can be taken on the 1st, 101st, 201st, etc., waveforms. Thus, there is no need to employ high-frequency techniques with respect to the sampling pulses.

12.6.2 Random Sampling

The need to pass the fast signals through a delay line, with possible consequent distortion, can be obviated by using random sampling. In this case the successive repetitions of the waveform are sampled at random times. With the sampled voltage stored in Y-memory, a Schmitt trigger, in response to the signal to be observed, generates a trigger pulse that initiates a sweep voltage (Figure 12.19).

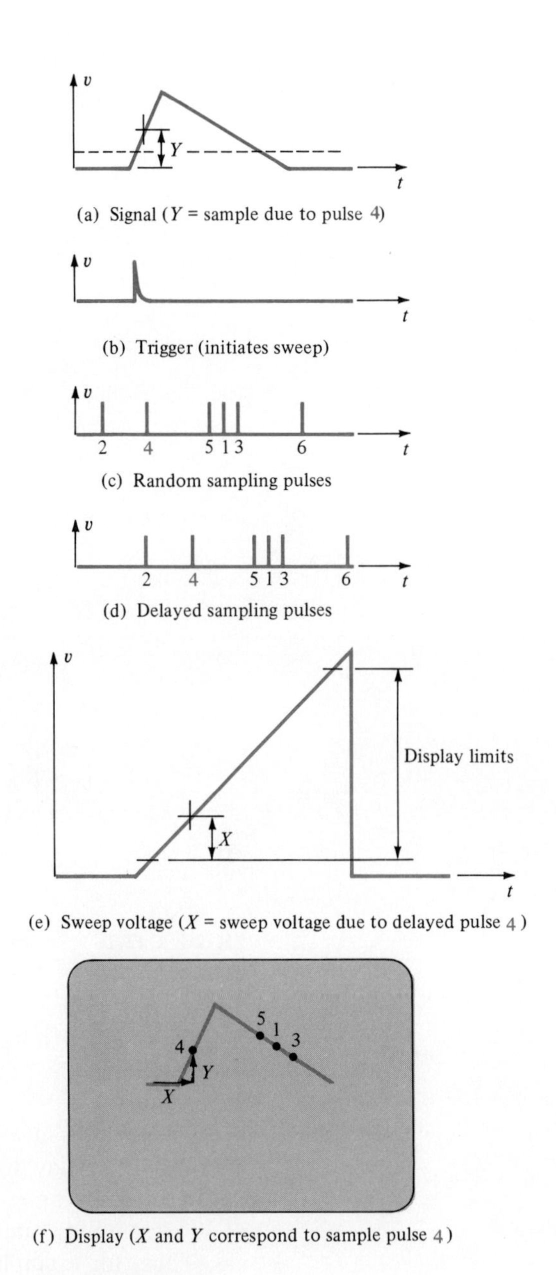

FIGURE 12.19. Random Sampling

The sampling pulses are each delayed by a fixed interval T and measure the sweep voltage after that same amount of delay, committing its value to X-memory. The x-y coordinates of the sample are then displayed on the CRT.

While we do not encounter the difficulty of maintaining the integrity of a short-duration signal in a delay line, random sampling requires a high trigger repetition rate. Why do we need random sampling? We need it because the start of the timing ramp is not connected with the sampling delay time, and random sampling assures coverage of the entire waveform desired.

12.7 ▪ STORAGE OSCILLOSCOPES

Conventional oscilloscopes (including sampling types) require repetitive signals. **Storage oscilloscopes** allow the observation of single events. Additionally, they can be used to advantage where the low repetition rate of repetitive signals would lead to low viewing intensities (for example, those associated with medical displays, mechanical motion, etc.).

When a particle is bombarded by very energetic electrons, such collisions may dislodge more electrons than were incident upon it, thereby leaving the originally charge-neutral material with a net positive charge. Subsequent low-energy electrons will be attracted to such positively charged particles. This process, by which the particles acquire positive charge, is termed **secondary emission.**

In one form of storage oscilloscope, a thin metallic layer is first deposited on the glass screen. This conductive layer will act as a controlling electrode. Atop this thin (transparent) conducting layer are scattered phosphor particles that serve as secondary emitters.

The storage CRT contains two types of electron guns. The **writing gun** produces high-energy electrons that trace out on the screen the signal pattern to be observed and that leave in their wake a series of positively charged phosphor particles. Because of their insulation properties, these charges do not migrate. When a second electron gun, constituting the **flood system**, directs a broad stream of low-energy electrons toward the screen, most of them will be repelled by a negative voltage on the controlling electrode. But, in those regions where the phosphor has been left positively charged, electrons will strike the phosphor particles and re-create the fluorescent path representing the original trace. The pattern may persist under the flood system illumination until erased by the control electrode. Many storage CRTs today utilize a more elaborate scheme than that just outlined—one that allows gradations of the trace intensity.

Storage oscilloscopes of the preceding form will have the display gradually fading with time. In a digital storage oscilloscope, the waveform is sampled and the data stored in a memory system, available to be displayed in a sequential manner, thereby reproducing a signal display of unlimited duration. Additionally, the digitized data can be processed before display, allowing for averaging, coordinate transformation, spectrum calculation, etc. The digital storage scope, however, will probably have a narrower bandwidth than a comparable analog storage system.

12.8 ■ LOGIC ANALYZERS

Although a **logic analyzer** superficially resembles an oscilloscope (both employ CRTs), it is really a different instrument, being used to analyze the correctness of assemblages of digital pulses such as might represent a binary number. An oscilloscope can display information only *after* a certain trigger instant. A logic analyzer can display signals not only after such a trigger but also *before*.

In a typical situation a computer might stop because of an improper sequence of pulses. As the various groupings of pulses (termed *words*) pass through the computer, they are also made to pass in sequence through the logic analyzer. If one has previously inserted into the analyzer a word (it might, for example, be the invalid one that is causing the trouble), when that word arises in the computer program, the analyzer will stop, having retained also the series of words that preceded it. (The number of such retentions depends on the memory storage capacity of the analyzer.) One may then examine the various steps in a leisurely fashion in order to determine the nature of the fault development. One could alternatively program the analyzer to display the sequence of words that follows the selected word—or any combination of such arrangements.

12.9 ■ SELECTING AN OSCILLOSCOPE

12.9.1 Manufacturers

Without exception the most important (in fact, indispensable) piece of equipment for anyone involved with electronic instrumentation is the oscilloscope. To aid in its selection, we will present some information on the various manufacturers as well as some specification details. The "big four" oscilloscope manufacturers are Gould, Hewlett-Packard, Philips, and Tektronix.

Gould, Inc. Gould, Inc. was originally the Gould Battery Company. By a series of acquisitions starting in the late 1960s, the company has become involved with motors, air conditioning equipment, medical instrumentation, electrically propelled cars, and sonar equipment. The head of the company during the times of acquisition was Wm. Ylvisaker (pronounced "ill-vis-soccer"), who was rated by *Fortune* magazine as among the 10 toughest bosses in the United States. His preoccupation with developing managerial talent has led to the creation of the company's Management Educational Center in Rolling Meadow, Illinois, outside Chicago.

The Instrument Division of Gould is located in Cleveland, Ohio, and includes a number of foreign manufacturing plants as well as worldwide service and distribution outlets. Its earliest entry into the data recording field was the development of the first electrocardiogram, in 1937. The company's original strong point was direct-writing pen and ink recorders. They introduced logic analyzers in 1973 and were one of the first to market a digital storage oscilloscope.

With digital storage a normal CRT is used to display signals that have been digitized and then stored in memory. Storage time is infinite, and the display is devoid of flicker.

Hewlett-Packard. David Packard and Wm. R. Hewlett were fellow benchwarmers on the Stanford football squad in 1931 where they became aware of their mutual interest in electronics. The company was started in a backyard garage in 1938. Their first product was the Model 200A Wien Bridge Oscillator (discussed in Section 8.4.2), and over the years Hewlett-Packard has probably produced more Wien bridge oscillators than the combined output of all other companies in the world. (They called it model 200 rather than model 1 in order to disguise their lack of maturity in manufacturing.)

Packard took over management functions and Hewlett, the company's design responsibilities. Throughout the years they have resisted rapid expansion, always limiting such moves to what could be financed out of their profits.

They are today the world's largest producer of electronics measuring and testing equipment. Although oscilloscopes form but a fraction of their manufacturing output, their quantity and caliber still place the company among the "big four."

Philips. Philips is a Dutch company, the biggest electrical manufacturer in Europe. Its strong point is its research laboratory and the ease with which fundamental discoveries are quickly translated into an engineered product. Such transitions from laboratory to manufacturing often prove to be a difficult process for most companies—and once was also at Philips.

Among the most notable names associated with the Philips Research Laboratories was that of Balthazar van der Pol. (Mechanical engineers might recognize his name as being associated with nonlinear mechanics in the form of the van der Pol equation.) In his early days at Philips, van der Pol failed in his attempts at developing significant research in AM, FM, and TV. He was also disappointed that his early suggestions for computer development were not accepted. These disappointments stemmed from Giles Holst, Director of Research, who was almost exclusively interested in gas discharge research. This interest made possible developments such as the fluorescent lamp, which Holst felt was more in line with the main objectives of the company as evidenced by its full name, Philips *Gloeilampenfabrieken* (Philips Lamp Company).

But such reluctance no longer endures, and Philips has been in the forefront of much in electronics development. For example, they were among the first to market digitized record players using lasers. An excellent book on oscilloscopes was authored by Rien van Erk, who is associated with Philips (see the Suggested Readings at the end of this chapter).

Tektronix. Unlike the three previous companies, Tektronix (founded in 1946) began with the purpose of developing a single instrument, the cathode ray oscilloscope, from which it has grown to become the largest producer of oscilloscopes in the world. The president, Howard Vollum, was a distinguished engineer and designer.

Prior to Tektronix, most instrument manufacturers bought ready-made components and merely assembled them into the final product. Tektronix, on the other hand, finding that available items were not capable of meeting their requirements, proceeded to design and produce their own improved versions. Such was the case of the cathode ray tube. They also made their own capacitors in order to achieve the demanded time base linearity, improved the connectors used, etc. The quality of their product was probably best acknowledged by a review of one of their instruments in the British publication *Wireless World* some years ago.

Wireless World was doing an evaluation of a newly marketed Tektronix scope. Many experienced observers, they reported, had to apply a "transatlantic factor" to the instrument specifications originating in the United States. It has sometimes even appeared, they said, that the U.S. output watt was larger than the British and the input watt smaller. But here was a Tektronix scope with a claimed 10 MHz bandwidth that actually measured 13.5 MHz.

For many years Tektronix retained its one-product resolve, but about a decade ago it started a degree of diversification and now produces a variety of instruments.

Other Manufacturers. There are other oscilloscope manufacturers, generally ones to fill the price range below the "big four." The most notable in this respect is probably Heathkit of Benton Harbor, Michigan.

This company started after World War II by buying up surplus government components and designing inexpensive electronic instruments that they sold in kit form—hence their name. The complexity and design of their products today is more sophisticated, and while some may still be purchased in kit form, many are purchased fully assembled.

There are many other oscilloscope manufacturers, such as B&K Precision. There are also some recent entries into the field including the Japanese firms of Leader Instruments and Hitachi.

12.9.2 Specifications

Bandwidth. Often, by using various forms of reactive compensation, the bandwidth of an amplifier can be significantly extended, as shown in Figure 12.20. The peaking of the response at high frequencies, however, can lead to pulse overshoot (Figure 12.21). Thus, merely specifying the bandwidth does not yield complete information on the expected performance. In addition to bandwidth the specifications should include a measure of the overshoot, such as $\leq 3\%$, with a six-division pulse having a rise time of 3 ns. A convenient approximate expression relating bandwidth and rise time is $BW \simeq 0.35/t_r$.

Phase Characteristics. Phase characteristics are particularly important in X-Y measurements. While the gain of the two channels need not necessarily be the same in many cases, a phase difference between them should be recognized. This often comes about because there is a delay line in the Y-channel but not in the X-channel. If the same signal is applied to both channels, the resultant phase shift is indicative of the channel unbalance in this regard.

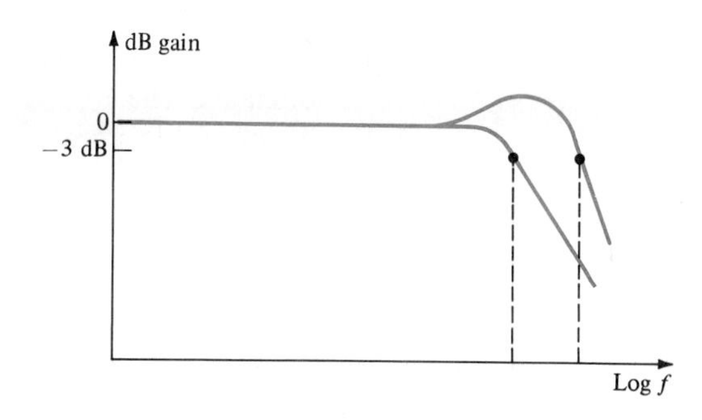

FIGURE 12.20. **Increasing Bandwidth by Reactive Compensation**

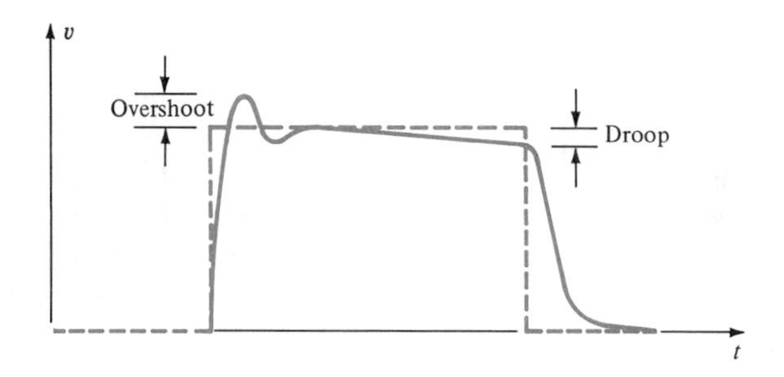

FIGURE 12.21. **Applied Signal (Dashed Curve) and Amplifier Response (Solid Curve)**

This unbalance appears as a specification in the *X*-channel deflection being listed as representing a maximum angle at a stated frequency.

Screen Phosphors. When the CRT's electron beam strikes the phosphor-coated screen, the immediate light emission is termed **fluorescence.** After the beam moves on, the remaining intensity constitutes **phosphorescence.** The rapidity of the phosphorescence decay is measured by the time taken for the intensity to drop to $1/e$ of its original value (e is the base of the natural logarithm; let us call the time constant t_p). This time constant leads to three categories of CRT screens: short persistence ($t_p < 1$ ns), medium persistence ($t_p < 2$ s), and long persistence ($t_p \sim 1$ min).

For radar applications long-persistence screens, which retain the image between very slow sweeps, are typified by the P19 phosphor. Both the fluorescence and phosphorescence in such CRTs are generally orange in color. For medical displays and the monitoring of mechanical motion, with their accompanying relatively slow sweep speeds, medium-persistence screens are used. P7 phosphors, with blue-white fluorescence and yellow-green phosphorescence, typify such applications. For high rep-rate sweeps, short persistence, such as provided by a P2 or a P11, is desirable. P2 yields a blue-green trace; P11 is blue and particularly suitable for photographing the trace. P4 phosphors leave a white trace and are used for black and white TV screens; P22 is the three-color dot pattern used for color TV.

A CRT screen subjected to an intense stationary spot for long time periods may end up having a hole burned in the phosphor. The various phosphors exhibit varying degrees of resistance to such damage. P19 is rather sensitive in this regard, P7 is less so, and P31 shows high resistance to such damage. The typical choice for a general-purpose screen material is P31, which yields a green trace (maximum eye sensitivity), has a short persistence (which eliminates multiple images as the trace moves), and possesses a high degree of resistance to burns. It is also rated as having a very high luminance level. An aluminized CRT (with an aluminum layer on the nonviewing side of the phosphor) also aids in reducing screen burns by acting as a heat sink. It does, however, require a higher beam-accelerating potential.

Often plastic filters are used between the screen face and the viewer. Thus, an amber filter might be used with P7 to eliminate short-term blue and increase the longer-persistence orange. A blue filter will reverse the process.

Plug-in Units. Many modern scopes employ a common wideband vertical amplifier in conjunction with one or more plug-in units that may be used to create a more versatile instrument than would otherwise be the case. The plug-in concept has been exploited to the greatest degree by Tektronix. Thus, if high sensitivity is needed and narrower bandwidths may be tolerated, this combination may be achieved by selecting an appropriate plug-in. Or if dual-channel operation is desired, there is probably a unit that is available. There may also be plug-ins available that will convert the scope into a completely different instrument—into a spectrum analyzer, a logic analyzer, a digital voltmeter, etc.

Is a plug-in scope always desirable? That depends on the use contemplated. They are more expensive than fixed-function units. For research laboratory purposes, where a variety of tasks is envisioned, they probably should be the instrument of choice. Where a scope is to be used for some sort of routine testing and the plug-in versatility is not needed, the fixed-function scope represents a more economical solution.

SUMMARY

12.1 The Basic Oscilloscope

- Signals to be observed are generally applied to the vertical deflection plates of an oscilloscope. The horizontal plates usually have a sawtooth applied to them.
- The trigger level control of an oscilloscope determines the magnitude of the pulse that initiates the oscilloscope sweep.

12.2 *X-Y* Plotting with an Oscilloscope

- Upon disconnnecting the sawtooth and applying to the *X*-deflection plates a signal voltage in its place, a cathode ray oscilloscope can be made to serve as an *X-Y* plotter.
- The use of *X-Y* plotting to determine either phase or relative frequencies makes use of Lissajous patterns.

12.3 Delayed Time Base

- A delayed sweep may be used to expand the trace of any selected segment of the main trace.

12.4 Dual Trace and Dual Beam

- Dual-beam scopes employ two separate electron beams for the simultaneous display of two different signals. A dual-trace scope, by periodic switching between the two signal sources, may be used to create the illusion of the simultaneous display of both signals using a single electron beam.

12.5 Signal Probes

- The display of signal currents may be accomplished by passing them through a small resistance, thereby converting them into proportional voltage waveforms. Alternatively, one may use a current probe.
- Signal probes attenuate the input signal but usually diminish the loading effect of the measuring instrument on the source being measured.
- Active signal probes include an FET amplifier within the probe.

12.6 Sampling Oscilloscopes

- Sampling scopes are used to reconstruct a waveform that is beyond the conventional bandwidth of the oscilloscope. The waveform must be repetitive, though not necessarily periodic.
- Sequential sampling takes place at modest rates. Random sampling requires the use of high-frequency pulse techniques.
- The rise time of an oscilloscope may be approximated by the expression 0.35/bandwidth.

12.7 Storage Oscilloscopes

- A storage oscilloscope may be used to observe single events.

12.9 Selecting an Oscilloscope

- The immediate light emission from a phosphor is termed fluorescence. The delayed emission is called phosphorescence.

KEY TERMS

alternate mode: the application of two different signals on alternate sweeps of a single-beam oscilloscope, creating the illusion of two beams

cathode ray oscilloscope (CRO): an instrument that provides a visual display of electrical signals including a CRT and the associated time base generators and signal amplifiers

cathode ray tube (CRT): a device that includes the electron beam source, the focusing electrodes, the deflection plates, and the fluorescent screen used to provide a visual display of electrical signals

chopped mode: the process of switching between two signals being applied to an oscilloscope, the switching taking place at a rate considerably exceeding the sweep rate

deflection plates, horizontal: parallel metallic plates whose normals lie on a horizontal axis; also termed the *X*-plates to which usually a sawtooth voltage is applied

deflection plates, vertical: parallel metallic plates whose normals lie along the vertical axis; also termed the *Y*-plates (The signals to be observed on an oscilloscope are generally applied to the *Y*-plates.)

delayed sweep: the sweep generated by the delayed time base circuit

delayed time base (DTB): the faster time base of a delayed time base oscilloscope

dual beam: the provision for observing two different signals simultaneously by employing two beams within an oscilloscope

dual trace: by means of electronic switching, the provision for observing two different signals with a single-beam CRT

flood gun: in a storage oscilloscope, the electron source that makes signal trace visible

fluorescence: the light emission resulting from direct excitation

Lissajous figure: the oscilloscope trace used in the determination of the phase and frequency relations between two signals

logic analyzer: an electronic device that records digital information on a CRT

main time base (MTB): the slower time base of a delayed time base oscilloscope

oscilloscope: an electronic instrument generally used to observe voltage waveforms as a function of time; may also be used for frequency comparison and phase determination

overshoot: the maximum excursion of a signal outside its final mean level

phosphor: the light-emitting coating of an oscilloscope screen

phosphorescence: the light emission that results after the direct source of excitation has been removed (The intensity diminishes in an exponential fashion.)

sampling, random: signal sampling at random times

sampling, sequential: signal sampling with time equality between samples

sampling oscilloscope: a means for observing short-duration signals by reconstruction from samples taken on successive repetitions of the waveform; analogous to the mechanical process of strobing

sawtooth: a voltage that increases linearly with time and then (ideally) drops back to the original voltage instantaneously

secondary emission: due to electron bombardment, the ejection from a target of more electrons than are incident upon the target

signal probe: a means for minimizing the disturbance of a signal source in the process of making a measurement

storage oscilloscope: signal retention for long time periods allowing the recording of single events, unlike the conventional oscilloscope practice that requires repetitive signals

sweep speed: the scanning rate of an oscilloscope, usually expressed in terms of divisions per (fractional) second

synchronization: the development of an integral relationship between signal and sawtooth periods

time base: the time scale employed in conjunction with the horizontal sweep of an oscilloscope

trigger level: the selected voltage level at which a particular circuit action is initiated; for example, the commencement of a sawtooth sweep

writing gun: in a storage oscilloscope, the source of the electron beam that records the signal

SUGGESTED READINGS

van Erk, Rien, OSCILLOSCOPES, FUNCTIONAL OPERATION AND MEASURING EXAMPLES. New York: McGraw-Hill, 1978. *An extensive glossary of oscilloscope terms, oscilloscope specifications, examples illustrating the use of oscilloscopes, and so on.* Pp. 98–102, *a brief introduction to logic analyzers.*

Sessions, Kendall W., and Walter A. Fischer, UNDERSTANDING OSCILLOSCOPES AND DISPLAY WAVEFORMS, pp. 155–175, 218–221. New York: Wiley, 1978. *A rather thorough discussion of oscilloscope probes, with numerous examples.*

Schlesinger, K., and E. G. Ramberg, "Beam Deflection and Photo Devices," PROC IRE, vol. 50, pp. 991–1005, May 1962. *History of cathode ray tubes from the first (gas) type by Braun (in 1897), through the fast (vacuum) one by Zworykin (1930), up to modern times.*

Oliver, B. M., and J. M. Cage, ELECTRONIC MEASUREMENTS AND INSTRUMENTATION, p. 361. New York: McGraw-Hill, 1971. *A listing of various phosphor types, with characteristics.*

Winningstad, C. N., PROCEEDINGS OF THE NATIONAL ELECTRONICS CONFERENCE, vol. 19, pp. 164–172, 1963. *Very thorough treatments of oscilloscope probes.*

EXERCISES

1. To what pair of deflection plates of an oscilloscope is the signal to be observed generally applied?

2. To appear stationary, what must be the relationship between signal periodicity and the sawtooth period?

3. If four sinusoid cycles are observed on an oscilloscope screen (10 cm width), with the sweep speed set at 1 ms/cm, what is the applied frequency?

4. What is the function of the trigger level control on an oscilloscope?

5. Circuitwise, what does the sweep speed control on an oscilloscope do?

6. Oscilloscopes allow for either dc or ac coupling of the input signal. In the latter case a coupling capacitor is inserted in series with the input. (See Figure 5.6.) Which input should be used to observe square waves with a low repetition rate? Which input should be used when observing small ac signals on a rather large dc level?

7. What is the function of an oscilloscope delay line?

8. What is the main function of a sampling oscilloscope?

9. What is the rise time of a 100 MHz oscilloscope?

10. What is the phase relation between two sine waves that lead to a circular Lissajous pattern? That lead to a straight

line with a 45° slope relative to the $+x$-axis? That lead to a slope of $+135°$ relative to the $+x$-axis?

11. With the same 100 kHz wave applied to both inputs of an oscilloscope, the ratio of the y-axis intercepts A/B is 1:40. What is the phase difference between channels?

12. Why should the display of an intense stationary spot on a CRT be avoided?

13. What is the source of the difficulty if a broken trace appears when using a dual-trace oscilloscope?

14. If four complete cycles of a waveform are displayed when the sweep speed is set at 20 μs/cm, what is the period of the waveform assuming an oscilloscope scale width of 10 cm? What is the frequency?

15. What is the function of a signal probe? Under what circumstances may a low-impedance (50 Ω) probe prove advantageous?

16. Indicate the meaning of the following: CRO, CRT, MTB, DTB.

17. What is generally the recommended demarcation line between the use of chopped vs. alternate modes with a dual-trace oscilloscope?

18. What is the distinction between dual trace and dual beam?

19. What must be the nature of a waveform if it is to be observed with a sampling oscilloscope? How does this situation differ from that encountered with a storage oscilloscope?

20. What is the advantage of random sampling over sequential sampling? What is a disadvantage?

21. In the case of storage oscilloscopes, distinguish between the writing gun and the flood gun.

22. What is a logic analyzer?

23. Name four major oscilloscope manufacturers.

24. Distinguish between phosphorescence and fluorescence.

25. What are the approximate time constants for the three phosphor categories? What phosphor is considered to be the typical choice for a general-purpose oscilloscope?

PROBLEMS

12.1

A neon bulb has the property that below some firing voltage (V_F) the bulb is an open circuit. But once the voltage reaches V_F, the bulb's gas will ionize, will lead to visible radiation, and represents a moderate resistance (about 1 kΩ) that can quickly discharge the capacitor. When the capacitor voltage drops below a value termed the extinguishing voltage of the bulb (V_E), the visible radiation ceases, and the bulb again assumes a high value of resistance, allowing the capacitor to recharge. By such means one may construct a flasher using a circuit such as the one shown Figure 12.22, which is a neon bulb relaxation oscillator. We have

$$v(t) = V_E + (V - V_E)(1 - e^{-t/RC})$$

At $t = T_C$ (termed the charging time), $v = V_F$. Solve for T_C. Compute the charging time for the circuit in Figure 12.22. If the decay time (T_D) is taken to be negligible compared to the charging time, what would be the repetition frequency? In terms of the firing voltage (V_F) and the resistance of the ionized bulb (R'), express the capacitor voltage as a function of time [$v(t)$]. Determine the expression for the decay time (T_D) and compute the value for the given circuit.

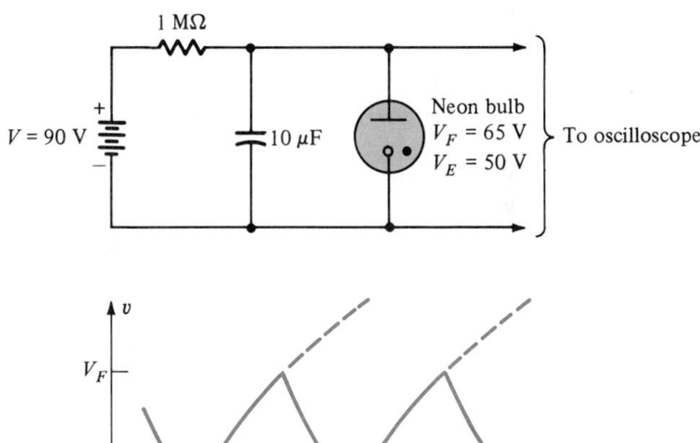

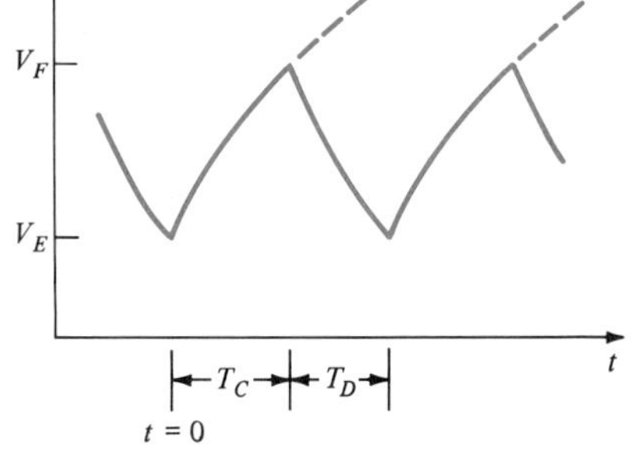

FIGURE 12.22

12.2

The neon bulb relaxation oscillator (as it is called) depicted in Figure 12.22 was the method used to generate sawtooth sweep voltages in the earliest oscilloscopes. Although no longer used, its simplicity and decided nonlinearity allows us an opportunity to study the influence of sweep nonlinearity on oscilloscope performance. The dc component of the charging curve can be removed by placing a blocking capacitor in series with one of the leads going to the CRO deflection plates. The charging voltage then simply becomes

$$v(t) = V(1 - e^{-t/RC})$$

with the sweep starting at $v = 0$ V and terminating at $v(t) = (V_F - V_E)$—that is, for the circuit in Figure 12.22, when $v(t) = 65 - 50$ V $= 15$ V.

a. Assume the desired sweep speed to be 1 ms/cm, with a total sweep width of 10 cm. With $V = 100$ V, compute the magnitude of the deflection voltages at 1 ms intervals using a time constant $\tau = RC = 6.153 \times 10^{-2}$ s. Compare the computed values with the ideal values, that is, 0 V, 1.5 V, 3.0 V, etc.

b. There is a variety of methods that are used to evaluate nonlinearity of which three are the following:

1. The maximum deviation from the ideal voltage, divided by the average interval
2. The largest deviation from the ideal, minus the smallest deviation, divided by the average interval
3. The same as method 2 except for division by the total sweep voltage interval rather than the average

Compute the percentage of nonlinearity by each method.

COMMENT: Often in method 3, rather than using the whole sweep voltage interval, the interval between 1 and 9 cm is used. In practice, rather than voltages, one applies a sequence of equally timed pulses and measures the nonlinearity in terms of graticule divisions.

12.3

a. By using a larger source voltage in conjunction with the sawtooth generator of Problem 12.2, the degree of nonlinearity will diminish since the initial charging curve will be more linear. Taking $V = 1000$ V and the same total sweep amplitude—that is, 15 V—compute the value of RC necessary to yield the same 1 ms/cm sweep speed.

b. Compute the percentage of nonlinearity by the same methods as outlined in Problem 12.2.

COMMENT: There are additional errors due to horizontal amplifier and CRT deflection nonlinearity. Unlike our consideration of the relaxation oscillator sweep nonlinearity in which all the errors were positive, the errors actually measured using CRT deflection will probably show both positive and negative values. This procedure will lead to method 2 showing the greatest percentage of nonlinearity, which is probably why it is not favored by scope manufacturers. Method 3 is most frequently used, but often rather than division by the full sweep voltage (or the total deflection length), it is the voltage (or distance) between 1 and 9 cm that is employed.

12.4

The signals shown in Figure 12.23 are applied to a NAND gate. Indicate the output trace that will be observed on an oscilloscope.

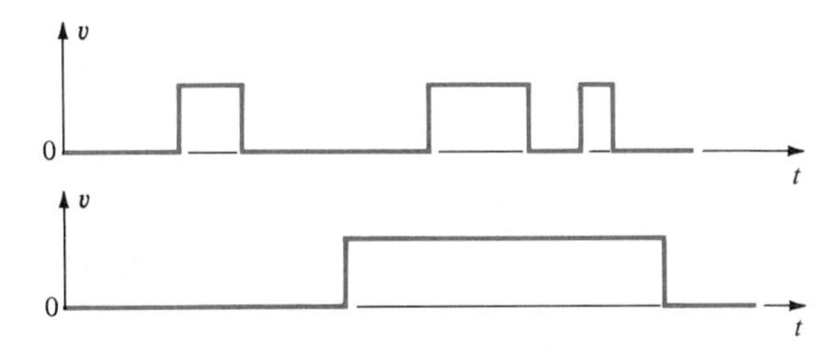

FIGURE 12.23

12.5

When rise time measurements approach those of the oscilloscope itself, one must compensate for the instrument's rise time by resorting to the equation

$$t_{rd} \simeq \sqrt{t_{rs}^2 + t_{ro}^2}$$

where t_{rd} = displayed rise time
t_{rs} = signal rise time
t_{ro} = oscilloscope rise time

If an oscilloscope is rated as having a rise time of 10 ns and the displayed rise time is 15 ns, what is the estimated signal rise time?

12.6

In terms of R and C, utilizing the expressions for rise time and amplifier turnover frequency (presuming a dominant single pole), compute the relation between rise time and the bandwidth of an oscilloscope.

12.7

If a significant fraction of the curved portion of a capacitor's charging curve is included in a sawtooth sweep, sketch the effect this will have on the appearance of a multiple, single-frequency, sine wave display.

12.8 (Z)

To demonstrate the influence of probe capacitance on signal response, with a 500 Ω source resistance, determine the percentage of signal remaining after loading when the following two probes are used to detect a 50 MHz signal and a 1 MHz signal:

a. 10 MΩ, 10 pF

b. 500 Ω, 0.7 pF

12.9 (Z)

The signal loss caused by loading impressed on the source by a probe can be represented as

$$\text{loss} = \frac{Z_s}{Z_s + Z_p}$$

where Z_s = source impedance

Z_p = probe impedance

Considering a 30 MHz source with a 1 V peak amplitude and source resistance of 500 Ω, compute the signal percentages remaining after loading and the signal magnitude after probe division for the following probes:

a. 500 Ω, 0.7 pF, 10:1 division

b. 5 kΩ, 0.7 pF, 100:1 division

c. 100 kΩ, 3 pF, 1:1 division

d. 1 MΩ, 2.5 pF, 10:1 division

e. 10 MΩ, 10 pF, 10:1 division

COMMENT: It will be noted that the lowest loading loss also leads to the smallest signal, and this in turn may limit the clarity of the display for small signals. The 500 Ω probe, on the other hand, while introducing a 50% loading loss, results in a larger display amplitude. The loss remains constant out to about 100 MHz leading to an accurately reproduced signal, and the loading loss may easily be taken into account by multiplying the observed values by a factor of 2.

12.10 △

Problems 12.8 and 12.9 concern the use of probes in conjunction with high-frequency signals. In this problem we consider the use of probes in conjunction with the measurement of fast rise time pulses. The observed rise time (t_o) is given by the equation

$$t_o = \sqrt{t_p^2 + t_s^2 + t_g^2 + t_{RC}^2}$$

- t_{RC} — due to probe-source interaction
- t_g — pulse generator specification
- t_s — oscilloscope specification
- t_p — probe specification

Three of these times are determined by equipment specifications. The fourth time (t_{RC}) is determined by the source resistance and the probe capacitance. Figure 12.24 shows a direct connection between the generator and the input to the measuring circuit. R_g and R_{in} in parallel lead to the equivalent series resistance R_c in Figure 12.24b. The rise time of the latter circuit is $2.2R_cC_{in}$. To diminish R_c, either R_g or R_{in} should be kept small. C_{in} should similarly be selected to have a small value. Again taking the same five probes considered in Problem 12.9, compute for each of them the value of t_{RC} as well as the percentage of loading, again taking the source resistance to be 500 Ω.

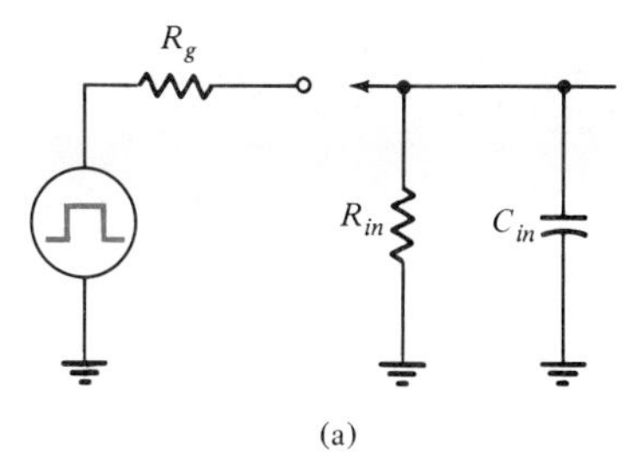

(a)

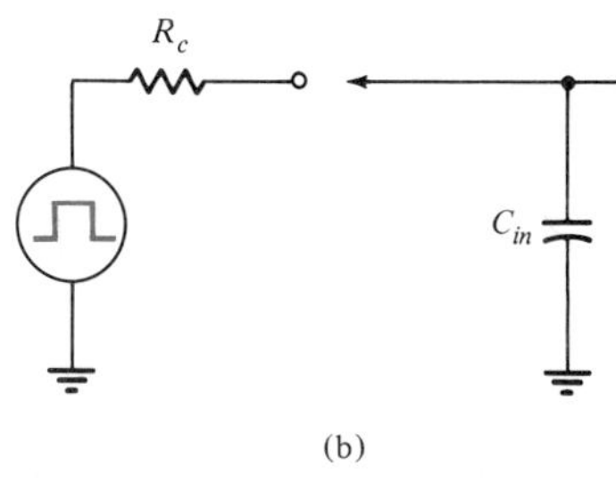

(b)

FIGURE 12.24

COMMENT: Generally, loading greater than 30% is considered to be too disruptive.

12.11

The impedance of a coaxial transmission line is given by the equation

$$Z_o \simeq \frac{138}{\sqrt{\kappa}} \log_{10}\left(\frac{D}{d}\right)$$

where κ = dielectric constant of the insulation

D = outer conductor diameter

d = inner conductor diameter

As one approaches high frequencies, one is faced with attempting to match the probe's coaxial cable to the source impedance. Taking a typical BJT output impedance to be 2 kΩ and desiring an equal characteristic impedance, we select d to be as small as possible. The smallest value might be the so-called diameter of the electron, $\sim 10^{-13}$ cm.

a. Compute the necessary value of D for an air dielectric—that is, $\kappa = 1$.

b. Repeat for insulation with $\kappa = 2.3$.

c. Repeat for $Z_o = 3000\ \Omega$ and $\kappa = 2.3$.

In all cases, reduce the answer to the largest practical unit.

12.12

a. A 30 MHz generator with a source impedance of 50 Ω is to be amplified by an amplifier whose output impedance is 2000 Ω. Assume two 10 MΩ, 10 pF, 10:1 divider probes are to be used to verify that the amplifier's phase shift is zero. Determine the measured phase shift.

b. Repeat the measurement with two 50 Ω, 0.7 pF probes.

c. At the output of the amplifier, what is the resistive loading in each of the two cases.

12.13

If the rise time of each contribution (probe specification, oscilloscope, source generator, and *RC* coupling; see Problem 12.10) is the same, what will be the observed rise time?

13 NOISE AND INTERFERENCE

OVERVIEW

The topic of electrical noise has previously entered into our considerations. In Chapter 4 it was in conjunction with the noise generated in composition resistors, in Chapter 8 it was the Fourier components of noise that initiated oscillation, in Chapter 11 it was noise that set the necessary width between the on and off states of the various logic categories, and in Chapter 12 it was the noise bandwidth that set the sensitivity limit of an oscilloscope.

Interference has also previously been considered in conjunction with selectivity and Q of resonant circuits (Chapter 2) and relative to the ability of differential amplifiers to discriminate against common-mode interference (Chapter 10).

In this chapter we investigate noise and interference in a more formal manner. Of particular importance is the manner in which the impedance level influences the degree of interference between signals passing along parallel conducting paths (Section 13.4.3).

These concepts will further be utilized in Chapter 15 (particularly Problem 15.10), in Chapter 16 (interference from silicon controlled rectifiers), and in Chapter 17 (minimizing 60 Hz interference in integrating-type digital voltmeters as well as in conjunction with the "classic" instrumentation amplifier).

OUTLINE

13.1 ■ INTRODUCTION

The earliest radio designers recognized that to develop large signals they needed large input resistances. But as this resistance was increased, they noted that the background noise level at the receiver's output also increased. Since this background level limits how small a signal may be amplified, an appreciation of the processes involved will aid in the intelligent use of oscilloscopes (and other electronic devices).

Such electrical *noise* arises because of the discreteness of the charged carriers and their statistical fluctuation. While the fluctuations are random, their average values are fairly predictable. Therein stems a difficulty.

One rather prominent electronics text points out that precisely because of such well-defined statistical properties, device noise gets the most attention in textbooks, despite overwhelming agreement among practitioners that it is **interference** rather than **noise** that usually constitutes the factor limiting circuit sensitivity. (The authors of the text then go on to say that they will bow to custom and also concentrate on the theoretical discussion of device *noise*.[1])

Interference can take on a variety of forms. It can be due to automotive ignition, it can be due to power-line surges, it can arise because of fluorescent lights, in a radio receiver it can be a disturbance created by stations other than the desired one, and, most ubiquitous of all, it can be due to 60 Hz pickup from power lines.

In Section 13.2 we will discuss electrical noise. In Section 13.3 we will consider some aspects of interference.

13.2 ■ ELECTRICAL NOISE

13.2.1 Johnson Noise

Valence electrons are free to move throughout the volume of a metal. This motion is of a random nature, and its violence increases with temperature. Consider a cylindrical metallic conductor. At any given moment there may be more electrons moving toward one end than toward the other; the next moment the magnitude of this difference (and even its direction) may change. Thus, given an idealized voltmeter placed across the ends of such a metallic sample, we would observe a random voltage variation. This variation constitutes the source of electrical noise in conductors and resistors. It is always present, and its magnitude depends on a number of parameters in addition to temperature. Such motion shows no periodicity and consists of a continuum of frequencies. We might then expect that the wider the bandwidth of the amplifier, the greater will be the noise content, as it is. Noise due to thermal motion is termed **Johnson noise**, being named after one of those early radio designers mentioned in the introduction.

[1]Stephen D. Senturia and Bruce D. Wedlock, *Electronic Circuits and Applications*. New York: Wiley, 1975, p. 550.

The magnitude of the noise voltage can be specified in terms of the effective (or rms) value. We have so far qualitatively established that

$$V_{rms\ noise} \simeq TR\ \Delta f \qquad (13.1)$$

where T = absolute temperature (in noise work generally taken to be 290 K)
R = resistance in ohms
Δf = *noise* bandwidth in hertz, generally being somewhat larger than the 3 dB bandwidth

To distinguish between bandwidths, henceforth we will designate the noise bandwidth as B:

$$V_{rms\ noise} = \sqrt{4kTRB} \qquad (13.2)$$

where, additionally, k is Boltzmann's constant ($k = 1.38 \times 10^{-23}$ J/K).

EXAMPLE 13.1

a. What is the rms value of the noise voltage contributions due to a 10^4 input resistance used with a 50 kHz (noise) bandwidth amplifier?
b. What would be the value if the resistance were increased to 1 MΩ?
c. What would be the noise contribution of a 1 MΩ input resistance representing the input of a 50 MHz (noise) bandwidth CRO amplifier?

Solution:

$$V_{rms\ noise} = \sqrt{4kTRB}$$

a. $$V_{rms\ noise} = \sqrt{4 \times 1.38 \times 10^{-23} \times 290}\ \sqrt{10^4 \times 5 \times 10^4}$$
$$= 1.27 \times 10^{-10} \times 2.24 \times 10^4 = 2.84\ \mu\text{V}$$

b. $$V_{rms\ noise} = 1.27 \times 10^{-10} \times \sqrt{10^6 \times 5 \times 10^4} = 28\ \mu\text{V}$$

c. $$V_{rms\ noise} = 1.27 \times 10^{-10} \times \sqrt{10^6 \times 50 \times 10^6}$$
$$= 898\ \mu\text{V} \simeq 0.90\ \text{mV}$$

13.2.2 Shot Noise

In addition to Johnson noise due to thermal fluctuations, there is also **shot noise**. Shot noise arises from the discreteness of electronic charges, analogous to the noise produced by rain falling on a tin roof. It is usually expressed in terms of the rms noise current superimposed on a dc current (I_{dc}):

$$I_{rms\ noise} = \sqrt{2eI_{dc}B} \qquad (13.3)$$

where e is the electronic charge, 1.6×10^{-19} C. Eq. 13.3 is called the *Schottky equation* after its originator, who happened to be aptly named.

13.2.3 Transistor Noise

The noise generated within a transistor is specified by the manufacturer as referred to the input (that is, the actual output noise divided by the gain) and appears as a noise generator (v_N) in series with the transistor input, as in

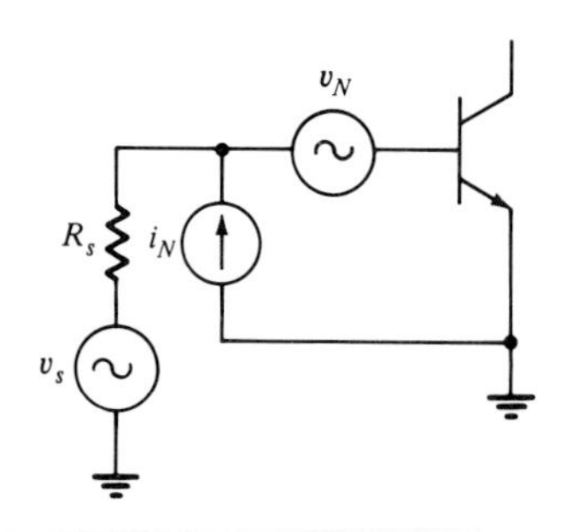

FIGURE 13.1. Input to a BJT Showing Johnson Noise Voltage Source (v_N) and Shot Noise Current Source (i_N)

Figure 13.1. Additionally, there is a current fluctuation, primarily due to the base current, that is represented by a noise generator i_N. This noise current, flowing through R_s (the source resistance), generates additional noise that adds to v_N. The component due to i_N flowing into the base has already been accounted for in v_N. In fact, *per unit bandwidth*, we have

$$v_N^2 = 4kTr_{bb'} + 2eI_C r_e^2 \tag{13.4}$$

where $r_{bb'}$ is the resistance of the base, and r_e can be recognized as the dynamic emitter resistance (that is, $r_e = kT/eI_C = 0.026/I_C$ [A]). We thereby have

$$v_N^2 = 4kTr_{bb'} + \frac{2(kT)^2}{eI_C} \tag{13.5}$$

The first term is a constant for a specific transistor, and the second term diminishes as the collector current increases. Therefore, v_N should diminish as the collector current increases.

On the other hand, i_N (Figure 13.1) comes about because of the base current through R_s (hence it is proportional to I_C) and leads to a term $(i_N R_s)^2$ that needs to be added to v_N^2. Since the former term increases with I_C while the latter diminishes, there must be some optimum value of collector current that minimizes the noise generated.

Given a particular source resistance (R_s) and the noise generated within it, in the process of amplifying the signal, there will always be some additional noise generated due to the amplifier. It is this additional noise that can be minimized by proper circuit design.

The signal-to-noise (S/N) ratio is defined as the ratio of mean square signal voltage to the mean square noise voltage (hence the ratio of signal power to noise power). For certainty of detection it is usually assumed that the minimum value of S/N should be 3. This value is a statistical result, predicting 98% probability of detection.

As mentioned, in the process of amplification, the amplifier itself introduces additional noise, and hence the signal-to-noise ratio is degraded. The ratio of *input* signal to noise to *output* signal to noise is termed the *noise* factor. Thus, we have

$$\text{noise factor} \equiv F = \frac{(S/N)_{in}}{(S/N)_{out}} \tag{13.6}$$

Designating the noise power of the source as N_s and that of the amplifier as N_a, with S the signal power, we have

$$(S/N)_{out} = \frac{S}{N_s + N_a} \qquad (S/N)_{in} = \frac{S}{N_s}$$

$$F = \frac{S}{N_s}\frac{N_s + N_a}{S} = \frac{N_s + N_a}{N_a} = 1 + \frac{N_a}{N_s} \tag{13.7}$$

For an ideal amplifier, $N_a = 0$, and hence an ideal noise factor is 1.

The signal-to-noise ratio (SNR) is often defined in terms of decibels:

$$\text{SNR [dB]} = 10 \log_{10}\left(\frac{V_s^2}{V_N^2}\right) \tag{13.8}$$

Also defining the **noise figure** (NF) in terms of decibels, we have

$$\text{NF [dB]} = 10 \log_{10}\left[1 + \frac{v_N^2 + (i_N R_s)^2}{4kTR_s}\right] \tag{13.9}$$

where $v_N^2 + (i_N R_s)^2$ is the mean squared noise voltage per hertz contributed by the amplifier. The noise figure is a measure of the signal-to-noise *degradation* due to amplification. The larger the noise figure, the greater the degree to which the signal-to-noise ratio suffers in the process of amplification.

The noise *factor* is the bracketed quantity in Eq. 13.9. That value of R_s that leads to a minimum noise factor can be obtained by differentiation and leads to

$$R_{s,opt} = \frac{v_N}{i_N} \tag{13.10}$$

and therefore,

$$F_{min} = 1 + \frac{v_N i_N}{2kT} \tag{13.11}$$

For a given V_s, the largest S/N arises when $R_s = 0$. Thus, the minimum noise factor does not necessarily represent the maximum S/N or the minimum value of noise. It does represent the smallest fractional contribution of the device.

EXAMPLE 13.2 Calculate the noise figure utilizing the following parameters appropriate to a 2N4250 *npn* transistor:

$R_s = 10\ \text{k}\Omega$ $\quad I_C = 14\ \mu\text{A}$

$f = 1\ \text{kHz}$ $\quad i_N = 0.35\ \text{pA}/\sqrt{\text{Hz}}$

$v_N = 0.0035\ \mu\text{V}/\sqrt{\text{Hz}}$

What percentage of the output noise is due to the amplifier?

Solution:

$$\text{NF [dB]} = 10 \log_{10}\left[1 + \frac{v_N^2 + (i_N R_s)^2}{4kTR_s}\right]$$

$$10 \log_{10}\left[1 + \frac{(3.5 \times 10^{-9})^2 + (3.5 \times 10^{-13} \times 10^4)^2}{4(1.38 \times 10^{-23})(290)(10^4)}\right] = 0.62\ \text{dB}$$

$$0.62 = 10 \log_{10}(1 + 0.15)$$

$$10^{0.62/10} = 1.15 = \frac{\text{total noise power}}{\text{source noise power}} = \frac{N_s + N_a}{N_s} = 1 + \frac{N_a}{N_s}$$

$$\frac{N_a}{N_s} = 0.15$$

Therefore,

$$N_a = 0.15\ N_s$$

The amplifier's contribution is $0.15/1.15 = 0.13$ or 13%.

Generally, one need consider only the noise figure of the first stage of amplification since the level of the amplified signal after the first stage usually makes negligible the noise contribution of the second and subsequent stages.

In looking at Eq. 13.9, *if* for a given R_s the noise voltage contribution is significantly greater than that due to i_N, it can be seen that the noise figure can be improved—that is, diminished—by increasing R_s. This increase should not be attempted by simply placing added resistance in series with R_s. You may get a lower value of NF, but remember that the noise figure merely tells you whether the amplifier is making a significant contribution to the output noise. By increasing R_s in this manner, you have increased the Johnson (thermal) noise due to the source so that the amplifier's percentage contribution has been made smaller, but you actually have a larger noise output. The way to increase R_s is to use a step-up transformer, thus making the source look to the transistor as having a larger value of resistance.

It can be seen that using a transistor's noise figure as a measure of performance can be rather questionable since it can be made small by using it in conjunction with a substantial source resistance. Much more meaningful is today's practice of specifying a transistor's v_N and i_N, which allows the computation of noise figures as a function of the source resistance.

EXAMPLE 13.3

For a transistor at 1 kHz and $I_C = 1$ mA, it is determined from the manufacturer's literature that $v_N = 2\ \text{nV}/\sqrt{\text{Hz}}$ and $i_N = 2\ \text{pA}/\sqrt{\text{Hz}}$.

a. Determine the value of the source resistance that will lead to a minimum noise figure.
b. What is the value of the minimum noise figure?
c. Show the results graphically.

Solution:

a. $$R_{s,opt} = \frac{v_N}{i_N} = \frac{2 \times 10^{-9}}{2 \times 10^{-12}} = 1\ \text{k}\Omega$$

$$F_{min} = 1 + \frac{v_N i_N}{2kT} = 1 + \frac{2 \times 10^{-9} \times 2 \times 10^{-12}}{2(1.38 \times 10^{-23})(290)} = 1.5$$

(Note that this computation is made on a "per unit bandwidth" basis.)

b. $\text{NF} = \log_{10} F = 10 \log_{10}(1.5) = 1.76\ \text{dB}$

c. Figure 13.2 shows the thermal noise voltage at 290 K. This plot is labeled "3 dB noise contour" because it corresponds to the case in which the device noise voltage equaled the thermal noise voltage. [NF = 3 dB; therefore, $F = 1 + 1$, where the second unity comes from (device voltage²/thermal voltage²).] Independent of R_s is v_N, and it appears as a horizontal line. The i_N contribution to noise voltage varies linearly with R_s and appears with a 45° slope. Where the two curves intersect, v_N and i_N are making equal contributions; therefore, the actual device noise is 3 dB above the intersection ($\sqrt{2}$ 2 nV = 2.83 nV). This value corresponds to the minimum noise figure and arises at R_s = 1 kΩ. At this R_s the thermal voltage is 4 nV/$\sqrt{\text{Hz}}$, the device voltage is 2.83 nV/$\sqrt{\text{Hz}}$, and the noise figure is 1.76 dB.

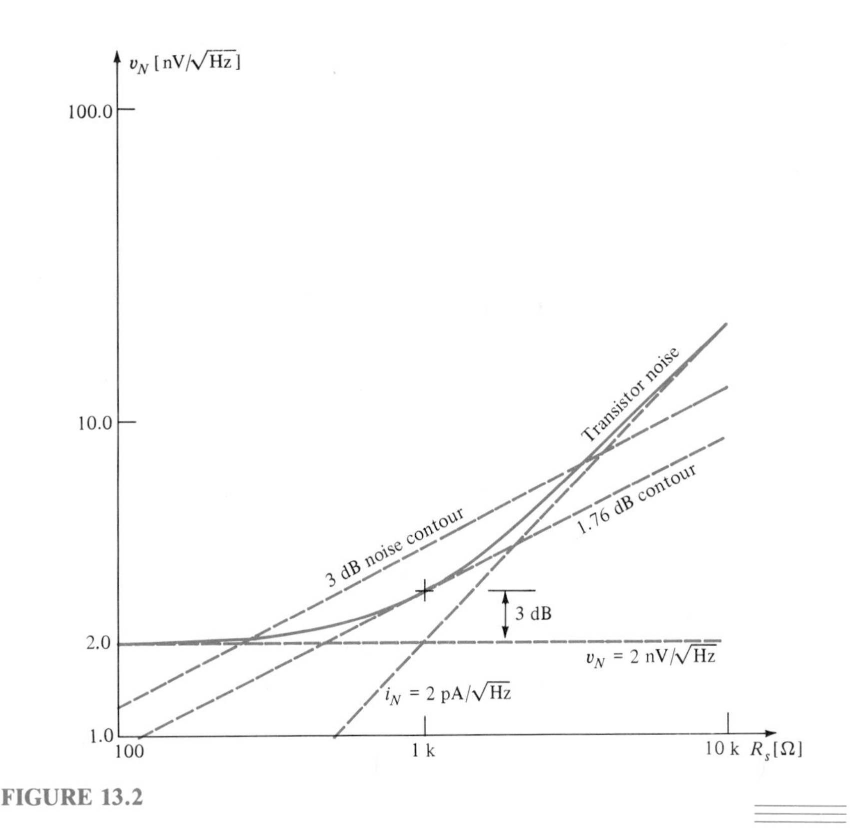

FIGURE 13.2

Thus far our consideration of noise in transistors has dealt with the BJT. Figure 13.3 may serve as a comparison between the noise generated with a BJT vs. that of a JFET. Being operated with a reverse bias on the drain, the i_N for a JFET tends to be rather small, and its influence is not felt until one reaches large values of R_s. On the other hand, at low values of R_s, v_N tends to be somewhat smaller for a BJT but with its significantly higher i_N will lead to more noise at large values of R_s.

Integrated circuit OPAMPs tend to be noisier than discrete transistors, and a typical curve is shown in Figure 13.3. Often if low noise performance is sought with an IC OPAMP, a discrete transistor is used as a preamplifier.

As a function of frequency the noise figure will rise at both high and low frequencies. In the former case this rise is due to the drop in gain at high frequencies. In the latter instance it is due in large part to $1/f$ noise, which we will now consider.

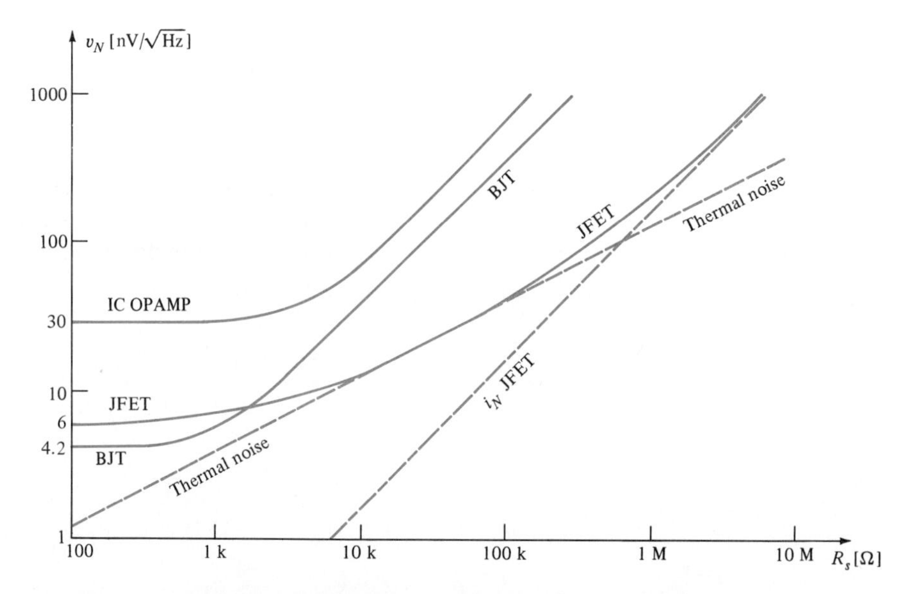

FIGURE 13.3. **Relative Comparison of Noise Voltages Due to BJT, JFET, and OPAMP Amplifiers of the IC Form**

13.2.4 1/*f* Noise

The noise power we have thus far considered has a uniform spectral density—that is, with the dimension of v_N^2 being square volts per hertz and noise power being proportional to v_N^2, the noise power per unit frequency interval is the same, irrespective of where in the frequency spectrum we find ourselves.

A noise source with a uniform power density is termed a **white noise** device. This expression stems from the optical fact that white light consists of a rather uniform power distribution.

In practical situations, particularly at low frequencies, noise will be found to exceed that due to Johnson and shot noise combined. This addition is termed **excess noise**, also called **l/*f* noise**, because the power density increases as the frequency is lowered. The physical causes for $1/f$ noise are varied and, in many instances, not fully understood. Such noise is referred to as **pink noise**, again the analogy coming from optics where a concentration of low-frequency (red) light would lead to a pinkish tinge.

For a $1/f$ source the noise density approaches infinity as the frequency approaches zero (that is, dc). And, of course, OPAMPs being dc-coupled do just that. How does one reconcile this difficulty?

For white noise the spectral density is constant—call it K. For a pink source it varies as K'/f (where K' is a constant). In calculating the mean square noise, we use

$$\overline{v_N^2} = \int_0^\infty \frac{K'\,df}{f[1 + (f/f_H)^2]} \rightarrow \infty \tag{13.12}$$

where f_H is the upper 3 dB point of the passband. The infinite result is, of course, unrealistic.

However, one does not make measurements of infinite duration. Assume the observation time to be T. Then any frequency component contained within $\overline{v_N^2}$ that has a period longer than T—that is, a frequency smaller than $1/T$—will not be observed. Thus, the difficulty is removed by making the lower limit of integration $1/T$. This finite limit then leads to a finite integral, as can be seen from the results of a problem at the conclusion of this chapter.

If there is no need to utilize an OPAMP down to dc, it is best to provide for a drop-off of gain at some low frequency. Such drop-off reduces the offset voltage fluctuation, which is a manifestation of $1/f$ noise.

13.2.5 Popcorn Noise

Another noise that arises at low frequencies (proportional to $1/f^2$, approximately) is popcorn noise, so named because of its sound when encountered in audio amplifiers, particularly so with background thermal noise providing an accompanying frying sound. It is generally considered to be the result of manufacturing defects, usually a result of metallic impurity in a semiconductor. The noise is burst-like, which, on an oscilloscope, looks like a rectangular pulse that pushes up through the background noise. The pulse width can vary from microseconds to seconds in duration, it is not periodic, and its amplitude for a given device is rather fixed, being a function of the junction defect.

13.3 ■ NOISE AND SOUND REPRODUCTION

We have seen that the noise content of an amplifier is proportional to the bandwidth, as the owner of an AM radio with a tone control can easily verify. Turning down the tone control diminishes the noise, but this decrease comes about at the expense of reducing the amplifier's high-frequency output. A better solution to this problem came into being with the development of FM radio.

13.3.1 Emphasis and De-emphasis

Rather than aiming for a flat frequency spectrum at the transmitter, the high audio frequencies are purposely amplified to a greater degree than are the low-band and midband frequencies (Figure 13.4a). Such unequal gain is termed **pre-emphasis.**

At the receiver the high-frequency response is purposely diminished, as shown in Figure 13.4b. The receiver's audio bandwidth, therefore, is much narrower than would be the case for maximally flat response, and with this narrower bandwidth there is much less noise in the output. The lowered high-frequency (hf) response constitutes **de-emphasis**, but when combined with the pre-emphasis, this de-emphasis gives a wide-band output. Figure 13.5a shows the noise content for a 10–15,000 Hz bandwidth. Figure 13.5b shows the noise content of the same receiver with pre-emphasis.

In addition to a broad frequency spectrum, a quality sound system should also be capable of accommodating a wide range of sound intensities. This quality is termed the system's **dynamic range** and can be defined as the

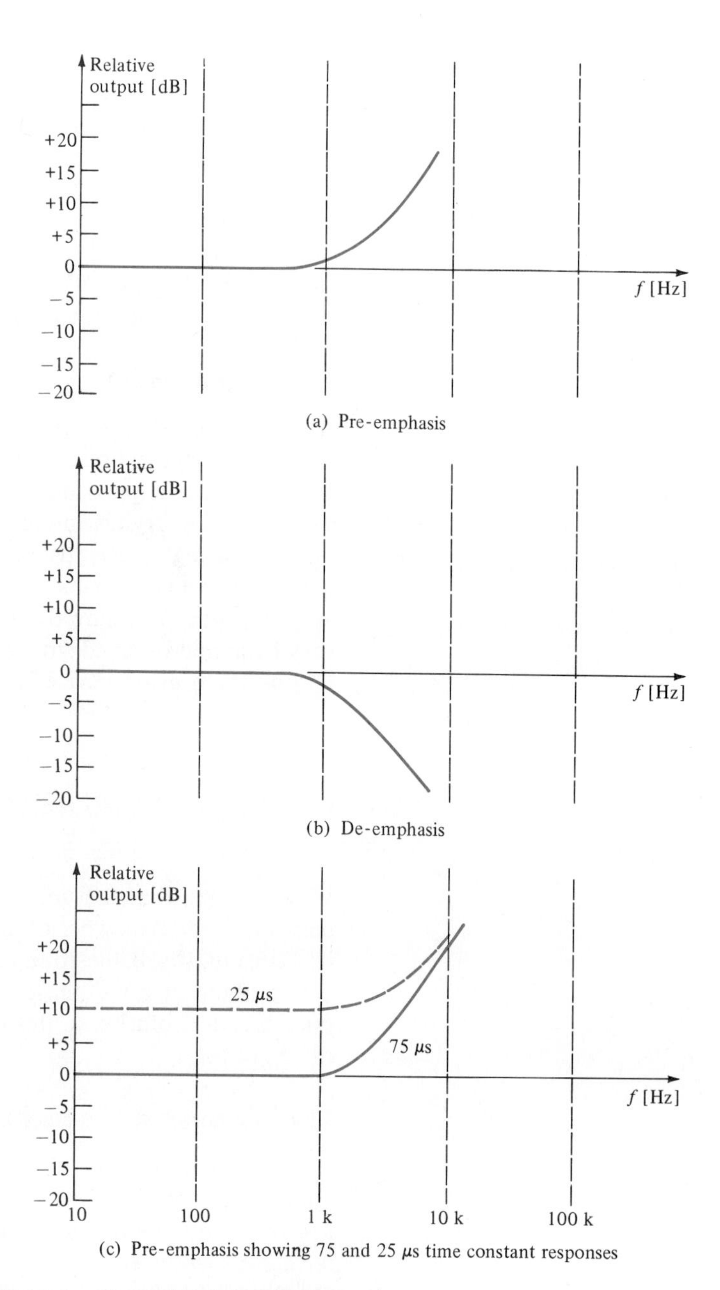

FIGURE 13.4. Control in Sound Systems

ratio of the largest to the smallest signal level. The upper limit may be set by such things as record-tape saturation or by the signal-handling capabilities of an amplifier. The lowest signal level that can be handled is generally determined by either the circuit noise, tape noise, or recording-disc noise. The dynamic range at concerts can be as great as 90–100 dB, but when recorded, it must be reduced to 60–70 dB; a broadcasting station has an even smaller range, 20–40 dB. Compaction of the dynamic range can be accomplished in a

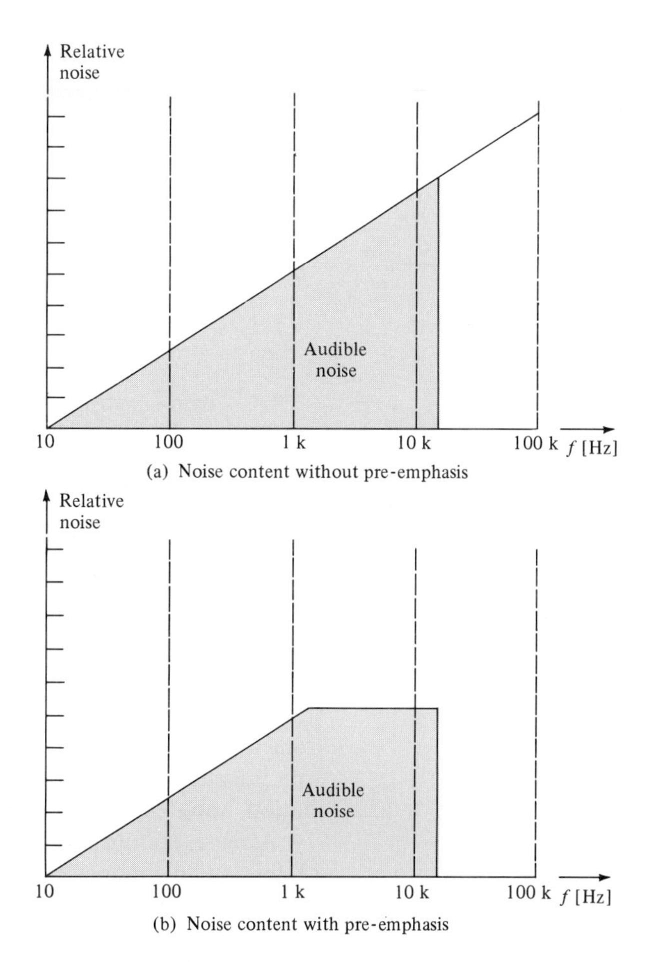

FIGURE 13.5. **Noise Limitation in Sound Systems**

variety of ways. Figure 13.6 shows two methods: A **limiter** prevents the output from exceeding some predetermined level (+10 dBm in the example) irrespective of the input signal amplitude. A **compressor**, having provided unity gain up to 0 dBm (again in the case illustrated), will exhibit a smaller gain beyond this point. (Beyond 0 dBm it can be seen that a 20 dB input change gives rise to only a 10 dB change in the output.) The action of the limiter is sudden, while the action of the compressor is more gradual. But sudden changes in gain may themselves give rise to extraneous noise. (Sharp amplitude changes contain numerous high-frequency components.) One may alternatively compress the low-level signals and not the higher-level ones or any combination thereof.

The reader will recall that in FM broadcasting the frequency of the audio signal is determined by the rate at which the varying radio frequency "swings" through the center (resting) frequency. The amplitude of the audio signal is determined by the extent of the deviation from this center frequency. To prevent interference between adjoining channels, the Federal Communications Commission (FCC) limits the extent of this deviation to ±75 kHz. With a

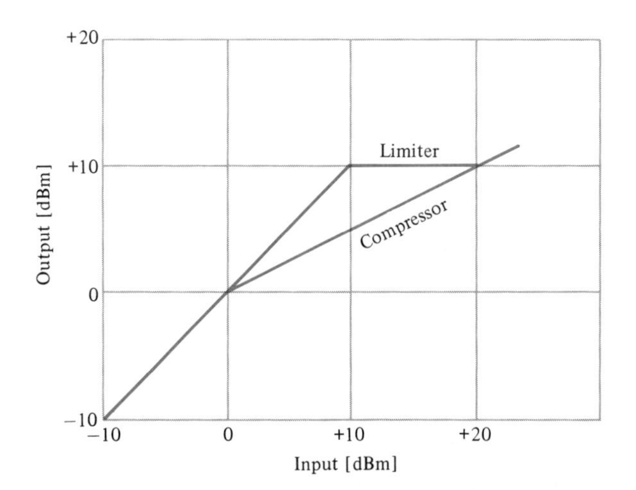

FIGURE 13.6. **Illustration of Limiter and Compressor Actions in Limiting Signal Amplitude**

flat audio spectrum, any audio frequency could utilize the full deviation and thereby offer maximum possible dynamic range to the sound. But if at the transmitter the higher frequencies are to be amplified to a greater degree (as with pre-emphasis), the lower frequencies cannot utilize the full ± 75 kHz deviation, and this restriction lowers their dynamic range.

A receiver's output, following a frequency curve such as that in Figure 13.4b, can be obtained by means of a simple *RC* filter. The roll-off (or turnover) point—that is, that frequency at which the output is down by 3 dB—depends on the filter's time constant ($\tau = RC$) corresponding to a roll-off frequency of $1/(2\pi RC)$. The same time constant is used at the transmitter for pre-emphasis, and since there must be agreement in this regard between the transmitter and receiver, the FCC has specified that the time constant should be 75 μs.

Figure 13.7 shows the output power capability of a 50 W receiver-amplifier using the 75 μs time constant. But experimental studies have shown that during peak levels of recorded music, arising between $\frac{1}{10}$% and 1% of the time, it is more desirable to have increased power at high frequencies. In Figure 13.7 these requirements are indicated by the dashed line. Such response can be achieved by using a 25 μs time constant for pre-emphasis and de-emphasis. While this value means that the noise reduction will now be smaller (since the receiver's bandwidth is greater), there are other means by which we can take care of this matter, and we will consider them in a moment. If now the full 75 kHz deviation is used for the highest audio frequencies, the lower ones can utilize a greater deviation than previously, thereby improving the loudness of the transmitted signal.

As a sound level varies, its average value constitutes its **loudness**. The broadcast of classical music leads to a loudness variation of about 22 dB, while pop music is generally compressed within the top 4–6 dB range. For this reason pop music appears much louder, though the maximum level attained by each may be the same.

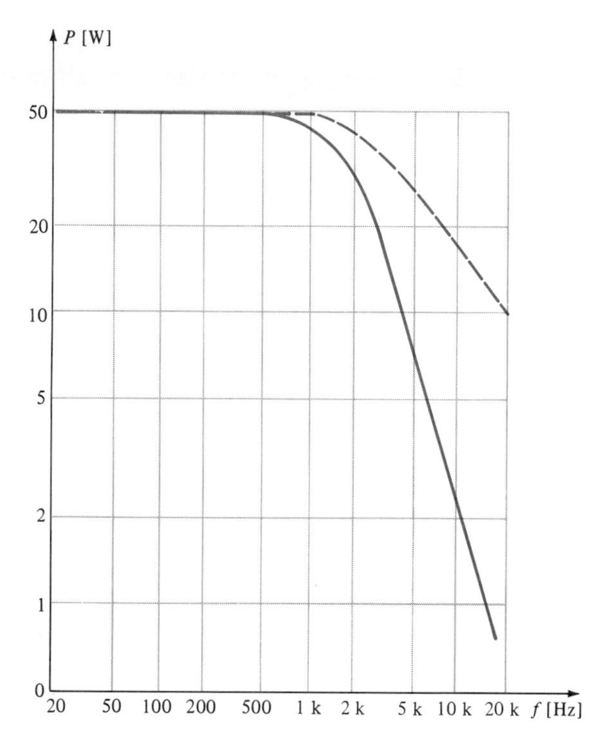

FIGURE 13.7. Amplifier Power Output The solid line corresponds to a 75 μs de-emphasis time constant; the dashed line shows the desired response characteristic with a 25 μs time constant.

13.3.2 Dolby Noise Suppression

Whether the 75 or the 25 μs constant is used, much of the time no pre-emphasis at all is needed. (It is only needed when there arises a low-level hf sound, for only then is noise reduction necessary.) The **Dolby noise reduction** system takes this probability into account by operating in a dynamic fashion—that is, pre-emphasis and de-emphasis are automatically introduced only when necessary and only to the extent necessary. While the Dolby system is used in FM transmission, it was originally developed to reduce recording-tape noise, and it is with that in mind that we will continue our discussion.

To understand the rationale behind the Dolby system, consider the noise spectrum of a cassette tape as shown in Figure 13.8. Rather than being a uniform broadband spectrum, the noise energy is concentrated at medium and high frequencies. As long as the sound level at such frequencies is fairly loud, the sound will mask the noise, but it will not do so during soft passages. Since the noise is introduced into the channel between the recording session and the replay—that is, it is the tape that is responsible for this noise—if the higher frequencies during the making of the recording are amplified to a greater degree than the low frequencies and during reproduction are amplified to a compensatingly smaller degree, the resulting spectrum should be uniform over the entire bandwidth, yet the accompanying noise will have been reduced.

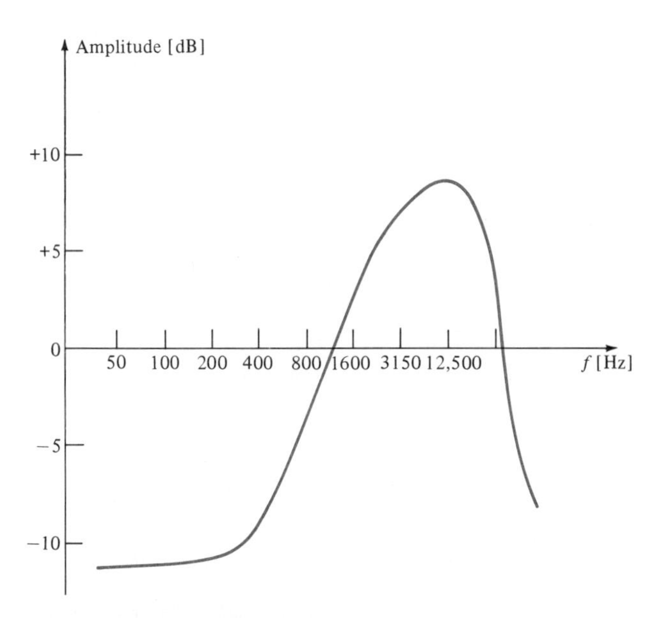

FIGURE 13.8. Noise Spectrum of a Cassette Tape

But if a particular high-frequency passage is already loud, subjecting it to such pre-emphasis will probably cause distortion since the magnetic tape may become saturated as a result, and there is no need for pre-emphasis in such cases since in the reproduction process these passages will be loud enough to cover the noise. (This subjective process is called **masking**.) Thus, the problem is to provide pre-emphasis at these frequencies only for soft passages and to diminish the amount of pre-emphasis as the sound level increases. This requirement is summarized in Figure 13.9. When the sound level at such frequencies is down by 10 dB, only a small amount of pre-emphasis is introduced, but the magnitude of the pre-emphasis increases as the sound level gets smaller. This change is accomplished by a circuit that monitors the sound level and produces a rectified control voltage that changes the amount of pre-emphasis accordingly. The higher the sound level, the greater the control voltage and the smaller the pre-emphasis. The controlling monitor is an FET voltage-controlled resistor, as discussed in Section 9.7.

But consider again Figure 13.9. Suppose we have an intense sound at 500 Hz. Do we want this signal to suppress the possible need for pre-emphasis at higher frequencies? Certainly not. Therefore, we want a circuit that will progressively increase the roll-off frequency of the boosted spectrum as the frequency of the strong signal approaches the boosted band. This action can be seen in Figure 13.10. In this case the boosted band will represent frequencies starting at about 200 Hz. But if we have a loud tone at 500 Hz, the boosted band will not start until about 1 kHz, while a loud tone at 2 kHz will lead to a boosted band starting above 2 kHz.

To summarize: Irrespective of the amplitude of the signal frequency below the hf passband and for low-level signals within the hf passband, pre-emphasis is provided and leads to noise reduction. For high-level signals within the hf passband, the FET diminishes the pre-emphasis since masking

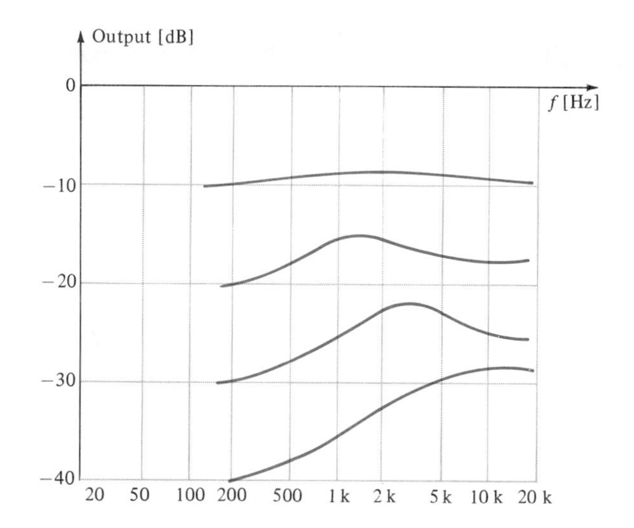

FIGURE 13.9. Degree of Pre-emphasis Showing Dependence on Output Signal Level The higher the signal level, the smaller the degree of pre-emphasis.

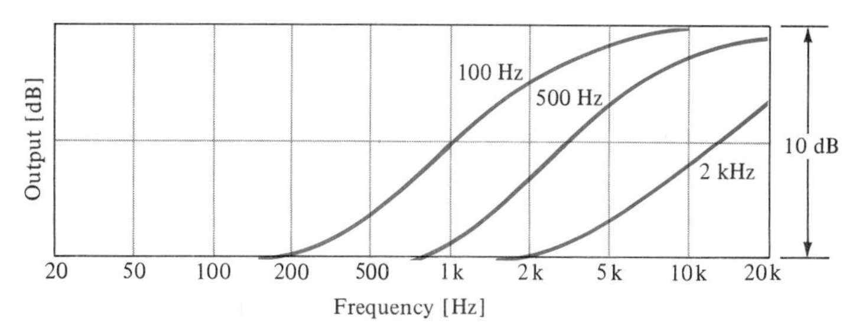

FIGURE 13.10. Variation of Pre-emphasis Turnover Frequency as a Function of the Input Signal Level The frequency labels on the three curves indicate pre-emphasis with a 0 dB tone applied at the indicated frequencies.

eliminates such need. To prevent high-level signals close to the roll-off frequency from diminishing necessary hf pre-emphasis, we wish to move the roll-off frequency upward.

In the reproduction process circumstances are inverted: The softer the passage, the smaller the gain to which the high frequencies will be subjected (since they have been made too loud originally), and this state of affairs will keep the noise low.

A major advantage of the Dolby system stems from allowing the same circuit to be used both to record and for playback. Figure 13.11 shows how this dual duty is accomplished: The signal is split into two paths. The direct path has all frequencies subjected to the same gain, and the side path contains a high-pass filter that passes only those frequencies that one may wish to pre-emphasize. The output from the latter is passed through a compressor that feeds a rectifier, and the resultant ac output voltage from this rectifier depends on the magnitude of the hf signal. If the hf signal is large, no *additional* significant hf gain is needed, and the dc voltage shuts down the

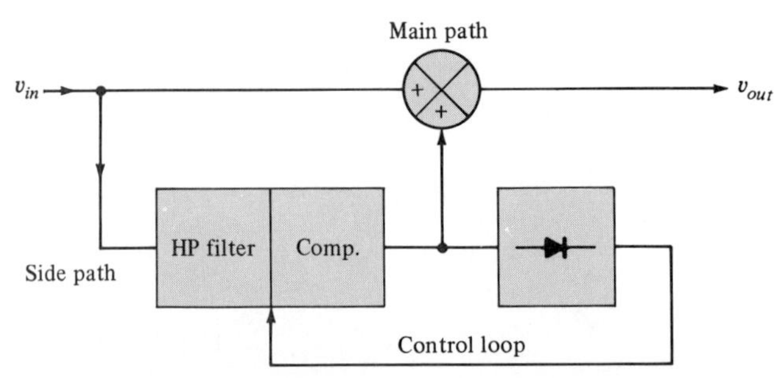

(a) Voltage-controlled filter and compressor with a positive feedback system that adds up to 10 dB of signal, resulting in pre-emphasis

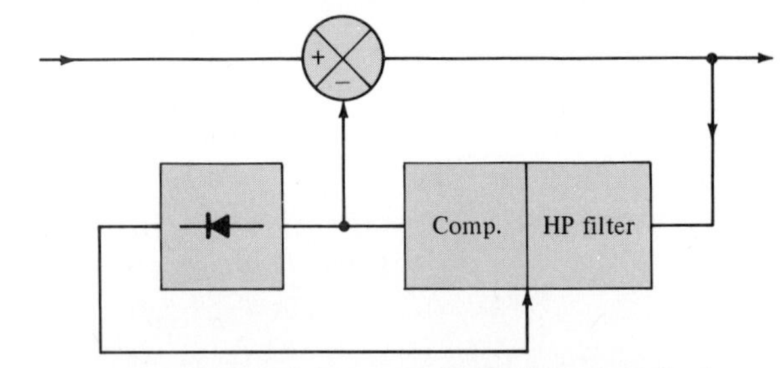

(b) Arrangement using negative feedback for de-emphasis

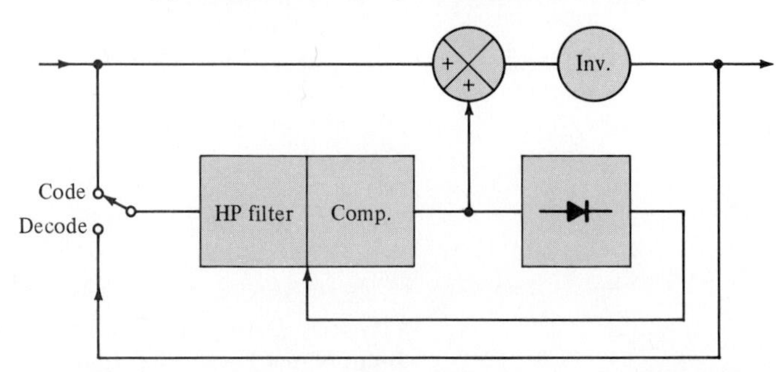

(c) Common network that can be used either for emphasis or de-emphasis

FIGURE 13.11. **Emphasis/De-emphasis Operation**

compressor. If, however, the hf signal is low, the dc voltage developed is insufficient to shut down the compressor, and additional hf amplitude from the side path is added to the main path, thereby providing pre-emphasis. Note that in recording (Figure 13.11a) this condition requires positive feedback from the side path (that is, hf from the side path is added to that in the main path). When used for playback (as indicated in Figure 13.11b), the input to the side path comes after the summation point and is to appear as negative feedback since, when needed, we wish to de-emphasize the high frequencies.

Looking at Figure 13.11c, note that if a phase inverter follows the summation point, both arrangements can employ positive feedback, and one may go from one condition to the other merely by flipping a switch (code-decode), allowing the same circuit to be used for both record and playback.

We now take a look at the appropriate circuitry. In Figure 13.12 we presume the input signal source to be of low impedance. Therefore, the 5.6 and 27 nF capacitors are effectively in parallel and with the 3.3 kΩ resistor form a high-pass filter (Figure 13.13). The turnover frequency in this case is 1.5 kHz. Disregarding the 4.7 nF capacitor for the moment, the high-pass filter feeds another voltage divider consisting of the 47 kΩ resistor and the FET. Without any control voltage to the FET, it constitutes a nominally infinite resistance so that the entire output of the high-pass filter appears at the output of the second voltage divider. There will be no dc control voltage for inputs of any magnitude at frequencies below the passband and for low-level inputs within the passband. With an hf signal in the passband (of sufficient intensity), a control voltage appears, the FET now has a finite resistance, and the second voltage divider (Figure 13.13) decreases the output within the passband—the greater the control voltage, the smaller the output within the passband, just what we want.

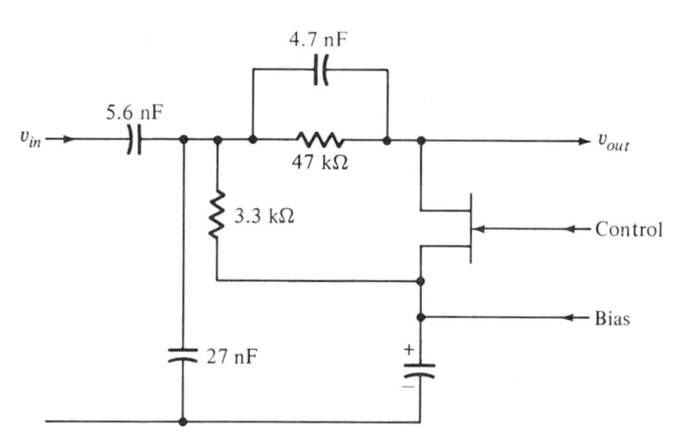

FIGURE 13.12. High-Pass Filter Whose Output Decreases in Response to a Gate Voltage

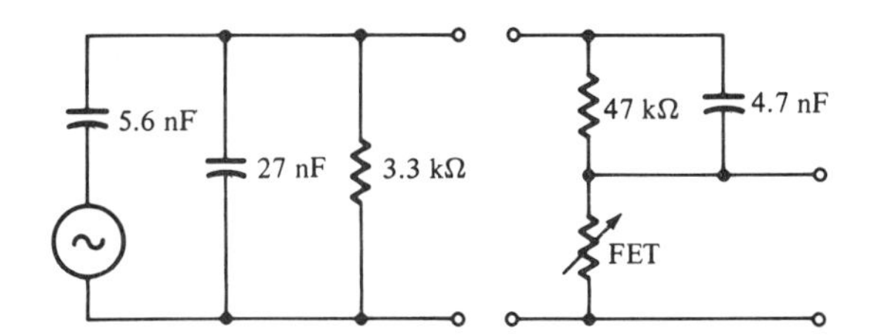

FIGURE 13.13. Circuit Equivalent to the High-Pass Filter Shown in Figure 13.12

But we also wanted the lower limit of the passband to move upward as the strong frequency responsible for the bias approached the passband. We can achieve such a response by having a variable turnover in the second voltage divider.

At frequencies well into the passband, the reactance of the 4.7 nF capacitor is small compared to the parallel 47 kΩ resistor and thereby effectively

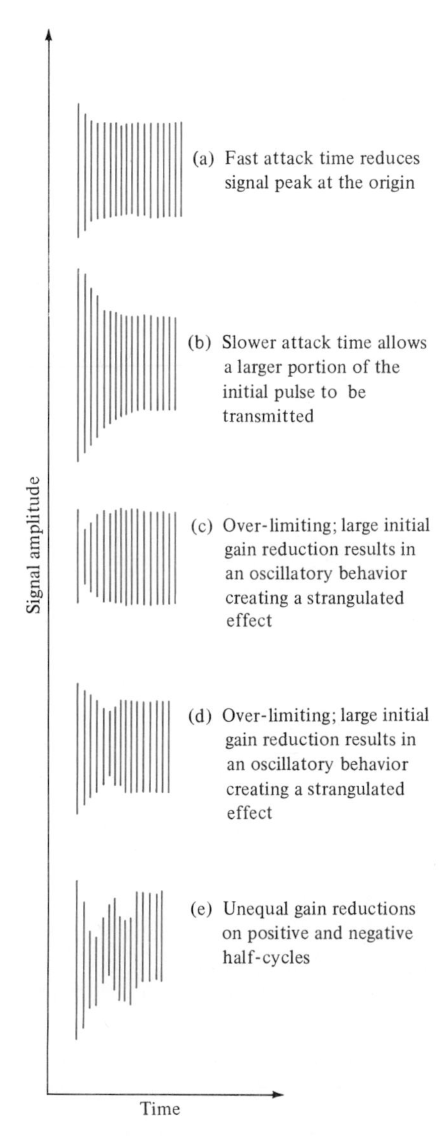

FIGURE 13.14. Effect of Noise Limiting in Sound Recording

places the FET in parallel with the 3.3 kΩ resistor. A strong signal lowers the output, and a weak signal increases the output. A comparably strong signal at a frequency along the skirt of the passband will be subjected to attenuation by both voltage dividers, lowering the magnitude of the output. In effect this process corresponds to moving the turnover point of the first voltage divider to higher frequencies.

Sudden changes in gain may themselves give rise to extraneous noise. (Remember that sharp amplitude changes contain high-frequency harmonics.) Such effects can be minimized by employing a long time constant in conjunction with the generation of the control voltage. The time taken by a limiter or a compressor to produce a gain change is called the **attack time**.

Figure 13.14 shows the influence of various attack times. That illustrated in Figure 13.14a results from fast attack times with the signal being smoothly reduced to a fixed level. In Figure 13.14b we have a slower attack time—more of the peak gets through before the amplitude is reduced to a fixed level. In Figures 13.14c and d, we see the effects of overlimiting; while the attack time is fast, the gain reduction is initially too large, resulting in a strangulated effect that the ear finds objectionable. Figure 13.14e is the result of unequal gain reductions for positive and negative half-cycles, leading to a shift of the dc level. This last response produces a thumping effect.

Fast attack times primarily cause transient distortion in conjunction with high frequencies. If a 10 kHz tone of excessive magnitude is to be brought back down to the threshold level within the first half-cycle (Figure 13.15), the attack time must be less than 50 μs. With tape recordings slower attack times are used, about 1 ms. Such response times takes care of the peaks of long duration, while saturation can be depended on to compress the short-duration peaks. Fortunately, the high frequencies generated in the latter case are beyond audibility.

The recovery or **release time**—that is, the time taken for the system to recover its normal gain after compression has taken place—is also an important parameter. In considering a complex sound wave with its succession of amplitude peaks and troughs, presume compression (hence a lowered gain) to arise during peak signals. When the amplitude drops into a trough, if the release time is long, the low amplitudes are also subjected to lower gain. With a fast release time, the gain is quickly increased, and gives rise to a heightened trough. Fast release times, therefore, tend to minimize amplitude variations, and thereby lead to a high mean (loudness) level.

Figure 13.16 shows how the control voltage is developed in the Dolby circuit and how the attack and release times are determined. To prevent rapid changes in gain, a long attack time for the FET is desired. But a short attack time is needed to minimize overshoots. In this circuit the time constant is determined by the rate of change of the signal.

D_4 will rectify the signal and thus furnish the dc control voltage. For signal frequencies the electrolytic capacitor can be considered a short, but a discharge path must be provided. This function is served by the 270 kΩ resistor. The time constant for the transistor amplifier thus consists of the 2.7 kΩ resistor and the 100 nF capacitor. Representing 0.27 ms (corresponding to a frequency of 600 Hz), this time constant allows slowly changing amplitudes to

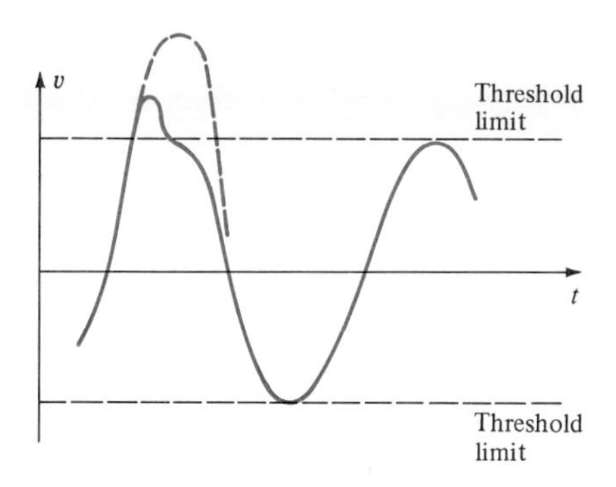

FIGURE 13.15. Influence of Fast Attack Time on an Increasing Signal Amplitude

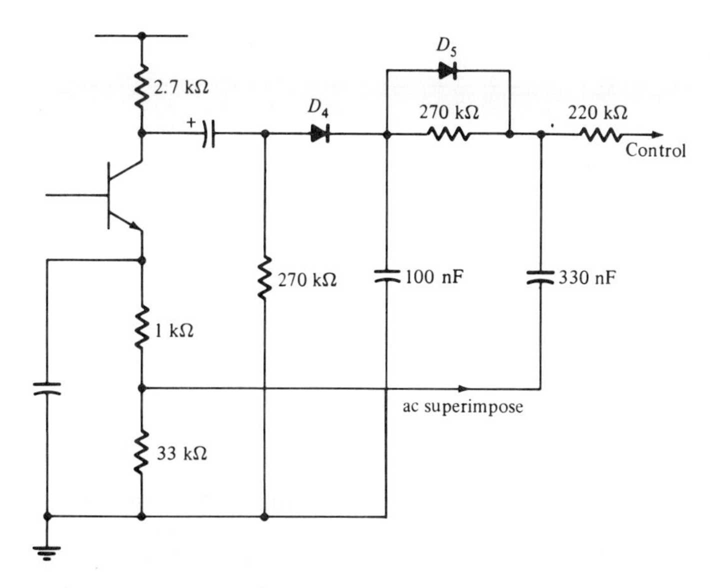

FIGURE 13.16. Dolby Control Circuit

be followed. But this RC combination feeds a longer time constant circuit (270 kΩ and 330 nF) that constitutes an attack time of 89 ms. If there is a large sudden change in the input signal, the potential across the 100 nF capacitor changes faster than that across the 330 nF capacitor. (Remember that $V = Q/C$; small C leads to large V.) The resulting voltage biases D_5 in the forward direction; it conducts and shorts the 270 kΩ resistor. 330 nF and 100 nF are now effectively in parallel, and the time constant is now about 1 ms. These two time constants represent extreme values. In general the charging of the 330 nF is shared by D_5 and 270 kΩ, leading to a variable attack time whose value is determined by the rapidity and extent of the amplitude change.

Very high-amplitude transients are handled by diode clippers elsewhere in the circuit. The normal release time is 100 ms, but for large, sharp reductions, this time constant is made smaller.

To keep distortion in the FET low, it is necessary that there be equality between the signals on the drain and gate. This equality is provided by superimposing some ac on the control voltage. (This procedure has previously been discussed in Section 9.7.)

The system we have discussed is termed the Dolby B. The Dolby A circuit is much more elaborate. Rather than providing dynamic noise reduction only at high frequencies, the A system splits the spectrum up into four parts, each of which is separately controlled. The band 80 Hz down takes care of hum and (turntable) rumble, the band 80 Hz to 3 kHz takes care of broadband noise and cross talk, and the bands 3 kHz and up and 9 kHz and up take care of hiss. The A system is very expensive and is used exclusively by recording studios.

13.4 ▪ INTERFERENCE

When first placed into operation, an electronic circuit often differs from its "paper" design because of the somewhat unpredictable presence of noise and interference (particularly interference arising from 60 Hz power lines). The usual attempt at resolving the interference problem involves the use of shielded wires and metallic enclosures, with a profusion of grounding connections. Often such procedures prove as ineffective as the procedure sometimes employed in evaluating the quality of a used car being considered for purchase, the kicking of its tires. In fact, indiscriminate grounding may actually aggravate the situation.

13.4.1 Chassis Grounding Procedures

To understand what follows, the reader should realize that the normal 120 V power outlet has one terminal grounded. Close inspection of the power plug on some appliances (such as TV sets) will often show one prong wider than the other, so that the plug must be inserted into the power outlet with a specific orientation (Figure 13.17a). In other cases a three-pronged plug ensures a specific orientation. It is common practice to connect the grounded side of the power leads to the metal chassis of the appliance. If no provision were made to assure a specific orientation of the plug in the wall (Figure 13.17b) and the chassis ground thereby got connected to the "hot" side of the outlet, a person who simultaneously came into contact with the chassis and a power-line ground (such as a radiator or a water pipe) would be exposed to the full line voltage. While such chassis grounding procedures are often employed with appliances for the sake of economy of manufacture, the grounding of an instrument chassis directly to a utility lead should be avoided. An isolation transformer between the instrument and the power line is strongly advisable, and reputable manufacturers will follow this procedure. In such cases usually one side of the secondary is made to serve as chassis ground, while both primary leads are left "floating."

The most common form of interference can be demonstrated with an oscilloscope. With no signal applied, but with the scope in a self-triggering mode, a fine horizontal (sweep) line is observed. Touching your finger to the input terminal will produce on the screen a 60 Hz signal of about 1 V magnitude. This display comes about because there is a capacitance between nearby 120 V power lines and your body, a capacitance that typically is about 0.2 pF (Figure 13.18). The capacitance of your body to ground is much greater (perhaps 2 pF) because there are many more grounds nearby than there are power sources, with each of these separate ground capacitances being additive. The input capacitance to the scope might be about 20 pF. Touching the scope input, therefore, gives rise to a voltage divider, as shown in Figure 13.19.

The measurement of electronic potentials and signals in most instances utilizes as a reference the metallic chassis on which the instrument is constructed. Thus, various dc voltages are either positive or negative relative to the chassis. The chassis is designated as being the *circuit ground.* Thus, the circuit ground is merely the reference terminal for the circuit on the chassis.

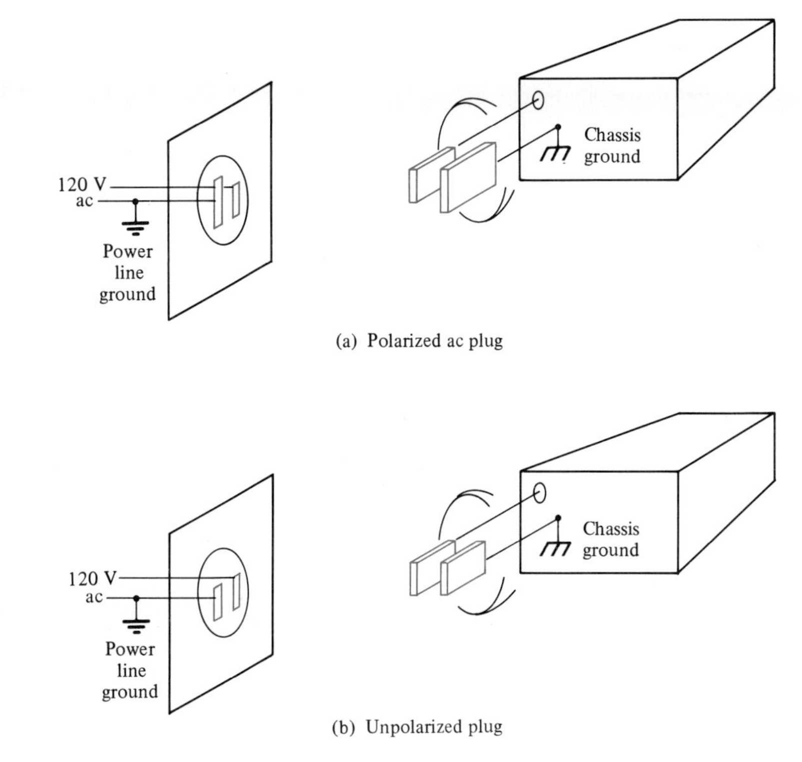

FIGURE 13.17. Two-Prong Utility Plugs

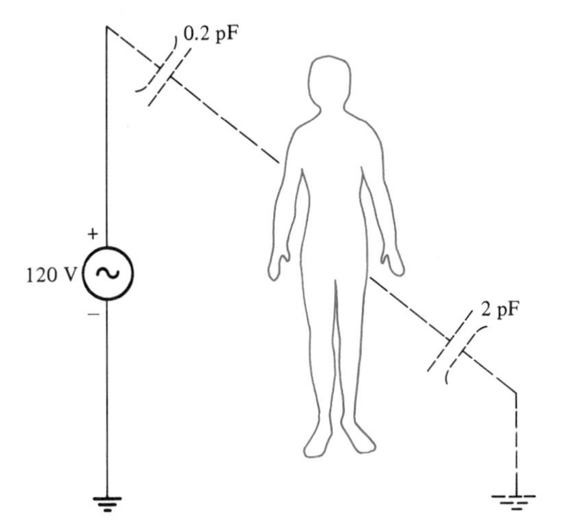

FIGURE 13.18. Capacitance Between Human Body and Nearby Power Lines and Between Body and the More Extensive "Grounds" in the Vicinity

Refer to Figure 13.20. Note that the isolated chassis behaves just like the human body depicted in Figure 13.18. Between the chassis and the power-line ground, there may exist a significant 60 Hz voltage. Of course if the power

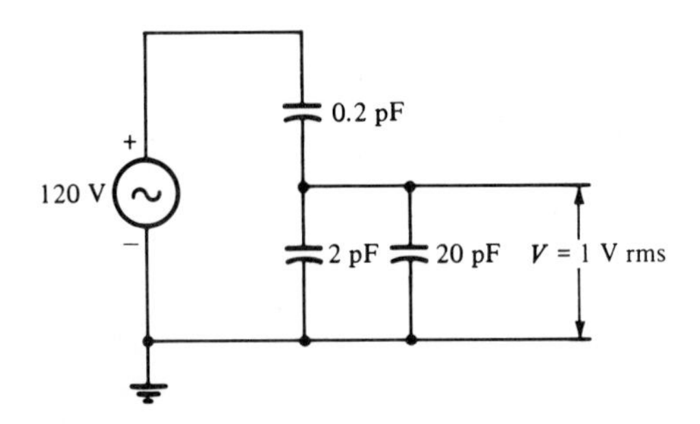

FIGURE 13.19. Electrical Equivalent of Situation Depicted in Figure 13.18 Plus Estimated (20 pF) Capacitance Due to Measuring Circuit

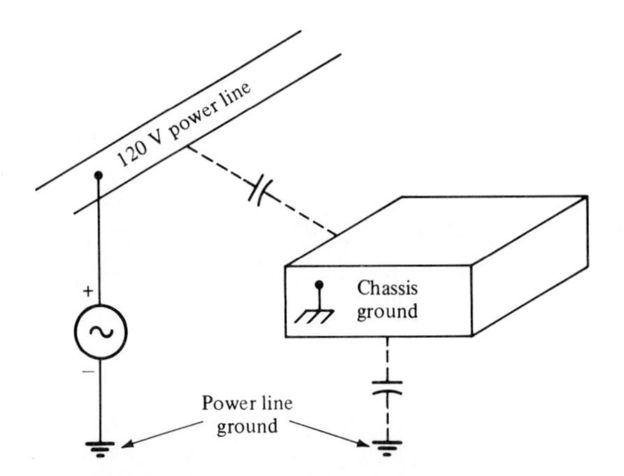

FIGURE 13.20. Manifestation of the Situation Depicted in Figure 13.18 Involving an Electronic Circuit Chassis

lead brought into the chassis has its grounded side attached to the chassis, the chassis ground is coincident with the power-line ground and the difficulty is eliminated.

13.4.2 Shielding

It is important to keep the area between the two input leads to a minimum since an interfering signal, particularly at power-line frequency, arises from a varying magnetic flux threading through this area. Obviously, such interference can be minimized by using either close-spaced parallel leads or twisted leads (Figure 13.21b). (In the case of the twisted leads, the induced voltage in successive loops will be in opposition and thereby lead to a minimal magnitude of interference.)

Regarding the further minimization of hum pickup, Figure 13.22a shows the signal source to be within a metallic enclosure and the emergent lead surrounded by a metallic shield. The capacitance between this shielded enclosure and nearby power lines might be about 1 pF. At 60 Hz this stray pickup causes about 4×10^{-8} A to flow through the shield to the chassis ground. If the shield were not present (Figure 13.22b), this current would flow through

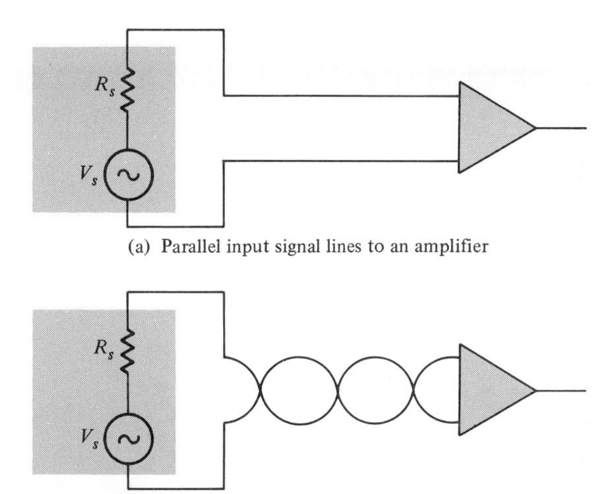

(a) Parallel input signal lines to an amplifier

(b) Twisted pair input leads that are subject to much less magnetic interference than in the situation depicted in part a

FIGURE 13.21. **Diminishing Magnetically Induced Interference in Amplifiers**

R_s and with 1 kΩ as a typical value of R_s would lead to a 40 μV hum superimposed on the signal source. If the source signals were of the order of microvolts also, the difficulty should be obvious. Such shields typically reduce the stray pickup by a factor of 100 or 1000.

Thus, we now have a metallic shield about the signal source, the amplifier sitting on a metallic chassis and maybe even within a metallic enclosure and a shielded wire transferring the signal between source and amplifier. It would seem natural that the wire shield should be attached to both metallic enclosures as shown in Figure 13.23a. But this is the wrong thing to do.

Insofar as the source enclosure is concerned, this situation corresponds to that depicted in Figure 13.20, and there will be a 60 Hz circulating current through the wire shield. This current is capacitively coupled to the signal wire and thus will introduce interference into the wire. Additionally, there is a large area enclosed between the two grounds, and any changing magnetic flux threading through this area will introduce additional interference. Such inadvertent conducting paths are termed **ground loops**. The arrangement shown in Figure 13.23b is much to be preferred. The shield is grounded at the source end; at the amplifier end it is connected to one of the amplifier's input leads, as it must be to properly introduce the signal into the amplifier. But the shield is not grounded to the amplifier's chassis, thereby making the ground loop resistance very high and reducing interference. One concludes that a shielded wire should be grounded at one end only.

Figure 13.24 introduces the comparable situation with a shielded two-wire lead. Figure 13.24b represents the preferred arrangement with the shield grounded only at one end. Figure 13.24c contains two ground loops, one involving the shield and the other, the lower of the two signal leads.

Our discussion has been limited to a consideration of interference arising from power lines. One may also encounter interference from nearby radio or TV stations, from motor ignition systems, and (within electronic circuits) between adjoining conductors.

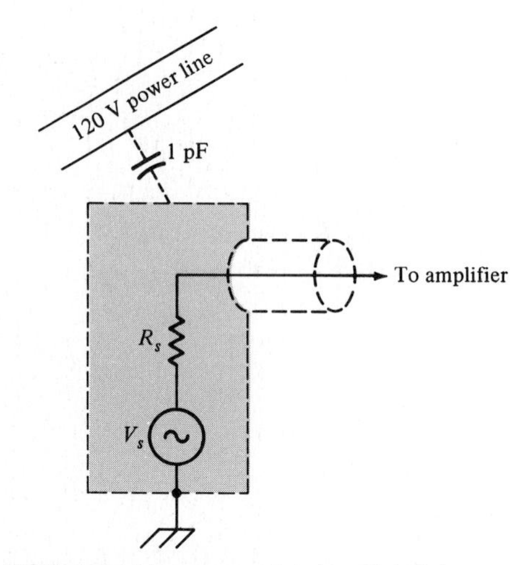

(a) Shielded signal source prevents induced interfering current from passing through the source resistance

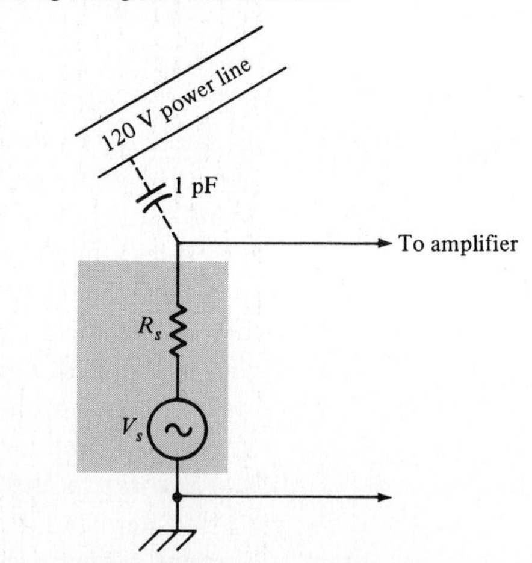

(b) Induced interfering current through a signal source resistance

FIGURE 13.22. Power-Line Interference in Amplifiers

13.4.3 Interference in Digital Circuits

The presence of interference pulses may cause erroneous logic transitions. To guard against such a likelihood, the various logic families have a designed gap introduced between the low and high states. The most likely sources of erroneous transitions are the capacitive coupling of an input circuit to adjoining signal lines, power leads containing transient fluctuations, and pulse reflections on improperly terminated lines. In TTL logic, for example, during a transition the current drawn from the power source is considerably larger than in a quiescent state. Any significant resistance in the power lead may cause the dc voltage to drop momentarily and lead to an erroneous triggering of other circuits powered by the same dc source. Such errors can be mini-

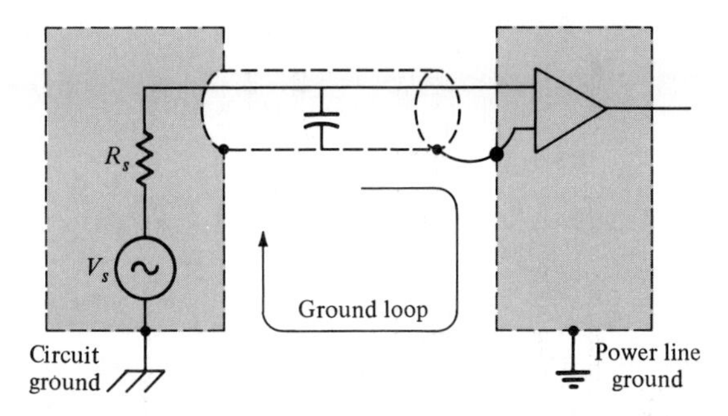

(a) Grounding the shield of a coax at both ends creates a ground loop through which a varying magnetic flux can create an interfering signal

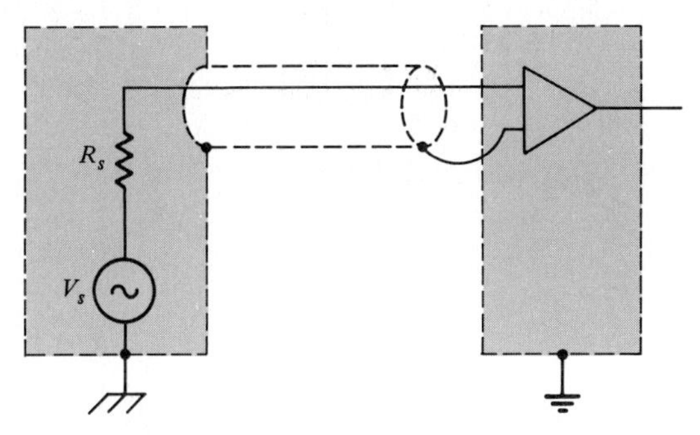

(b) Grounding of a coax at one end only eliminates the ground loop

FIGURE 13.23. **Ground-Loop-Induced Interference**

mized by the judicious placement of capacitors where their charged condition may be used to furnish the necessary current transient.

There is a significant variation in the degree to which the various logic families are susceptible to interference. Interference sensitivity for ECL is the smallest, necessitating a spread of only 0.2 V between high and low states (called the noise margin, though the author would prefer "interference margin" since signal levels are generally much above the background noise level). By using balanced differential inputs, the total current drawn by an ECL gate remains essentially constant irrespective of its status. TTL, on the other hand, is 10 times more susceptible, and this sensitivity is reflected in its value for the noise margin, 1.0 V.

The interference between adjoining conductors is illustrated in Figure 13.25, together with a simplified equivalent. If the ratio of separation distance (D) to wire diameter (d) is greater than 3, the coupling capacitance can be approximated by the following equation:

$$C_{12} = \frac{\pi\epsilon}{\ln(2D/d)} \text{ [F/m]} = \frac{2.78 \times 10^{-11}}{\ln(2D/d)} \tag{13.13}$$

where we presume air separates the lines; therefore $\epsilon = 8.854 \times 10^{-12}$ [F/m].

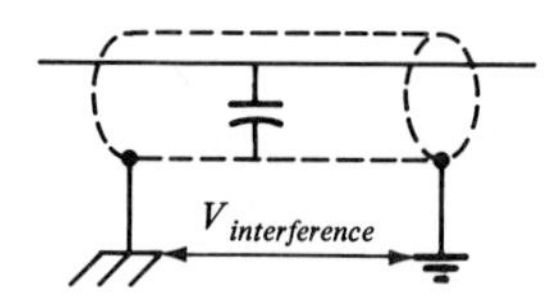

(a) Grounded at both ends, an induced current in the ground lead can be capacitively coupled to the shielded conductor

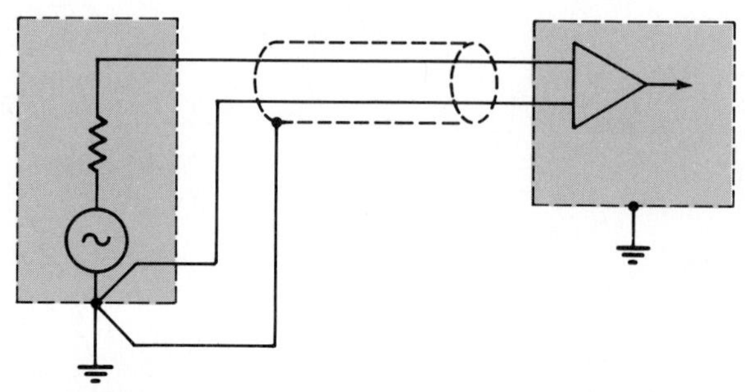

(b) Preferred arrangement for shielded pair of signal leads

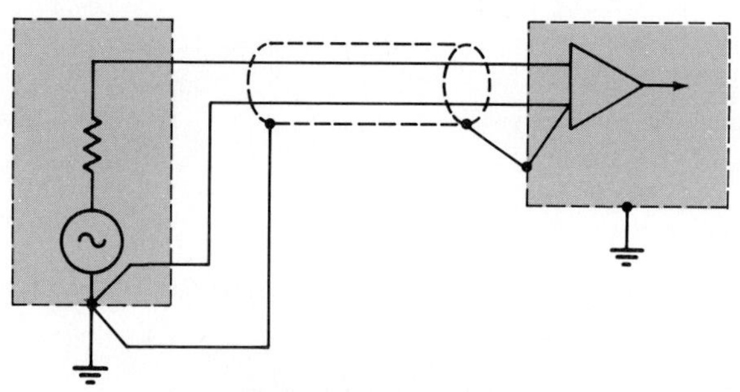

(c) Ground loops involving shielded pair

FIGURE 13.24. Shielding Paired Leads

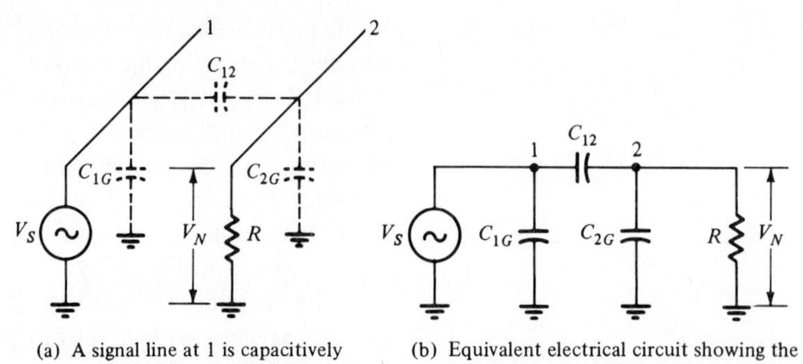

(a) A signal line at 1 is capacitively coupled to line 2, which has a resistance R to ground

(b) Equivalent electrical circuit showing the interfering signal V_N appearing across R

FIGURE 13.25. Interference Induced Between Parallel Conductors

The value of C_{1G} does not affect the result. By solving the remainder as a voltage divider problem, with $R << 1/[j\omega(C_{12} + C_{2G})]$, the interfering voltage (V_N) in terms of the interfering source voltage (V_S) is

$$V_N = j\omega RC_{12}V_S \tag{13.14}$$

If $R >> 1/[j\omega(C_{12} + C_{2G})]$, we have

$$V_N = \left(\frac{C_{12}}{C_{12} + C_{2G}}\right)V_S \tag{13.15}$$

which is independent of frequency. (This latter equation represents the same situation as was encountered when we first considered the automotive ignition probe in Section 3.4.) In this case the interference is independent of the frequency and larger in magnitude than would be the case with smaller R.

With regard to Eq. 13.14, V_S and ω are generally fixed so that only the values of R and C_{12} are available to minimize interference. In using Eq. 13.14, what frequency do we use in conjunction with digital pulses? In terms of pulse rise time (t_r), the equivalent maximum frequency is

$$f_{max} \simeq \frac{1}{2\pi t_r} \tag{13.16}$$

Nonrepetitive pulses include all frequencies from zero to f_{max}. For repetitive pulses the content includes frequencies from the repetition frequency up to f_{max}.

The function of a shielded wire is illustrated in Figure 13.26. The voltage induced on the shield is

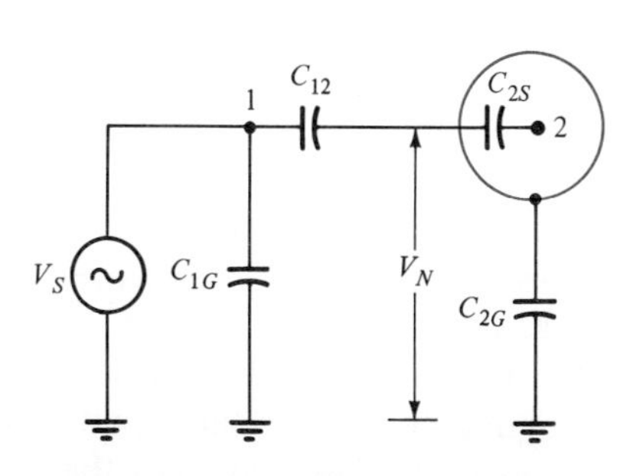

FIGURE 13.26. Effectiveness of Shielded Wire If wire 2 and the surrounding shield are both "floating," the interference induced into the shield appears on line 2 since no current flows through C_{2S}. If the shield is grounded, the voltage appearing at the shield is zero, as is the interference on line 2, provided line 2 is fully enclosed by the shield.

$$V_N = \left(\frac{C_{12}}{C_{12} + C_{2G}}\right)V_S$$

If wire 2 is ungrounded, there is no current through C_{2S}, and the interference induced on the shield appears also on wire 2. If the shield is grounded, the shield voltage is zero, as is the interference on line 2, provided the wire does not extend beyond the shield. If there is an extension, the situation with regard to the unshielded portion is treated in a manner similar to that first considered.

EXAMPLE 13.4

With a 4 MHz digital pulse rate having a 5 V amplitude, when applied to one of two 2.5 in. long parallel wire conductors (0.025 in. in diameter) separated by 0.1 in. and placed 0.1 in. above the ground plane, at 4 MHz, what is the induced voltage in the second wire when

a. The second wire has a 1 MΩ resistance to ground?
b. The resistance to ground is 50 Ω?

Solution: The value of C_{12} is 0.85 pF, and the value of C_{2G} is 1.7 pF.

a. The combined reactance is 15.6 kΩ, and since R is much larger, we have

$$V_N = \left(\frac{C_{12}}{C_{12} + C_{2G}}\right)V_S = \left(\frac{0.85}{0.85 + 1.7}\right) \times 5\text{ V} = 1.67\text{ V}$$

b. In this case $R << 15.6$ kΩ, and one uses $|V_N| = \omega RC_{12}V_S$. We have

$$V_N = 2\pi(4 \times 10^6)(50)(0.85 \times 10^{-12}) \times 5\text{ V} = 5.35\text{ mV}$$

SUMMARY

13.1 Introduction

- Noise and interference represent the two most unpredictable aspects of electronic performance.
- Noise arises from the discreteness of electrical charge. Interference generally arises from man-made sources.

13.2 Electrical Noise

- Noise arising from thermal agitation of charged carriers is termed Johnson noise.
- Shot noise is due to charge discreteness.
- Noise voltage increases as the square root of the bandwidth.
- $1/f$ noise increases inversely with frequency and is generally of significance only at very low frequencies.
- Transistor noise is due to both shot noise and Johnson noise, and optimum performance will arise when, by design, there is an equality between the two sources.
- For certainty of detection, a signal-to-noise ratio of 3:1 is generally assumed.
- Noise factor is a measure of the degradation of a signal-to-noise ratio in the process of amplification. The larger the noise factor, the greater is the amplifier's contribution to the noise level.

13.3 Noise and Sound Reproduction

- Pre-emphasis represents the purposeful introduction of frequency distortion to minimize noise in high-fidelity reproduction systems.
- The range of intensities accommodated by a sound system represents its dynamic range.
- A limiter prevents the output signal from exceeding some predetermined level. A compressor diminishes the gain in proportion to the signal level.
- Masking represents the ability of a strong audio signal to diminish the recognition of noise presence.
- Loudness is the average value of the sound level.
- A Dolby system is primarily intended to improve the dynamic range of a sound system.
- In frequency modulation (FM) the rate at which the radio frequency is varied constitutes the signal frequency to be transmitted. The magnitude of the radio-frequency deviation determines the signal amplitude.
- Noise and static in FM are reduced by the use of a noise limiter.

13.4 Interference

- One lead of an ac outlet is grounded, and in many appliances this lead is connected to the chassis. High-quality instruments should avoid such procedures and utilize isolation transformers between the apparatus and the utility outlet.
- To minimize 60 Hz interference, the grounding at both ends of shielded leads, thereby creating a ground loop, should be avoided.

KEY TERMS

attack time: the time taken by either a limiter or a compressor to produce a gain change

compressor: a device that progressively diminishes the gain as the signal amplitude increases

de-emphasis: a compensating decrease in the gain to which higher audio frequencies are subjected in order to diminish the background noise level

Dolby noise suppression: a means to automatically alter the degree of pre-emphasis (and de-emphasis) in order to make use of the maximum allowable dynamic range

dynamic range: the ratio of largest to smallest signal level

electrical noise: a random signal fluctuation arising from the discreteness of electronic charges and their thermal agitation

excess noise: an increase in the noise level at very low frequencies

ground loop: an unintentional connection that creates a closed conducting path through which a varying magnetic field will give rise to an induced interfering voltage

interference: signal disruption arising from extraneous sources such as hum, utility line impulses, motor ignition, and so forth

Johnson noise: arises from the thermal agitation of

charged carriers within conductors; also called thermal noise

limiter: a device to prevent an output signal from exceeding a predetermined level

loudness: the average sound level

masking: the inability to subjectively distinguish background noise in the presence of strong audio signals

noise figure: a measure of the additional noise introduced in the process of amplification (A noise figure of unity indicates an amplifier that makes zero contribution to the background noise; it is a measure of signal-to-noise degradation in the process of amplification.)

pink noise: an increase in noise density with diminishing frequency

pre-emphasis: subjecting the higher audio frequencies to a proportionally larger gain than that employed at lower frequencies

release time: the time taken for a system to recover its normal state after being subject to either limiting or compressing action; also called recovery time

shot noise: noise arising from the discreteness of electronic charges

white noise: noise that exhibits a uniform spectral density—that is, the noise level does not vary with frequency

$1/f$ noise: noise that increases inversely with frequency

SUGGESTED READINGS

Johnson, J. B., "Thermal Agitation of Electricity in Conductors," PHYSICAL REVIEW, vol. 32, pp. 97–109, July 1928. *The original paper reporting on the experimental investigation of thermal (Johnson) noise.*

Nyquist, H., "Thermal Agitation of Electric Charge in Conductors," PHYSICAL REVIEW, vol. 32, pp. 110–113, July 1928. *A companion paper reporting on a theoretical derivation of the Johnson noise formula.*

Skomal, Edw. N., MAN-MADE RADIO NOISE. New York: Van Nostrand Reinhold, 1978. *A consideration of radiated noise (but not noise conducted through power lines). Apropos of a continuing thread in our text, however, Chapter 2 is devoted to automotive noise considerations.*

Cohen, Arnon, "Noise and Interference in Electrical Networks—an Undergraduate Experiment," IEEE TRANSACTIONS ON EDUCATION, vol. E-16, pp. 161–166, Aug. 1973. *Although designed as an undergraduate laboratory experiment, very informative with regard to "cross talk" between parallel conductors and shielded cables.*

Zepler, E. E., THE TECHNIQUE OF RADIO DESIGN, 2nd ed., pp. 240–264 and pp. 305–322. New York: Wiley, 1949. *Although an old (and on the face of it perhaps an unsuspected) reference, among the best practical suggestions for some simple guidelines concerning shielding practices.*

Ott, Henry W., NOISE REDUCTION TECHNIQUES IN ELECTRONIC SYSTEMS. New York: Wiley, 1976. *If you desire a modern comprehensive treatment of noise and interference with ample analytical considerations and a number of problems (with answers), this work is highly recommended.*

Lee, Y. W., and J. B. Wiesner, "Correlation Functions and Communications Applications," ELECTRONICS, vol. 23, pp. 86–92, June 1950. *With regard to noise correlation, here we have another one of those old references that, in terms of clarity and conciseness, the author finds superior to more modern and sophisticated presentations.*

EXERCISES

1. Why does a large input resistance improve the signal voltage? Why does it increase the background noise level?

2. State the Johnson noise equation and identify the various factors.

3. State the Schottky noise equation and identify the various factors.

4. Distinguish between noise and interference.

5. In what manner does the noise voltage of a bipolar junction transistor vary with collector current?

6. What is generally considered to be the minimal signal-to-noise ratio for signal detection with certainty?

7. How is *noise factor* defined?

8. What is the smallest possible value for the noise factor? What is the smallest value for the noise figure?

9. What would happen to the noise figure if a resistor were inserted in series with a source lead? Why?

10. Why can one usually disregard the noise contributions of voltage amplifier stages beyond the initial one? When would this procedure not be justified?

11. What is white noise? Pink noise?

12. What is pre-emphasis? De-emphasis? What is the function of such processes?

13. What is dynamic range?

14. Indicate the dynamic range of the following: concerts, recordings, broadcasting stations.

15. Distinguish between a limiter and a compressor.

16. In FM, what determines the audio frequency received? What determines the amplitude of the received audio frequency?

17. In FM, what is the result of using a 25 μs time constant for pre-emphasis in contrast to a 75 μs time constant? What does this smaller time constant do to the overall dynamic range?

18. What is loudness?

19. What is masking?

20. What is the prime function of a Dolby system?

21. For pre-emphasis, what kind of feedback is needed in the Dolby system? For de-emphasis?

22. What is the meaning of attack time? Of release time?

23. What is the danger that comes about from a direct connection between chassis ground and a utility power ground?

24. What is the purpose of twisting the input lead wires of an amplifier?

25. What is a ground loop?

PROBLEMS

13.1
The rms noise generated in a resistance can be represented by its Thevenin equivalent, a generator ($v_N = \sqrt{4kTRB}$) in series with an ideal resistor R.

a. Under what circumstances will the resistor deliver the maximum noise power into a connecting load?

b. What is the value of this maximum deliverable power?

13.2
Noise power (particularly in conjunction with microwaves) is often stated in terms of *noise temperature.* If the source temperature is T_s and the noise power generated within the amplifier is such as to be represented by a temperature T_a, what is the expression for the noise factor? Microwave amplifiers employing thermionic electron sources tend to be very noisy. Compute the noise factor of a microwave receiver with a noise temperature of 900 K. Compute the noise factor of a maser, the microwave equivalent of a laser, with a noise temperature of 10 K.

13.3
Verify that Eq. 13.11 is dimensionally correct.

13.4
Given v_N, i_N, and R_s, determine the value of R_s that leads to the minimum value of F.

13.5
The LM394 is a matched pair BJT known for its low noise properties. To illustrate the influence of collector current on the noise, calculate the noise figure at 1 kHz with $R_s = 1\text{k}\Omega$ and

a. $I_C = 1\ \mu\text{A}$

b. $I_C = 1\ \text{mA}$

Use the following specifications

At 1 μA, $i_N = 0.03\ \text{pA/Hz}^{1/2}$, and $v_N = 13\ \text{nV/Hz}^{1/2}$.

At 1 mA, $i_N = 0.9\ \text{pA/Hz}^{1/2}$, and $v_N = 1\ \text{nV/Hz}^{1/2}$.

13.6
A low-noise BJT transistor has the following noise parameters:

$$v_N = 2.0\ \text{nV}/\sqrt{\text{Hz}}$$
$$i_N = 0.2\ \text{pA}/\sqrt{\text{Hz}}$$

As a function of source resistance, over the range 100 Ω to 10 MΩ, construct a log-log plot consisting of

a. The 3 dB thermal noise contour

b. v_N

c. i_N

d. The 3 dB point above the intersection of the i_N and v_N curves

What is the minimum noise figure for this transistor?

13.7
The 2N6483 is an n-channel matched pair JFET with a low noise rating. To illustrate the influence of R_s on JFET noise, calculate the noise figures at $R_s = 1\ \text{k}\Omega$ and $100\ \text{k}\Omega$ using the following applicable parameters: $v_N = 3.2\ \text{nV}/\sqrt{\text{Hz}}$, and $i_N =$

5.6×10^{-3} pA/$\sqrt{\text{Hz}}$. Why is the noise figure so much lower at the larger value of R_s?

13.8

A TTL circuit draws 5 mA in its on state and 1 mA when off. If the transition time is 5 ns and the power supply lead has an inductance of 0.5 μH, what is the interference voltage generated across the power supply wiring when the TTL changes state?

13.9

The original definition of noise factor was in terms of

$$\frac{(S/N)_{in}}{(S/N)_{out}} = F$$

Show that the following expression is an equivalent one:

$$F = \frac{\text{noise power output of the actual device}}{\text{noise power output of the ideal device}}$$

13.10 △

The rms noise voltage from two sources can be represented by two noise voltage generators in series with an ideal—that is, noiseless—input resistance. If the respective rms noise voltages are $v_{N1} = 10\ \mu$V and $v_{N2} = 15\ \mu$V, with $R_{in} = 100$ kΩ, compute the noise power

a. By first adding the voltages and then determining the noise power

b. By computing the individual power contributions and then summing the result

COMMENT: Which procedure should be used in practice? By way of an answer, consider the following: If x and y are fluctuating quantities with $\overline{x} = \overline{y} = 0$ (where the bars indicate a time average), the *correlation coefficient* is

$$r \equiv \frac{\overline{xy}}{\sqrt{\overline{x^2}\,\overline{y^2}}}$$

If the correlation is zero—that is, the two noise sources are independent—we add them quadratically:

$$\overline{(x + y)^2} = \overline{x^2} + \overline{y^2}$$

If there is a correlation—that is, they are not independent—they are to be added linearly:

$$\overline{(x + y)^2} = \overline{x^2} + 2\overline{xy} + \overline{y^2} = \overline{x^2} + 2r\sqrt{\overline{x^2}\,\overline{y^2}} + \overline{y^2}$$

where the degree of correlation can vary between 0 (uncorrelated) and 1 (fully correlated). One instance in electronic circuits where correlation between noise sources arises is in feedback circuits of OPAMPs. Disregarding correlation may lead to an error factor as large as 1.4—that is, $\sqrt{2}$—in the noise voltage.

13.11

Refer to Figure 13.25. Derive the general expression for V_N in terms of V_S. Show that in the limits of small and large R this equation reduces to Eqs. 13.14 and 13.15, respectively.

13.12

On linear graph paper, sketch v_N vs. ω for Eqs. 13.14 and 13.15. At what radian frequency, expressed in terms of R and $C_{12} + C_{2G}$, do the two approximate expressions lead to an equality? What is the actual value of v_N at this frequency relative to the value predicted by Eq. 13.14?

13.13

Figure 13.27 illustrates the recommended utility connections for a 120 V system and a combination 120 V/240 V system. Why are separate grounds used to connect a metal enclosure to the service entrance ground rather than simply connecting them to the neutral wire at the load?

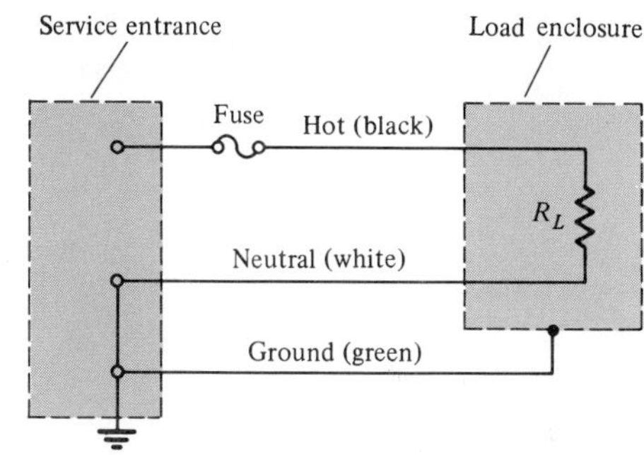

(a) Three-pronged connector for 120 V system

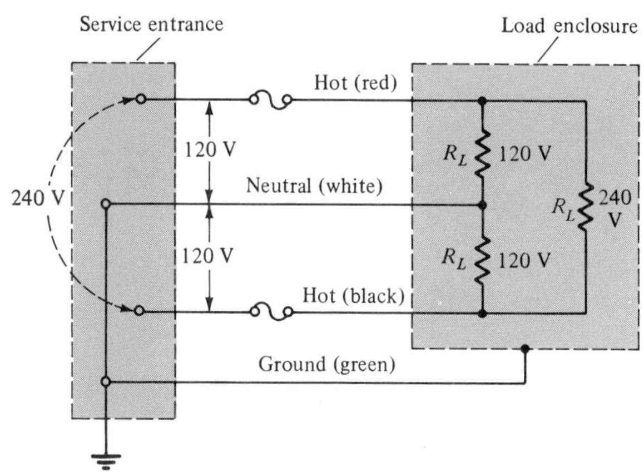

(b) Four-pronged connector for 120/240 V system

FIGURE 13.27

14 MICROCOMPUTERS

OVERVIEW

Basic digital circuits were discussed in Chapter 11, together with the necessary elements of binary manipulation. Our objective has been their utilization in the creation of digital computers. This application we undertake to consider in the present chapter. The 8-bit Z80 microprocessor, whose operation is considered in some detail, is simple enough to demonstrate the principles of microcomputer operation without the encumbrance of the 16 data bits utilized by later processors such as the Z80 successor, the Z8000, and the 32 bits of the Motorola 68000, which is upwardly incompatible with previous machines and hence not typical.

The material in this chapter will be used to illustrate some basic principles of electronic automotive controls in Chapter 15.

OUTLINE

14.1 ■ INTRODUCTION

In keeping with the philosophy of this text—that is, to introduce the engineering student succinctly, but effectively, to various aspects of electronics—rather than make any attempt at the detailed treatment of the rather broad topic of microcomputers, it will be presented in the context of a specific problem, that of an electronically controlled automobile engine.

We have, of course, already made a brief acquaintance with electronic ignition systems. Now we consider the use to be made of a computer in controlling, for example, the emission gases from an internal combustion engine.

In the conventional engine the function of the carburetor is to furnish a vaporized fuel-air mixture to the various cylinders. The optimum ratio of fuel-to-air mixture is about 1:14.7, which constitutes the **stoichiometric ratio** and results in complete combustion of the fuel. Combustion in an engine can be sustained only within a limited range of fuel-to-air mixture—80% of the stoichiometric ratio on the "lean" side (about 1:18) and 10% above on the "rich" side (1:13). Maximum *power* arises when a rich mixture is used, such as when starting the engine. The engine *efficiency*, however, should improve as the mixture is made leaner (the power output per unit mass of fuel increases). But in practice too lean a mixture causes slow combustion and a rapid drop in efficiency. As a result the optimum efficiency arises at a value that is somewhat smaller than the stoichiometric ratio.

In terms of pollutants, rich mixtures provide insufficient air for total combustion, giving rise to exhaust carbon monoxide (CO) and hydrocarbons (HC). On the other hand, an excess of air, while reducing HC and CO (by resulting in the creation of rather inert carbon dioxide, CO_2), increases the emission of nitrogen oxides (NO_x)—it is nitric oxide that reacts with sunlight, producing smog as a result of a photochemical reaction.

Emissions may be reduced by using a three-way catalyst, one for each component, but for its proper operation, the fuel-air ratio must be held within a very narrow tolerance. Such reduction can be accomplished by means of a closed-loop control system that continually samples the oxygen in the exhaust gases and adjusts the engine's air intake accordingly.

The sensor might be a thimble-shaped piece of zirconium dioxide that has both surfaces coated with porous platinum electrodes, this combination acting like an electromotive cell. As the mixture changes from a few percent rich to a few percent weak, the potential difference drops sharply from 800 to 50 mV. Such signals can be used to adjust the fuel-air ratio, but this simple relationship may not suffice. For example, when the engine is first turned on, a richer mixture is needed. Although the sensor may have a fast response time (20 μs), the response time of the mechanical controller may be as much as 0.5 s. The optimum ratio also shows a dependence on engine temperature, ambient pressure, etc. The need for a computer that will take all these matters into account should be obvious. There are two aspects to computer operation:

1. The **hardware**, consisting of the electronics, the wiring, the input/output devices, and so forth

2. The **software**, comprising the detailed instructions for the arithmetic and/or logic operations to be executed by the computer

14.2 ▪ THE MICROPROCESSOR

All computers have need for a unit that actually performs the needed arithmetic computations (termed the ALU, the arithmetic/logic unit). Additionally, there will be registers that temporarily store information, a sequencer that controls the order in which various program steps are serviced, and the associated wiring between these units (termed buses). The sum of these components constitutes a central processing unit (CPU) and represents the first step toward constructing a compact computer using integrated circuit techniques. When combined with external memory storage, input and output devices, etc., it constituted a microcomputer. When the first CPU was produced using IC techniques (in 1971), it was termed a "microprogrammable computer on a chip." In 1972 this expression was shortened to "microprocessor." At that time IC techniques were limited to the production of 30 or fewer gates per chip [small-scale integration (SSI)] and 30 to 300 per chip [medium-scale integration (MSI)].

With the advent of large-scale integration (LSI—300 to 3000 gates per chip) and very large-scale integration (VLSI—more than 3000 gates per chip), it also became possible to incorporate memory and other functions on a single chip. The name **microprocessor** is now applied to such devices. The first complete computer on a single chip was the 8048, introduced in 1976. A typical microprocessor unit (MPU), incorporating a CPU and associated memory, input/output circuits, etc., is contained within a 40-pin, $2 \times \frac{1}{2}$ in. package.

A major consideration in the selection of a microprocessor is the number of bits that it can handle. Eight bits will handle numbers 0 to 255—that is, 0 to $2^8 - 1$. Since the signals from the transducers (for example, engine temperature) will have to be converted into digital form, the number of data bits will determine how fine the data resolution will be. More importantly, however, the bit capacity also determines the number of memory locations that can be addressed.

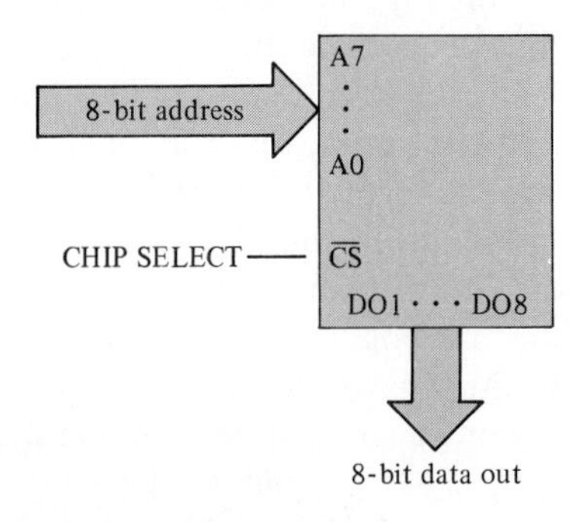

FIGURE 14.1. Read-Only Memory With $\overline{CS}$ applied, an 8-bit address (A0···A7) applied to the read-only memory (ROM) will cause the contents of the stated address to appear on the 8-bit output lines (D01···D08).

The program to be followed by the computer is generally available in *storage*. Since in the case of an automotive system there will be no need to change this program (as such it is called a **dedicated computer**), it will be available as a read-only memory (ROM), such as depicted in Figure 14.1. The combination of leads used to convey digital information is the *bus*. In Figure 14.1 we have an *address bus* and a *data bus*.

Terminals that have a bar over a designation, such as $\overline{CS}$, imply that they are activated upon application of a 0 to the terminal; otherwise, a terminal is activated by a 1. Refer again to Figure 14.1. When $\overline{CS}$ (termed *chip select*) is enabled, the ROM delivers at its DATA OUT terminals (D01–D08) the contents of the 8-bit address it has received at terminals A0–A7. The desired addresses and the enabling pulse—that is, $\overline{CS}$—come from the microprocessor, which, in turn, accepts the data from storage.

Figure 14.2 depicts the microprocessor (μP), sometimes called the MPU and sometimes, the CPU. It responds to timing pulses it receives from an oscillator that can be as simple as a multivibrator flip-flop.

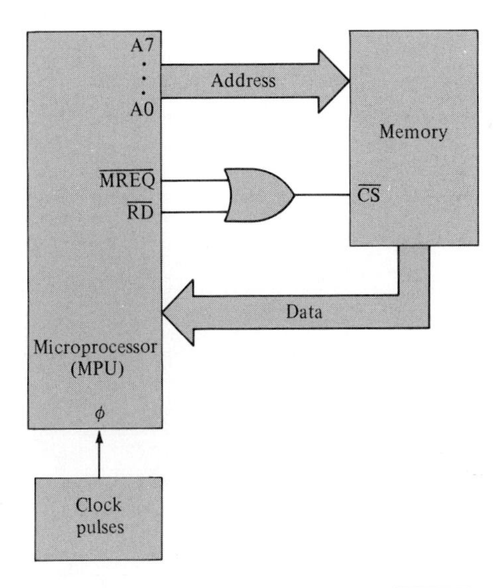

FIGURE 14.2. Memory Activation Memory request ($\overline{MREQ}$) and read ($\overline{RD}$) from the microprocessor, applied to the OR Gate, yields chip select ($\overline{CS}$) signal, allowing the memory to insert onto the data bus the contents of the memory location requested on the address bus.

When the MPU requires something from memory, $\overline{MREQ}$ is enabled (*memory request*). Since it is possible to either request something from memory or to wish to write something into memory, for a read operation the $\overline{RD}$ terminal on the MPU must be enabled. Applying $\overline{MREQ}$ and $\overline{RD}$ to an OR gate furnishes the necessary $\overline{CS}$ signal.

In executing the program, data from the outside are sometimes needed. [In our automotive system such data will be provided by sensors via an A/D (analog-to-digital) converter.] This data the MPU can request via an input port (Figure 14.3). To activate this input, two signals are required: $\overline{RD}$ (read) and $\overline{IORQ}$ (input/output request). Since the device select (DS2) requires a 1, an inverter is interposed between $\overline{IORQ}$ and DS2; the $\overline{RD}$ 0 is applied to $\overline{DS1}$. The same data bus used when the memory was queried is used here also but now the memory is disengaged.

To get data out of the MPU, we may now use an output buffer (Figure 14.4). This buffer must be a **latch device.** Since the data furnished by the MPU are available for only a few microseconds before the MPU goes on to other chores and the output data (D08–D01) must remain for a considerable length of time to allow the slow external devices to use them. Such is the function performed by the latch circuitry.

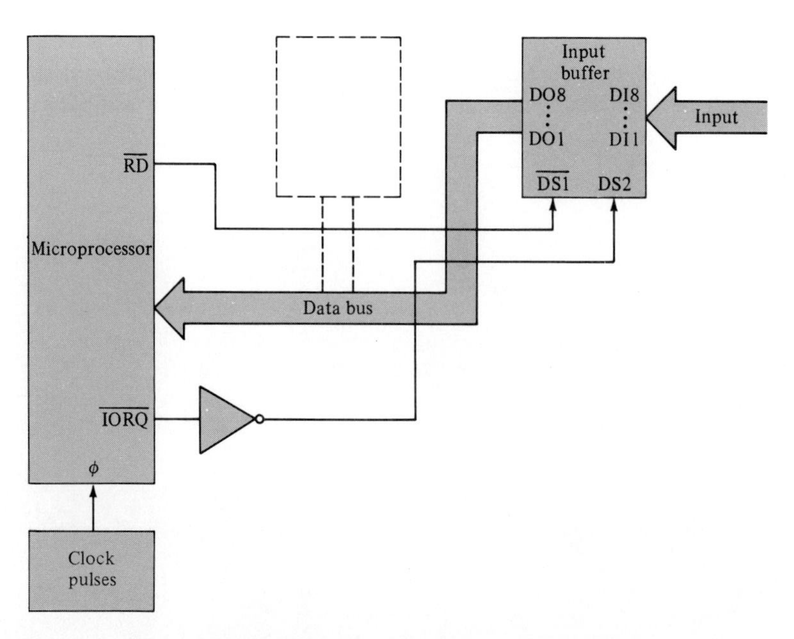

FIGURE 14.3. **Input Signal Activation** Input signals from outside the microcomputer are applied to a buffer to mediate between the fast response time of the microprocessor and the slow input rate of the signal. The input/output request ($\overline{IORQ}$) and read ($\overline{RD}$) signals activate the buffer and connecting bus.

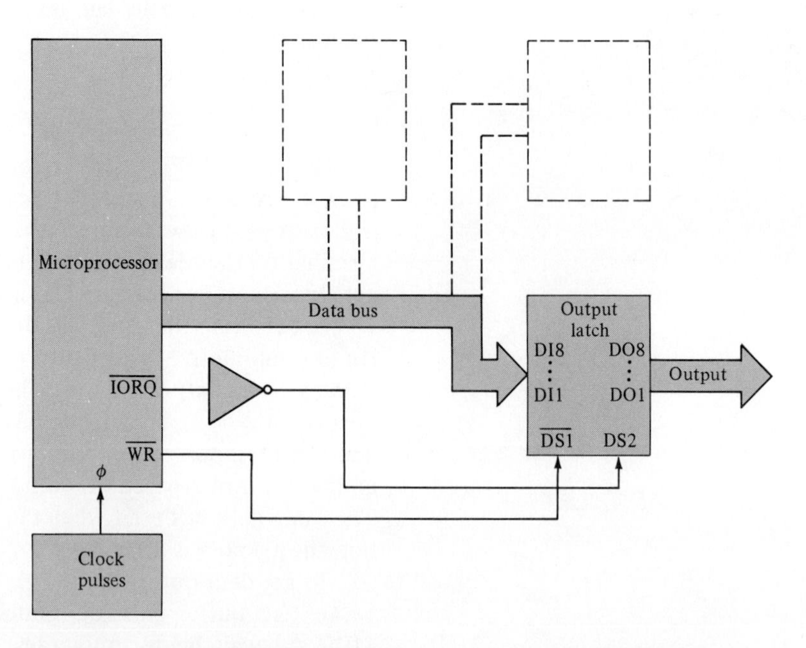

FIGURE 14.4. **Output Signal Activation** The high signal rate from the microprocessor must pass through an output latch to match the slower output rate demanded by printers, and so forth. The input/output request ($\overline{IORQ}$) and write ($\overline{WR}$) signals activate the buffer and bus.

14.3 ■ MEMORY

The simplest memory to visualize is a sequence of flip-flops, the Q output of each representing either the 0 or the 1 bit of a binary number. We may easily *read* the status of such a sequence, thereby determining the number *stored*. The stored number may be changed by applying to the flip-flop inputs the desired sequences and pulsing the clock input. This procedure constitutes a *write* operation. An 8-bit byte would require eight flip-flops. By proper connection the 8 binary bits could be made to enter (write) or leave (read) in either a sequential fashion (1 bit at a time) or in a parallel format (all 8 bits at once). By increasing the number of flip-flops, we could increase the number of stored bits. The desired combination of flip-flops is identified by an **address.**

In addition to flip-flops, the earliest memories included magnetized toroids, with a 0 and a 1 being distinguished by whether a toroid was magnetized in a clockwise or a counterclockwise direction. The advantage of such a system is the retention of the memory contents should the computer shut down, either purposely or unintentionally (for example, due to a power failure). Such a storage method is said to be **nonvolatile**. Such is not the case with flip-flops; their information would be lost, and they would be termed **volatile**.

Flip-flops, of course, can take on a variety of forms by utilizing the various logic families: MOS, CMOS, ECL, etc. The **cycle time** of a memory is the time taken to deliver the contents after application of the address. ECL memories are fastest, but they require large amounts of power. MOS flip-flops are fast, have a high packing density (you can fabricate a large number of cells per unit), and have low power consumption. Bipolar flip-flops (TTL) are fast but require more space and are less energy-efficient than MOS units.

Memory units whose contents can be easily altered are termed RAMs, the designation standing for **random access memory**, although now often termed read-and-write memory, indicative of their usage. Flip-flops make up a *static* RAM. In a *dynamic* RAM the bits are stored as charges on capacitors. Since such charges would disappear in less than a second, such memories must be continually "refreshed," typically every 2 ms or so. The computer is programmed to automatically refresh the memory. The seeming inconvenience of a dynamic RAM is justified because it utilizes less space than a static unit, allowing a larger packing density.

By way of contrast to the RAM, the ROM **(read-only memory)** represents a nonvolatile form of information storage. Most computers will include at least one ROM that contains all the start-up information needed by the computer when first turned on. Often the information content of a ROM cannot be changed once entered in the manufacturing process.

There are, however, **programmable read-only memories** (PROMs) that allow the user to set his or her own program—but, again, once entered, the program cannot be changed. In one form each bit addressed consists of a small wire fuse. If the fuse is maintained intact, it constitutes a 1 bit, and if (purposely) blown in the process of programming, it represents a 0. Obviously, once blown, it cannot be changed back to a 1.

An EPROM is an **erasable programmable read-only memory.** It can take the form of a MOS transistor with an isolated gate. A charge deposited on the gate will be retained and interpreted as a 1 bit and no charge, as a 0. It has been estimated that the charge will drop by less than 30% over a 100-year period (certainly a nonvolatile form of storage). To erase the charges, the memory cells are exposed to ultraviolet light for the order of 30 min to an hour, the light discharging them by means of the photoelectric effect, thereby creating a blank that can be reprogrammed. EAROMs are electrically alterable ROMs that can be programmed and erased electrically.

Memories that are located on the microprocessor chip itself provide much more rapid access time than does an offboard memory. Such memories are often termed **registers**, and microprocessors are divided into two broad categories depending on their reliance on registers. Register-oriented microprocessors (such as the 8080 and the Z80, which we will consider in some detail) have a large number of internal registers that can be manipulated by the programmer. They tend to require more complex timing and control. Memory-oriented microprocessors (such as the 6800 and 6500) have few internal registers but have instruction sets that allow easy manipulation of memory content.

14.4 ■ Z80 MICROPROCESSOR

8 bits	8 bits
A register	F
B	C
D	E
H	L
Index register IX	
Index register IY	
Program counter (PC)	
Stack pointer (SP)	
I register	R register

FIGURE 14.5. Registers of the Z80 Microprocessor The I register is involved with interrupts. The R register is used to refresh dynamic memories.

The Zilog Z80 microprocessor has been among the more popular microprocessors, having been used in the Radio Shack TRS80 computer. This microprocessor occupies an area about $2\frac{1}{2} \times \frac{1}{2}$ in. on a standard 40-pin DIP (dual-in-line package). It normally handles 8-bit computation, but 16 bits is also possible at a somewhat slower speed. The clock pulses used with the Z80 are square waves whose repetition rate can be as high as 4 MHz, in which case the timing intervals are 0.25 μs. The instructions executed by the Z80 require from 4 to 20 clock pulses—that is, from 1 to 5 μs.

Let's take a closer look at what is inside the Z80. Most importantly, there is an ALU (arithmetic/logic unit), and there are registers. The A **(accumulator)** register (Figure 14.5) in the Z80 is the main data-handling register. It is used to perform arithmetic and other operations. Additionally, there are six other 8-bit registers that, however, may be paired (B&C, D&E, H&L) when 16-bit data are to be handled. The advantage of these registers is that data can be called up from them in about 4 μs, versus 13 μs if called up from memory.

The eighth register, generally associated with A, is the F or **flag register.** The various bits within the F register—termed flags—may be used to check on the status of various devices and/or to alter the operating sequence of the memory instruction set. Some of these flags are shown in Figure 14.6. By way of illustrating their use, if the result of an arithmetic operation is 0, the Z flag is "set," and being made aware of this eventuality, the programmer should have made provision to have subsequent programming altered taking such an eventuality into account. The S (sign) flag is set when a negative number arises. If as a result of an arithmetic operation there is a carry, the C flag will

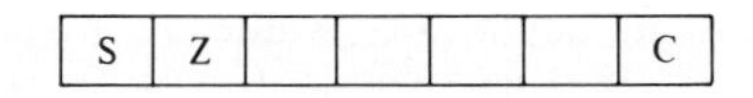

FIGURE 14.6. Some of the Z80 Flag Registers S is the sign flag (1 if an operational result is negative and 0, otherwise). Z is the zero flag (1 if an operational result is zero). C is the carry flag (1 if there is a carry from the higher-order accumulator bit).

be set. At appropriate points in the program, such flags are checked and may alter the program ordering.

There are also two *index registers* (IX and IY) in the Z80. These registers are used for table look-up operations. One might, for example, need to transform a voltage reading by a sensor into a physical parameter such as the rpm of an engine. The voltage reading could be made part of the address at which would be found the corresponding numerical rpm value.

There is a **program counter** (PC) that points to the next byte of instructions available to the computer. And there is also the **stack pointer** (SP), which, if there is an interruption in the normal sequence of the program, indicates the address to which the program should return after acting upon the interruption.

While the computer is carrying on a routine operation, an **interrupt** requiring immediate attention may arise. In the Z80 such an interrupt can take on one of two forms:

1. An $\overline{\text{INT}}$ signal, such as might come about when one is ready to type in information from the terminal keyboard. The CPU would ordinarily honor the request at the end of the present instruction cycle. However, it first checks that an internal CPU flip-flop is set. This FF is set by a software command, and thus action on such interrupts can either be delayed or entirely ignored (if desired). (If you don't have enough money in the bank, the automatic money dispenser can be programmed to ignore your request.)
2. An $\overline{\text{NMI}}$ **(nonmaskable interrupt)** signal, which *must* be serviced at the end of the current instruction cycle. Such might be the case of a "three-mile episode" being imminent or an emergency situation involving a medical monitor.

When acting upon an interrupt request, prior to the introduction of the Z80, all the information in the MPU would have been pushed into memory, the interrupt would have been processed, and then the original contents of the MPU would have been recalled and the MPU would have continued on with its interrupted routine. In the Z80 there is a duplicate set of registers, labeled A′F′, B′C′, D′E′, and H′L′. When an interrupt arises, the MPU leaves everything in place and switches over to the primed registers. Having completed the interrupt, the MPU returns to the unprimed registers and continues on.

The execution of a typical arithmetic operation is illustrated in Figure 14.7. The quantity in the A register (operand 1) is to be added (or subtracted or whatever) to another quantity from an MPU register (or from memory). The result will appear at the output of the ALU, and this new result will replace what had been in the A register. The other MPU register (or a memory) will still retain the operand 2. The former is called **destructive readout** and the latter, **nondestructive readout.**

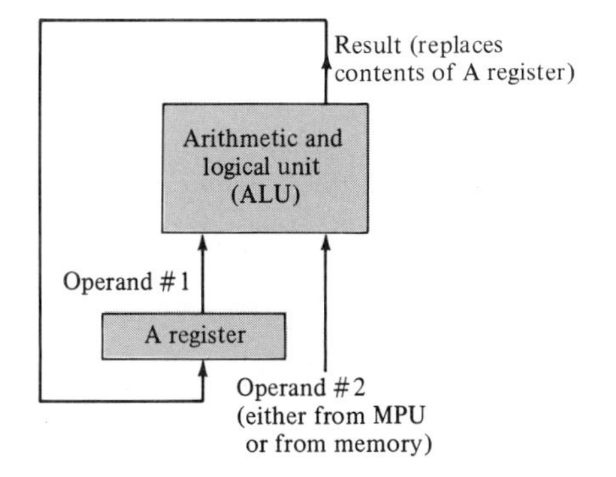

FIGURE 14.7. Signal Paths Involved in the Use of the Arithmetic/Logic Unit (ALU)

The reader may have wondered how the computer gets started. The PC register plays a role in start-up. When the computer is first turned on, we must allow time for the clock MV to start. At the RESET terminal of the Z80, there is a series *RC* circuit, as shown in Figure 14.8. It takes time for the voltage on the capacitor to build up toward its asymptotic value of +5 V. When the RESET reaches a value of about +3 V, the PC register is set to zero, and a short while later the MPU interprets this PC 0 as the first instruction that it obtains from memory location 0000 0000. Before actually responding to the operation code, the PC is incremented by 1. The reason for such incrementing is that some instructions in the Z80 may require as much as four 8-bit sequences—that is, 4 bytes—to be executed. Having obtained the first byte from 0000 0000, if that instruction is recognized as a multibyte form, the MPU will go back to the PC, get the next byte (located at 0000 0001), and increment the PC, in case another byte of instruction (or data) is needed. If not, the PC is ready with the location of the next task to be performed by the MPU. Once the instruction is completely assembled, the MPU performs the required operation.

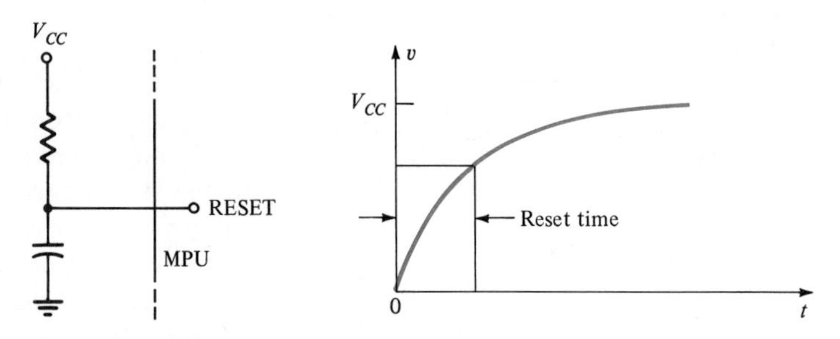

FIGURE 14.8. **Start-up of Computer Involves a Charging Capacitor** When the capacitor reaches a voltage within the range that constitutes the high state of a trigger circuit, the computer resets itself.

The program normally proceeds to follow the instructions in sequence. However, should a JUMP be called for to some location out of sequence, the PC would be loaded with that address. The JUMP may arise as a result of a particular operation—a zero result, for example, rather than a finite value. By querying the Z flag, the normal sequence may be interrupted as a result.

When a JUMP location has been entered into the PC and made, the program would not return to the takeoff point since there is no record as to the location from which the JUMP took place. If after executing the JUMP operation it is desired to come back to the original program point, one should use the expression CALL (rather than JUMP). In this case the next address in the normal sequence is inserted into the SP, and when the CALL is completed, a RET (return) will cause the SP entry to be placed into the PC, and the program will resume its normal sequence.

14.5 ■ PROGRAMMING THE COMPUTER

The same combination of binary digits can be interpreted by the computer as either of two different things: as an instruction or as a number. Thus, in the Z80,

0111 0110

may be interpreted as either a HALT or as the number 118_{10}. The computer distinguishes between interpretations by location. Thus, the first byte of a program step always represents an operation, and depending on the nature of the operation, the bytes that follow will be recognized either as additional required operations or as numbers.

In the process of devising a program for a microprocessor, one utilizes **mnemonics**—that is, short combinations of letters (and possibly numbers) whose meaning has been agreed upon by the computer designer and user. Thus,

ADD A,35

for the Z80 means "add 35_{10} to the contents of the A register." In the computer the instruction to add to the A register appears as

1100 0110 [that is, C6(H)]

while the number 35_{10} appears in the program as

0010 0011 [that is, 23(H)]

An expression such as ADD A,35 represents a statement in *assembler language*. To be used in the computer, there must be an **assembler** that converts this statement into the corresponding binary instruction (which constitutes **machine language**). Thus, the assembler represents a fixed dictionary of mnemonics, together with their machine language translations.

If one wished to evaluate something like the square root of a number, when using an assembler language, the computer would require detailed programming of each step in the process. When using a higher-level language like FORTRAN, only one instruction would be necessary (SQRT). But the computer would have another dictionary (termed a **compiler**) that would translate that one instruction into a sequence of binary instructions. Such high-level languages simplify programming, but they use so much memory space that they are too costly to be employed in conjunction with microprocessors.[1]

Let us follow through on the detailed procedure of a single program instruction. Consider placing the decimal numerical value 16 into register C. The source program would read

LD C,16

In the operational listing of Z80 codes, we find

LD r,n =	00 r 110	n

where r represents the register number (C ≡ 001) and n is the immediate address (in this case the decimal number 16). If

[1]One may, however, use a main frame computer terminal with languages such as PL/M or BASIC to obtain the requisite microcomputer programming.

LD r,n =	0000 1110	0001 0000
	C register	16 = 10(H)

were the first step in the program, it might be assigned location 0 in memory storage. The computer would recognize the operational byte (LD) as requiring a second byte, a number. The start of the next instruction in memory would be assigned location 2 (locations 0 and 1 having been used for the first step).

Such long binary series are subject to considerable error in transcription, and their conversion to hexadecimal eases that possibility. (Each byte consists of two hex digits.) Thus, we have

0000 1110 0001 0000 = 0E 10 (H)

If an assembler is available, we insert (type) LD C,16, and the assembler converts this sequence into the primary binary form, inserting the result, for example, into an EPROM's 0 and 1 locations. The next step typed in will appear at EPROM's memory location 2. If that instruction requires 3 bytes, for example, the next step in the program will be entered into location 5, etc. Once the program has been entered into the EPROM, that chip is inserted into the Z80 microcomputer and is ready for use.

A simple assembler may be constructed from a series of switches that may be used to address locations in the EPROM (switch on = 1, switch off = 0) and by another set of switches to insert the entries on a one-by-one basis. As each is entered, the voltage on the EPROM is momentarily raised from its normal level of 5 V to the 25 V level, which causes the memory location to be loaded.

EXAMPLE 14.1

a. It is desired to place the decimal number 100 into the A register of a Z80 microprocessor. Indicate the statement in assembler form and in binary form.

b. Now presume that it is the number in *memory location* 100 that is to be placed into the A register. Indicate the statement in assembler form and in binary form. (The number of a memory location is placed in parentheses.)

Solution:

a. Looking in the code listings for the Z80, we find

LD r,n =	00 r 110	n

where r is the register coding (for A it is 7_{10}—that is, 111_2) and n is the number to be inserted. In assembler language we have LD A,100.

In machine language we then have

0011 1110	0110 0100	= 3E64(H)

b. From the code listing we get

LD A,(nn) = | 0011 1010 | n | n |

That it is a memory call is reflected in the different operational code (the first byte). The 2-byte representation of 100_{10} is 0000 0000 0110 0100. A peculiarity of the Z80 is the transposition of the lower- and higher-order numerical bytes when they are assembled in machine language. Thus, we have

LD A,(100) = | 0011 1010 | 0110 0100 | 0000 0000 |

= 3A0064(H)

The first byte is the operational code, the second byte is the *least* significant bits of n, and the third byte is the *most* significant bits of n.

EXAMPLE 14.2 It is desired to place the decimal number 8400 into a Z80 register. Indicate the statement in assembler form and in binary form.

Solution: Since this decimal number requires more than 1 byte, a pair of registers must be selected. The A register, being limited to 1 byte, adds and subtracts numbers as large as 8400 by utilizing a pair of registers (BC, DE, or HL), with HL as the (assumed) destination address of the resulting operation in the present example. (Index registers IX and IY could also be used.)

The appropriate operation code listing would be

LD dd,nn = | 00dd0001 | n | n |

where dd is the register pair (0 = BC, 1 = DE, 2 = HL, 3 = SP) and nn is the immediate address value (in this case the decimal constant).

In the present case the coding becomes (presuming HL to be selected)

| 0010 0001 | 1101 0000 | 0010 0000 | = 2120D0(H) = LD HL,8400

HL = 2

where again the assembler reverses the ordering of the lower- and higher-order bytes representing the number.

EXAMPLE 14.3 The addition of 8-bit numbers in the Z80 makes use of the A register. Indicate the statement that will add a number in the A register to that of memory location 1023.

Solution: This operation must be performed in two steps:

1. The HL register is loaded with the memory location number (1023 in this case).
2. Then the addition between register A and the contents of the memory location *as indicated by HL* is performed.

Thus,

LD dd,nn =	00dd0001	n	n

ADD A,(HL) =	10000110

1023_{10} = 0000 0011 1111 1111 = 03FF (H)

Since HL = 2_{10} = 10_2, the required steps are

0010 0001 1111 1111 0000 0011 = 2103FF
1000 0110 = 86

where, again, in binary form the high and low numerical bytes are inverted.

EXAMPLE 14.4

In our discussion of automotive control systems, we will employ a tachometer that measures engine rpm in terms of a proportional voltage. That voltage will then be used as a memory address that will contain the corresponding numerical value of the rpm in the binary form necessary for computer use. Such table "lookups" make use of the indexing registers.

Assuming the initial table entry to be at memory location 2112_{10} = 0000 1000 0100 0000 = 0840(H), for an 8-bit system and rpm range from 0 to 6000 (actually we'll use 0 to 100 rps), indicate the call to be made to load the accumulator with the binary value corresponding to 3000 rpm (that is, 50 rps).

Solution: Consulting the operational code listings, we find

LD IX,nn =	1101 1101	0010 0001	n	n

LD r,(IX + d) =	1101 1101	01 r 110	d

where d represents the displacement from the beginning of the table. A 0 V reading will correspond to 0 rps; a +5 V reading, to 100 rps; and a +2.5 V reading, to 50 rps. Register IX is to be loaded with the number 2112_{10}. Since r = 7_{10} (111_2) for the A register, d = 50_{10} = 0011 0010_2 = 32(H), and we have

LD IX,2112 = 1101 1101 0010 0001 0100 0000 0000 1000
= DD210840

LD r,(IX + d) = 1101 1101 0111 1110 0011 0010 = DD7E32

EXAMPLE 14.5[2]

Indicate by means of a few program steps the distinction between a JUMP procedure and a CALL procedure.

[2]The text material following the semicolon (;) will appear in a printout of the assembler program but will not be coded into the microcomputer. It proves useful in refreshing one's thinking concerning the design of the program.

Solution:

```
0820  ADD  A,-25     ;-25 is added to the contents of A
0822  JP   M,1510    ;if the result is negative (i.e., minus) go
                      to instruction 1510
0825  INC  A         ;increment A
       .
       .
       .
1510  LD   DE,800H   ;DE is loaded with hex number 800
1513  LD   A,(DE)    ;A is loaded with the contents of the
                      memory location specified by DE (that
                      is, 800H)
1514  ADD  A,B       ;contents of B added to A
       .
       .
       .
```

Once the JUMP has been made to 1510, there is no record from where it came, and the program will continue on from 1514. If we subsequently want to come back to step 0820, we could do this by giving that step a "name." For example, we may have

```
LOOP  0820  ADD  A,-25
      0822  JP   M,1510
             .
             .
             .
      1510  LD   DE,800H
      1513  LD   A,(DE)
             .
             .
             .
      1580  JP   LOOP
```

Anytime a particular step is given a name (for example, LOOP in the preceding case), the computer will equate the name with the location (0820 above). The danger here is that if on subsequent passes the program always reaches step 1580 it will get caught in an endless loop. Somehow in subsequent passes the program, having accomplished its task, must be made to bypass step 1580 if it is to continue or to end.

As an example of a call procedure, consider the following:

```
0820  ADD   A,-25
0822  CALL  M,1510
0825  INC   A
```

.
.
.

```
1510  LD   DE,800H
1513  LD   A,(DE)
1514  RET
```

In this case if step 0820 leads to a negative result, there will again be a transfer to 1510; otherwise, it will continue on to step 0825. If a transfer to 1510 has taken place, the next instruction (location 0825) will be stored in the stack pointer (SP), the subroutine (steps 1510 and 1513) will be executed, and at step 1514 the program will return to continue step 0825 on.

Will the last series of steps not also lead to an endless loop each time the program reaches 1514? Subroutines (such as represented by steps 1510 to 1514 in the above) are generally gathered together at locations that are outside the program's normal sequence. Reference can be made to them whenever desired, and an infinite loop should not arise.

14.6 ■ DATA TRANSMISSION

14.6.1 Bus Transceivers

We have noted that data can be transferred in two directions along the same lines. How is this done, and how do the data know between what two units they are to go? Refer to Figure 14.9. Gates 1 and 2 will be operative only if their respective enable inputs are made to be 1. If the signal is to be from A_0 to B_0 (RECEIVE), gate 1 is enabled, and if the signal is to be from B_0 to A_0 (TRANSMIT), gate 2 is enabled. The simultaneous disabling of both gates isolates A_0 from B_0.

To disable both gates, each of their enabling pulses has to be a 0. This condition will arise when a 1 is applied to the CHIP DISABLE (CD) gate. It can then be verified that irrespective of whether the ($T/\overline{R}$) is a 1 or a 0 gates 1 and 2 remain disabled.

When a 0 is applied to ($\overline{CD}$), that same 0 appears as one of the inputs to ($T/\overline{R}$). If the other input to ($T/\overline{R}$) is also a 0, this leads to a 0 on gate 2 (disabling it), while at the same time the resulting 0–0 input to (CD) leads to a 1 output, enabling gate 1 (that is, we receive from A_0 to B_0). With a 0 applied to (CD), that same 0 at the ($T/\overline{R}$) input when combined with a 1 input to ($T/\overline{R}$) leads to a 1 output, which activates gate 2. This 1 combined with the 0 input at (CD) leads to a 0 output, which disables gate 1, and we get transmission from B_0 to A_0.

A single 20-pin IC package (designated as 74LS245) contains eight such bidirectional transceivers that are simultaneously controlled by a single (CD) and a single ($T/\overline{R}$) gate. Interposing the 74LS245 between, say, a memory device and a common set of bus lines, the memory can be attached to the lines and signals sent in the required direction as dictated by circumstances. Or, in

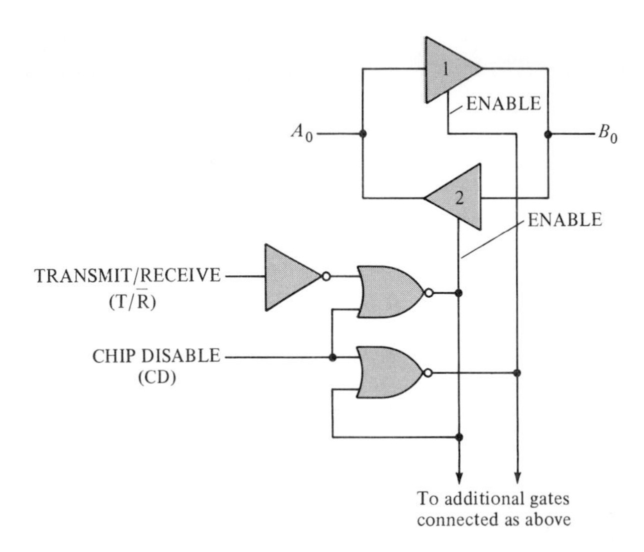

FIGURE 14.9. **Bidirectional Gate** Depending on whether gate 1 or gate 2 is activated, signals pass either from A_0 to B_0 or from B_0 to A_0, respectively. This figure shows how the same bus may be used for two-way communication.

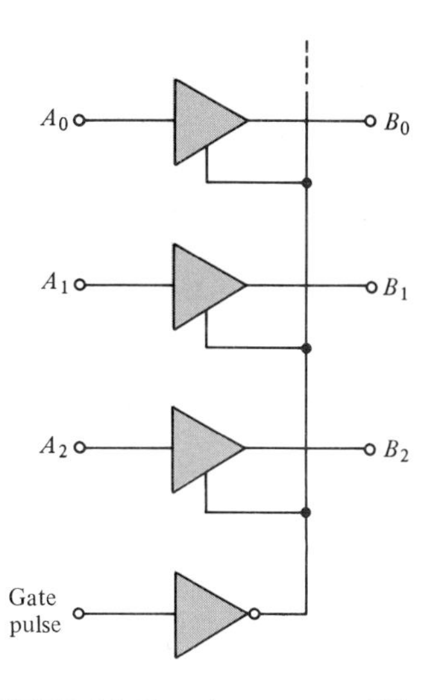

FIGURE 14.10. **Arrangement Showing How One-Way Data Bus May Be Made to Be Either Active or Inactive**

like fashion, input or output devices can be connected to the lines as desired if a 74LS245 is associated with each of them.

With regard to the *address* lines associated with an MPU, it is generally only required that the MPU *send* addresses (to memories, etc.) and not receive them. But if we want to send addresses between, say, an input device and memory directly [called direct memory access (DMA)], we may want to detach the MPU from the lines. This task can be accomplished using a comparable IC that has gates with only one direction of transmission (Figure 14.10). A typical unit of this sort is the 74LS244. If the address lines are 16-bit form, two such 74LS244s may be employed, activated by a single enabling unit.

14.6.2 Bus Standards

In 1975, following an article that appeared in *Popular Electronics* magazine, an Albuquerque manufacturer named MITS began to market a home computer kit called the Altair 8800. The required stack of printed circuit (PC) boards were screwed together using spacers and interconnected by individually soldered wires. But soon after came a modification consisting of a "motherboard" containing a number of 100-pin card-edge connectors. The leads from each PC board were brought out to a 100-pin edge plug and inserted into one of the 100-pin connectors on the motherboard. The functional purpose of each pin was standardized. Known originally as the Altair **bus**, after some modification it became the S100 bus—that is, standard, 100 pins.

The S100 includes 4 power-supply lines that deliver unregulated dc to the cards each of which contains its own voltage regulator to furnish voltages

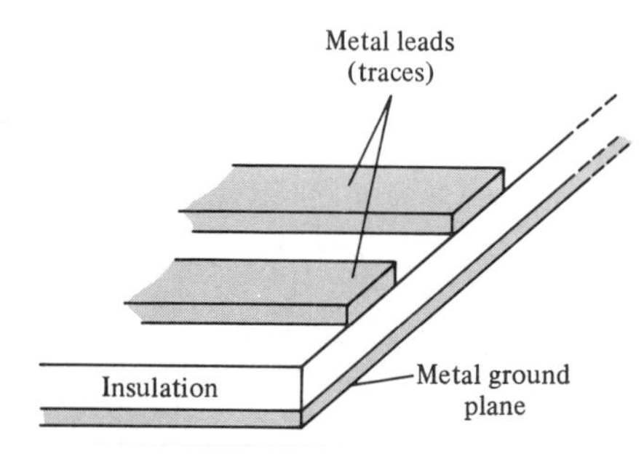

FIGURE 14.11. Physical Aspect of a Signal Bus

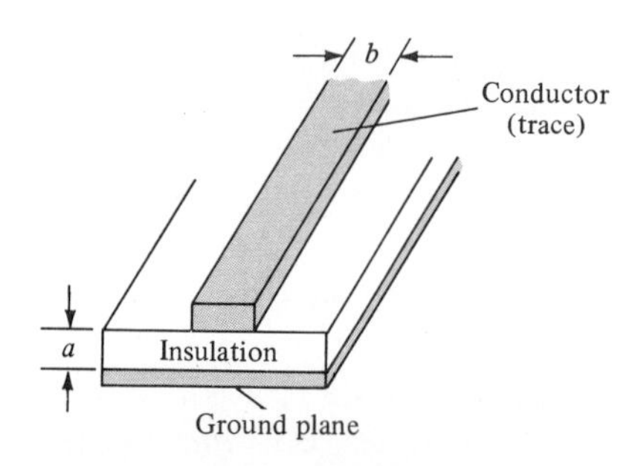

FIGURE 14.12. Metallic Trace Considered as a Transmission Line

such as +5 and ±12 V. There are also 5 ground lines, 24 address lines, two sets of 8-bit data lines that may be used as a 16-bit bidirectional line, various status lines, clock pulse lines, etc. (Incidentally, the separate lines are termed *traces*.)

Somebody had given MITS a good deal on 100-pin connectors, which is why they appear in the S100 bus, but they didn't know what to do with all those pins. Unfortunately, other manufacturers found lots of things to do with them, and chaos rather than standardization was the result.

Eventually, the IEEE (Institute of Electrical and Electronic Engineers) intervened and established the IEEE488 General-Purpose Interface Bus. Today the S100 has been extended into a 16-bit bus and appears as a standard, IEEE696. To handle expansion to 32-bit systems, the IEEE896 has been developed. There are also other standard buses in use, proposed by various manufacturers, such as the HPIB of Hewlett-Packard and Intel's Multibus-II.

One should not be misled into thinking that these bus standards merely involve assignments of particular traces to specific common functions. In particular, one must treat signal traces as transmission lines. With L_0 and C_0 being the inductance and capacitance per unit length, we have

$$Z_0 = \sqrt{\frac{L_0}{C_0}} \tag{14.1}$$

while the propagation delay per unit length, T_{pd}, is

$$T_{pd} = \sqrt{L_0 C_0} \tag{14.2}$$

The individual traces in conjunction with the ground plane (Figure 14.11) can ideally be considered as transmission lines that take the form shown in Figure 14.12.

The individual impedances are

$$Z_0 = \frac{377}{\sqrt{\epsilon_R}}\left(\frac{a}{b}\right) \tag{14.3}$$

where ϵ_R is the relative permittivity of the insulation material.

Typical values for a *motherboard* (as the assemblage of traces is called) are $Z_0 = 100\ \Omega$ and $T_{pd} = 1.7$ ns/ft. But these values are for *unloaded* boards. When the plug-in *daughterboards* are inserted, the loaded values become

$$Z_L = \frac{Z_0}{\sqrt{1 + (C_L/C)}} \tag{14.4}$$

$$T_{pdL} = T_{pd0}\sqrt{1 + \left(\frac{C_L}{C}\right)} \tag{14.5}$$

An unloaded board might have $C = 20$ pF/ft. The loading capacitance of the plug-in card can be 12–20 pF for TTL devices. If there are 15 cards/ft, Z_L can be 20 Ω and T_{pdL} can be 8.25 ns/ft.

The reduced impedance means large signal current demand from the driver. With a 3 V pulse and 20 Ω impedance, the current is 150 mA. The increased delay lowers the rate at which signal pulses can be transmitted (that is, a lower *throughput*).

At the point where the board plugs in, there is a change in the value of C, and hence the transmission line cannot be considered to be a uniform one. Where such discontinuities arise, there are reflections. Also, the terminating resistances cannot be accurately matched. As a result pulses on such buses lead to all kinds of "hash." With synchronous lines involving clocked pulses, one simply waits until the transients have settled out before responding to the signals.

It is not the objective of this section to provide a detailed analysis of bus problems but merely to provide an appreciation for their design and to emphasize that they do not represent as simple a matter as it might appear at first glance.

14.6.3 Keyboards

The introduction of the program and data into the computer may be accomplished via a keyboard (typewriter). Let us confine our attention to the introduction of decimal integers; letters may be introduced by a simple expansion of the system employing numbers. Dealing with decimals (0–9), we will use the arrangement shown in Figure 14.13.

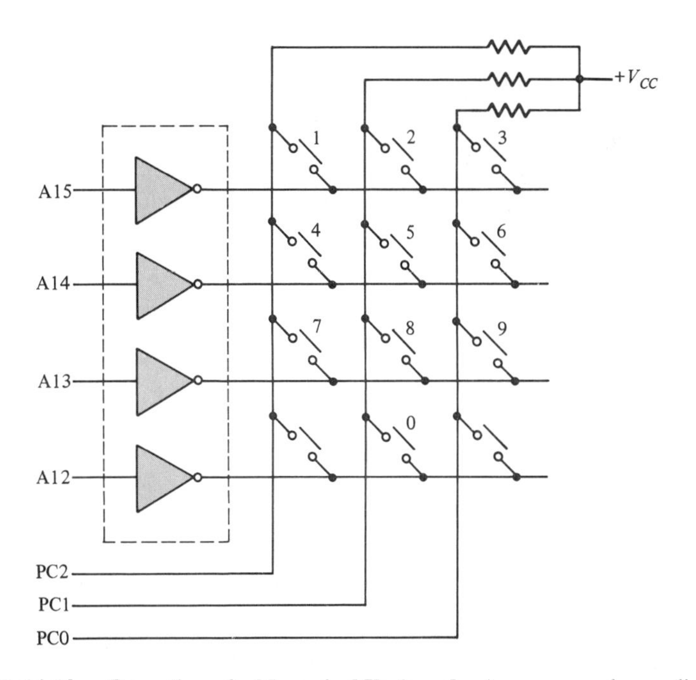

FIGURE 14.13. Operation of a Numerical Keyboard A scanner pulse applied successively to A12, A13, A14, and A15 will determine the coding outputs from PC0, PC1, and PC2, depending on the key depressed. The combined 7 bits represent a unique address in memory where the numerical value of the key depressed may be found.

Refer to Figure 14.13. With no keys depressed, all three outputs (PC2, PC1, PC0) will be 1 because each is connected to $+V_{CC}$. If any key is depressed *and* the output of the connecting row is zero at that time, this depression makes the column intersecting that row zero also. The objective is to make only one row zero at any given time. Such a condition is assured by applying 1s successively to A15, then A14, then A13, and finally A12. When passed through the inverters, the applied 1 causes only one horizontal line to be zero at any given time.

EXAMPLE 14.6 Suppose key 8 on a numerical keyboard is depressed. Indicate the nature of the PC outputs as the unit pulse moves through the A inputs. (Refer to Figure 14.13.)

Solution: The contents of the various lines will be as follows:

	PC2	PC1	PC0	A15	A14	A13	A12
1 at A15	1	1	1	1	0	0	0
1 at A14	1	1	1	0	1	0	0
1 at A13	1	0	1	0	0	1	0
1 at A12	1	1	1	0	0	0	1

When a 0 appears among the PC2, PC1, PC0 outputs, it indicates that a key has been depressed in the row being interrogated at that time, and the PC column that is 0 indicates the column concerned, thus uniquely identifying the number. The particular combination of PC and A bits can be used to specify a memory location that yields the corresponding numerical value.

The scanning process, of course, is much faster than the successive key depressions by the operator. How does the processor determine whether it is the same key depression or two successive depressions of the same key? On successive sweeps, if it keeps coming up with the same number, it takes no further action. Only if a "no key depression" intervenes will it interpret the same value as constituting a subsequent key depression of the same value.

An expansion of the process depicted in Figure 14.13 to more columns can be used to create an alphabetic series of keys in addition to the numerics we have been considering. The binary coding for alphanumerics has been standardized into the ASCII (American Standard Code), pronounced "ass-key."

The ASCII code utilizes 7 bits, making it suitable for use with 8-bit processors. With the most significant bit (MSB) appearing at the left,[3] the first three MSB digits are arranged to designate columns:

[3]Actually, the most significant bit, b_8, is either set equal to zero or used as an error check, termed a *parity bit*. Parity can be either even or odd. With odd parity b_8 is adjusted so that the number of 1 bits in a word is an odd number. If in transmission a bit is subsequently missed, a parity checker will indicate an error.

$b_7 \rightarrow$	0	0	0	0	1	1	1	1
$b_6 \rightarrow$	0	0	1	1	0	0	1	1
$b_5 \rightarrow$	0	1	0	1	0	1	0	1
$b_4b_3b_2b_1$								
0 0 0 0	·	·	·	0	·	P	·	p
0 0 0 1	·	·	·	1	A	Q	a	q
0 0 1 0	STX	·	·	2	B	R	b	r
·	·	·	·	·	·	·	·	·
·	·	·	·	·	·	·	·	·
·	·	·	·	·	·	·	·	·

Each column is termed a *stick*. The remaining bits (the least significant ones) are used to designate rows 0–15. Within stick 011, from row 0 to row 9, one finds the decimal integers. Sticks 100 and 101 primarily incorporate capital letters. Sticks 110 and 111 have the corresponding lowercase letters. Note that one can switch between upper- and lowercase letters by simply altering bit 6. The remaining ASC characters represent various punctuation marks and nonprinting codes. As an example, 0000010 (STX) is used to indicate "Start of Text." Others are CR (carriage return), BS (no! not that; backspace), and so forth. A full listing of the code may be found in Horowitz and Hill, *The Art of Electronics*, p. 476, together with additional elaboration (see the Suggested Readings at the end of this chapter).

14.6.4 Programmable Peripheral Interface

Rather than a simple input buffer (Figure 14.3) or an input latch (Figure 14.4) between the microcomputer and the outside world, one often employs a **programmable peripheral interface** (PPI), which performs a multitude of functions. As one example, consider its use in driving an LED display (Figure 14.14).

Each of the LED indicators used in conjunction with the frequency counter discussed in Section 11.8.4 was continually energized (except when, momentarily, new values were being introduced). This procedure leads to considerable power expenditure, which may be reduced by alternative circuitry. Because of the eye's retentivity, we may rapidly switch from one digit to the next, every $\frac{1}{100}$ s, for example, thus displaying each digit for $\frac{1}{25}$ s. To the eye, however, this display appears to be continuous.

The four LED cathodes are driven by a 7437, a high-current inverter. The PPI by itself provides insufficient current to do the job directly. Only one cathode is grounded at any given time, depending on which of the PPI lines are low (PA7, PA6, PA5, or PA4). With a cathode grounded and a combination of anode lines (a–g) excited, one digit will be displayed. The PA3, PA2, PA1, PA0 output of the PPI is in BCD and determines the digit displayed.

The PPI is an Intel development. A comparable functional unit by Motorola is termed the parallel interface adapter (PIA). Other manufacturers issue comparable devices.

The transmission of information using a parallel format is limited to

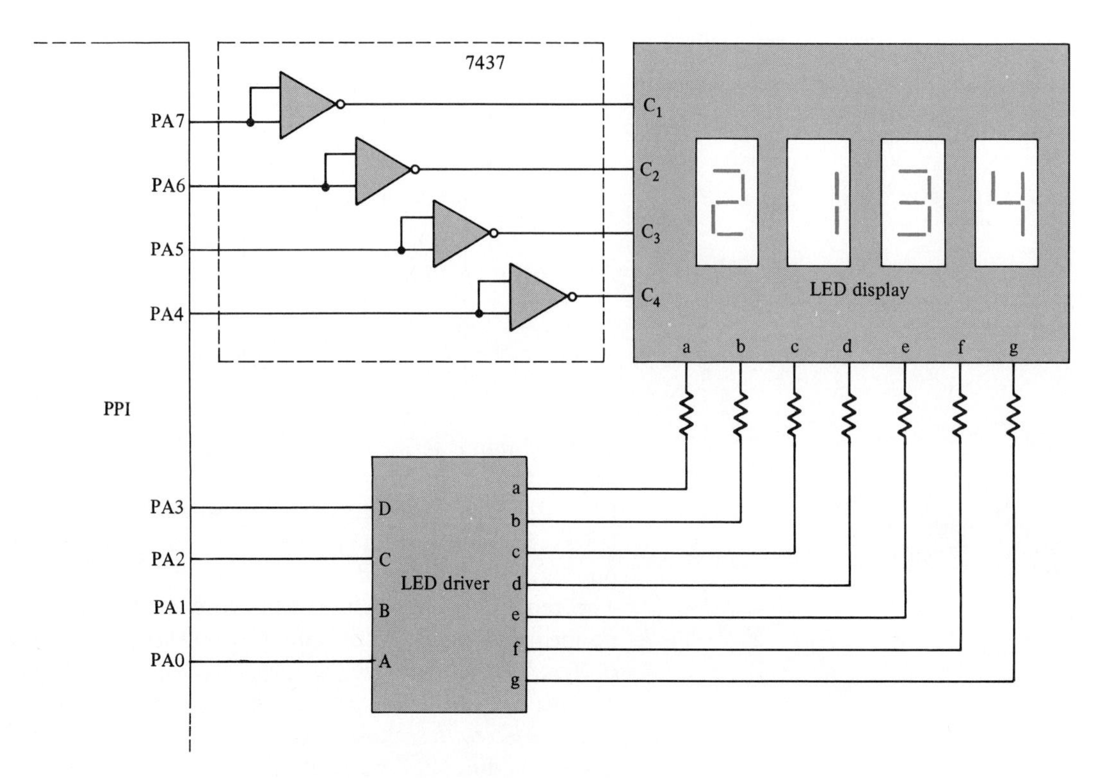

FIGURE 14.14. **The Use of a PPI to Provide a Scanning Display** Outputs PA7···PA4 activate one digit position at a time, but the rapidity of the scanning operation creates the impression of a continuous display.

distances of, perhaps, a few feet. Beyond that distance the wiring costs become prohibitive. For longer distances a serial format is used, with the bits succeeding one another along a line, much in the manner of telegraph signaling. In such cases there must be some method provided to distinguish the beginning and end of a binary word.

Serial transmission appears deceptively simple. Complications arise because the standard employed (termed RS-232C) was promulgated before the advent of TTL circuits. As a result the logic levels are not 5 and 0 V but rather somewhere between +5 to +15 and −5 to −15 V, respectively. Place such a 25–30 V swing on a TTL gate input, and you've got problems. The subsequent advent of more involved standards further complicates the situation. For a detailed discussion, see the S. Leibson reference in the Suggested Readings at the conclusion of this chapter.

14.7 ■ POSTSCRIPT

The Z80 microprocessor, considered in this chapter in some detail, made its first appearance in 1976. About 2 years later a number of manufacturers introduced a 16-bit microprocessor, including Zilog, which came out with the

Z8000; 32-bit microprocessors are now also becoming available. As already mentioned, the main advantage of such units is not necessarily the larger numbers they can handle (after all, for a century or so, engineering got along very nicely with three significant figure calculations). Their advantage lies in the much greater memory storage capacity that can be addressed.

The first microprocessor appeared as recently as 1971. It was the 4004, made by Intel. Intel had been approached in 1968 by a Japanese calculator manufacturer (now defunct) to produce a set of 12 chips that would print, display, constitute ROMs, etc. Each chip was to contain 600–1000 transistors. This proposed design did not prove cost-effective, but Intel went on to settle for a three-chip design—a CPU, a ROM, and a RAM. (A shift register was later added.) In final form the CPU was to have about 2300 transistors. To a large degree the development was due to Federico Faggin, who later became the founder of Zilog. (Zilog in turn was later bought out by Exxon.) A member of the Japanese concern who also worked on the project was Masatoshi Shima, who later developed the Z80 and Z8000 for Zilog. The four-chip Intel system was dubbed the MCS-4, consisting of the 4001 (ROM), 4002 (RAM), 4003 (output register), and 4004 (CPU), in an arrangement closely approximating what we have considered in Section 14.2.

The introduction of the 8-bit 8008 in 1972 by Intel was followed two years later by the much improved 8080 (also 8-bit). The Zilog Z80 then appeared, followed by Motorola's 6800 (in 1974), the first to use a single +5 V power supply. (Previously, multiple voltage sources were necessary.) The 6800 represented a significant advance because the elimination of multiple power supplies lowers the product cost.

The 8086 by Intel, introduced in 1978, represented the first real move toward a fast 16-bit processor. By a new *N*-channel MOS process, it incorporated 29,000 transistors into a single unit. This accomplishment was soon to be exceeded by Motorola's 68000, with 68,000 transistors. Rather than the usual 40-pin chip, it utilizes 64 pins.

A characteristic of the Intel microprocessors has been their **upward compatibility**—that is, the instruction set developed for one chip (for example, the 8008) is incorporated into subsequent developments (for example, the 8080). (The Z80 that we looked at in some detail also used programs compatible with the 8080.)

The *N*-channel MOS came into general use with the 8080, 6800, and Z80. While bipolars provide speed and the CMOSs, low power, the high density of the *N*MOS technique still provides sufficient speed. It was the 8080 that was used in the MITS Altair home computer (see Section 14.6.1).

SUMMARY

14.2 The Microprocessor

- A microprocessor performs arithmetic and logical manipulation. When combined with memories, power supplies, and input/output devices, it constitutes a microcomputer.
- A latch is used to mediate between the high-speed computations of a computer and the much slower action of input and output devices.

14.3 Memory

- The varieties of memory ICs differ from each other by the ease of accessibility and alterability of their data.

14.5 Programming the Computer

- Mnemonics are short combinations of characters whose meaning has been agreed upon by the computer designer and user.
- An assembler converts the program mnemonics into a machine language operation—that is, a binary instruction.
- High-level languages, like FORTRAN, utilize a complier to convert a single instruction into a series of binary instructions.

14.6 Data Transmission

- Bus lines are used for data transmission between the various units of a computer.
- A bus transceiver determines the direction in which signals will travel along a set of data lines—that is, a bus—and the units between which such transfer will take place.

KEY ACRONYMS

ALU: arithmetic-logic unit
ASCII: American standard code
CMOS: complementary metal-oxide semiconductor
CPU: central processing unit
DIP: dual-in-line package
EAROM: electrically alterable memory
ECL: emitter-coupled logic
EPROM: erasable programmable read-only memory
LCD: liquid crystal display
LED: light-emitting diode
LSI: large-scale integration
MOS: metal-oxide semiconductor
MPU: microprocessor unit
PC: program counter
PC board: printed circuit board
PIA: parallel interface adapter
PPI: programmable peripheral interface
PROM: programmable read-only memory
RAM: random access memory (also known as read-and-write memory)
ROM: read-only memory
SP: stack pointer
TTL: transistor-transistor logic (also T^2L)
VLSI: very large-scale integration

KEY TERMS

accumulator: the basic work register of a computer
address: a numerical identification indicating the location of a particular piece of data
assembler: a device that converts operational mnemonics and numerics into machine language
bus: the wires used to carry signals, power, and so forth, between various computer units (*Bus* implies communication between more than two units; if only two units are involved, one generally refers to the connections as wires, traces, or signals.)
compiler: a device that converts a single command into the sequence of machine language commands necessary to carry out a specific task
cycle time: the time interval between a request for data and their availability
dedicated computer: a computer designed to perform a specific task, such as controlling the operation of an automotive engine; contrasts with a general-purpose computer that may perform a variety of tasks
destructive readout: the destruction of the contents of either a register or a memory location as a consequence of a readout operation
erasable programmable read-only memory (EPROM): memories whose contents can originate with the programmer and can subsequently be changed but not as part of a computer program
flag register: a register used to record the occurrence of certain computational results such as a zero, a negative, or a positive result; the specific result may then be used to alter the sequence of the program execution
hardware, computer: the electronic devices and circuits comprising a computer
interrupt: a signal indicating a request to stop a program routine and act instead upon a prearranged program whose procedure depends on the nature of the externally generated request
keyboard: in essence a typewriter whose output pulses

serve to identify to the computer the particular key that has been depressed and the character to be introduced into the computer

latch: an electronic device used to handle data between two segments of a computer each operating at different speeds

machine language: the binary representation of alphabetic and numeric characters recognized by a computer as constituting either a program's instruction or data

maskable interrupt: an interrupt signal whose request may be either delayed or ignored depending on prerequisite conditions having been satisfied

memory: stored data and instructions necessary for computer operation

microcomputer: a combination of a microprocessor with various memories, power supplies, and input and output devices

microprocessor: a miniature electronic device capable of performing arithmetic and logic operations

mnemonic: a short combination of letters (and possibly numbers) whose operational meaning has been agreed upon by the computer designer and programmer alike

motherboard: a set of sockets whose corresponding terminals are connected in parallel, allowing various hardware units to be serviced with power, data, etc., irrespective of the specific socket into which they are plugged

nondestructive readout: retention of the information in a register or memory location following a readout determination

nonmaskable interrupt: an interrupt signal that *must* be acted upon

nonvolatile storage: data that will be retained in a computer even if the power is removed

program counter (PC): a register that indicates the next instruction to be executed

programmable peripheral interface (PPI): a connecting link between a computer and various input and output devices

programmable read-only memory (PROM): memories whose contents can originate with the programmer but cannot be subsequently changed

random access memory (RAM): memory locations whose contents may be determined (read) or easily replaced (write operation); also termed read-and-write memory

read-only memory (ROM): memories whose contents cannot be changed

readout: a determination of the nature of the contents of a register or a memory location

register: a memory location within the CPU; generally, it provides a more rapid cycle time than would an offboard memory

software, computer: the detailed instructions for the arithmetic and/or logic operations to be executed by a computer

stack pointer (SP): indicates the address to which a program should return after acting upon an interrupt request

stoichiometric ratio: air-to-fuel mixture in the ratio of 1:14.7; results in complete combustion

transceiver: a device that allows a set of data lines—that is, a bus—to be used to send data in either direction

upward compatibility: the ability of an advanced computer to accommodate the program mnemonics developed for use with an earlier design

volatile storage: data that are lost when power is removed from the computer

SUGGESTED READINGS

Easton, John, "Multibus II—An Architecture for 32-Bit Systems," ELECTRONIC ENGINEERING, vol. 56, pp. 47–51, March 1984. *An advanced signal bus.*

Waite, Mitchell, and Michael Pardee, MICROCOMPUTER PRIMER, 2nd ed. Indianapolis: Howard W. Sams, Publ., 1980. *An introduction to the circuitry and hardware of microcomputers.*

Horowitz, P., and W. Hill, THE ART OF ELECTRONICS, Chap. 11. New York: Cambridge University Press, 1980. *A detailed discussion of the 8085 microprocessor and peripheral equipment.*

Noyce, Robt. N., and Marcian E. Hoff, Jr., "A History of Microprocessor Development at Intel," IEEE MICRO, vol. 1, pp. 8–21, Feb. 1981. *A history of microprocessor development.*

Barden, William, Jr., Z-80 MICROCOMPUTER DESIGN PROJECTS. Indianapolis: Howard W. Sams, Publ., 1980. *Constructing your own computer, including detailed programming instructions, fabricating a simple EPROM programmer, and even making your own numeric keyboard.*

Gustavson, David B., "Computer Buses—A Tutorial,"

IEEE MICRO, vol. 4, pp. 7–22, Aug. 1984. *A detailed treatment of bus design.*

Ditlea, Steve (ed.), DIGITAL DELI. New York: Workman Publ. Co., 1984. *A "fun" book that, among many other interesting aspects concerning computer development, will tell you how the MITS* Altair *got its name and about the "rise and fall of Altair," how Apple Computer came into being, about Atari, and about Silicon Valley (California), Silicon Alley (Boston), and Silicon Prairie (Texas).*

Leibson, Steve, THE HANDBOOK OF MICROCOMPUTER INTERFACING. Blue Ridge Summit, Penn.: Tab Books, 1983. *Highly recommended for the serious user of microprocessors.*

A CLOSER LOOK: Microprocessors

Resume of Major Microprocessor Developments

Type	Year	Mfg.	Pins	Data Word	Instruction Set	Remarks
4004	1971	Intel	16	4-bit	45	First microprocessor
4040		Intel	24	4-bit	60 (45 compatible with 4004)	First introduction of interrupt
TMS 1000	1974	Texas Instruments		4-bit		Very rudimentary as microcomputer but used widely to control ovens, tuners, etc.
8008	1972	Intel	18	8-bit	48	Allowed program compiling on a main frame computer terminal using high-level PL/M language
8080	1973	Intel	40	8-bit	78	Program compiled using either PL/M or BASIC language; used in MITS Altair 8800 computer
6800	1974	Motorola	40	8-bit	72	First to use single (+5 V) power supply; first to supply complete set of support ICs
6502		MOS Technology	40	8-bit	56	Formed basis of Commodore 6502 and Apple II computers
PACE	1976	National Semiconductor	40	16-bit	45	Primarily designed for manufacturing process control—hence the name, processing and control element; output is not compatible with TTL logic

8048	1976	Intel	40	8-bit	96	First complete computer on a single chip
8748		Intel	40	8-bit	96	Identical to 8048 but with an on-board EPROM
8085	1976	Intel	40	8-bit	80	Discussed in detail in *The Art of Electronics* by Horowitz and Hill
Z80	1976	Zilog	40	8-bit	158 (compatible with 8080 instructions)	Used in Radio Shack TRS80 computer
8088		Intel	40	16-bit		8-bit bus version of 8086; used in IBM PC and PC Jr. computers
8086	1978	Intel	40	16-bit		New *N*-channel process allowed 29,000 transistors in the unit
68000	1980	Motorola	64	16-bit		68,000 transistors per unit
Z8001		Zilog		16-bit		

Abridged Listing of Operational Codes for the Z80

Mnemonic	Format	Description
ADD A,(HL)	[1000 0110]	A←A + (HL): contents of HL added to contents of A and placed in A
ADD A,n	[1100 0110] [n]	A←A + n: number n added to contents of A and placed in A
AND n	[1110 0110] [n]	A←A·n: logical AND performed between A and number n (in binary) and result placed in A
CALL cc,nn	[11 cc 100] [n] [n]	Depending on condition code (cc), either start subroutine at location nn or continue on to the next step in the program
CCF	[0011 1111]	The carry flag will be complemented—that is, if 1, then 0; if 0, then 1
CP n	[1111 1110] [n]	Compare A to n—that is, A–n: S flag is set if negative, reset otherwise; Z flag is set if zero, reset if 1; C flag for no borrow, reset otherwise, etc.
CPL	[0010 1111]	Complement A (1's complement)
HALT	[0111 0110]	Halt

Abridged Listing of Operational Codes for the Z80 *(cont.)*

Mnemonic	Format	Description
INC r	00 r 100	Increment register r by 1
INC dd	00dd 0011	Increment register pair dd by 1
JP cc,nn	11cc 010 | n | n	Depending on condition code (cc), either jump to location nn or proceed on to the next step
LD A,(nn)	0011 1010 | n | n	Load A with contents of location nn
LD dd,nn	00dd 0001 | n | n	Load register pair dd with nn
LD IX,nn	1101 1101 | 0010 0001 | n | n	Load register IX with nn
LD r,(IX + d)	1101 1101 | 01 r 110 | d	Load register r with contents of location specified by IX plus indexing displacement d = +127 to −128
LD r,n	00 r 110 | n	Load register r with n
RET	1100 1001	Return from subroutine
SCF	0011 0111	Set carry flag to 1
XOR r	1010 1 r	$A \leftarrow A \oplus r$: exclusive OR between A and contents of register r to be placed in A

r register designations: B = 0, C = 1, D = 2, E = 3, H = 4, L = 5, A = 7.
dd register pair designations: BC = 0, DE = 1, HL = 2, SP = 3.
cc condition codes: NZ = 0, Z = 1, NC = 2, C = 3, P = 6, M = 7.

EXERCISES

1. What is the meaning of stoichiometric ratio as applied to fuel combustion?

2. What is the distinction between a microprocessor and a microcomputer?

3. What is a dedicated computer?

4. What is the meaning of a bar over a terminal designation, such as $\overline{CS}$?

5. Distinguish among ROM, RAM, PROM, and EPROM.

6. What is the difference between a register and a storage location?

7. What is the function of a flag register?

8. What is the meaning and purpose of the PC and SP registers?

9. Distinguish between destructive and nondestructive readout.

10. What is the function of an assembler? A compiler?

11. What is the function of a bus line?

12. Distinguish between volatile and nonvolatile memory.

13. What does the cycle time represent?

14. What is an ALU?

15. Distinguish between maskable and nonmaskable interrupts.

16. Identify the following microprocessors: 4004, 8008, 8080, Z80, 68000.

17. Name three major microprocessor manufacturers.

PROBLEMS

14.1

See Figure 14.15. The first EPROM employs 10 address inputs, A0–A9, allowing 1024 addresses to be called. But the second EPROM also uses the same lines, while the RAM uses lines A0–A6 to address its 128 memory locations. If the CPU addressed location 100, all three memories would believe it was their 101st location that was being interrogated. (Remember that there is a 0 location.) How can one indicate a distinction if 16 bits of address line are available? (*Suggestion*: Write out in binary form the limits for each address and note the distinction between the call for location 100 in each.)

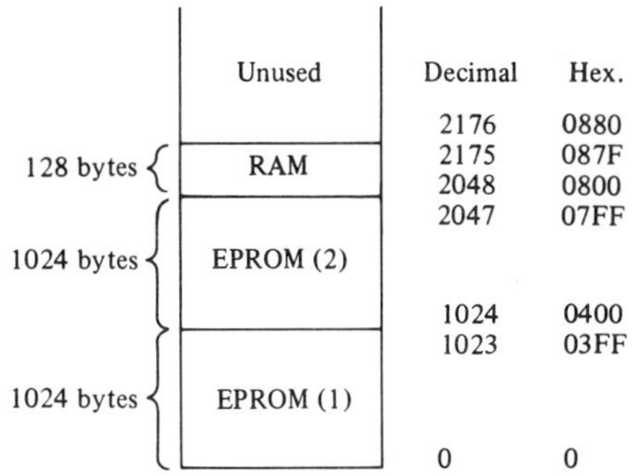

Size	Block	Decimal	Hex.
	Unused		
		2176	0880
128 bytes	RAM	2175	087F
		2048	0800
1024 bytes	EPROM (2)	2047	07FF
		1024	0400
1024 bytes	EPROM (1)	1023	03FF
		0	0

FIGURE 14.15

14.2

It is desired to place the decimal number 8400 into one of the registers of the Z80. Indicate the operational statement entered into the assembler and the binary form that will result. (Since the decimal number requires more than 1 byte, a pair of registers must be used. Select H and L. The operational code for this pair is 2.)

14.3

Starting at memory location 810(H) and running through 81E(H), there are 15 positive numbers (each different), with a maximum value of +127. Follow through the following program and indicate its objective. Anytime a particular step is given a "name" (for example, LOOP and NEXT in what follows), the computer will equate the name with the location number. One may thereafter refer to the location by name rather than by number.

```
       LD    C,15
       LD    HL,810H
       LD    B,127
LOOP   LD    A,(HL)
       CP    B
       JP    P,NEXT
       LD    B,A
NEXT   INC   HL
       DEC   C
       JP    NZ,LOOP
       LD    A,B
```

14.4

In addition to arithmetic operations, the ALU may also be used to perform logical manipulation. During the course of a program, it is frequently desired to clear the accumulator—that is, set the register equal to zero. This may easily be done by using

```
XOR r
```

which represents an EXCLUSIVE OR between the contents of the accumulator and the contents of register r.

a. If r is designated as A, this operation amounts to an EXCLUSIVE OR of the contents of A with itself. Show why this procedure resets A to zero.

b. Show by a specific example that one may also use SUB A to accomplish the same result.

c. What does

```
CPL A
INC A
```

accomplish?

14.5

What do the steps

```
AND  04
JP   NZ,LED05
```

accomplish? LED05 is the "name" of some location in the program.

14.6

Refer to Figure 14.14. If it is desired to display the hundreds digit with the number 3, what should be the binary signal applied to lines PA0–PA7?

14.7

Indicate what the following program segment accomplishes:

```
       LD    HL,2000H
       LD    DE,3000H
       LD    BC,-1000H
LOOP   LD    A,(HL)
```

```
LD      (DE),A
INC     HL
INC     DE
INC     BC
JP      NZ,LOOP
HALT
```

14.8

To clear a carry flag, one could use the sequence

```
SCF
CCF
```

Why is the CCF by itself insufficient?

15 PHASE DETECTORS AND PHASE-LOCKED LOOPS

OVERVIEW

The first treatment of automotive ignition systems arose in Chapter 3 as an example of the interaction among resistance, inductance, and capacitance. In Chapter 7 the automotive ignition system was used as a vehicle to introduce transistor operation. In this chapter microcomputer control of automotive ignition and fuel injection is considered. It utilizes many of the digital concepts introduced in Chapter 11 and the microcomputer concepts of Chapter 14.

OUTLINE

15.1 ■ INTRODUCTION

There are many instances in electronics when it is necessary to generate a signal that accurately duplicates the phase and frequency of another signal. Consider two instances: In digital instruments the scanning frequency is often 61,440 Hz. This frequency is 1024 times the 60 Hz power-line frequency that is used as a style reference. In another instance a practical FM detector must be able to accurately follow the frequency variations of the transmitter. A basic circuit that can perform these (and a multitude of other) tasks is the **phase-locked loop** (PLL).

The block diagram of a PLL is shown in Figure 15.1. The **VCO (voltage-controlled oscillator)** is presumed to emit a signal whose frequency is ω_o. The incoming frequency is ω_i, and it (as well as ω_o) is applied to the phase detector. The output consists of the frequencies $\omega_i \pm \omega_o$, where the high-frequency component is eliminated by a low-pass filter. This selection gives rise to an error voltage proportional to the frequency difference, $\omega_i - \omega_o$. The error voltage, applied to the VCO, alters ω_o in such a manner as to bring it into synchronism with ω_i—that is, $\omega_i - \omega_o$ becomes equal to zero. Once the VCO frequency is identical with ω_i, there remains a phase difference between the two, and it is this phase difference that is responsible for creating a dc voltage that maintains the VCO at ω_i.

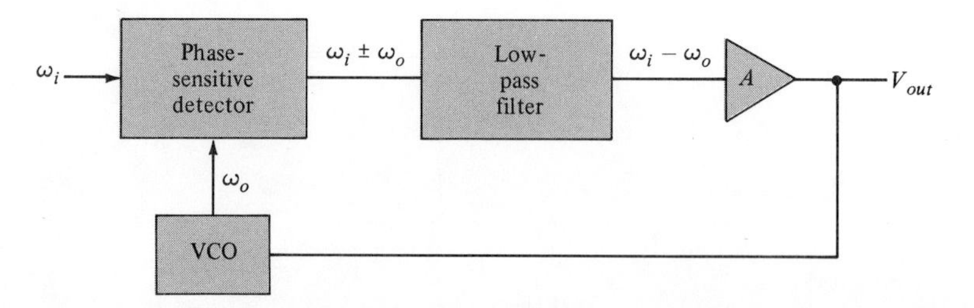

FIGURE 15.1. **Phase-Locked Loop** The frequency difference ($\omega_i - \omega_o$) applied to the phase-sensitive detector generates sum and difference frequencies ($\omega_i \pm \omega_o$), with the sum being filtered out by the low-pass filter. The difference signal is amplified and applied to a voltage-controlled oscillator whose output varies in such a manner as to bring it into synchronism with ω_i.

15.2 ■ PHASE DETECTORS

We turn first to a detailed consideration of phase detection. In Figure 15.2 the signal is applied to the primary winding of the transformer. If we short out the terminals marked "reference signal," the circuit closely resembles that of a full-wave rectifier. The major distinction lies in the use of a single, common resistor in the full-wave rectifier, whereas in the phase detector two resistors in series are used.

Rather than obtaining a pulsating dc output as in a full-wave rectifier

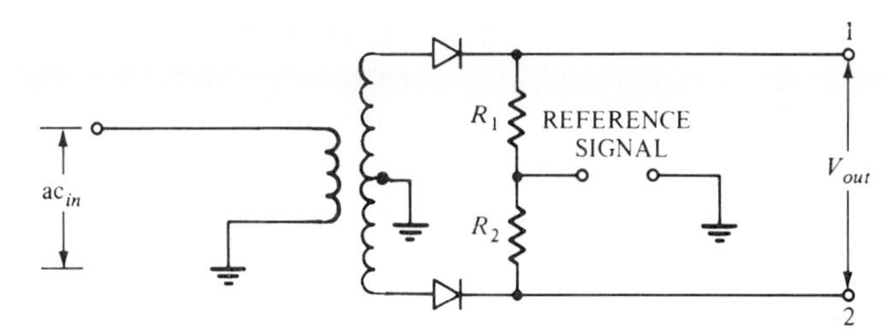

FIGURE 15.2. **Phase Detector** The output voltage depends on the phase difference between the ac input and the reference signal.

(Figure 15.3), in Figure 15.4 we obtain an ac output of the same frequency as the input. A dc voltmeter across the full-wave rectifier output might measure the rms voltage. The same voltmeter would read zero when placed across the output terminals 1 and 2 in Figure 15.4. (It is assumed that the frequency variations are too fast to be followed by the lethargic meter movement.)

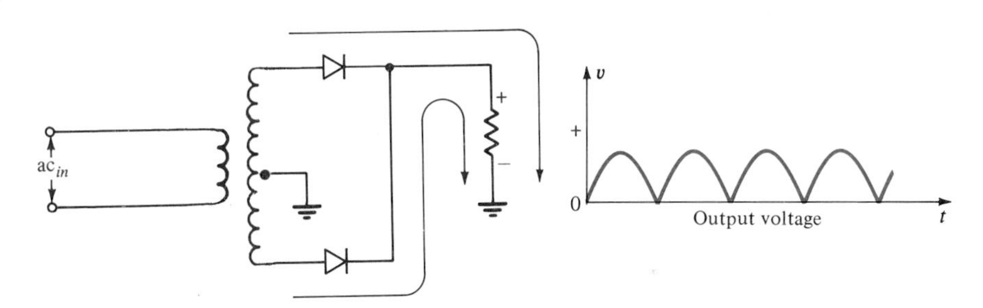

FIGURE 15.3. **Full-Wave Rectifier** The output is a rectified version of the input.

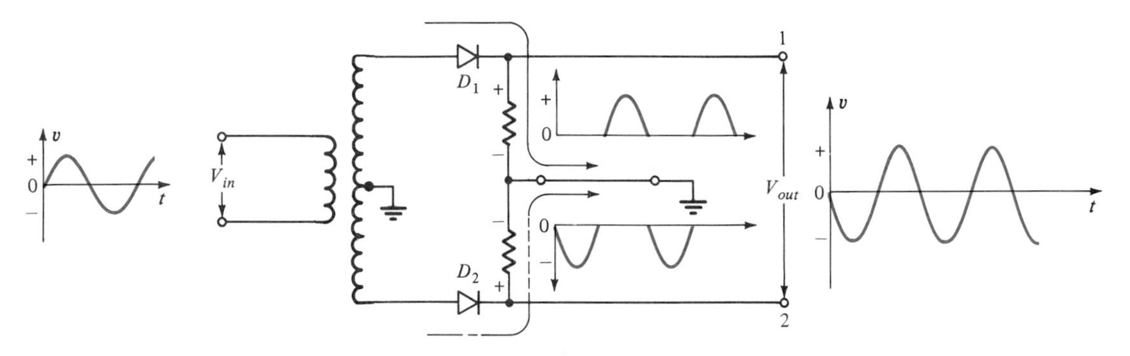

FIGURE 15.4. **Phase Detector with Reference Terminals Shorted** Output and input signals are of the same form.

Let us now eliminate the short across the terminals marked "reference signal" and apply between them an ac signal whose frequency is the same as the one we wish to detect. All ac signals will be measured with ground as the reference point.

Figure 15.5 shows the variations of the voltage at anode 1 and anode 2 as well as the reference voltage. Note that the reference voltage is being applied simultaneously to both cathodes.

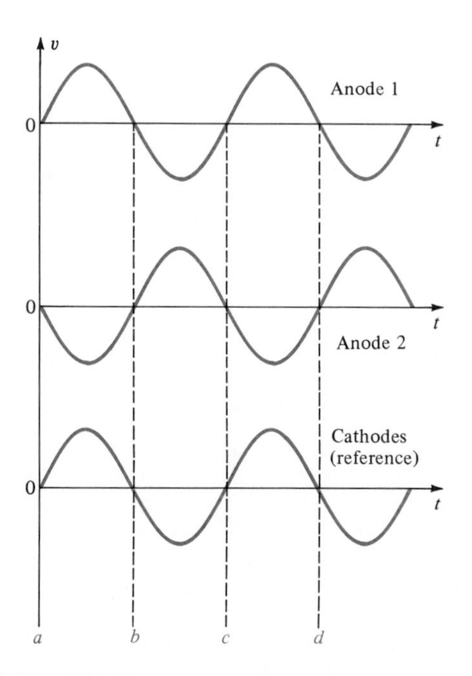

FIGURE 15.5. **PLL Signals (in Phase)** With input and reference (cathode) signals in phase, anode voltage 1 is presumed in phase with the common cathodes; voltage at anode 2 is 180° out of phase.

Between *a* and *b*, neither diode will conduct, for anode 2 is negative with respect to its cathode, while the cathode of diode 1 follows its anode voltage, and hence no potential *difference* exists across this diode. Between *b* and *c*, diode 1 does not conduct, and its anode again retains the same potential as its cathode. Diode 2, however, does conduct, and causes terminal 1 to assume a varying negative voltage with respect to terminal 2, as indicated in Figure 15.6. During the interval *c* to *d*, no conduction takes place, to be followed by a negative pulse during the next half-cycle.

If substantial capacitors are placed across R_1 and R_2, just as in the case of a rectifier, current can be caused to flow continually through the resistors even when the diodes are nonconducting. A dc voltmeter placed between terminals 1 and 2 would now give a finite reading proportional to the magnitude of the pulsating dc.

Now assume that there arises a change in the phase relationship between the input signal and the reference signal, say a 90° phase shift. The voltages at the two anodes now assume (with respect to the reference voltage) the relationships shown in Figure 15.7. During the interval *a* to *b*, anode 1 is positive with respect to the cathode, and hence conduction takes place through R_1. During the interval *b* to *c*, the cathodes remain more positive than either of the anodes, and no conduction takes place. During the interval *c* to *d*, anode

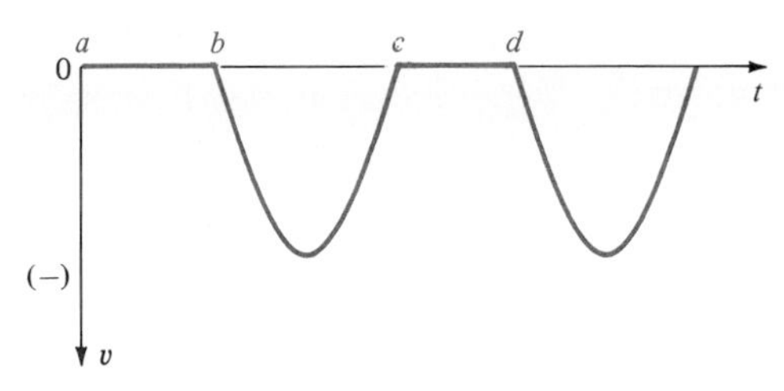

(a) Output from phase detector with input and reference signals in phase

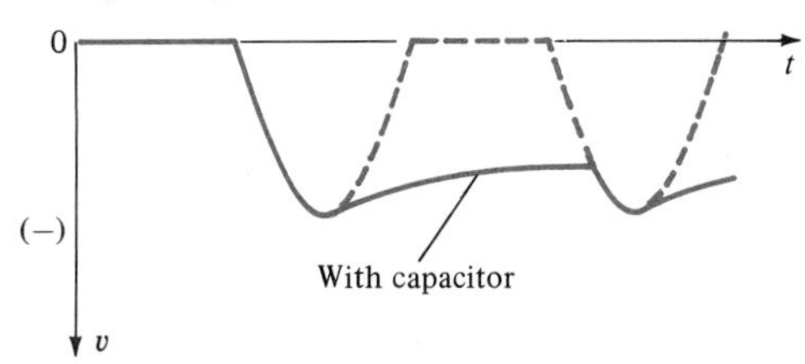

(b) Negative pulsations are reduced with a capacitor across the output terminals

FIGURE 15.6. Phase Detector Output

2 is more positive than the cathode, and conduction takes place through R_2. During the interval d to e, both anodes are more positive than the cathodes, and both diodes conduct.

The succession of events leads to the conduction pulses shown in Figure 15.8, where positive pulses refer to diode 1 and negative pulses, to diode 2. The positive and negative pulses average out to a mean voltage of zero across R_1 plus R_2, and hence a dc voltmeter between terminals 1 and 2 would read zero.

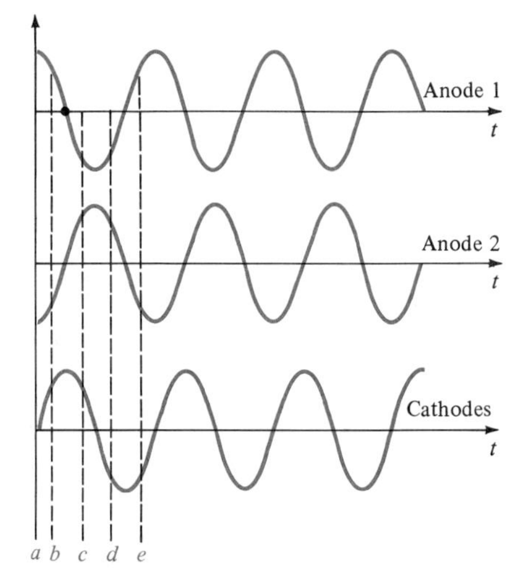

FIGURE 15.7. PLL Signals (90° Phase Difference) With input and reference (cathode) signals 90° out of phase, voltage at anode 1 leads reference voltage by 90° and voltage at anode 2 lags by 90°.

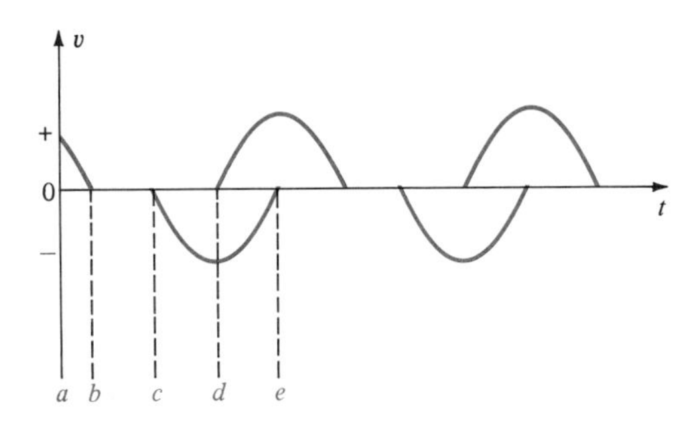

FIGURE 15.8. PLL Output (90° Phase Difference) Output voltage from phase detector averages to zero when signal and reference are 90° out of phase.

If we were to cause a 180° phase shift between the signal and reference voltages, the corresponding variations would look as in Figure 15.9. The dc reading would be equal in magnitude (but of opposite polarity) to that originally encountered.

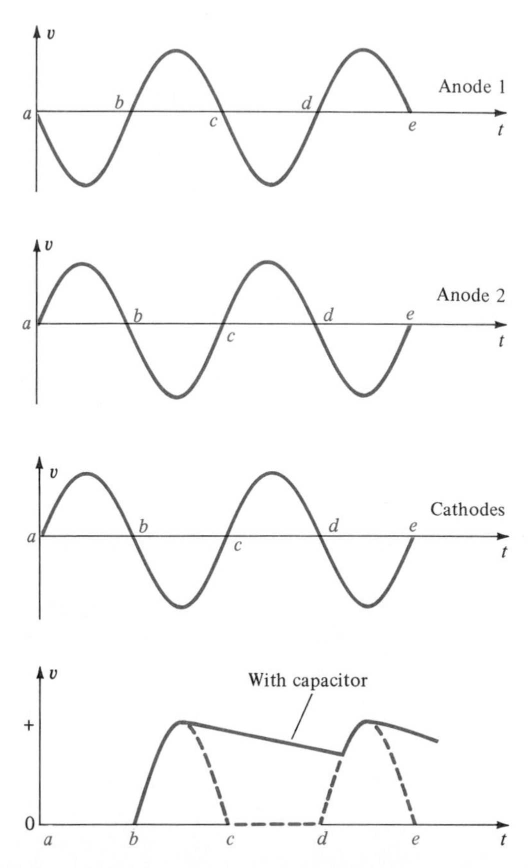

FIGURE 15.9. PLL Signals and Output (180° Phase Difference) With input and reference signals 180° out of phase, anode 1 is out of phase with the cathode (reference) and anode 2 is in phase. Positive pulsations are reduced with the capacitor across the load.

Let us recapitulate: Starting with the input and reference signals in phase, the output voltage would be at a maximum. As the two get out of phase, the magnitude of the dc voltage diminishes, becoming zero when the two are 90° out of phase. Further increases in the phase difference will cause the voltage to increase, but with opposite polarity, finally reaching a maximum when the two are 180° out of phase. We can, therefore, if we wish, calibrate the dc voltmeter in terms of phase angle rather than voltage, and we then have a **phase detector**.

15.3 ■ THE PHASE-LOCKED LOOP

Now assume that the input and reference signals are not quite the same in terms of frequency. They will, therefore, periodically fall into and out of phase. Insofar as the phase detector is concerned, this sequence appears as a signal that is constantly shifting phase with respect to the reference signal.

The farther apart the two frequencies are, the more rapid the apparent phase shift. This continual falling into and out of synchronism with the reference signal gives rise to an alternating voltage at the output terminals of the phase detector. This we term a **phase-sensitive detector**.

In the IC version of the phase-locked loop, such as the unit designated by the number 565, the output of the VCO is in the form of square waves, and the reference signal applied to the phase detector, therefore, is a square wave. Figure 15.10 shows a simplified version of the phase detector used in the 565. By presuming first that there is no signal input, Q_1 and Q_2 each pass one half of the current furnished by the **constant current generator**.[1] Presume further that the instantaneous polarity of the square wave is such that the bases of Q_4 and Q_5 are both positive, implying in turn that the bases of Q_3 and Q_6 are negative. With Q_4 and Q_5 forward-biased, I_1 passes through R_{c2}, and I_2 passes through R_{c1}. On the next half-cycle of the square wave, Q_3 and Q_6 are forward-biased, I_1 passes through R_{c1}, and I_2 passes through R_{c2}. The output voltage (V_d) is determined by the voltage drop across R_{c2}. If Q_1 and Q_2 are passing equal amounts of current—that is, $I_1 = I_2$—there is no change in the voltage drop across R_{c2}, and $\Delta V_d = 0$.

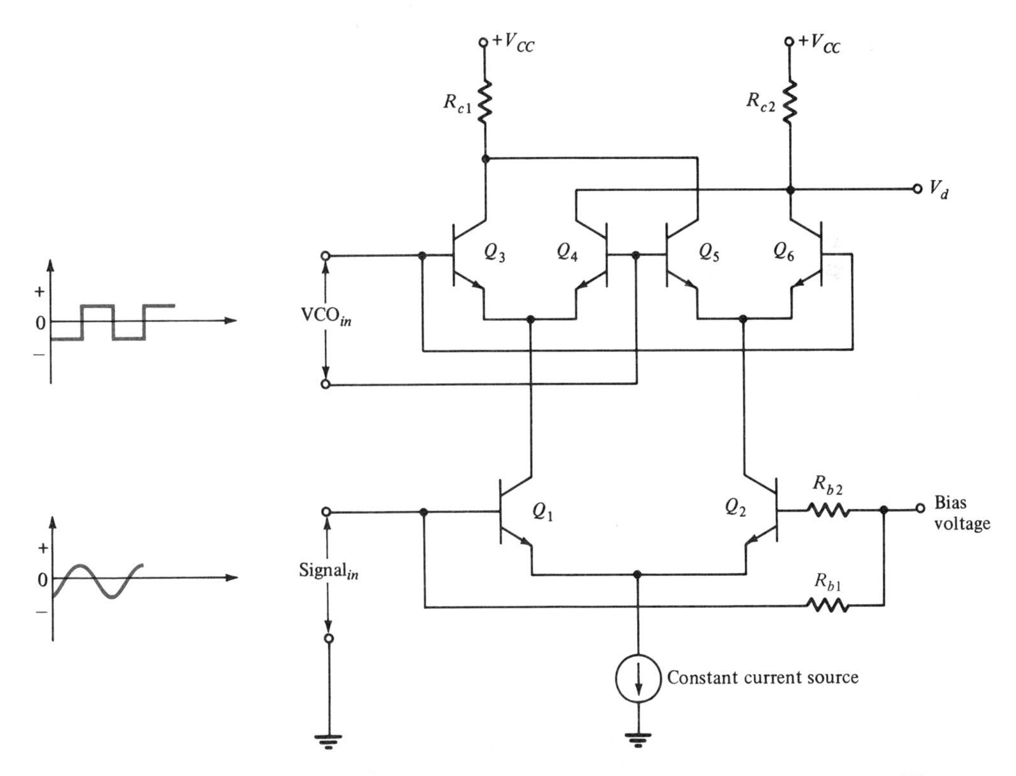

FIGURE 15.10. Simplified Version of 565 Phase-Locked Loop Showing VCO Square Wave Input (ω_o), Signal Input (ω_i), and Output Voltage (V_d)

[1]See "A Closer Look" at the end of this chapter for a discussion of constant current generators.

Now apply signal ω_i (Figure 15.11a). Recall that the total current through Q_1 plus that through Q_2 must always be constant (because their emitters are tied to a constant current generator). With the square wave and input signals of the same frequency and in phase, when the input signal is on the positive portion of its cycle, I_2 is passing through R_{c2}. As the input signal starts on its negative half-cycle, I_1 is diminishing, but I_2 is increasing. But at this time it is the current I_1 that is passing through R_{c2}. The output voltage (V_d) looks like an amplified full-wave-rectified version of the input signal. If this pulsating dc is now put through a low-pass filter that removes the ac components, we obtain a dc output voltage that may be used to control the output frequency of the VCO.

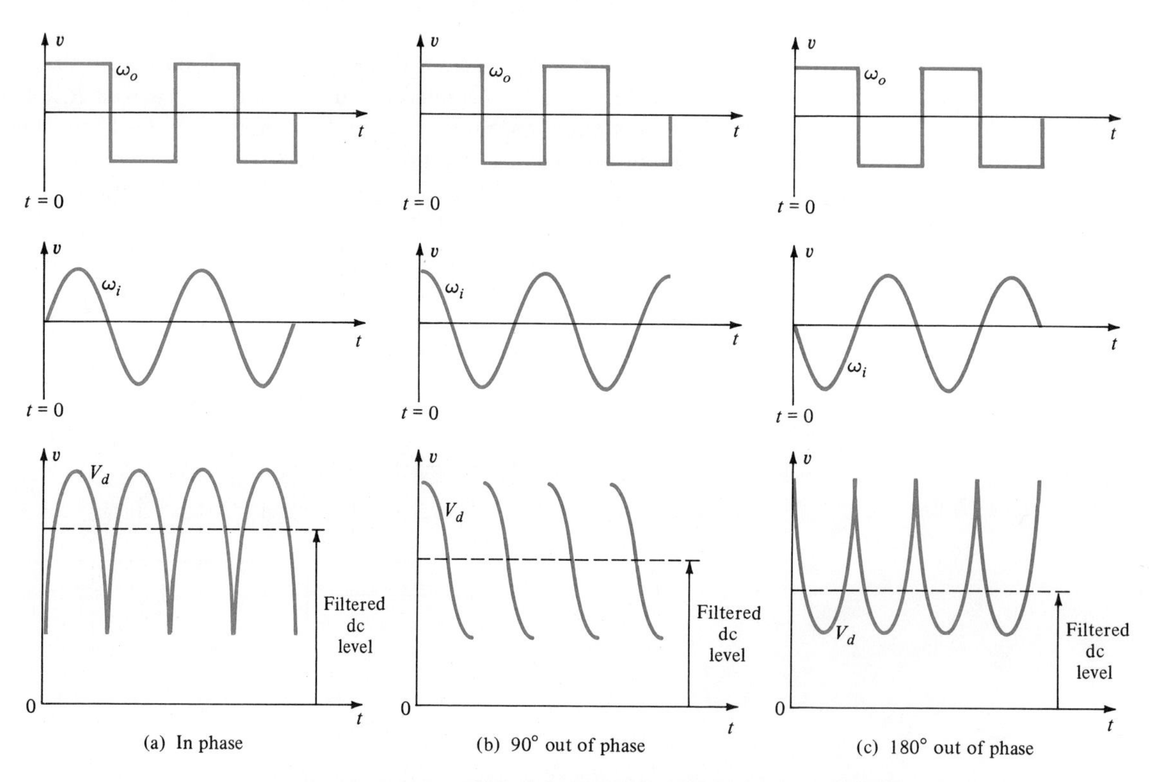

FIGURE 15.11. PLL with Square Wave Output Some assumed phase relations between square wave VCO output and sinusoidal signal input into a 565 PLL. The dc level changes with the phase difference between the two waveforms.

We have presumed in the preceding that the square wave and signal have the same frequency and phase. Let their frequencies remain the same, but consider now what happens if they differ in phase by 90° (Figure 15.11b). In this case V_d looks as shown in Figure 15.11b. Filtering out the ac components, we are left with a lower value for the dc voltage.

Continuing to presume the signal and square wave are still of the same frequency but now 180° out of phase (Figure 15.11c), we will again obtain a full-wave-rectified type of output but inverted and with a still lower dc level.

Now presume that the signal frequency differs slightly from the resting frequency of the VCO. The dc level will then keep changing at a rate equal to the frequency difference. As a result the VCO frequency will keep changing in response to this dc voltage. When the voltage reaches a value such that $\omega_o = \omega_i$, the two frequencies will "lock."

We may profitably turn at this time to a mathematical consideration of the operation of the PLL's phase detector. We again assume a sinusoidal signal frequency, ω_i:

$$v_i = V_i \sin(\omega_i t + \theta_i) \tag{15.1}$$

The square wave consists of a fundamental and odd harmonics of the VCO frequency, ω_o:

$$v_o = \sum_{n=0}^{\infty} \frac{4}{\pi(2n+1)} \sin[(2n+1)\omega_o t] \tag{15.2}$$

Multiplying v_i by v_o (which is what the phase detector does) and presuming a differential stage gain of A_d, after using the appropriate trigonometric relations, we obtain

$$v_d = \frac{2A_d}{\pi}\left\{\sum_{n=0}^{\infty} \frac{V_i}{2n+1} \cos[(2n+1)\omega_o t - \omega_i t - \theta_i] - \sum_{n=0}^{\infty} \frac{V_i}{2n+1} \cos[(2n+1)\omega_o t + \omega_i t + \theta_i]\right\} \tag{15.3}$$

If ω_i is close to the square wave fundamental—that is, $n = 0$—the first term constitutes a beat frequency that feeds around the loop. The varying value of V_d causes the VCO frequency to change, finally reaching a value that causes a *lock* to occur between ω_i and the VCO. When this happens, we have

$$V_d = \frac{2A_d V_i}{\pi} \cos \theta_i \tag{15.4}$$

This equation yields the voltage ncessary to keep the VCO at ω_i. It is possible that ω_o may equal ω_i momentarily during the lock-up process and yet have the phase be incorrect, so that ω_o passes through ω_i. Then the developed voltage will bring the VCO frequency back down. This condition may arise a number of times before the two frequencies coincide with the proper phase relationship.

Considering the second term in Eq. 15.3, note that during lock $\omega_o + \omega_i = 2\omega_i$. This frequency represents the pulsations that we graphically identified as looking similar to those of a full-wave rectifier.

V_d also depends on V_i, the amplitude of the signal frequency. If the input signal is amplitude-modulated, V_d will keep changing value (because of a variation in V_i), and hence monitoring the variation of V_d presents us with a method of extracting amplitude variations of the input signal. In other words, the PLL can serve as an amplitude detector.

If the input signal frequency is varying, as is the case with frequency modulation, θ_i will keep changing to keep a frequency lock, and the quent variation of V_d allows us to obtain the demodulation of an FM signal as well.

Turning finally to a consideration of how the voltage-controlled oscillator functions, consider Figure 15.12. I_1 is a charging current that originates from a constant current generator whose value is controlled by the voltage V_d from the phase detector. Q_3 is off so that the only path to ground is through C_1. Therefore, C_1 charges through diode D_2.

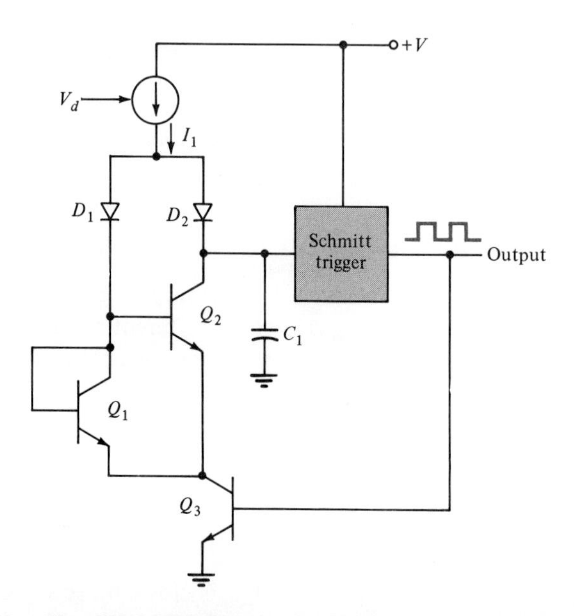

FIGURE 15.12. Simplified VCO Circuit Input voltage V_d controls the charging current of C_1, which, in turn, determines the switching rate of the Schmitt trigger. The output of the Schmitt trigger represents the VCO's output.

When the voltage across C_1 reaches the triggering level of the Schmitt, that circuit activates Q_3, which grounds the emitters of Q_1 and Q_2. The charging current I_1 now flows through D_1, Q_1, and Q_3 to ground. This condition arises because this path represents a much lower impedance to ground than does C_1.

Once the current switches to this path, both Q_1 and Q_2 have the same base-emitter bias. While I_1 flows through Q_1, a like current passes through Q_2, the latter being the discharge current from capacitor C_1.

When the voltage across C_1 reaches a value equal to the lower threshold for the Schmitt trigger, the latter switches Q_3, causing D_1 to stop conducting. I_1 switches back through D_2, and the capacitor again starts to recharge.

The charging of a capacitor from a constant current source will give rise to a linear increase in capacitor voltage. [Since $Q = CV$, $dQ/dt = I = C(dV/dt)$. Since I is constant, so is dV/dt.] Similarly, the discharging voltage across C will also drop linearly. Therefore, across the capacitor a symmetrical triangular pulse is developed. Across the output of the Schmitt trigger, there is a square wave whose period is the same as that of the triangular wave. The frequency of both is determined by the control voltage V_d from the phase detector. Voltage-controlled oscillators are widely used and are available not only as part of a PLL (such as the 565) but also as separate IC units (for example, the 566).

Finally, there are two types of PLLs. Our discussion has been confined to type I, which employs either analog signals or square waves. There is also a type II, which is driven by edge transitions—that is, it is sensitive to any difference in the relative timing between the edges of input pulses and the pulses generated by a VCO.

15.4 ■ PLL STABILITY

The use of negative feedback in conjunction with amplifiers can lead to instability at some high frequency when the phase shift through the amplifier assumes a value of 180°, thereby negating the −180° introduced by the negative feedback. At that frequency the feedback is positive, and if, additionally, the loop gain (βA) is unity, the gain theoretically becomes infinite according to the equation

$$A' = \frac{A}{1 - \beta A} \tag{15.5}$$

Phase-locked loops have a reputation for instability. This potential instability arises in the following manner: The PLL is also a feedback system, with the VCO behaving like an integrator and thereby introducing a 90° phase shift. Therefore, the phase change accompanying the drop-off in loop gain with increasing frequency needs to introduce an additional shift of only 90° to satisfy the instability condition that, in a straight-forward feedback amplifier, required a total shift of 180°. The possibility of instability in a PLL can easily be eliminated if one is aware of its cause.

Refer to Figure 15.11. The desired VCO voltage is represented by the dashed (average) value of V_d, which moves up and down as the system searches for the proper value to cause a lock-in condition to prevail. The rapid pulsations constitute high-frequency components whose phase shift in going around the feedback loop can lead to the trouble that we just mentioned. The magnitude of these oscillations can be minimized by employing a low-pass filter, as shown in block diagram form in Figure 15.1.

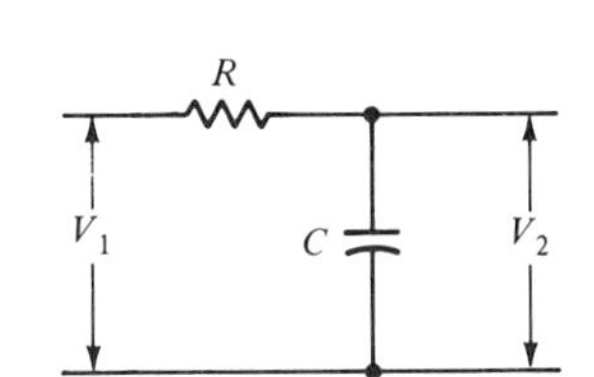

FIGURE 15.13. **Low-Pass (Lag) Filter**

The simplest form of low-pass filter is shown in Figure 15.13, termed a **lag filter** since it introduces a delay phase angle. Much more satisfactory is the **lag-lead filter** shown in Figure 15.14.

For stable operation we need a design that assures us that at that frequency that leads to a 180° phase shift around the feedback loop the gain will be less than unity. We proceed with the loop gain calculation (Figure 15.15).

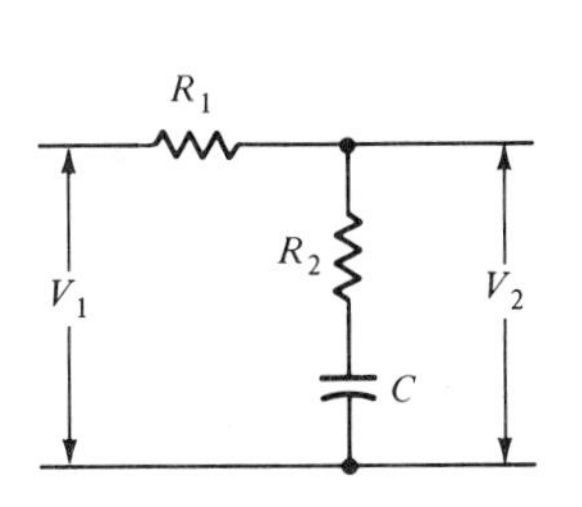

FIGURE 15.14. **Low-Pass (Lag-Lead) Filter**

The voltage output from the phase detector (V_1) is proportional to the phase difference ($\Delta\theta = \theta_i - \theta_o$). The proportionality factor represents the phase detector "gain" (K_P):

$$V_1 = K_P \, \Delta\theta \tag{15.6}$$

The loop gain equals $\theta_o/(\theta_i - \theta_o) = \theta_o/\Delta\theta \equiv K$.

The filter's output (V_2) is proportional to the input voltage (V_1), and we

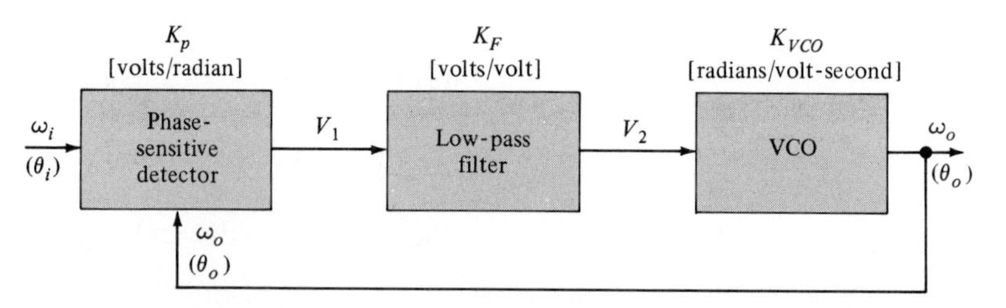

FIGURE 15.15. Basic Phase-Locked Loop Showing Stage Gains and Their Respective Dimensional Units

will call the transfer function (V_2/V_1) for the filter K_F. The analytical form of K_F will depend on the type of filter used. We have

$$V_2 = K_F V_1 \tag{15.7}$$

The VCO generates a frequency that is proportional to V_2. V_2, however, is an oscillatory function as long as the PLL is searching for a lock-in condition $(V_2 = V_2' e^{j\omega t})$. Also, frequency ω_o is the time rate of change of phase. Thus, we have

$$\omega_o = \frac{d\theta_o}{dt} = K_{VCO}(V_2' e^{j\omega t}) \tag{15.8}$$

To obtain θ_o (which is applied to the phase detector), we integrate:

$$\theta_o = \frac{K_{VCO} V_2' e^{j\omega t}}{j\omega} = \frac{K_{VCO} V_2}{j\omega} \tag{15.9}$$

(The j in the denominator verifies that there is a 90° phase shift associated with the VCO.)

By selecting the simple low-pass filter (Figure 15.13), the complete expression for the loop gain (K) becomes

$$K = \frac{\theta_o}{\Delta\theta} = \frac{K_{VCO} K_P}{j\omega} \frac{V_2}{V_1} = \frac{K_{VCO} K_P}{j\omega} \frac{1}{1 + j\omega RC} \tag{15.10}$$

where for the low-pass filter $K_F = V_2/V_1 = (1/j\omega C)/[R + (1/j\omega C)] = 1/(1 + j\omega RC)$. The phase margin—that is, the closeness to $-180°$ at unity gain—is small with the lag filter, and any slight additional phase change, such as from the amplifier shown in Figure 15.1, can lead to instability. One generally makes use of the lag-lead filter of Figure 15.14. This selection assures avoidance of the critical instability condition by a considerable margin.

EXAMPLE 15.1

A phase-sensitive detector generates a 61,440 Hz frequency using a 60 Hz power line as a reference. The output dc voltage has a range of 0–10 V for respective phase changes of 0–360°.[2] The corresponding VCO frequency vari-

[2]This specification indicates a type II PLL since the VCO phase range is 360°. For a type I, it would be limited to 180°. Being a type II, the 60 Hz power-line sine waves must first be converted into a 60 Hz square wave.

ation will be 20–200 kHz, which, when passed through a (÷ 1024) circuit, would allow lock-in operation for input signal variations of 19.5–195 Hz. If a simple lag filter with $R = 1\ \text{M}\Omega$ and $C = 1\ \mu\text{F}$ is used, determine the frequency at which the loop gain will be unity and the phase shift close to $-180°$.

Solution:

$$K_P = \frac{10\ \text{V}}{2\pi\ \text{rad}} = 1.59\ \text{V/rad}$$

$$K_F = \frac{1}{1 + j\omega CR} = \frac{1}{1 + j\omega(1)}$$

$$K_{VCO} = \frac{2\pi(200 - 20) \times 10^3\ \text{rad/s}}{10\ \text{V}} = 113 \times 10^3\ \text{rad/V}\cdot\text{s}$$

$$K_{DIV} = \frac{1}{n} = \frac{1}{1024}$$

$$K = 1.59\left(\frac{113}{j\omega}\right)\left(\frac{1}{1 + j\omega}\right)\left(\frac{1}{1.024}\right)$$

$$|K| \equiv 1 = \frac{175}{\omega\sqrt{1 + \omega^2}}$$

$$\omega^4 + \omega^2 - 3.06 \times 10^4 = 0$$

$$\omega^2 = -0.5 \pm \sqrt{3.06 \times 10^4}$$

$$\omega = \sqrt{174} = 13.2\ \text{rad/s} \qquad (2.1\ \text{Hz})$$

$$K = \frac{175}{j(13.2)(1 + j13.2)} = \frac{175}{(13.2)(13.2)} \frac{1}{\angle 90° \ \angle 85.7°}$$

$$= 1.00 \angle -176°$$

It will be noted that in Example 15.1 the frequency at which the loop gain passes through unity (2.1 Hz) is well *below* the lower limit of the phase-sensitive detector's frequency range (19.5–195 Hz). In this type of application, the objective is to maintain a constant output frequency (61,440 Hz) in synchronism with a 60 Hz reference. Any variations of the power-line frequency are expected to be very slow, certainly less than 2 Hz, and the band limit of the filter is correspondingly small. If the PLL were being used in an application such as the demodulation of an FM signal, the filter passband would have to be wide enough to accommodate the entire modulation bandwidth.

When using a lag-lead filter (Figure 15.14), the ratio R_1/R_2 determines the stability. Small (or zero) R_2 diminishes stability. If R_2 is too large, however, it will take a long time for the system to adjust to frequency changes. This delay might be an advantage if the objective of the PLL is to generate a constant frequency, and the reference signal is noisy, or even "drops out" occasionally. The voltage on the capacitor would remain rather constant, as would the frequency. If the objective is to follow rapid changes, such as in FM detection, R_2 needs to be small and the bandwidth of the filter, large.

EXAMPLE 15.2 A 565 PLL is used to demodulate an FM signal. The lag filter used has the value $R = 3.6\ \text{k}\Omega$, and $C = 330$ pF. What is the bandwidth of the filter?

Solution:

$$f = \frac{1}{2\pi RC} = \frac{1}{2\pi(3.6 \times 10^3 \times 330 \times 10^{-12})} = 134\ \text{kHz}$$

15.5 ■ TOUCH-TONE® TELEPHONE: A PLL EXAMPLE

Depressing the button of a Touch-Tone® telephone generates two tones, a different combination for each digit. In the range 700–1700 Hz, eight different frequencies are used. The low band includes 697, 770, 852, and 941 Hz. A high band utilizes 1209, 1336, 1477, and 1633 Hz (Figure 15.16). These specific frequencies have been selected to avoid harmonic relationship.

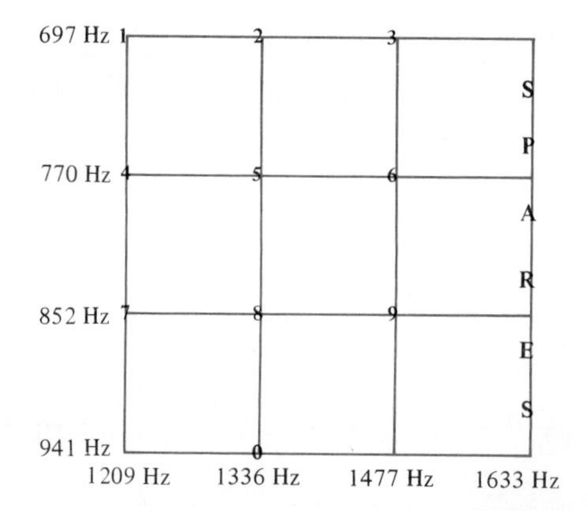

FIGURE 15.16. Frequency Combinations Associated with Each Digit of a Touch-Tone® Telephone

The basic circuit used in the dialing system is the 566, which serves as a voltage-controlled oscillator (VCO); it is essentially the VCO portion of a 565 PLL. A constant current generator is used to charge a capacitor C_1 through a resistor R_1. The frequency of the unit may be changed in any of three ways:

1. By changing the value of C_1
2. By changing the control voltage
3. By changing the value of R_1

In this application it is the change in R_1 that is used. Each dialing unit needs two 566s since two simultaneous frequencies are required. Figure 15.17 shows the arrangement.

In both instances $C_1 = 0.022\ \mu$F. In one unit R_1 is 12.1 kΩ; in the other it is 6.8 kΩ. These values determine the highest frequency generated. The VCO frequency is obtained from the equation

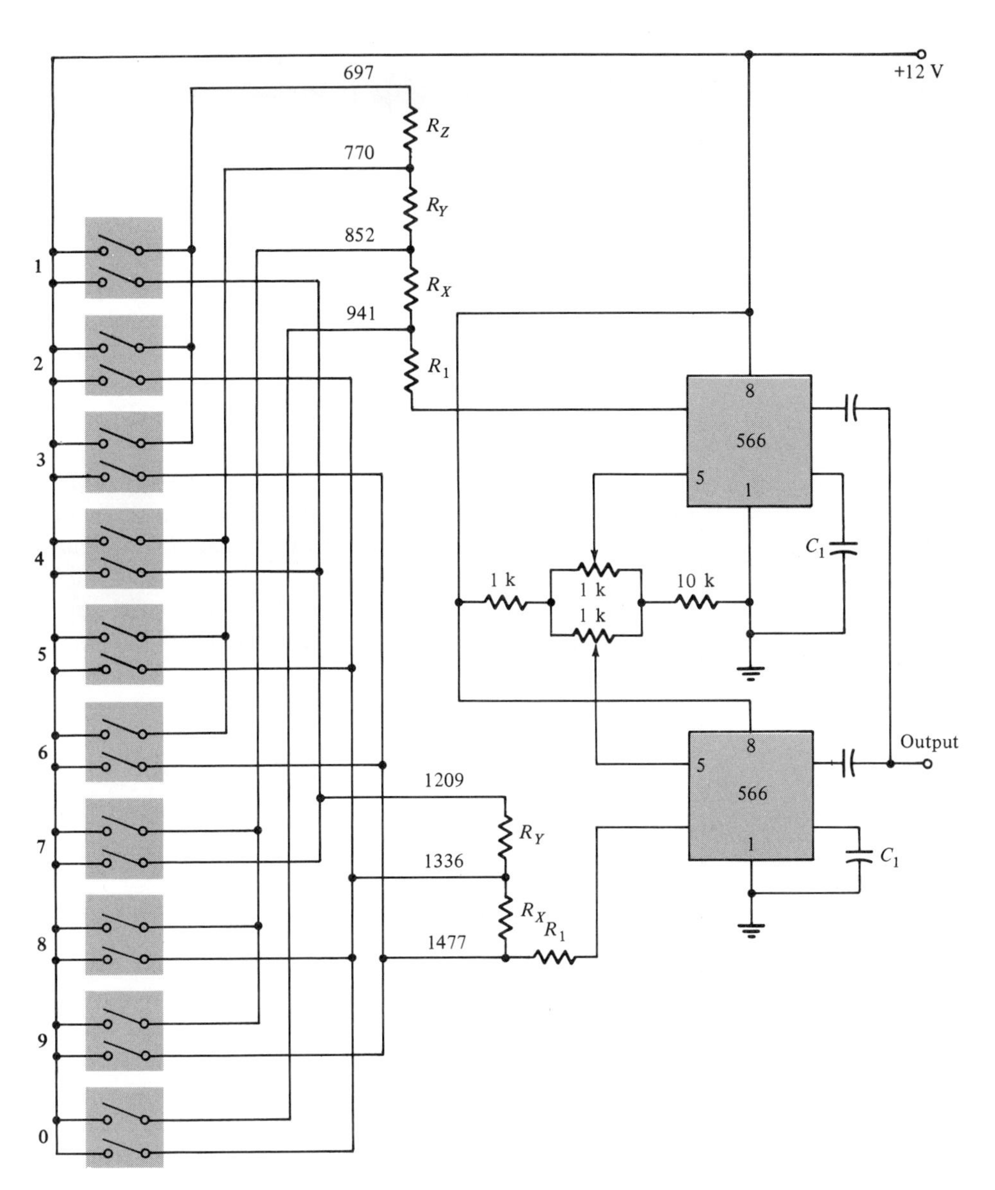

FIGURE 15.17. **Each Digit of a Touch-Tone® Telephone Is Associated with Two Switches, Each of Which Controls a 566 Tone Generator** Tones depend on the resistances (R_Z, R_Y, R_X, R_1) inserted into each 566 by the push button depressed.

$$f_o \simeq \frac{2}{R_1 C_1}\left(\frac{V_8 - V_5}{V_8 - V_1}\right) \qquad (15.11)$$

where V_i represents the dc voltage at the ith pin of the 566 unit. Typically, $V_8 = 12$ V, V_1 is at ground potential, and V_5 is obtained from an adjustable *pot* that is used to calibrate the unit. Additional resistances together with R_1 are used to produce the other needed frequencies:

$$\frac{R_X + R_1}{R_1} = \frac{f_1}{f_2} \tag{15.12}$$

$$\frac{R_Y + (R_1 + R_X)}{R_1 + R_X} = \frac{f_2}{f_3} \tag{15.13}$$

$$\frac{R_Z + (R_1 + R_X + R_Y)}{R_1 + R_X + R_Y} = \frac{f_3}{f_4} \tag{15.14}$$

For decoding the Touch-Tone® signals, a 567 unit is employed (Figure 15.18). This unit contains a current-controlled oscillator and a phase detector, rather similar to the 565. But unlike the 565, which is usually made to respond to a wide range of frequencies, each 567 needs to respond only to a specific frequency.

Seven 567 decoders are needed (only four are shown in Figure 15.18). Their inputs are connected to the phone line (or any other source of tone signals). Each is tuned to a specific frequency by proper selection of R_1 and C_1. The other capacitors and resistors are adjusted to prevent overlaps in the response of the individual units. The logical circuitry responds to the two tones, which are simultaneously present and give the proper decimal output indication.

15.6 ■ TRANSISTORIZED ENGINE CONTROLS: ANOTHER PLL EXAMPLE[3]

15.6.1 Operating Modes

The maximum power of an internal combustion engine is delivered when the piston is approximately 15° past **top dead center** (TDC). Since complete burn-through of the fuel-air mixture takes about 0.003 s, we see the necessity to fire the spark plug at a significantly earlier time. Additionally, as the revolutions per minute increase, this time interval constitutes a larger fraction of the angular rotation of the crankshaft, and the ignition must be applied at an even earlier time—that is, the spark must be *advanced*.

In the following we presume we are dealing with a four-cylinder, four-stroke engine, whose rotational speed may vary between 300 and 6000 rpm. Attached to the rotating engine is a transducer that provides a synchronizing pulse each time one of its pistons reaches a position 60° before TDC. There are, therefore, two synchronizing pulses per engine revolution, being one for every 180° of rotation.

The system can work in any of three modes:

1. Under normal running conditions a computer controls the firing time and the amount of fuel injected in response to the engine speed,

[3]This section is based on a paper by R. J. Freimark, "Engine Controls Become More Cost Effective," *Automotive Engineering*, vol. 89, 1981, p. 28.

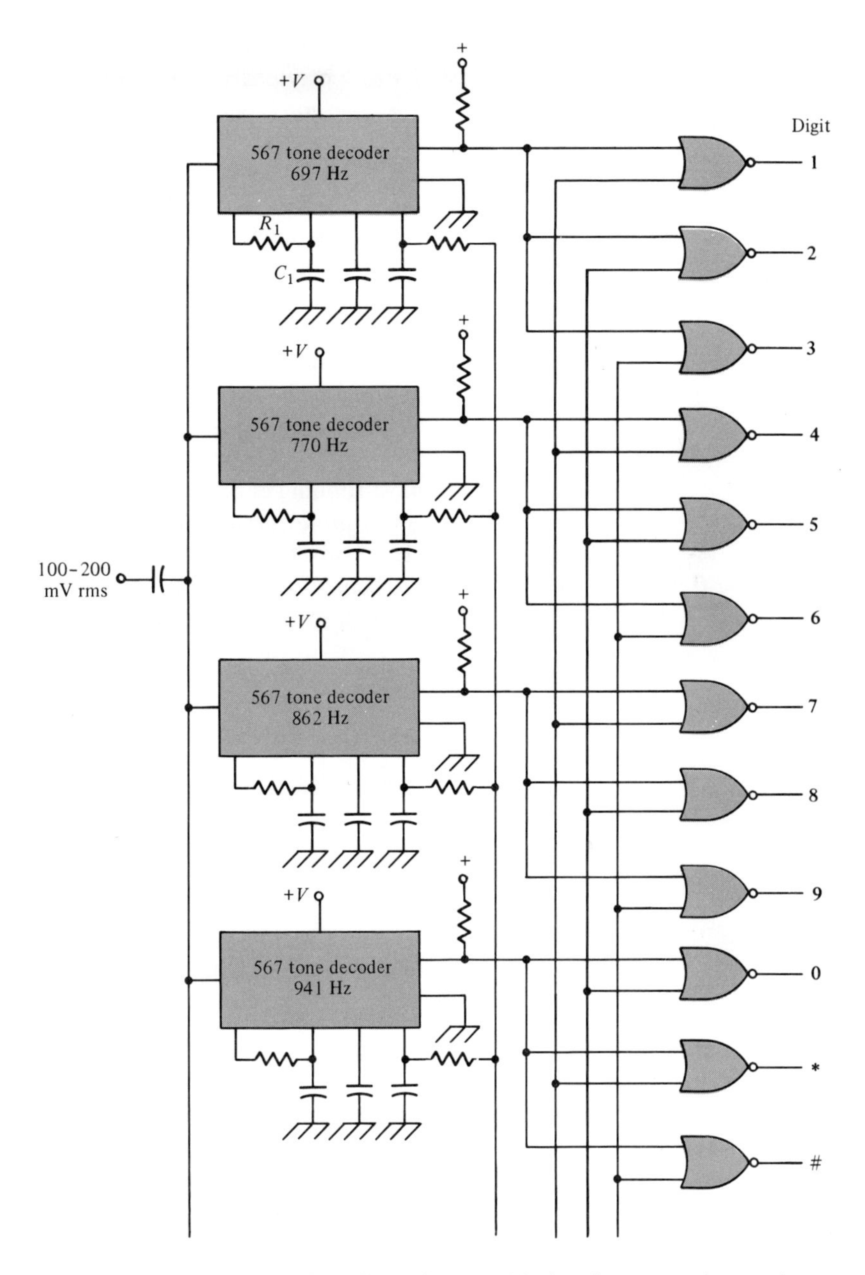

FIGURE 15.18. Telephone Decoders Separate 567 decoders respond to two impressed frequencies and activate logic digital circuitry appropriate to the incoming tone combinations. Four of the required seven decoder units are shown.

exhaust oxygen content, and so forth. This mode constitutes the *closed-loop* operation.

2. The exhaust oxygen detector, which represents the main element in the pollution control system, has to be brought up to temperature before its output readings become effective. Therefore, initially the

engine's running parameters are determined in response to numbers previously stored in the computer. This mode constitutes the *open-loop* fuel control.

3. Should the computer fail to operate properly, as well as when the engine is being started, provision is made to operate without the computer. These conditions give rise to the *backup mode.* Although parameters in the last case are far from optimum, they do allow operation under starting conditions as well as in emergencies.

Backup Mode. Consider first the operation under the backup mode. The engine synchronizing pulses are of the form shown in Figure 15.19. The spark is fired on the rising edge of the reference pulse. This edge occurs at 15° **before top dead center** (BTDC) and would constitute the proper advance angle at about 1600 rpm and thus represents a somewhat retarded situation for most running conditions. The falling edge of the pulse (at +60° BTDC) triggers a monostable MV that generates a 1 ms pulse, and this pulse is used to electrically actuate fuel ejectors twice per engine revolution. These serve as the backup fuel pulses. This value is rich for idle and slow running and lean for high speed, but it serves to get the driver back in an emergency.

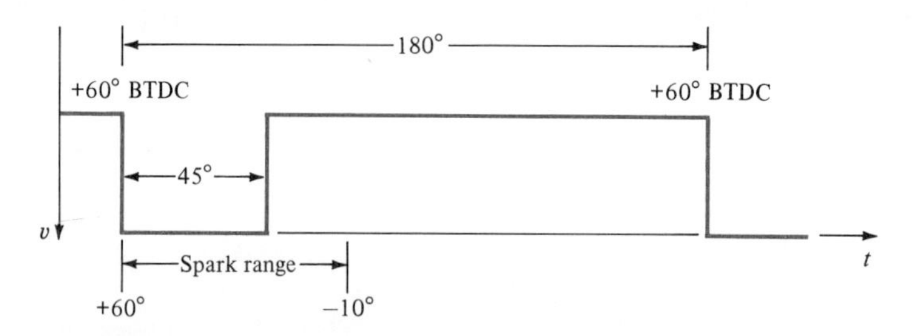

FIGURE 15.19. Spark Timing Diagram

The fuel ejectors are mechanical devices, of course, and there is about a 500 μs delay between pulse application and delivery. The ejector operations are battery-sensitive, and the 1 ms pulse is made to trigger another monostable MV whose pulse length can be varied to compensate for battery voltage variations.

Open-Loop Operation. We turn now to a consideration of open-loop operation. The pulse that occurs at +60° BTDC begins a countdown within the microcomputer, and when the count reaches zero, the spark will fire. This countdown time is not a fixed quantity but rather varies according to engine conditions. For example, at 1000 rpm the spark might fire at about 3° BTDC; at 2000 rpm it could be advanced to about 21° BTDC. It might be desired to have other conditions influence the firing time, and these conditions can be fed into the computer to effect a variation.

The angular position of the engine is determined through the use of a PLL in the following manner: The VCO (Figure 15.20) associated with the PLL has a frequency range of 5120–102,400 Hz. This signal is fed into a

÷512 counter, yielding output frequencies in the range 10–200 Hz. Engine speeds in the range of 300–6000 rpm correspond to 5–100 rps, and since two synchronizing pulses arise per revolution, the signal from the crankshaft is 10–200 Hz. Thus, as the engine speed changes, the phase comparator will generate a dc voltage such as to make the output of the VCO match that of the engine (÷512, of course). The undivided VCO output (5120–102,400 Hz) serves as a source of clock pulses for the computer. Each undivided pulse represents 0.35° of the crankshaft rotation. If at 300 rpm, for example, it is desired to fire the plug at TDC, the countdown starts at 171, and after a crankshaft rotation of 60°, the plug will fire.

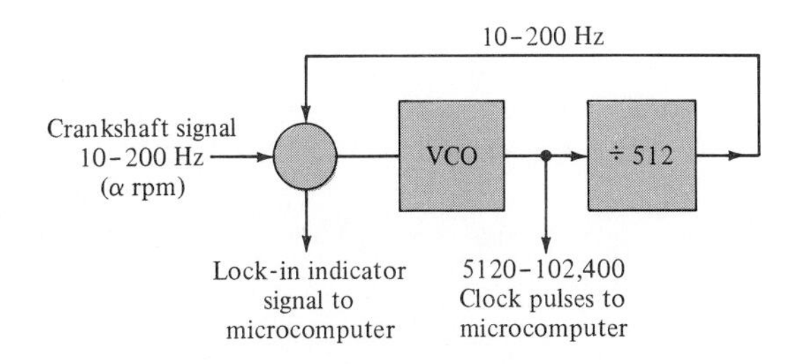

FIGURE 15.20. **Fuel System Phase-Locked Loop**

Tachometer. So far our timing has been only in terms of engine degrees, but some functions, such as the duration of the fuel injection and spark duration, demand real time determinations depending on rpm. Such functions are handled by having each incoming reference pulse fire another monostable MV that generates a pulse of 5 ms duration (Figure 15.21). By feeding the succession of such pulses through an *RC* filter, the resulting dc voltage will be proportional to the engine rpm. We thereby have the makings of a **tachometer.** Voltages between 0 and 5 V correspond to engine speeds between 0 and 6000 rpm, respectively. (Above 6000 rpm the dc level remains at 5 V.) The dc level is read by an 8-bit analog-to-digital converter (A/D converter) (see Chapter 18), and in the computer this 8-bit binary value is used as a table entry that yields the numerical value of rpm.

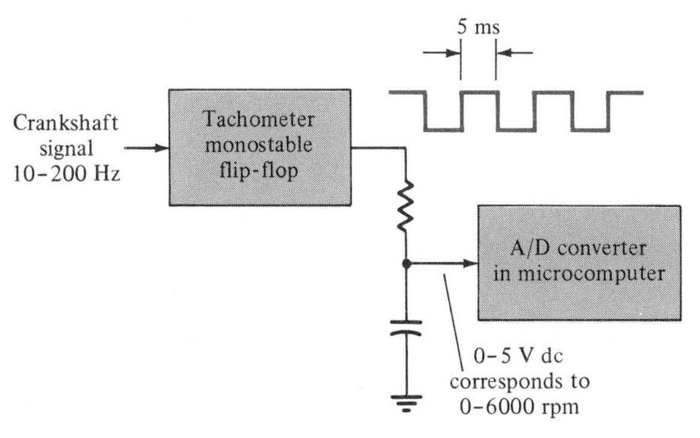

FIGURE 15.21. **Tachometer Detail**

15.6.2 Operational Sequence

Consider Figure 15.22. Since the cranking speed is below 600 rpm, the engine always starts in the backup mode. If the engine is not fully warmed, the 1 ms fuel pulse is lengthened in proportion to the inverse of the temperature. This enrichment process is enabled only upon starting—it is activated by the presence of a voltage across the starter relay. After the engine has started, the PLL should become fully locked below 600 rpm (it should, in fact, lock at 300 rpm). The microprocessor checks whether the tachometer voltage corresponds to that above 600 rpm and then whether there is a lock-in at the PLL. If both signals are satisfactory, the computer control takes over, and the data selector changes from the backup mode to the open-loop computer control.

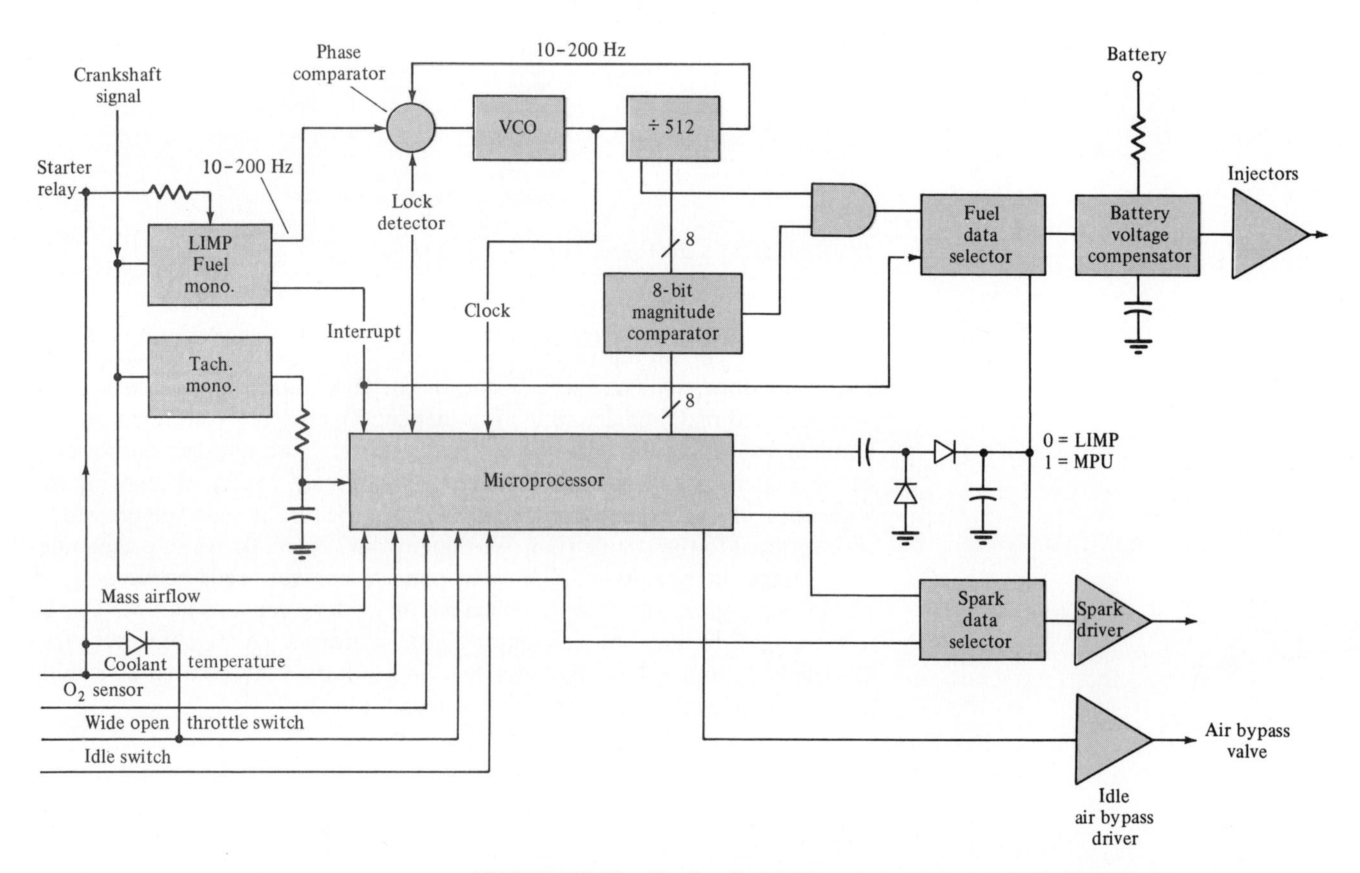

FIGURE 15.22. Engine Control Unit Using Motorola MC6805R2 Microprocessor

The duty cycle of the fuel injection during the open-loop operation—that is, the number of degrees of injection per rotation—is proportional to the mass airflow needed to maintain a constant air-fuel ratio. The airflow sensor provides a voltage that (by means of another A/D converter) allows the computer to consult a table of duty cycle values. This voltage is an 8-bit number, but to conserve memory space in the computer, only the most significant 4 bits—that is, the upper *nibble*—are listed in the table. More precise values are calculated by the microcomputer using interpolation. This procedure results in a table that occupies only 16 bytes.

Before the engine is up to temperature, the mixture is enriched beyond the stoichiometric ratio. The coolant temperature enters as a dc voltage and is converted during the calculation period into a proportional 8-bit binary number by the on-board A/D converter. A temperature enrichment algorithm increases the duty cycle linearly with lower coolant temperature. Additional enrichment arises if the accelerator is fully depressed (WOT—wide open throttle). This enrichment is a fixed percentage of the calculated duty cycle based on mass airflow. The WOT signal enters the computer through one of the I/O lines.

The open-loop calculations serve as fuel loop control only while the O_2 sensor is coming up to temperature. After the output voltage of this sensor is judged appropriate, a rich indication will cause a RAM byte to be decremented every eight engine revolutions. (If lean, the byte will be incremented.) This error byte is thus a record of the difference between the proper stoichiometric operating point (as determined by the O_2 sensor) and that predicted by the open-loop calculation. It is the corrected value of the calculation that is presented to the magnitude comparator that determines the fuel injection duty cycle. However, there are limits placed on the closed-loop operation. It cannot deviate more than 50% in either direction from the predicted values since such a large deviation from the expected operation would indicate a probable malfunction. In fact, it is suggested that any attempt at operation with such a large deviation might be used as a diagnostic trap, indicating to the driver the need for engine servicing.

In the idling condition, with the throttle valve closed, the duty cycle of a 180 Hz signal controls the amount of air through a throttle plate bypass. This air does flow past the airflow sensor, though. The 180 Hz signal corresponds to a 5.5 ms interval. It is not necessary that this signal be synchronized with the engine rotation, and, therefore, the duty cycle is decided by computer programming.

Computation of normal engine control requires about 1 ms. At the end of the program loop, the processor counts down an 8-bit idle number stored in RAM. When this number reaches zero, the bypass shuts off. Since the maximum idle count of 256 requires about 3 ms, the idle number can extend from 1 up to 4 ms. Every other time through the computation loop it is the complement of the idle number that is decremented and constitutes the other half of the duty cycle.

With such an arrangement the idle duty cycle range is 20–80%. The idle byte is incremented or decremented depending on the difference between the engine speed and the desired speed, reduced by the rate of engine speed change between samples. This procedure leads to stable operation under most conditions. The idle speed control is engaged only when the throttle plate switch is in the fully closed position as determined by the "idle" switch connected to the throttle plate.

15.6.3 MC6805R2 Microprocessor

It should prove profitable to take a closer look at the microcomputer used in conjunction with the automotive controls we have been discussing since it represents in some ways a considerable technical advance over what could be done with the Z80. It is the 6805R2 made by Motorola.

Figure 15.23 shows the 6805R2 in block diagram form. The unit includes an ALU, the CPU control, the accumulator, an index register, the stack pointer, and the program counter, although the bit capacity of some of these elements may be less than in the Z80. For example, in the Z80 the stack pointer accommodated 16 bits but in the 6805, only 5 bits. In the former case the index registers also held 16 bits, and now we have only 8. The program counter was 16 bits, and now we have 12 (split into 4 bits of high PC—that is, PCH—and 8 low—that is, PCL). But the most notable advance in the 6805 is the on-board memories—64 8-bit RAM locations, 2048 8-bit ROM.

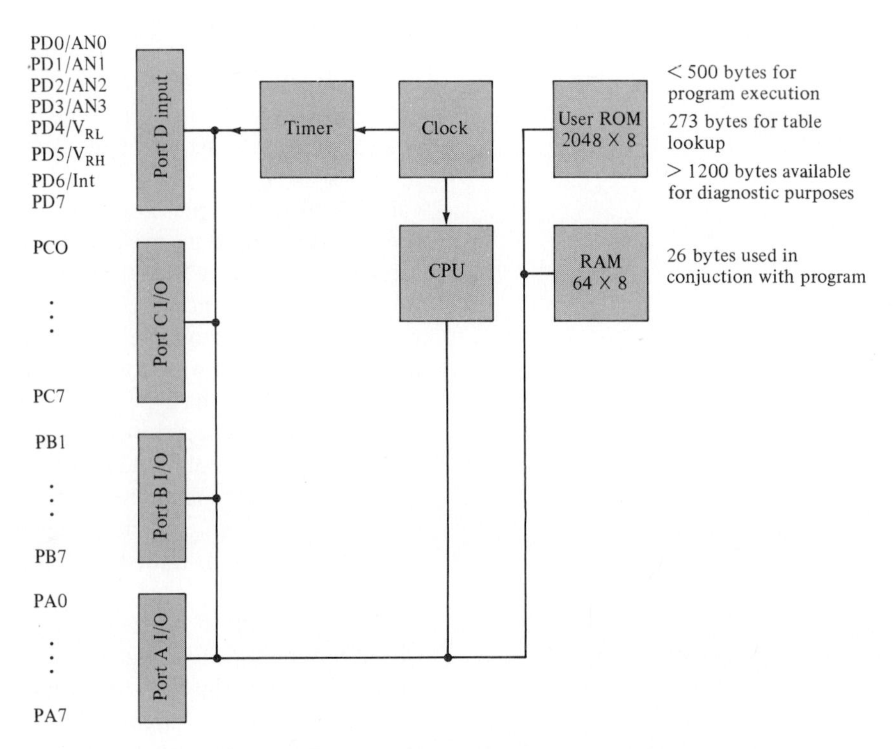

FIGURE 15.23. Simplified Block Diagram of Motorola MC6805R2 Microprocessor as Used in Conjunction with Automotive Control System

Also conspicuous in the 6805 are the sets of data lines, four of them. Ports A, B, and C each consist of 8-bit lines that can be used for either data input or output, their function being controlled by software programming of the data direction registers (DDRs). The D set can be used as a conventional set of 8-bit input lines, but, more importantly for our automotive system, these lines may also be used to convert analog signals into digitized form.

Analog lines AN0–AN3 may be directed into the analog **multiplexer** (MUX). A multiplexer allows individual lines to be addressed. To lines PD4 and PD5 are connected dc voltages that represent the reference limits of the analog signals (V_{RH} and V_{RL}, being the high and low limits, respectively). The A/D converter will proportionally convert the analog signals into corresponding digital forms. Memory location 14 in the RAM is the A/D control register, and bits 0, 1, and 2 control which line is to be addressed. The converter operates continuously, and the completed conversion, in digitized form, is placed in location 15 of RAM.

15.7 ▪ PLL SPECIFICATIONS

15.7.1 The 565

With f_o the free-running center frequency, θ_d the maximum phase error under lock-in conditions, and A_d the gain of the internal amplifier, the following characteristics apply to the 565 IC PLL:

$$K_{VCO} = \frac{50 f_o}{V_{CC}} \quad \text{[rad/V·s]}$$

$$K_P \simeq \frac{1.4}{\pi} \quad \text{[V/rad]}$$

$$\theta_d = \pm \frac{\pi}{2} \quad \text{[rad]}$$

$$A_d = 1.4$$

The center frequency depends on externally selected values of R and C by means of the following equation:

$$f_o \simeq \frac{1.2}{4RC} \tag{15.15}$$

The radian frequency limit on each side of f_o (ω_L, where $2\omega_L$ is termed the **lock range**) is derived from the equation

$$\omega_L = K_{VCO}\, K_P\, A_d\, \theta_d = \frac{50 f_o}{V_{CC}} \frac{1.4}{\pi} (1.4) \frac{\pi}{2} \simeq \frac{49 f_o}{V_{CC}} \tag{15.16}$$

The **capture range**, on the other hand—that is, $2\omega_C$—the frequency range within which the lock-in operation is initiated, depends on the time constant of the filter, τ:

$$\omega_C \simeq \sqrt{\frac{\omega_L}{\tau}} \tag{15.17}$$

It can be seen that the capture range is always less than the lock-in range. This approximation (Eq. 15.17) is valid for $f_C \gtrsim \frac{1}{3} f_L$. The 565 is a general-purpose PLL designed to operate at frequencies below 1 MHz.

EXAMPLE 15.3

Determine the lock range and capture range of a 565 PLL with $V_{CC} = 5$ V and $f_o = 1$ kHz if the low-pass filter has a time constant of 1 ms.

Solution:

$$\omega_L = 2\pi f_L = \frac{49 f_o}{V_{CC}}$$

$$2f_L = \frac{49(10^3)}{5\pi} \simeq 3 \text{ kHz}$$

$$2f_C \simeq \frac{1}{\pi}\sqrt{\frac{\omega_L}{\tau}} = \frac{1}{\pi}\sqrt{\frac{2\pi(1.5 \times 10^3)}{10^{-3}}} = 977 \text{ Hz}$$

15.7.2 The 4046

The 4046 PLL includes two comparators, one for type I operation and a second for type II. It employs CMOS circuitry, which leads to very low power consumption, typically 70 μW.

Type I Operation. The comparator in this case is an EXCLUSIVE OR circuit, and the signal and comparator input frequencies should be 50% duty cycle square waves. With no signal present, the comparator's output voltage is $V_{DD}/2$ (V_{DD} being the applied dc voltage), and the VCO will then oscillate at its center frequency f_o. With this comparator, lock-in operation should remain even in the presence of considerable noise.

At the center frequency there is a 90° phase difference between the signal and VCO (control voltage at $V_{DD}/2$), which diminishes to 0° (when the control voltage is approximately zero) and 180° (where it is V_{DD}).

The center frequency (without offset; see Figure 15.24a) is obtained from the equation

$$f_o = \frac{K_1}{R_1 C_1} \tag{15.18}$$

and the lock-in range is $f_L = \pm f_o$. K_1 is a constant that depends on the dc voltage V_{DD}. For $V_{DD} = 10$ V, $K_1 \simeq 0.7$, and for $V_{DD} = 5$ V, $K_1 \simeq 0.5$. R_1 and C_1 represent the charging circuit in the VCO.

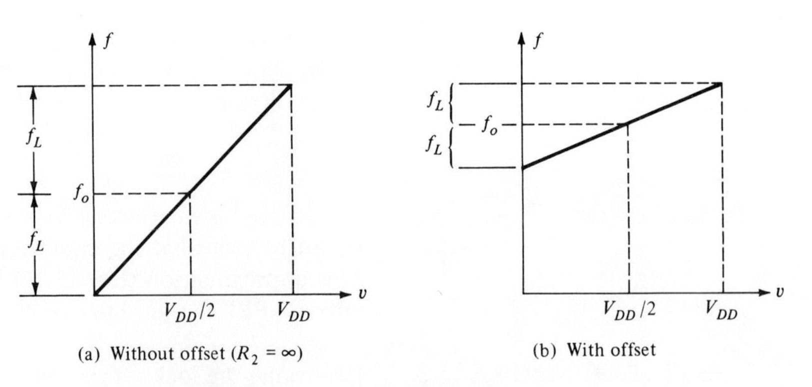

FIGURE 15.24. VCO Frequency vs. Control Voltage

For offset operation (Figure 15.24b), we have

$$R_1 = \frac{K_1}{f_L C_1} \quad \text{and} \quad R_2 = \frac{2K_1}{(f_o - f_L)C_1} \tag{15.19}$$

where R_2 is another external resistor in the VCO circuit. We also have

$$f_o \simeq \frac{K_1}{R_1 C_1} + \frac{2K_1}{R_2 C_1} \tag{15.20}$$

$$f_L = \pm \frac{f_o}{1 + (2R_1/R_2)} \tag{15.21}$$

$$f_c \simeq \pm \frac{1}{2\pi}\sqrt{\frac{2\pi f_L}{R_3 C_2}} \tag{15.22}$$

where R_3C_2 represents the low-pass filter.

Type II Operation. The comparator in this case reacts to the time interval between the leading edges of the signal and the VCO output (Figure 15.25). In a locked condition the leading edges of both the VCO and the signal will be coincident, as will both the repetition rates, but the duty cycle of each is irrelevant.

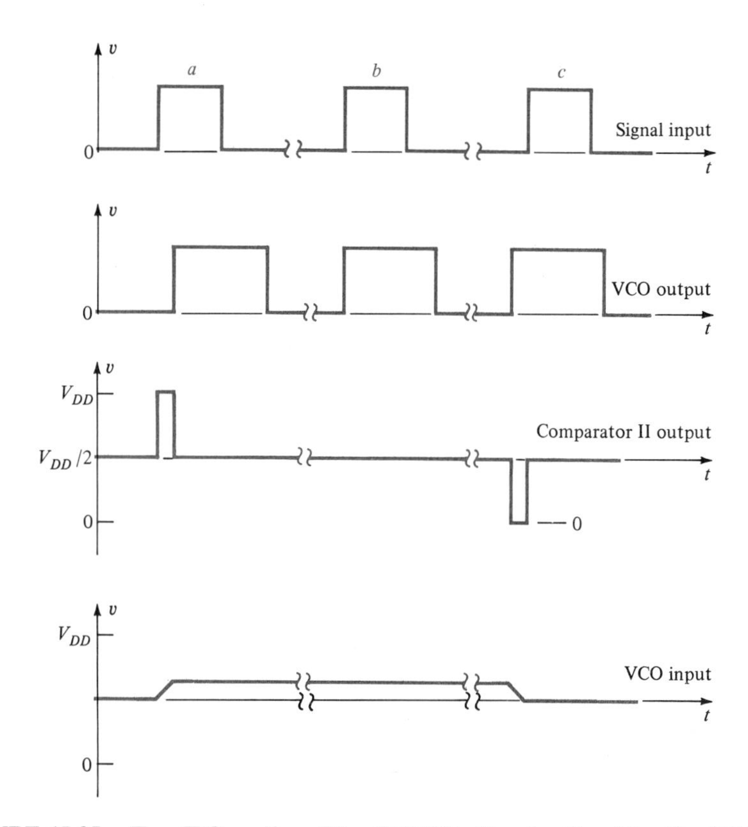

FIGURE 15.25. **Type II Operation of the 4046 PLL, Showing Signal Leading VCO at *a*, Signal in Phase with VCO at *b*, and Signal Lagging VCO at *c***

In the absence of a signal, the VCO in this type of operation emits its lowest frequency. If the signal loses a pulse or an extra one appears, this type of operation interprets such an event as a large error. As a result this type of operation is rather intolerant of noise.

SUMMARY

15.1 Introduction

- A phase-locked loop generates a frequency that rather accurately maintains its frequency and phase relative to some reference frequency.
- A VCO is an oscillator whose output frequency is determined by the magnitude of the applied dc voltage.

15.2 Phase Detectors

- A phase detector can be used to measure the phase difference between two signals whose frequency is identical.

15.3 The Phase-Locked Loop

- A phase-sensitive detector yields an output frequency that represents the frequency difference between two signals. (It also has an output equal to their sum, but this frequency is usually discarded.)
- A phase-locked loop makes use of a phase-sensitive detector and a VCO to bring a reference frequency into synchronism with an input signal frequency.

15.5 Touch-Tone®: Another PLL Example

- Depression of each button on a Touch-Tone® telephone generates two, unharmonically related, tones.
- Each dialing unit needs two VCOs, one for each of the two generated frequencies.
- Seven decoders are needed for each receiving unit.

15.6 Transistorized Engine Controls: Another PLL Example

- With the burn time of a fuel-air mixture remaining approximately constant, the ignition must be applied at progressively earlier times as an internal combustion engine increases its rpm—that is, the spark must be advanced.
- A byte (8 bits) may be split into four most significant bits (MSBs) and four least significant bits (LSBs), being, respectively, the upper and lower nibbles.
- Electronic automotive controls operate in three modes: open-loop, closed-loop, and backup mode.

KEY TERMS

beat frequency: when two frequencies are applied to a nonlinear device, it gives rise to a frequency equal to their difference

capture range: that span of frequencies within which a PLL can accomplish a locked condition, usually smaller than the lock range

constant current generator: a current source that is constant under load variations

lag filter: the phase of the output lags behind the input in a progressively increasing fashion as the frequency increases

lag-lead filter: with increasing frequency, an initial increase in the phase delay subsequently gives rise to a diminishing phase delay

lean mixture: a fuel-air mixture ratio less than the stoichiometric ratio

lock range: that span of frequencies within which a PLL remains in a locked condition

multiplexer: a unit that allows for the selection of one among a number of data lines

phase detector: as used in this text, a device that produces a dc voltage proportional to the difference between two signals of equal frequency

phase-locked loop (PLL): a combination of phase-sensitive detector and VCO that generates a frequency equal to that of an incoming signal frequency

phase-sensitive detector: as used in this text, a phase detector to which the applied frequencies are not necessarily the same and whose output consists of their sum and difference frequencies

rich mixture: a fuel-air mixture ratio greater than the stoichiometric ratio

stoichiometric ratio: a fuel-to-air mixture of 1:14.7; this is the mixture that leads to complete combustion

tachometer: an instrument used to determine the rotational speed of an engine

top dead center (TDC): the condition in an internal combustion engine when the cylinder volume assumes its smallest value—that is, when the connecting segment of the crankshaft is at its greatest upward extension

voltage-controlled oscillator (VCO): a signal generator whose output frequency is determined by the value of dc control voltage applied

KEY DEVICES

565: popular phase detector designed to be driven by either an analog signal or digital square waves (type I)

566: linear VCO with either square wave or rectangular output

4044: TTL phase detector designed to be driven by edge transitions (type II)

4046: CMOS phase detector containing both type I and type II detectors

SUGGESTED READINGS

"Phase-Locked Loop Applications," SIGNETICS DIGITAL, LINEAR, AND MOS APPLICATIONS, pp. 6-1 through 6-64. Sunnyvale, Calif.: Signetics Corp., 1974. *Detailed circuitry of the 565, 566, 567, etc., together with numerous applications of phase-locked loops.*

Kinley, Harold, THE PLL SYNTHESIZER COOKBOOK. Blue Ridge Summit, Penn.: Tab Books, 1980. *The use of PLL for frequency generation, particularly in CB radio applications.*

Blanchard, Alain, PHASE-LOCKED LOOPS, APPLICATION TO COHERENT RECEIVER DESIGN. New York: Wiley-Interscience, 1976. *A thorough treatment of PLL, with many numerical examples.*

Gardner, Floyd M., PHASELOCK TECHNIQUES, 2nd ed. New York: Wiley-Interscience, 1979. *A book somewhat less detailed than the Blanchard text and therefore probably more readable; however, it is rather devoid of numerical examples.*

Boylestad, Robt., and Louis Nashelsky, ELECTRONIC DEVICES AND CIRCUIT THEORY, 3rd ed., pp. 630–635. Englewood Cliffs, N.J.: Prentice-Hall, 1982. *Illustrative examples of PLL as used in frequency demodulation, frequency synthesis, and FSKs (frequency shift decoders—such as used for reception of teletype transmission).*

Lancaster, Don, CMOS COOKBOOK, pp. 361–371. Indianapolis: Howard W. Sams, Publ., 1977. *A consideration of the various applications of the 4046 PLL, which allows one to select either type I or type II operation.*

A CLOSER LOOK: Constant Current Source and Pollution Control

Constant Current Source

The ideal constant current source should provide the stated current independent of the value of the load. In practice a constant current source provides the stated current only over a limited range of load resistance. The output voltage range over which it behaves in a satisfactory manner is termed its output *compliance*.

The basic circuit is illustrated in Figure 1. The load current I_L is the same as that through R_1. Since the voltage drop across R_1 is $V_1 - V_L$, if $V_L \ll V_1$, we have

$$I_L = \frac{V_1 - V_L}{R_1} \simeq \frac{V_1}{R_1}$$

That is, if V_L is small compared to V_1, variations of the load resistance will have little influence on the current.

The departure from constant current is given by the relation $[100(V_L/V_1)]\%$. If V_L is required to be several volts and the current variation is to be held to within a few percent, V_1 must be of the order of hundreds of volts, and this condition often proves to be inconvenient to satisfy.

A somewhat more practical constant current source

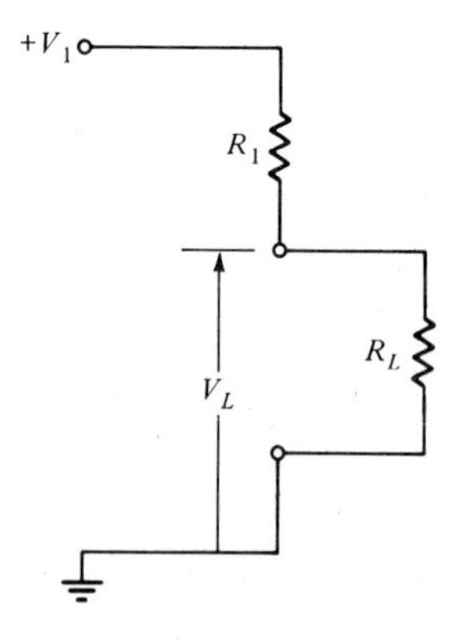

FIGURE 1

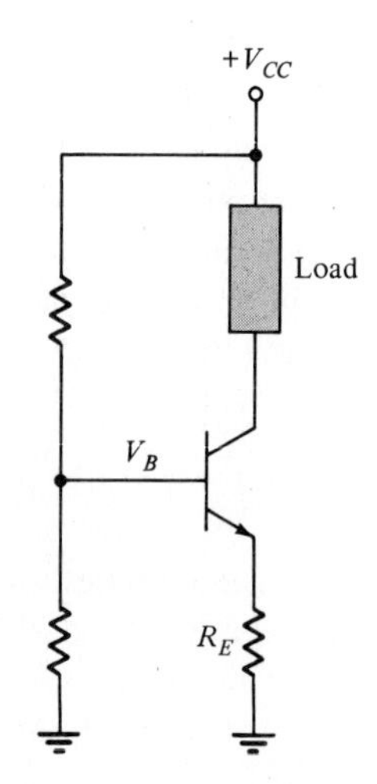

(a) BJT constant current generator circuit

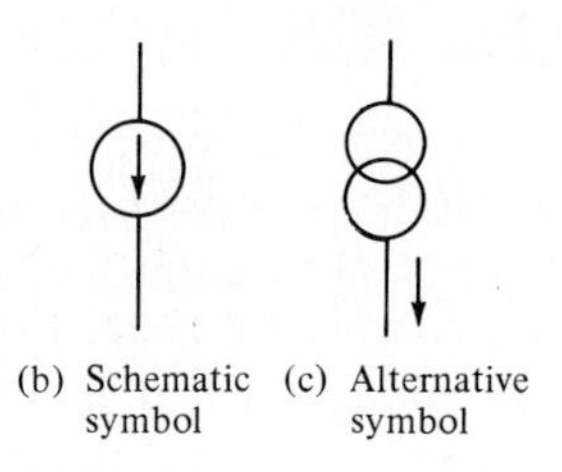

(b) Schematic symbol (c) Alternative symbol

FIGURE 2

makes use of a BJT operating with a constant base voltage and in the active operating region. The active region corresponds to the flat portion of the i_C vs. v_{CE} characteristic curve (see Figure 7.24). With $V_{BE} = 0.7$ V, we have

$$I_E = \frac{V_E}{R_E} = \frac{V_B - 0.7}{R_E}$$

For large values of h_{FE}, $I_C \simeq I_E$ and

$$I_C \simeq \frac{V_B - 0.7}{R_E}$$

which is independent of V_C provided the transistor is not saturated—that is, $V_C \gtrsim V_B + 0.2$ V. Figure 2 shows the arrangement. Constant current transistor circuits can become more elaborate, particularly in terms of maintaining a constant base voltage. An FET makes an even better current source since its characteristic curves can show an even smaller slope.

In Figure 15.10 what constitutes the load is the complete circuitry involving transistors Q_1 through Q_6.

Pollution Control

The three main automotive exhaust components that contribute to atmospheric pollution are hydrocarbons (HC), carbon monoxide (CO), and oxides of nitrogen (NO_x). Figure 3 indicates that the CO emissions decrease with increasing fuel-air ratios. This increase is to be expected, of course, since increased oxygen tends to guarantee more complete combustion. Theoretically, an air-fuel ratio of about 15:1 (termed the **stoichiometric mixture**) should lead to complete combustion of all the carbon and hydrogen in the fuel. In practice, because of imperfect air-fuel mixing as well as the short combustion period, air in excess of the stoichiometric value is needed for complete combustion.

In addition to adequate oxygen, air-fuel mixing, and combustion duration, the combustion chamber temperature also affects the emissions. Thus, as more air is fed into the combustion chamber there will be a cooling effect, and such cooling slows the combustion process. Beyond about 18.5:1 the HC emissions begin to increase again. Thus, 18:5 is about the upper limit for most engines.

If the lean mixtures by themselves do not diminish the HC and CO to acceptable levels, the next step requires that the excess be burned in the exhaust system itself. This burning can be accomplished by means of a catalytic reactor. Since the reactors operate best at elevated temperatures, engine operation with a somewhat retarded spark often proves helpful. While a retarded spark reduces the peak combustion chamber temperature, the temperature of the exhaust gas is actually increased. Of course, a retarded spark will probably lower the engine's cycle efficiency.

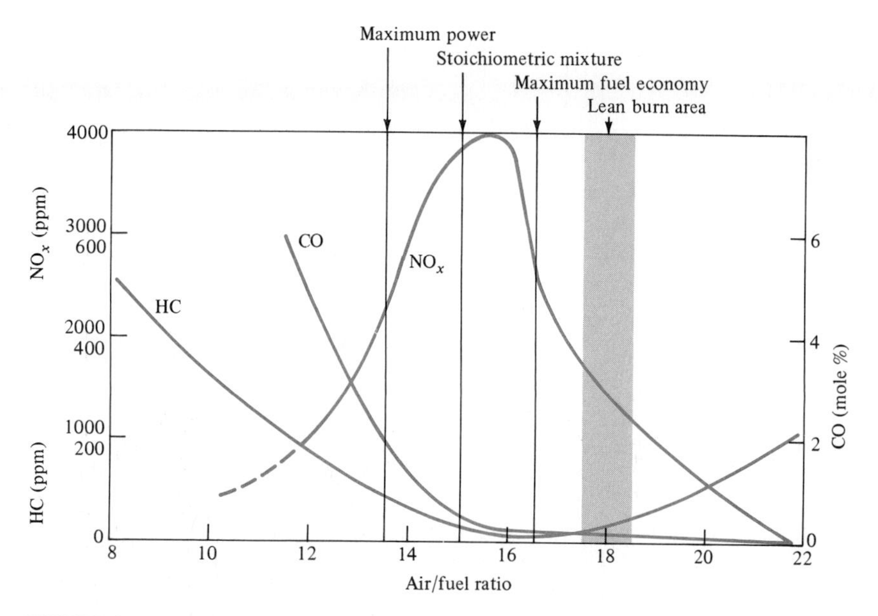

FIGURE 3

Let us turn now to a consideration of the NO_x emissions. Unfortunately, most of the methods that tend to reduce CO and HC tend to increase NO_x content. The optimum combustion chamber temperature is in excess of 3500°F. At such temperatures oxygen and nitrogen combine to form nitric oxide, a component that (through photochemical atmospheric reactions) forms ozone and other oxidants.

Nitric oxide production may be diminished by lowering the combustion temperature through the addition of either water, more air, or exhaust gas recirculation, but the HC content will go up. There will also be an accompanying loss in fuel economy. Probably the best solution stems from the use of a three-way catalyst, one that takes care of NO, HC, and CO.

The NO_x emissions are reduced by using a rhodium catalyst. This material may be combined with an oxidation catalyst (such as platinum) to take care of CO and HC. For effective operation such three-way catalysts can tolerate only very small amounts of oxygen. It is necessary, therefore, to maintain very accurate control of the air-fuel mixture. Figure 4 shows this narrow range in terms of CO concentration. Additional removal of CO and HC may be effected by using a downstream oxidation catalyst.

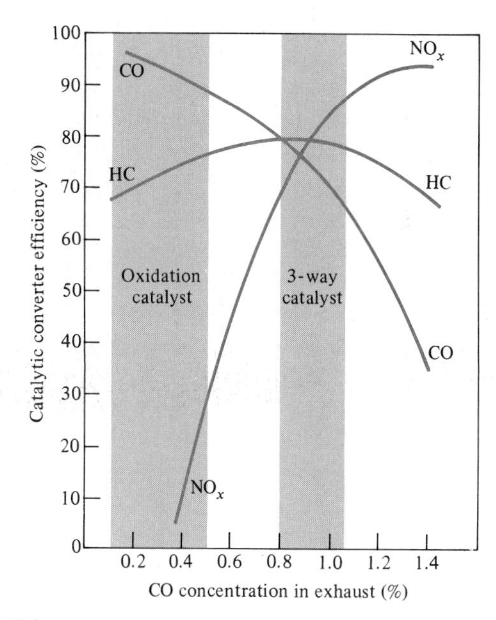

FIGURE 4

EXERCISES

1. What is the meaning of PLL?
2. In block diagram form, sketch the basic units of a PLL.
3. What is a VCO?
4. Why does the PLL have a reputation for instability?
5. What is the simplest form of filter that can be used with a PLL? What is a preferred filter in this respect?
6. Distinguish between PLL types I and II.
7. What determines the time constant of the filter used in conjunction with a PLL?
8. Distinguish among 565, 566, and 567 units.
9. Indicate the dimensions associated with each of the following gains: K_P, K_F, K_{VCO}.
10. Why are the tones used in a Touch-Tone® telephone purposely made to avoid harmonic relationshps?
11. What is the meaning of TDC?
12. What is the time interval for complete burn-through of a fuel-air mixture?
13. What does it mean to have the spark advanced as the engine speed increases?
14. What are the three operating modes of an electronic engine control?
15. Verify that the backup pulse at $+15°$ BTDC corresponds to the proper advance angle for 1600 rpm (assuming a 3 ms burn time).
16. Assuming a 3 ms burn time, what would be the proper spark plug advance at 1500 rpm for the engine we have considered?
17. With maximum power at 15° past TDC, for the engine under consideration, with a 3 ms burn time, what would be the rpm corresponding to 60° BTDC, the longest time available? For engine speeds of the order of 6000 rpm, what modification of this calculation would seem to be required?
18. At 2400 rpm, what is the undivided output frequency of the VCO?
19. In conjunction with the controls discussed, describe the operation of the tachometer.
20. While some aspects of the 6805 are inferior to the Z80, what is a major advance?
21. What are the three main atmospheric pollutants encountered in automotive engine exhaust?
22. In relation to the stoichiometric ratio, what generally is the air-fuel mixture required? What is the upper limit for most engines?
23. Why might a somewhat retarded spark aid in pollution control? What does such retardation do to the engine efficiency?

PROBLEMS

15.1
It is desired to have a digital tachometer measure the 300 rpm of a four-cylinder, four-stroke engine. What is the frequency of the digital pulses obtained from the cylinder firings? If an accuracy of 0.1% is desired, what should be the measuring interval?

15.2
The measuring time in Problem 15.1 is impractically long. If we decide 0.5 s to be a *practical* maximum for the measuring interval and for convenience decide to have a count of 600 pulses represent 600 rpm, this count will represent a pulse frequency of

$$\frac{600 \text{ pulses}}{0.5 \text{ s}} = 1200 \text{ Hz}$$

which is 120 times the input frequency. We can handle this problem by having an oscillator that runs at 120 times the input frequency, and the oscillator output is what we count during the 0.5 s interval. We also *divide* the oscillator frequency by 120 and compare that output with the pulse input from the engine. Any difference between these two frequencies can be made to develop a voltage proportional to the difference. This voltage is applied to a voltage-controlled oscillator that will alter the oscillator so as to bring the error voltage to zero and thereby have the oscillator frequency precisely $\frac{1}{120}$ times that of the pulse frequency. Sketch the following in block diagram form: the time reference, sequencer, VCO, divider circuit, phase-sensitive detector (with speed input), error voltage amplifier, and digital counter.

15.3
In the operation of a phase-locked loop, a sinusoidal frequency variation

$$v_i = V_i \sin(\omega_i t + \theta_i)$$

is multiplied by a square wave oscillator signal:

$$v_o = \sum_n \frac{4}{\pi(2n + 1)} \sin[(2n + 1)\omega_o t]$$

Show that the resulting input signal is of the form

$$V_d = \frac{2}{\pi}\left\{ \sum_{n=0}^{\infty} \frac{V_i}{2n + 1} \cos[(2n + 1)\omega_o t - \omega_i t - \theta_i] - \sum_{n=0}^{\infty} \frac{V_i}{2n + 1} \cos[(2n + 1)\omega_o t + \omega_i t + \theta_i] \right\}$$

15.4
Plot the transfer function (V_2/V_1) in decibels and the phase variation as a function of radian frequency for a lag filter with $R = 1\ M\Omega$ and $C = 1\ \mu F$ (Figure 15.13). After plotting the results, justify the expression *lag* filter.

15.5
a. Plot the transfer function (in decibels) for the lag-lead filter shown in Figure 15.14 with $R_1 = 1\ M\Omega$, $R_2 = 0.5\ M\Omega$, and $C = 1\ \mu F$. Also plot the phase variation.

b. Do the same if R_2 is decreased to $0.1\ M\Omega$, plotting the results on the same graph as developed in part a.

c. Analytically determine the crossing point between the transfer function curves of parts a and b. Compare the result with your graphical solution.

15.6
A lag-lead filter, with $R = 1\ M\Omega$, $R_2 = 100\ k\Omega$, and $C = 1\ \mu F$, is to be used in conjunction with the PLL considered in Example 15.1. Determine the unity loop gain frequency and the phase shift at that frequency. Plot the decibel gain and the phase.

15.7
a. Show graphically that two 50% duty cycle square waves of equal frequency but displaced in phase when applied to an EXCLUSIVE OR circuit may be used to generate a dc voltage suitable for PLL operation.

b. If the signal frequency increases, show graphically what happens to the EXCLUSIVE OR voltage.

c. Sketch the average dc voltage variation of the EXCLUSIVE OR phase detector as the signal square wave moves in and out of phase with the reference square wave.

15.8
Justify the statement that a VCO acts as an integrator.

15.9
In terms of the two time constants, what is the expression for the phase of a lag-lead network?

15.10 △

a. As a radiotelescope "looks out" into space, it receives background noise that may be characterized by a noise temperature (see Chapter 13). For a background noise temperature of 120°K and with an antenna representing a characteristic impedance of 50 Ω, what is the rms background noise voltage delivered by the antenna into a 1 MHz noise bandwidth?

b.
1. A typical radio star emits noise power of about $10^{-26}\ W/m^2 \cdot Hz$. Using an antenna with a cross-sectional area of 10,000 m^2 and a receiver noise bandwidth of 1 MHz, what is the "signal" power received?
2. What change in temperature above the background noise level does this signal represent?
3. What is the magnitude of the increase in received noise voltage due to the radio star?
4. What is the input signal-to-noise ratio?

A simple low-pass filter at the output of a phase-sensitive detector may be used to extract signals that are much smaller than the noise level. By presuming the "signal" to be a low frequency (or even dc), as one increases the output time constant, the noise bandwidth is progressively restricted to such an extent that the signal eventually will stand out above the remaining noise. Such is the method by which radio astronomers, for example, extract signals that may be 1000 (or even 10,000) times smaller than the accompanying noise level.

c. With V_n representing the noise voltage and $\Delta\omega$ the bandwidth, we have

$$V_n \propto \sqrt{\Delta\omega}$$

If the input bandwidth to the phase-sensitive detector is $\Delta\omega_i$ and $\Delta\omega_o$ is the output from the low-pass filter, we have

$$\frac{V_{n,in}}{V_{n,out}} = \sqrt{\frac{\Delta\omega_i}{\Delta\omega_o}}$$

Designating S as the desired signal, we have

$$\frac{(S/V_n)_{out}}{(S/V_n)_{in}} = \sqrt{\frac{\Delta\omega_i}{\Delta\omega_o}}$$

If the input signal-to-noise ratio were that computed in part b, what would be the time constant of the output filter, taking $(S/V_n)_{out} = 3$ as the requirement for certainty of detection?

d. What would be the time constant if the signal-to-noise ratio were $\frac{1}{1000}$? What would this time constant imply concerning the detection of signal variation?

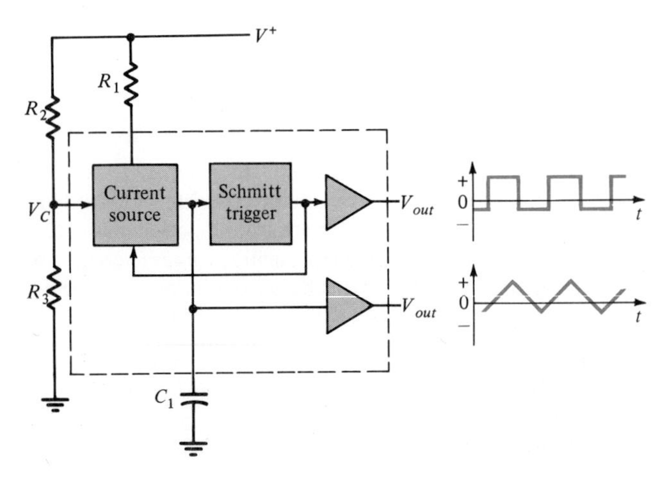

FIGURE 15.26

15.11

For a 566 VCO (Figure 15.26), with $V^+ = 12$ V, $R_1 = 10$ kΩ, and $C_1 = 820$ pF, what should be the values of R_2 and R_3 in order to obtain a 10 kHz square wave? Since the available resistance values will often not agree with the computed values, what should be done to adjust the value of V_C?

$$f_o = \frac{2}{R_1C_1}\left(\frac{V^+ - V_C}{V^+}\right)$$

$2\text{ k}\Omega \leq R_1 \leq 20\text{ k}\Omega$

$0.75\, V^+ \leq V_C \leq V^+$

$f_o \leq 1\text{ MHz}$

$10\text{ V} \leq V^+ \leq 24\text{ V}$

PART IV

MEASUREMENT, CONTROL, AND POWER

16 POWER SUPPLIES AND POWER CONTROL

OVERVIEW

The analytical consequences of converting ac into (pulsating) dc were first considered in conjunction with Fourier analysis (Chapter 5). The introduction of the diode in Chapter 6 allowed us to examine a device by means of which this conversion could be accomplished. Precision rectifiers employing OPAMPs appeared in Problems 10.7 and 10.8.

To maintain constant circuit voltages that are subject to minimal variations due to line voltage and load fluctuations, rather elaborate circuits must be employed. They are introduced in this chapter. Again, this fuller treatment of the topic had to be deferred until the necessary preliminary concepts had been considered, specifically the transistor (Chapter 7), negative feedback (Chapter 8), and operational amplifiers (Chapter 10).

OUTLINE

16.1 ■ INTRODUCTION

We have previously considered the conversion of ac into dc and a few methods by means of which the resulting dc pulsations may be made more nearly equal to a constant voltage. There are two major causes that may lead to departures from this constancy:

1. The value of the load current being drawn from the supply may change. Various activated relays may suddenly impose additional power demands. An audio amplifier may be called upon to reproduce a loud musical passage. An electronic instrument is taken from a standby condition into an active state.
2. The power-line ac voltage may vary. The power demanded of the utility company does not remain constant, and, although the utility makes adjustments to compensate for such variations, the compensation is not immediate.

Both of these instances—an electronic amplifier and a utility company power line—we can consider to be ideal generators in series with internal resistances. As the current demand varies, so does the voltage drop across the internal resistance, and this varying current leads to a variation in the delivered voltage.

While the functions performed by a power supply are simple in principle, the complexity of the circuitry that may be employed can be truly surprising. This complexity arises because of the large number of eventualities for which provision must be made, such as the accidental short-circuiting of the output. Power supply design is a rather specialized branch of electronics, and one of the objectives of this chapter is to acquaint the reader with some of the problems and some of the solutions and to justify the substantial price that is sometimes demanded for a good power supply, although the advent of IC regulators has significantly lowered such prices. The second part of this chapter is concerned with the control of power devices.

16.2 ■ VOLTAGE AND CURRENT REGULATION

16.2.1 Zener Diodes

In Figure 16.1 is shown the current vs. voltage curve for a Zener diode. Normally, one would apply a positive voltage to the anode of the diode and a negative (or less positive) voltage to the cathode. The right-hand side of the characteristic curve (Figure 16.1) illustrates a typical voltage-current (*V-I*) variation. If, however, one reverses the polarity of a diode, it will conduct very little under such reverse bias conditions unless this voltage is made sufficiently large, in which case there will be a sudden increase in the current (termed a **Zener breakdown**), sufficient to damage the device unless this current is limited in some manner. A resistor placed in series with the device provides such limitation. This series resistance determines the operating point, and from Figure 16.1 it can be seen that for a considerable variation of the "reverse" current the voltage across the diode shows little variation. The fabrication of a Zener diode is such as to accentuate this vertical portion of the *V-I* curve.

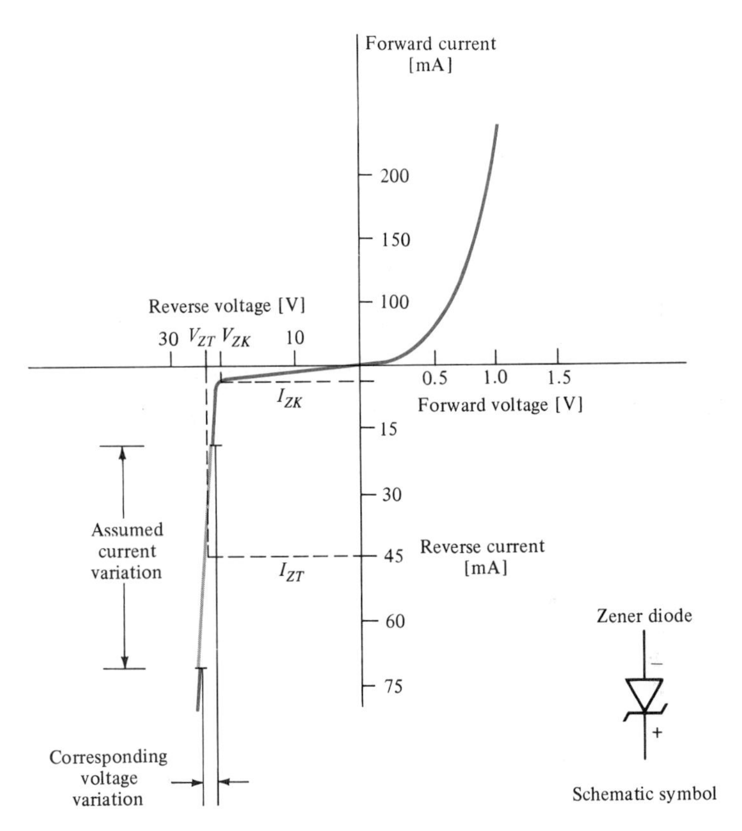

FIGURE 16.1. **Diode Current vs. Voltage, Showing the Zener Characteristic in the Third Quadrant**

The current drawn by the diode at its nominal Zener voltage is called the **test current** I_{ZT}. The maximum power rating of the diode is usually taken to be four times the power dissipation at the Zener voltage—that is, $4I_{ZT}V_{ZT}$. The **knee current** I_{ZK} represents the minimum Zener current commensurate with stable operation.

EXAMPLE 16.1

a. A Zener diode (Figure 16.2) with $V_Z = 51$ V is used in conjunction with a 200 V source. The allowable Zener current extends from 5 to 40 mA. Calculate the value of R that allows voltage regulation with the load varying from $I_L = 0$ to $I_{L(max)}$. What is the value of $I_{L(max)}$?
b. With $I_L = 25$ mA and the resistance as calculated in part a, what is the input voltage variation that can be tolerated?
c. Assume the design is based on an R prescribed by $I_{ZT} = 25$ mA.

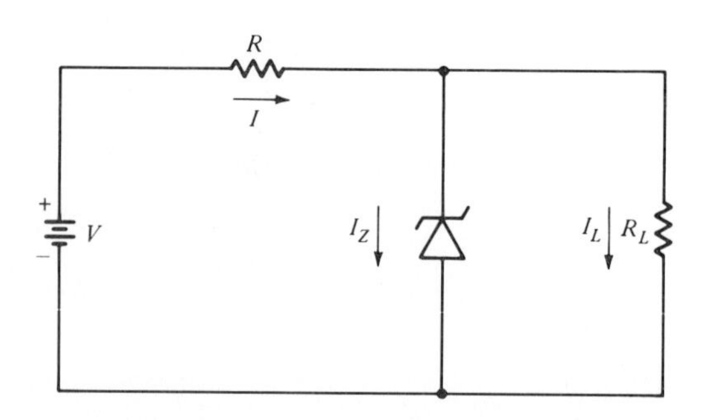

FIGURE 16.2. **Zener Voltage Regulator Circuit**

How does this design alter the range of allowable load current variation?

d. What should be the power rating of the Zener in each case?

Solution:

a. $I_Z = I - I_L$

As R_L is varied, $I_L = V_Z/R_L$, but I remains constant. $I_Z = 40$ mA when $I_L = 0$. We have

$$R = \frac{V - V_Z}{I} = \frac{200 - 51}{40 \times 10^{-3}} = 3.725 \text{ k}\Omega$$

With minimum Zener current—that is, I_{ZK}—of 5 mA, the maximum load current is

$$I_{L(max)} = I - I_{ZK} = 40 - 5 = 35 \text{ mA}$$

b. At minimum diode current, $I = 5 + 25 = 30$ mA. We have

$$V = IR + V_Z = 30(3.725) + 51 = 162.8 \text{ V}$$

At maximum Zener current, $I = 40 + 25 = 65$ mA. We have

$$V = 65(3.725) + 51 = 293.1 \text{ V}$$

c. $$R = \frac{V - V_{ZT}}{I_{ZT}} = \frac{200 - 51}{25 \times 10^{-3}} = 5.96 \text{ k}\Omega$$

I stays fixed. When $I_Z = 5$ mA, $I_L = 20$ mA. When $I_Z = 25$ mA, $I_L = 0$ mA.

d. $P = V_{ZT}I_{ZT}$

For part a, $P = 51\,(40 \times 10^{-3}) = 2.04$ W. By applying the generally used safety factor of 4, the rated power should be 8.16 W. (The nearest standard unit is rated at 10 W.)

For part c, $P = 51\,(25 \times 10^{-3}) = 1.275$ W. Therefore, a 5 W unit will do.

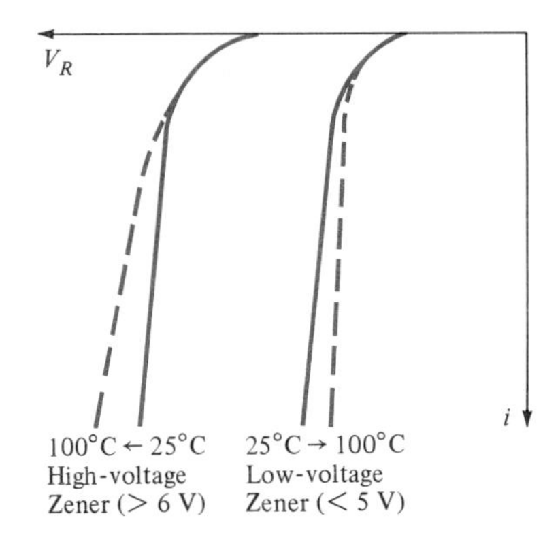

FIGURE 16.3. Variation of Zener Voltage with Positive Temperature Coefficients ($V_Z > 6$ V) and Negative Coefficients ($V_Z < 5$ V)

Zener diodes are available in values from 2 to 200 V, in the same series of values as 5% resistors, with power ratings from a fraction of a watt to 50 W. They are current- and temperature-sensitive, but there is a variety of ways to improve their performance.

A Zener diode, made to serve as a source of current, is termed a **voltage regulator**. Others are designed to serve as **voltage references**. As voltage regulators they tend to be noisy, and the Zener voltage variation with temperature and current can be as much as 0.1%/°C and 1%, respectively, the latter when the current changes by a factor of about 5. With regard to temperature and current variations, however, there are fabrication methods that can lead to vast improvements in performance, making them suitable as voltage references.

There are two mechanisms responsible for Zener breakdown. One operates below 5 V (termed Zener breakdown) and exhibits a negative **temperature coefficient** (termed tempco, or TC). The other (termed avalanche breakdown) operates above 6 V and exhibits a positive temperature coefficient. For Zeners between 5 and 6 V, both mechanisms are operative, and between 5.1 and 5.6 V, Zeners can be fabricated with a temperature coefficient in the vicinity of zero. Since the temperature coefficient also depends on the current, it is subject to some adjustment in this manner (Figure 16.3).

Temperature compensation at higher voltages may be obtained by a series combination of a conventional diode with a Zener. Since conventional diodes have a tempco of about −2.5 mV/°C, by selecting a zener with an equal *positive* coefficient, the two in series may be made to compensate for one another. Such a diode combination is commercially available, for example, in the 1N821 series. In contrast to the 3600 ppm/°C variation for a normal diode, one has

- 1N821: 100 ppm/°C
- 1N823: 50 ppm/°C
- 1N825: 20 ppm/°C
- 1N827: 10 ppm/°C
- 1N829: 5 ppm/°C

Their temperature range extends from −55°C to +100°C. They are rated at 6.2 V, with a range of 5.9–6.5 V at 7.5 mA, and have a dynamic impedance of 15 Ω. They are also available with a 10 Ω impedance, being in such cases distinguished by a suffix A appended to the designation. A 1N821 costs about $1.00 and a 1N829A, about $25.00.

But to maintain these small temperature coefficients, the Zener must be operated at a constant current. Figure 16.4 shows how this current may be controlled using an OPAMP. The circuit operates in the following manner: When first turned on, the 1N829 has a high resistance (that is, an open circuit), and its voltage divider action leads to positive feedback. As the output voltage of the OPAMP rises, the Zener breaks down and thereby places the noninverting input at the voltage V_{ZT}. Now the negative feedback through R_2 and R_3 dominates, and a stable condition is established, with the amplifier amplifying the V_{ZT} voltage. R_2 and R_3 are chosen to set the OPAMP output voltage level, and R_1 is chosen for the desired current in the diode.

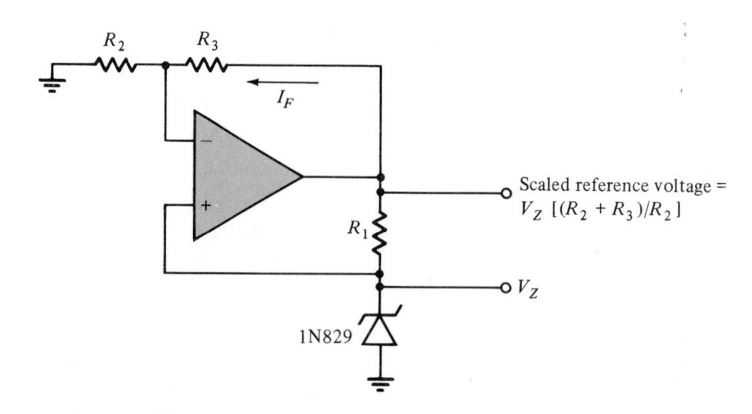

FIGURE 16.4. **Utilizing an OPAMP to Provide Constant Current for a Reference Zener as Well as a Scaled-up Reference Voltage**

EXAMPLE 16.2

a. A 1N829 compensated Zener, rated as 6.2 V ± 5% at 7.5 mA, has an incremental resistance of 15 Ω and a temperature coefficient of 5 ppm/°C. Compare the change in the reference voltage accompanying a 1 mA current variation with the variation accompanying a temperature change from −55°C to +100°C (the rated change for this Zener).

b. The 1N829A has an incremental resistance of 10 Ω. Compute the voltage change accompanying the same 1 mA current change.

c. What should be the minimum power rating of the 1N829?

Solution:

a. $\Delta V = r_Z\,\Delta I = 15(10^{-3}) = 15\text{ mV}$

$$\frac{5\text{ ppm}}{°\text{C}} \times 155°\text{C} = 775\text{ ppm} = \frac{775}{10^6}$$

$$\frac{775}{10^6} \times 6.2\text{ V} = 4.8\text{ mV}$$

Therefore, the current change is about three times more significant than the rather extensive temperature change.

b. $\Delta V = 10(10^{-3}) = 10\text{ mV}$

In this case the respective sensitivity is 2 to 1.

c. $P = V_{ZT}I_{ZT} = 6.2(7.5 \times 10^{-3}) = 46.5\text{ mW}$

Applying the customary safety factor of 4 yields 186 mW. The units are actually rated at 400 mW.

EXAMPLE 16.3

Calculate the values of R_1, R_2, and R_3 if the 1N829 Zener in Figure 16.4 is to be operated at $I_{ZT} = 7.5$ mA and $V_{ZT} = 6.2$ V and the desired scaled reference voltage is to be +10.0 V.

Solution: With $V_0 = +10.0$ V, we have

$$R_1 = \frac{V_0 - V_{ZT}}{I_{ZT}} = \frac{10 - 6.2}{7.5 \times 10^{-3}} = 5.07 \times 10^2\ \Omega$$

Use $R_1 = 511 \pm 1\%\ \Omega$. With $(R_2 + R_3)/R_2$ being the gain of the OPAMP, we have

$$\frac{R_2 + R_3}{R_2} V_{ZT} = V_0$$

$$\left(\frac{R_2 + R_3}{R_2}\right) 6.2 = 10$$

$$\frac{R_2 + R_3}{R_2} = \frac{10}{6.2} = 1.613, \quad \text{which represents the gain of the OPAMP}$$

$$\frac{R_3}{R_2} = 0.613$$

Since the current (I_F) through R_2 and R_3 detracts from that available for the power supply load, one might typically make this current 1 mA, at most. We have

$$\frac{V_0}{I_F} = \frac{10}{10^{-3}} = 10\ \text{k}\Omega = R_2 + R_3 = R_2(1 + 0.613)$$

$$R_2 = \frac{R_2 + R_3}{1.613} = \frac{10}{1.613} = 6.20\ \text{k}\Omega$$

$$R_3 = 0.613(6.2) = 3.80\ \text{k}\Omega$$

16.2.2 Regulated Power Supplies

When one must accommodate larger variations in line and load voltages than is possible with Zener diodes, a more elaborate electronic system must be utilized. The basic principle of the series voltage regulator is shown in Figure 16.5.

Refer to Figure 16.5. There is inserted in series with one of the dc leads an element whose resistance is variable. Typically, this element is a transistor. Across the unregulated voltage terminals, there is placed a voltage reference (for example, a Zener diode), and this reference provides a rather precise, constant voltage to one of the inputs of a differential amplifier (the error amplifier). Across the regulated output there is a voltage sampler—which can be as simple as a voltage divider. This divider furnishes the other input to the error amplifier. Under normal circumstances the reference signal and the sampled voltage are the same, there is no signal input to the error amplifier, and the series element represents a fixed resistance across which there is developed a voltage drop appropriate to the current being drawn by the load.

Should the line voltage suddenly rise, the voltage reference will be affected little (if at all), but the voltage at the output sampler will rise. This increase provides an error signal to the amplifier, which, in turn, alters the bias on the series transistor in such a manner as to represent an increased resistance. There will then be an added drop across the series element equal to

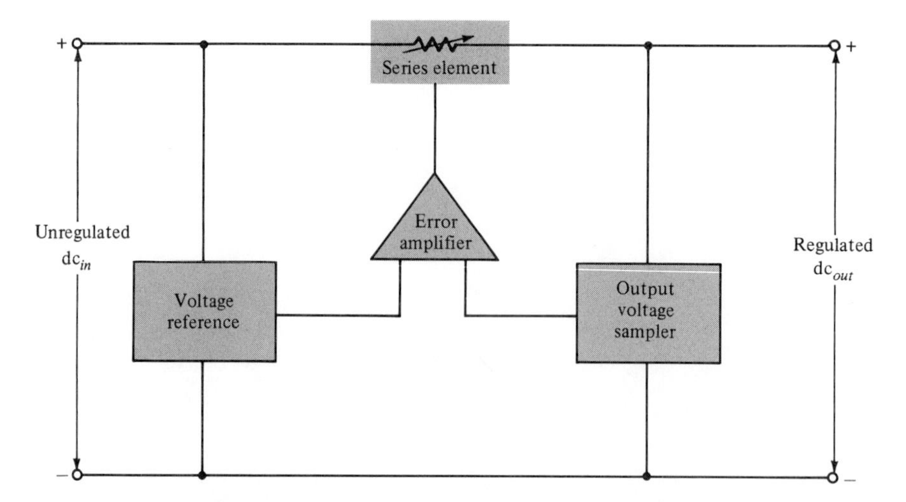

FIGURE 16.5. **Block Diagram of a Voltage-Regulated Power Supply**

the source voltage increase, thereby maintaining the output voltage at a constant value.

Of course, if the line voltage drops or the load demand changes up or down, the system will operate to compensate for such variations. Since often the entire load current passes through the series element, these transistors must be rugged and capable of sufficient heat dissipation. A number of them may be put in parallel to provide for added current capacity.

With regard to the voltage reference, it does not handle any of the load current as would be the case for a straightforward Zener regulator. Only a small maintaining current passes through, and it can easily accommodate the current variations due to line and load fluctuations. Of course, it would be a reference Zener that would be used in such applications.

Figure 16.6 shows more actual details of such a series voltage regulator. The error amplifier can be an OPAMP operating in a noninverting mode, and from Eq. 10.14 we have the expression for the output voltage in terms of the input (reference) voltage:

$$V_{out} = \left(\frac{R_1 + R_2}{R_1}\right)V_R = \left(1 + \frac{R_2}{R_1}\right)V_R \tag{16.1}$$

If r_Z represents the dynamical (incremental) resistance of the Zener, any change in the unregulated input voltage will cause a change in the Zener voltage, small though this variation may hopefully be, and the change will be proportional to $r_Z/(R + r_Z)$. This variation will bring about a change in the regulated output, which will be proportional to $R_1/(R_1 + R_2)$. The ratio of the output change to the input change constitutes a measure of the regulation:

$$\text{load regulation} = \frac{r_Z}{R + r_Z}\,\frac{R_1 + R_2}{R_1} \tag{16.2}$$

EXAMPLE 16.4 The regulated power supply shown in Figure 16.6 has $V_{in} = 20$ V, $R = 1$ kΩ, $V_Z = 6.5$ V, $r_Z = 25$ Ω, $V_{out} = 9.75$ V, and $R_2/R_1 = 4.7$ kΩ/10 kΩ. Compute the percentage of load regulation.

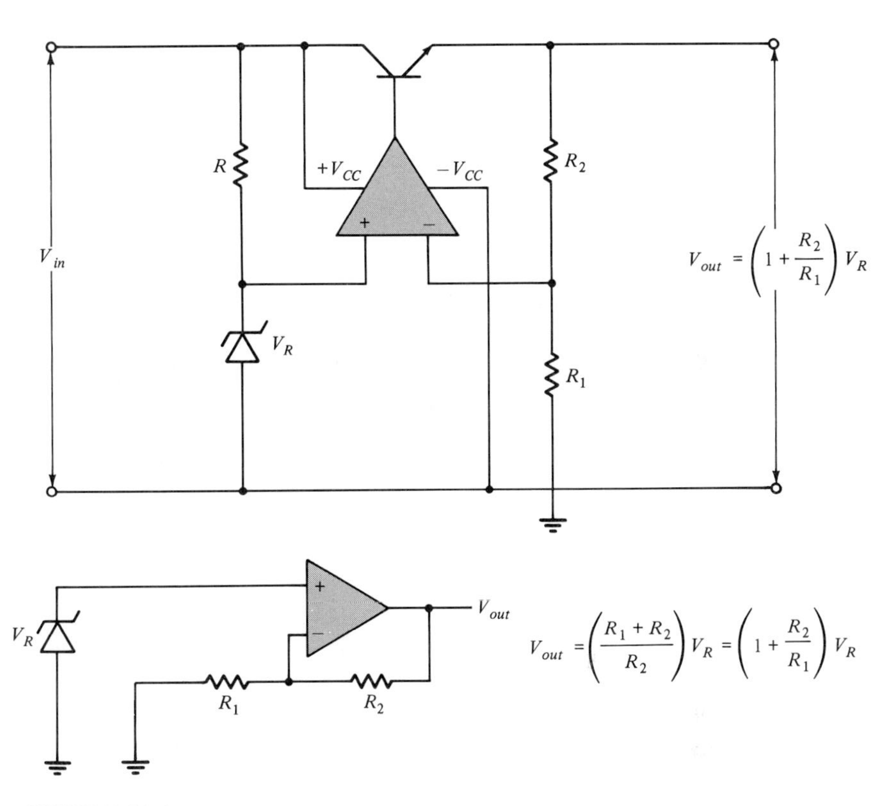

FIGURE 16.6

Solution:

$$\text{load regulation} = \frac{r_Z}{R + r_Z}\frac{R_1 + R_2}{R_1} = \frac{25}{10^3 + 25}\frac{14.7}{10}$$
$$= 3.58 \times 10^{-2} \simeq 3.6\%$$

In the example we have seen how the constancy of the Zener diode voltage influences the regulation. The load regulation may be further improved by operating the Zener from a separate constant current source.

With such power supplies one may also make provision by appropriate circuitry for current limiting—that is, the power supply may be made to furnish a constant voltage up to some limiting (usually adjustable) value of current. Irrespective of the nature of the load, the system will not furnish a current greater than the set limiting value, even if the output terminals are short-circuited. This feature is particularly helpful if the load is defective and demands an abnormally high current that might otherwise damage the power supply.

It can be seen that regulated supplies represent rather complex electronic circuitry as one proceeds to demand greater degrees of regulation, power delivered, short-circuit provisions, etc. These various features are selectively available in commercial power units.

16.2.3 The 723 Regulator and Other IC Regulators

Like the 741 OPAMP, the 723 IC regulator constitutes a classic circuit and is worth looking at in some detail since more modern units are based on its design. It includes a temperature-compensated Zener (Figure 16.7) operated under constant current conditions delivering an output voltage V_{REF}. There is also an error amplifier, a series pass transistor, and a protective circuit to limit the current. Used with a few external components, it constitutes a regulated power supply capable of line and load regulations of 0.01% and 0.03%, respectively, with a temperature coefficient of 0.003%/°C. The latter is due to the Zener being a compensated unit.

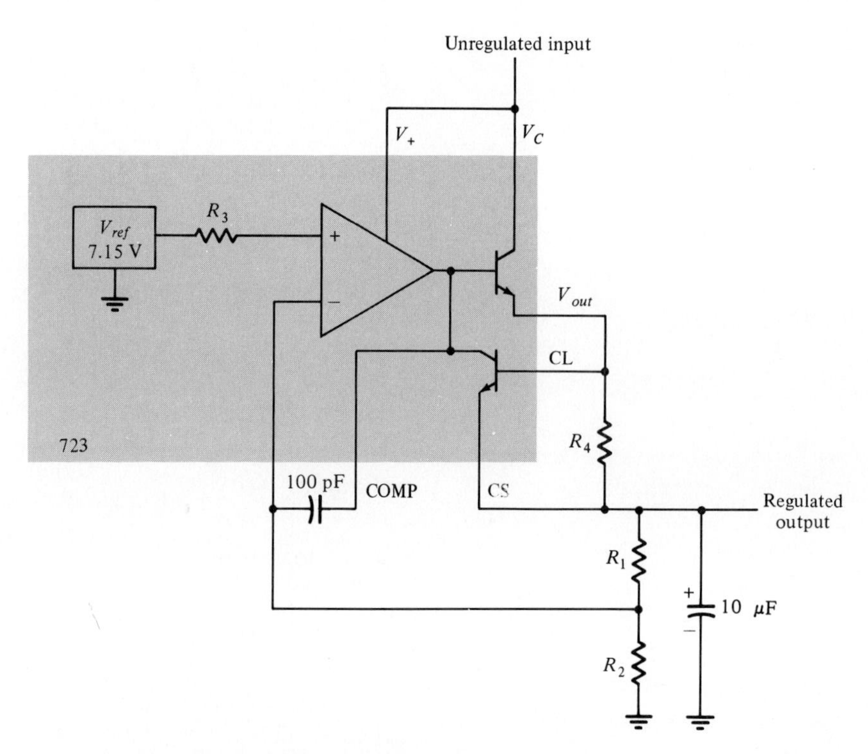

FIGURE 16.7. **723 IC Voltage Regulator Unit**

The voltage divider R_1 and R_2 constitutes the voltage sampler. Resistor R_4 is given a value such that at the maximum desired output current the voltage drop across it is about 0.5 V. Since this voltage represents the bias on the current-limiting transistor, when the latter is turned on (by excessive current flow), the drive voltage on the pass transistor is terminated, and the system shuts down. R_3 is used to balance the input impedance on the OPAMP and thereby to minimize the effect of current offset. The 100 pF capacitor stabilizes the feedback loop. The amount by which the input voltage must exceed the output voltage—that is, the **dropout voltage**—is 3 V. The values of R_1 and R_2 can be adjusted to yield the precise regulated voltage desired.

Newer IC voltage regulators for most noncritical applications are exemplified by three-terminal devices such as the 7800 series (Figure 16.8). These devices are specified as 78xx, where the last two digits take on the values 05, 06, 08, 10, 12, 15, 18, or 24 and respectively represent the output voltage. They can provide up to 1 A of output current, but there are also low-power versions available (78Lxx) as well as a negative voltage series (7900 series). Similar four-terminal regulators are also available; the extra terminal allows one to adjust the output voltage. Neither the three-terminal nor the four-terminal units are precision regulators, typically having temperature stabilities of 2%/°C, but their performance suffices in many instances.

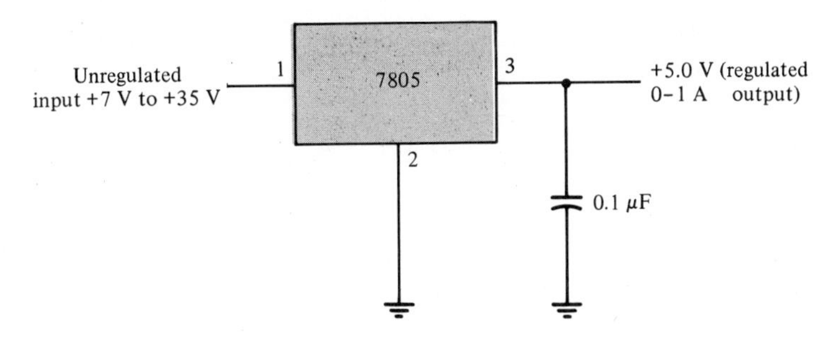

FIGURE 16.8. **Three-Terminal 7805 IC Regulator**

16.2.4 Current Mirrors

Many IC circuits utilize a **current mirror** to set the biasing currents for the various transistor stages. Figure 16.9 shows the circuit in its simplest form. Because of the short between the base and collector of Q_1, this transistor acts simply as a forward-biased diode and develops V_{BE}, a voltage appropriate to the forward current and a value determined by the R placed in the collector circuit. Q_2 is impressed with the same value of V_{BE}, and hence its collector—that is, the load—current "mirrors" that of Q_1.

A deficiency of this simple mirror circuit is the slight dependence of V_{BE} on the value of V_{CE}. It is this dependence that results in the V_{CE} vs. I_C curves not being completely flat (Figure 7.24). The circuit shown in Figure 16.10 corrects for this deficiency. The function of Q_3 is to keep the V_{CE}s of Q_1 and Q_2 at constant values (equal to $2V_{BE}$ and V_{BE}, respectively) irrespective of the collector currents. The collector of Q_1 is the programming terminal, and Q_2 is still the source of the output current.

Current mirrors can also be used to yield load currents that are fractionally related to the programming current by adding a resistor to the emitter of Q_2 (Figure 16.11). The insertion of resistor R_2 in series with the emitter of Q_2 leads to a difference in the V_{BE}s for Q_1 and Q_2, given by the expression

$$\Delta V_{BE} = V_T \ln\left(\frac{I_{C1}}{I_{C2}}\right) \qquad (16.3)$$

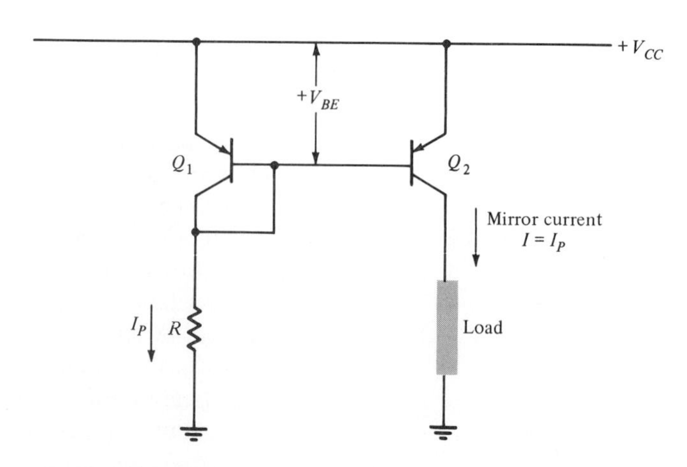

FIGURE 16.9. Simple Current Mirror R determines the programmed current I_P through diode-connected transistor Q_1. The resultant V_{BE} is also the bias on Q_2, which should pass an identical current.

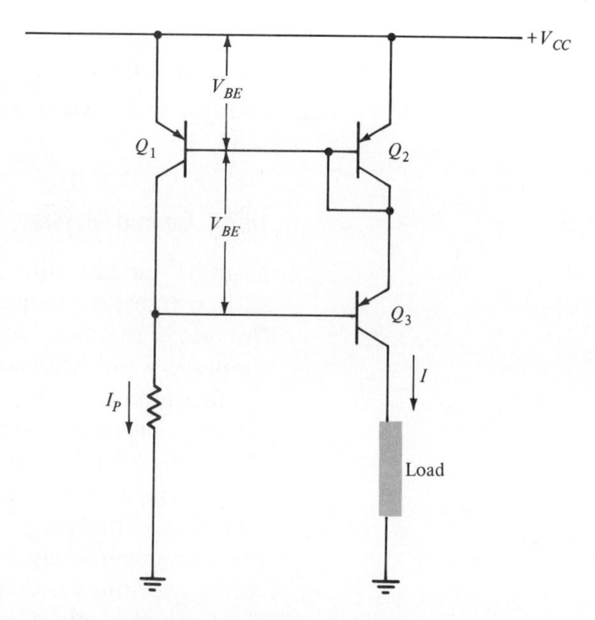

FIGURE 16.10. Current Mirror with Stable Programmed Current

where I_{C1} and I_{C2} are the respective collector currents of Q_1 and Q_2 and V_T = $T/11{,}600$, T being the absolute temperature.

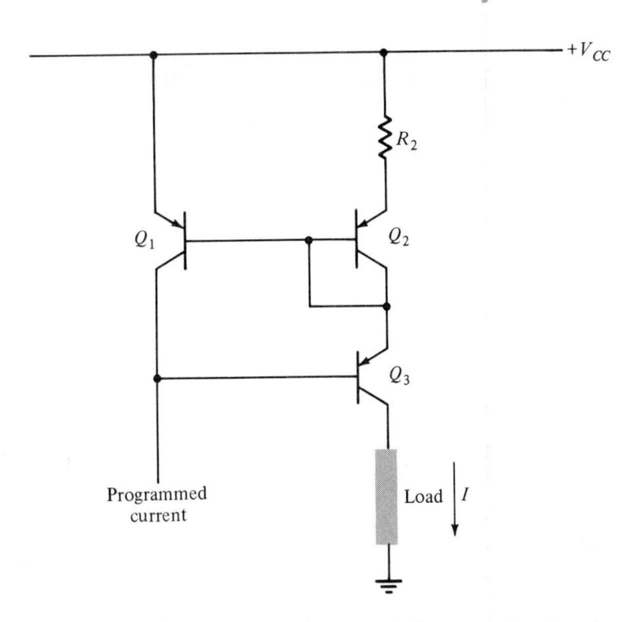

FIGURE 16.11. **Current Mirror** Load current is fractionally related to programmed current through the value of R_2.

EXAMPLE 16.5

Since the region between an emitter and the shorted base-collector junction of a transistor constitutes a diode, using the approximate diode equation

$$I_C = I_0 e^{V_{BE}/V_T}$$

where V_{BE} is the base-emitter voltage, I_0 the reverse saturation current, and $V_T = 0.026$ V, verify that for a current mirror with emitter resistors R_1 and R_2 (in the program and mirror circuits, respectively) the ratio of programmed current to mirror current is approximately equal to the inverse of the resistance ratio, that is,

$$\frac{I_{C1}}{I_{C2}} \simeq \frac{R_2}{R_1}$$

(See Figure 16.12.)

Solution:

$$I_{C1} = I_0 e^{V_{BE1}/V_T}$$

$$I_{C2} = I_0 e^{V_{BE2}/V_T}$$

$$\ln\left(\frac{I_{C1}}{I_{C2}}\right) = \frac{V_{BE1} - V_{BE2}}{V_T}$$

$$I_{C1}R_1 + V_{BE1} = I_{C2}R_2 + V_{BE2}$$

$$V_{BE1} - V_{BE2} = I_{C2}R_2 - I_{C1}R_1$$

$$V_T \ln\left(\frac{I_{C1}}{I_{C2}}\right) = I_{C2}R_2 - I_{C1}R_1$$

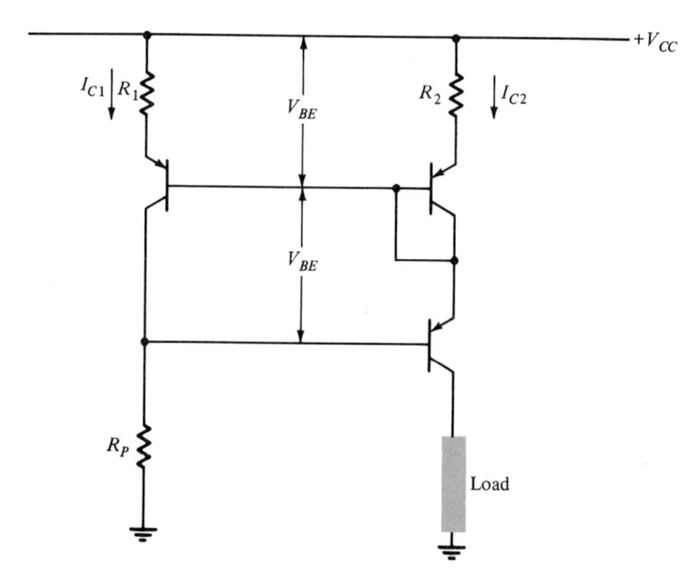

FIGURE 16.12

Since $V_T \ln (I_{C1}/I_{C2})$ is generally a small quantity, we have

$$\frac{I_{C1}}{I_{C2}} \simeq \frac{R_2}{R_1}$$

16.3 ■ SILICON CONTROLLED RECTIFIERS

The silicon controlled rectifier (SCR) represents a most useful device in controlling large currents such as encountered in many industrial applications. Its construction is similar to that of an *npn* transistor, with an additional *p*-segment thrown in, as shown in Figure 16.13.

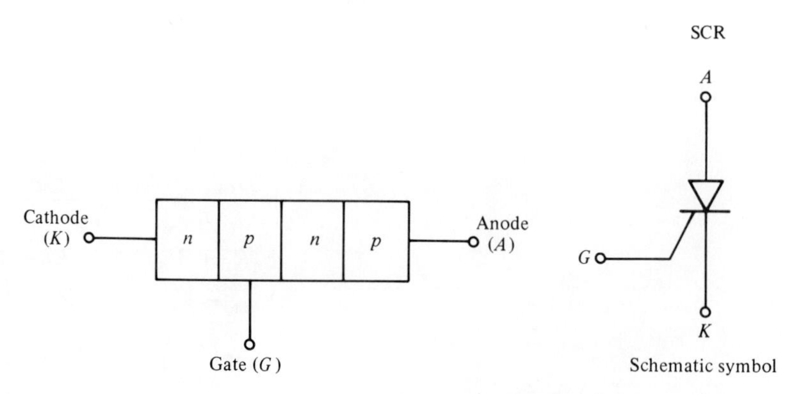

FIGURE 16.13. **Physical Structure of a Silicon Controlled Rectifier and Schematic Symbol**

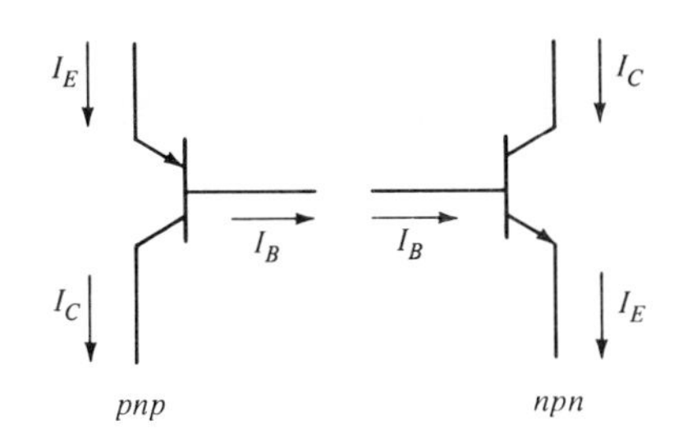

FIGURE 16.14. Current Flow in *pnp* and *npn* BJT Transistors

A typical application would have a resistive load, through which one wishes to switch a considerable amount of current, connected to the power source in series with the anode. Despite what might be a very large source voltage, no current will flow until a relatively small gate current is applied at G. Thus, the SCR serves as an electric switch but without having a very large current pass through switch contacts.

16.3.1 Operating Principle

To understand the operation, let us review again the basic operations of *npn* and *pnp* transistors (Figure 16.14). In each case a small base current $I_{B1,2}$ controls a much larger collector current $I_{C1,2}$. Let us now connect the two transistors in such a way that the collector current I_{C2} becomes the base current I_{B1} (plus I_G) and the base current I_{B2} becomes the collector current I_{C1}. The current I_{B1} is amplified by Q_1, and the resulting collector current I_{C1} is amplified by Q_2, becoming I_{C2} (hence I_{B1}) (Figure 16.15).

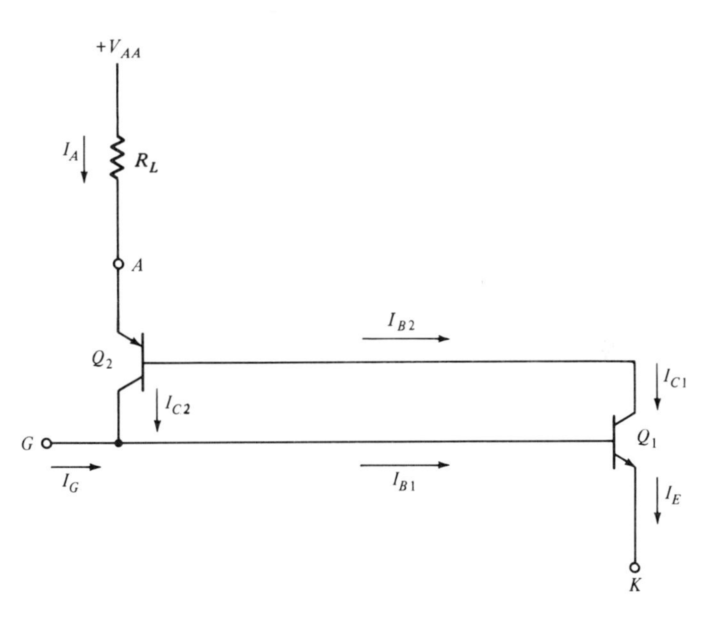

FIGURE 16.15. *pnp* and *npn* Equivalents of an SCR

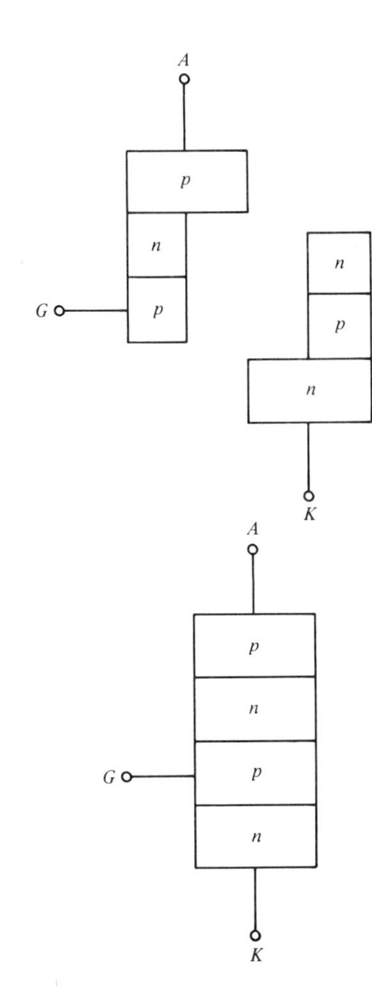

FIGURE 16.16. Physical Combination of *pnp* and *npn* Leading to the Creation of an SCR

If, to start with, the gate voltage at G is insufficient to cause conduction in Q_1, I_{C1} is also insufficient to cause conduction through Q_2, point A of the load resistor is at a potential V_{AA} (the source voltage), and no current passes through the load. If G is brought up to a point sufficient to cause Q_1 to conduct, Q_2 will do likewise, and a large current will flow through R_L.

Once conduction starts, the gate no longer can control matters since the gate current I_G is but a small fraction of I_{B1}, which is now mainly I_{C2}. Either removal of the voltage V_{AA} or diminishing it to a very low value (while maintaining I_G below the trigger level) will cause the conduction to stop.

The two transistors need not be individual units. Figure 16.16 shows that our original sandwich, consisting of *npnp*, is equivalent to Figure 16.15.

16.3.2 Small Motor Control

One example of the use of an SCR is its control of the speed of a dc motor (Figure 16.17). The *conduction angle* (θ) may be varied from 0° to 180° by controlling the *firing angle* (ϕ). The average current will be

$$I_{dc} = \frac{1}{2\pi}\int_{\phi}^{\pi} I_m \sin \omega t \, d(\omega t) = \frac{I_m}{2\pi}(1 + \cos \phi)$$

$$= \frac{I_m}{2\pi}(1 - \cos \theta) \qquad (16.4)$$

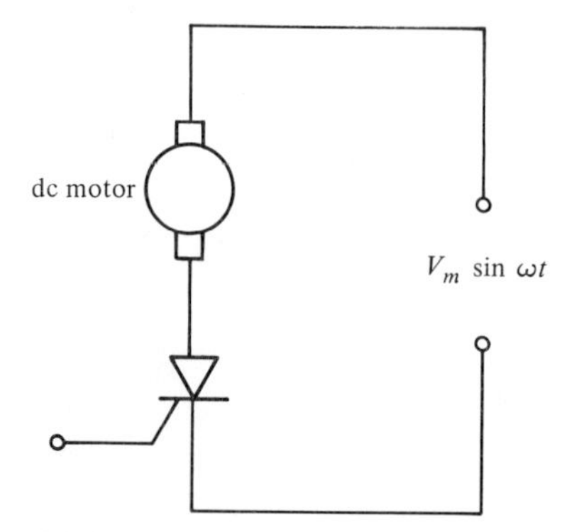

(a) dc motor speed control employing an SCR

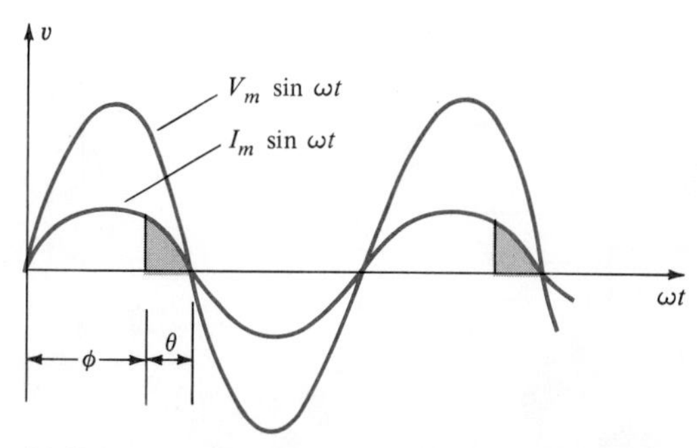

(b) Voltage and current waveforms showing firing angle (ϕ) and conduction angle (θ)

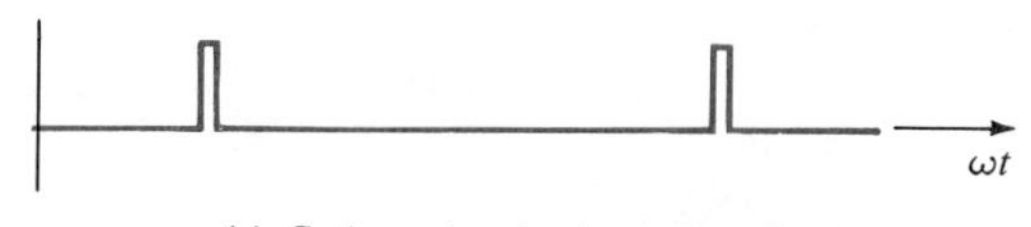

(c) Gating pulses I_G that initiate firing

FIGURE 16.17. SCR Application

Again, note that once the SCR has been turned on it cannot be turned off except by reducing the supply voltage.

To show a practical form in which the SCR can be used, we assume we have the problem of controlling a $\frac{1}{4}$ in. electric drill that usually runs only at a single high speed. Figure 16.18 shows the circuit for a variable speed controller.

The black dot within the schematic symbol for a neon bulb represents a

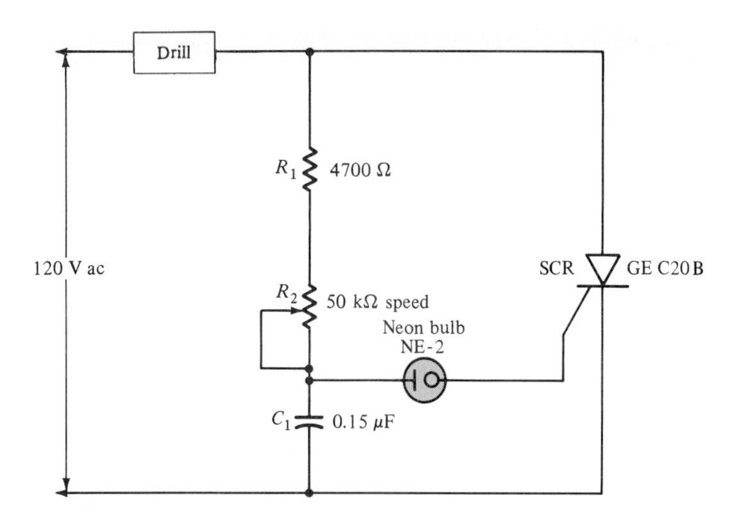

FIGURE 16.18. **Speed Control for $\frac{1}{4}$ in. Drill**

convention to indicate a gas-filled envelope. The gas in the neon bulb, when subjected to an appropriate firing voltage, will become luminous and electrically highly conductive. The capacitor C_1 will charge through R_1 and R_2 and will fire the neon bulb upon reaching the latter's ignition voltage. For an NE-2 bulb the firing voltage is about 55 V. The neon bulb ignition provides the trigger pulse for the SCR.

Since we are applying ac to the *RC* circuit, the neon bulb could fire on both positive and negative half-cycles. But the SCR will only conduct during the positive half-cycles since only then is its anode positive with respect to the cathode.

The setting of R_2 determines the charging rate of C_1. Decreasing R_2 increases the speed of the drill (Figure 16.19).

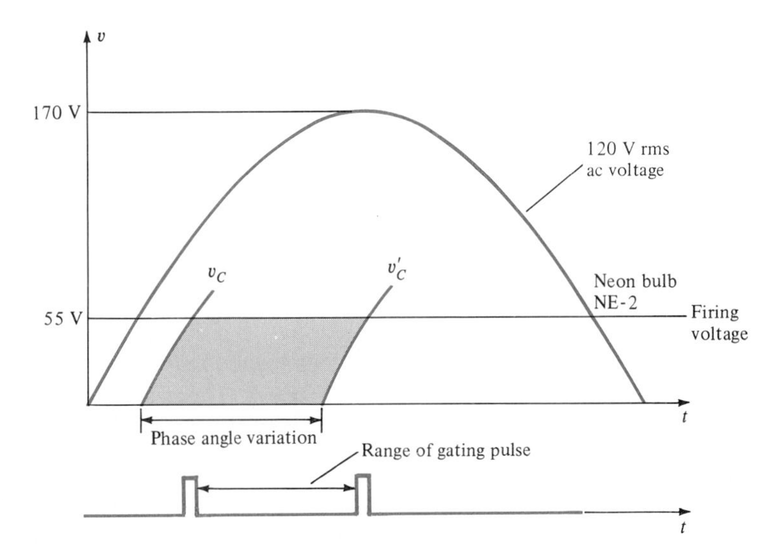

FIGURE 16.19. **Neon Bulb Phase Angle Variation When Used in Conjunction with an SCR**

EXAMPLE 16.6

See Figure 16.20.

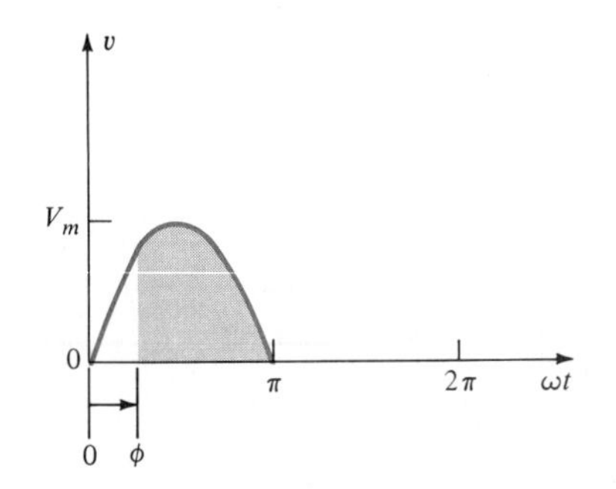

FIGURE 16.20

a. Compute V_{rms} for the half-wave output of an SCR with firing angle ϕ and peak voltage V_m.
b. Disregarding the voltage across the SCR, what is the power delivered to a load resistance R?

Solution:

a. $v = V_m \sin \omega t$

$$V_{rms} = \left[\frac{1}{2\pi}\int_{\phi}^{\pi} v^2\, d(\omega t)\right]^{1/2} = \left[\frac{1}{2\pi}\int_{\phi}^{\pi} V_m^2 \sin^2 \omega t\, d(\omega t)\right]^{1/2}$$

$$= \left[\frac{V_m^2}{2\pi}\left(\frac{\omega t}{2} - \frac{\sin 2\omega t}{4}\right)\Big|_{\phi}^{\pi}\right]^{1/2}$$

$$= \frac{V_m}{\sqrt{2\pi}}\left(\frac{\pi}{2} - \frac{\sin 2\pi}{4} - \frac{\phi}{2} + \frac{\sin 2\phi}{4}\right)^{1/2}$$

b. Disregarding the voltage drop across the SCR, we have

$$P = \frac{V_{rms}^2}{R}$$

16.3.3 The Triac

Like the half-wave rectifier that conducts only on the positive half-cycles of the ac waveform and in contrast to the more efficient full-wave rectifier that conducts during both half-cycles, a disadvantage of the SCR is its half-wave behavior. The **Triac** [standing for *tri*ode (three-electrode) *ac* semiconductor switch] can be triggered by positive and negative gate signals and conducts during both halves of the ac cycle.

Figure 16.21 shows the schematic symbol for the Triac, its symbolism being obvious—that is, a two-way diode—and a simple circuit making use of the Triac. In position 1 of the switch, there is no conduction through the load since there is no gate connection. In position 3 triggering pulses take place during both positive and negative half-cycles and in position 2, only during positive half-cycles. Thus, moving from position 3 to position 2 cuts the power level by one half.

Another useful circuit is shown in Figure 16.22. Across the secondary of the step-down transformer (a so-called *filament* transformer from vacuum tube days) is a switch that, when open, represents a high (ideally, infinite) impedance across the secondary, which also means a high impedance across the primary. Therefore, the trigger voltage that appears across R_1 is too low to activate the Triac. But with the closing of the switch, a short appears across the secondary, which lowers the impedance across the primary, and across R_1 a voltage sufficient to activate the Triac is developed. Power pulses will continue to pass through the load as long as the switch is closed.

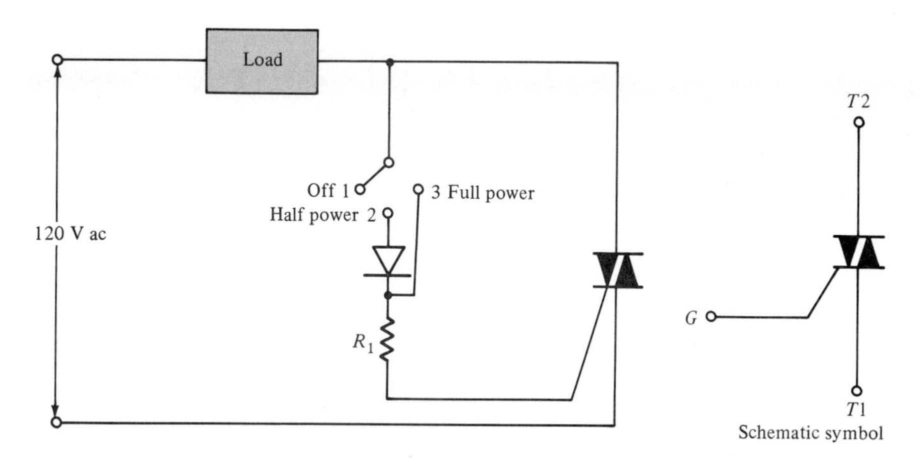

FIGURE 16.21. Simple Triac Control Circuit and Schematic Symbol

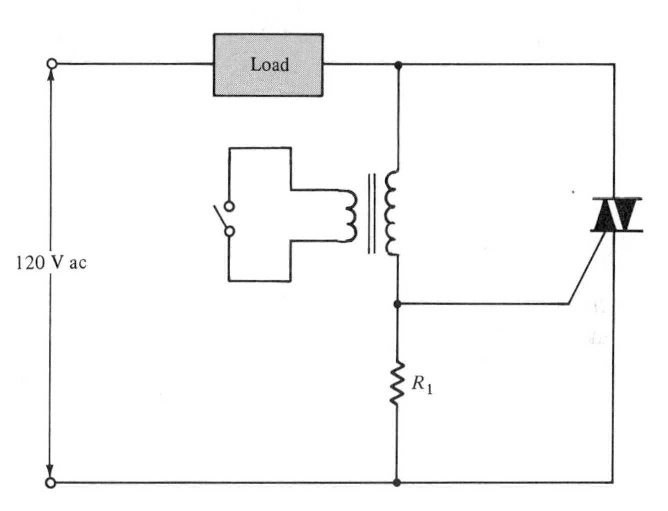

FIGURE 16.22. Triac Control of Power The controlling switch is electrically isolated from the power line, thereby constituting a safety feature.

Figure 16.23 illustrates a practical use of this arrangement as a means of fluid-level control. With the level contact engaged, the fluid conducts and bypasses the gate current that would otherwise activate the SCR. If contact is lost, the SCR gate voltage appears, the activated SCR acts as a short, and the action that follows is identical to that of the closed switch in Figure 16.22. The pump (presumed to be the load) will operate until the level rises sufficiently to allow contact to again be made.

One may also use the Triac in an arrangement similar to that employed as a speed control for a $\frac{1}{4}$ in. drill (Figure 16.24). In that conjunction it might be

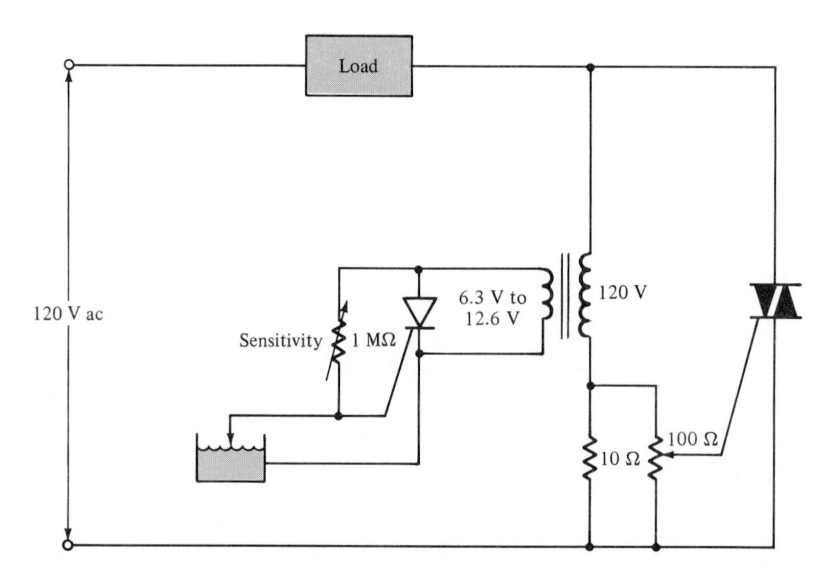

FIGURE 16.23. **Liquid-Level Control Employing an SCR and a Triac**

mentioned that there is a semiconductor equivalent to the neon bulb we used as a gate trigger. This device is a Diac [*di*ode (two-electrode) *ac* switch], made specifically to be used to trigger Triacs but also useful in conjunction with SCRs. The schematic symbol for the Diac is as shown in Figure 16.24, and it will trigger with either positive or negative pulses.

16.3.4 Some Complications

Refer again to a half-wave phase control (Figure 16.25). When the SCR fires, there is a residual charge left on the capacitor, V_0 (Figure 16.26). If the charge

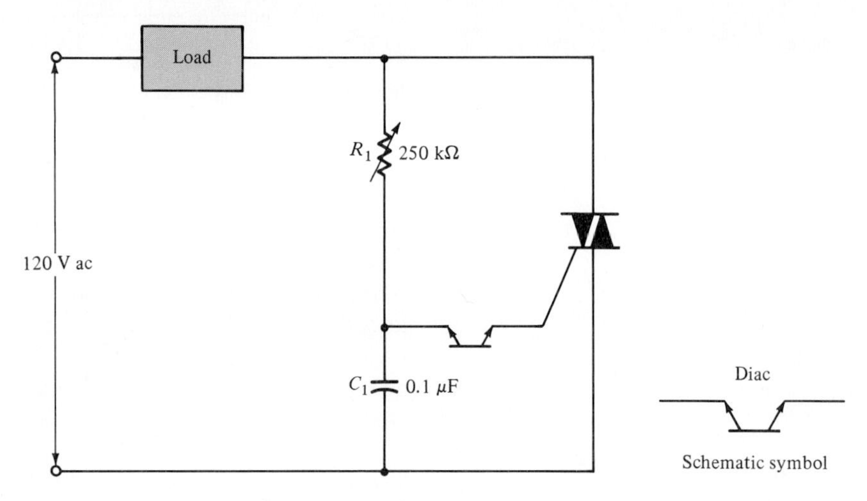

FIGURE 16.24. **Simple Diac Control Circuit of a Triac and Schematic Symbol**

deposited during the next positive cycle is *in*sufficient to fire the SCR (for example, because R_1 is too large), the capacitor is now left with a large residual voltage (V_0') that can lead to a relatively small firing angle at the next positive half-cycle. The result is *cycle skipping*, as shown in Figure 16.27.

There are a couple of solutions to this problem, one being to limit the range of R_1 so that such a situation cannot arise; the second is to automatically reset the capacitor to a known voltage after each half-cycle so that the system always starts from a common voltage. In the latter case the time constant is such that the system either works properly or does not work at all. A simple way to accomplish this automatic reset is to replace diode D_1 (Figure 16.25) with a resistor. The resistor will cause the capacitor to start at a fixed negative value. This solution presumes that the trigger does not fire on nega-

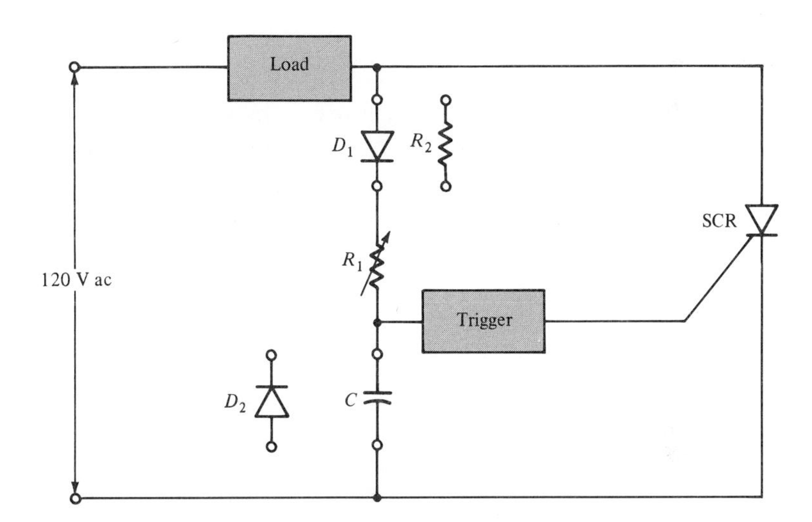

FIGURE 16.25. ***RC* Control of an SCR** R_2 may be used to replace D_1 in order to prevent cycle skipping. D_2 may be placed across C to prevent negative trigger operation.

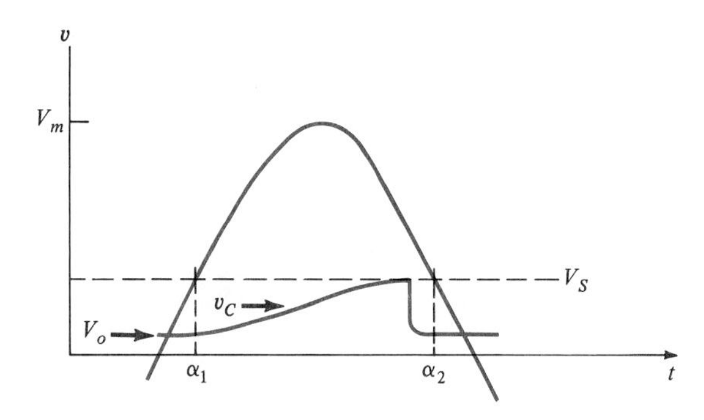

FIGURE 16.26. Capacitor Charging Voltage Showing Residual Voltage (V_o) After Firing of SCR

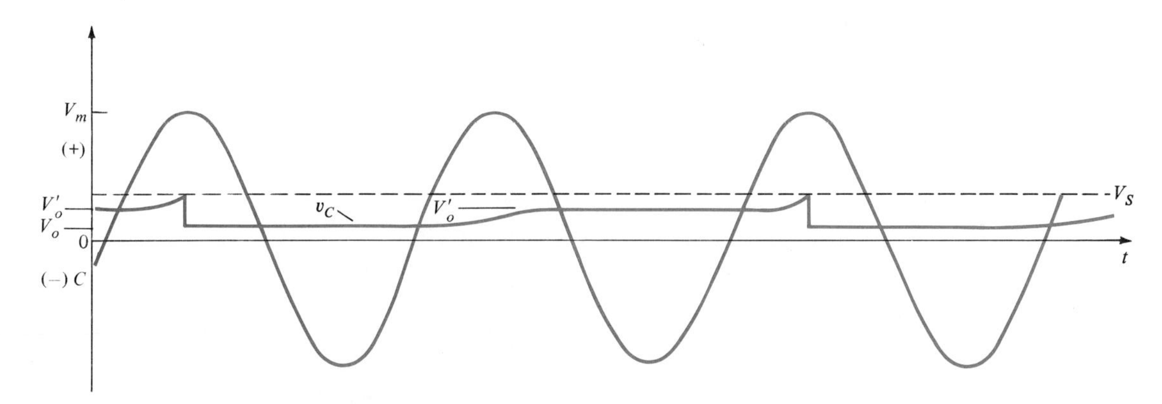

FIGURE 16.27. **Insufficient Capacitor Charging Rate Causes Cycle Skipping in an SCR**

tive half-cycles. If it does, a diode D_2 will fix V_0 at about -0.7 V on the negative half-cycle, a value presumed insufficient to fire the trigger.

Full-wave control may also pose problems: It may be subject to "snap-on" effects. Assume the system has been operating near its upper firing angle limit (Figure 16.28). R_1 is then increased, thereby carrying the system out of conduction. R_1 is then decreased slightly in the hopes of bringing the system back into operation. But when it does, it is with a much earlier firing angle than existed when control was lost, for the following reason: The solid curve for v_C represents a properly operating system. With a full-wave operation, the capacitor starts out with a residual voltage $-V_0$ on the positive half-cycle and $+V_0$ on the negative one. If R_1 is now advanced so that the capacitor voltage is insufficient (dotted curve in Figure 16.28), the capacitor is left with a residual charge almost equal to $+V_s$, the breakover voltage. As a result, on the following negative half-cycle, it is far short of the required breakover voltage.

Reducing R_1 will increase v_C, but when V_s is again reached, the system fires and suddenly the residual charge (V_0) is drastically reduced. This reduction causes an increase in v_C on the next half-cycle and leads to triggering at a firing angle substantially earlier than the threshold value. This early firing causes the load to snap from a near zero value to some intermediate value from which it may then be smoothly controlled over the full range, right up to the lower threshold.[1]

The addition of a second phase-shift network may be used to diminish

[1]J. H. Galloway, in his General Electric Semiconductor Application Note 200.35 ("Using the Triac for Control of AC Power," 1970), presents an interesting analogy to this hysteresis effect. A kerosene lantern cannot be lit with the wick in the last position in which it was used; it must be advanced. It then lights at some intermediate brightness, from which it can be turned down to the desired brightness. This interesting analogy has been left out of the GE *SCR Manual*; perhaps not too many electrical engineers these days are acquainted with the oddities of a kerosene lamp.

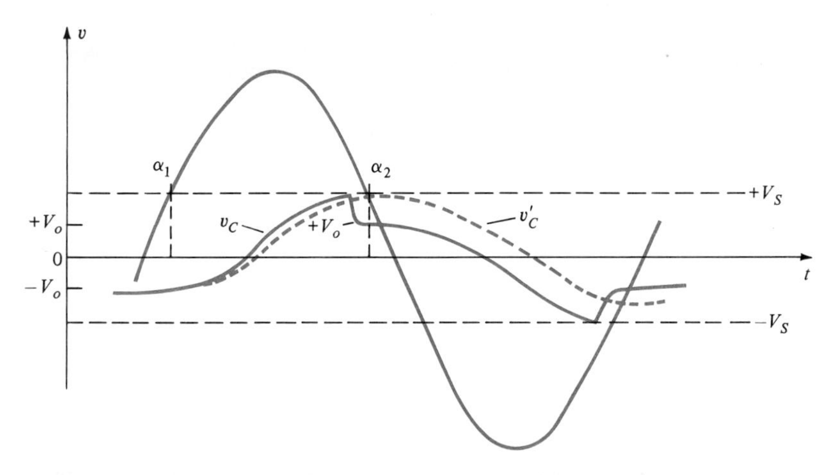

FIGURE 16.28. **Hysteresis Effect with Triac** The solid curve represents proper operation near zero load current conditions. The dashed curve arises when the phase control advances beyond the trigger point. $\pm V_o$ show residual voltages under proper operating conditions.

this hysteresis effect. See Figure 16.29. The power through the load diminishes as R_1 is increased. But, as we have seen, hysteresis does not allow the power level to be gradually brought up from zero. With the second network, at large values of R_1, C_1 will primarily be recharged through R_3 from the voltage across C_2, thereby allowing a wider range of control, between 5% and 95%.

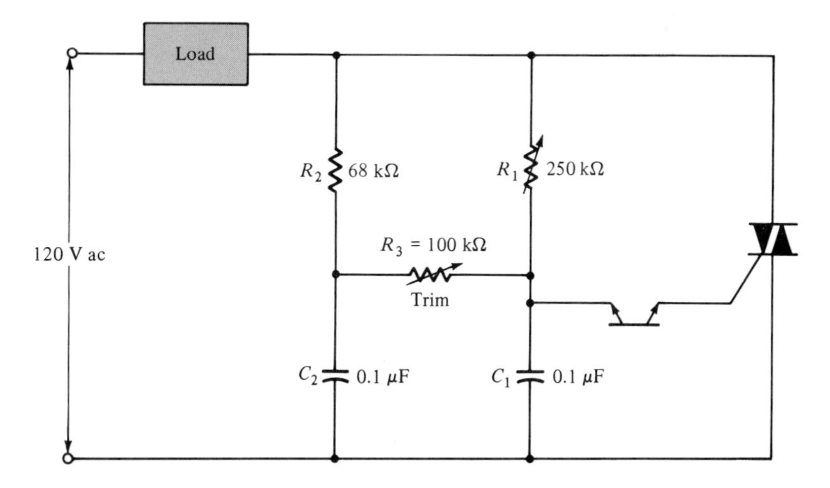

FIGURE 16.29. **Two-Network Shift Control to Minimize Hysteresis Effect**

There are more elaborate circuits employing various arrangements of diodes that can eliminate the hysteresis entirely, but we will terminate our discussion at this point.

EXAMPLE 16.7 Devise an automotive battery-charging circuit utilizing SCRs operating from a 120 V ac source.

Solution: See Figure 16.30. D_1 and D_2 constitute a full-wave rectifier. R_1, R_2, and SCR2 form a potential divider. When SCR2 is in a nonconducting state, the anode voltage of D_3 is such as to cause conduction and fire SCR1 early in each half-cycle of the full-wave output. The conduction angle of SCR1 is thus large and provides a substantial charging current to the battery. By presuming the battery voltage to be low, the voltage V_R is insufficient to cause a breakdown of the Zener diode, and this condition prevents SCR2 from triggering. As the battery voltage builds up through charging, the Zener will start to break down, initially at the voltage peaks, thereby leading to a 90° conduction angle in SCR2. Since SCR1 has already fired during that half-cycle, such conduction does not influence the charging through SCR1. But as the battery voltage builds up, the firing angle of SCR2 advances until, eventually, SCR2 is triggering before the ac waveform has reached sufficient magnitude to fire SCR1. With SCR2 conducting, D_3 is back-biased, SCR1 does not conduct, and the battery charging ceases.

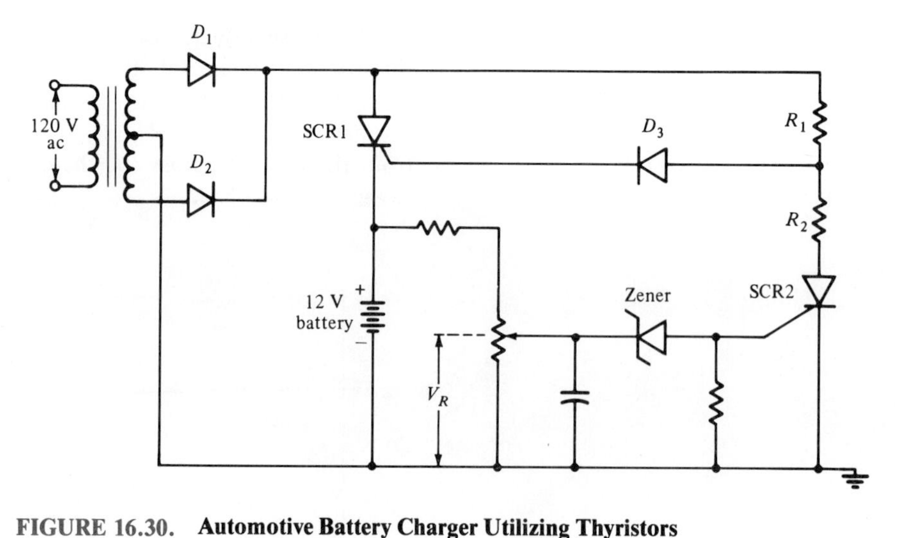

FIGURE 16.30. **Automotive Battery Charger Utilizing Thyristors**

16.3.5 Some Concluding Comments

We have confined our consideration of SCRs and Triacs[2] to loads that are resistive. In many instances the loads include reactive components (particularly inductive) that, expectedly, complicate the matter of phase control.

Another complication in the use of phase-controlled thyristors arises from the radio interference that they can create. Upon being triggered, a thyristor's load current goes from zero to the load limit within a few microseconds. A Fourier analysis of the resulting waveform would show it to be a

[2]As a category, *pnpn* devices such as SCRs and Triacs are called *thyristors*.

prolific producer of harmonics. While their intensity at FM and TV frequencies has probably diminished sufficiently so as not to lead to a bothersome situation, such is often not the case at AM radio frequencies. Additionally, in an industrial environment where a number of thyristor control circuits may be in use simultaneously, through such harmonics they may be made to adversely interact with one another.

The generated interference leaves by two methods:

1. As electromagnetic radiation
2. By conduction along the power lines

Interference in the first instance may be diminished by proper shielding and in the second, by appropriate filtering.

Another solution to the interference problem lies in the use of **zero-voltage switching**. Rather than phase control, the thyristors are turned on as the ac voltage is passing through zero, and the power control is achieved by altering the number of complete cycles that follow each such triggering operation.

And let us close this chapter with a few words about some "classic" SCRs and their specification. For handling currents of the order of 3 A, the GE C106 series provides a versatile assortment. In addition to the current-handling capacity, the *blocking voltage* is also an important specification. This voltage represents the maximum (either forward or reverse) peak voltage that may be applied. For the C106 series these various values are coded by postscript letters: Q = 15 V, Y = 30 V, F = 50 V, A = 100 V, B = 200 V, C = 300 V, and D = 400 V. The trigger requirements for the C106 series are 800 mV at 200 μA.

SUMMARY

16.2 Voltage and Current Regulation

- A Zener diode may be used to stabilize dc load voltages against modest changes in ac line voltage and/or changes in load resistance.
- Zener diodes are available with voltage ratings that have the same series of values as 5% tolerance resistors.
- Zener diodes that have voltages in the vicinity of 5.3 V exhibit the greatest degree of temperature stability.
- In order to maintain a high degree of voltage stability, Zener diodes must be operated at constant currents.
- Electronically regulated power supplies are needed where large load and line voltage variations are expected as well as under conditions of large power demand.

16.3 Silicon Controlled Rectifier

- The silicon controlled rectifier (SCR) serves as an electric switch but without the potentially large load current passing through switch contacts.
- Relative to the start of the ac power cycle, the firing angle (ϕ) represents the angle at which the SCR begins to conduct, and the conduction angle (θ) represents the duration of the cycle during which conduction takes place. The conduction angle of an SCR is limited to 180°.
- A Triac can conduct on both positive and negative half-cycles and hence has a conduction angle that can extend up to approximately 360°.
- A Diac is the semiconductor equivalent of a neon bulb trigger.
- SCRs and Triacs are grouped together in a class of semiconductor devices called thyristors.

KEY TERMS

current mirror: parallel conduction paths with the current in one being determined by the adjusted value in the other; widely used to set biasing currents in IC transistor amplifiers

Diac (*di*ode *ac* switch): a triggering device with initiation taking place by means of either positive or negative pulses

diode temperature coefficient (tempco, TC): variation of diode voltage with temperature (A conventional diode has a negative TC, with the voltage diminishing with increasing temperature; a Zener may have either a positive or a negative TC, with a transition from one to the other as the Zener voltage is altered.)

dropout voltage: in a regulated power supply, the minimal required difference between the unregulated input voltage and the regulated output

phase control: altering the fraction of a sinusoidal waveform during which conduction takes place in a thyristor device

regulated power supply: a voltage source with provisions for reducing the output voltage variations in the face of load and line voltage fluctuations

silicon controlled rectifier (SCR): a solid-state switch, used to control power devices (In effect it is a half-wave rectifier with the capability of altering the fraction of the half-cycle during which conduction takes place.)

thyristor: a semiconductor switch; an SCR and a Triac are two examples

Triac (*tri*ode *ac* semiconductor switch): effectively, a full-wave version of an SCR

Zener breakdown: that value of reverse bias at which the Zener diode exhibits a rather sudden, large increase in conduction current

Zener diode: a diode operated under reverse bias conditions that, over a limited range, serves as a voltage source relatively independent of the current

Zener knee current, I_{ZK}: the minimum Zener current commensurate with stable operation

Zener test current, I_{ZT}: the value of conduction current in a Zener diode at its nominal operating voltage

Zener voltage reference: a Zener diode whose function is to provide a rather precise voltage but under negligible load conditions

Zener voltage regulator: a Zener diode that, within a limited range, is used to minimize line and voltage fluctuations

zero-voltage switching: initiation of thyristor conduction only when the applied ac is passing through its zero value (The power is controlled by the number of conduction cycles that immediately follow; the interference created by such switching is much less than that encountered with phase control systems.)

KEY DEVICES

1N821–1N829: a series of temperature-compensated Zener diodes with 6.2 V at 7.5 mA; those with a terminal A in the designation have 10 Ω dynamic resistance in contrast to the usual 15 Ω

723: a classic adjustable voltage regulator; 2–37 V output voltage at 150 mA; 0.03% load regulation; 0.1% line regulation

78xx: a series of three-terminal regulators with positive output voltage where xx = 05, 06, 08, 10, 12, 15, 18, 24 V at 1 A; 0.4% long-term stability

78Lxx: low-power equivalent of 78xx; 0.1 A output current; 0.25% long-term stability

79xx series: same as 78xx but with negative output voltage; also available in low-power version, 79Lxx

C106: a popular 3 A SCR available for use with the following voltages: 15, 30, 50, 100, 200, 300, 400 V

SUGGESTED READINGS

Hemingway, T. K., ELECTRONIC DESIGNER'S HANDBOOK, 2nd ed., pp. 19–26. Blue Ridge, Penn.: Tab Books, 1970. *Zener diode applications.*

Millman, Jacob, MICROELECTRONICS, pp. 676–689. New York: McGraw-Hill, 1979. *Regulated power supplies, including the 7800 series regulator circuitry.*

Morris, Noel, ADVANCED INDUSTRIAL ELECTRONICS, Chap. 10. Maidenhead, England: McGraw-Hill, 1974. *A variety of power supply regulator schemes.*

Jung, Walter G., IC OPAMP COOKBOOK, Chap. 4. Indianapolis: Howard W. Sams, Publ., 1974. *Voltage and current regulator circuits.*

General Electric Co., SCR MANUAL, 5th ed. Syracuse, N.Y.: General Electric Co., 1972. *The SCR "bible."*

EXERCISES

1. With regard to power supplies, what are two major causes that may lead to a departure from voltage constancy?

2. Relative to a Zener diode, what is the test current I_{ZT}? The knee current I_{ZK}?

3. In terms of test current, what is generally taken to be the maximum allowable power dissipation in a Zener diode?

4. Why is it preferable to use a Zener voltage in the vicinity of 5.3 V?

5. What is the dominant characteristic of the 1N821 series of diodes?

6. Sketch the OPAMP circuit that may be used to maintain a constant current through a Zener. What are the respective functions of the three resistors used in this circuit?

7. Briefly describe the function of a 723 IC regulator.

8. What is the output voltage of a 7815 IC regulator? What is the magnitude of the current output?

9. What exemplifies the 78Lxx version of IC regulator? The 7900 series?

10. With the base of a transistor shorted to the collector, how may the value of V_{BE} be adjusted?

11. Sketch the basis for a simple current mirror and explain its mode of operation. What is its main deficiency?

12. In the active region of the transistor's I_C vs. V_{CE} curves, what causes the finite slope of the base current lines?

13. What is the electrical equivalent of an SCR?

14. Once conduction starts in an SCR, how may it be terminated?

15. Distinguish between the conduction angle (θ) and the firing angle (ϕ) of an SCR.

16. Sketch the SCR circuit used to control the speed of a $\frac{1}{4}$ in. drill.

17. Functionally, how does the Triac differ from an SCR? What is a Diac?

18. What is the cause of cycle skipping in an SCR?

19. What are thyristors?

PROBLEMS

16.1

a. For small changes in supply voltage (ΔV), in analyzing a Zener diode circuit, the Zener may be replaced by an emf equal to V_Z in series with an incremental resistance r_Z. Assuming that $r_Z \ll R_L$ and therefore that $V_R \simeq V_Z$, show that the change in the reference voltage (ΔV_Z) is related to the series resistance (R) and the load resistance (R_L) by the following equation:

$$\Delta V_Z = \frac{\Delta V}{1 + (R/R_L) + (R/r_Z)}$$

b. In the circuit shown in Figure 16.31, what should be the value of R?

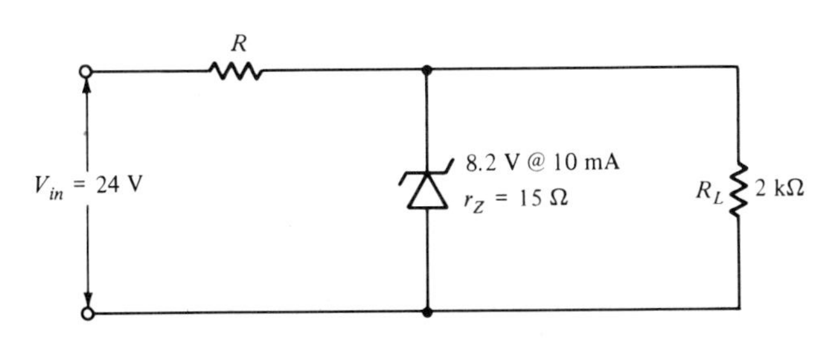

FIGURE 16.31

c. If V_Z is to be held to within 0.001%, to within what range of voltage variation should the input be maintained?

16.2

Compute the required power rating for the Zener diode in the circuit shown in Figure 16.32. Take the incremental resistance to be 15 Ω. Let the applied voltage change by 10%. What will be the percentage variation of the Zener voltage? What would be the variation due to a temperature change from 20°C to 80°C?

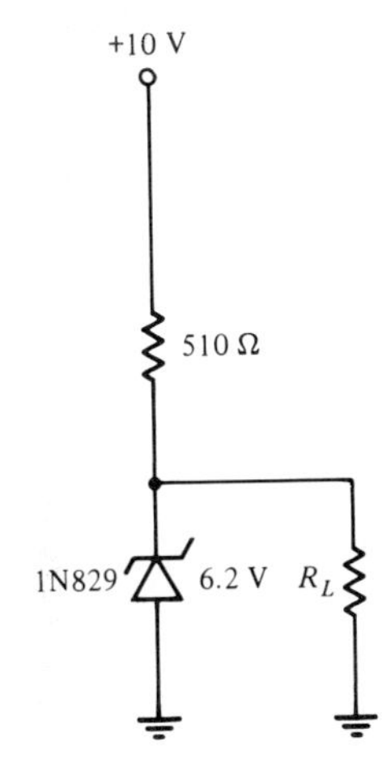

FIGURE 16.32

16.3

a. What is the smallest value of R_L that can be used in the circuit shown in Figure 16.33?

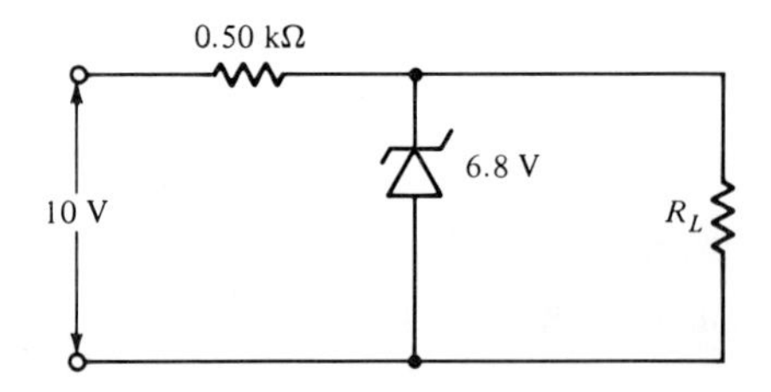

FIGURE 16.33

b. If $R_L = \infty$, what is the change in the regulated voltage if the input voltage changes by 1 V, with $r_Z = 50\ \Omega$? In terms of percentages, what are the input and output changes?

c. What should be the dissipation rating of the zener?

16.4

For the circuit shown in Figure 16.11, it is desired to have a programmed current of 20 μA and a mirror current of 10 μA. What should be the value of the resistor inserted in series with the programmed lead if $V_{CC} = 10$ V, and what should be the value of the emitter resistor R_2?

16.5

Draw a simple current mirror using an *npn* transistor.

16.6

Draw the *npn* equivalent of Figure 16.10.

16.7

With a simple mirror circuit, if $V_{CC} = +12$ V, what should be the value of R_p to produce a load current of 1 mA?

16.8

a. In using the simple mirror circuit of Problem 16.6, if $V_{CC} = 12$ V and $I_L = 10\ \mu$A, what should be the value of the programming resistor?

b. The value in part a turns out to be unreasonably large for IC fabrication. One may then resort to using a proportional mirror circuit by using an emitter resistor in conjunction with the transistor containing the load. What should be its value if the programmed current is 1 mA?

16.9

For an SCR, verify the relationship among I_{dc}, the firing angle ϕ, and the conduction angle θ, as expressed by Eq. 16.4.

16.10

Using the results of Example 16.6, what is the power delivered into a 10 Ω load by an SCR operated from a 120 V rms ac source with a *firing* angle of 120°?

16.11

See Figure 16.34. The conduction of an SCR control can be extended from a minimum of 180° to values in excess of 180° by using a conventional diode in conjunction with an SCR, although the firing angle will still be limited to 180°. Show how this might be done.

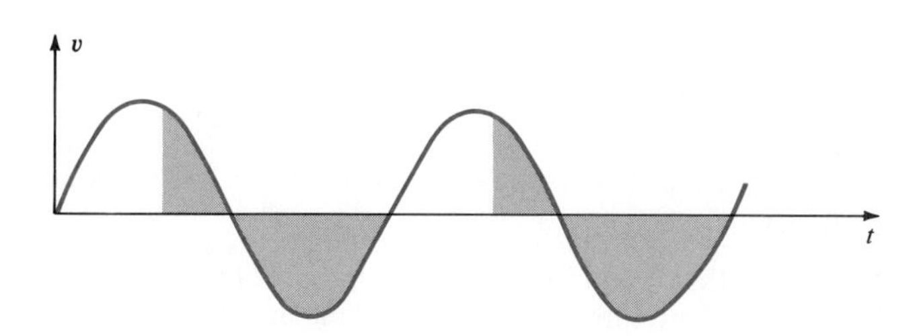

FIGURE 16.34

16.12

Compare the power content of a half-wave rectified sinusoidal wave to the power content of the same wave with a firing angle of 30°. With a firing angle of 150°. What do you conclude concerning the power content of the first and last 30° of each half-cycle? For full-wave power control, what is the percentage of power content in the initial and final 30°?

16.13

While the average value of current may be well within the specifications of a particular SCR, for low duty cycles with high peak values the rms current rating is also important to prevent excessive heating of joints, interfaces, leads, etc.

a. Compute the rms values for half-wave pulses of V_m peak magnitude with a firing angle of 45°. With a firing angle of 135°.

b. What are the average values for the same firing angles?

16.14

A simple SCR trigger circuit can be constructed as in Figure 16.35.

a. What is the range of phase angles over which the circuit may be made to operate?

b. With SCR control, relative to the power that would be delivered to the load when connected directly across the ac line, what would the power be with the SCR full on?

c. What would be the power percentage when the SCR is half on?

d. What should be the value of the *pot* if the trigger voltage is 0.8 V?

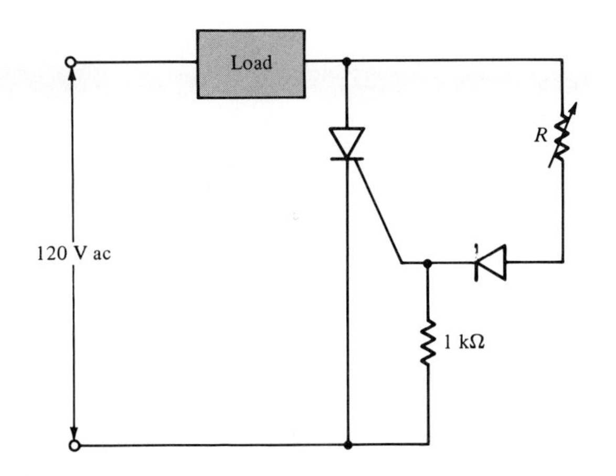

FIGURE 16.35

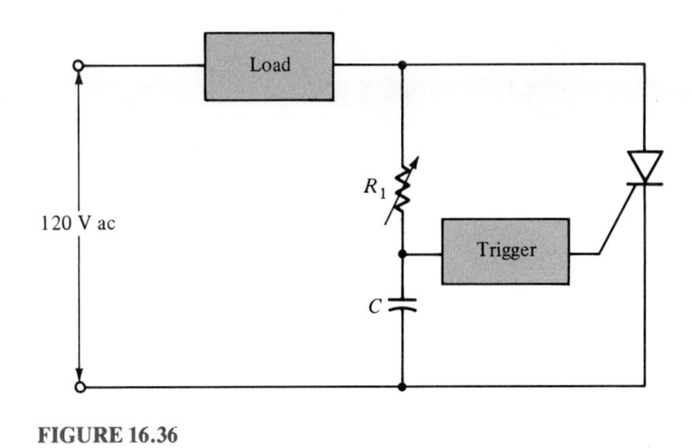

FIGURE 16.36

16.15
For an *RC* trigger of the form shown in Figure 16.36, with a residual voltage on the capacitor of V_0 and a breakover voltage of V_s, show that the value of R_1 leading to $\phi = \pi$ is

$$R_1 = \frac{2V_m}{\omega C(V_s - V_0)}$$

17 MEASUREMENTS

OVERVIEW

The measurement of dc was discussed in Chapter 1. A discussion of ac measurement requires information developed in subsequent chapters: the Fourier analysis of Chapter 5 and the diodes of Chapter 6. Additionally, digitized measurement of ac and dc requires the intervening development of FET theory (Chapter 9), the OPAMP (Chapter 10), digital electronics (particularly the digital readout discussed in Section 11.7), and the reduction of 60 Hz interference considered in Chapter 13.

In this chapter, many of the concepts introduced in the earlier chapters are applied to two illustrative measuring techniques, those for strain and temperature.

OUTLINE

17.1 ■ INTRODUCTION

This chapter is basically divided into two parts. The first discusses a long-deferred topic, ac voltage measurements. Allied with this discussion is a consideration of some of the methods used to convert analog signals into digitized form, thus making them suitable for digital display.

In the last half of the chapter, to illustrate what may be accomplished upon applying electronics to physical measurements, we select two areas for a somewhat detailed treatment: strain gages and the measurement of temperature. In particular, the latter shows how an old technique, the measurement of temperature using thermocouples, has been updated through the use of electronics.

17.2 ■ MEASUREMENTS

17.2.1 Peak- and Average-Reading Instruments

Since sinusoidal alternating currents spend as much time traveling in one direction as in the other, impressing such a current on a dc meter would probably yield a zero reading since the meter movement would not be able to follow the rapid current fluctuations of other than a very low-frequency current. There is one type of ac meter that involves but a slight modification of the basic d'Arsonval movement, the electrodynamometer mentioned in Section 1.5.4; rather than a pivoted coil mounted between permanent magnets, the coil is mounted between electromagnets whose coils are connected in series with the indicator coil. As the current reverses direction, both the fixed and rotatable magnetic fields reverse polarity, and the torque direction is unaffected.

But the majority of measuring devices make use of a rectifier in combination with a dc meter. This rectifier can be either the half- or full-wave form. After this rectification a dc voltmeter will indicate the *average* voltage (or current), which in the two cases is $0.318 V_{peak}$ and $0.636 V_{peak}$, respectively, presuming the waveform to be sinusoidal. But what one usually desires is the *effective* (that is, rms) value of the voltage (or current).

As long as the waveform is sinusoidal, this requirement poses no problems since there is a unique numerical relationship between V_{eff} and V_{avg}:

$$\frac{V_{eff}}{V_{avg}} = 1.1 \qquad \text{(full wave)} \tag{17.1}$$

$$\frac{V_{eff}}{V_{avg}} = 2.2 \qquad \text{(half wave)} \tag{17.2}$$

Therefore, we merely relabel the meter face to obtain the rms value. Thus, if an ac meter is using a full-wave rectifier and the dc value is 90.9 V, the meter scale is labeled 100 V rms—that is, the effective value—[90.9 V (average) $\times$ 1.1 = 100 V (rms)].

Suppose, however, the input signal is not truly sinusoidal but a very distorted sinusoid with small positive peaks, as in Figure 17.1. In this case the negative parts of the waveform are rejected by the rectifier (assuming half-wave rectification), and the average value of the positive peaks is quite small. The average-reading type of ac voltmeter, whose scale, however, implies that this value is an effective value, would give a rather small reading. The heating effect of such a signal might be substantial and would cause us to wonder how such a small signal (as indicated by the meter reading) could be causing such an appreciable amount of heating.

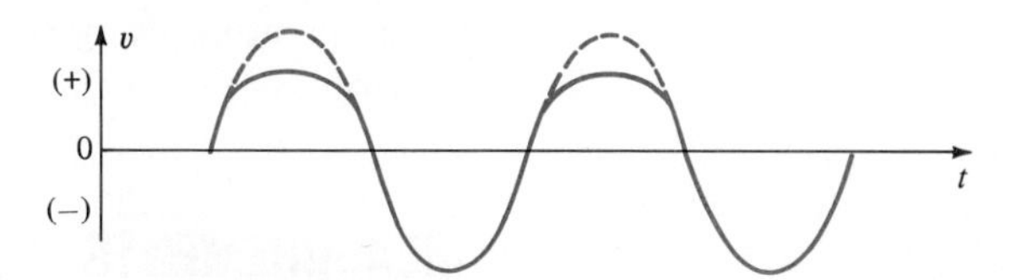

FIGURE 17.1. Distorted Sine Wave

"Well," you say, "let's reverse the input probes." This time, however, we would obtain a heating effect smaller than that implied by the meter reading since what are now the negative pulses would not be making the assumed contribution.

If we used a full-wave rectifier with an averaging type of meter, we would still get an erroneous reading because the numerical factor relating average and rms values would not be 1.1. The moral of this discussion is that average-reading meters indicate true rms only for sinusoidal waveforms.

There is another large group of ac voltmeters that *read* peak values but also *indicate* rms values by a scale change. The basic circuit might look as in Figure 17.2. As R_b is made larger, less charge "leaks off" the capacitor during the nonconducting periods of the diode, and the voltage peak is converted into the rms value by the factor $V_{rms} = 0.707 V_{peak}$. Again, this relationship holds only for truly sinusoidal waveforms.

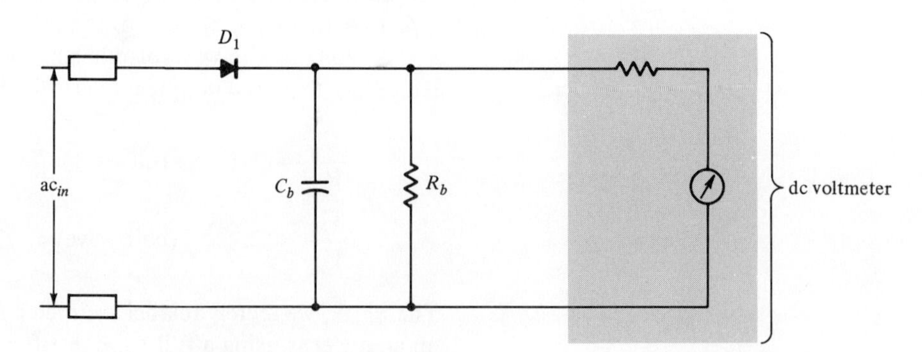

FIGURE 17.2. Peak-Reading ac Voltmeter

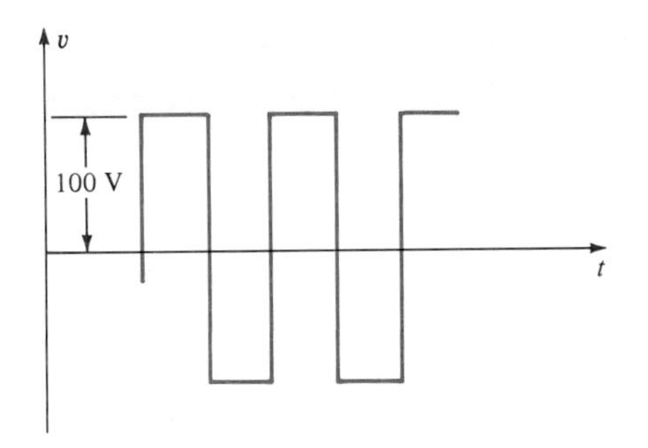

FIGURE 17.3. **Square Wave with 100 V Peak Amplitude**

How erroneous such readings can be if the waveform is not sinusoidal may be illustrated by considering the 100 V square wave shown in Figure 17.3. To find the rms value, we square the instantaneous values: This operation yields 10,000 V^2. Then take the square root, yielding 100 V. This result is also the mean value. Therefore, the rms value of the square wave is 100 V. (This result may easily be realized since inverting the negative pulses gives us a constant voltage of 100 V. By definition the rms value is equal to the dc value that yields the same heating effect.) But a peak-reading meter would apply a factor 0.707 and indicate (erroneously) an rms value of 70.7 V.

Are there any meters that can read rms directly? Yes. By using a thermocouple, a meter might measure the heating effect of current passing through a resistance. A major difficulty with such an instrument used to be the long time constant; any rapid variation of the rms value would go undetected. Also, in many instances, the signals one attempted to measure might be so small that the meter lacked the necessary sensitivity. In recent years, however, there has been a considerable improvement in such procedures.

There are some unique electronic circuits that can accomplish the task, however. One of the earliest was the Ballantine 300 voltmeter, which rather closely indicated a true rms value for waveforms containing harmonics as high as the tenth.

With the advent of integrated circuits, a new resolution to the problem has become available. The rms value of a complex waveform can be computed. Explicit computing of the rms value of a waveform requires three mathematical operations. With v_{in}, squaring yields v_{in}^2/V_R, where $1/V_R$ is a scaling factor, averaging yields $\overline{v_{in}^2}/V_R$, and taking the square root leads to $\sqrt{\overline{v_{in}^2}}$. (The result of an electronic square root operation yields $\sqrt{V_R \cdot \overline{v_{in}^2}}$.) A major disadvantage of such direct computation is the limited dynamical range that is available. With an input range of 1:100 (say 0.1–10 V), the output of the squaring circuit (a multiplier) will have a 1:10,000 dynamic range [$(0.1)^2/10$ to $10^2/10$ = 1 mV to 10 V]. Since multiplier errors can significantly exceed 1 mV, an accurate overall dynamic input range will be limited to 1:100 or even 1:10. A much simpler solution, providing a much wider dynamic range, uses a divider circuit that can also be made to furnish a square root.

Consider an input signal x applied to a divider where the divisor is the circuit's output y:

$$\begin{aligned} y &= \frac{x}{y} \\ y^2 &= x \qquad (17.3) \\ y &= \sqrt{x} \end{aligned}$$

If $x = v_{in}^2$, averaging yields $\bar{x} = \overline{v_{in}^2}$ and $(\bar{x}/y) = y$, $y = \sqrt{\bar{x}} = \sqrt{\overline{v_{in}^2}}$, and we can identify y as the rms value of the input voltage.

Consider the circuit in Figure 17.4. By following the multiplier-divider with a low-pass filter, we obtain V_{out} = average (v_{in}^2/V_{out}). This circuit in effect performs the squaring, averaging, and dividing by the output in succession, but since the output remains essentially constant over the period of the

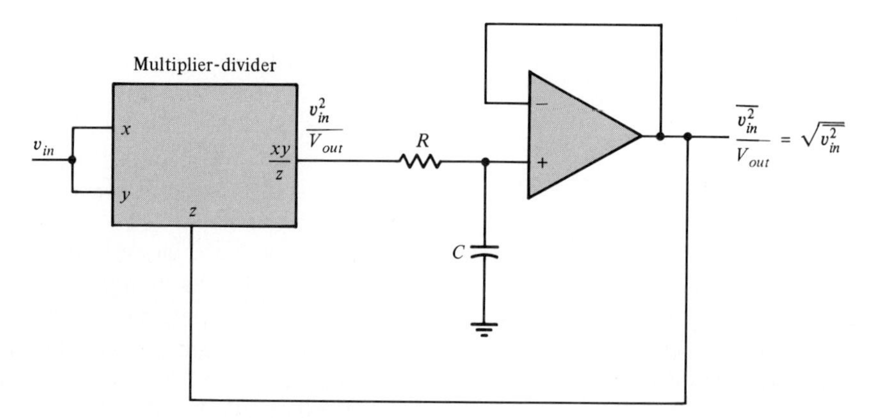

FIGURE 17.4. **rms Computation**

signal, we may divide *before* taking the average. We thereby identify V_{out} as the rms value of the input signal:

$$V_{out} = \overline{v_{in}^2}/V_{out}$$

$$V_{out} = \sqrt{\overline{v_{in}^2}} = V_{rms}$$

Such devices are commercially available. The Analog Devices AD536, for example, computes the rms values of ac and dc signals within a 100 kHz bandwidth for signal levels above 100 mV. An important parameter in such devices is the **crest factor**, being the ratio of peak signal swing to the rms value. The 536 accommodates a crest factor of 6 with a 1% error. The value of an external capacitor sets the low-frequency ac accuracy.

The AD442 may be used for conversion from dc to 8 MHz and is ideally suited for measuring thermal noise and transistor noise. It may also be used for measuring mechanical phenomena such as strain, stress, vibration, and shock when such signals are noisy, nonperiodic, and superimposed on dc levels.

17.2.2 Analog-to-Digital Conversion

The modern trend in measuring instruments is to employ a digital rather than an analog display. There are a number of methods that can be used for this purpose: parallel (flash) encoder, ratiometric or dual-slope, single-ramp counting, and successive approximation. The unit central to all of these methods is the comparator. Early on, this was the 741 IC, which we have used for so many other purposes.

The comparator provides information concerning the relative state of two input voltages. If one input is a reference voltage, the comparator allows a determination of whether the other (unknown) input is above or below the reference voltage. Taking a look at Figure 17.5, one can see that if $V_{in} > V_{ref}$, the output takes on the value of $+V_{sat}$; if $V_{in} < V_{ref}$, the output is at $-V_{sat}$. As used in a digital voltmeter, the voltage to be measured (unknown) is applied to one input while the voltage applied to the other is swept through a

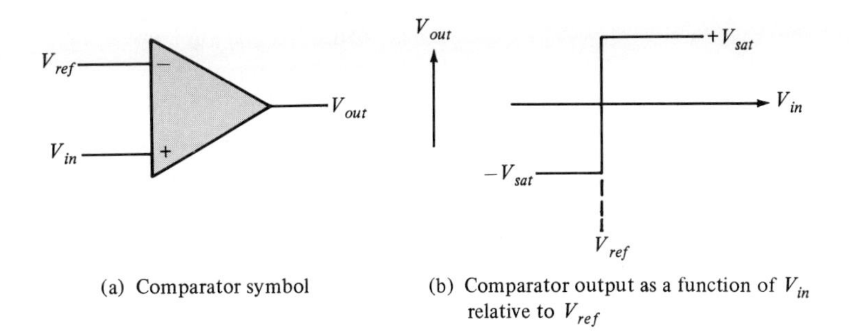

(a) Comparator symbol

(b) Comparator output as a function of V_{in} relative to V_{ref}

FIGURE 17.5. **Voltage Comparator**

series of values in some controlled manner. When the two inputs arrive at coincidence, the comparator output changes state, and this change can be used to activate a numerical display related to the value of the "unknown" voltage.

Parallel (Flash) Encoder. Most generally, physical measurements appear in the form of analog signals (temperature, pressure, displacement, voltage) and it is desired to convert these signals into digital form. A stack of comparators may be made to serve in such an analog-to-digital converter (A/D converter, or ADC). In simple form this process is illustrated in Figure 17.6. This circuit

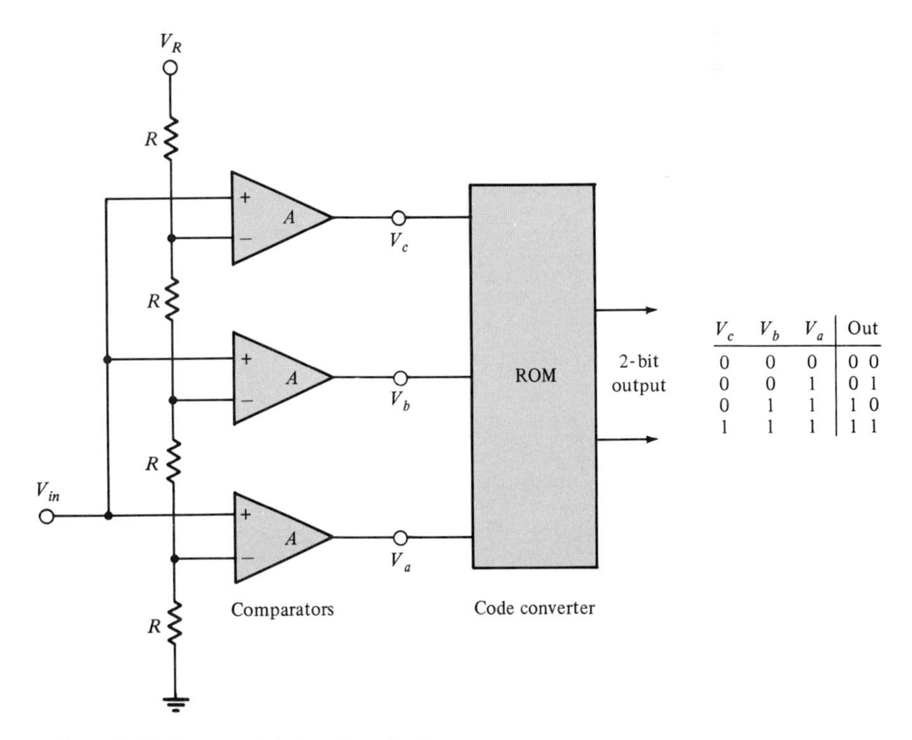

V_c	V_b	V_a	Out
0	0	0	0 0
0	0	1	0 1
0	1	1	1 0
1	1	1	1 1

FIGURE 17.6. **Parallel (Flash) A/D Converter**

is termed a **parallel encoder**, also known as a **flash encoder** because of its high operating speed.

The analog signal (V_{in}) is applied to the input terminal. At the inverting terminals of the comparators, one has (proceeding upwards) the following voltages: $\frac{1}{4}V_R$, $\frac{2}{4}V_R$, and $\frac{3}{4}V_R$. If $V_{in} < \frac{1}{4}V_R$, all three comparators are "down"—that is, $V_a = V_b = V_c \equiv 0$. If V_i lies between $\frac{1}{4}V_R$ and $\frac{2}{4}V_R$, V_a is "up" and V_b and V_c are down. V_i between $\frac{2}{4}V_R$ and $\frac{3}{4}V_R$ leads to both V_a and V_b being up and V_c being down, etc.

The read-only memory (ROM) interprets these results as addresses and yields a 2-bit output consisting of digital values previously stored at those locations. The comparator stack may be expanded to accommodate more extensive displays. For an n-bit display, $2^n - 1$ comparators are needed. Commercial encoders with 4-bit and 8-bit outputs are readily available (16 to 256 levels, respectively), but they become very expensive at the higher values. A 256-level unit, for example, can cost hundreds of dollars. Flash converters may achieve rates of from 10 to 100 million conversions per second.

Ratiometric, or Dual-Slope, Conversion. Another rather popular form of conversion makes use of the **ratiometric A/D converter** shown in Figure 17.7. Presume that $V_a > 0$ and $V_R < 0$. With S_1 open and S_2 closed, the counter is set to zero. V_a is applied via S_1, S_2 is opened, and V_a is integrated for a fixed number of clock pulses, n_1. If the clock period is T, the integration time is $T_1 = n_1T$. (See Figure 17.8.) For an N-stage ripple counter, with $n_1 = 2^N$, at time t_2 all counters will reset to zero. This event can be used to connect S_1 to V_R. Since V_R is negative, v now has a positive slope. Also since we assume that $|V_R| > V_a$, the second integration time is less than the first—that is, $T_2 < T_1$. As long as v is negative, the output of the comparator is positive, and the AND gate allows pulses to be counted. At $v = 0$ (which occurs at t_3) the gate is inhibited, and no further clock pulses are counted.

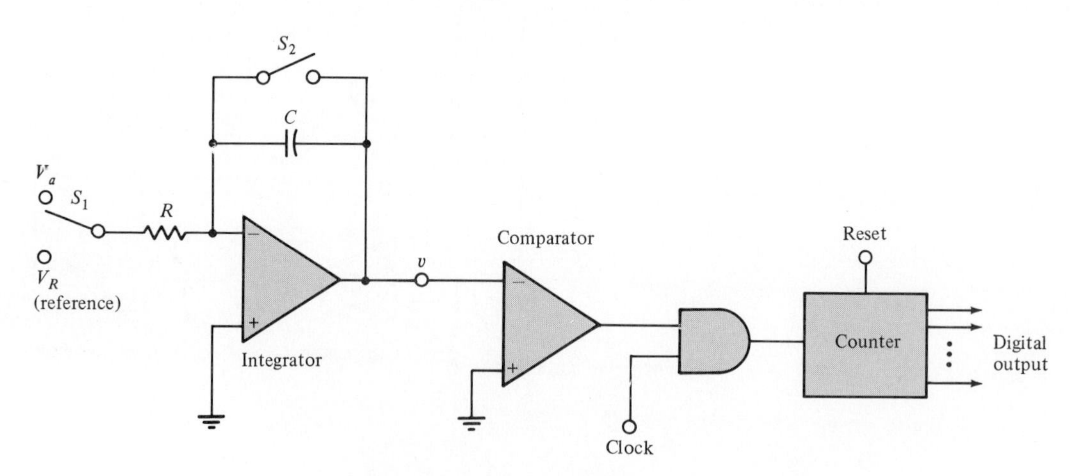

FIGURE 17.7. **Ratiometric A/D Converter**

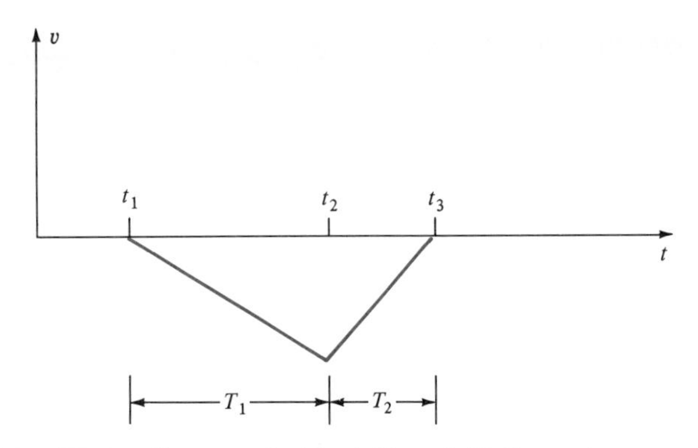

FIGURE 17.8. **Timing Sequence for Dual-Slope A/D Converter**

The counts at t_3 are proportional to the analog voltage. The value of v at t_3 is

$$v = -\frac{1}{RC}\int_{t_1}^{t_2} V_a\,dt - \left(-\frac{1}{RC}\int_{t_2}^{t_3} V_R\,dt\right) = 0 \qquad (17.4)$$

(Remember, $V_R < 0$.) Since V_a and V_R are constants, we have

$$V_a(t_2 - t_1) + V_R(t_3 - t_2) = 0 \qquad (17.5)$$

$$V_aT_1 + V_RT_2 = 0 \qquad (17.6)$$

$$T_2 = n_2T \qquad (17.7)$$

$$T_1 = n_1T = 2^NT \qquad (17.8)$$

$$V_a = \frac{T_2\,|V_R|}{T_1} = \frac{n_2\,|V_R|}{n_1} = n_2\frac{|V_R|}{2^N} \qquad (17.9)$$

Since $|V_R|$ and N are constants, $V_a \propto n_2$. The result is independent of the time constant RC.

The digital output of a **dual-slope converter** represents an average of the input signal during the first integration. This procedure allows the signal to change during conversion. The dual-slope method is the most widely used technique in conjunction with digital meters and can be made fairly immune from 60 Hz interference by choosing an integration time that is an integral number times the power-line period. We shall say more about this process in a succeeding section.

The conversion time of the dual-slope method requires n_1—that is, 2^N—clock periods for the first integration and n_2 for the second. The conversion time is variable, therefore, with a maximum value of $2 \cdot 2^N$. A 10-bit dual ramp with a 1 MHz clock may require as much as 2.048 ms/conversion, with less than 500 conversions/second. Although relatively slow, its noise rejection capability and linearity make it an attractive ADC.

EXAMPLE 17.1

See Figure 17.9. If f represents the pulse frequency and n_1 the number of counts during the signal integration time of the dual-slope ADC, determine the total charge on the integrator due to a sample voltage V_a.

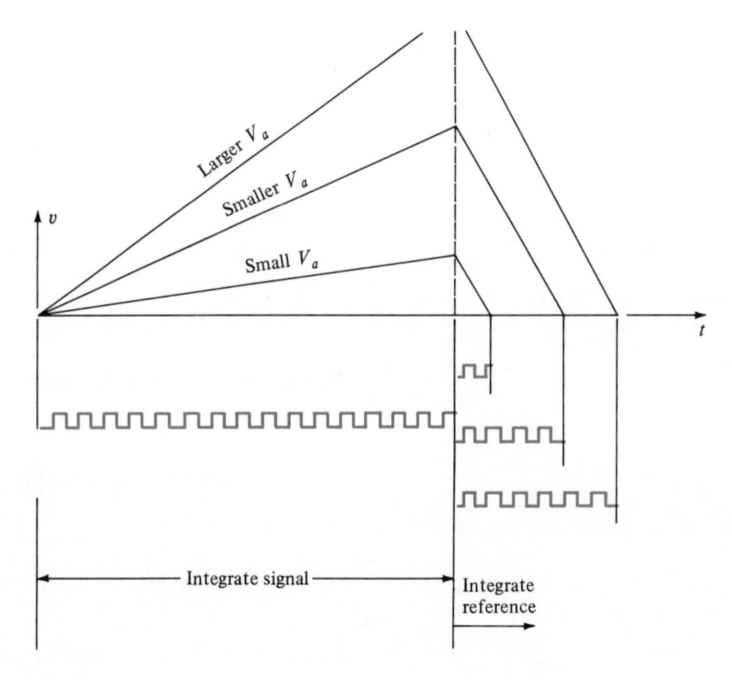

FIGURE 17.9

If the discharge count is n_2, what is the expression for n_2 in terms of the sample voltage V_a, the reference voltage V_R, and n_1?

Solution:

$$q_C = \frac{V_a}{R}\Delta t = \frac{V_a}{R}\frac{n_1}{f}$$

In the discharge process, we have

$$q_C = \frac{V_R}{R}\frac{n_2}{f}$$

Therefore, we have

$$\frac{V_a}{R}\frac{n_1}{f} = \frac{V_R}{R}\frac{n_2}{f}$$

and

$$n_2 = \frac{V_a}{V_R}n_1$$

A Counting (Single-Ramp) ADC. Figure 17.10 illustrates a counter form of ADC in the form of a **single-ramp** digital voltmeter. Let us assume that the voltage to be measured is +1.5 V. A clock is started when the comparator recognizes the first coincidence, and it is shut off when the sweep voltage passes through zero. If the slope has been properly selected, the number of clock pulses (except for the decimal placement) is numerically equal to the unknown voltage. The polarity of the voltage can easily be determined. It depends on whether the zero crossing occurs before or after the coincidence.

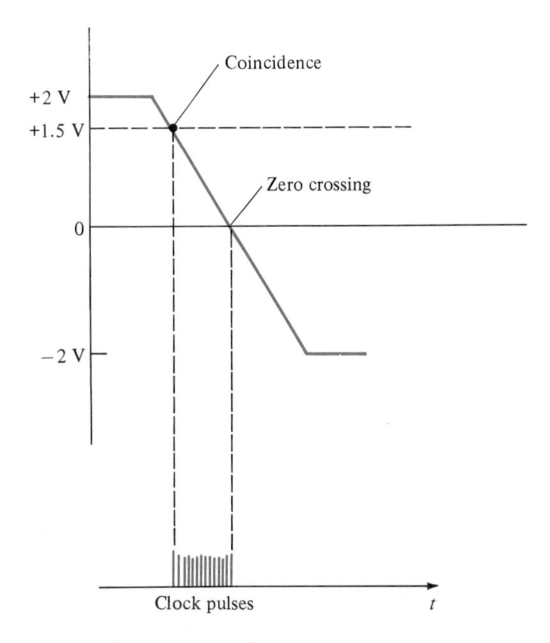

FIGURE 17.10. Single-Ramp Digital Voltmeter

In a more general sense, the counter can be of the binary form, and parallel format output at the termination of the clock pulses constitutes the digital form of the analog input.

The slope of the sweep voltage is usually determined by an *RC* time constant, which constitutes a major disadvantage of this type of ADC. Over long periods of time it is difficult to maintain the time constant, particularly in the presence of temperature variations.

Rather than a linear sweep voltage, one may employ a **staircase** sweep. In this case, again, a counter is used, the number of steps needed to reach the signal level constituting the output. Like the linear sweep, each conversion may require considerable time, thus limiting the conversion rate.

If the analog signal is varying with time, one needs to employ a sample-and-hold system. A **tracking converter** constitutes an improvement in this regard.

In the tracking converter an up-down counter is used. When the stepped sweep exceeds the signal level, the counter reverses and starts to count "down." But after one down count it again reverses direction, for one count. Therefore, as long as the input level remains constant, it will bounce between ± 1 LSB. Should the input signal be subject to variation (provided the variation is not too rapid), this type of ADC will follow the variation.

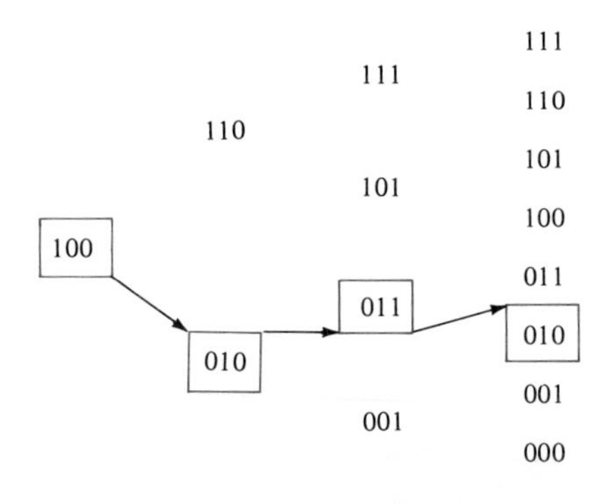

FIGURE 17.11. Three-Bit A/D Converter Using Successive Approximation

The staircase ramp is somewhat less susceptible to noise interference in contrast to the linear ramp where a noise pulse may give a false indication of ramp termination. But the integrating type of digital meter is the most secure against such noise interference.

Successive Approximation ADC. Rather than a binary counter, in this form of ADC a programmer is used. The latter sets the MSB to 1 and all other bits to 0. If the comparator indicates the ADC output to be larger than the signal, the 1 is removed, and the next bit is tried as being the most significant. If the analog input is larger, the 1 remains in that bit.

The 1 is tried for each bit until, at the end, the binary equivalent of the analog is obtained. Rather than a worst case of 2^N pulses needed with a counting ADC, the **successive approximation** method needs only N clock periods (Figure 17.11). A 10-bit converter with a 1 MHz clock can perform 100,000 conversions/second.

Analog vs. Digital Instruments. The advantages of digital instruments are probably obvious: fast and accurate readings, repeatability, automatic ranging, automatic polarity determination, etc.

Are there any disadvantages? Yes. If one is making circuit adjustments and using a digital instrument as an indicator, it is difficult to determine the rate of change that is needed to guide one to a peak or a null in the response. Digital instruments are necessarily averaging instruments with the attendant limitations we have previously discussed in conjunction with averaging analog instruments. One may also encounter display difficulties under high ambient light conditions, such as sunlight.

With respect to some of these shortcomings, mention might be made of the Simpson 360 digital VOM (volt-ohm-meter). It provides both digital and analog displays. We have spoken primarily of digital voltmeters, but like the analog multimeter, many of the digital instruments also can be made to perform these various functions, such as the Simpson 360.

Many digital voltmeters are classified as having a $\frac{1}{2}$-digit capability. For instance, a 5-digit readout display might be classed as a $4\frac{1}{2}$-digit instrument and one with 4 digits, as a $3\frac{1}{2}$-digit device. Such specifications arise because in these instruments all the digits except the most significant (the extreme left one) have a 0–9 capability; the most significant one can read only either 0 or 1, making it a "half-digit." Since the instruments are generally divided into decade ranges, this restriction does not prove to be a handicap.

On a given range, however, these instruments generally allow for **"overranging."** Thus, a voltmeter might be made to read as high as 19.99 on the 0–10 V setting, thereby representing virtually a 100% overrange; others may make provision for only a 10% overrange. If the overrange is exceeded, some sort of warning indicator advises of the overload and the need to switch to a higher scale.

EXAMPLE 17.2

If a $3\frac{1}{2}$-digit BCD counter utilizing a dual-slope converter is to be made to use an integration period equal to one period of 60 Hz, what should be the clock frequency if a 10% overrange is desired? What should be the clock frequency for a $5\frac{1}{2}$-digit instrument with a 10% overrange? What is the integration time?

Solution: The full range count is 1000. With a 10% overrange the count should be 1100. Therefore, we have

$$\frac{1100}{1/60\ \text{s}} = 66{,}000\ \text{Hz} = 66\ \text{kHz}$$

The desired count is 11,000. We have

$$\frac{11{,}000}{1/60\ \text{s}} = 660\ \text{kHz} = 660\ \text{kHz}$$

The integration time is

$$\frac{1}{60} = 16.7\ \text{ms}$$

EXAMPLE 17.3

The resolution of a DVM refers to its dynamic range. A 5-digit unit can resolve 1 part in 100,000 or 0.001%, while a 4-digit unit has a resolution of 0.01%.

What, respectively, are the resolution capabilities of a $4\frac{1}{2}$-digit and a $3\frac{1}{2}$-digit instrument?

Solution: For a $4\frac{1}{2}$-digit unit, since readings beyond 10,000 generally constitute an overrange, we have

$$\frac{1}{10{,}000} = 10^{-4} = 0.01\%$$

that is, about the same as a 4-digit instrument.

For a $3\frac{1}{2}$-digit instrument, we have

$$\frac{1}{1000} = 10^{-3} = 0.1\%$$

that is, about the same as a 3-digit instrument.

17.2.3 Normal-Mode Rejection

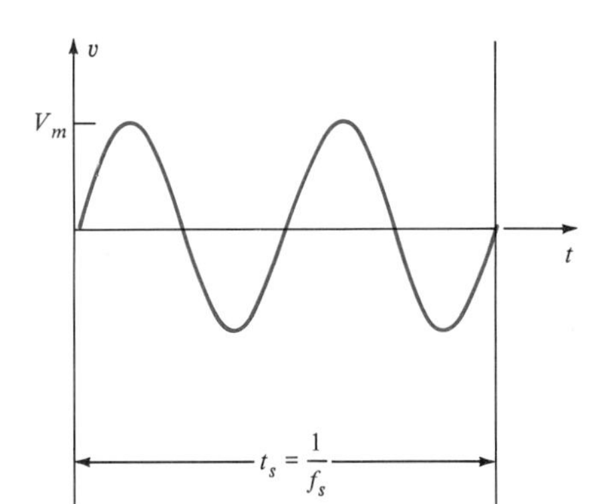

FIGURE 17.12. Interfering Wave, Integrally Related to the Sampling Interval

Even when using a battery-powered instrument, one is inevitably faced with the problem of 60 Hz interference because of the ubiquitous presence of power lines that induce such interference into the instruments. (See Chapter 13.) Integrating procedures, by a proper choice of integrating interval, permit considerable discrimination against interference.

The effect of such interference can be completely negated if a whole number of cycles is integrated over a given time t_s (Figure 17.12). If the time t_s does not encompass a whole number of cycles, there will remain a noncancelling voltage that will add to the dc level, the time remaining after a given number of whole cycles being t_r (Figure 17.13). The maximum amount of error is approximately obtained from

$$v_{error} = \frac{(1/2)v_{max}t_r}{t_s} \tag{17.10}$$

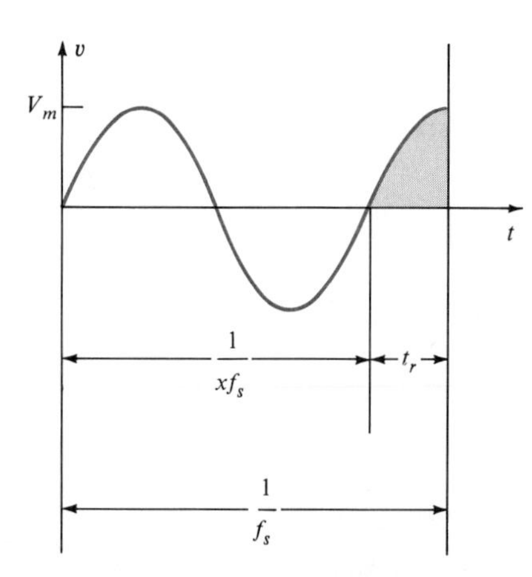

FIGURE 17.13. Interfering Sine Wave, Not Integrally Related to the Sampling Frequency There is increasing contribution to the error as the interfering frequency is increasing.

Such residual signals represent a normal mode, and the relationship between t_r and t_s represents the **normal-mode rejection** of a device. The error is proportional to t_r/t_s.

Consider the repetition rate of sampling to be f_s, corresponding to an integration time of $1/f_s$. The interfering signal has a frequency x times as large, and its period is $1/xf_s$. The residual time (t_r) is then

$$t_r = \frac{1}{f_s} - \frac{1}{xf_s} \tag{17.11}$$

$$\frac{t_r}{t_s} = \frac{x - 1}{x} \tag{17.12}$$

The integrated error will reach a maximum when $t_r = 1/2xf_s$—that is, when $x = \frac{3}{2}$. Thereafter the error will diminish as t_s is made larger relative to the period of the interfering signal. We then have (Figure 17.14)

$$t_r = \frac{1}{2xf_s} - \frac{1}{f_s} + \frac{1}{xf_s} + \frac{1}{2xf_s} = \frac{1}{f_s}\left(\frac{2 - x}{x}\right) \tag{17.13}$$

and in this range

$$\frac{t_r}{t_s} = \frac{2 - x}{x} \tag{17.14}$$

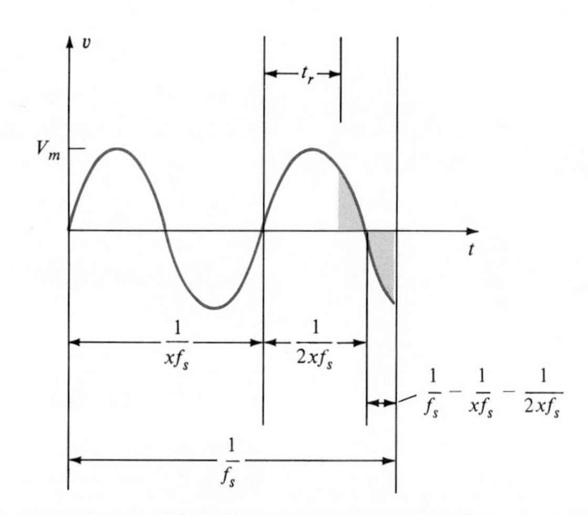

FIGURE 17.14. Interfering Sine Wave, Not Integrally Related to the Sampling Period There is diminishing contribution to error as the interfering frequency is increasing.

At frequencies below the sampling repetition rate, the appropriate equation is

$$\frac{t_r}{t_s} = \frac{1 - x}{x} \tag{17.15}$$

Summarizing, we have the following: At frequencies below f_s, we have

$$\frac{t_r}{t_s} = \frac{1 - x}{x} \tag{17.16}$$

In the interval from f_s to $1.5f_s$, we have

$$\frac{t_r}{t_s} = \frac{x - 1}{x} \tag{17.17}$$

From $1.5 f_s$ to $2f_s$, we have

$$\frac{t_r}{t_s} = \frac{2 - x}{x} \tag{17.18}$$

From $2f_s$ to $2.5f_s$, we have

$$\frac{t_r}{t_s} = \frac{x - 2}{x} \tag{17.19}$$

From $2.5f_s$ to $3f_s$, we have

$$\frac{t_r}{t_s} = \frac{3 - x}{x} \tag{17.20}$$

and so forth.

The normal-mode rejection, in decibels, is

$$\text{NMR [dB]} = 20 \log_{10}\left(\frac{t_r}{t_s}\right) \tag{17.21}$$

where t_r = residual time

t_s = integration period

The amplitude of the interfering signal does not enter this equation since the NMR merely indicates the degree of diminution whatever the signal amplitude might be.

EXAMPLE 17.4 With an integration time of 0.1 s, on semilog paper, plot the NMR (in decibels) as a function of the interfering frequency. Why is the NMR equal to zero below 5 Hz since there certainly is a normal mode error in this range?

Solution: An integration time of 0.1 s corresponds to a sampling frequency of 10 Hz. Taking a specific interfering frequency to be xf_s, in the range from 5 to 10(−) Hz, we use

$$\text{NMR} = 20 \log_{10}\left(\frac{1 - x}{x}\right)$$

Typically, at 9 Hz, $x = 0.9$ and NMR $= 20 \log_{10}[(1 - 0.9)/0.9] = -19$ dB.

In the range from 10(+) to 15 Hz, we use

$$\text{NMR} = 20 \log_{10}\left(\frac{x - 1}{x}\right)$$

Typically, at 13 Hz, $x = 1.3$ and NMR $= 20 \log_{10}[(1.3 - 1)/1.3] = -12.7$ dB.

In the range from 15 to 20(−) Hz, we use

$$\text{NMR} = 20 \log_{10}\left(\frac{2 - x}{x}\right)$$

Typically, at 18 Hz, $x = 1.8$ and NMR $= 20 \log_{10}[(2 - 1.8)/1.8] = -19$ dB. (See Figure 17.15.)

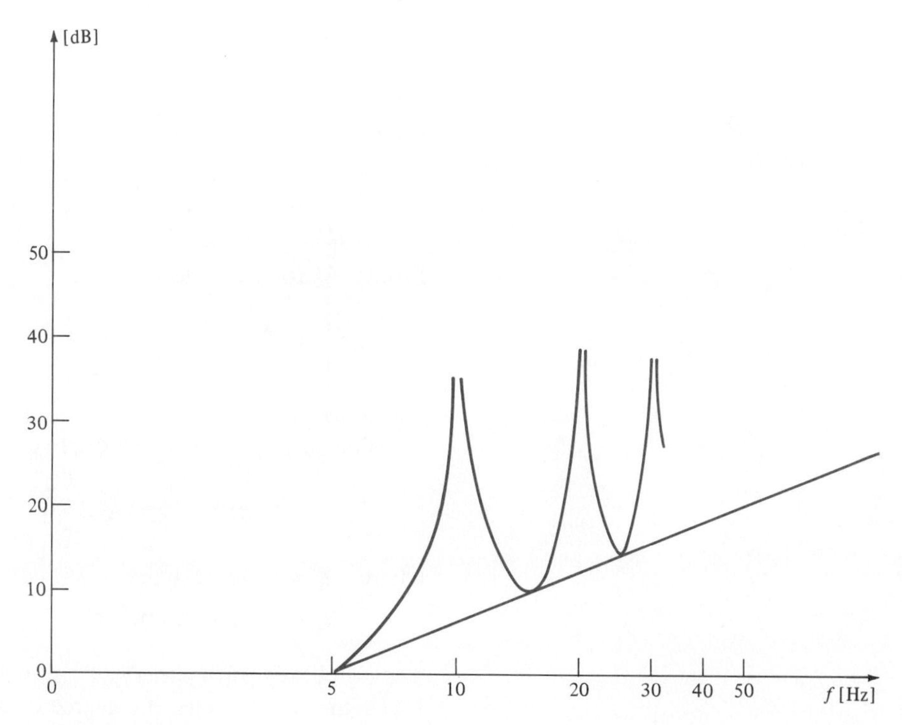

FIGURE 17.15

The NMR indicates the degree of rejection of the normal mode. Below 5 Hz there is a residual error, but it persists in undiminished fashion. Therefore, there is no attenuation, and the NMR = 0 dB.

17.2.4 Digital-to-Analog Conversion

The most basic circuit in digital-to-analog conversion (DAC) schemes makes use of the **summing amplifier** that in Figure 17.16 is shown in perhaps its simplest form. Since point X is at virtual ground, the input current (I) is

$$I = \frac{V_1}{R} + \frac{V_2}{R} + \frac{V_3}{R} + \frac{V_4}{R} \tag{17.22}$$

With $R_F = R$, we have

$$V_{out} = -IR_F = -(V_1 + V_2 + V_3 + V_4) \tag{17.23}$$

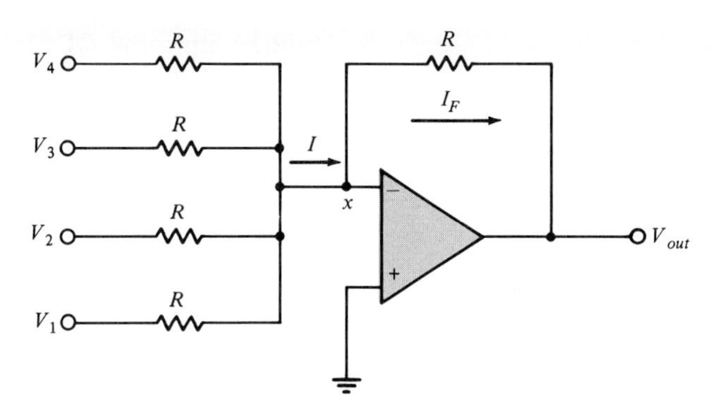

FIGURE 17.16. **Summing Amplifier**

The inputs can be either positive or negative. If the input resistances are not equal, the output is equal to the weighted sum.

EXAMPLE 17.5

See Figure 17.17. The inputs of the summing amplifier are to represent binary inputs corresponding to 1, 2, 4, and 8. With R_F = 10 kΩ, select the respective input resistances to yield an output equal to the binary input.

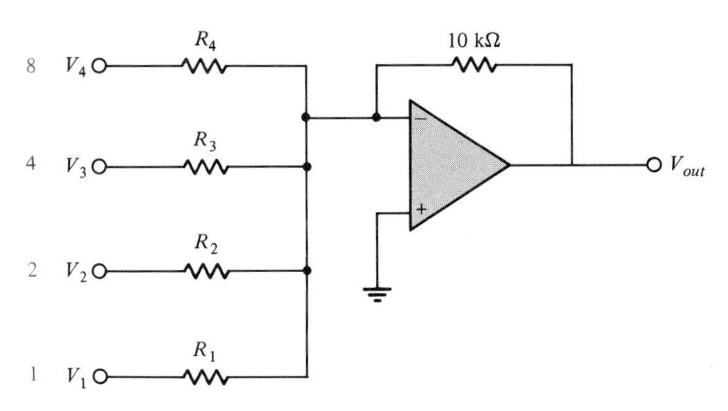

FIGURE 17.17

Solution: To weight the output proportional to the binary input, v_1 might be subjected to unity gain; v_2, to a gain of 2; v_3, to a gain of 4; and v_4, to a gain of 8. Therefore, we have

$$R_1 = 10\ \text{k}\Omega \qquad R_2 = 5\ \text{k}\Omega,$$
$$R_3 = 2.5\ \text{k}\Omega \qquad R_4 = 1.25\ \text{k}\Omega$$

If the number 15_{10}—that is, 1111_2— is applied, the output voltage will be

$$V_{out} = -10\left(\frac{1}{1.25} + \frac{1}{2.5} + \frac{1}{5} + \frac{1}{10}\right) = -15\ \text{V}$$

Presuming the OPAMP to operate with V_{CC} = ± 15 V, generally one should not allow the output to swing beyond 13 V. Therefore, the output should be scaled down, perhaps by increasing each input resistance by a factor of 10.

Approaching the problem from a more general viewpoint, consider Figure 17.18. The switches are electronically controlled in response to binary bits. If a particular bit is a 1, that switch is connected to V_R and if 0, to ground. The output voltage will be

$$V_{out} = (2^{N-1}a_{N-1} + 2^{N-2}a_{N-2} + \cdots + 2^2 a_2 + 2^1 a_1 + a_0)V \quad (17.24)$$

where V is a proportionality factor that we will evaluate in a moment.

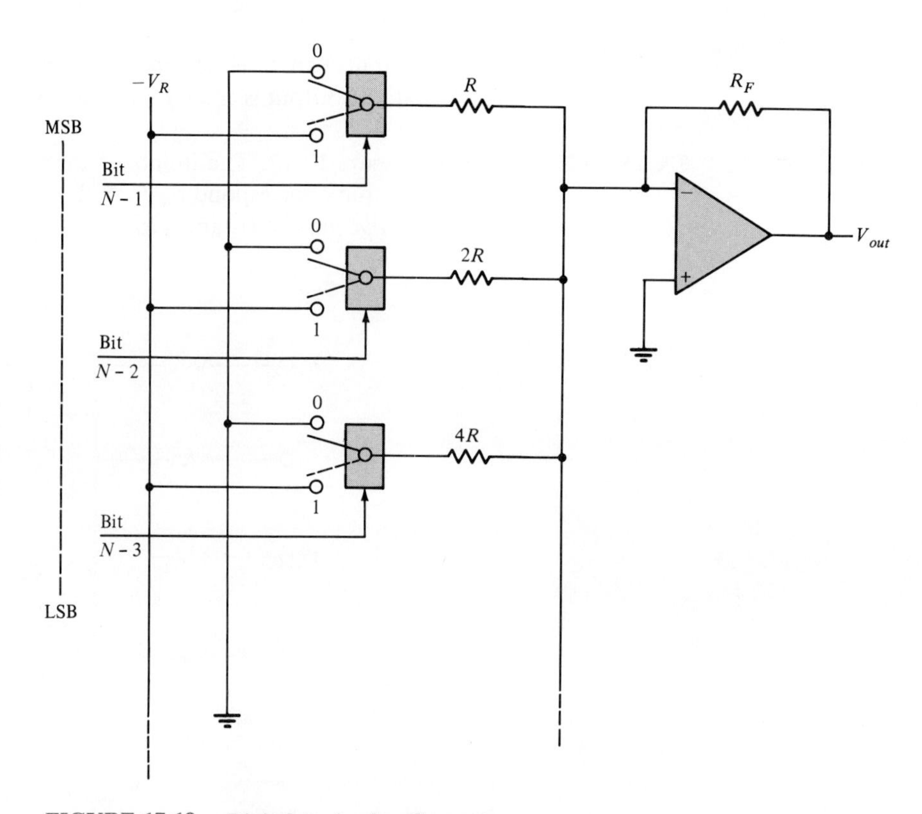

FIGURE 17.18. **Digital-to-Analog Converter**

The *as* assume values that are either 0 or 1, depending on the binary number. MSB is the most significant bit and LSB, the least significant. In turn, we have

$$V_{out} = \left(a_{N-1} + \frac{1}{2}a_{N-2} + \ldots + \frac{1}{2^{N-3}}a_2 + \frac{1}{2^{N-2}}a_1 + \frac{1}{2^{N-1}}a_0\right)2^{N-1}V \quad (17.25)$$

If all the *as* are zero except for the MSB, which is a 1, the current through resistor R is $(-V_R/R)$, and the output will be $V_R R_F/R$. Comparing Eqs. 17.24 and 17.25, we see that the proportionality factor is

$$V = \frac{V_R R_F}{2^{N-1} R} \tag{17.26}$$

EXAMPLE 17.6

Consider a 4-bit DAC of the form shown in Figure 17.18 with $V_R = -1$ V and $R = 10$ kΩ. Compute the output with

a. Only MSB = 1
b. Only LSB = 1
c. All bits equal to 1

Solution:

a. For MSB = 1, all others equal to zero, the current through R is $(-V_R/R)$, and the output is

$$V_{out} = \frac{V_R R_F}{R}$$

b. For LSB = 1, all others equal to zero, the output is

$$V_{out} = \frac{V_R R_F}{8R}$$

c. With all bits at 1, the output is

$$V_{out} = \left(1 + \frac{1}{2} + \frac{1}{4} + \frac{1}{8}\right)\frac{V_R R_F}{R} = (8 + 4 + 2 + 1)\frac{V_R R_F}{8R}$$

The accuracy and stability of such DACs depend on resistor accuracy and the closeness with which the resistances track each other as the temperature rises. The largest resistance—that is, $2^{N-1}R$—can become excessively large. Such large resistors are expensive and difficult to maintain in terms of stability and precision. On the other hand, if the largest resistance assumes a reasonable value, the smallest may become comparable to the switch resistance, and this selection again affects the accuracy.

EXAMPLE 17.7

For a 10-digit DAC (Figure 17.18), with $R_F = 10$ kΩ,

a. What are the smallest and largest values of input resistance, presuming $V_R = -1$ V?
b. If the largest resistor is set equal to 100 kΩ, what is the smallest value? If 10 kΩ?

Solution:

a. The smallest value of resistance is

$$\frac{V_R R_F}{R} = \frac{1 \times 10}{R} \equiv 1$$

and

$$R = 10 \text{ k}\Omega$$

The largest resistance is $2^{N-1}R = 2^{10-1}R = 5.120 \text{ M}\Omega$.

b. With 100 kΩ, we have

$$\frac{100 \text{ k}\Omega}{2^{N-1}} = \frac{100 \text{ k}\Omega}{2^9} = 0.195 \text{ k}\Omega \quad \text{or} \quad 195 \ \Omega$$

With 10 kΩ it is 19.5 Ω.

An R-$2R$ **ladder-type converter** can minimize the preceding difficulties. Refer to Figure 17.19. Consider S_{N-1} to be connected to V_R and all the other switches to be grounded. What *resistance* is seen by V_R? At node 0, $2R$ in parallel with $2R$ equals R. This combination added to R brings us to node 1 where the resistance looking to the left is $2R$—that is, $(2R \parallel 2R) + R$. Moving on to node 2, we would similarly see a resistance $2R$, and so it would be for every node. Thus V_R would see a resistance equal to $3R$—that is, $(2R \parallel 2R) + 2R$.

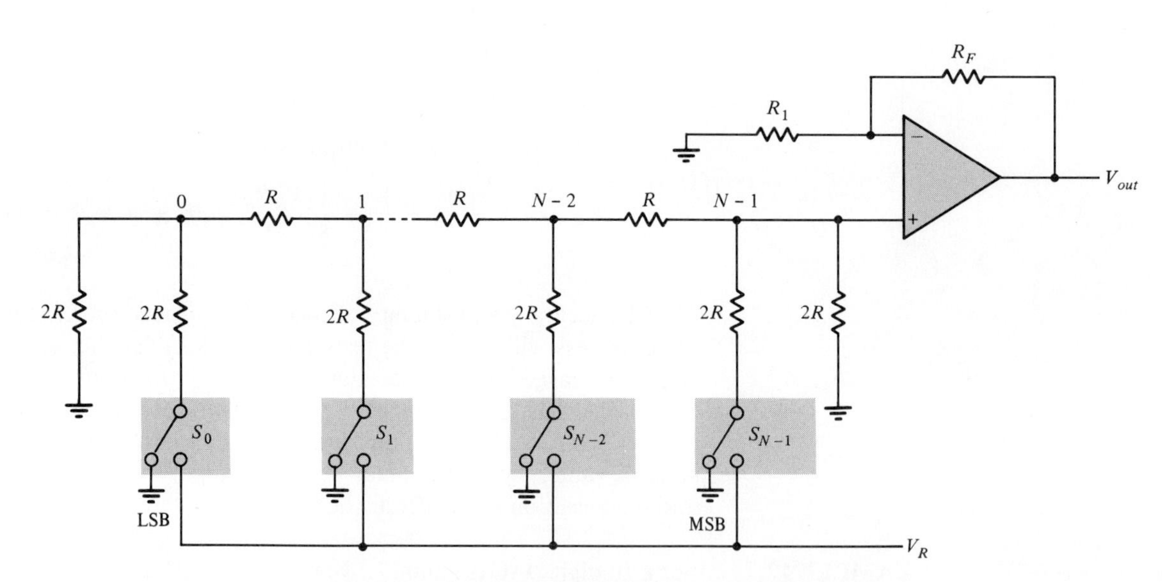

FIGURE 17.19. ***R*-2*R* Ladder Digital-to-Analog Converter**

With S_{N-1} attached to V_R—that is, a digital 1— and all other switches grounded (that is, all others digitally 0), what is the *voltage* at node $N - 1$? At node $N - 1$, looking to the right, we have $2R$; looking to the left, we also have $2R$. The resistance to ground at $N - 1$ is then R—that is, $2R \parallel 2R$. This parallel combination forms a voltage divider with a $2R$ in series with S_{N-1}, and the input voltage to the OPAMP is $\frac{1}{3} V_R$, leading to an output voltage of

$$V_{out} = \frac{V_R}{3} \frac{R_1 + R_F}{R_1} \tag{17.27}$$

With only S_{N-2} activated, node $N - 2$ has a voltage $V_R/3$, which is attenuated by a factor of $\frac{1}{2}$ before being applied to the OPAMP so that the input is $V_R/6$. From node $N - 3$ we would get an input voltage of $V_R/12$. In general we would again obtain Eq. 17.25. However, we would have accomplished the task by using only two values of resistance and with a resistance spread of only R—that is, $2R - R$.

A disadvantage of the R-$2R$ ladder DAC is the presence of capacitances between the nodal points and ground. Such capacitances lead to time-varying delays, depending on which bit is energized. (When the LSB switch closes, it takes the signal longer to reach the OPAMP than if MSB were energized.)

With regard to the **inverted ladder** (see Figure 17.20), since the OPAMP's (−) input is at virtual ground, the resistors are grounded in both positions of the switches. The only transients that arise are those due to the time it takes to go from a 0 to a 1 or vice versa.

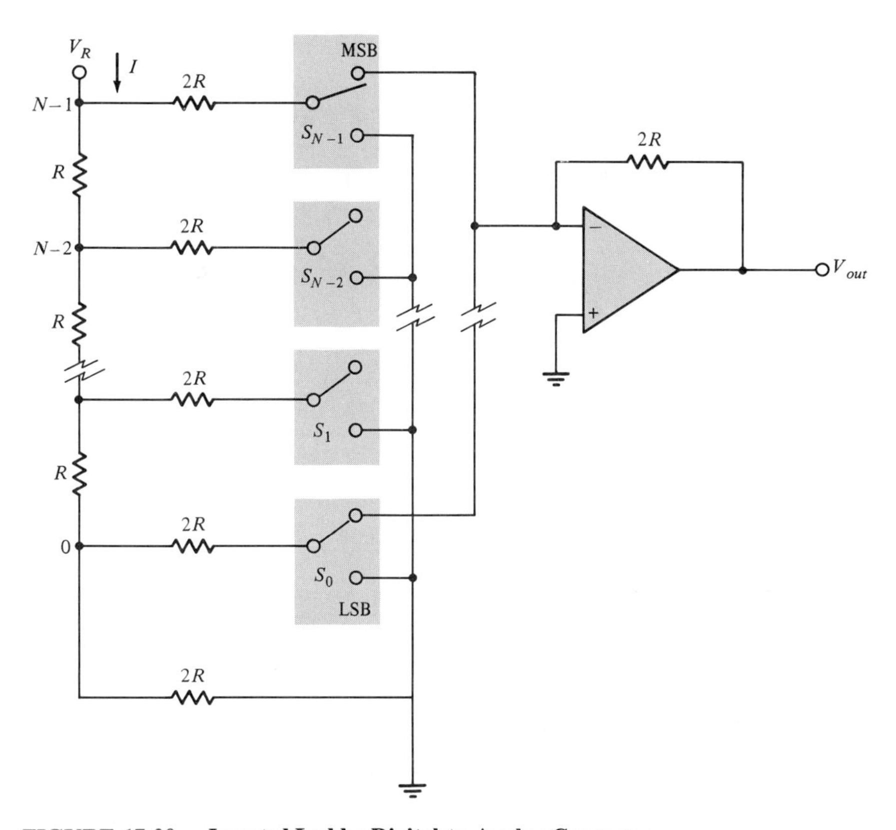

FIGURE 17.20. Inverted Ladder Digital-to-Analog Converter

As the input code of a DAC changes, it passes through major and minor transitions. Particularly troublesome are major transitions such as arise with changes at midscale. With a switch in the MSB, all the switches are changing state, for example, in going from 0111 1111 to 1000 0000. If the switching-off time is faster than the switching-on time, the DAC will momentarily give a zero output before assuming the intended value (1000 0000). (The same sort of condition can arise when going from 1000 0000 to 0111 111, as a counterexample.) The resulting large transient spike is known as a **glitch** and can prove to be quite troublesome, constituting an extraneous pulse. Normally, this difficulty can be handled by using a sample-and-hold circuit, which holds the output constant until the switches all reach equilibrium. Another possibility is the use of a low-pass filter to minimize the effect of a glitch.

We have concentrated our attention on D/A converters wherein the output represents a *voltage* proportional to the digital input. One may alternatively have a device where the output is a proportional *current*. One popular IC unit of this type is the 1408. But the analog current can be converted into a proportional voltage simply by passing it through a resistor. Since the output of the 1408 represents a current sink, the proportional output voltage will be negative.

Digital-to-analog circuits are often used to achieve analog-to-digital conversion. By using an up-down counter such as a 74LS191 (Problems 11.27 and 11.28), its digital *output* is connected to the digital *input* of a 1408. The 74LS191 has a clock input that can be counted either up or down depending on its enabling input $\overline{\text{U}}$/D. A comparator compares a negative analog signal (+) input to that of the (negative) DAC output (−) input. If the former is larger, the comparator output is down, and the 74LS191 continues to count up until the 1408 output becomes the larger. The comparator output then goes up, and the counter starts down, but only for one count, after which it again starts up (for one count). The net result is a continuous switching back and forth by one count in the LSB as long as the analog signal remains constant. A digital readout device across the parallel lines between counter and DAC yields the signal voltage. Should the analog signal change, the system will follow the variation, provided it is not too rapid. We have, therefore, a tracking ADC.

17.3 ■ STRAIN GAGES

17.3.1 Basis of Operation

The electrical resistance of a cylindrical sample of material of length l and cross-sectional area A is given by

$$R = \rho \frac{l}{A} \tag{17.28}$$

where ρ is the material resistivity measured in ohm-meters, presuming l and A are given in terms of meters and square meters, respectively—although in many engineering circumstances, if feet and square feet are used, corresponding changes should be made.

If this sample material is extended by applying to it a longitudinal stress, the consequent increase in l and diminished cross-sectional area will give rise to a change in resistance. A measurement of this change can, in turn, be related back to provide a determination of the stress that caused it. Such is the basis of operation of a **strain gage.**

To increase the sensitivity of such a determination, the gage wire may be fabricated in the form shown in Figure 17.21. In their commercial version such strain gages are furnished mounted on a base that is to be bonded to the mechanical element within which the strain is to be measured. Such bonded strain gages were initially manufactured using fine wires, although now a folded metallic foil is often used since this configuration can be accurately photoetched—but the principle of operation remains the same.

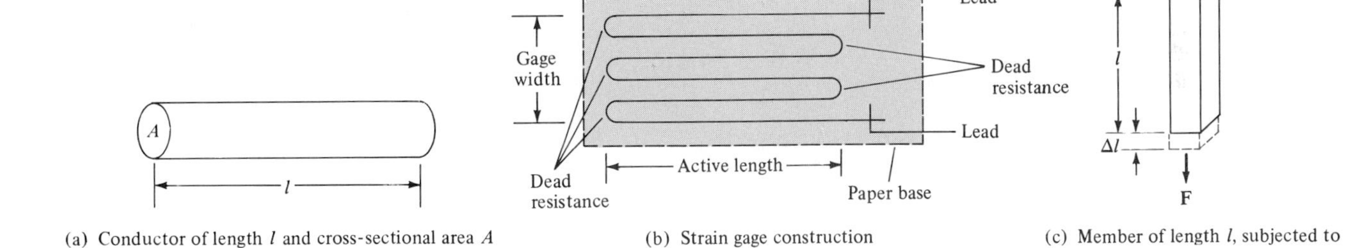

(a) Conductor of length l and cross-sectional area A

(b) Strain gage construction

(c) Member of length l, subjected to a force **F** that produces an extension Δl

FIGURE 17.21. Strain Gage Fundamentals

When a bar of length l and area A is subjected to a force F, the ratio of elongation Δl to the original length l is, of course, termed the strain ϵ, and the ratio of the applied force (F) to the cross-sectional area (A) is the stress:

$$\sigma = \frac{F}{A} \equiv \text{stress} \qquad (\text{typically, } \text{lb}_f/\text{in.}^2) \tag{17.29}$$

$$\epsilon = \frac{\Delta l}{l} \equiv \text{strain} \qquad (\text{a dimensionless quantity}) \tag{17.30}$$

The ratio of these two quantities represents the **modulus of elasticity**, also called **Young's modulus**, E:

$$E = \frac{\sigma}{\epsilon} \tag{17.31}$$

σ is often measured in microinches per inch, designated as **microstrains.**

This simple relationship among stress, strain, and modulus of elasticity exists only in the special case of uniaxial stress and only in the direction of maximum stress. In the case of a uniaxial stress, accompanying any elongation will be another strain at right angles to the axis, the *lateral* strain that is always of opposite sign to the primary strain. This lateral strain constitutes the *Poisson effect*, and the ratio of transverse to primary strain is called the *Poisson ratio* ν. Since the lateral strain is always the negative of the primary strain, ν is a negative number.

Let us return to the strain gage of Figure 17.21. The small curved segments of the wire are not affected by the tension (or compression) strains. They are, therefore, termed the **dead resistance**. These segments, however, lie in line with the Poisson strains and numerically diminish the magnitude of the Poisson strain. The effect is called transverse sensitivity and diminishes the normal gage sensitivity. The strain gage manufacturer's calibration has generally taken this effect into account in the stated value of the **gage factor**. The most commonly available strain gages have a resistance of either 120 or 350 Ω and gage lengths from 0.008 to 4 in. The gage factor is usually about 2, where

$$\text{gage factor} \equiv \text{GF} = \frac{\Delta R/R}{\Delta l/l} \tag{17.32}$$

Differentiating Eq. 17.28, one obtains

$$\frac{dR}{R} = \frac{d\rho}{\rho} + \frac{dl}{l} - \frac{dA}{A} \tag{17.33}$$

With the transverse strain being $[-\nu\,(dl/l)]$, where again ν is the Poisson ratio, if the original diameter of the wire were d_0, under strain it would be

$$d_f = d_0\left(1 - \nu\frac{dl}{l}\right) \tag{17.34}$$

Then we have

$$\frac{dR}{R} = \frac{d\rho}{\rho} + \frac{dl}{l}(1 + 2\nu) \tag{17.35}$$

Defining **material sensitivity** as $S_A = (dR/R)/\epsilon$—that is, the fractional change in resistance per unit change in normal strain—we have

$$S_A = 1 + 2\nu + \frac{dl/l}{\epsilon} \tag{17.36}$$

The value of S_A for metallic alloys used in the construction of strain gages varies between 2 and 4. The most commonly used material is constantan (also called Advance alloy), consisting of 45% nickel and 55% copper. It has an S_A of 2.1, and among its numerous advantages are its linear strain sensitivity over a wide range of strain and good temperature stability. Another useful material is isoelastic alloy (36% Ni, 8% Cr, 0.5% Mo, 55.5% Fe). Its S_A value is 3.6, and therefore it is used in dynamic applications where larger signal magnitudes and a high degree of fatigue strength are desired. But it has a high degree of temperature sensitivity and limited linearity.

Semiconductor gages have values in the order of 50–200 but, again, are very sensitive to temperature variations.

17.3.2 Gage Readout

A direct measure of the resistive change in a strain gage is not usually attempted for a number of reasons: The change is generally small, and its magnitude must be amplified before being used to drive any indicating device. A resistance variation also arises due to temperature change, and it is helpful to make provisions to compensate for this variation. If the strains are either impulsive or cyclic in nature, some means of signal processing is usually attempted.

EXAMPLE 17.8

With a gage factor of 2, a strain gage having a resistance of 120 Ω exhibits a strain of 1000 μin./in. What is the change in gage resistance? What value of calibrating resistance, placed in parallel with the gage, will yield the same reading?

Solution:

$$\text{GF} = \frac{\Delta R/R}{\Delta l/l}$$

$$\Delta R = R(\text{GF})\left(\frac{\Delta l}{l}\right) = 120(2)(1000) \times 10^{-6} = 240 \times 10^{-3} = 0.24\ \Omega$$

With the calibrating resistor in place, we have

$$\Delta R = R - \frac{RR_s}{R + R_s}$$

$$\frac{\Delta R}{R} = 1 - \frac{R_s}{R + R_s} = \frac{R + R_s - R_s}{R + R_s} = \frac{R}{R + R_s}$$

Since $R_s >> R$, we have

$$\frac{\Delta R}{R} \simeq \frac{R}{R_s}$$

$$R_s = \frac{R^2}{\Delta R} = \frac{R}{\Delta R/R} = \frac{120}{2 \times 10^{-3}} = 60\ \text{k}\Omega$$

The simplest circuit arrangement employed in conjunction with strain gage readout is the voltage divider (Figure 17.22). R_g is the strain gage resistance, and R_b is the resistance of the fixed ballast resistor. The **circuit sensitivity** is defined as

$$S_c = \frac{\Delta V}{\epsilon} \tag{17.37}$$

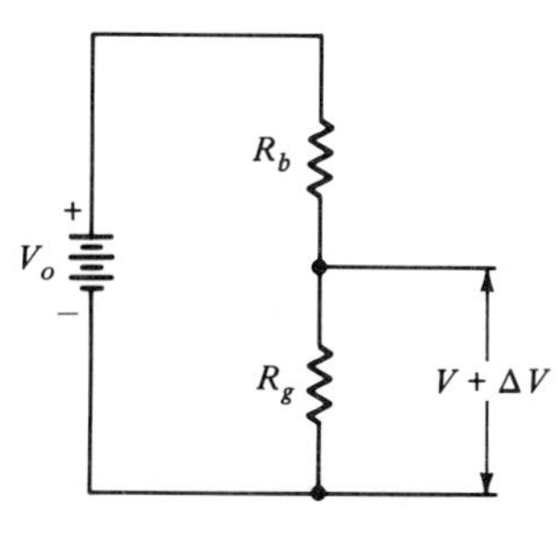

FIGURE 17.22. **Potentiometric Strain Gage Circuit** R_b is the ballast resistor, R_g is the strain gage, and V_o is the dc voltage source.

Here ΔV represents the output signal of the voltage divider, and ϵ is the corresponding strain. We have

$$V = \left(\frac{R_g}{R_g + R_b}\right)V_o = \frac{1}{1 + r}V_o \tag{17.38}$$

where $r \equiv R_b/R_g$

V_o = source voltage

For an incremental change in strain gage resistance ΔR_g, one obtains the following voltage variation:

$$\Delta V = \frac{r}{(1 + r)^2} \frac{\Delta R_g}{R_g} V_o \tag{17.39}$$

The circuit sensitivity thus becomes

$$S_c = \left[\frac{r}{(1 + r)^2} \frac{\Delta R_g}{R_g}\right] \frac{V_o}{\epsilon} = \frac{r}{(1 + r)^2} S_g V_o \tag{17.40}$$

where S_g is the gage factor. For a given gage the sensitivity is controlled by two factors: One depends on the circuit resistance ratio, and the other depends on the voltage source V_o. There is a limit as to the value of V_o that may be employed as a means of increasing the sensitivity, and that limit is generally set by the power dissipation capability of the gage. This limit depends, among other factors, on the size of the gage and the heat conductivity of the material on which the gage is mounted. One may determine the limiting value of usable V_o by noting the maximum gage dissipation (P_g) capability of the gage as specified by the manufacturer. With I_g as the gage current, we have

$$P_g = I_g^2 R_g \tag{17.41}$$

$$I_g = \frac{V_o}{R_b + R_g} \tag{17.42}$$

$$V_o = (1 + r)\sqrt{P_g R_g} \tag{17.43}$$

$$S_c = \frac{r}{1 + r} S_g \sqrt{R_g P_g} \tag{17.44}$$

The factor $r/(1 + r)$ represents the circuit efficiency, which increases with increasing r. But the large values of r mean that a large V_o is needed, and the source voltage could become excessively large. Generally, an r of about 9 is selected, leading to a circuit efficiency of 90%. The circuit sensitivity also strictly depends on gage factors $S_g\sqrt{P_g R_g}$ whose value may range from 3 to about 700, depending on the gage selected.

The output of the voltage divider consists of $V + \Delta V$, with V typically being of the order of a few volts and ΔV being of the order of millivolts. If this circuit is being used to measure a static strain, the determination of a millivolt variation atop a dc level of the order of a volt is very difficult. If the measured strain is a dynamic variation, this small voltage poses little difficulty since the output variation may be coupled to the measuring circuit via a capacitor that removes the dc level, leaving the small signal to be amplified by the usual means.

Thus, the voltage divider is not employed for static strain measurements. The use of a Wheatstone bridge allows both static and dynamic measurements to be made.

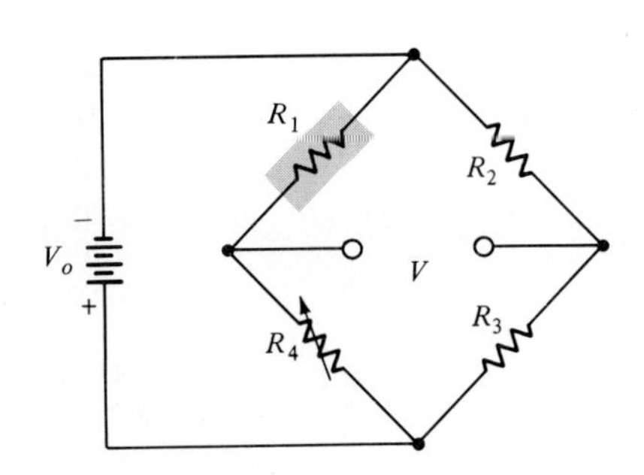

FIGURE 17.23. Wheatstone Bridge Strain Gage Circuit R_1 is the strain gage, R_4 is the balancing resistor, R_2 and R_4 are fixed resistors.

Consider first a static measurement using the Wheatstone bridge (Figure 17.23). With dc excitation and R_1 presumed to be the strain gage, the variable resistance R_4 is adjusted for a null condition, that is, no potential difference existing between the junctions of (R_1,R_4) and (R_2,R_3). Then we have

$$R_1 = \frac{R_2}{R_3} R_4 \tag{17.45}$$

If the strain gage is subjected to a change ΔR_1, this change is related to the measured change (ΔR_4):

$$\Delta R_1 = \frac{R_2}{R_3}\Delta R_4 \qquad (17.46)$$

Under the usually justified assumption that $\Delta R << R$ and with the initial resistance of each leg being the same (R), the highest output voltage will be

$$V = \frac{V_o \Delta R}{4R} \qquad (17.47)$$

This result assumes that the detector has an infinite input impedance. To the extent that it does not, the output voltage will diminish.

Rather than having three fixed resistors (R_2, R_3, and R_4) and one strain gage, it often proves convenient to have strain gages comprising two legs—that is, R_1 and R_2. But before considering that case, let us presume that all four legs consist of transducers. By assigning a ΔR_i variation to each leg (where $i = 1, 2, 3, 4$), the output voltage will be

$$V = V_o \frac{r}{(1 + r)^2}\left(\frac{\Delta R_1}{R_1} - \frac{\Delta R_2}{R_2} + \frac{\Delta R_3}{R_3} - \frac{\Delta R_4}{R_4}\right) \qquad (17.48)$$

where $r = R_4/R_1$.

If all four resistances are initially the same—that is, $r = 1$—and if R_2 and R_4 change their values in a direction opposite to R_1 and R_3 but with the same magnitude, we have

$$V = \frac{V_o}{4}\left(\frac{4\,\Delta R}{R}\right) = \frac{V_o\,\Delta R}{R} \qquad (17.49)$$

Thus, the output would be four times as great as would be the case with one transducer. If only R_3 and R_4 are fixed resistors and R_1 and R_2 are like transducers, we have

$$V = \frac{V_o\,\Delta R}{2R} \qquad (17.50)$$

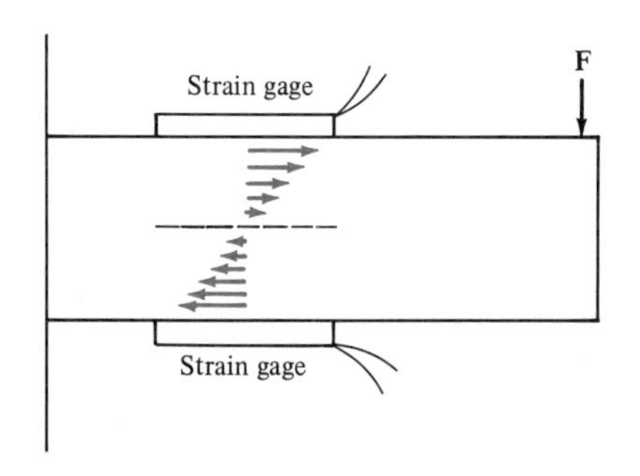

FIGURE 17.24. **Use of Two Strain Gages to Compensate for Temperature-Related Linear Expansion (Contraction)**

Consider the cantilever deflection shown in Figure 17.24. With R_1 placed above and R_2 below, the two changes will be of opposite sign, and the output will be twice that obtained from a single gage. Additionally, any temperature change that causes the beam to expand (or contract) will affect the two gages equally and not lead to an erroneous indication of a change in the bending strain.

Often one gage, a dummy, is placed so that it shares the same environment as the active gage but is not subject to the strain to be measured. It thereby compensates for any temperature variation.

The expression for circuit sensitivity of the Wheatstone bridge is the same as that for the voltage divider, namely, Eq. 17.44. If a dc source is used and a static strain is being measured, the amplifying device that follows the bridge must be capable of amplifying dc. This requirement means that any multistage amplifier cannot be capacitor-coupled, and before the advent of solid-state circuitry meeting this condition presented considerable difficulty. The main problem arose because of dc drift in vacuum tube circuits, and a variety of compensation methods had to be employed to minimize the effect.

Today, of course, differential amplifiers in IC form complete such tasks very nicely, and their common-mode rejection helps to minimize the 60 Hz interference so easily picked up with high-impedance inputs. (Common-mode rejection, CMRR, was discussed in Section 10.3.) Alternatively, one may apply an ac source to the bridge, and any unbalance of the bridge will appear as modulation of the applied ac, the modulation being recoverable by any of the usual means of detection. (See Section 17.3.4.)

17.3.3 Temperature Effects

A change in ambient temperature (ΔT) can give rise to three important effects that may alter strain gage accuracy:

1. The gage itself may elongate:

$$\frac{\Delta l}{l} = \alpha \, \Delta T \tag{17.51}$$

2. The material on which the gage is mounted may elongate:

$$\frac{\Delta l}{l} = \beta \, \Delta T \tag{17.52}$$

3. The gage resistance may change because of the temperature coefficient of resistivity of the gage material:

$$\frac{\Delta R}{R} = \gamma \, \Delta T \tag{17.53}$$

The three effects may be combined into a single equation:

$$\left(\frac{\Delta R}{R}\right)_{\Delta T} = (\beta - \alpha)S_g \, \Delta T + \gamma \, \Delta T \tag{17.54}$$

α is the thermal expansion coefficient of the gage material, β is the comparable coefficient for the base material, and γ is the temperature coefficient of resistivity of the gage material.

One of the earliest methods used to produce STC (**self-temperature compensation**) gages consisted of making the gage from two different wire materials and connecting them in series in such a manner that their temperature effects cancelled, usually over a rather limited temperature range. More recently, the following technique has been employed: There is a statistical variation in the values of α and γ for different batches of the same gage material. The resulting gages are then selected so that their α matches that of the base material on which they are to be mounted. Such temperature gages are available to match the β coefficients of a variety of materials such as quartz, titanium, mild steel, stainless steel, etc., but, again, the compensation holds only over a limited range, and there still is a variation if γ is not equal to zero.

EXAMPLE 17.9 An STC gage is connected to a bridge via two 10 ft lengths of #18 copper wire (Figure 17.25). If the wires are subjected to a temperature change of 30°F, what is the change in resistance?

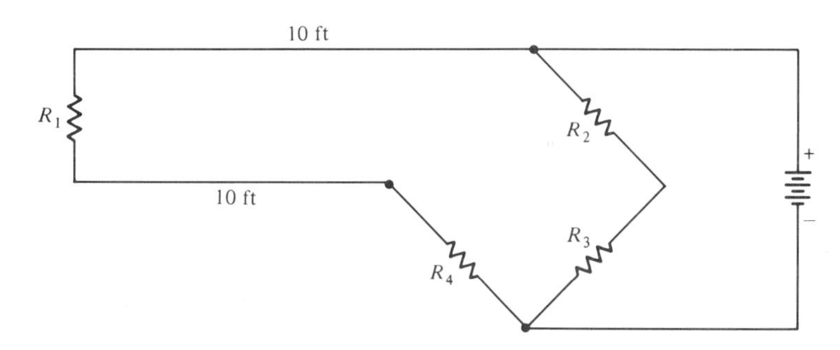

FIGURE 17.25

If the GF is 2.0 and the bridge resistances are 120 Ω, what is the apparent stress if the gage is mounted on a steel beam whose E = 30 million psi?

Solution: #18 copper wire has a resistance of 5.88 Ω/1000 ft. We have

$$\frac{5.88}{1000}(20) = 0.118\ \Omega$$

If the temperature coefficient of resistance is 0.0022/°F, we have

$$\Delta R = 0.0022(30)0.118 = 7.79 \times 10^{-3}\ \Omega$$

$$\epsilon = \text{apparent strain} = \frac{\Delta R}{R(\text{GF})} = \frac{7.79 \times 10^{-3}}{120(2)} = 3.24 \times 10^{-5}$$

$$E\epsilon = 3 \times 10^{7} \times 3.24 \times 10^{-5} = 972\text{ psi}$$

17.3.4 ac-Operated Strain Gages

We now consider a strain gage being subjected to a time-varying stress resulting in a resistance

$$R(t) = R_0 + r \sin \omega_0 t \tag{17.55}$$

where R_0 = its nominal resistance
r = maximum change in resistance
ω_0 = angular frequency of vibration (See Figure 17.26.)

For the moment, consider the applied voltage to be constant with a value V_{max}. The bridge output is

$$v = \frac{V_{max}}{2} - \frac{V_{max}R(t)}{R + R(t)} = \frac{V_{max}(R_0 + r \sin \omega_0 t)}{R + R_0 + r \sin \omega_0 t} \tag{17.56}$$

If $R = R_0$, with no stress applied, we have

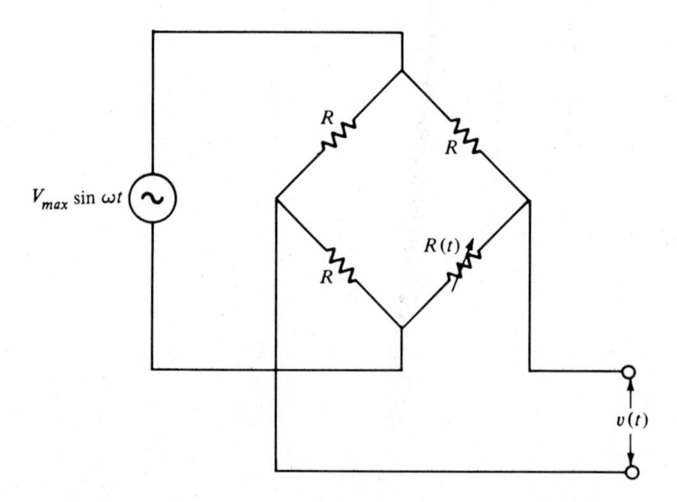

FIGURE 17.26. **ac-Operated Wheatstone Bridge Strain Gage Circuit**

$$v = V_{max}\left(\frac{1}{2} - \frac{R_0 + r \sin \omega_0 t}{2R_0 + r \sin \omega_0 t}\right) \tag{17.57}$$

The denominator in the second term can be put in the form

$$\text{denominator} = \left(1 + \frac{r}{2R_0} \sin \omega_0 t\right)2R_0 \tag{17.58}$$

thereby leading to the equation

$$v \simeq V_{max}\left[\frac{1}{2} - \frac{R_0 + r \sin \omega_0 t}{2R_0}\left(1 - \frac{r}{2R_0} \sin \omega_0 t\right)\right] \tag{17.59}$$

The approximation arises from recognizing that $[1/(1 \pm x)] \simeq 1 \mp x$ if $x << 1$. Retaining only the first order term in r that results after multiplication in Eq. 17.59, we obtain the final result

$$v = V_{max}\left(-\frac{r}{4R_0} \sin \omega_0 t\right) \tag{17.60}$$

Now, introducing the ac source variation, we have

$$v = -\frac{V_{max} r}{4R_0} \sin \omega t \sin \omega_0 t \tag{17.61}$$

$$v = \frac{V_{max} r}{8R_0} [\cos(\omega_0 + \omega)t - \cos(\omega - \omega_0)t] \tag{17.62}$$

The result is AM modulation, giving rise to a carrier frequency ω and two sidebands, $\omega + \omega_0$ and $\omega - \omega_0$. The usual methods of demodulation may be used to extract ω_0.

17.4 ■ THE "CLASSIC" INSTRUMENTATION AMPLIFIER

The amplification of strain gage signals presents an example of a task to be performed by an instrumentation amplifier, characterized as a high-gain dc-coupled differential amplifier with single-ended output, high input impedance, and a high CMRR. Such amplifiers are used to amplify transducer signals in the presence of large common-mode interference.

The "classic" arrangement consists of three operational amplifiers (Figure 17.27) with two of them (A_1 and A_2) providing a gain of $1 + (R_2/R_1)$. The third (A_3) has unity differential gain and serves as a differential subtractor.

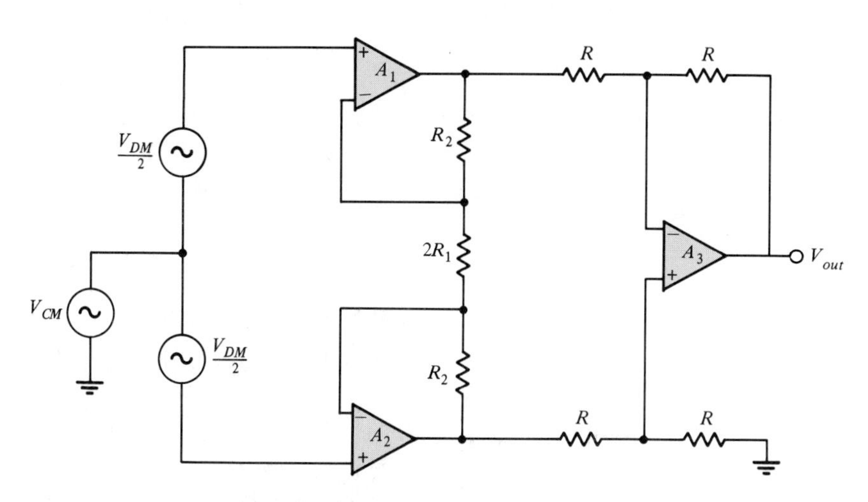

FIGURE 17.27. **"Classic" Instrumentation Amplifier**

The symmetrical aspect of the input amplifiers is shown in Figure 17.28. For differential-mode signals point A is at virtual ground, and A_1 and A_2 each assume the form shown in Figure 17.29a. In the differential mode the circuit performs like a single noninverting amplifier (Eq. 10.14).

For common-mode signals (Figure 17.29b) point A can be considered an open circuit resulting in no current through the resistors. With common-mode signals, A_1 and A_2 act as voltage followers. The respective total outputs of A_1 and A_2 are

$$A_1 = V_{CM} + \left(1 + \frac{R_2}{R_1}\right)\frac{V_{DM}}{2}$$
$$A_2 = V_{CM} - \left(1 + \frac{R_2}{R_1}\right)\frac{V_{DM}}{2} \tag{17.63}$$

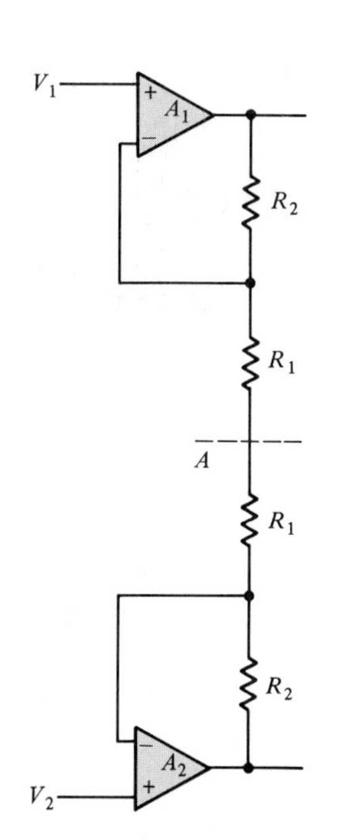

FIGURE 17.28. **Differential Input Amplifiers of a "Classic" Instrumentation Amplifier Showing Symmetrical Resistor Arrangement**

A_3 provides an output that is the difference between the two inputs:

$$A_3 = \left(1 + \frac{R_2}{R_1}\right)V_{DM} \tag{17.64}$$

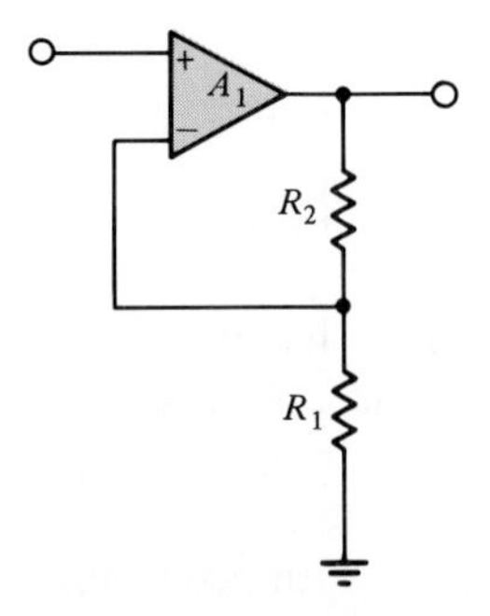

(a) Input stage appearance seen by differential signal

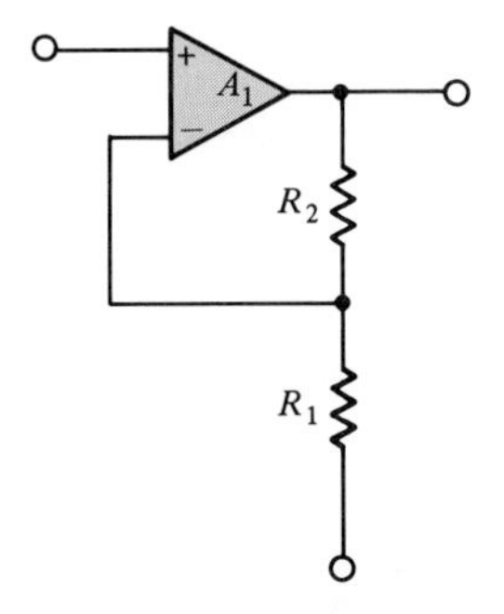

(b) Input stage appearance seen by common-mode signal

FIGURE 17.29. **Circuit Appearance, Depending on the Nature of the Signal**

By designating the CMRR of the first stage as λ_1 and that of the second as λ_2, the overall CMRR is

$$\Lambda = \lambda_1 + \frac{\lambda_2}{A_D} \tag{17.65}$$

where A_D is the differential gain of the first stages (the gain of A_3 should be unity). λ_2 is primarily determined by the resistor mismatch, being

$$\lambda_2 \simeq 2\frac{\Delta R}{R} \tag{17.66}$$

The gain of the "classic" amplifier may easily be varied by either "switching in" different values for $2R_1$ or making this resistance variable.

EXAMPLE 17.10

The common-mode signal need not necessarily be an ac interference. It can be a common dc level. Consider a strain gage bridge circuit consisting of 350 Ω resistors excited by a 10 V dc source. If the full-scale reading is to be 20 mV and an accuracy of 0.05% is desired, what is the needed value of CMRR?

Solution:

$$0.05\% \times 20 \text{ mV} = 0.01 \text{ mV}$$

This signal "rides" atop a 5 V dc level—that is, 5000 mV. The needed rejection is

$$20 \log_{10}\left(\frac{0.01}{5000}\right) = 114 \text{ dB}$$

EXAMPLE 17.11

The common-mode gain of a differential amplifier is given by (Eq. 10.22)

$$A_{CM} = -\frac{R_F}{R_1} + \frac{R_F}{R_1}\frac{R_3}{R_3 + R_2} + \frac{R_3}{R_3 + R_2}$$

$$= -\frac{R_F}{R_1} + \frac{R_3}{R_1}\left(\frac{R_F + R_1}{R_3 + R_2}\right)$$

For the subtractor stage of a "classic" instrumentation amplifier, show that an estimate of the common-mode gain is approximately equal to $2(\Delta R/R)$, where $(\Delta R/R)$ is the fractional error f of the resistors.

Solution: Since, typically, in a worst-case situation, we have

$$A_{CM} = -\frac{R_F(1 + f)}{R_1(1 - f)} + \frac{R_3(1 + f)}{R_1(1 - f)}\left[\frac{R_F(1 + f) + R_1(1 - f)}{R_3(1 + f) + R_2(1 + f)}\right]$$

with all Rs having the same nominal values, we have

$$\begin{aligned} A_{CM} &= -\frac{1 + f}{1 - f} + \frac{1 + f}{1 - f}\left(\frac{1 + f + 1 - f}{1 + f + 1 + f}\right) \\ &\simeq -(1 + f)(1 + f) + (1 + f)(1 + f)(1 - 2f) \\ &= -(1 + 2f) + (1 + 2f)(1 - 2f) = -1 - 2f + 1 + 2f \\ &\quad - 2f = -2f \end{aligned}$$

where, with $f \ll 1$, the approximation $1/(1 \pm f) = 1 \mp f$ has been used and second-order terms (involving f^2) have been disregarded.

The preceding estimate is that due to resistor mismatch and assumes that the CMRR of the amplifier itself is considerably greater. A_{CM} due to resistor mismatch may be minimized by making one of the resistors adjustable.

17.5 ▪ TEMPERATURE MEASUREMENTS

Thermistor temperature sensors have previously been discussed (Chapter 6). They offer the convenience of simplicity, accuracy, and low cost. However, they are fragile, have a rather restricted temperature range (−50°C to +300°C) and, most important of all, suffer from self-heating.

A much older device for temperature measurements, the thermocouple, overcomes many of these deficiencies and now has been made into a much more useful instrument as a result of electronic developments. The basic circuit is shown in Figure 17.30. It consists of two junctions between dissimilar metals (iron and constantan are illustrated). One juncture constitutes the sensing junction, and the other (the reference) needs to be kept at a reference temperature. In the past the latter was an ice-water mixture representing 0°C. The thermocouple develops an emf proportional to the difference between the two temperatures—for iron and constantan, about 50 μV/°C.

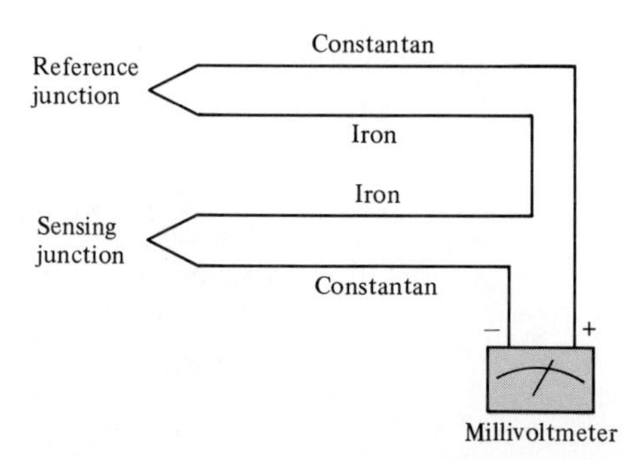

FIGURE 17.30. **Typical Thermocouple Pair with Readout Meter**

Today modern techniques allow one to do away with the inconvenience of ice baths. Instead, one generates a current that (in microamperes) equals the absolute temperature (Figure 17.31). The Analog Devices AD590 represents such a unit, being a current source producing 1.0 μA/K. The reference junction is placed in thermal contact with this current source.

In series with the thermocouple circuit and readout amplifier, one places a resistance (R_1) whose value is equal to the temperature coefficient of the thermocouple pair (51.5 Ω for the iron-constantan shown in Figure 17.31).

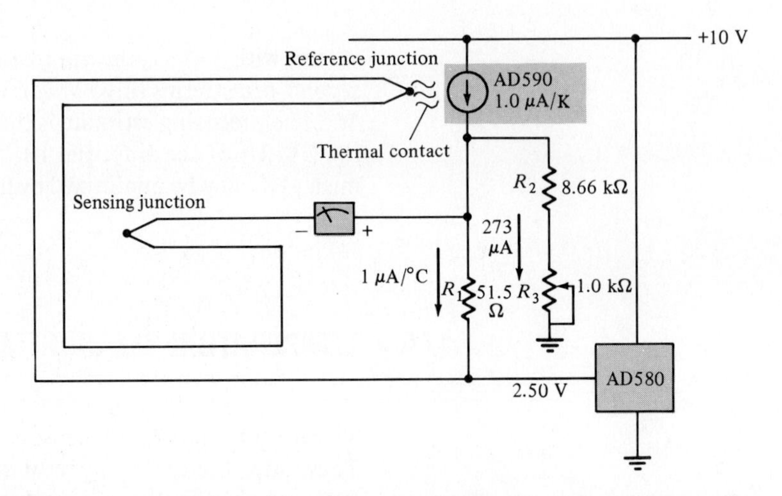

FIGURE 17.31. **Thermocouple Circuit with Compensation for Changes in Reference Temperature**

The Analog Devices AD580 is a low-drift voltage reference whose output provides 2.50 V at one end of the 51.5 Ω resistor. At 0°C the resistance of $R_2 + R_3$ is adjusted so that the other end of the 51.5 Ω resistor is also at 2.50 V, and hence no voltage is introduced in series with the thermocouples.

Should the reference junction temperature rise, for example, to an ambient value, there will be a corresponding increase in the AD590 current output at a 1 μA/°C rate. Through the 51.5 Ω resistor, this will give rise to a voltage in series with the thermocouple circuit, a voltage that compensates for the change in reference temperature. The readout of the sensing junction temperature should remain uninfluenced by the change in reference temperature.

SUMMARY

17.2 Measurements

- Many electronic voltmeters, while giving an indication of rms readings, in actuality are measuring either the average value or the peak value. Their calibrations, therefore, are accurate only when used in conjunction with sinusoidal waveforms.
- The most popular method of digital voltage measurement is the dual-slope method.
- The dual-slope method of voltage measurement can be made relatively immune to 60 Hz interference by proper selection of the integration time.
- The MSB of half-digit instruments is limited to values that are either 0 or 1. The remaining digits can assume values between 0 and 9.
- Analog-to-digital conversion may be accomplished by a variety of methods, including parallel (flash) encoding, ratiometric (dual-slope) conversion, single-ramp counting, or successive approximation.
- Digital-to-analog conversion may be accomplished using summing amplifiers, *R*-2*R* ladder networks, and inverted ladders.

17.3 Strain Gages

- The resistance change of a strain gage can be related to the stress that induced the change through deformation of the gage.
- Stress is often measured in microstrains, being the stress in microinches per inch.
- The gage factor of a strain gage is the fractional change in resistance, divided by the fractional change in length. A typical value is 2.
- The fractional change in resistance of a material divided by the strain constitutes the material's sensitivity. Typical values are between 2 and 4 but may be as high as 50 or 200 for semiconductors.
- Direct reading of the resistance change of strain gages is not generally attempted because the changes are small, there are often temperature changes for which correction is usually attempted, and signal processing is often needed to discriminate against interfering signals.
- Strain gage sensitivity may be increased by increasing the source voltage, but the resulting increase in heating may lead to significant temperature effects.
- Changes in ambient temperature may alter strain gage accuracy due to gage elongation, elongation of the mounting material, and resistance temperature effects.
- Self-temperature compensation (STC) gages compensate for temperature variations by being constructed of dissimilar materials.

17.4 The "Classic" Instrumentation Amplifier

- The "classic" instrumentation amplifier features high differential gain that is easily variable, high input resistance, and high common-mode rejection.

17.5 Temperature Measurements

- The reference temperature of thermocouples may be maintained by electronic means without resorting to such procedures as ice baths.

KEY TERMS

analog-to-digital conversion (ADC): replacing a continually varying signal by its digital equivalent

parallel (flash) encoder: a string of comparators each of which responds to a different level of a voltage divider connected between a reference voltage and the analog voltage (It is the most rapid of the various A/D conversion methods.)

ratiometric (dual-slope) method: determination of the duration of a sweep voltage, whose slope is proportional to the analog magnitude, via the elapsed number of clock pulses and compared to the number accompanying a fixed reference voltage sweep (The analog voltage is a function of this ratio; the method is particularly advantageous in terms of the discrimination it provides against 60 Hz interference.)

single-ramp method: relating the number of clock pulses between analog signal coincidence and the zero crossing of a sweep voltage to the analog magnitude (Its main disadvantages are susceptibility to noise and the temperature variation of the sweep voltage slope.)

staircase ramp: a sweep voltage in the form of discrete steps rather than a linear sweep; tends to minimize the noise errors to which the latter is particularly susceptible

successive approximation: method for determining digital signals by successive tests (To a comparator, a 1 is applied to the most significant bit to determine whether the analog or the digital signal is greater and then to each lower bit in turn; conversion speed is considerably greater than that for a dual-slope but slower than that for a flash encoder.)

tracking converter: method which, within a limited range of rate variation, the digital "output" follows the analog variations rather than requiring a redetermination of each output value *ab initio*

circuit sensitivity: change in the output voltage per unit strain

common-mode rejection: the ability of a differential amplifier to discriminate against in-phase signals at its two inputs [Expressed as a ratio of differential gain to common-mode gain, it is the common-mode rejection ratio (CMRR); the designation CMR is sometimes used when the rejection ratio is expressed in decibels.]

common-mode signal: a simultaneous change at both inputs of a differential amplifier

crest factor: ratio of peak to rms value

current-sinking terminal: one into which current will flow; for example, the output terminal of the 1408 ADC

dead resistance: that portion of a strain gage resistance that does not respond to the applied stress

digital-to-analog conversion (DAC): replacing a digital signal by its continually varying equivalent

inverted ladder network: minimizes the capacitance-induced variable delay times encountered with a ladder DAC

ladder network: a DAC whose many resistors assume only two different (modest) values easily reproduced in integrated circuits, in contrast to the wide (and often impractical) range needed with summing methods

summing method: a multi-input OPAMP with input resistance values adjusted to provide gains that vary in the ratio of 1, 2, 4, and 8

gage factor: the fractional change in resistance divided by the fractional change in the length of the strain gage

glitch: extraneous pulses introduced by unequal turn-on and -off times of switching circuits

material sensitivity: the fractional change in a material's resistance, per unit change in normal strain

microstrain: microinch per inch, being the dimensional change used as a measure of strain

normal-mode rejection (NMR): a noninteger relationship between a sampling rate and a common-mode interference frequency can result in the creation of normal-mode interference (A measure of the circuit's ability to discriminate against such interference is termed its NMR, expressed in decibels)

normal-mode signal: a differential change at the two inputs of a differential amplifier

overranging: the ability of a digital meter to provide a reading in excess of that indicated by the range scale

self-temperature compensation (STC): the incorporation of temperature compensation into a gage rather than through external circuitry

strain gage: a resistance whose value is altered by the stress to which it is subjected

1408: an 8-bit D/A converter using an R-$2R$ ladder with current-sinking output

SUGGESTED READINGS

Jaeger, Richard C., IEEE MICRO: "Part I. Digital-to-Analog Conversion," vol. 2, pp. 20–37, May 1982; "Part II. Analog-to-Digital Conversion," vol. 2, pp. 46–56, Aug. 1982; "Part III. Sample-and-Holds, Instrumentation Amplifiers, and Analog Multiplexers," vol. 2, pp. 20–35, Nov. 1982; "Part IV. System Design, Analysis, and Performance," vol. 3, pp. 52–61, Feb. 1983. *A series of tutorial papers devoted to A/D conversion.*

Topping, J., ERRORS OF OBSERVATION AND THEIR TREATMENT. London: Chapman & Hall, 1957. *When considering the overall effect of tolerances (for example, those of resistors in differential amplifiers), an excellent concise treatment regarding how one should estimate cumulative errors when subjecting the various quantities to arithmetic manipulation.*

Perry, C. C., and Lissner, H. R., STRAIN GAGE PRIMER, 2nd ed. New York: McGraw-Hill, 1962. *The "bible" with regard to strain gage practice.*

EXERCISES

1. While the scale of many ac meters using a dc meter in conjunction with a rectifier is labeled *rms*, what are they truly measuring?

2. What is the crest factor?

3. Distinguish among the following ADC methods: flash encoder, single-ramp, dual-ramp, successive approximation.

4. How can the dual-slope method of voltage measurement be made immune to 60 Hz interference?

5. What is the expression for the normal-mode rejection?

6. For a sampling time of 0.45 s, what is the normal-mode rejection of 60 Hz interference?

7. What are some of the advantages of digital meters? What are their disadvantages?

8. In what way does a $4\frac{1}{2}$-digit instrument differ from a 5-digit one?

9. Distinguish among the following DAC methods: summing amplifier, *R*-2*R* ladder, inverted ladder.

10. What is the main advantage of an *R*-2*R* ladder in a DAC in contrast to other schemes? What is its disadvantage?

11. Strain is often measured in units of microstrain. To what is this quantity equivalent dimensionally?

12. What represents the dead resistance of a strain gage?

13. What is the mathematical expression for gage factor?

14. What is the mathematical expression for material sensitivity?

15. What class of materials has the highest values of material sensitivity? What is their major disadvantage?

16. Why does one not usually attempt a direct measure of the resistance change from a strain gage?

17. How is circuit sensitivity defined relative to strain gage measuring circuits?

18. What two factors control gage sensitivity?

19. What generally sets the limit on the V_o (the source voltage) to be used with a strain gage?

20. What is usually selected as the ratio of ballast resistance to gage resistance?

21. How is the measurement of a small voltage variation due to a cyclic resistance change, atop a relatively large dc voltage, accomplished?

22. Why is the simple voltage divider not used for static strain gage measurements?

23. With identical initial resistances (R) in a Wheatstone bridge,

a. What is the highest output voltage obtainable in terms of the source voltage V_o and the resistance change ΔR?

b. What is the maximum value if all four branch elements are resistance gages, with one pair subject to a change opposite to that of the other pair? With two fixed resistors and two like strain gages?

24. What three important temperature effects can alter strain gage accuracy?

25. What is an STC gage?

26. In a "classic" instrumentation amplifier, what is the common-mode gain of the input stages? What is the differential-mode gain of the differential subtractor?

27. What is the AD580? The AD590?

PROBLEMS

17.1

A Heathkit multirange meter has ac full-scale values that are multiples of 1.5 and 5.0. (Their terminal points at the high ends do not coincide.) On the 0–15 V scale, the manufacturer specifies that 0 dB represents 7.74 V. The manufacturer's instructions state that when an ac range lower or higher than 0–15 V is used, 10 dB should be subtracted or added, respectively, for each change in range position.

a. To what voltage does 0 dB (scale reading) correspond on the 0–50 V scale?

b. To what voltage does 0 dB correspond on the 0–5 V scale?

17.2

A Simpson 260 multimeter is rated as having an accuracy of 3% of full-scale reading on the ac voltage scales. On this meter the full-range scales are 2.5, 10, 50, and 250 V. The dB scale is referenced to 1 mW across 600 Ω—that is, 0.775 V across 600 Ω represents 0 dB. The manufacturer specifies that on the 10 V range one should add +12 dB to the dB reading, +26 dB on the 50 V range, and +40 dB on the 250 V range.

a. If 7.6 V is read on the 10 V scale, corresponding to 8 dB on the decibel scale, what is the actual dB reading? Check your answer analytically.

b. On the same 10 V full scale, a reading of 2.6 V corresponds to 0 dB. What is the dB reading according to the manufacturer's instructions? Check your answer analytically. Explain any significant discrepancy.

c. On this same meter 2.42 and 9.60 V both correspond to +10 dB. What are the values in decibels according to the manufacturer's instructions and according to an analytical calculation?

17.3

The following apply to a dual-slope voltmeter:

$$n_1 = 1000$$
$$R = 10\ \text{k}\Omega$$
$$V_R = 1.00\ \text{V}$$
$$f = 100\ \text{kHz}$$
$$C = 0.1\ \mu\text{F}$$

If $V_A = 350$ mV, what is the readout count? What is the charge on C at the end of the integration period? What is V_c at the end of the integration period? Why must $|V_A|$ be less than $|V_R|$?

17.4

The oscillator used in conjunction with an integrating voltmeter has an accuracy of ±0.01%. The normal-mode rejection, taking into account the oscillator stability and the line voltage fluctuation, leads to a value of −48 dB. In parts per million, what is the assumed stability of the line voltage?

17.5

An oscillator, used to provide the integration time for a digital voltmeter, has an accuracy of 0.01%.

a. Assuming an exact 60 Hz power frequency, what would be the normal-mode rejection in decibels at 60 Hz?

b. Upon monitoring the power-line frequency, it is seen to vary by as much as ±500 ppm over an 8 min period. What might be the expected NMR under the circumstances?

17.6

The Simpson 360 meter used a 40 kHz clock with a count of 8000 per sample period. Calculate the 60 Hz normal-mode rejection in decibels.

17.7

Verify the validity of Eq. 17.16:

$$\frac{t_r}{t_s} = \frac{1 - x}{x}$$

17.8

Verify the validity of Eq. 17.19.

$$\frac{t_r}{t_s} = \frac{x - 2}{x}$$

17.9

Using a 60 Hz sampling rate, on semilog paper, plot the NMR (in decibels) as a function of the interfering frequency.

17.10

a. If the full-scale voltage of a staircase sweep is 10 V and a 10-bit counter is employed, what is the conversion resolution?

b. If a 1 MHz clock is used, what is the maximum conversion time for a counter ramp-converter?

c. What is the minimum number of conversions that would be carried out per second?

17.11

Considering a tracking ADC, we wish to determine the maximum rate of change in the analog signal that can accurately be followed given the bit capacity of the counter (n) and the clock frequency (f_c). This rate may be estimated by considering a sinusoidal signal whose peak-to-peak value is equal to the full-scale voltage of the converter (V_{FS}). The maximum rate of change of the DAC is one LSB per clock period.

a. Determine the expression for the highest frequency that can accurately be followed.

b. What is the expression for the maximum rate of change of the analog voltage?

c. For a 10-bit converter with a 1 MHz clock frequency, what is the maximum rate of change that can be followed if $V_{FS} = 5$ V?

17.12

The successive approximation method of ADC demands that the input signal change, at most, by $\frac{1}{2}$ LSB during the conversion period—that is, $V_{FS}/(2 \cdot 2^n)$ during the time $T_c = n/f_c$. (If it does exceed this value, a sample-and-hold circuit must precede it.)

a. For a 10-bit counter using a 1 MHz clock, compute the upper limit on the sinusoidal frequency change that may be tolerated.

b. To what does this frequency correspond in terms of time rate of change of input signal voltage, taking $V_{FS} = 10$ V?

c. Compare this result with part c of Problem 17.11.

17.13
A 350 Ω strain gage with a gage factor of 2 is fastened to a steel member subjected to a stress of 15,000 psi. Taking the modulus of elasticity to be 30,000,000 psi, calculate the voltage change if the gage is used in a Wheatstone bridge circuit. What is the power dissipation in the gage?

17.14
Derive Eq. 17.35 using Eqs. 17.28 and 17.34.

17.15
A potentiometric strain gage circuit (Figure 17.22) may be used for dynamic measurement if the output across R_g is taken via a series capacitor. If the input resistance to the measuring device is 1 MΩ, what should be the value of the series capacitor in order to provide *negligible* frequency distortion down to 5 Hz?

17.16
a. From Eq. 17.40 it is seen that the sensitivity of a voltage divider strain gage circuit depends on the ratio of ballast to strain gage resistances ($r = R_b/R_g$). For a fixed value of source voltage and for $r = 1$ used as a reference, plot the relative output voltage as a function of r using a semilog plot extending from $r = 0.1$ to $r = 10.0$.

b. What condition leads to a maximum output?

c. In what manner does the output differ if obtained from across R_b rather than R_g?

17.17
a. The output signal of a voltage divider strain gage circuit may be increased if the value of the ballast resistor is increased, together with an increase in battery voltage such that the gage current is maintained. Using a value of $r = 1$ as a reference, plot the relative output voltage.

b. Plot the supply voltage required to maintain the constant gage current

$$I = \frac{V_o}{2R_g}$$

17.18
As a means of calibrating a Wheatstone bridge strain gage circuit, one may, by means of a switch, introduce a resistance R_{cal} that will unbalance the bridge in a known manner. What is the strain (in microinches per inch) resulting from a stress of 15,000 psi applied to a steel sample with a modulus of elasticity of 30×10^6 psi? If the gage has a resistance of 120 Ω and a gage factor of 2, what is the resulting change in resistance? What value must R_{cal} assume to simulate this change?

17.19
Five foot lengths of wire are used to connect a strain gage to a Wheatstone bridge, as in Example 17.9. The leads are made of copper wire with a total resistance of 0.0206 Ω and are subjected to a temperature change of 150°F. What stress corresponds to the resulting change in lead resistance? The gage resistance is 120 Ω, and the gage factor is 2.

17.20
Why does the arrangement in Figure 17.32 eliminate lead errors due to a temperature difference between the gage location and the bridge location?

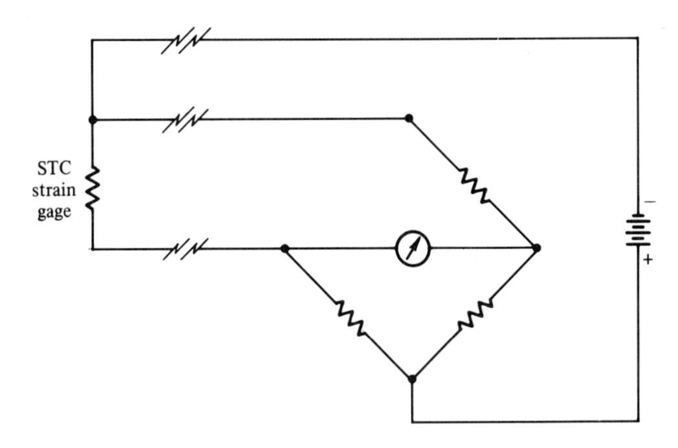

FIGURE 17.32

17.21
Verify Eq. 17.39:

$$\Delta V = \frac{r}{(1 + r)^2} \frac{\Delta R_g}{R_g} V_o$$

17.22
Verify Eq. 17.60:

$$v = V_{max} \left(\frac{r}{4R_0} \sin \omega_0 t \right)$$

17.23
With regard to a classic instrumentation amplifier, what is the worst-case value of CMRR due to resistor mismatch in the A_3

stage using 0.01% resistors? What would it be for 5% resistors?

17.24

With regard to Figure 17.31, compute the change in the voltage developed across R_1 if the reference temperature changes from 0°C to 35°C, with R_2 and R_3 having initially been adjusted to provide a zero correction voltage at 0°C. What is the corresponding error in the injected voltage through R_1, expressed in terms of °C? (The specifications for the AD590 indicate a compensation accuracy within ± 0.5°C.)

18 MAGNETIC CIRCUITS

OVERVIEW

Many aspects of electrical conduction are analogous to the conduction of magnetic flux. Thus, there is a magnetic circuit equivalent to Ohm's law. This chapter is primarily devoted to a consideration of those fundamental magnetic concepts necessary to understand the operation of electric motors, which constitute the last topic to be considered in this text.

OUTLINE

18.1 ■ INTRODUCTION

There is a close analogy between current flow and magnetic flux (Figure 18.1). While it is an (emf) electromotive force (V) that creates a current (I), it is a **magnetomotive force** (mmf)—the product of amperes times the number of turns and specified as *ampere-turns* ($\mathfrak{F}$)—that gives rise to the **magnetic flux** (Φ). Corresponding to Ohm's law for the electrical case, which is

$$V = IR \tag{18.1}$$

there is the magnetic's case,

$$\mathfrak{F} \text{ [ampere-turns]} = \Phi \text{ [webers]} \times \mathcal{R} \text{ [ampere-turns per weber]} \tag{18.2}$$

where $\mathcal{R}$ is the **reluctance** of the magnetic circuit.

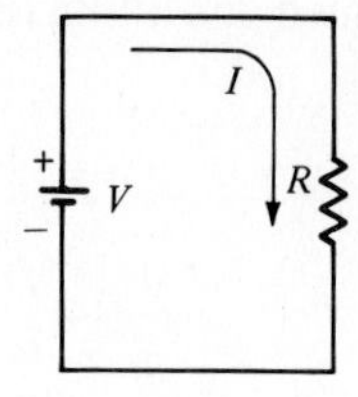

(a) Electromotive force (V) produces (I) through a resistance (R)

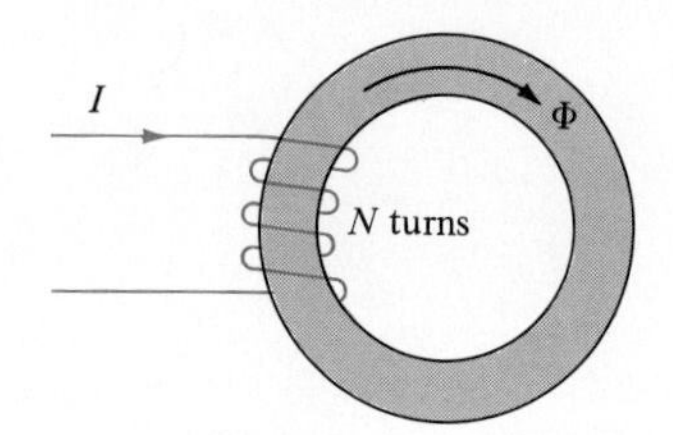

(b) Magnetomotive force ($\mathfrak{F}$) produces flux (Φ) through a magnetic circuit constituting a reluctance ($\mathcal{R}$)

FIGURE 18.1. Electric and Magnetic Analogs

The resistance of a conductor of length l and cross-sectional area A is

$$R = \frac{l}{\sigma A} \tag{18.3}$$

where σ is the conductivity of the circuit material. In a magnetic circuit, we have

$$\mathcal{R} = \frac{l}{\mu A} \tag{18.4}$$

where μ is termed the **permeability** of the magnetic path. If the magnetic path is in air (or other nonmagnetizable material), the permeability has the value $\mu_0 = 4\pi \times 10^{-7}$ [henry per meter]. For magnetizable material, $\mu = \mu_r \mu_0$, where μ_r is the *relative permeability*—that is, the permeability measured with respect to that for air. Typically, for iron, $\mu_r = 2000$, meaning that an iron magnetic circuit offers a reluctance to the flow of magnetic flux that is 2000 times smaller than that for a like circuit made up of a nonmagnetizable material.

This "magnification" of a magnetic flux in an iron circuit, for example, arises because the iron contains numerous small magnetic domains that ordinarily are randomly oriented. As a magnetic field is established by means of a current passing through a coil, these magnetic domains progressively line up with the externally established field, thereby adding to it. The externally established field is designated as H, the **magnetic field intensity** [ampere-turns per meter], and when multiplied by μ determines the **flux density** B [weber per square meter, now called a tesla]:

$$B = \mu H \tag{18.5}$$

Figure 18.2 shows how the flux density (B) varies as a function of H. Note that at large values of H this relationship departs from linearity. The reason for this departure is easy to understand: Since the number of magnetic domains is finite, the material will ultimately reach saturation—there are no more domains left to be aligned.

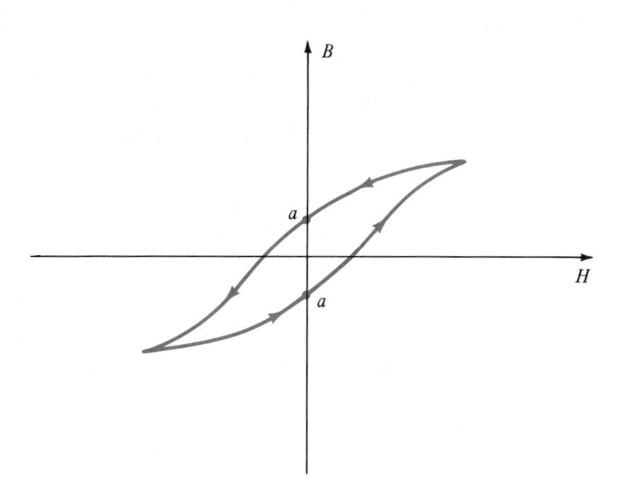

FIGURE 18.2. **B-H Curve Showing Hysteresis Loop** Points a represent residual magnetism—that is, remanence—when magnetic intensity (H) is reduced to zero.

The relationship between magnetomotive force ($\mathfrak{F}$) and magnetic field intensity (H) is

$$\mathfrak{F} = Hl \tag{18.6}$$

where l is the path length of the magnetic circuit.

A rather common form of magnetic circuit consists of a steel core such as shown in Figure 18.3. Current I flows through N turns, creating a magnetic flux (Φ) whose path length can be taken as representing the mean dimensions of the core. Often a gap exists in the core, and it is desired to know the flux within this gap. This gap is the region into which are inserted devices that require a magnetic field for their proper operation. In the case of a motor (or generator), the rotor would be inserted into such a gap.

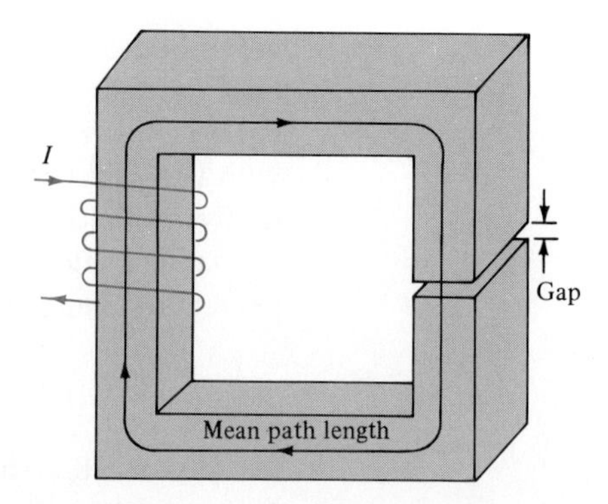

FIGURE 18.3. **Laminated Iron Core** Shown are the mean path length of magnetic flux created by current I through a coil and an air gap in the core.

18.2 ■ HYSTERESIS

In magnetizing ferromagnetic materials, energy is required to rotate the magnetic domains. Also, when the external magnetic field is removed—that is, H reduced to zero—the magnetic domains may not all return to their original random orientation, thereby leaving the material in a permanently magnetized state. We then have a permanent magnet, with flux density indicated by points a in Figure 18.2.

The magnitude of this residual magnetization is termed the **remanence.** If it is an alternating current that creates the magnetizing force H, the flux density will trace out the curves shown in Figure 18.2. It is the area within these curves that represents a measure of the work done in orienting the magnetic domains of a material. Figure 18.4 shows the shape of two such curves, one for silicon steel and the other for alnico. Obviously, for a transformer or a motor, we would prefer a core made of silicon steel in preference to alnico since the **hysteresis** loss (area within the curve) would be less. The remanence of alnico, however, is higher than that of silicon steel and would therefore be preferred as a permanent magnet material.

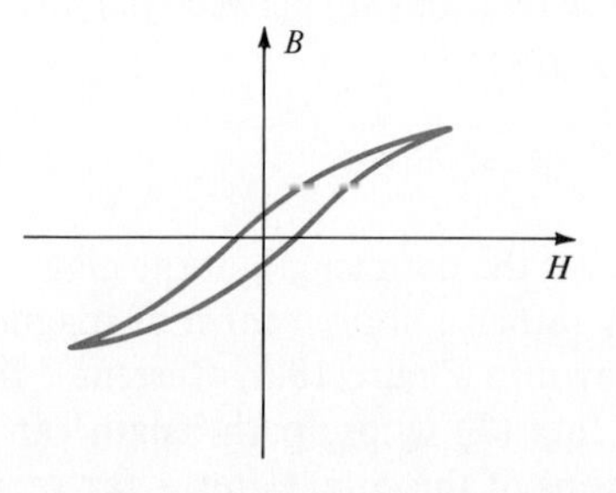

(a) Silicon steel hysteresis loop

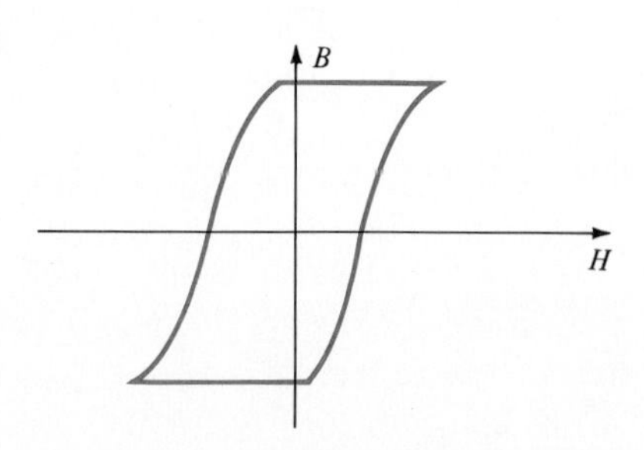

(b) Alnico hysteresis loop

FIGURE 18.4. **Hysteresis Loops**

Empirically, it is found that hysteresis loss in power (per unit volume) (P_h) can be expressed approximately by the equation

$$P_h \cong K_h f B_{max}^{1.6} \tag{18.7}$$

where f = applied frequency

B_{max} = peak flux density

K_h = a constant

18.3 ■ EDDY CURRENTS

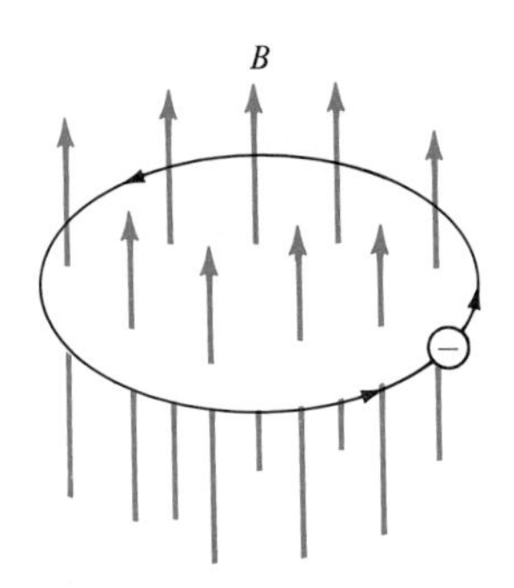

FIGURE 18.5. Influence of a Changing Magnetic Field on a Charge Carrier A free (metallic) electron exposed to a perpendicular changing magnetic field will execute circular motion in a plane perpendicular to the magnetic direction.

A moving charged particle finding itself in a magnetic field will execute circular motion (Figure 18.5). If the magnetic field is an oscillatory one, the charge will continually reverse directions.

Since a metal has numerous free electrons that can respond to such an oscillating magnetic field, such circular motion of charges constitutes a current whose passage through the material represents a power loss. Such motions are termed **eddy currents**, and the use of solid iron cores in transformers, motors, and so on would lead to a large resulting power loss. The emf giving rise to the eddy currents depends on the magnitude of the flux threading through the circular loops. The losses may be decreased by diminishing the loop area.

An induced emf in an N-turn coil is equal to $N(d\Phi/dt)$. These eddy current loops can be considered to be single-turn coils, and therefore the induced emf is equal to $d\Phi/dt$. With Φ varying in a sinusoidal fashion, the induced emf will also depend on the frequency. If t is the thickness of the conductor and ρ the material resistivity, the eddy current power loss per unit volume is given by the expression

$$P_e = \frac{\pi^2}{6\rho} t^2 f^2 B_{max}^2 \equiv K_e f^2 B_{max}^2 \tag{18.8}$$

it being recognized that power loss is equal to V^2/R; K_e is a constant. Thus, silicon steel, having a resistivity that is three times that of ordinary sheet steel, is preferred as a core material.

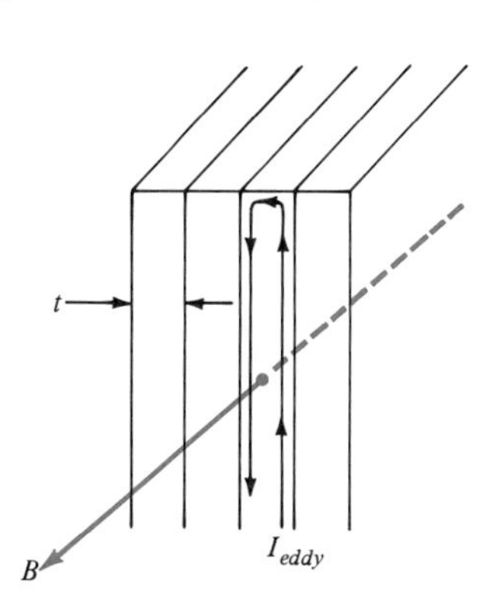

FIGURE 18.6. Current Path in Laminated Magnetic Core Free charged particles within metallic laminations subjected to a changing magnetic field will give rise to eddy currents.

The dependence on t^2 also explains why laminated sheets (Figure 18.6) are used in place of solid core materials. As the frequency of the magnetic field variations increases, the power loss may be reduced by diminishing t. But at high frequencies (such as radio waves represent) the laminations do little good – they cannot be made thin enough. At such frequencies *powdered iron* particles of small dimension are embedded in a binder. Thus, at radio frequencies magnetic cores are made of powdered iron.

Ultimately, at microwave frequencies, for example, even the dimensions of the iron powder are too great, and one must use magnetic particles of molecular dimensions. Such magnetic materials are called **ferrites.**

EXAMPLE 18.1 The cast steel core in Figure 18.3 has a mean path length of 40 cm, with an air gap of 1 mm. With 2 A passing through a 100-turn coil, compute the flux density in the gap.

Solution: Because of the large reluctance of air (compared to that of steel), as a first approximation we can consider all the magnetomotive force to be "dropped" across the air gap (Figure 18.7):

$$B = \mu_0 H = \frac{\mu_0 \mathfrak{F}}{l_{gap}} = \frac{\mu_0 NI}{l_{gap}} = \frac{4\pi \times 10^{-7} \times 100 \times 2}{10^{-3}}$$
$$= 0.25 \text{ [weber per square meter or tesla]}$$

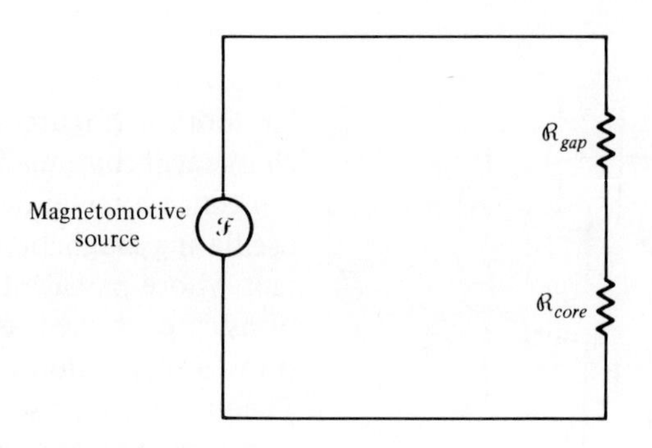

FIGURE 18.7

The actual B is slightly less because of the drop in the core. Assume B_{gap} = 0.22 [weber per square meter] or 0.22 [tesla, T]. We have

$$\mathfrak{F}_{gap} = H_{gap} l_{gap} = \frac{Bl_{gap}}{\mu_0} = \frac{0.22 \times 10^{-3}}{4\pi \times 10^{-7}} = 175 \text{ A.t. [ampere-turn]}$$

For a flux density of 0.22 T, the required field intensity (from Figure 18.8) for cast steel is about 160 A.t./m. We have

$$\mathfrak{F}_{cast\ steel} = H_{cs} l_{cs} = (160)(0.40) = 64 \text{ A.t.}$$

The total mmf required is 175 + 64 = 239 A.t. We have only 200 A.t. available, representing about a 20% error in the result.

As a second approximation, take B_{gap} to be 0.2 T:

$$\mathfrak{F}_{gap} = \frac{0.2 \times 10^{-3}}{4\pi \times 10^{-7}} = 159 \text{ A.t.}$$

For B = 0.2 T, from the magnetization curve, H = 150 A.t./m. We have

$$\mathfrak{F}_{cs} = (150)(0.40) = 60 \text{ A.t.}$$

The total mmf required is 159 A.t. + 60 A.t. = 219 A.t., about a 9.5% error.

We can make a still closer approximation if desired.

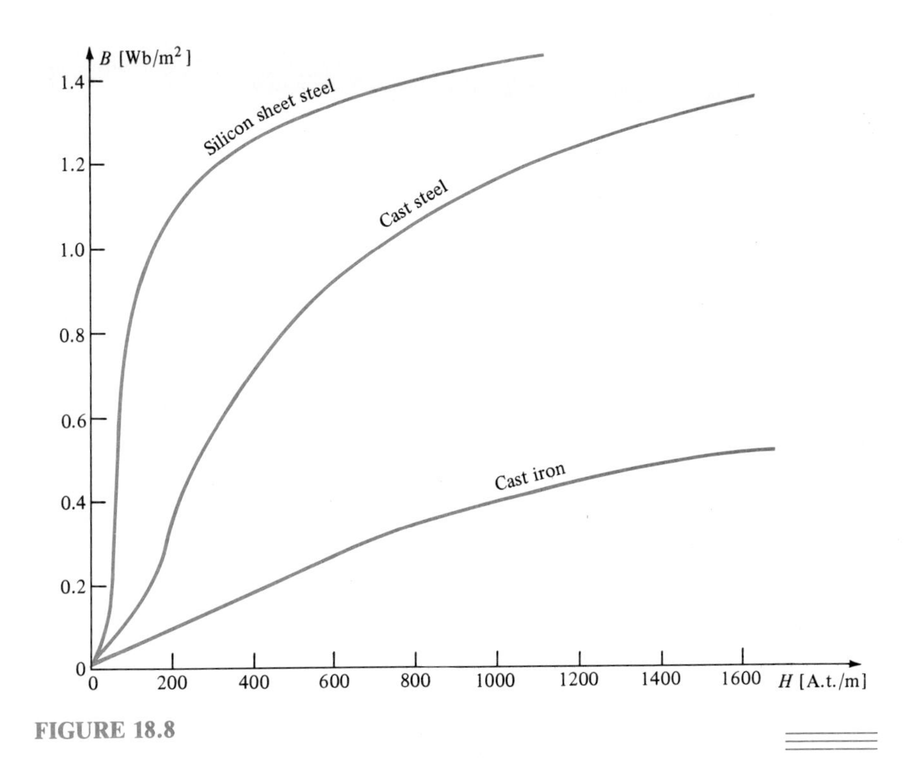

FIGURE 18.8

18.4 ■ ELECTROMAGNETS

A rather common industrial tool is the lifting magnet (Figure 18.9). Magnetic flux lines may be looked upon as rubber bands in tension, always attempting to shorten their lengths. The existence of an air gap in a magnetic circuit, therefore, will give rise to a force tending to diminish the gap.

Considering the circuit shown in Figure 18.10 as having a cross-sectional area A and a gap width h, with the gap width taken to be small compared to the area, the flux density can be taken to remain uniform within the entire circuit. The energy (W) within a magnetic field of volume V and flux density B is

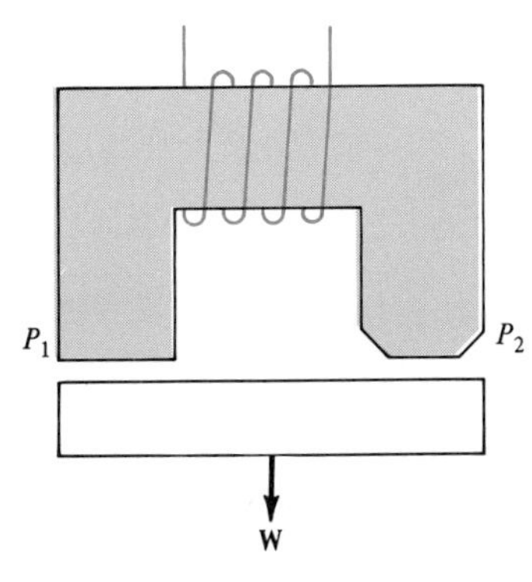

FIGURE 18.9. **Electromagnet** P_1 has a squared pole face and P_2, a tapered pole.

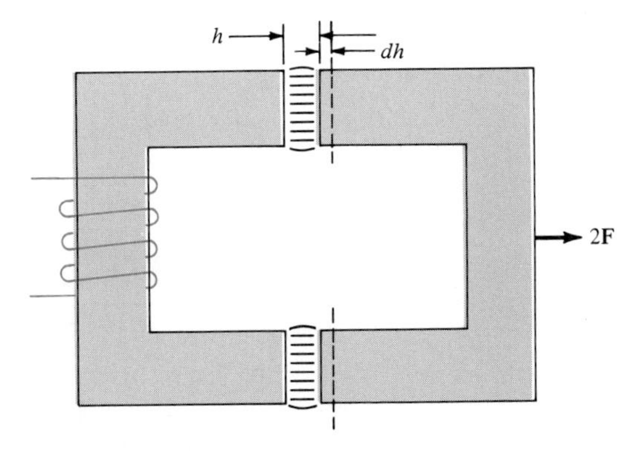

FIGURE 18.10. **Magnetic Circuit Subjected to a Force 2*F*, Leading to an Increase in Gap Widths, from *h* to *h* + *dh***

$$W = \frac{B^2}{2\mu}V \tag{18.9}$$

that within each air gap of Figure 18.10 will be

$$W = \frac{B^2 A}{2\mu}h \tag{18.10}$$

Drawing apart the two portions in Figure 18.10 is presumed to require a force $2F$. A gap change of dh results in additional energy storage within a gap of

$$dW = \frac{B^2 A}{2\mu}\,dh \tag{18.11}$$

The work done per gap is

$$F\,dh = \frac{B^2 A}{2\mu}\,dh \tag{18.12}$$

and the force is

$$F = \frac{B^2 A}{2\mu} \tag{18.13}$$

Since the force is proportional to B^2, but only to the first power of A, it is advantageous to round off the corners of the pole pieces. While this procedure diminishes A, it increases B. Too much rounding off, however, will increase the circuit reluctance and diminish Φ. The reader should note that in lifting magnets these forces can be considerably reduced by the presence of small air gaps between the magnet and the object being lifted.

EXAMPLE 18.2

Refer to Figure 18.9. The flux density at P_1 is taken to be 1.0 Wb/m², while at P_2 it is 2 Wb/m². Compute the force exerted if the cross-sectional area at P_1 is 10 cm² and that at P_2 is 5 cm².

Solution:

$$F = \frac{B^2 A}{2\mu}$$

$$F_1 = \frac{1^2 \times 10^{-3}}{2(4\pi \times 10^{-7})} = 398 \text{ N} = 89.5 \text{ lb}$$

$$F_2 = \frac{2^2(5 \times 10^{-4})}{2(4\pi \times 10^{-7})} = 796 \text{ N} = 179 \text{ lb}$$

18.5 ■ INDUCTANCE – AGAIN

Thus far in this text, we have relied on a somewhat intuitive concept of inductance—that it was a quantity that determined the magnitude of inductive reactance, that it depended on the number of turns making up a coil, and that it was increased by inserting a magnetic core within the coil. At this point, having considered the nature of magnetic fields in a bit more detail, we should place the concept on a firmer analytical basis.

When the magnetic flux threading through a coil is changing at a rate $d\Phi/dt$, there is induced into the coil an emf (v) given by the equation

$$v = N\frac{d\Phi}{dt} \qquad (18.14)$$

The direction of the induced voltage (according to Lenz's law) is such as to produce a current whose accompanying magnetic field opposes the flux change. In turn, we have

$$v = N\frac{d\Phi}{dt} = N\frac{d\Phi}{di}\frac{di}{dt} \equiv L\frac{di}{dt} \qquad (18.15)$$

where the quantity $N(d\Phi/di)$ constitutes the *inductance* measured in terms of the unit *henry* (H). Thus, we see that if the induced emf is 1 V, created by a current change of 1 A/s, the inductance is 1 H.

When the flux is directly proportional to current, the permeability is constant. This constant permeability implies that no part of the magnetic circuit has reached magnetic saturation. Under such conditions, we have

$$L = \frac{N\Phi}{I} \qquad (18.16)$$

With Φ being directly proportional to the current, the inductance is independent of the current and depends only on the geometry and permeability of the circuit. Also, since Φ itself is proportional to N, $L \propto N^2$.

There may also exist an *erroneous* intuitive concept concerning inductance that probably should be corrected at this time, that is, that an inductance necessarily implies a coil of some sort. That a straight piece of wire also constitutes an inductance is an important concept that manifests itself when considering high-frequency and fast pulse circuits. This statement may qualitatively be justified by recognizing that a magnetic field surrounds a current-carrying wire and that the surrounding field represents a magnetic energy density:

$$W = \tfrac{1}{2}LI^2 \qquad (18.17)$$

Alternatively, we have

$$L = \frac{2W}{I^2} \qquad (18.18)$$

The flux lines linking a conductor may originate either externally or from within the conductor itself. In the latter instance one has a case of *self-inductance*. In cases where the flux from two inductors interacts, the flux lines linking both units contribute to the *mutual inductance*.

SUMMARY

18.1 Introduction

- Magnetomotive force ($\mathfrak{F}$) in magnetics corresponds to voltage (V) in electricity; magnetic flux (Φ) corresponds to electrical current (I) and reluctance ($\mathfrak{R}$), to resistance (R).
- The unit of $\mathfrak{F}$ is the ampere-turn; of Φ, the weber; and of $\mathfrak{R}$, the ampere-turn per weber.
- Reluctance is given by $\mathfrak{R} = l/\mu A$, where l is the magnetic circuit path length, A is the cross-sectional area of the circuit, and μ (the permeability) is a characteristic of the material making up the magnetic path.
- The permeability of air, μ_0, is $4\pi \times 10^{-7}$ [henry per meter]. Most materials are nonmagnetic and may also be considered as having a permeability equal to μ_0.
- Magnetic materials have significantly larger permeabilities than μ_0 and are represented by μ_R, their permeability being measured relative to μ_0.

18.2 Hysteresis

- Hysteresis is directly proportional to f, the applied frequency; eddy current losses are proportional to f^2.

KEY TERMS

eddy current: circulating currents due to induced emf arising from a time-varying magnetic field passing through an electrically conducting medium

ferrites: magnetic materials whose eddy currents may be sufficiently low to allow their magnetic properties to persist at high frequencies (into the microwave region) (Not all ferrites have low electrical conductivities.)

hysteresis: magnetic loss incurred because of growth and rotation of magnetic domains

magnetic field intensity (H): the magnetic equivalent of voltage drop per unit path length (Unit is the ampere-turn per meter.)

magnetic flux (Φ): the magnetic equivalent of electrical current (Unit is the weber.)

magnetic flux density (B): flux per unit area (Unit is the tesla, which equals 1 Wb/m^2; still in rather wide usage is the gauss, which equals 10^{-4} T.)

magnetomotive force ($\mathfrak{F}$): the magnetic equivalent of the electromotive force (voltage) of electrical circuits (Unit is the ampere-turn.)

permeability (μ): a measure of the ease with which magnetic flux is conducted through a medium (Unit is the henry per meter.)

reluctance ($\mathfrak{R}$): the magnetic equivalent of electrical resistance (Unit is the ampere-turn per weber.)

remanence: residual magnetism remaining after the magnetizing source is removed

SUGGESTED READING

Kolm, Henry, et al., eds., HIGH MAGNETIC FIELDS. New York: Wiley, 1962. *There are conductors (termed superconductors) that, under very low temperature conditions, exhibit zero resistance, making them ideal for magnetic generation. On the eve of their widespread adoption, a conference, held at the Massachusetts Institute of Technology, provided an excellent review of magnetic generation, both conventional and superconducting. This book is the written record of this conference.*

EXERCISES

1. Identify and state the appropriate SI unit corresponding to each element of the *magnetic* "Ohm's law."

2. In terms of length, area, and material properties, write the equations of electrical resistance and of the equivalent magnetic quality.

3. What is the numerical value (and the SI unit) of μ_0?

4. What is the dimensional unit associated with relative permeability?

5. **a.** What is the name and associated unit of H?

   b. What is the name and associated unit of B?

6. At a large value of H, why is B no longer a linear function?

7. What quantity does H times length equal? What is the appropriate SI unit?

8. In terms of hysteresis, what kind of magnetic material is preferred for such applications as permanent magnets? For use as a transformer core material?

9. In what manner does the hysteresis loss vary with frequency?

10. What is the cause of eddy currents in metals?

11. How does the eddy current loss depend on the thickness of the magnetic conductor in a plane perpendicular to the flux direction? On the frequency of the flux variation?

12. Why is it advantageous to round off the edges of a lifting magnet? What happens if too much of the cross-sectional area is removed?

13. $H = 40$ A.t./m in air. What is the corresponding flux density? (The figure is an average value of that due to the earth's magnetic field.)

14. What is the flux passing through a rectangular area 1×1.25 in. if $B = 0.5 \times 10^{-4}$ Wb/m^2?

15. **a.** For a path length of 0.25 in. and an area of 1 in.^{2}, what is the reluctance in air?

b. What would be the required NI for a flux of 0.1 Wb?

c. How many ampere-turns would be required if the gap were iron with $\mu_R = 1000$?

16. **a.** If one attempted to use a power transformer intended for 60 Hz operation in an aircraft installation where the frequency is 400 Hz, what would be the factor by which the hysteresis losses would increase?

b. By what factor would the eddy current losses be increased?

17. **a.** The core of a transformer has a 1 in.^{2} cross-sectional area through which a $\frac{1}{32}$ in. gap is cut at right angles to the flux lines. If the flux density in the gap is 1 Wb/m^2, what is the force acting to close the gap?

b. Does the force depend on the gap width?

c. What is the magnetic energy stored in the air gap?

PROBLEMS

18.1
A steel core with a permeability of 2500 has a 5 cm^2 cross section and a mean length of 20 cm. When energized by 100 A.t., what is the circuit reluctance?

18.2
Given the silicon sheet steel core as depicted in Figure 18.3, with a mean path length of 80 cm, a core thickness of 5 cm, and a core width also of 5 cm but *without* any gap, find the total magnetic flux produced by a current of 2 A through a 200-turn coil. (Use the data shown in Figure 18.8.)

18.3
Upon cutting a 1 mm gap in the silicon core of Figure 18.1, and with the same parameters that apply in Problem 18.2, compute the flux density within the gap.

18.4
In SI units the flux density is given in terms of tesla units (1 T = 1 Wb/m^2). It is still common practice in magnetics to express flux density in terms of gauss, the CGS unit. The large magnet at the National Magnet Laboratory at M.I.T. produces a flux density of 250,000 G. Express this value in terms of tesla units.

18.5
Using Figure 18.8, estimate the (average) permeability for

a. Cast iron at a field intensity of 400 A.t./m

b. Cast steel at $H = 400$ A.t./m

c. Silicon sheet steel at $H = 400$ A.t./m

d. Silicon sheet steel at $H = 800$ A.t./m

How do you account for the diminished permeability in the last instance?

18.6
The eddy current power loss per unit volume is often expressed in the form

$$P_{avg} = K_e f^2 B_{max}^2$$

a. What are the dimensional units associated with K_e?

b. Using dimensional analysis, set

$$[K_e] = [\rho]^a [t]^b$$

where ρ is the material resistivity (the reciprocal of conductivity) and t is the lamination thickness. Evaluate a and b.

18.7
What is the magnetic energy density within the gap of Problem 18.3? What is the force tending to close the gap?

18.8
A flux density of 0.025 Wb/m^2 is to be generated within an air gap of 1 mm in an iron core using a current of 10 mA. With an iron permeability of 1000 leading to a core reluctance negligible compared to that of the gap, estimate the number of turns required.

18.9
See Figure 18.11. Stored magnetic energy comes from an electric circuit, and the total electric input energy is

$$W = \int_0^t vi\,dt$$

Since $v = N(d\Phi/dt)$ and $Ni = \mathfrak{F} = Hl$ while $\Phi = BA$, show that the area between the magnetization curve and the B-axis represents the energy stored per unit volume. (Within a closed magnetization curve, such as represented by a hysteresis loop, the enclosed area represents the energy density per cycle.)

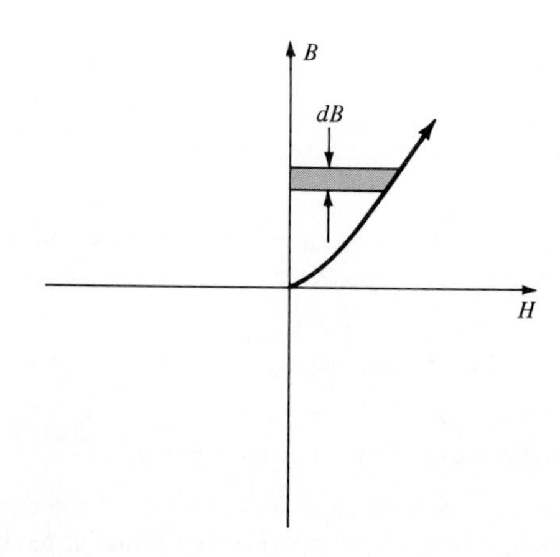

FIGURE 18.11

19 ELECTRIC MOTORS

OVERVIEW

This final chapter, perhaps more than any other, considers a realm in which the disciplines of electricity and mechanics most closely interact. The electric motor converts electrical power into mechanical power.

Despite its long history and simple physical basis, to many the electric motor retains a degree of mystery. This condition stems from a most fundamental failure on the part of most texts—that is, to provide an explanation as to how a rotating magnetic field is created by stationary coils. This topic will represent our first objective, to be followed by a consideration of a few of the varieties of electric motors.

OUTLINE

19.1 ■ INTRODUCTION

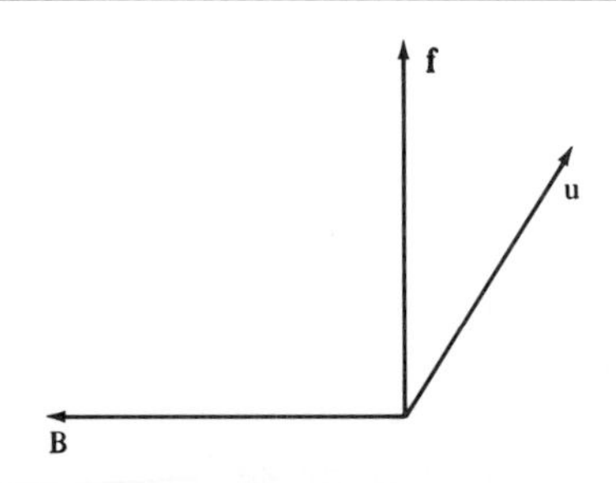

FIGURE 19.1. **Right-Hand Rule Applied to a Magnetic Field** Vector relationship between force (**f**) exerted on a (positively) charged particle having velocity **u** and passing through a magnetic field of flux density **B**.

A positive electric charge q moving with velocity **u** in a direction perpendicular to a flux density **B** will experience a force **f** in the third (orthogonal) direction (Figure 19.1). The *right-hand rule* may be conveniently used to determine these relationships: The forefinger is pointed in the direction of **u** and the middle finger in the direction of **B**, and the thumb will then point in the direction of the force **f** acting on the charge. Vectorially, we have

$$\mathbf{f} = q\mathbf{u} \times \mathbf{B} \tag{19.1}$$

Alternatively, if a collection of free charges, such as exist in a metal, have a force imparted to them in a direction perpendicular to a magnetic field **B**, they will move in the third direction, thereby giving rise to an electric current in a completed circuit. These two processes, respectively, constitute the basis of operation of an electric motor and an electric current generator.

Consider a segment of metal as in Figure 19.2a. Due to a force $\mathbf{f}_a$, this wire is made to move to the right with velocity **u** through a magnetic field directed downward (into the paper). From the right-hand rule the force on the positive charges can be seen to be such as to establish a potential difference (that is, an electromotive force, emf)[1] between the ends of the metal segment. If this wire has a length l, we have

$$\text{force} = quB = \left[\text{charge} \cdot \frac{\text{distance}}{\text{time}} \cdot \text{flux density}\right]$$
$$= \left[\frac{\text{charge}}{\text{time}} \cdot \text{distance} \cdot \text{flux density}\right]$$
$$\text{work} = \text{force} \times \text{distance} = quBl$$
$$\text{emf} = \text{work/unit charge} = \frac{quBl}{q} = uBl \equiv e$$

If an external circuit with resistance R connects the wire ends to a battery with voltage V, the magnitude of the current flow will be

$$i = \frac{e - V}{R} \tag{19.2}$$

If $e > V$, the current is positive, and the system constitutes an electric current generator. The work done in moving the wire is converted into an electrical current.

But with current i flowing there is a force opposing $\mathbf{f}_d$ (Figure 19.2b). [Again use the right-hand rule. Now determine the force ($\mathbf{f}_d$) due to current flow in a magnetic field.] $\mathbf{f}_d$ and $\mathbf{f}_a$ must, in fact, be equal and opposite if a constant velocity **u** is to be maintained.

Let us turn now to Figure 19.2c. Here we assume $\mathbf{u} = 0$. Now a current will flow from the battery, and this i' will cause the metal to accelerate to the

[1] To distinguish between a generated emf and a voltage source (such as a battery), we use the designation E in the former case and V in the latter.

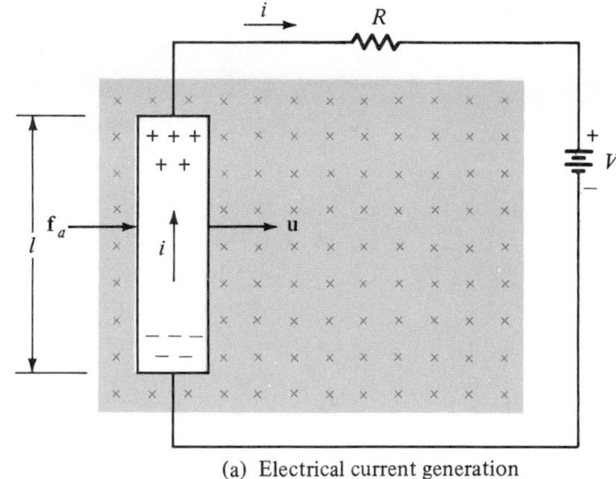

(a) Electrical current generation

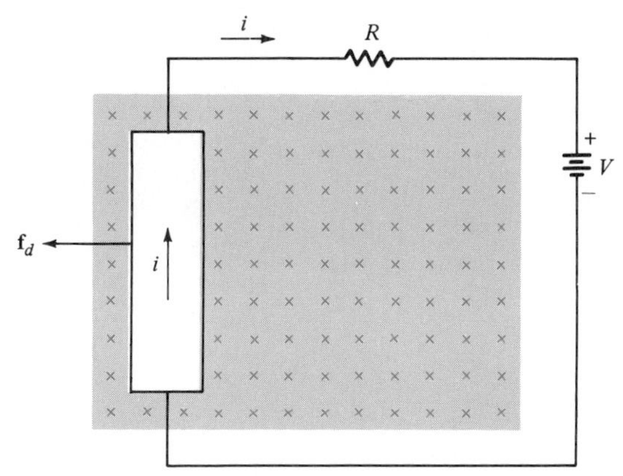

(b) Resistive force due to current in a moving conductor

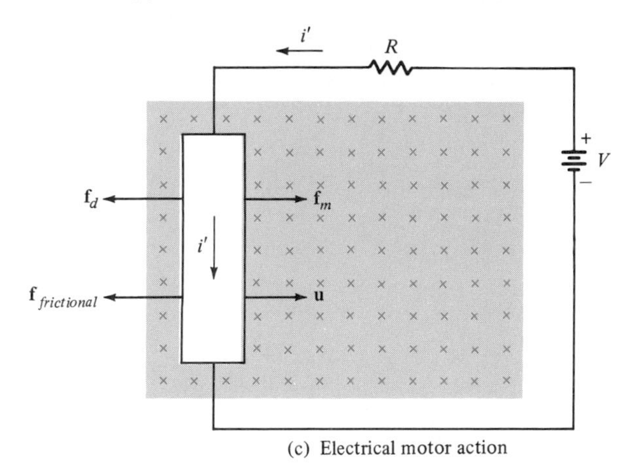

(c) Electrical motor action

FIGURE 19.2. **Electro-Mechanical Conversion**

right. But such a movement will give rise to a current i that opposes i'. (The actual current that will flow will be $i' - i$.)

Associated with i' is a force $\mathbf{f}_m$ to the right; associated with i is a force $\mathbf{f}_d$ to the left. There would also be an additional force acting to the left due to mechanical resistance. Upon reaching a balance between these forces, the wire segment would be moving to the right at a constant velocity, and such a system constitutes an electric motor.

In terms of i and i', we have an explanation of why there is a surge in current when a motor first starts (often visibly evident by the room lighting dimming momentarily). Initially, i' is large, and a large accelerating force (or torque) is developed, but as **u** increases, this velocity creates a counter-emf associated with i. The actual current drawn $(i' - i)$ reaches a steady-state value appropriate to the force balance.

With the type of geometrical arrangements we have been considering, there arise two practical difficulties:

1. We must provide for a large linear distance through which the metal segment can move freely.
2. We require a uniform magnetic field spread over a large area.

If, rather than a single length of wire (l), we were to consider two parallel lengths that can rotate about a common parallel axis, we would have the arrangement depicted in Figure 19.3. With a downwardly directed magnetic field **B** and an applied current i flowing in opposing directions through two segments, we would develop a torque. In this case the field **B** would only have to extend over the area traced out by the rotary motion. Here we have the basis for the conventional rotating motor.

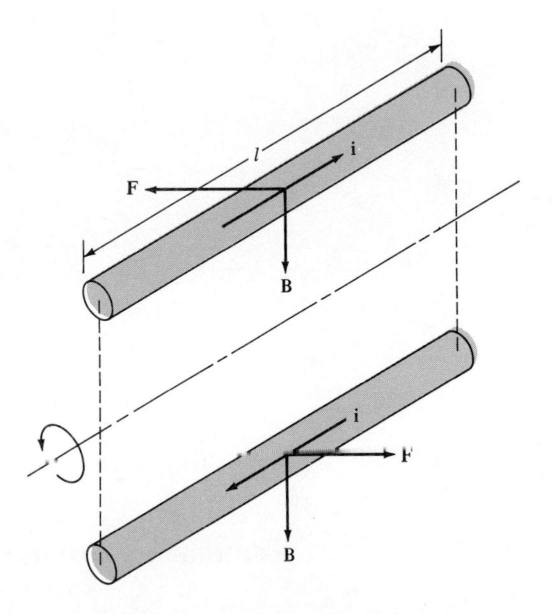

FIGURE 19.3. **Current Interaction with a Magnetic Field** Two metallic current elements pivoted about a centerline and subjected to a downward magnetic flux density **B** will be subject to a counterclockwise torque.

The two wires we have been considering might actually be two segments of a wire loop such as shown in Figure 19.4, with the supplied current furnished via slip rings. To maintain the unidirectional current flow, it is necessary to reverse the applied current after each half-turn; such reversal may be accomplished by the use of a split slip-ring **commutator.**

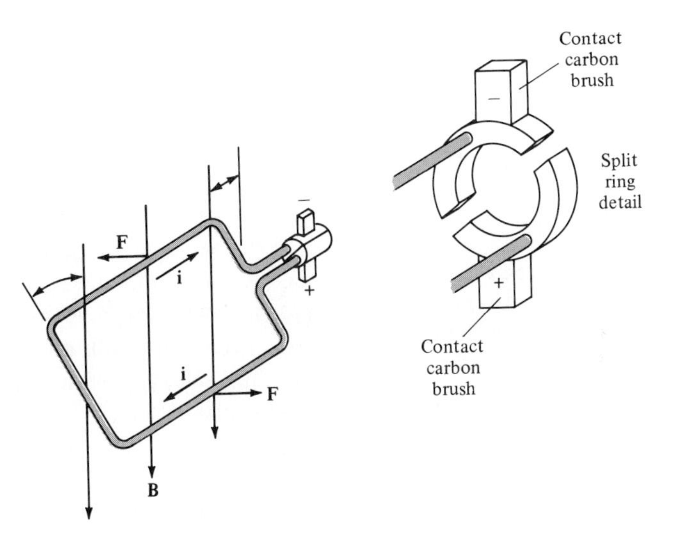

FIGURE 19.4. **Basic Electric Motor Configuration** To maintain unidirectional current flow in a wire loop rotating in a magnetic field, the polarity of the applied voltage is reversed after each half-turn by applying it through a split-ring commutator. The brush contacts slide along the split-ring commutator.

One desirable characteristic of a dc-excited motor is the ease with which one may vary its speed over a wide range. An undesirable feature arises from the need to use slip rings that are subject to wear, and from the power loss and electrical interference arising from the arcing between the commutator and the contact brushes.

In ac machines one may do away with commutators, use being made of the inherent periodic change in current direction to maintain a constant torque directivity. But the use of ac means that there is a relationship between the frequency of the ac and the rotational speed of the motor.

The simplest of rotors is a cylinder magnetized along its diameter. Surrounding this cylinder are magnetic pole pieces that, upon rotation, will cause the magnetized rotor to follow (Figure 19.5). This arrangement does not constitute a practical motor, however, since the rotational energy imparted to the rotating pole pieces must necessarily be greater than the energy that might be extracted from the rotor. In a practical motor the external rotating magnetic field is created by electrically exciting stationary coils. Let us see how such a rotating magnetic field might be created.

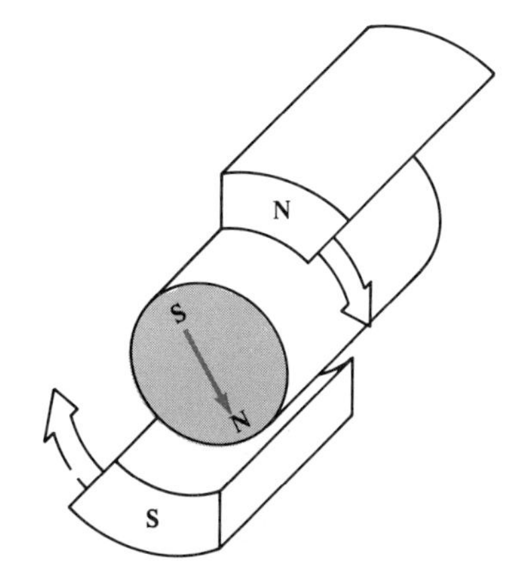

FIGURE 19.5. **A Simple Motor** A magnetized cylinder will rotate in response to the coaxially rotating magnetic poles, N and S.

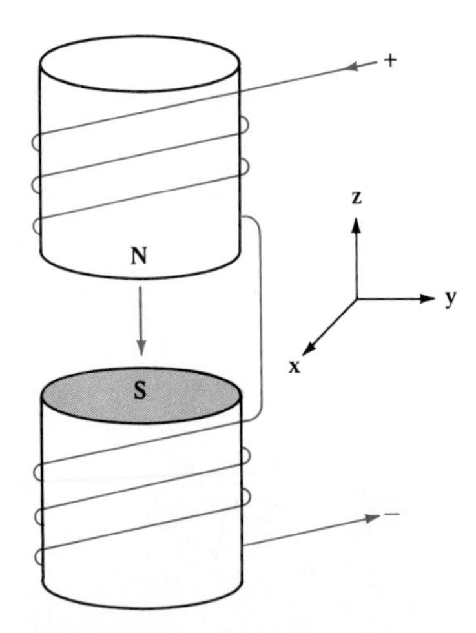

FIGURE 19.6. Electromagnets Created by Coil Current An applied ac creates an oscillating magnetic field along a vertical axis.

Two electromagnetic pole pieces arranged as in Figure 19.6, when excited by an alternating current, create an alternating magnetic field along the z-direction. Since the flux density can be taken to vary as $\cos \omega t$, we have

$$\cos \omega t = \tfrac{1}{2}(e^{j\omega t} + e^{-j\omega t}) \tag{19.3}$$

One can see the appearance of positive and negative frequencies ($+\omega$ and $-\omega$) represented by vectors rotating in opposite directions each with a frequency equal to that of the impressed alternating current. The successive vector additions of these two counterrotating vectors are shown in Figure 19.7a for one half-cycle of ac. A permanent magnet (PM) rotor inserted between the ac-excited pole pieces will attempt to follow these rotating vectors.

But which one will it follow, clockwise or counterclockwise? All other factors being equal, it will turn in whatever direction it is given an initial impulse.

But one usually desires that a motor turn in a unique direction when energized. A unique direction can be achieved by using *two* pairs of stator poles (Figure 19.8). Each of these orthogonal oscillating fields consists of two vectors, thus four in all, and if the ac is applied in phase simultaneously to both winding sets, the vector addition will lead to an alternating magnetic field that is oriented at 45° relative to the z-axis. It changes its magnitude but not its direction (Figure 19.7b).

If, however, the two coils are excited 90° out of phase with each other, the four rotating vectors will add to produce a single equivalent magnetic field that rotates in the clockwise direction, as shown in Figure 19.7c.

And what is an easy way to produce a 90° phase shift between the two excitations? Connect one winding directly to the single-phase utility source, and connect the other winding to the same source but through a capacitor. Thus, one winding has current (and magnetic field) in phase with the applied voltage, and the other has current (and magnetic field) leading the applied voltage. The resulting motor, with a PM rotor and orthogonally excited windings, constitutes a **synchronous motor.**

EXAMPLE 19.1 Verify analytically the results depicted in Figures 19.7a, b, and c.

Solution:

a. Since both vectors commence at $t = 0$ with vertical components at maximum values, taking each to have a unit amplitude, we have

$$j\cos(-\omega t) + j\cos(+\omega t) = j\left(\frac{e^{-j\omega t} + e^{j\omega t}}{2}\,\frac{e^{-j\omega t} + e^{j\omega t}}{2}\right)$$

$$= j\left(e^{j\omega t} + e^{j\omega t}\right) = 2j\cos\omega t$$

that is, a vector of magnitude 2 executing a cosine variation along the vertical axis.

b. As shown in part a, the vertical pair can be represented as $2j\cos\omega t$. The horizontal pair can be represented as $2\cos\omega t$. We have

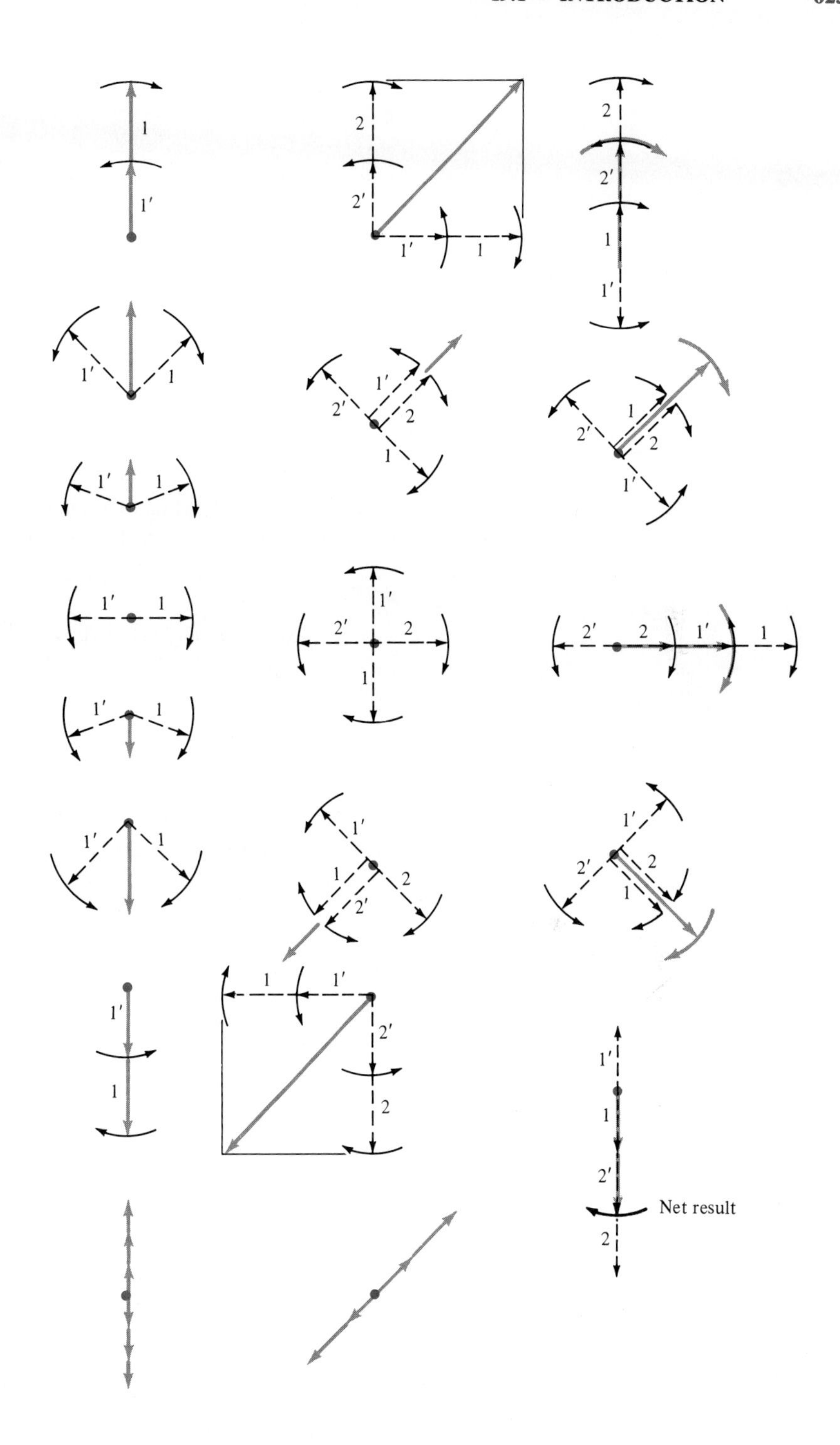

FIGURE 19.7. **Resolution of an Oscillating Magnetic Field**

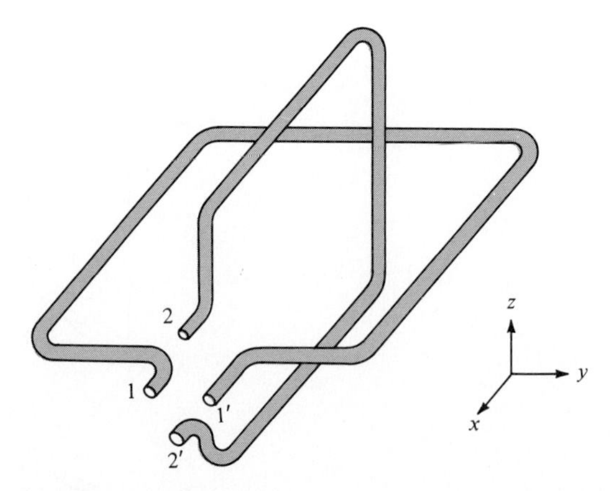

FIGURE 19.8. **Rotating Magnetic Field** Two stationary coils, when excited by currents that are 90° out of phase, will create a rotating magnetic field.

$$
\begin{aligned}
2(\cos \omega t + j \cos \omega t) &= 2\left(\frac{e^{j\omega t} + e^{-j\omega t}}{2}\right) + 2j\left(\frac{e^{j\omega t} + e^{-j\omega t}}{2}\right) \\
&= (1 + j)(e^{j\omega t} + e^{-j\omega t}) = 2(1 + j)\cos \omega t \\
&= 2\sqrt{2}\ \angle 45^\circ \cos \omega t
\end{aligned}
$$

that is, a vector of magnitude $2\sqrt{2}$, at a fixed 45° angle, executing a $\cos \Omega t$ variation.

c. In this case we have

$$
\begin{aligned}
2j \cos \omega t + je^{-j\omega t} - je^{j\omega t} &= 2j\left(\frac{e^{j\omega t} + e^{-j\omega t}}{2}\right) \\
&\quad + je^{-j\omega t} - je^{j\omega t} = 2je^{-j\omega t}
\end{aligned}
$$

that is, a vector that, at $t = 0$, has a magnitude of 2 and $+j$ orientation and subsequently rotates in the negative angular direction with frequency ω.

19.2 ■ SYNCHRONOUS MOTORS

Consider Figure 19.9. Here are shown two diametral windings, 1–1′ and 2–2′. These windings are shown as occupying the gap between the permanent magnet rotor and the (outer) stator. In practice the windings can be inserted into stator slots. The permanent magnet rotor provides a radial magnetic flux that can interact with the current passing through the diametral windings and subject them to a tangential torque. But since these windings are firmly fixed, they will not move and will, therefore, give rise to a reaction torque in the opposite direction. This reaction constitutes the motive drive for the rotor.

The flux density around the permanent magnet rotor will show a variation with angle, being maximum at the north and south poles and diminishing about the periphery (Figure 19.10). We can approximate this variation by

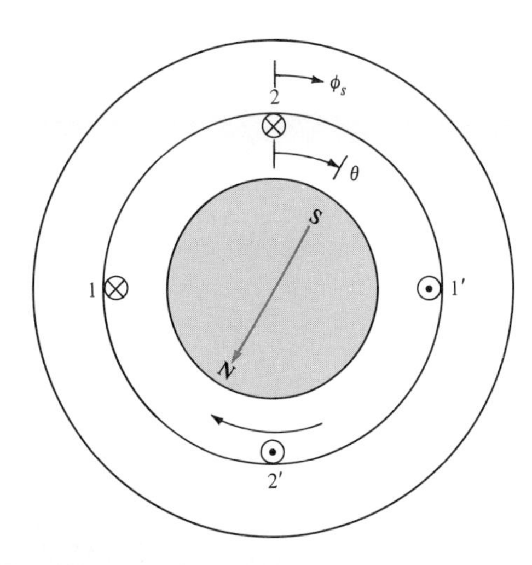

FIGURE 19.9. Two Stationary Coils (1 and 1′, 2 and 2′) with a Magnetized Rotor Between Them By using the vertical axis as a reference, the location of conductor elements is designated by ϕ_s and that of rotor orientation, by θ.

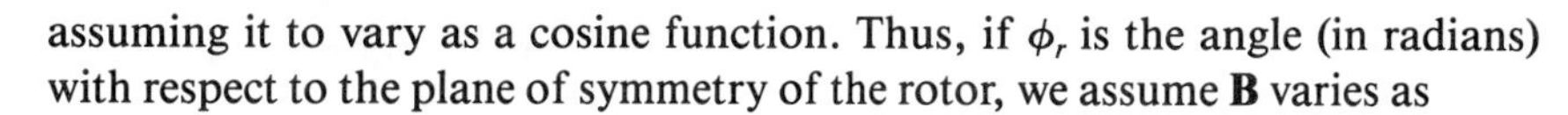

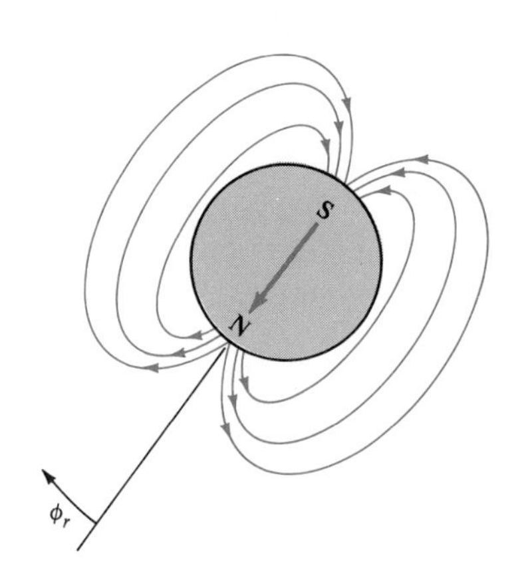

FIGURE 19.10. Magnetic Flux Distribution Around a Magnetized Rotor Variation of flux density about a permanently magnetized rotor is specified in terms of the angle ϕ_r.

assuming it to vary as a cosine function. Thus, if ϕ_r is the angle (in radians) with respect to the plane of symmetry of the rotor, we assume **B** varies as

$$B = \hat{B} \cos \phi_r \tag{19.4}$$

The force (F) on winding 1 is

$$F = I_{s(1)} l B \tag{19.5}$$

where $I_{s(1)}$ is the current through the element whose length is l and that is subjected to a gap flux density B.

For an abritrary angle (θ) of the rotor with respect to the stator (Figure 19.11), the flux density at a stator element is given by the equation

$$B = \hat{B} \cos (\phi_s - \theta) \tag{19.6}$$

Rotation in the direction of increasing θ, as shown in Figure 19.9, is taken to be positive. This is in keeping with the convention adopted by Kamerbeek. (See the Suggested Reading list at the end of this chapter.) Since ϕ_s for segment 1 is $-\pi/2$, trigonometric expansion yields

$$F = I_{s(1)} l \hat{B} \sin \theta \tag{19.7}$$

The force on the other side of this winding (1′), where $\phi_s = +\pi/2$, is equal but opposite in its sense. The torque on the complete winding is $2a\mathbf{F}$, where $2a$ is the width of the coil. The reaction torque on the rotor is in the opposite direction. Thus, we have

$$T_{e(1)} = -2alI_{s(1)} \hat{B} \sin \theta \tag{19.8}$$

We have previously shown that a unique sense of rotation is established if the two coils are excited with alternating currents differing in phase by 90°:

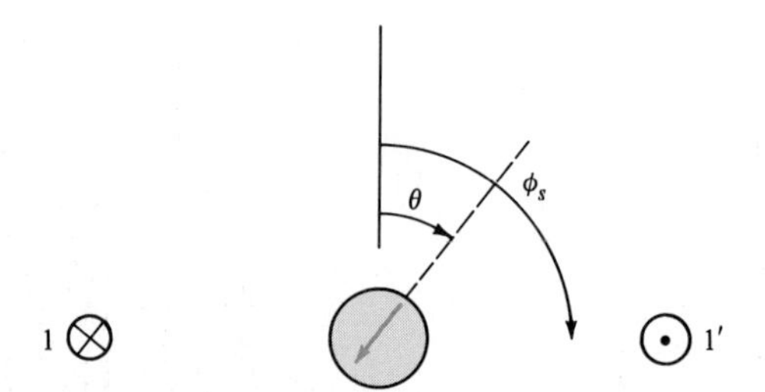

FIGURE 19.11. **Rotor Position θ** The angular location of conductor element 1′ is shown as $\phi_s = +\pi/2$. Element 1 is at $\phi_s = -\pi/2$.

$$i_{s(1)} = \hat{\imath}_s \cos \omega t \qquad i_{s(2)} = \hat{\imath}_s \sin \omega t \tag{19.9}$$

Summing up the two torques, and with a rotor revolving with angular frequency ω_r so that $\theta = (\omega_r t + \theta_0)$, we find that the total torque produced by the two coils is

$$T_e = -2al\hat{\imath}_s\hat{B} \sin[(\omega_r - \omega)t + \theta_0] \tag{19.10}$$

The time average of this torque can differ from zero only if $\omega_r = \omega$, that is, only if the motor runs at synchronous speed. Two consequences follow:

1. This type of motor will not operate at other than synchronous speed.
2. It will not start of its own accord since $\overline{T}_e = 0$ at $\omega_r = 0$. To avoid this difficulty, synchronous motors are sometimes provided with a second rotor attached to the same shaft, this second (starting) rotor operating on a different principle.

At synchronous speeds the torque is constant and given by

$$T_e = -2al\hat{\imath}_s\hat{B} \sin \theta_0 \tag{19.11}$$

If the motor is running at constant (synchronous) speed, the accelerating torque must be zero. Zero torque means that the electromagnetic torque developed must exactly equal that of the load and frictional torque being imposed on the shaft. For a given current the only variable in the expression for T_e is θ_0, the angle between the rotor and stator fields, termed the **torque angle**. For motor operation θ_0 is negative, and the rotor lags behind the stator field, falling further and further back as the load increases. When $\theta_0 = -\pi/2$, the electromagnetic torque has reached its maximum deliverable value, and any further increase in the loading will bring the rotor to rest.

Synchronous machines may also be used as electrical generators, and since utility companies furnish ac within rather close tolerances in terms of frequency (60 Hz), synchronous generators delivering hundreds of megawatts have been constructed. No permanent magnet rotor, however, could deliver such power. In large units externally supplied dc is passed to rotor windings through slip rings and is used to create the rotor's magnetic field.

Since the same synchronous machine can be used as either a motor or a generator, it is instructive to look at how the phasor diagram is altered as the unit is shifted from one type of operation to the other. A simple model for an alternator is shown in Figure 19.12. The left-hand side represents the mechanical input and the right-hand side, the electrical aspects of the machine. I_F represents the dc excitation to the rotor winding, R_F. Considered as a generator, an external source provides torque T at a rotational velocity Ω. Some of this torque is used to overcome losses so that only a fraction (T_D) is converted into electrical energy. The generated voltage is **E**, but because of electrical losses and phase shift due to winding reactance X_s,[2] the delivered voltage (**V**) differs from **E**. Over a wide range the generated voltage **E** is directly proportional to the field current I_F. This variation is qualitatively depicted in Figure 19.13. We have

$$\mathbf{V} = \mathbf{E} - \mathbf{I}(R_A + jX_s) \simeq \mathbf{E} - j\mathbf{I}X_s \tag{19.12}$$

where we have taken $R_A << X_s$.

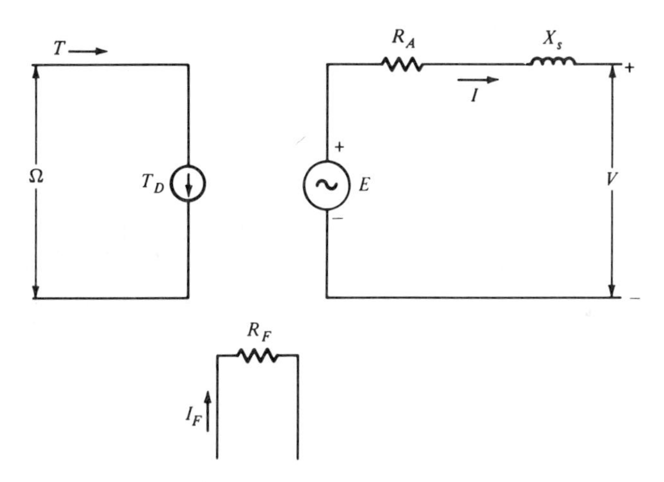

FIGURE 19.12. **Schematic Model of an Electromechanical Machine** Part of the applied torque (T_D) at angular frequency Ω is converted into electromotive force **E**, which, because of the voltage drop in R_A and X_s, is delivered as output voltage V. Variation of current I_F in the rotor coil (R_F) may be used to control the magnitude of the emf **E**.

Consider an alternator as being driven at synchronous speed by a gasoline engine. The input mechanical power ($T_D\Omega$) is converted into electrical power ($VI\cos\theta$). If the electrical output is connected to the utility power lines and I_F adjusted so that $E = V$, the generator furnishes *no* current—that is, $I = 0$. There is no transfer of power between the alternator and the line (Figure 19.14a).

Now try to speed up the driving engine. The rotor will now lead the

[2]Reactance X_s has two sources: In the rotor there is a component of induced voltage that is in quadrature with the current, and this component contributes inductive reactance. Also, that portion of the stator flux that does not react with the rotor represents the other contribution.

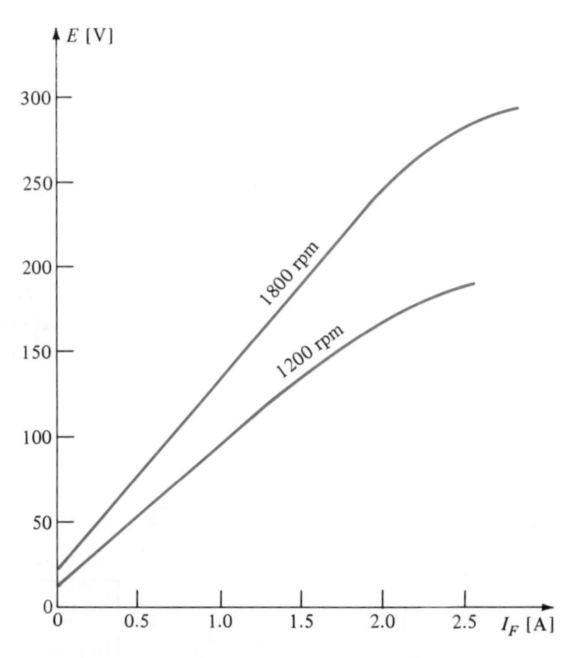

FIGURE 19.13. Variation of emf Generated (E) as a Function of the Rotor Field Current (I_F) for Two Different Synchronous Rotational Speeds

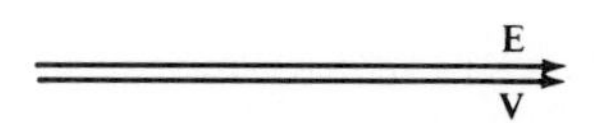

(a) Field current (I_F) adjusted so that $\mathbf{E} = \mathbf{V}$.

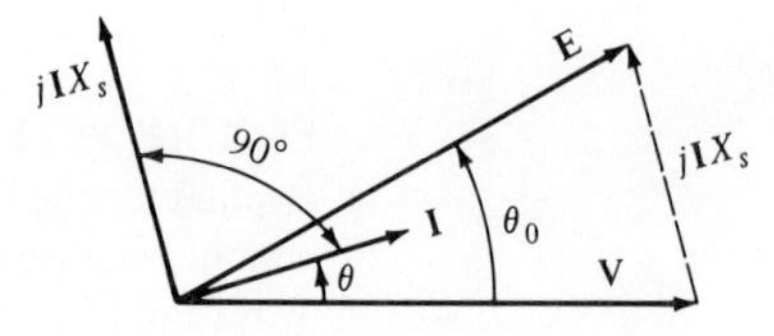

(b) Attempting to accelerate the rotor ahead of the stator field gives rise to an output current **I**, and the system acts as an electrical generator.

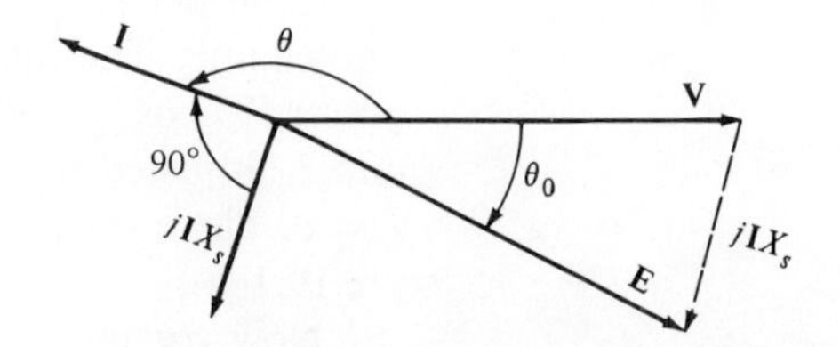

(c) Throttling the rotor drive back from synchronous speed causes current **I** to flow *into* the machine and the system acts as a motor, its electrical input supplementing the mechanical input and thus maintaining the rotor at synchronous speed.

FIGURE 19.14. Current-Voltage Relationship in Synchronous Electric Machines

stator field. (This angular increase can be seen by the use of a strobe light.) The two, however, will remain in synchronism, with their angular difference being termed the *torque angle* θ_0 (Figure 19.14b). Since the mechanical input has been increased (T_D is larger, although Ω stays the same), the excess work appears in the form of an output current into the power lines, and there is a voltage drop ($j\mathbf{I}X_s$) across the stator winding:

$$\mathbf{V} + j\mathbf{I}X_s = \mathbf{E} \tag{19.13}$$

The j indicates that the voltage drop across the inductance leads the current by 90°, while the phase angle θ shows the output current leading the output voltage V. The increase in the torque angle acts as an added load, thereby maintaining the synchronous speed.

If the driving engine is throttled back to its original state, the system goes back to that depicted in Figure 19.14a. Now try to diminish the driving speed by slowing down the driving motor. The torque angle θ_0 will now become negative (again, this angular change can be seen with a strobe light), and current will be withdrawn from the power lines—that is, $\mathbf{I}$ is negative (Figure 19.14c). With power being drawn from the lines, this contribution acts as a "makeup" torque, again resulting in the maintenance of synchronous speed.

When dealing with motors, rather than the depiction in Figure 19.12, one may draw the electrical circuit at the left (with the current entering) and the mechanical portion at the right (with the torque exiting). In this case positive power represents mechanical output (Figure 19.15).

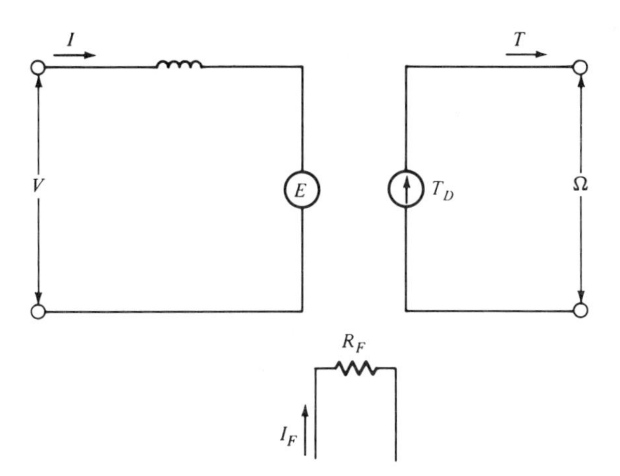

FIGURE 19.15. **Schematic Model of an Electrical Motor** Input electrical power at the left is converted into mechanical output power at the right.

There is a special type of synchronous motor in which the torque can differ from zero at all speeds and not just at synchronous speed. This synchronous machine has the advantage that it can be started without any special provisions being required. As long as the load torque remains below some designated maximum, like the ordinary synchronous motor, this type (called a **hysteresis motor**) will also run at synchronous speed. This motor will be treated in more detail in a separate section.

EXAMPLE 19.2 Recalling that in a capacitor the current leads the voltage—that is, we have a leading power factor—and in an inductor the opposite holds (we have a lagging power factor), show that a synchronous motor may be made to behave like a capacitor and its capacitive reactance used to compensate for a reactive load with inductive characteristics.

Solution: Refer to the motor depiction in Figure 19.15. We have

$$\mathbf{V} = \mathbf{E} + j\mathbf{I}X_s$$

$$\mathbf{I} = \frac{\mathbf{V} - \mathbf{E}}{jX_s} = -j\frac{\mathbf{V} - \mathbf{E}}{X_s}$$

A synchronous machine is made to act like a motor by having the torque angle negative (**E** lagging **V**), but the current can be made to either lead or lag the line voltage by, respectively, making **E** either larger or smaller than **V** (through the adjustment of the magnitude of field excitation). In the former case this condition amounts to a leading power factor, and such a motor, in addition to developing positive power (**VI** cos θ), will also exhibit a negative reactance that may be used to compensate for an inductive load. Such a field condition is termed **overexcitation**.

Figure 19.16 shows the apparent, real, and reactive power components of an inductive load, such as presented by a resistance furnace. When placed in parallel with an overexcited synchronous motor, the two reactive components may be made to compensate for each other. The advantage of this procedure over that of a simple capacitor compensator lies in the additional availability of useful work from the synchronous machine.

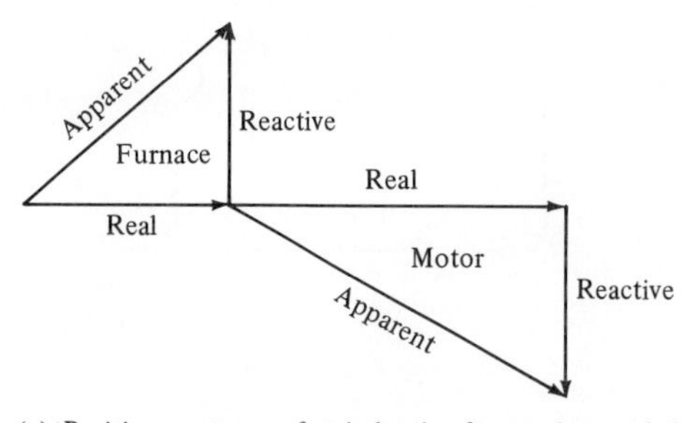

(a) Positive reactance of an inductive furnace is canceled by the negative reactance of an overexcited synchronous motor; in addition to furnace heat, one also obtains mechanical output from the motor

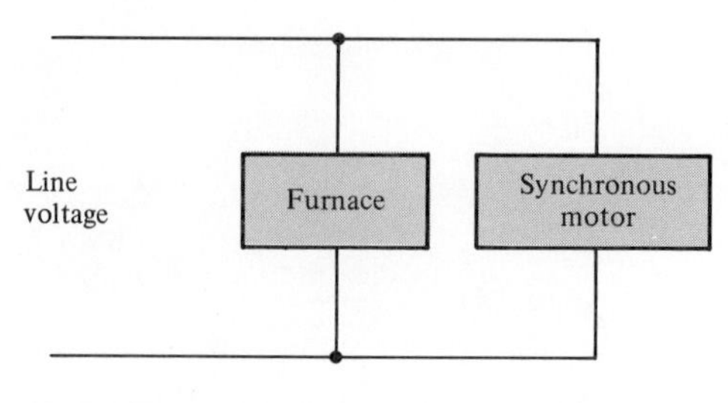

(b) Synchronous motor and furnace connected across input power line

FIGURE 19.16

The watt unit is used only in conjunction with *real power. Reactive power* and *apparent power* are designated in terms of volt-ampere. When dealing with large motors or generators, the corresponding units are kilowatt and kilovolt-ampere, abbreviated as kW and kVA, respectively. Reactive power is generally designated as kVAR, kilovolt-ampere, reactive.

19.3 ▪ INDUCTION MOTORS

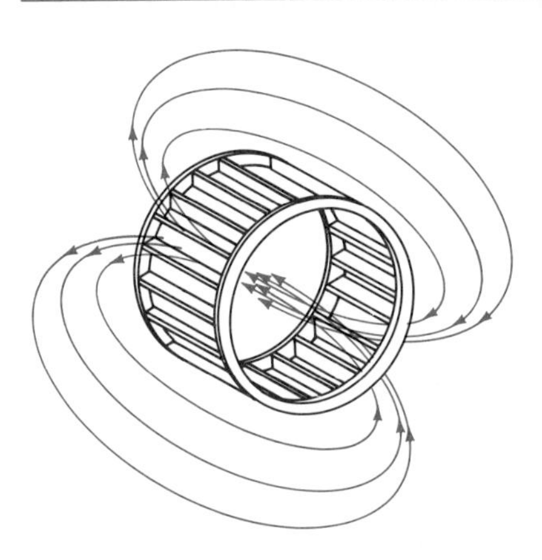

FIGURE 19.17. Squirrel-Cage Rotor The rotating stator field induces voltage into rotor conductors, giving rise to rotor current whose associated magnetic field interacts with the stator field.

Among the most widely used electric machines that furnish mechanical power is the **induction motor.** In such motors the rotor includes a wire winding, but this winding does not necessarily have to come out to slip rings. The rotor turns can be (and often are) short-circuited. From the configuration that results (Figure 19.17), the reason for their being called *squirrel-cage rotors* should be obvious. As the stator coils are energized and give rise to a rotating magnetic field, the flux cutting across these conductors gives rise to an induced voltage. The resulting rotor current creates a magnetic field that can interact with the rotating stator field.

When the rotor is first energized, the motion of the stator field across the rotor windings is at its maximum, and hence a starting torque is developed. But since these conductors are embedded in iron, they represent a considerable inductance. (Remember, $L \propto \mu$.) In an inductive circuit the current lags behind the voltage (the induced voltage in this case). Because of this lag, the torque at standstill is generally not at its maximum value. We could improve upon this torque by increasing the circuit resistance. This increase may be accomplished by diminishing the size of the rotor conductor.

Further clarification of the induction process may be obtained by referring to Figure 19.18. In the position illustrated, the stator field ($\mathbf{B}_s$) induces the maximum emf into the rotor windings. The rotor field would then assume the direction shown by the dotted vector, and the torque would be at a maximum—that is, $\sin \theta_0 = 1$. But this condition would prevail only if the induced voltage and resulting current in the rotor windings were in phase. Because of rotor winding inductance, the current lags behind the induced voltage, and the rotor field ($\mathbf{B}_r$) lags $\mathbf{B}_s$ by more than 90°—that is, $\sin \theta_0 < 1$. A diminished phase angle (θ) (accompanied by a desired decrease in the torque angle θ_0) may be achieved by increasing the resistive component of the rotor voltage.

As the motor speeds up, the difference between the rotor frequency and the stator field rotation diminishes—that is, there is a **slip** between them—and so does the induced rotor voltage. If the motor is running at synchronous speed, there is no flux cutting across the rotor bars, and hence no torque is developed.[3] In use the motor will slow down below synchronous speed to a value such that the slip leads to the torque required to overcome the load (Figure 19.19).

But the higher the rotor resistance (for a given load), the greater must be the extent to which the motor must slow down in order to develop sufficient voltage for the current to be forced through the cage. Thus, increasing the rotor resistance reduces the full-load speed and causes an increase in the I^2R losses. A motor with 50% slip could have 50% of its rotor power lost as heat.

If the rotor windings are brought out through slip rings, one can use a large resistance for starting and a small one for running. Figure 19.20 shows the effect of progressive increases in rotor resistance. Note that for a constant

[3]The slip disappears when synchronization prevails.

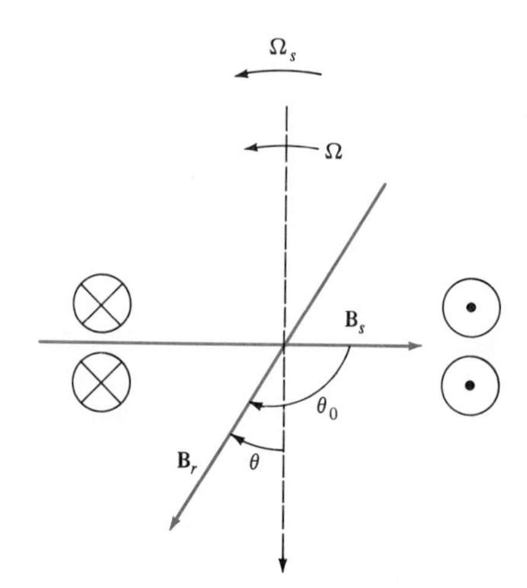

(a) Vector relationship between rotor and stator fields

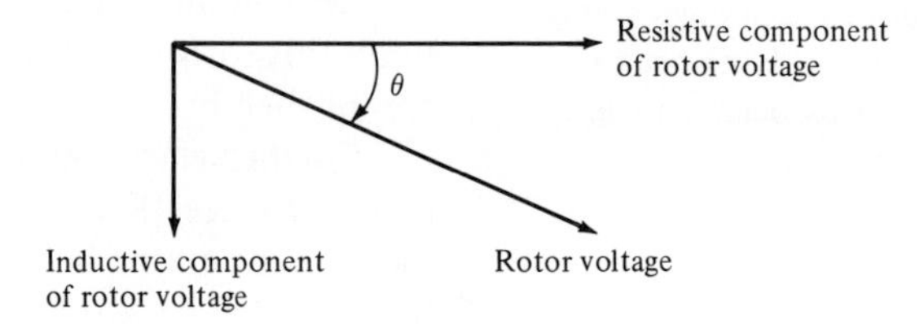

(b) Resistive and reactive rotor voltage components

FIGURE 19.18. Induced Torque in an Induction Motor

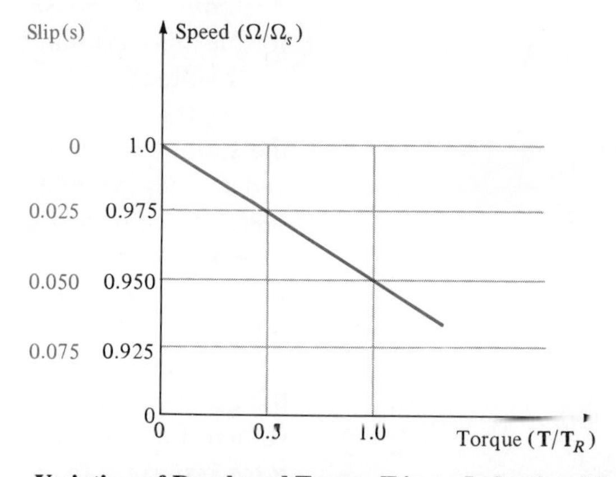

FIGURE 19.19. Variation of Developed Torque T in an Induction Motor as a Function of Slip *s* T_R is the rated torque.

delivered torque the speed can be changed by varying the external resistance. Also, the heating due to this additional slip occurs in the external circuit and not in the rotor.

Since motors are generally produced in high volume, the savings of a few

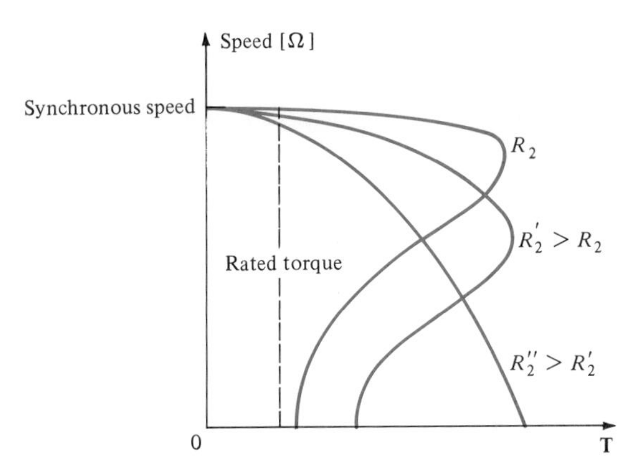

FIGURE 19.20. Rotor Speed vs. Developed Torque for an Induction Motor, with Increasing Rotor Resistance (R_2) as a Parameter

cents on each item, cumulatively, may represent a considerable amount of money. Induction motors are made of laminated steel (for the same reason that transformer cores are—to reduce eddy current losses), and the earliest practice had copper wires being inserted into the rotor slots. Today many rotors consist of molded aluminum squirrel cages, a much cheaper production process. Because aluminum has a lower conductivity, cast aluminum rotors are larger than those of equivalent copper construction. Being molded from hot aluminum, each element of the rotor cage can easily be made to assume complicated forms, such as shown in Figure 19.21a. With such an arrangement one may automatically achieve high rotor resistance when starting and lower resistance in a running condition. To see how this arises, consider the following: When current flows through region 1 (Figure 19.21b), a significant portion of the surrounding magnetic field that is generated by this current lies in air. Because of the smaller permeability, the corresponding inductance is smaller than if the "wire" were more deeply embedded in the rotor core. Region 2, for that reason, represents a higher inductance.

The rotor current depends on the squirrel-cage impedance, which consists of resistance and reactance. Upon starting the motor, the lower part of the cage (region 2) has a large reactance since the flux-cutting frequency is large. (Remember, $X_L = 2\pi fL$.) As the motor speeds up, the effective cutting frequency diminishes, and we have a large current conduction area and hence a lower rotor resistance at the operating speed.

To conclude this section, and before turning to some more theoretical considerations of the induction motor operation, let us consider a facet of comparison between synchronous and induction motor performance. So far we have concentrated on the two-(stator) pole motor. It is also possible to arrange a larger number of poles around the stator periphery (but always in pairs). For a synchronous motor, in terms of a 60 Hz applied frequency and the number of poles, we have

$$\text{synchronous rpm} = \frac{60 \times 120}{\text{poles}} \tag{19.14}$$

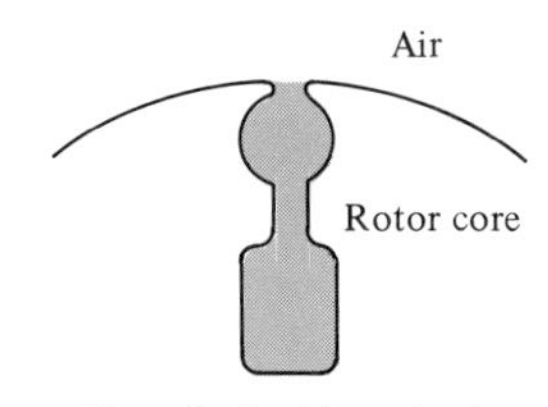

(a) Cross section of a double squirrel cage element

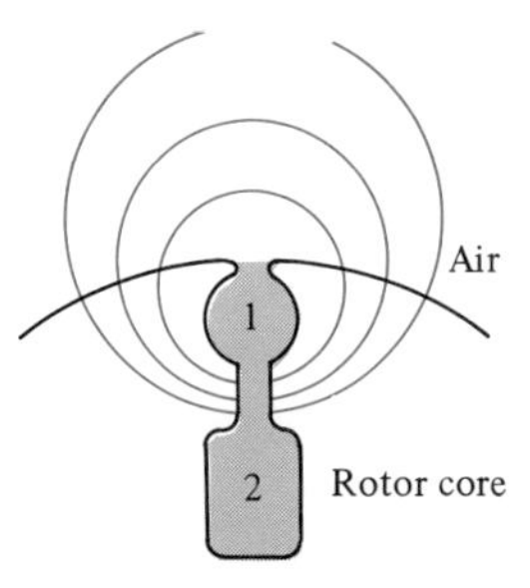

(b) Due to lower permeability, smaller inductance relative to region 1 than to the more deeply embedded region 2

FIGURE 19.21. Automatic Adjustment of Induction Rotor Resistance

A typical induction motor might have a 5% slip. A comparison of motor speeds yields the following:

	Synchronous Speed (rpm)	Nominal Rated Speed of Induction Motor with 5% Slip (rpm)
Two-pole motor	3600	3420
Four-pole motor	1800	1710
Six-pole motor	1200	1140

Note that in either case the rotational speed is diminished by increasing the number of poles.

19.4 ■ INDUCTION MOTOR THEORY

If we call the emf induced into the rotor of an induction motor $\mathbf{E}_2$, the rotor current is

$$\mathbf{I}_2 = \frac{\mathbf{E}_2}{R_2 + j2\pi f L_2} = \frac{\mathbf{E}_2}{R_2 + jX_2} \tag{19.15}$$

If slip is taken into account, the induced voltage is $s\mathbf{E}_2$, and the frequency "seen" by the rotor is sf. (Note that at synchronous frequency $s = 0$ and $sf = 0$, and the induced voltage is zero.) The current is

$$\mathbf{I}_2 = \frac{s\mathbf{E}_2}{R_2 + jsX_2} = \frac{\mathbf{E}_2}{(R_2/s) + jX_2}$$

$$= \frac{\mathbf{E}_2}{R_2 + jX_2 + R_2(1 - s)/s} \tag{19.16}$$

$$\mathbf{E}_2 = \mathbf{I}_2(R_2 + jX_2) + \mathbf{I}_2 R_2(1 - s)/s = \mathbf{I}_2 Z_2 + \mathbf{E}_g \tag{19.17}$$

Alternative representations of the rotor circuit are shown in Figure 19.22. Since $\mathbf{E}_2$ is the "secondary" voltage, the rotor of an induction motor looks either like a transformer whose load varies depending on the motor speed or, alternatively, like a counter-emf in series with resistance and inductance. The power transferred to the secondary load is $E_2 I_2 \cos\theta_2$. The loss is $I_2^2 R_2$. Conversion into mechanical power is represented by the quantity $E_g I_2$. If Ω is the rotor speed and T_D the developed torque, we have

$$T_D\Omega = E_g I_2 \tag{19.18}$$

$$T_D = \frac{E_g I_2}{\Omega} \tag{19.19}$$

If Ω_s is the stator frequency, the total power transferred from stator to rotor is

$$P_2 = E_2 I_2 \cos\theta_2 \tag{19.20}$$

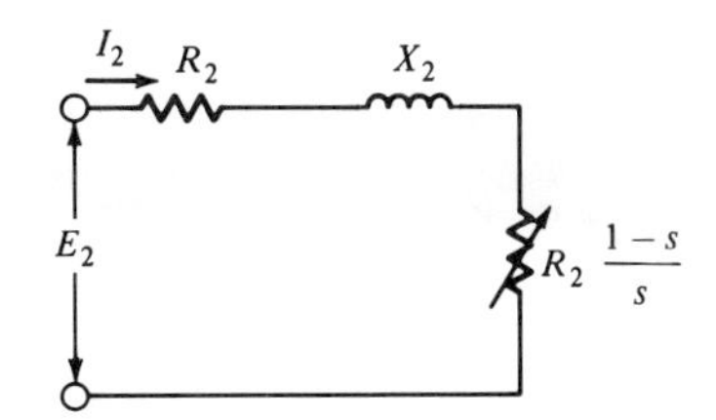

(a) Electrical equivalent of an induction rotor represented in terms of a variable load dependent on the slip

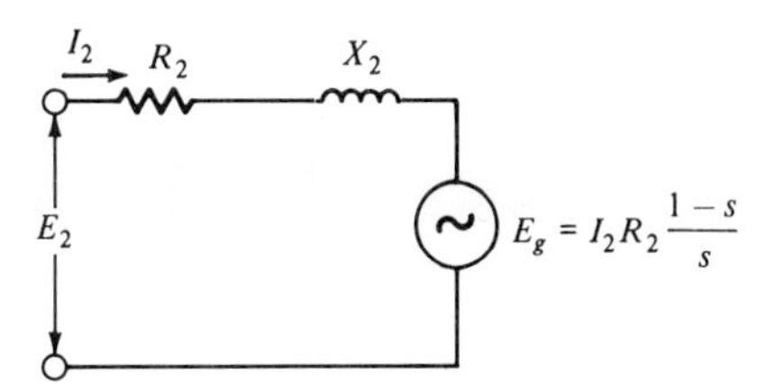

(b) Load of an induction rotor represented in terms of the counter-emf E_g

FIGURE 19.22. **Alternative Representations of Induction Rotor**

$$T_2 = \frac{P_2}{\Omega_s} = \frac{E_2 I_2 \cos \theta_2}{\Omega_s} \tag{19.21}$$

The developed torque T_D is less than T_2 by the amount of the power loss $I_2^2 R_2$. The complete equivalent circuit for the induction motor might be represented as in Figure 19.23a. For the sake of simplicity, assuming the turns ratio to be unity, we get Figure 19.23b. R_m and X_m represent both rotor and stator losses. The sum of these losses may be conveniently determined under no-load conditions, with I_0 being the no-load current.

Let us continue to refer to Figure 19.23b. When the no-load current is drawn, there is a drop across R_1 and X_1. Analytical consideration is simplified

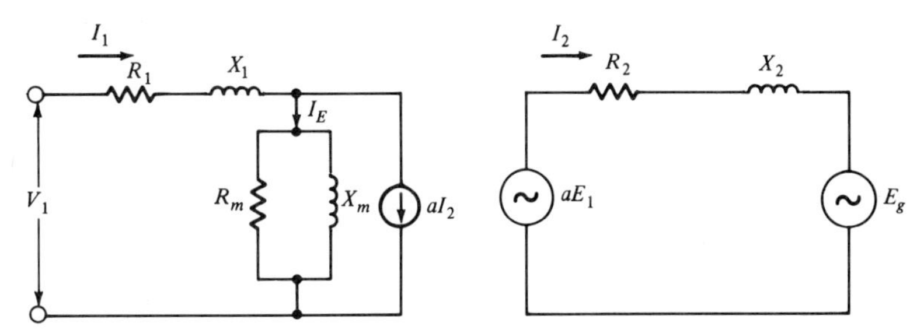

(a) Transformer equivalent of an induction machine with a representing the turns ratio between stator winding (at left, subscript 1) and rotor (at right, subscript 2)

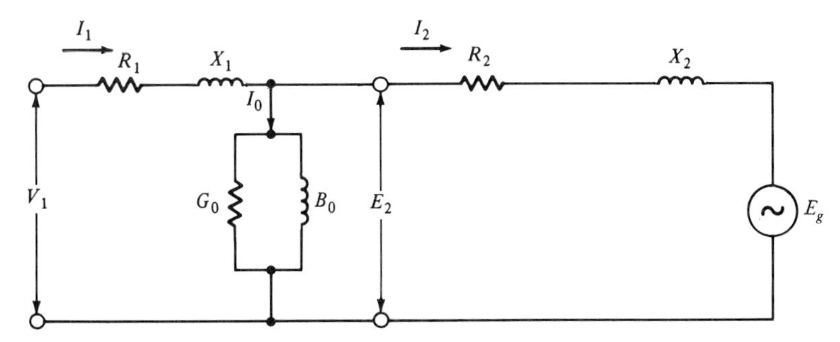

(b) Induction machine equivalent obtained by setting $a = 1$, with G_0, B_0 constituting unloaded loss leading to current I_0 and rotor induced voltage (E_2) equal to applied voltage V_1 less unloaded current drops across R_1 and X_1

FIGURE 19.23. **Transformer Equivalent of an Induction Machine**

by adjusting the terminal voltage to some hypothetical value that takes this drop into account. The hypothetical voltage is fairly accurately given by

$$V_A = V_1 - I_0\sqrt{R_1^2 + X_1^2} \tag{19.22}$$

$$\begin{aligned} T_D &= \frac{E_g I_2}{\Omega} = \frac{I_2^2 R_2(1-s)/s}{\Omega_s(1-s)} = \frac{I_2^2 R_2}{s\Omega_s} \\ &= \frac{V_A^2}{\Omega_s} \frac{R_2/s}{(R_1 + R_2/s)^2 + X^2} \end{aligned} \tag{19.23}$$

Since $X_1 \simeq X_2$, we assume that $X = 2X_1$, and the total resistance is

$$R_1 + R_2 + R_2(1-s)/s = R_1 + R_2/s$$

If s is small, R_2/s is large compared to X, and

$$T_{normal} \simeq \frac{V_A^2 s}{\Omega_s R_2} \tag{19.24}$$

It can be seen that in the normal operating range the torque is proportional to s. Also, $T \propto 1/R_2$, meaning that the use of larger rotor conductors leads to larger induced currents and greater torque for a given slip. When the speed is low, $s \simeq 1, X >> R$, and

$$T_{low\ speed} \simeq \frac{V_A^2 R_2}{s\Omega_s X^2} \tag{19.25}$$

and the starting torque is seen to be proportional to R_2.

One other point of interest is the maximum torque developed. Differentiating T_D with respect to s and setting the derivative equal to zero, we have

$$s = \frac{R_2}{\sqrt{R_1^2 + X^2}} \tag{19.26}$$

$$T_{max} = \frac{V_A^2}{2\Omega_s(R_1 + \sqrt{R_1^2 + X^2})} \tag{19.27}$$

Incidentally, small fractional horsepower motors, such as are used in hand drills, sewing machines, and electric fans, are usually *universal motors.* The stator windings are recessed in such a way that the magnetic polarity of each successive slot is reversed. They are also excited in series with a rotor commutator in such a way that all polarities simultaneously reverse. Such motors, therefore, may be used with either ac or dc.

19.5 ■ INDUCTION MOTOR PRACTICE

19.5.1 Three-Phase Systems

We have seen that for a single cycle of alternating current the power delivered appears in pulsating form (Figure 19.24a). In a three-phase system the power delivery is constant and hence more efficient (Figure 19.24b).

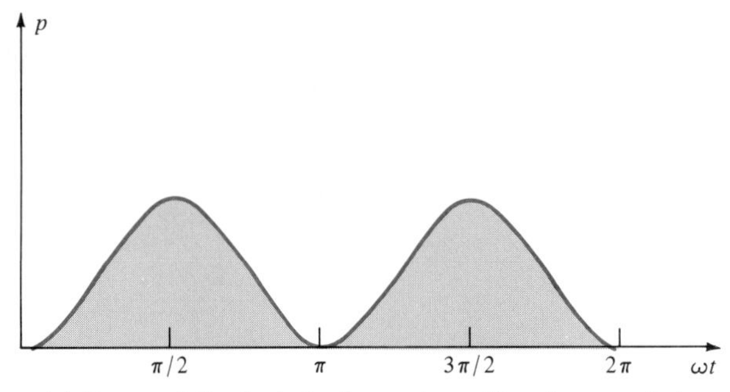

(a) Power variation in single-phase system, with unity power factor

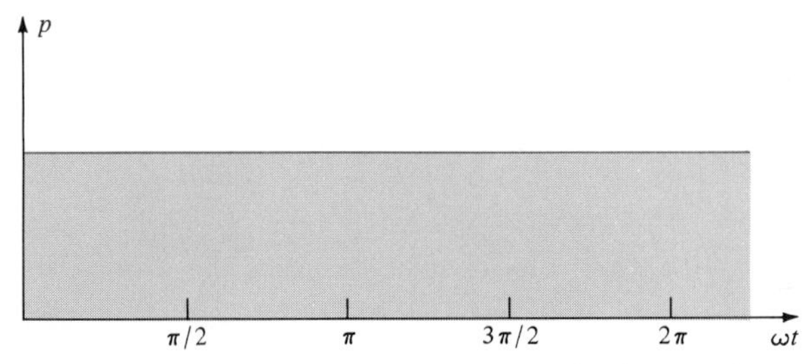

(b) (Constant) power in three-phase system, with unity power factor

FIGURE 19.24. **Power Input Variation of Single-Phase and Three-Phase Systems**

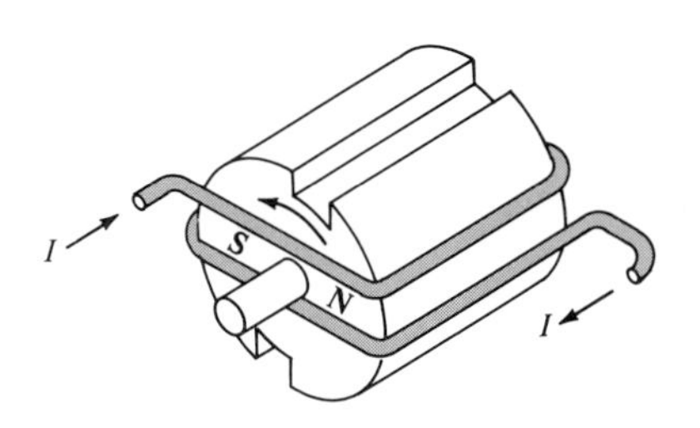

FIGURE 19.25. **Alternating emf Induced by a Rotating Permanent Magnet**

Figure 19.25 shows a single winding. As a magnetized rotor rotates through one cycle, it will give rise to a single cycle of alternating voltage. If three such coils are wound about the rotor, with their respective axes separated by 60°, a three-phase voltage can be generated. Voltages in such coils will reach their maximums 60° apart (Figure 19.26a). However, when connected to the transmission lines, one phase (phase C) is deliberately inverted, thereby creating the 120° difference between them (Figure 19.26b).

There are two alternative ways that are usually used to interconnect the coils; one, the Y (or *wye*) connected, is shown in Figure 19.27. The primed terminals form a common junction, and the unprimed ones are connected to the output lines. The voltage between any line and the common (or neutral) wire constitutes a **phase voltage**, and the voltage between the lines is the line-to-line voltage—or, more simply, just the **line voltage.**

The subscripts in Figure 19.28 indicate the direction of voltage measurement. Thus, $\mathbf{V}_{aa'} = -\mathbf{V}_{a'a}$. The voltage between any two lines is obtained by a phasor sum. Thus, we have

$$\mathbf{V}_{a\ to\ b} = \mathbf{V}_{a\ to\ a'} + \mathbf{V}_{b'\ to\ b} \tag{19.28}$$

This phasor addition is shown in Figure 19.29.

Thus, for a wye connection we have

$$\mathbf{V}_{line} = \sqrt{3}\,\mathbf{V}_{phase} \angle +30° \tag{19.29}$$

It should be evident, however, that in the wye connection the phase and line currents are identical:

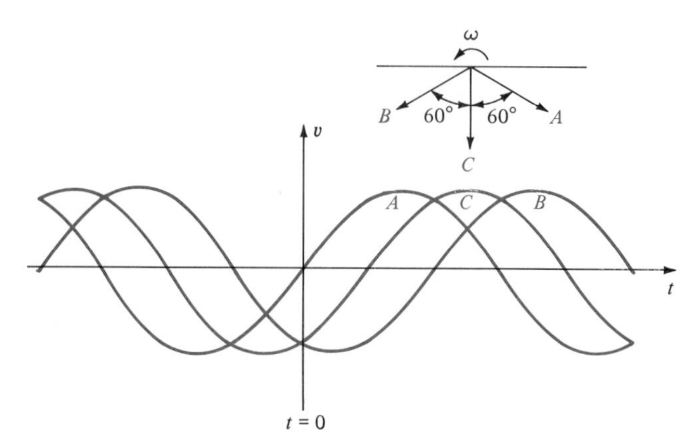

(a) Output voltage variation due to a magnet rotating within three coils oriented with 60° between their axes

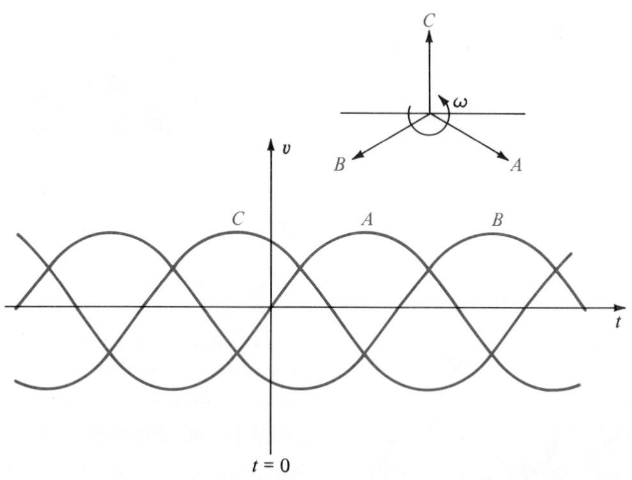

(b) Output voltage variations available under comparable conditions with leads of coil c reversed relative to a and b coils

FIGURE 19.26. Voltage Induced by a Rotating Magnet into Three Coils The coils are oriented at 60° relative to each other, with a variation dependent on their interconnection.

$$\mathbf{I}_{line} = \mathbf{I}_{phase} \tag{19.30}$$

The alternative way of connecting the coils is in a *delta* configuration (Figure 19.30). In this case the line voltages and phase voltage are the same, and the phasor diagram is as shown in Figure 19.31. The phasor sum around the delta loop equals zero. Thus, each phase supplies current to its own load, with no current circulating around the delta loop (but see below).

For a delta connection, we have

$$\mathbf{V}_{line} = \mathbf{V}_{phase} \tag{19.31}$$

but the line currents are not equal to the phase currents. This situation is illustrated in Figure 19.32. We have

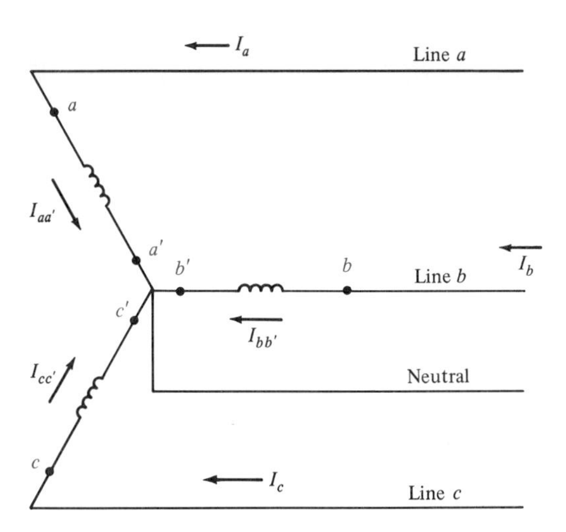

FIGURE 19.27. **Wye Three-Phase System**

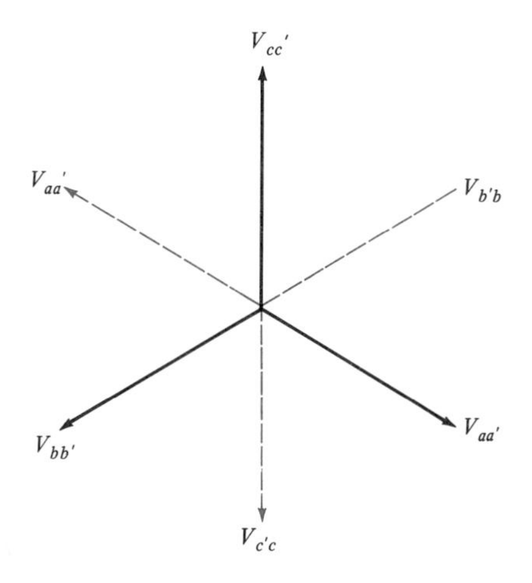

FIGURE 19.28. **Phase Voltages of a Wye System** Reversing the subscript ordering reverses the voltage polarity.

$$\mathbf{I}_{line} = \sqrt{3}\mathbf{I}_{phase} \angle -30° \tag{19.32}$$

A most fundamental question should now arise: When do we use a delta configuration and when a wye? In motors and generators the wye configuration is preferred since the smaller voltages across the coils (close to 75% smaller because of $\sqrt{3}$) ease insulation problems; however, it requires an additional line (the neutral). Thus, for transmission purposes the delta is preferred since the need for this fourth line is obviated. In terms of power transmission, the delta also has another advantage, the maintenance of a sinusoidal waveform.

Iron, of course, is used in electric machines, and the *B-H* curves are not linear. Sinusoidal *H* variations will lead to *B* variations with symmetrical depressions of their maximum values. Sinusoidal *B* variations will lead to symmetrical peaking of the *H* values. In our prior discussion of Fourier analysis, we have seen that such distorted sinusoids constitute third harmonic (and other odd harmonic) distortion. Therefore, the prior (cautious) statement that there are no circulating currents in a delta connection is not true if there

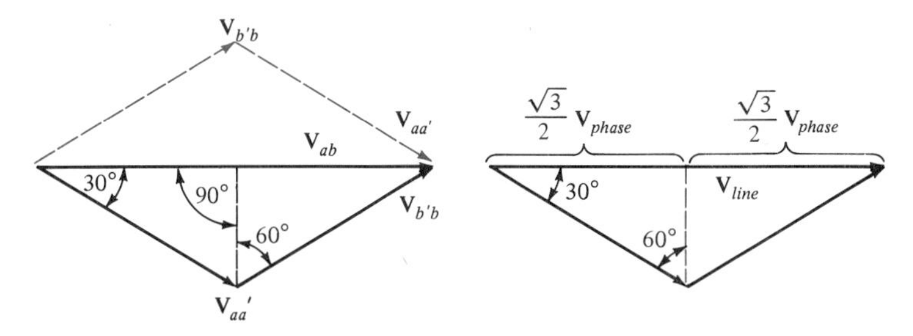

FIGURE 19.29. **Phasor Addition of Wye Phase Voltages ($\mathbf{V}_{aa'} + \mathbf{V}_{b'b}$) to Yield Line Voltage $\mathbf{V}_{ab}$**

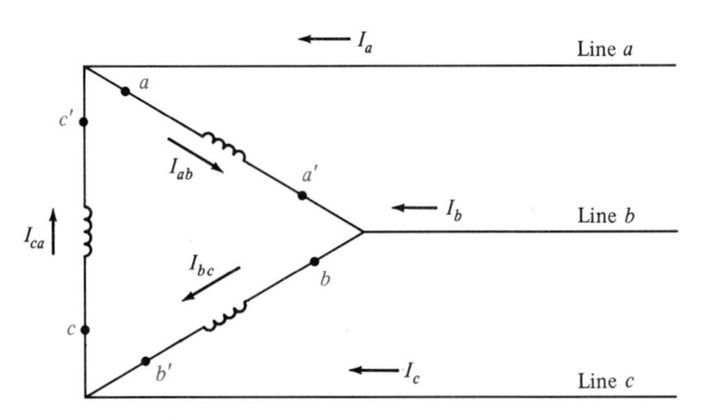

FIGURE 19.30. **Delta Three-Phase System**

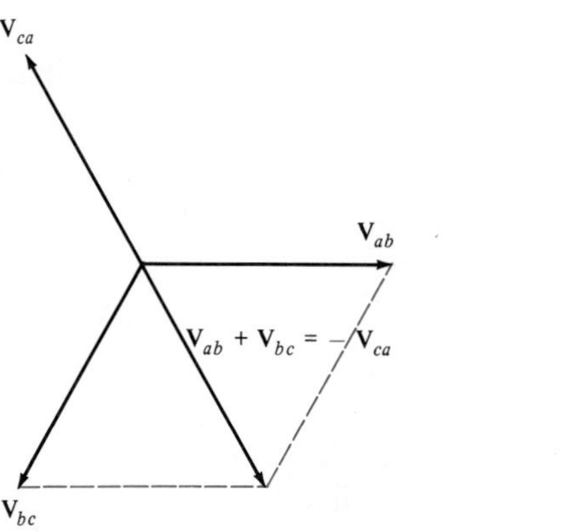

FIGURE 19.31. **Phasor Voltages for a Delta System** $\mathbf{V}_{ab} + \mathbf{V}_{bc} + \mathbf{V}_{ca} = 0$.

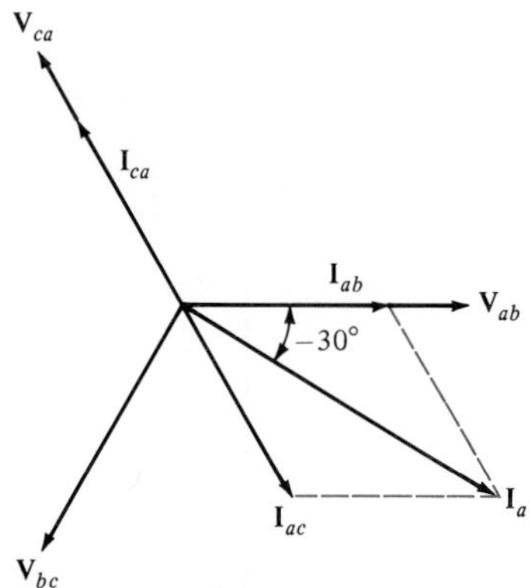

FIGURE 19.32. **Addition of Delta Currents ($\mathbf{I}_{ab} + \mathbf{I}_{ac}$) Yielding Line Current $\mathbf{I}_a$ for Resistive Phase Loads**

is distortion. These harmonic currents circulate around the delta, which acts like a trap, minimizing their appearance on the lines. Additionally, unbalanced loading of the individual phases tends to be equalized by a delta configuration. Thus, a common procedure is to use a wye generator, transform the power into a delta configuration for transmission, and then via a transformer return to a wye configuration at the destination.

EXAMPLE 19.3

A 220 V, 25 hp, 60 Hz, three-phase, six-pole induction motor (Figure 19.33) is characterized by the following parameters: $R_1 = 0.04\ \Omega$, $R_2 = 0.06\ \Omega$, $X_1 = 0.25\ \Omega$, $X_2 = 0.25\ \Omega$. The no-load power input is 700 W at a line voltage of 220 V, with an input current of 21 A. Calculate the adjusted voltage V_A. Repeat the calculations in phasor terms and show the difference to be small.

Solution: The phase voltage $V_P = 220\text{ V}/\sqrt{3} = 127$ V. Under no-load conditions, we have

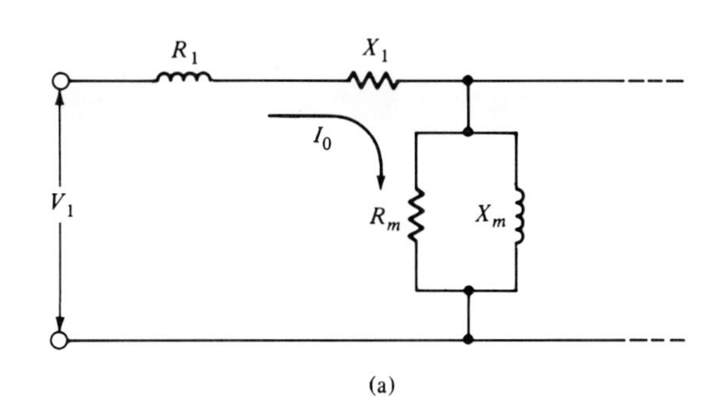

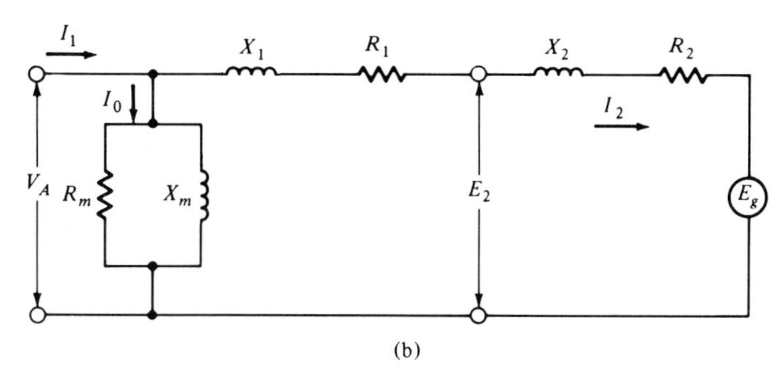

FIGURE 19.33

$$700\text{ W} = 3V_1 I_0 \cos\theta = 3\frac{220}{\sqrt{3}}21\cos\theta$$

$$\theta = \cos^{-1}\left(\frac{700}{3}\frac{\sqrt{3}}{220}\frac{1}{21}\right) = 85°$$

$$V_A = V_1 - I_0\sqrt{R_1^2 + X_1^2} = 127 - 21\sqrt{(0.04)^2 + (0.25)^2} = 122\text{ V}$$

In phasor terms, we have

$$R_1 + jX_1 = 0.04 + j0.25 \simeq 0.25 \angle 81°\ \Omega$$

$$\mathbf{V}_A = \mathbf{V}_1 - \mathbf{I}_0(R_1 + jX_1) = 127 \angle 0° - 21 \angle -85°(0.25 \angle 81°)$$
$$= 127 - 5.24 + j0.37 \simeq 122\text{ V}$$

EXAMPLE 19.4 For the motor in Example 19.3, estimate the starting torque as well as the maximum torque. What value of R_2 leads to a maximum starting torque?

Solution:

$$T_{start} \simeq \frac{3V_A^2 R_2}{\Omega_s X^2} = \frac{3(122)^2(0.06)}{40\pi(0.5)^2} = 85.3\text{ N}\cdot\text{m}$$

$$T_{max} = \frac{3V_A^2}{2\Omega_s(R_1 + \sqrt{R_1^2 + X^2})} = \frac{3(122)^2}{2 \times 40\pi(0.04 + \sqrt{0.04^2 + 0.5^2})}$$
$$= 328\text{ N}\cdot\text{m}$$

Using Eq. 19.26, with $s = 1$, we have

$$1 = \frac{R_2}{\sqrt{R_1^2 + X^2}}$$

$$R_2 = \sqrt{0.04^2 + 0.5^2} = 0.5\ \Omega$$

EXAMPLE 19.5 With a 4% slip, what are the output power and the input power of the motor in Example 19.2? What is the efficiency?

Solution:

$$\mathbf{I}_2 = \frac{V_A}{R_1 + (R_2/s) + jX} = \frac{122 \angle 0°}{0.04 + (0.06/0.04) + j0.5}$$

$$= \frac{122 \angle 0°}{1.54 + j0.5} = \frac{122 \angle 0°}{1.62 \angle 18°} = 75.3 \angle -18°\ \text{A}$$

$$\mathbf{I}_1 = \mathbf{I}_0 + \mathbf{I}_2 = 21 \angle -85° + 75.3 \angle -18°$$

$$= 1.83 + 71.6 - j20.9 - j23.3 = 85.7 \angle -31°\ \text{A}$$

The total power input is

$$3V_1 I_1 \cos\theta = 3(127)(85.7)(0.857) = 28.0\ \text{kW}$$

From Eq. 19.23, the total developed torque is

$$T_D = \frac{3I_2^2 R_s}{s\Omega_s} = \frac{3(75.3)^2(0.06)}{0.04(40\pi)} \simeq 203\ \text{N}\cdot\text{m}$$

The output power is

$$P_{hp} = \frac{T_D\Omega}{746} = \frac{203(1 - 0.04)40\pi}{746} = 32.8\ \text{hp}$$

$$\eta = \frac{\text{output}}{\text{input}} = \frac{32.8(0.746)}{28.0} = 0.87 \quad \text{or} \quad 87\%$$

19.5.2 Single-Phase Systems

Many small induction motors are classed as single-phase units. But we have indicated that the stator field of two-pole machines consists of two oppositely rotating vectors for which some method of establishing a unique sense of rotation must be provided. Previously, this unique direction was established by means of two additional poles, making it into a four-pole machine. In many instances this second pair of poles is not left permanently connected, and such motors can then be considered to be of single-phase form. A variety of starting methods is available in this conjunction:

1. *Split-phase motor.* The main winding has a low resistance and a high reactance. The starting winding has a high resistance and a low reactance. This combination gives rise to a phase angle of some 30°–40°, and this phase difference is sufficient to provide a starting torque. When the rotor reaches about 75% of its rated speed, a centrifugal

switch disconnects the starter winding. Such motors find use in fans, blowers, etc., and are rated up to about $\frac{1}{2}$ hp.

2. *Capacitor-start motor.* Here a second (starter) coil is energized through a capacitor, but once started, a centrifugal switch opens up the starter winding. Such motors can have ratings up to about 10 hp. The starter capacitance may be of the order of hundreds of microfarads and gives rise to a starting torque up to four times the rated torque. In some motors two different capacitors are used. The one providing the large starting torque is switched out after starting, leaving a smaller capacitor permanently connected for efficient operation under running conditions.
3. *Shaded-pole motor.* Very small, inexpensive single-phase induction motors truly use only one pair of coils. As shown in Figure 19.34, each of the **salient poles** (that's what those rounded pole ends are called) is slotted at opposite corners, and single turns of copper conductor are placed into these slots. As the magnetic field builds up in the gap, the resulting current flow in these single turns creates an opposing magnetic field. Therefore, the fields are greatest at the right of the upper pole and at the left of the lower one. But as the flux reaches its peak (and momentarily constant) intensity, with no flux change there is no longer any opposition field. The net effect is the production of a stator field that moves in a counterclockwise direction (as depicted in Figure 19.34) and gives rise to a unique starting direction. Such shaded-pole motors are rated up to $\frac{1}{20}$ hp.

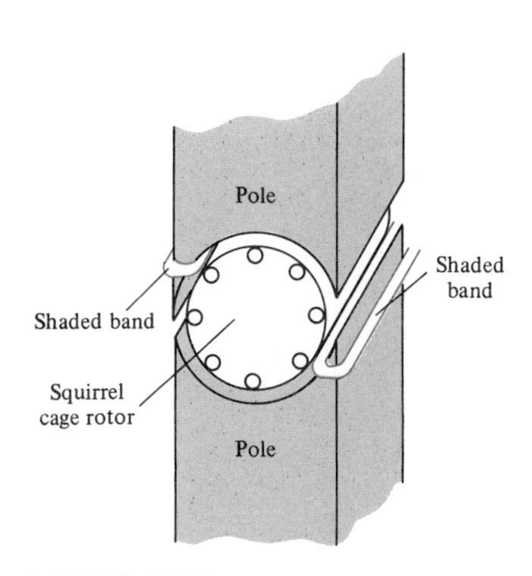

FIGURE 19.34. **Shaded-Pole Motor**

19.6 ▪ HYSTERESIS MOTORS

Assume we have a stator field rotating at velocity ω and a solid rotor with slip s (thus rotating at velocity $s\omega$). If there were no hysteresis, the rotor magnetization would be in phase with the exciting field that creates it. Thus, in Figure 19.35a the magnetic axis of the rotor (*A-C*) would coincide with that of the stator field (*B-D*).

But if the stator field has a higher velocity than that of the rotor, how can the two fields remain synchronized? Note that nothing was said about the *magnetization* of the rotor having to be synchronized with the rotor's rotation. The direction of the rotor magnetization can move relative to the rotor itself.

If there is hysteresis present, the rotor magnetization will lag behind the magnetizing force (the stator field), and thus *A-C* (the axis of the rotor magnetization) lags behind the stator field by an angle θ_0 (Figure 19.35b). That a torque is thereby exerted on the rotor can easily be visualized by again thinking of the magnetic flux lines as rubber bands that attempt to assume linear extension. Such a visualization leads to the conclusion that no torque would prevail in the case illustrated in Figure 19.35a.

It has previously been noted that the area of the hysteresis curve represents the energy expended in altering the magnetic domains during one cycle. The rotor's magnetization, cutting across the stator coils, gives rise to an induced counter-emf E:

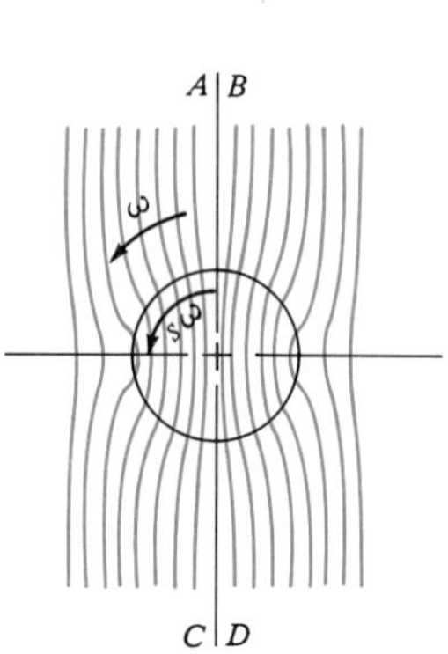

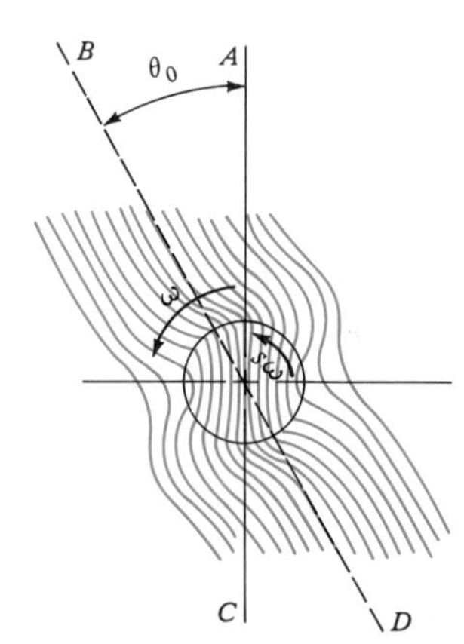

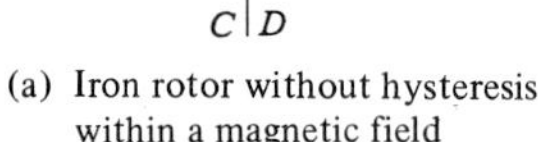

(a) Iron rotor without hysteresis within a magnetic field

(b) Iron rotor with hysteresis within a magnetic field, with θ_0 representing the angular lag between the stator field and the rotor magnetization

FIGURE 19.35. **Field and Rotor Magnetization in Hysteresis Motors**

$$E = N\left(\frac{d\Phi}{dt}\right) \tag{19.33}$$

where Φ = magnetic flux

N = number of turns

If we consider the flux passing through a unit area of the rotor surface, that constitutes the flux density (B), and with the flux varying in sinusoidal fashion, the emf induced into one turn of the stator winding will be

$$e = \frac{d}{dt}(B_m \sin \omega t) = \omega B_m \cos \omega t \tag{19.34}$$

The rms value of the induced emf will be $\omega B_m/\sqrt{2} = 2\pi f B_m/\sqrt{2}$.

The magnetic intensity H_f can be split into two parts: H_f'' is in phase with B, and H_f' is the out-of-phase component. It is this out-of-phase component that gives rise to the torque.

The current i passing through the stator coils in opposition to the counter-emf gives rise to a magnetomotive force (mmf), a portion of which appears as a mmf "drop" ($\mathfrak{F}$) across the rotor. We have

$$\mathfrak{F} = Ni = H_f l \tag{19.35}$$

where H_f is the magnetic intensity (per unit path length in the rotor). Its rms value is $H_f/\sqrt{2}$. Thus, the electrical power converted into mechanical power (per cubic inch of rotor material)[4] is

[4]There is still considerable opposition in magnetics toward the utilization of the meter as a unit of length. Thus, in the case of hysteresis rotors, one will encounter power densities specified as being of the order of 30 W/in.^{3}; translated into strictly SI units, this value amounts to 1.8 MW/m^3! In terms of practical rotors, the former would appear to be the more convenient number to use. It constitutes an example of *English* units. In other cases CGS units, such as the gauss and the oersted, are still used.

$$P = \left(\frac{2\pi f B_m}{\sqrt{2}}\right)\left(\frac{H_f}{\sqrt{2}}\right) \sin \theta_0 \quad (19.36)$$

Thus, the first factor represents voltage (counter-emf), the second represents a current, and the third is the phase angle between them. (See Figure 19.36.)

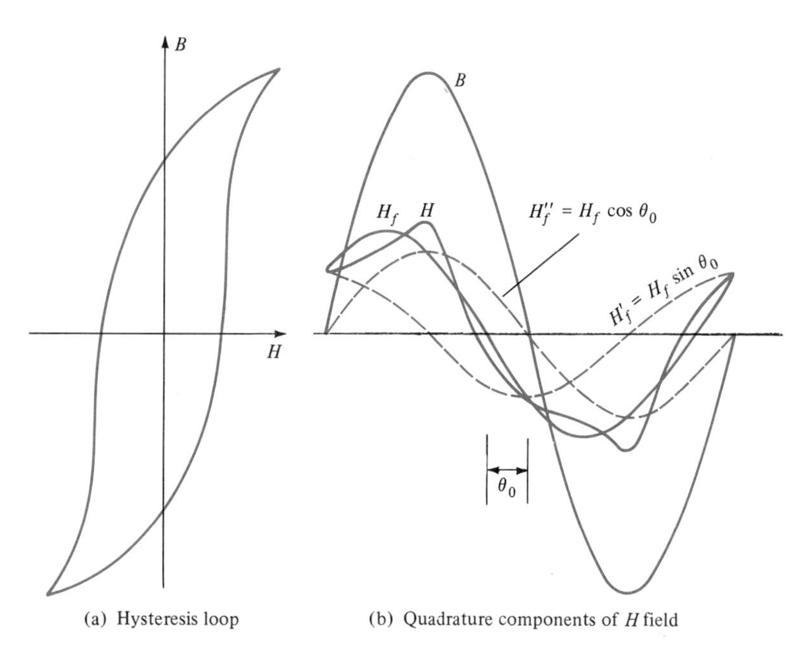

FIGURE 19.36. **Quadrature Magnetic Components in Hysteresis Motors**

Alternatively, since the area within a hysteresis loop constitutes energy, for which a value is generally known for a specific magnetic material, we have

$$P = fVW_h = 2\pi nT \quad (19.37)$$

where W_h = energy in joules per cubic inch per hertz
V = rotor volume in cubic inches
f = applied frequency

This power is set equal to the torque (T) times the rotational velocity of the field ($2\pi n$). Solving for the torque, we have

$$T = \frac{fVW_h}{2\pi n} \times 8.86 = \frac{1.41 fVW_h}{n} \text{ [inch-pounds]} \quad (19.38)$$

$$= \frac{22.6 fVW_h}{n} \text{ [inch-ounces]}$$

Of this hysteresis power transmitted to the rotor, some, proportional to the slip, appears as heat in the rotor. The remainder, proportional to the rotor speed, appears at the shaft as mechanical work. But among the reasons why

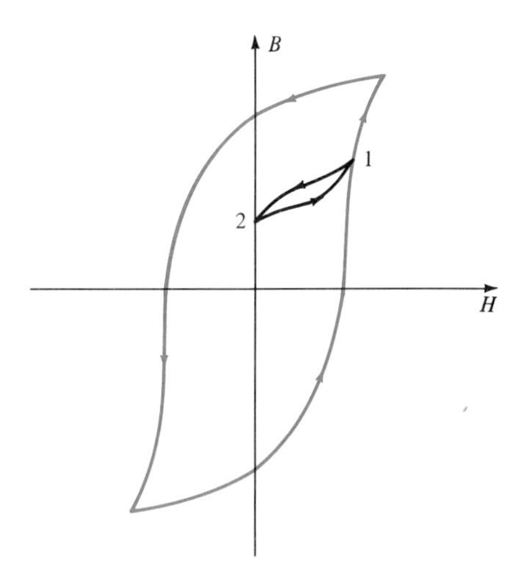

FIGURE 19.37. **Hysteresis Loop** The superimposed minor loop is due to an assumed variation of flux density as the rotor passes an open stator slot.

the performance may deviate from this ideal arises because of the effect of slot openings.

With the stator windings inserted into slot openings, while the separation between the rotor and the solid teeth of the stator may be small, it is considerably increased between the rotor and the open slot gap, and this widening results in a minor hysteresis loop being superimposed on the major loop. Such minor loops constitute rotor heat loss and show up as a mechanical drag on the rotor (Figure 19.37). Under synchronous conditions there will be as many such minor loops per power supply cycle as there are teeth per pair of poles, and the power loss due to slot openings will be proportional to the minor loop area times the number of teeth per pair of poles.

In some cases where the slot openings are wide and the air gap small, the total area of the small loops may exceed that of the major loop. When this condition prevails, the rotor will slip below synchronism. (Equilibrium below synchronism is possible with a solid rotor due to an increase in the torque produced by eddy currents in the rotor.) One way of minimizing such a condition is to use a wider air gap between rotor and stator. The wider gap smooths out the flux density variation, again allowing synchronous operation, but this improvement is achieved at the expense of increased copper losses and reduced efficiency. A better way is to close the slots. Such closed slots in conventional induction and synchronous motors would not be desirable because of a reduction in the power factor caused by shunting the magnetic field to adjoining teeth, away from the rotor gap. In the case of the hysteresis motor, the advantage of slot closing can be made to outweigh the disadvantage.

Another interesting aspect of hysteresis motors involves their behavior under load. Started without a load, $\theta_0 \simeq 0$, and the rotor will acquire its maximum magnetization under the prevailing excitation conditions. Subsequently loaded, it can carry a greater load before falling out of synchronization than it can if started under load. For this reason, accurate quantitative results, depending as they do on the state of rotor magnetization, are difficult to predict when operating in a synchronous manner.

To summarize, the hysteresis motor is self-starting and can reach synchronous speed within a few cycles of the power-line frequency. Such performance makes it particularly useful in applications such as phonograph turntables, magnetic tape drives, etc. At synchronous speeds, by proper design, a considerable fraction of the hysteresis power may be converted into mechanical power. As the mechanical load is increased, the lag angle increases until the sine of the angle reaches unity. Further loading will cause the rotor to slip, but such slip will create eddy currents along the surface of the rotor, thereby giving rise to additional magnetic flux and torque. The motor thus continues to rotate, but at less than synchronous speed. Further loading creates greater slip, more eddy currents, and greater torque to take up the added loading. The hysteresis torque is independent of speed. The eddy current torque is proportional to the slip.

EXAMPLE 19.6

Show that both hysteresis and eddy currents make contributions to the acceleration of a hysteresis motor but that only the hysteresis effect makes a contribution at synchronous speeds.

Solution: Eddy current losses are expressible in the form

$$P_e = K_e f_2^2 B_m^2$$

where f_2 = eddy current frequency
B_m = maximum flux density
K_e = a constant

The rotor frequency (f_2) is related to the stator frequency (f_1) through the slip. We have

$$f_2 = sf_1$$

Therefore, we have

$$P_e = K_e s^2 f_1^2 B_m^2$$

Setting eddy current power equal to mechanical power, where ω_2 is the rotor velocity and T_e the eddy current torque, and since $\omega_2 = s\omega_1$, we have

$$T_e = \frac{P_e}{\omega_2} = sK'$$

where $K' = K_e f_1^2 B_m^2 / \omega_1$.

For the hysteresis loss, we have

$$P_h = K_h f_2 B_m^{1.6} / \omega_1$$

and the corresponding torque can be expressed as

$$T_h = K''$$

where $K'' = K_h f_1 B_m^{1.6} / \omega_1$.

Thus, we see that T_e is proportional to slip, and its contribution diminishes as the rotor accelerates, becoming zero at synchronous speeds. T_h, on the other hand, remains constant at all speeds and is the only torque-producing effect at synchronous speeds.

19.7 ■ MINIATURE dc MOTORS

There has recently been a resurgence of interest in dc motors, primarily those of the miniature variety that are finding wide use in toys, portable electric shavers, tape recorders and players, and Polaroid cameras. Their supply voltages may lie in the range from 1.5 V (a single-cell battery) to 12 V (a car battery). The range of delivered power may vary from a few tenths of a watt (toys and tape players) to a few watts (shavers). The required motor life may be as much as several thousand hours (for tape units).

In the design of such motors, the two most important aspects are their efficiency and their lifetime. Their lifetime is primarily determined by the lifetimes of the commutator and brushes.

For use in toys, low unit cost is most important and efficiency, secondary. Because batteries represent a high cost per kilowatt hour, in equipment such as tape units and shavers, efficiency is most important, and greater unit cost can be tolerated. Where equipment operates from rechargeable nickel-cadmium cells, efficiencies may also be important if long operation per charging cycle is desired.

Figure 19.38 shows the cross section of such a dc motor. The shaded portion is a magnetized plastic ring—that is, magnetic material embedded in a plastic binder. The rotor is a three-slot laminated iron core that will contain three coils each wound about a rotor tooth that has a width b. The outermost ring is steel and conducts magnetic flux between the magnetic poles. The three windings have their leads brought out to a three-segment commutator (Figure 19.39). At any given time the applied voltage appears across one coil as well as the other two in series (Figure 19.40).

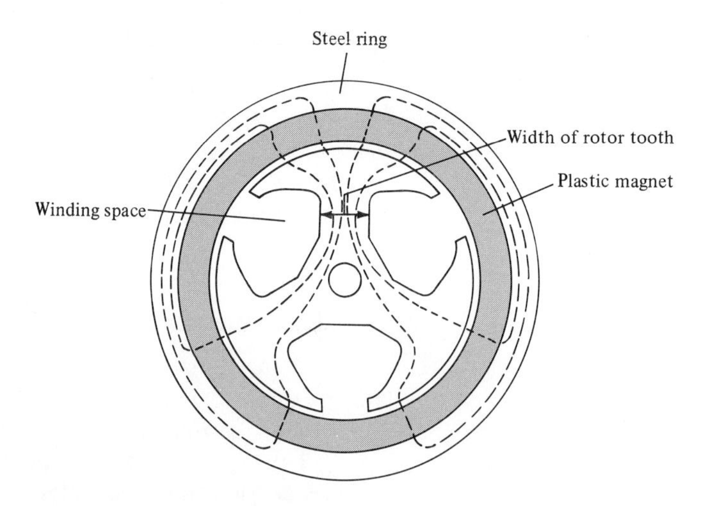

FIGURE 19.38. Cross Section Through a Small dc Motor Showing Magnetic Flux Lines

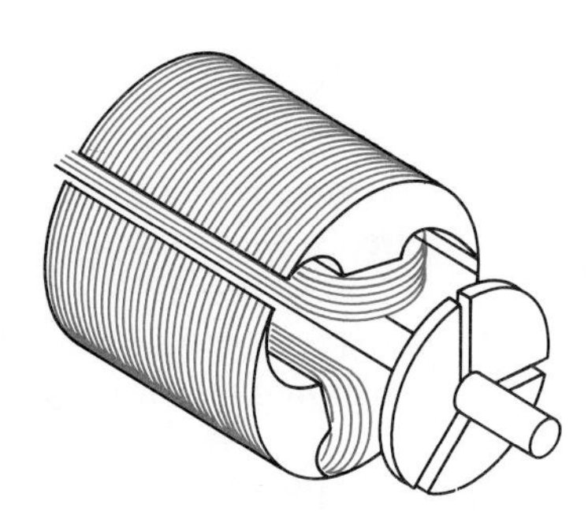

FIGURE 19.39. Sketch of a Small dc Motor Showing the Trisectioned Commutator

There are two means by which motor losses can be classified:

1. They can be classified as electrical and mechanical. The electrical losses arise from rotor and commutator resistance as well as eddy currents and magnetic hysteresis. Mechanical losses arise because of brush friction, bearings, and air resistance.
2. An alternative classification of losses makes use of those that prevail when the motor is energized but prevented from turning—in other words, stationary losses—and those that arise when the motor is driven by an external drive in such a manner that the generated counter-emf (E) and the applied voltage (V) are equal. This classification splits the losses into those that are due to conduction and those due to iron losses and friction. This second method of classifying losses is particularly convenient in calculating efficiency since the loss is approximately proportional to motor speed.

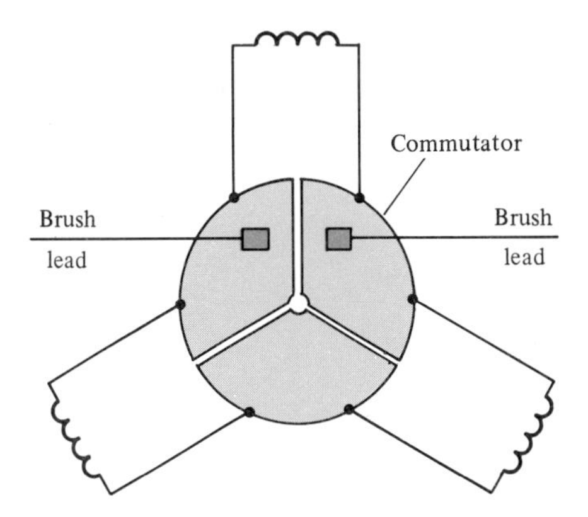

FIGURE 19.40. **Small dc Motor Windings, Their Connection to the Trisectioned Commutator, and Brush Locations**

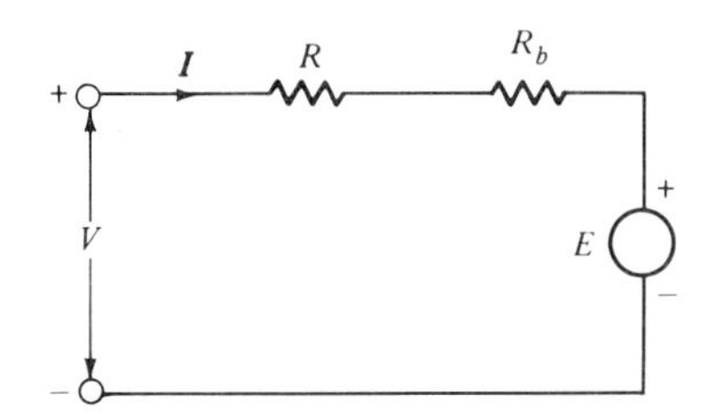

FIGURE 19.41. **Electrical Equivalent of dc Motor** *R* is the winding resistance, R_b is the brush resistance, and *E* is the developed counter-emf.

Figure 19.41 shows the equivalent circuit for the dc motor. V is the supply voltage, I is the supply current, R is the diametral rotor resistance (that is, the resistance of one coil in parallel with the other two that are in series), R_b is the commutator resistance, and E is the back-emf induced into the coil rotors. It is E that represents the mechanical output. We have

$$V = E + I(R + R_b) \tag{19.39}$$

The power P_i supplied is VI, so that

$$P_i = EI + I^2(R + R_b) \tag{19.40}$$

The conduction losses are $I^2(R + R_b)$, EI is the electrical power that is converted into mechanical power, and P_e is termed the *electromechanical power* (also being equal to the angular velocity times the torque).

The electrical efficiency η_e is

$$\eta_e = \frac{P_e}{P_i} = \frac{V - I(R + R_b)}{V} = \frac{E}{V} \tag{19.41}$$

It is shown in "A Closer Look" at the end of this chapter that the counter-emf is

$$E = 2nN\Phi_{max} \tag{19.42}$$

where n = revolutions per second

N = total number of turns on the three coils

Φ_{max} = maximum flux through a rotor tooth

The electromechanical power is EI in electrical terms and ΩT_e in mechanical terms, where Ω is the angular rotational velocity and T_e the generated torque. We have

$$P_e = 2\pi n T_e = EI = 2nNI\Phi_{max} \tag{19.43}$$

$$T_e = \frac{NI\Phi_{max}}{\pi} \tag{19.44}$$

Not all of this torque is available at the shaft, there being a loss torque T_l due to friction, etc. This braking torque may be considered to remain constant over a limited range of speeds. We have

$$T_d = T_e - T_l \tag{19.45}$$

The mechanical power is

$$P = 2\pi n T_d \tag{19.46}$$

The mechanical efficiency is

$$\eta_{mech} = \frac{T_d}{T_d + T_l} \tag{19.47}$$

and the overall efficiency is

$$\eta = \frac{T_d}{T_d + T_l} \frac{V - I(R + R_b)}{V} \tag{19.48}$$

The efficiency varies with the delivered torque and the supply voltage V.

Proceeding to eliminate I in Eq. 19.48 by substitution, we obtain

$$\eta = \frac{T_d}{T_d + T_l} - \frac{\pi T_d (R + R_b)}{N\Phi_{max} V} \tag{19.49}$$

This equation may be simplified by introducing a number of parameters: The motor can be driven externally at a speed n_0 such that the induced voltage E equals the applied voltage V. We have

$$V = 2n_0 N\Phi_{max} \tag{19.50}$$

This condition can be recognized by the absence of any current being drawn by the motor—all the mechanical power being put in by the driving system (P_0) is being consumed in iron and frictional losses. We have

$$P_0 = 2\pi n_0 T_l \tag{19.51}$$

If, on the other hand, the shaft is held stationary and a voltage V applied, all the power P_s supplied electrically will be dissipated in the resistance. We have

$$P_s = \frac{V^2}{R + R_b} \tag{19.52}$$

We then define G as follows:

$$G \equiv \frac{P_0}{P_s} = \frac{\text{rotational losses}}{\text{electrical losses}} = \frac{2\pi n_0 T_l (R + R_b)}{V^2} \tag{19.53}$$

By using the results of Eqs. 19.50 and 19.53, the expression for efficiency becomes

$$\eta = \frac{T_d}{T_d + T_l} - G\frac{T_d}{T_l} \tag{19.54}$$

Introducing a second dimensionless parameter, we have

$$a \equiv \frac{T_d}{T_l} = \frac{\text{delivered shaft torque}}{\text{torque loss}} \qquad (19.55)$$

$$\eta = \frac{a}{1 + a} - Ga \qquad (19.56)$$

Figure 19.42 gives the efficiency as a function of a, with G as a parameter. The larger the value of G, the greater the possible efficiency. If Eq. 19.56 is differentiated with respect to a, for any given value of G the maximum efficiency can be shown to prevail at a load such that

$$a_{opt} = \frac{1}{\sqrt{G}} - 1 \qquad (19.57)$$

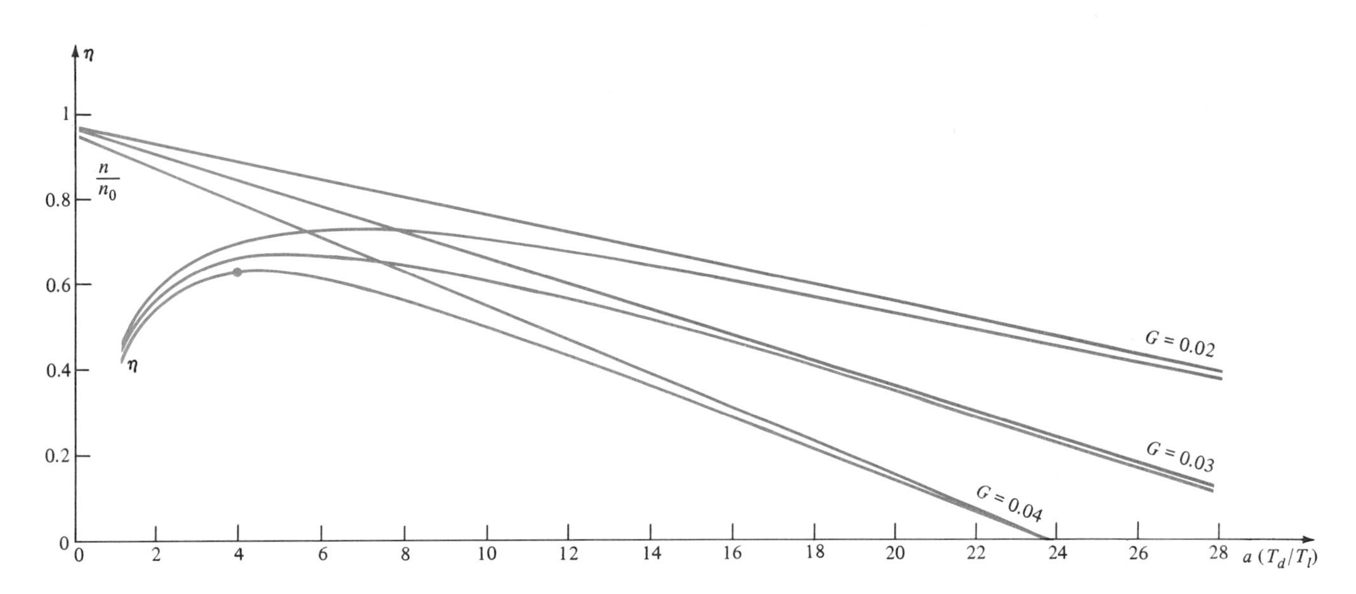

Curved lines: efficiency (η) as a function of a (T_d/T_l); the parameter values on these curves is the value of G

Straight lines: speed-torque characteristic; n = speed, n_0 = reference speed

FIGURE 19.42. **dc Motor Characteristics**

The maximum efficiency is then

$$\eta_{max} = (1 - \sqrt{G})^2 \qquad (19.58)$$

With that optimum load the electrical and mechanical efficiencies are equal.

What sort of efficiency should we strive to achieve? Figure 19.43 shows the power to be supplied for an output of 1 W, as a function of η. At higher efficiencies the proportional increase in power per unit change in η is much smaller than at low values of η. Additionally, with the present state of the art, it is difficult to develop $\eta > 0.8$ even if one is willing to undertake the higher cost. As a result an efficiency of 80% should be considered a practical maximum. Aside from energy conservation reasons, in favor of a high efficiency is the observation that the efficiency remains high over a wide range of a. This insensitivity allows the motor manufacturer to satisfy a wide range of requirements with a fewer number of designs.

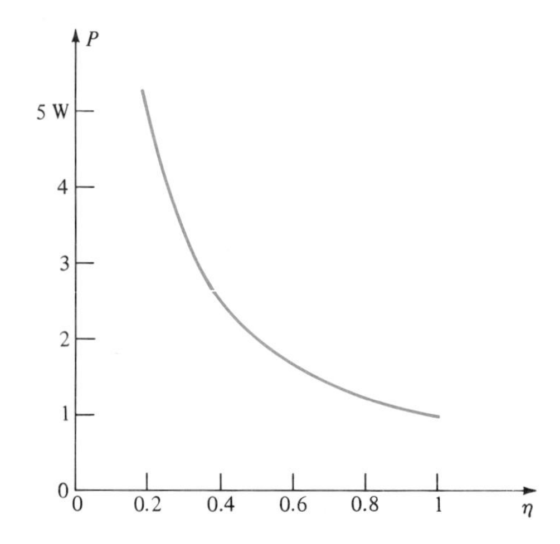

FIGURE 19.43. **Power to Be Supplied to Achieve 1 W of Output Power for Various Values of Efficiency for a Small dc Motor**

Like the efficiency, the motor speed (or, more accurately, the speed relative to the reference speed n_0) can be expressed as a function of a and G. We have

$$\frac{n}{n_0} = \frac{E}{V} = \frac{V - I(R + R_b)}{V} = \eta_e \tag{19.59}$$

Eliminating I as before, we have

$$\frac{n}{n_0} = 1 - G(a + 1) \tag{19.60}$$

This function is also plotted in Figure 19.42 (the straight lines). These lines are the speed-torque characteristics. It can be seen that the speed falls off faster as a function of the load as G increases.

EXAMPLE 19.7 Discuss a practical example of small dc motor design.

Solution: We consider an electric toothbrush design due to R. H. Dijken of the Philips Co., Eindhoven, Netherlands. (See his "Designing a Small dc Motor," *Philips Technical Review*, vol. 35, pp. 96–103, 1975.)

The practical limit to the external motor diameter was about 26 mm (about 1 in.). It was desirable to magnetize the rotor teeth to saturation, which amounted to 1.5 T. To produce this flux with optimum dimensions of the rotor teeth required a remanence of at least 0.4 T within the plastic magnetic material. The best that could be done was 0.25 T, and this limitation led to a less than optimum design since the additional flux had to be provided by a coil volume larger than that dictated by the optimum dimensions.

To accommodate the necessary magnetic flux, the outer steel band of the motor had to be at least 1.5 mm thick, the magnetic ring had to be at least 2.8 mm thick (for mechanical strength), and the air gap was 0.3 mm (limited by machining tolerances). Thus, the maximum diameter available for the rotor was 16.8 mm.

The most important dimension is the rotor diameter d. Conveniently, the rotor tooth width (b) and the motor length (l) are expressed in terms of d:

$$\beta = \frac{b}{d} \qquad \lambda = \frac{l}{d}$$

The variation of rotor volume can be studied in terms of β and λ. For a three-toothed rotor, the volume shows a minimum at $\beta \simeq 0.4$. (In most existing small motors, this value is considerably smaller, $\beta = 0.15$ being typical.) The optimum value for λ is 0.2–0.3. (For existing dc motors the value of λ is generally between 0.5 and 2.0.)

To squeeze in the required number of turns using a wire gage sufficient to handle the power, the best that Dijken could do with respect to the tooth width was $\beta = 0.22$. With this limitation he had to go to a rotor greater in length than optimum to get the required flux. The rotor length was 10 mm, making $\lambda = 0.60$.

The motor, designed for $V = 4.8$ V and delivering a power of 1 W at 6500 rpm, resulted in value of $G = 0.04$, $a = 4$, and an efficiency of 64%.

SUMMARY

19.1 Introduction

- Current in a conductor, the direction of the magnetic field, and the magnetic force acting on the conductor act in mutually perpendicular directions.
- Alternating current applied in series with two mutually orthogonal paired coils will create an oscillating magnetic field at 45° to the coil normals. If applied to one pair 90° out of phase relative to the other, a rotating magnetic field is created.

19.2 Synchronous Motors

- A stator field (created by applying ac to stationary coils) rotating at the same velocity as the magnetized rotor constitutes a synchronous motor. As the motor load increases, the angle between the stator and rotor fields (termed the torque angle) increases, up to a maximum of 90°.

19.3 Induction Motors

- In an induction motor the rotor coil has an induced voltage that creates a current that leads to a rotor field that interacts with the rotating stator field. The rotor rpm of an induction motor is always less than the rpm of the rotating stator field.

19.4 Induction Motor Theory

- For starting an induction motor, a high rotor resistance is desirable. A low rotor resistance is desirable under normal operating conditions.

19.5 Induction Motor Practice

- Three-phase systems result in a more efficient transfer of power than that obtained from single-phase systems.
- Two common three-phase connections are the wye system and the delta system.
- To establish a unique sense of rotation in single-phase motors, one may resort to using either a split-phase winding, a capacitor-start, or a shaded-pole.

19.6 Hysteresis Motors

- Hysteresis motors are synchronous motors that are self-starting and, under heavy load, continue to operate at below synchronous speeds, unlike the conventional synchronous motors that are not self-starting and stall under excessive loads.

19.7 Miniature dc Motors

- The design of miniature dc motors is primarily dictated by either the desired operating efficiency or the desired low cost.
- The lifetime of a miniature dc motor is primarily determined by the ruggedness of the commutator and brushes.
- Motor losses may alternatively be classed as either mechanical and electrical or as pertaining to stationary and rotating performance.

KEY TERMS

alternator: the mechanical rotation of a magnetized rotor induces into the stator windings an ac voltage that constitutes the electrical output

asynchronous machine: one whose rotational speed is slower than that of the stator field

commutator: a segmented slip ring that converts the ac output of a rotating machine into a unidirectional output current or, alternatively, when used with a motor converts dc excitation current into ac, thereby leading to a unidirectional torque

generator: by means of a commutator, a machine that converts rotary mechanical motion into a (pulsating) dc output voltage

hysteresis motor: a motor whose rotor magnetization at synchronous speeds arises from a lag between the peak of rotor magnetization and the peak inducing stator field (Under high load conditions it may continue to develop torque but in an asynchronous manner, with the additional torque arising from eddy current magnetization.)

induction motor: one whose rotor magnetization is (either fully or partially) created by the rotating stator field

kilovolt-ampere (kVA): power unit used in conjunction with apparent power

kilovolt-ampere, reactive (kVAR): power unit used in conjunction with reactive power

kilowatt (kW): power unit used in conjunction with real power

line voltage: the voltage between any pair of power lines

overexcitation: a rotor field magnitude that gives rise to an output with a leading power factor

phase voltage: the voltage between any power line and the neutral wire

salient pole: a stator pole that is rounded in such a manner as to encompass a significant portion of the rotor surface

slip (s): $1 - (\Omega/\Omega_s)$, where Ω_s is the stator rotational velocity and Ω the rotor velocity (At synchronous speeds $\Omega = \Omega_s$ and $s = 0$; under starting conditions $\Omega = 0$ and $s = 1$.)

slip ring: a rotating contact that allows electrical contact with a rotor winding

synchronous machine: one whose rotational speed coincides with that of the stator field

torque angle (θ_0): the angle between the stator field direction and the direction of the rotor field (The torque varies as $\sin \theta_0$.)

SUGGESTED READINGS

Schweitzer, Gerald, FRACTIONAL HORSEPOWER MOTORS AND REPAIR. Rochelle Park, N.J.: Hayden Book Co., 1960. *For practical considerations of various small motors (shaded pole, universal, repulsion, etc.).*

Nasser, S. A., and L. E. Unnewehr, ELECTROMECHANICS

AND ELECTRIC MACHINES. New York: Wiley, 1979. *Basic concepts and laws governing rotating machines.*

Smith, Ralph J., CIRCUITS, DEVICES AND SYSTEMS, 4th ed., Chaps. 23 and 24. New York: Wiley, 1984. *dc and ac machines.*

Kamerbeek, E. M. H., "Electric Motors," PHILIPS TECHNICAL REVIEW, vol. 33, pp. 215–234, 1973. *Our treatment of synchronous motor theory follows that found here.*

Dijken, R. H., "Designing a Small dc Motor," PHILIPS TECHNICAL REVIEW, vol. 35, pp. 96–103, 1975. *Our treatment of small dc motors follows that found here.*

Roters, Herbert C., "The Hysteresis Motor—Advances Which Permit Economical Fractional Horsepower Ratings," AIEE TRANSACTIONS, vol. 66, pp. 1419–1427, 1947. *Detailed designs of hysteresis motors.*

A CLOSER LOOK: Counter-emf in Miniature Motors

Assuming the flux through one rotor tooth varies in a sinusoidal manner as it passed by a magnetic pole, the average emf is

$$E = \frac{3}{\pi}\int_{\pi/3}^{2\pi/3} e_{max} \sin\theta \, d\theta = \frac{3}{\pi} e_{max}$$

θ is the position of the rotor relative to the stator. The voltage induced into *one* rotor coil is

$$e = -\frac{N}{3}\frac{d}{dt}(\Phi_{max}\cos\theta)$$

$$= -\frac{N}{3}\frac{d}{dt}(\Phi_{max}\cos 2\pi nt)$$

$$e_{max} = 2\pi n\frac{N}{3}\Phi_{max}$$

from which we get the overall speed voltage (see Figure 1).

$$E = 2nN\Phi_{max}$$

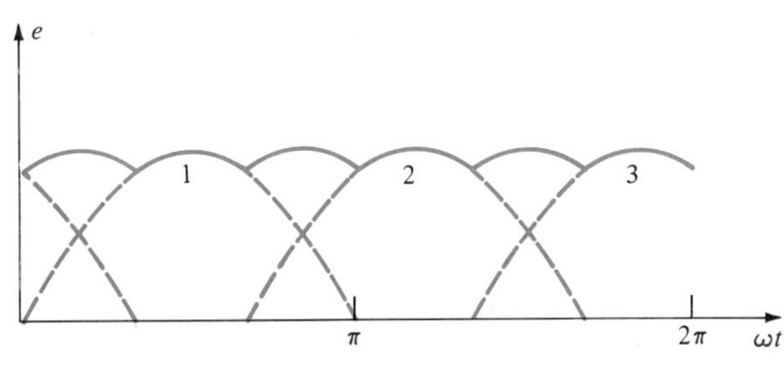

FIGURE 1

EXERCISES

1. **a.** Since we have two expressions for induced emf, $L(di/dt)$ and $N(d\Phi/dt)$, how is L related to flux and current?

   b. Dimensionally, to what does 1 H correspond?

   c. What is the proportional relation between inductance and the number of coil turns?

2. If, as a result of an excessive load, an electric motor is stalled, why is there a large increase in the current drawn?

3. What is the nature of the magnetic field created by two sets of coils with perpendicular orientation if both are excited by the same ac? If the applied ac of one is 90° out of phase with that applied to the other?

4. For a synchronous motor, what is the relationship between the rotor and stator field positions? What is their difference called? What happens to this parameter as the load increases?

5. Why is a synchronous motor not self-starting?

6. Why would there be no torque developed by an induction motor running at synchronous speed? In a practical case, does an induction motor ever run at synchronous speed?

7. Explain why the squirrel-cage rotor of the form shown in Figure 19.21a exhibits large starting resistance and low running resistance.

8. What are the two alternative ways of connecting coils in a three-phase system?

9. What constitutes the phase voltage of three-phase systems?

10. In a wye-connected 440 V three-phase system, what is the voltage across each coil? In a delta system?

11. What is the nature of the phase and line currents in a wye system? In a delta system?

12. Why is a delta three-phase system preferred over a wye system?

13. Discuss the operational methods of the following single-phase motors: split-phase, capacitor-start, shaded-pole.

14. When the stator field of a hysteresis motor has a higher rotational velocity than the rotor, how can the rotor field be synchronized with the stator field?

15. The magnetic intensity of a hysteresis motor can be split into two parts, one in phase with **B** and the other out of phase. Which component is responsible for the developed torque?

16. In Eq. 19.38, distinguish between f and n. Why does the torque increase with diminished n?

17. What are two important parameters in the design of small dc motors?

18. What primarily determines the lifetime of small dc motors?

19. Name two means by which motor losses may be classified?

20. What is a practical limit on the efficiency of small dc motors?

PROBLEMS

19.1
Calculate the force on a wire 1 m in length carrying 1.0 A at right angles to a flux density of 0.5 Wb/m^2. (For comparison purposes, 0.5 Wb/m^2 is about 10,000 times greater than the flux density due to the earth's magnetic field.)

19.2

a. Justify that the weber is equal to a volt-second.

b. Justify the permeability unit being henry per meter.

19.3
We have considered an alternating magnetic field as equivalent to two magnetic fields rotating in opposite directions. Mathematically, we have

$$B(\theta,t) = B_m \cos\theta \sin\omega t$$

where ω represents the frequency of the applied current and θ the starting reference angle. Using a trigonometric expansion, show that the preceding equation *does* lead to two oppositely rotating vectors each of magnitudc $B_m/2$.

19.4
In terms of **E**, **V**, and the voltage drop $j\mathbf{I}X_s$, show by means of a phasor diagram the existence of leading and lagging power factors as the excitation voltage of a synchronous machine is varied. (For the sake of clarity, assume an idling condition, with **E** and **V** in synchronism.)

19.5
Show by means of phasor diagram sketches that a synchronous generator may be made to exhibit leading and lagging power factors as $E < V$ and $E > V$, respectively. Show that for synchronous motors the reverse situations prevail.

19.6
Since $E \propto d\Phi_R/dt$ and $V \propto d\Phi_s/dt$, Φ_R and Φ_s being, respectively, the rotor flux and the air gap flux, show that the angle between Φ_s and Φ_R is the same as that between E and V—that is, the torque angle.

19.7
A three-phase, Y-connected, six-pole alternator is rated at 20 kVA, 220 V, 60 Hz. With negligible series resistance and X_s = 1.0 Ω, what is the necessary generated emf E for an 0.9 lagging power factor. Sketch the phasor diagram. By what means can the value of E be adjusted?

19.8
Consider a 750 kVA induction furnace with a lagging power factor of 0.7 to be used in parallel with a 1000 kVA synchronous motor. If the combined power factor is unity, what useful mechanical load can be carried by the machine?

19.9
The relationship between the torque angle (θ_0) and the phase angle (θ) can be seen from Figure 19.44 to be

$$E \sin\theta_0 = IX_s \cos\theta$$

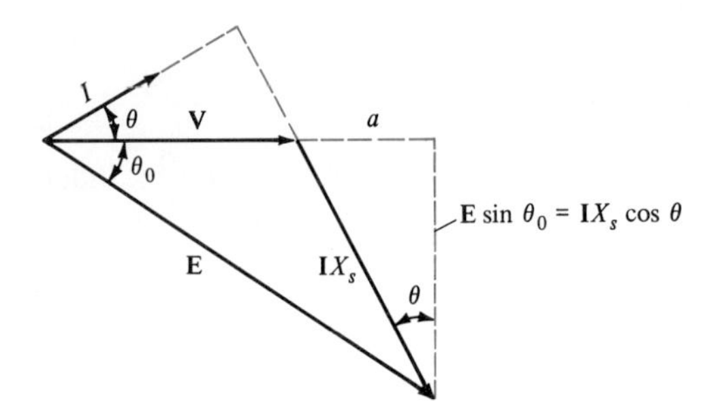

FIGURE 19.44

a. A three-phase, Y-connected synchronous motor with X_s = 1.27 Ω/phase develops 30 kW of power at a torque angle of 25°. Neglecting resistance, calculate the reactive kVA of the motor. The line voltage is 220 V.

b. Across the synchronous motor of part a there is a Y-connected load at a lagging power factor of 0.707, drawing 50 A. What will be the overall power factor of the motor and load?

19.10

How many poles would be needed for an induction motor to operate at 1000 rpm if the applied frequency were 2000 Hz?

19.11

If $R_1 = 0.1\ \Omega$ and $X_1 = 0.15\ \Omega$, for a 220 V, three-phase, Y-connected motor, with a no-load input power of 1000 W and a no-load line current of 20 A, what value should be used for the adjusted voltage V_A? (Check your results in phasor terms.) Why does the approximation hold?

19.12

A 220 V, 60 Hz, three-phase, six-pole, Y-connected induction motor is characterized by the following parameters: $R_1 = 0.05\ \Omega$, $R_2 = 0.06\ \Omega$, $X_1 = 0.3\ \Omega$, $X_2 = 0.3\ \Omega$. The no-load input is 850 W at a line voltage of 220 V, with an input current of 25 A. Calculate the following:

a. The adjusted voltage V_A. (Consider both the algebraic approximation and the more accurate phasor method.)

b. With 5% slip, the input power and the output power and the efficiency.

c. T_{start} and T_{max}.

19.13

To illustrate the incongruity of SI units as applied to magnetic problems, compute the values of hysteresis power density in SI units for the following two cases:

a. 17% cobalt steel: 35.4 W/in.^{3}

b. Alnico V: 162 W/in.^{3}

19.14

In conjunction with the discussion of the hysteresis motor, we used English units. One weber of SI flux is equal to 10^8 *lines* in English units. (Note that this quantity is the same as 10^8 maxwells.) Find the numerical equivalents for the following:

	SI Units	English Units
B	Tesla	_____ [lines/in.^{2}]
H	Ampere-turn per meter	_____ [A.t./in.]

19.15

For a peak value of B_m equal to 1 T, with 17% cobalt steel, the area of the hysteresis loop, W_h, is 0.59 J/in.^{3}·cycle. What power per cubic inch of steel can be obtained from a hysteresis rotor operating at 60 Hz? If the corresponding area of Alnico V is 2.7, what is the *idealized* power developed per cubic inch of material?

19.16

For a 1.75 in. diameter rotor and a 3 in. length, compute the torque in inch-ounces developed in each of the cases in Problem 19.15 for a four-pole unit. (These figures will also be idealized. In practice, with closed stator slots, the actual torque may be one order of magnitude smaller. With open slots the actual torque may be two orders of magnitude smaller.)

19.17

Convert the idealized hysteresis power of Alnico V and 17% cobalt steel motors of Problem 19.16 into horsepower.

19.18

In Eq. 19.38, justify the numerical factor 8.86 if f is in hertz, V is in cubic inches, and W_h is in joules per cubic inch-cycle.

19.19

Derive Eq. 19.49 from Eqs. 19.43 and 19.48.

19.20

Determine the efficiency of Dijken's toothbrush motor cited in Example 19.7. What was the value of n_0 in his design?

19.21

It was Nikola Tesla who established 60 Hz as the power frequency standard, representing a compromise between the amount of necessary iron and the effectiveness of the electromagnetic induction process. Generally, in Europe, 50 Hz is used as a standard. In aircraft applications, 400 Hz has been commonly used. Indicate the relative advantages and disadvantages of the latter two frequencies as compared to the standard 60 Hz.

ANSWERS TO SELECTED PROBLEMS

CHAPTER 1

1.1 **a.** 12×10^{-3} A, 12 mA
b. 7.5×10^{16} charges per second
1.3 **a.** 1.08×10^{-2} A
b. 0.108 V, 10.8 V, 1.08 V
c. $0.108 + 1.08 + 10.8 \simeq 12$ V
1.6 **a.** 0.119 A
b. 0.108 A, 0.0108 A, 0.119 A
c. 10.8 V, 10.8 V, 1.19 V
1.10 **a.** +22 V **b.** +2 V **c.** 0 V **d.** +1 V
e. +11 V
1.15 **a.** 339.5 V **b.** 622.4 V
1.17 3990 Ω
1.22 $V_{max}/2$
1.26 0.25
1.28 **a.** 458 W **b.** 114.5 W **c.** 480 W, 120 W
1.31 1700 W
1.35 **a.** 100 mV **b.** 50 mV
1.37 4.44 Ω
1.41 50 Ω

CHAPTER 2

2.1 **a.** 3.77 Ω **b.** 62.8 MΩ, 6.28 GΩ
2.3 $4700 \angle 0°$ Ω, $62.8 \angle +90°$ MΩ
2.5 $0.392 \angle +78.7°$ A
2.7 $2 \angle 0°$ A
2.9 **a.** 2.5 C/s **b.** 0.21 F
2.11 50 Ω
2.13 **a.** 112 Ω **b.** 63.4° **c.** 11.15 V, 22.3 V
d. 63.4°
2.15 25.83 kVA
2.17 **a.** 3.18 kHz **b.** $0.1 \angle -90°$ V **c.** $0.1 \angle +90°$ V
2.20 **a.** 2.76 kHz **b.** 200 Ω
2.23 **a.** $L' = C(R^2 + X_C^2)$ **b.** 25% **c.** 53.7 Hz
2.29 10.6 μV
2.31 **a.** 40 mS, −20 mS
b. $G = 13.8$ mS, $B = -34.5$ mS

CHAPTER 3

3.1 **a.** 4.4 ms **b.** 2.75 ms
3.3 22.5 mW·s, 16.8 mW·s
3.5 $i = I_o e^{-Rt/L}$, $i = I_o e^{-Rt/2L} \cos \omega t$
3.7 **a.** 15 V **b.** 10 V **c.** 1.5 mA **d.** 2.0 mA
3.9 **a.** $V_T = 14$ V, $R_T = (50/15)$ Ω
b. $I_N = 4.2$ A, $G_N = 0.3$ S
3.11 (6/55) A
3.13 **a.** 5:1 **b.** 0.4 V **c.** 0.02 W
3.15 0.0938 V (−20.5 dB), 0.185 V (−14.6 dB), 0.426 V (−7.4 dB), 0.686 V (−3.3 dB), 0.883 V (−1.08 dB), 0.994 V (−0.052 dB), 0.999 V (−0.0087 dB)
3.19 **a.** ≃3 dB **b.** ≃3 dB
3.21 **a.** 137 dB
b. 120 dB (resistances are not the same)
3.23 63
3.25 **a.** 10^9 **b.** 3.16×10^6 **c.** 10,000
3.27 $\pi\omega C^2 V^2 R$
3.32 $f = 1.414 f_c$
3.37 Both start at a value equal to unity, but while case 1 has a crossover point between the damped and undamped cases, case 2 does not.
3.40 4.5 in.

CHAPTER 4

4.1 **a.** 10 kΩ ± 20% **b.** 47 Ω ±20%
c. 4.7 kΩ ±5% **d.** 1 MΩ ±10% **e.** 102 Ω
f. 442 kΩ **g.** 11.8 Ω **h.** 10.0 Ω
4.3 **a.** 1 W (2 W preferred) **b.** 0.0047 W (use 1/8 W)
4.5 33.4 pF
4.8 **a.** 0.0025 μF **b.** 0.04 μF
4.10 1.2 mC
4.12 4.7 kΩ, 117 Ω
4.14 $C_{eff} = C/(1 - \omega^2 LC)$

CHAPTER 5

5.1 **a.** $f(t) = (2V_m/\pi)(1 - \frac{2}{3}\cos 2\omega t - \frac{2}{15}\cos 4\omega t - \frac{2}{35}\cos 6\omega t - \cdots)$
b. $f(t) = (2V_m/\pi)(1 + \frac{2}{3}\cos 2\omega t + \frac{2}{15}\cos 4\omega t + \frac{2}{35}\cos 6\omega t + \cdots)$

5.4 **a.** $D_2 = 0, D_3 = 1.7\%, D_4 = 0$
b. Asymmetric distortion primarily affects the second harmonic; symmetric distortion primarily affects the third harmonic.

5.6 **a.** $v = \sum_{n,odd} \left(\frac{4V_m}{n\pi}\right) \sin n\omega t$

b. $v = -\sum_{n,odd} \left(\frac{4V_m}{n^2\pi^2}\right) \cos n\omega t$

The amplitude of the harmonics for the triangular wave drops much more rapidly with frequency than does that for the square wave. The sharp leading and trailing edges of the square wave have large harmonic content; the gradual slopes of the triangular waves do not.

5.9 It will be a sine wave of frequency f. The impressed signal will contain the fundamental f and its various harmonics.

5.11 $-1.045, -1.443, -1.15, -0.667, -0.442, -0.443, -0.337, 0$ (a sawtooth)

5.13 **a.** Addition yields $(v_L + v_R)/2$; subtraction yields $(v_L - v_R)[(2/\pi)\cos \omega t]$
b. a low pass filter with cutoff below ωt will pass $(L + R)$. A high pass filter passes $(L - R)$
c. $(L + R) + (L - R) = 2L$
$(L + R) - (L - R) = 2R$
having utilized the Fourier expansions
$v_L[\frac{1}{2} + (2/\pi)\cos \omega t - (2/3\pi)\cos 3\omega t + \cdots]$ and
$v_R[\frac{1}{2} - (2/\pi)\cos \omega t + (2/3\pi)\cos 3\omega t - \cdots]$

5.15 $f(t) = (2V_m/\pi)(\sin \omega t + \frac{1}{2}\sin 2\omega t + \frac{1}{3}\sin 3\omega t + \frac{1}{4}\sin 4\omega t + \cdots)$
For a square wave:
$f(t) = (4V_m/\pi)(\sin \omega t + \frac{1}{3}\sin 3\omega t + \frac{1}{5}\sin 5\omega t + \cdots)$

CHAPTER 6

6.1 1.6×10^{-19} J, 10^{-12} ergs
6.3 **a.** 3.47×10^8 W/ft **b.** 28.8 ft **c.** 1.19×10^4 K
6.5 6.67×10^{-5} m/s
6.7 -0.028 Ω/Ω°C This is an order of magnitude greater than the value for copper.
6.11 **a.** 54 V **b.** 108 V
6.13 **a.** 2.03×10^{-14} A, 2.95×10^{-16} A
b. 1.42×10^{-8} A, 1.72×10^{-9} A
6.15 approximately 10°C
6.17 1.21 or 121%, 0.48 or 48%
6.19 960 kHz ± 3 kHz = 957 kHz and 963 kHz

6.23 **a.** $v = A\cos \omega_c t + \frac{mA}{2}[\cos(\omega_c + \omega_m)t] + \frac{mA}{2}[\cos(\omega_c - \omega_m)t]$
b. 1000 kHz ± 1 kHz = 999 kHz and 1001 kHz
c. 10 kHz
6.25 The harmonic content would be considerable and probably exceed the allowed bandwidth.
6.27 **a.** 3,1,30 **b.** 5

CHAPTER 7

7.1 0.9933
7.4 **a.** 200 Ω **b.** 4.17 Ω
7.6 4 mA
7.8 40 kΩ
7.10 **a.** 9 V, −8.3 V **b.** 0.7 V, 0 V
c. the transistor is in saturation
7.12 1.5 kΩ, 50, 2.65 V rms
7.14 −20 dB∠+84.3°, −0.043 dB∠+5.71°
7.16 $R_1 = 1.71$ MΩ, $R_2 = 2.46$ MΩ, 83%
7.17 34%
7.19 22.5 mW·s, 40 mW·s
7.21 Place a diode between collector and emitter.
7.23 **a.** 60 mA, 0.7 mA, 6 V **b.** 2 V, 10 V **c.** 100 Ω
d. 16.14 kΩ, 15 kΩ
7.25 $R_E = 5.5$ Ω, $R_1 = 79.4$ Ω, $R_2 = 17.4$ Ω, $C_E \simeq 100{,}000$ μF, $C_B = 11$ μF
7.27 at 1, 180° out of phase with the input; at 2, in phase with the input
7.29 **a.** 33.3 W **b.** 38.9 W

CHAPTER 8

8.1 **a.** 0.99% **b.** The tolerance value of the resistor might be 10%
8.3 **a.** 0.0356 (3.56%) **b.** 21.9
8.6 1.78 MHz
8.9 **a.** −3 dB **b.** −9 dB **c.** −135° **d.** −180°
e. −18 dB
8.11 5%, −26 dB
8.13 **a.** 7.55 kΩ **b.** that is a characteristic of voltage amplifiers
8.15 34.6 Ω(emitter follower), 25.1 Ω(Darlington)
For $R_s = 0$, 24.75 Ω (emitter follower), 25.00 Ω (Darlington)
8.17 **a.** 1.28 mA, 28% **b.** 1.05 mA, 5%
8.19 **a.** 5.2% ($\Delta\beta$), 37.5% (ΔV_{BE}), 28% (ΔI_{CO})
b. 2.45% ($\Delta\beta$), 15% (ΔV_{BE}), 12.8% (ΔI_{CO})
8.21 Input resistance is diminished; output resistance is diminished.
8.23 **a.** $\omega = 1/(\sqrt{6}RC)$ **b.** $\omega = 1/\{RC[6 + 4(R_c/R)]^{1/2}\}$

c. $h_{fe} = 4(R_c/R) + 29(R/R_c) + 23$ **d.** 2.7
e. 44.5

CHAPTER 9

9.2 $2\sqrt{kI_{DS}}$
9.4 ≈ 16.5 V
9.5 16.7 V
9.7 2 kΩ
9.9 1.8 kΩ
9.11 7.6 kΩ
9.13 400 Ω, 7.2 kΩ
9.15 **a.** 2.2 V **b.** 11.6 V **c.** 2.37 V
9.17 **a.** 5 kΩ
b. 1.126 mA and 2.274 mA, 6.37 V, 12.74 V
c. 7.66 kΩ, 96.5 μF
d. In Figure 9.38, the R_1 connection, additionally, introduces negative feedback.
9.19 9.35 V, 22.7 mA
9.24 **a.** 4 V(dc), $2V_m$(peak signal), 0.25 V_m^2 (peak distortion), 2 (gain)
b. 1.5, same distortion; 2.5, same distortion
9.26 $D = 19.4\%$

CHAPTER 10

10.1 10.0001 V, 1, 10^{10} Ω
10.3 6.67 V, 4.45 mW, 10 V, 10 mW
10.5 0.1 V
10.9 32 dB, 63.1, 0.0158 (1.58%)
10.11 18.2 mV, 180 mV, 1 mV, 10 mV
10.13 90°/decade, 45°, 31.6 MHz, −40 dB/decade, 40 dB, 0.1%
10.15 1.13 V/μs
10.17 **a.** $A_V = A_o/[1 + (R/R_F)(1 - A_o) + (R_1/R_i)]$
b. 5%, 0.5%, 0.05%
10.19 The loop gain is taken to be negative.
10.22 **a.** $(R_3R_4 + R_2R_3 + R_2R_4)/R_4$
b. $-(R_3R_4 + R_2R_3 + R_2R_4)/R_1R_4$
c. 10 MΩ

CHAPTER 11

11.1 **a.** 2.03 V, 5.18 V, 7.78 V **b.** −10 V, −7.97 V, −2.79 V, +5 V **c.** +1.89 V, −7.41 V **d.** This second transistor is cut off. **e.** 2.29×10^{-4} A, 1.91×10^{-3} A
11.4 **a.** $V_{ref} < (R_1/R_2)V_o$ **b.** $V_{ref} = 0$
11.6 Reverse the diode.
11.8 1.58×10^3/s, 4.84×10^{-5} s
11.10 **a.** 000
010
100
111
b. 0→0, 1→1 (i.e., maintains the *status quo*)
c. 0→1, 1→0 (i.e., inversion)
11.12 $\overline{\overline{A} + \overline{B}} = A \cdot B$
$\overline{\overline{A} \cdot \overline{B}} = A + B$
11.15 exclusive OR
11.17 **a.** positive-edge triggers **b.** negative-edge triggers
c. Master triggers on leading edge; transfer takes place on trailing edge.
11.20 **a.** 0001 1000 0111
1000 1000 0111
0010 0011 1000
0001 0001 0010
0101 0110 0110
b. 555, 677, 959, not BCD, 271
c. 0010 0010 1011
0010 1010 0101
0011 1011 1111
0001 0000 1111
11.22 **a.** 0001 0000 0111 0001
b. 0 to 9999 **c.** 16 bits **d.** 0 to 65,536
11.25 **a.** 47.5 kHz **b.** 250 Hz/degree

CHAPTER 12

12.1 $T_c = RC \ln[(V - V_E)/(V - V_F)]$
4.7 s, 0.21 Hz, $v(t) = V_F e^{-t/R'C}$
$T_D = R'C \ln(V_F/V_E)$
2.62 ms
12.5 11.2 ns
12.6 bandwidth $= 0.35/t_r$
12.9 **a.** 51.6%, 48.4 mV **b.** 14.2%, 8.58 mV
c. 22.3%, 777 mV **d.** 19.1%, 80.9 mV
e. 48.5%, 51.5 mV
12.11 **a.** 21.2 inches **b.** 5930 miles **c.** 106 light years
12.13 twice the value of an individual contribution

CHAPTER 13

13.1 **a.** matched load **b.** kTB
13.4 $R_s = v_N/i_N$
13.7 1.64 (2.15 dB), 1.007 (0.028 dB) The small noise current for a JFET makes $i_N R_s$ much smaller than v_N, while thermal noise increases with R_s. Thus the amplifier's percentage contribution has diminished.
13.10 **a.** 6.25×10^{-15} W **b.** 3.25×10^{-15} W
13.13 Any current drawn through the powerline grounds will cause the load enclosure to assume a potential difference relative to the utility ground.

CHAPTER 14

14.2 LD HL, 8400

14.4 **a.** For example,
```
  1101 0011
  1101 0011
  ---------
  0000 0000 = XOR
```
b. For example,
```
  1010 1010
- 1010 1010
  ---------
```
In terms of two's complement this is
```
  1010 1010
  0101 0110
  ---------
 10000 0000
```
c. It leads to the two's complement of the quantity in the accumulator

14.6 0100 0011
↑ ↑
PA7 PA0

14.8 CCF complements the flag, so if it was 0 to start with it would end up being set. The SCF assures us that the flag *is* set, and CCF clears it.

CHAPTER 15

15.1 10 Hz, 1.67 minutes
15.8 $\theta = -jKv_ct/\omega$, where the angle associated with θ is $-90°$
15.10 **a.** 2.03×10^{-7} V
b. 10^{-16} W, 3.62 K, 5×10^{-8} V, $\frac{1}{4}$ **c.** 23 μs
d. 1.4 s Only variations slower than about 1.4 s would be observable.

CHAPTER 16

16.1 **b.** 1580 Ω **c.** 0.1%
16.3 **a.** 1.0625 kΩ **b.** 91 mV, 10% (input), 1.3% (output)
c. 174 mW (250 mW is a standard rating)
16.10 141 W
16.13 **a.** $0.477\ I_m$, $0.151\ I_m$ **b.** $0.272\ I_m$, $0.0466\ I_m$

CHAPTER 17

17.1 **a.** 24.5 V **b.** 2.45 V
17.3 350, 0.350 μC, 3.5 V If $V_R < V_A$, $n_2 > n_1$ which is not allowed since n_1 is the maximum possible count.
17.5 **a.** −80 dB **b.** −64.4 dB
17.10 **a.** 9.77 mV **b.** 1.024 ms **c.** 977/s
17.12 **a.** 15.5 Hz **b.** 0.49 V/ms
c. an order of magnitude smaller than for a single slope
17.15 3.18 μF
17.18 500 μin./in., 0.1200 Ω, 119.88 Ω
17.23 −74 dB, −20 dB

CHAPTER 18

18.1 1.27×10^5 A.t./Wb
18.3 $\simeq 0.45\ T$
18.5 **a.** 400 **b.** 1500 **c.** 2500 **d.** 1390 (saturation takes place)
18.7 80 J/m^2, 45 lb$_f$

CHAPTER 19

19.1 0.5 N (0.1124 lb$_f$)
19.3 $B_m \cos\theta \sin\omega t = \frac{1}{2} B_m \sin(\omega t - \theta) + \frac{1}{2} B_m \sin(\omega t + \theta)$
19.9 **a.** 26.3 kVAR **b.** 0.959
19.11 $\simeq 123$ V The reactive component is small.
19.13 **a.** 2.16 MW/in.^{3} **b.** 9.89 MW/in.^{3}
19.15 35.4 W/in.^{3}, 162 W/in.^{3}
19.17 0.34 hp, 1.57 hp
19.20 64%, 8125 rpm

INDEX

TK146 .S84 1987

Suprynowicz, V. A.
(Vincent A.)
Electrical and
electronics
c1987.

SCHEMATIC SYMBOLS

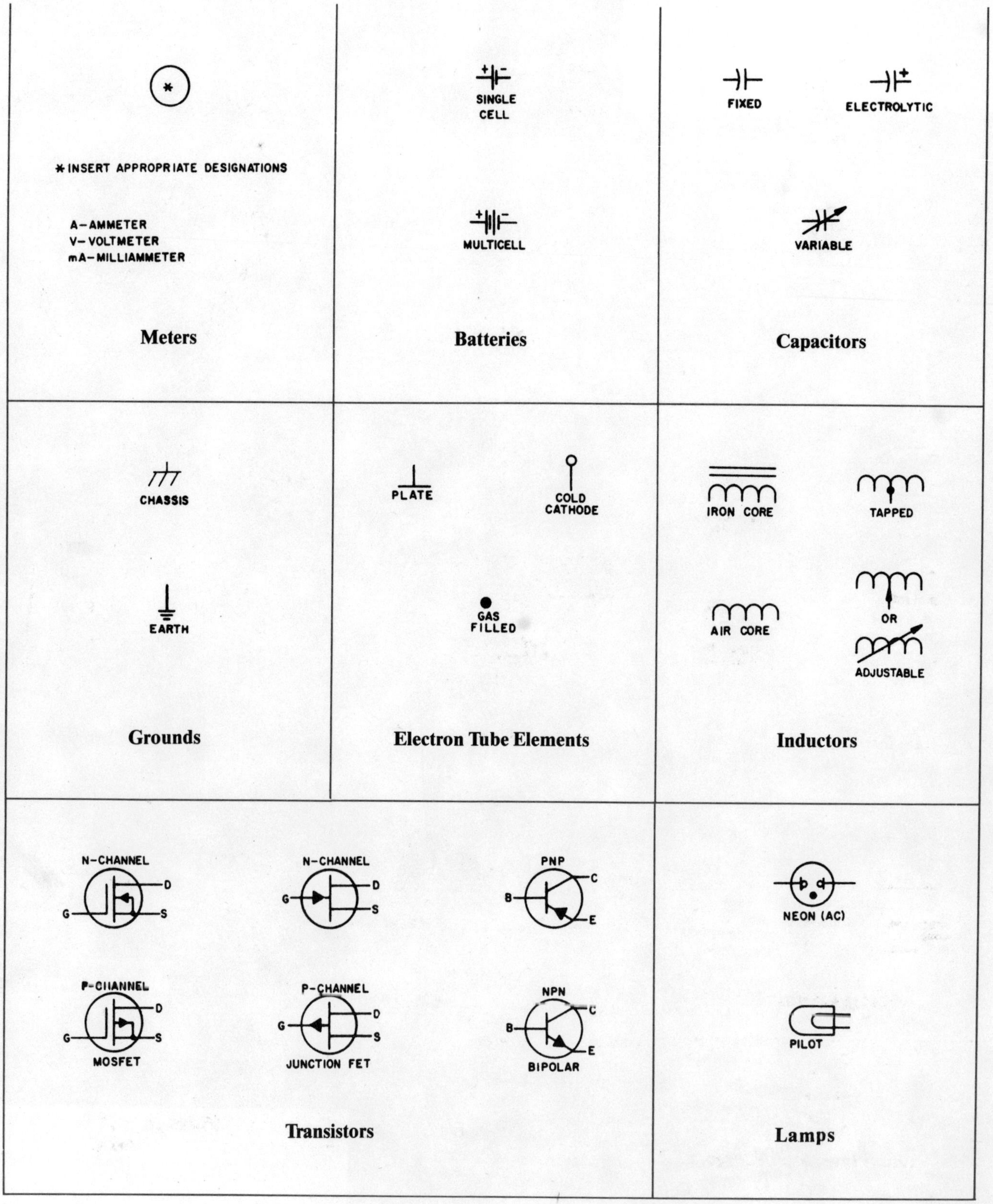

*INSERT APPROPRIATE DESIGNATIONS

A–AMMETER
V–VOLTMETER
mA–MILLIAMMETER

Meters

SINGLE CELL

MULTICELL

Batteries

FIXED

ELECTROLYTIC

VARIABLE

Capacitors

CHASSIS

EARTH

Grounds

PLATE

COLD CATHODE

GAS FILLED

Electron Tube Elements

IRON CORE

TAPPED

AIR CORE

OR

ADJUSTABLE

Inductors

N-CHANNEL
D
G
S

P-CHANNEL
D
G
S

MOSFET

N-CHANNEL
G
D
S

P-CHANNEL
G
D
S

JUNCTION FET

PNP
B
C
E

NPN
B
C
E

BIPOLAR

Transistors

NEON (AC)

PILOT

Lamps